ADVANCES IN NITROGEN FIXATION RESEARCH

ADVANCES IN AGRICULTURAL BIOTECHNOLOGY

Related titles previously published

Akazawa T., et al., eds: The New Frontiers in Plant Biochemistry. 1983. ISBN 90-247-2829-0

Gottschalk W. and Müller H.P., eds: Seed Proteins: Biochemistry, Genetics, Nutritive Value. 1983. ISBN 90-247-2789-8

Marcelle R., Clijsters H. and Van Poucke M., eds: Effects of Stress on Photosynthesis. 1983. ISBN 90-247-2799-5

Advances in Nitrogen Fixation Research

Proceedings of the 5th International Symposium on Nitrogen Fixation, Noordwijkerhout, The Netherlands, August 28 – September 3, 1983

edited by

C. VEEGER

Department of Biochemistry
Agricultural University
Wageningen
The Netherlands

and

W.E. NEWTON

C.F. Kettering Foundation
Research Laboratories
Yellow Springs
Ohio 45387
USA

1984

MARTINUS NIJHOFF / DR W. JUNK PUBLISHERS
THE HAGUE / BOSTON / LANCASTER

PUDOC
WAGENINGEN

Distributors

for the United States and Canada: Kluwer Boston, Inc., 190 Old Derby Street, Hingham, MA 02043, USA
for all other countries: Kluwer Academic Publishers Group, Distribution Center, P.O.Box 322, 3300 AH Dordrecht, The Netherlands

Library of Congress Card Number: 83-23755

ISBN 978-90-247-2906-7 ISBN 978-94-009-6923-0 (eBook)
DOI 10.1007/978-94-009-6923-0

PREFACE

As for the preceding four International Symposia on Nitrogen Fixation, held in Pullman, Washington USA (1974); Salamanca, Spain (1976); Madison, Wisconsin, USA (1978); and Canberra, Australia (1980), the 5th Symposium held from August 28 - September 3, 1983 in Noordwijkerhout, The Netherlands, received the generous support of the Charles F. Kettering Foundation Research Laboratory and the Tennessee Valley Authority. This support has helped research progress in this broad field of science by offering a forum both for the exchange of ideas and for scientific summary and discussion as captured over the last 10 years in each of the four books published previously. Although all previous meetings were well attended, the present conference was the largest so far. 550 scientists from 60 different countries attended the "Leeuwenhorst Conference", representing the many different disciplines actively involved in research in this field: chemists, biochemists, molecular biologists, geneticists, microbiologists, plant physiologists, agriculturalists. A large number of them had to go through a difficult period to raise the necessary funds to attend.
In addition, a parallel meeting of "policymakers" from Southeast Asia, Africa and South America was held under the auspices of Crosscurrents International Institute, Dayton, OH, USA and the United Nations University, Tokyo, Japan. These participants attended some of the scientific sessions to benefit from the vision of a number of scientists at the symposium. Visits to several agricultural institutions in the Netherlands, including an introduction to the Dutch views on agricultural research, were also provided by the Netherlands Ministry of Agriculture and Fisheries.
The organizers of this meeting adopted the trend which has guided all the previous meetings: to promote the interaction between the various disciplines involved in nitrogen fixation. But not only to have interaction between overlapping disciplines like chemistry and enzymology, but also between more disparate interests, such as chemistry and agriculture and with genetics.
To give all participants the opportunity to take notice of developments in other disciplines - particularly through contributions from many "2nd-generation" scientists - no parallel plenary sessions were held. Unfortunately, these good intentions could not be fully maintained for the poster sessions. All 350 posters, the majority of which were presented by the younger scientists, were on display throughout the whole meeting, with the most interesting and promising developments discussed in parallel sessions moderated by poster discussion convenors. The convenors were asked to capture the highlights and to add, as experts, their own ideas about future trends and present omissions. Although the organizers realize that such views can be no more than indicative, it is our hope that these contributions might lead to better understanding of the progress and possibly help generate new ideas.
It was the aim of the organizers to focus somewhat more on the agricultural possibilities offered by present and future applications of nitrogen fixation, while realizing that it is difficult to predict the future - as the Dutch say - from coffee grounds (tea leaves in English). Moreover, applications were emphasized by the post-symposium workshop on *Frankia* and the UNDP/IRI-sponsored training course for scientists from developing countries in various institutes in Wageningen, Leiden, Amsterdam and Groningen.

A meeting of this size, followed by the rapid publication of its proceedings, cannot be successful without the support and financial and technical assistance of many.

- Without the generous support of our sponsors, listed on page xxii, this meeting could never have been organized in this time of world-wide economic difficulties. We express our sincere gratitude for their generous donations.

- A great advantage of the "Leeuwenhorst" conference was that 85% of the participants lived for seven days in one building - to participate, discuss, present, eat, drink and sleep. We express our gratitude to the management and staff of the Leeuwenhorst Congress Centre in Noordwijkerhout for their splendid help and cooperation before, during and after the meeting.

- We wish to thank the Chairpersons of the sessions for successfully carrying out the "hard" job of maintaining the strict time schedule demanded by the overloaded scientific program.

- The poster convenors deserve our gratitude for their willingness not only to promote discussions - often much longer than planned on the schedule - but to finish their last-minute written contributions on time, making it possible to include them in these Proceedings.

- Although a beautifully printed and edited book is a treasure in any study, the editors decided not to aim for perfection in publication, but for rapidity of publication of the Proceedings, with its actuality prevailing over its layout. Thus, small mistakes and inconsistencies are inevitable in our attempts to make the many interesting scientific achievements available to all in the shortest possible time. The flexibility of the publisher and his willingness to provide free copies to all participants, which is, in our opinion, especially beneficial for the lesser-developed countries, is gratefully acknowledged.

- We thank the members of the International Advisory Board for suggestions with respect to the program and organization of this meeting. Their views, which in many cases were based on experience with previous meetings, were most valuable.

- Although the Organizing Committee worked very hard for this meeting, especially in the last few months, two members did the "dirty work". Dr. Rommert van den Bos, as Treasurer, needed to take hard decisions sometimes in order to stop the enthusiasm of his fellow committee members. Dr. Huub Haaker, as Secretary, not only wrote many pleasant letters to us all, but also organized the poster displays. The problems of the late (and even the late, late) contributions were solved by Huub in an excellent way, making it possible for every participant to take part in the discussions. The trans-Atlantic telephone line between Wageningen and the Yellow Springs Lab operated properly. Interruptions in the contact between the Dutch Organizing Committee and its American co-chairperson was mainly due, on both sides, to absence by travelling.

- We hope that the initiative taken at this meeting to present awards to the three best posters will be continued by the organizers of future meetings to highlight the importance not only of this opportunity to present recent results but to do so in a way that is clearly and easily understood.

- Being a member of an Organizing Committee of a symposium like this does not mean that one has knowledge of running a meeting. Registration, housing and abstracts are topics one hears about, but for which one needs professional assistance. The way that Mr. Jan Drijver of the International Agricultural Centre, Wageningen and his three helpful and always laughing assistants (Ellen P. v.d. Wetering, Mariette Th. Burgers and Elisabeth J.L. Hotke-Staal) solved many small and big problems before and during the Symposium, all to the benefit of the participants enabling them to do science properly, was masterful. The "for he's a jolly good fellow" sung by the participants to Jan Drijver extends to all of them.

- Almost last, but certainly not least, we extend our sincere thanks to the three lovely, enjoyable, hard-working ladies who enabled us to get the Proceedings ready in time: Jenny Toppenberg-Fang, Lyda Verstege and Vicki Newton. Without their devoted help, accuracy and mutual interaction and understanding as well as great working capacity, the publication schedule of this book would have been much extended. We sometimes had to stop them from working! We, the editors, ate out of their hands! To quote the remark of a participant: "Which Newton is the editor of this book?"

- Finally, we thank all participants for their contributions, presentations, and discussion remarks. Without their presence, this symposium would have been a monologue.

C. Veeger — Wageningen, The Netherlands

W.E. Newton — Yellow Springs, Ohio, USA

09-09-1983

CONCLUDING REMARKS

J. R. POSTGATE
A.R.C. UNIT OF NITROGEN FIXATION, UNIVERSITY OF SUSSEX, BRIGHTON BN1 9RQ, UNITED KINGDOM

So we have come to the end of the 5th International Symposium, sponsored primarily by the Charles F. Kettering Foundation Research Laboratory and the Tennessee Valley Authority, and I thought I'd start by reminding you of the purpose behind these meetings when they were first started. The intention was to stimulate research at the interface of the disciplines bearing on nitrogen fixation. Chemists were supposed to interact with biochemists, biochemists with physiologists, both of them to interact with geneticists, and ecologists and agriculturalists to draw from geneticists and physiologists and perhaps, rather more distantly from chemists and biochemists. And underlying this intention was the confidence that our deliberations would lead to improvements in food production and agriculture and we would, thereby, improve the human condition.
Well, how far have we progressed since Canberra? The deluge of information, posters, highly condensed lectures and the presence of 550 of you here-or at least there were 550 of you here-tells us that the subject should be moving very fast at all levels. Cees Veeger said I should take five or ten minutes to give you a summary overview of what happened. Well, for heavens sake, is that some kind of Dutch joke? I am sure you will agree with me that I can't possibly give you any serious scientific overview in that time and indeed to be honest, I think, like many of you, I am still dazed by much of the information that I have been exposed to. In fact, I'd rather like to digress for a moment and congratulate our Dutch hosts upon achieving what seemed to be impossible-of presenting something of interest to everyone at this meeting, even if we had to make agonizing choices about the discussion groups that we would choose to attend or which lecuture we would miss in order to be able to talk to our colleagues. It is daunting to think that a further increment in the exponential growth of this subject will have taken place by the time we meet again in Corvallis.
As I said, I find it too early to be able to assess in detail what we have achieved since Canberra, but I can discern some encouraging trends. For example, the chemistry has continued its shift into mechanisms and our understanding of the protonation of bound dinitrogen is beginning to parallel our understanding of the actual world of nitrogenase. We ought very soon to understand this enzyme at the molecular level. The awkard thing is still the structure of that entity we call FeMoco. Not much new has happened in that area. It is significant I think that our main handle, or our main way into this subject at the moment, seems to be through molecular biology, through a mutant. Some of our Dutch colleagues have challenged the perceived dogma on the structure of the iron protein and on the specific activity of nitrogenase. It is always good to challenge perceived dogmas, but I have the impression that we are already getting hints from molecular genetics of what the solutions to these differences will be; particularly, when one thinks about the repetition of *nif* genes that are being found in certain organisms other than *Klebsiella*.
On the molecular biology front, *Klebsiella* remains the cornerstone of molecular genetics, but it is gratifying to see the surge of interest in other microbes, including *Rhizobium*, phototrophic bacteria, *Azotobacters*, *Azospirillum* and so on. Cloning and probing has led to a ground swell of information, which is already changing our views on *nif* expression and which will probably feed back into the biochemistry and physiology of these organisms. I said probably. In fact it must feed back, because I have the impression that there is a danger of the molecular biology of the rhizobial system, for example, outstripping its physiological basis.

At the more ecological level, things progress rather more slowly. But it is gratifying to see how quickly the exciting innovations reported at Canberra, such as the stem nodule rhizobia or the free-living diazotrophic Frankias, have become assimilated into our thinking and are already targets for cloning, plasmid hunting and physiological study. Efficiency and effectiveness are still with us. They remain burning questions, even though we now compose them at more subtle levels. People are still asking what is the role of leghemoglobin in the nodule, although we continue to gain deeper insights into that. The physiological status of the bacteroid, I am rather surprised to say, is still a subject for disagreement. We still don't know really what underlies the role of hydrogenase, if there is one, in nodule efficiency. And I am really pleased to see that work is still continuing on the question of the contribution of diazotrophy to the rhizocoenoses, those associations between Azotobacter or Azospirillum and cereals or between the gram-negative rods and rice. These studies had a period of disappointment a few years ago, yet I am entirely convinced that there is a great deal to be done in these areas.
These are all questions where molecular genetics is already throwing up the mutants and genetic constraints, which are enabling us at present to get answers to these problems. And as you all well know, just as soon as we have answers to these questions, entirely new and even more difficult questions will have risen, which will concern to a greater extent the plant genome and the plant physiology. So, it is very gratifying to note that plant molecular biologists are already getting into position to handle these questions as and when they arise.
Well, I could go on like this, but I am sure you are all getting rather tired of generalities. This has been a very condensed and intensive meeting. Although most of you have, no doubt, been absorbed in your own specialities, talking to colleagues who share your interests and perhaps superficially, seeming to defeat the objectives of these symposia. Yet I am sure, when you get home and are able to reflect in tranquility on the week's proceedings, you will be impressed, as I have been, by the way in which the pure sciences-chemistry and biochemistry, genetics and molecular biology-are reinforcing each other, feeding back into physiology and ecology, and telling us more about how the diazotroph works and therefore, about how we can make it work for us more usefully and more effectively And that, after all, is why we are here.
Finally, I know that everyone will wish me to convey to our hosts, through you Cees, our thanks for the splendid organization which has supported this meeting. Running a meeting of this size is no small operation. And I know from the expression of anxiety on your face when you have occasionally rushed up and down these corridors, that it may not run as smoothly all the time for you as it has for us. But as far as we are concerned, this has been a most efficiently-run meeting and we are extremely grateful to you and your back-up team, including Vicki Newton, and the staff of this congress center for making it all so easy for us. Scientifically, in the words of one of my young colleagues, it has been a fantastic meeting!

REPORT FROM THE POLICY MEETING ON NITROGEN FIXATION RESEARCH

R. KOKKE[+] AND W.P. SHAW[++]
+ UNITED NATION UNIVERSITY, TOMO SEIMEI BLD. 15-1, SMIBUYA 2-CHOME, SHIBUYA-KU, TOKYO 150, JAPAN
++CROSSCURRENTS INTERNATIONAL INSTITUTE, 5335 FAR HILLS AVENUE-SUITE 300, DAYTON, OH 45429-2382, U.S.A.

In a meeting run parallel to the 5th International Symposium on Nitrogen Fixation, a group of policymakers, farmers and industrialists, from all continents, were invited to discuss the policy implications of the ongoing and future research.

The parallel meeting was sponsored by the United Nations University in Tokyo and by Crosscurrents International Institute in the USA. Eighteen participants and observers from countries such as Czechoslovakia, Australia, Sri Lanka, India, Holland, Switzerland, Zimbabwe, Colombia, Federal Republic of Germany, USA, Vietnam and Venezuela, some of them with little or no expertise on nitrogen fixation research, were exposed to lectures from scientists present at the Symposium and from specially invited guest speakers. The objectives of such sessions were not only to instruct the participants on the current nitrogen fixation research in the world, but to equip them in order to analyze its policy implications. Discussions after each lecture, informal conversations with many of the more than 500 scientists at the Symposium, visits to the poster exhibitions and to Dutch and International Research Centers complemented the participants' view on biological nitrogen fixation research (BNF).

Nitrogen fixation research is an example of agricultural research with significant potential for solving the world food problem. A large proportion of the registered increase in world agricultural output has been linked to higher fertilizer usage. Therefore, BNF becomes a policy issue.

It was within such a context that the participants at the policymakers' meeting discussed a range of policy questions. What is the potential of BNF research in promoting increased agricultural output? How long will it take before major research results lead to field application? Are there specific lines or types of BNF research that should presumably lead to prior or faster applied results? Have the needs of the LDC's farmer been taken into account? Should the BNF research continue to be undertaken mainly in the developed countries of the western world and, if so, what should be the role of the third world scientists? In the event that major breakthroughs effectively take place, which countries and segments of its populations would be most benefited by its results? How important should BNF funding be vis-a-vis other types of agricultural research? Which countries and what types of institutions (public or private) should be financing such research? Is there a wide communication network of BNF research that involves international research institutes and national government systems?

The participants at the policymaking meeting did not pretend to presume that their coverage of these questions was adequate and, even less, that the answers given are correct. They, nevertheless, considered it pertinent to transmit some of their discussion points to this wider audience in order to provoke some discussion or at least some thought.

A first, perhaps obvious but important, conclusion reached is the fact that the priority of BNF research and the funds allocated to it have to be evaluated, not in isolation but from an integrated and long-term perspective approach that involves the whole agricultural research program and the agricultural policies of each country. This is so since, in the first place, there is no clear evidence that the payoff from BNF research would be greater than that of alternative research which might utilize agricultural research funds. And, even if the payoff

was higher, the solution of the nitrogen problem would immediately bring forth other limiting factors (water, phosphorus, marketing, etc.). Compare with the green revolution rice - which brought about new problems.
A second conclusion had to do with the topic of basic or pure research vis-a-vis the more applied. The feeling of the group was the there seems to be a growing gap between the fundamental understanding of BNF and the farmer. The recommendation of participants, as expected from policymakers, leans in favor of increasing efforts in field research work and suggesting to donors such priority in their resource allocation, not as a substitute, but as a complement to basic research funding. It was suggested that more attention be paid to bringing into practice more fully the existing BNF knowhow as, for example, in the traditional and newer intercropping systems and the biomass-soil fertility interaction.
Moreover, from the discussions held, it was evident that developing countries must improve their technology transfer and extension capacities as well as the training in order to benefit more fully from BNF research. It was suggested that international institutions and donor agencies should induce international cooperation in order to link more adequately the pure and applied research efforts.
Finally, participants stressed the importance and convenience of improving world wide information networks of BNF research results.

TABLE OF CONTENTS

ORGANIZING COMMITTEE:

C. Veeger, Chairperson

W. E. Newton, Co-Chairperson

A. van Kammen, Co-Chairperson

H. Haaker, Scientific Secretary

J. Drijver, Administrative Secretary

R. C. van den Bos, Treasurer

A. Quispel, Member

J. Reedijk, Member

A. H. Stouthamer, Member

INTERNATIONAL ADVISORY BOARD:

W. E. Newton, Chairperson, Yellow Springs, OH, U.S.A.

J. E. Beringer, Harpenden, England

R. H. Burris, Madison, WI, U.S.A.

P. J. Dart, Patanchero, Po., India

Y. R. Dommergues, Dakar, Sénégal

A. H. Gibson, Canberra, ACT, Australia

W. H. Orme-Johnson, Cambridge, MA, U.S.A.

J. R. Postgate, Brighton, England

C. Rodriguez-Barrueco, Salamanca, Spain

A. E. Shilov, Moscow, U.S.S.R.

ADMINISTRATIVE ORGANIZATION:

International Agricultural Centre, Wageningen

Sponsors of the Symposium:

This symposium has been sponsored by the following committees, companies and institutions, for which the Organizing Committee wishes to express its sincere appreciation.

- Agricultural University Wageningen
- Biogen S.A.
- Charles F. Kettering Foundation Research Laboratory
- Dutch State Mines (DSM)
- Faircraft Engineering
- KLM Royal Dutch Airlines
- Ministry of Agriculture and Fisheries
- Netherlands Programme Committee on Biotechnology
- Royal Netherlands Academy of Sciences (KNWA)
- Royal Netherlands Chemical Society (KNCV)
- Skalar Analytical
- Tennessee Valley Authority
- United Nations Development Programme

APPLICATION OF DINITROGEN FIXATION IN AGRICULTURE AND FORESTRY

POSTER DISCUSSION 1A
MEASURING NITROGEN FIXATION

POSTER DISCUSSION 1B
APPLICATION OF BIOLOGICAL NITROGEN FIXATION FOR INCREASING CROP PRODUCTION

THE APPLICATION OF MOLECULAR GENETICS IN AGRICULTURE

JOHN E. BERINGER
ROTHAMSTED EXPERIMENTAL STATION, HARPENDEN, HERTFORDSIRE. AL5 2JQ ENGLAND.

Over the last few years there have been major advances in our ability to manipulate genes in microorganisms and to isolate genes from higher organisms and maintain and manipulate them in bacteria. Work in a number of laboratories with *Agrobacterium tumefaciens* and *A. rhizogenes* has shown that when these bacteria infect plants DNA is transferred to the plants where it becomes stably integrated within the nuclear DNA. Foreign genes which are inserted within the DNA sequences which are transferred from *Agrobacterium* to plants (T-DNA and R-DNA) are also inherited.

Recently it has been shown that a drug resistance gene present in T-DNA was expressed after transfer of the T-DNA to plant cells. Work of this nature suggests that in the near future it will be possible to introduce foreign DNA into a range of plant species as part of a normal breeding programme, though as yet, the methods for utilising *Agrobacterium* DNA are not feasible for monocotyledonous plants. Intense research is also being concentrated on methods for introducing foreign DNA into animal cells.

The isolation of genes, the manipulation of them in microorganisms and attempts to introduce them into novel genetic backgrounds is called molecular genetics, which is now recognised to be an extremely exciting and important aspect of biotechnology. The aim of this brief review is to point out the potential for gene manipulation in agriculture and to stress the limitations of existing technology.

Because agriculture is concerned with the production of plants and animals, and their conversion into foodstuffs, it has been an area where genetics has been exploited for centuries, both in the selection of superior plants and animals for future breeding stock and in the crossing of different strains of plants and animals to produce progeny carrying desirable attributes derived from both parents. The success of commercial breeding is obvious: there have been constant improvements in yield, and breeders have been able to introduce genes conferring resistance to important diseases and to select material that has desirable marketing properties, such as shape and colour. The intense interest in biotechnology is based on the hope that gene manipulation can provide a method for introducing novel genes and for speeding-up breeding.

While it is true that gene manipulation techniques, such as those based on the use of *Agrobacterium* DNA have enormous potential for producing entirely new assortments of genes, it should be recognised that a major problem associated with conventional plant breeding, the length of time taken to produce and test new varieties, is unlikely to be reduced significantly. This is because once a novel gene has been introduced into a variety seed will need to be bulked-up and then subjected to variety testing in exactly the same way as is done with conventional material. However, extensive backcrossing to eliminate unwanted genes which are inherited in normal crosses will be avoided.

It is the potential for introducing genes into organisms from quite different genetic backgrounds, and being able to make specific alterations to the genetic makeup of organisms that is so attractive. Obvious examples are introducing gene coding for disease resistance or for the production of novel compounds, in which case a desired objective is known and specific methods for isolating genes and handling them can be envisaged. It is perhaps worth considering an existing problem in plant breeding which could be solved by genetic manipulation, but which may be much more readily solved by traditional breeding techniques.

My example is the kiwifruit (Actinidia chinensis), which is a major agricultural export from New Zealand. It is a fruit which is sold largely on the basis that it is novel and attractive. A problem with kiwifruits is that they have a hairy skin which detracts from their appearance and makes them unpleasant for some people to handle. The obvious solution to this cosmetic, but commercially important, problem is to breed hairless varieties. This could be done by traditional methods, which would involve screening plants from many different sources for hairless fruits, or by utilising molecular genetics to inactivate the offending gene or genes.

Conceptually the problem is a simple one. The gene or genes involved in producing hairs need to be identified and inactivated, or another gene could be inserted into the plant whose function would be to prevent the formation of hairs. In practice this problem exemplifies a major difficulty demonstrated by following a logical sequence of steps to specifically inactivate the gene(s) involved.

1. Identify the function of the gene or genes in question.
2. Devise method for selecting for the gene.
3. Find the gene.
4. Clone the gene in a suitable vector.
5. Inactivate the gene.
6. Insert the gene into a suitable vector to return it to the original host.
7. Select for transfer of the gene to the new host.
8. Select for the stabilisation of the gene in the host.

Existing techniques for handling DNA are so well developed that steps 4-6 are routine in many laboratories working with bacteria and those plants which are susceptible to Agrobacterium tumefaciens. Step 4 usually depends upon steps 1 and 2 and sometimes preceeds step 3. Nearly all the work that has been done on the cloning and manipulation of DNA has involved genes whose function is well known and for which there are well established methods for assaying the function and selecting for expression. Genes for resistance to antibiotics are classic examples. An ability to select for the inheritance of a gene is extremely important when manipulation procedures yield only a few cells among many that do not contain the new gene. If such cells cannot be selected directly our ability to obtain progeny carrying an introduced gene (or genes) will be much lower than when the recipient expresses a function that will enable it and/or its progeny to survive and grow under conditions that are not suitable for the majority not carrying the new gene(s).

Lack of hair on a fruit is an obvious example of a property that cannot be selected directly, especially at the level of single cells growing in tissue culture. In the absence of a good method for selecting specific genes, steps 2, 3, 7 and 8 become extremely difficult. Indeed one of the few ways around the problem would be to insert a gene whose function can be selected (e.g. a drug resistance) beside the desired gene and then proceed to select for and manipulate DNA carrying the introduced function. Without a knowledge of the location of the gene that one wishes to handle the prospects for obtaining insertion of another gene beside it are remote. Furthermore for most plants and probably all animals methods for introducing foreign DNA to mark the position of native genes, or to inactivate them, do not yet exist.

Another way to inactivate hair production would be to introduce foreign genes which prevent hair formation. In principal this would involve introducing the genes into T-DNA or R-DNA and allowing Agrobacterium to introduce the novel genes into the host plant. However, we know almost nothing about the biochemistry of hair formation on kiwifruits and thus have no rational basis for selecting genes whose function could interfere with their production, if indeed such genes exist. It would be essential to have a very much better knowledge of step 1 than is required simply to clone DNA to be able to use this approach.

The list of problems relating to the potential for utilising molecular genetic techniques to produce a hairless kiwifruit show clearly that there is still a role for traditional genetics in breeding new varieties of organism and that a good knowledge of biochemistry and physiology of an organism will be needed in many cases where altered gene function are desired. They also show that where genes coding for known selectable functions are concerned and where a host is susceptible to gene transfer by Agrobacterium, there is real potential for making novel plants. A physical method for removing hairs from fruits is another form of biotechnological response to the problem.

Of most interest to those involved in research on biological nitrogen fixation is the potential for moving nitrogen fixation (nif) genes around to produce new nitrogen-fixing species. We know that in Klebsiella pneumoniae nif genes are clustered together and can be transferred to other bacterial species, such as Escherichia coli, in which they can function. We also know that these genes do not function in all the bacterial species into which they have been introducted, presumably because other gene functions are required or because the organisms are not able to decode the genes to produce functional proteins. The extreme sensitivity of nitrogenase to oxygen also limits the range of species and environments in which nif genes can function. Another extremely important constraint on the possibility of transferring nif genes to different organisms is the requirement that the function of these genes be regulated so that sufficient nitrogen is fixed for the needs of the organism. Too much nitrogen fixation would not only cost too much in wasted energy, but could also lead to ammonia toxicity problems. Indeed an important component of regulation is that the organism must have an efficient system for assimilating the fixed nitrogen.

Much effort is being expended in studying the genetics of nitrogen fixation in a range of microorganisms (see other chapters in this Proceedings) and there is no logical reason to believe that the genes involved will not be transferred to some species of plants in the near future. Whether it will ever be possible to obtain plants that are self-sufficient in nitrogen nutrition and still yield as well as nitrogen fertilised varieties not carrying *nif* genes is debatable, as also in the question whether the solution to the problem lies in extending the host range of existing symbiotic nitrogen-fixing microorganisms such as *Rhizobium* or *Frankia*.

It is not my intention to take sides in this arguement, either method, even if the resulting plants are relatively inefficient when compared to well-fertilised cultivars growing in fertile soils, offers the potential for increasing crop yields in developing countries when N fertiliser is unavailable and is growth limiting. The question that I feel should be raised is what are we doing about animals? Ruminants in particular are extremely inefficient in converting plant protein into meat. Should we not be considering the potential for altering microorganisms which already thrive in anaerobic environments present in animal digestive systems? Theoretically it should be much easier to introduce *nif* genes into bacteria and reintroduce the manipulated bacteria into animal guts than to make nitrogen-fixing plants.

In summary it should be stressed that the genetic manipulation of organisms will provide a way of increasing the rate at which new characteristics can be introduced into plants and animals as well as enabling us to introduce genes which could not be crossed into an organism by conventional procedures. In the long term it is the introduction of novel functions, such as nitrogen fixation, that promises to revolutionise agriculture. However our optimism should be tempered with the realisation that the eventual success in applying our new genetic skills will depend to a large extent on a greatly expanded knowledge of the biochemistry and physiology of the plants and animals that we are interested in.

NITROGEN-FIXING TREES IN THE TROPICS: POTENTIALITIES AND LIMITATIONS

Y.R. DOMMERGUES[1], H.G. DIEM[1], D.L. GAUTHIER[1], B.L. DREYFUS[1], F. CORNET[2].

1 CNRS/ORSTOM, BP 1386, DAKAR, Sénégal
2 ISRA-CNRF/CTFT, BP 2312, DAKAR, Sénégal

1. INTRODUCTION

Shifting cultivation, logging, ranching, mining and road construction are rapidly reducing the surface of tropical forests. Estimates of losses vary widely (7.5 to 20 million ha each year), but even if the lower figures are taken into account, the losses are staggering. The slash-and-burn condition which is a major cause of forest destruction should no longer continue, but there are no practical means to stop it, especially when the forest is converted to crops for human subsistance. The only solution to the coming wood crisis is to set up an intensive program of reforestation: (1) to provide the people in developing countries with the wood they require for their own needs; (2) to restore and, if possible, to increase the production of timber and other ligno-cellulosic materials that will be requested in larger amounts in the near future. Since good quality soils will be mostly devoted to agriculture, leaving only land of low fertility to reforestation, it will be necessary to use tree species with limited requirements in nutrients and especially N. Fortunately N_2-fixing trees (NFT) (legumes or actinorhizal trees) exhibit the most desirable characteristic of thriving in soils devoid of N and also in P-deficient ones, if they are adequately inoculated with effective endo- or ectomycorrhizae. NFT do not only provide wood or ligno-cellulosic material and a number of useful products (fodder for animals, fruit, gums, etc.), they can also maintain or restore soil fertility by improving the N status of the soil, protecting vulnerable soils against wind and water erosion. It is not yet possible to enumerate the NFT species belonging to the Leguminoseae, since only a part of these species has been tested for their nodulating ability. Tentative lists of the most important genera of the leguminous NFT have already been published (e.g.: Döbereiner, Campelo, 1977; National Academy of Sciences, 1979; Brewbaker et al., 1982).
Compared with the vast group of N_2-fixing Leguminoseae, the actinorhizal non-legumes form a much less numerous body of N_2-fixing species. In spite of this numeric inferiority actinorhizal plants are of considerable fundamental and practical significance for at least two reasons: (1) the ability of actinorhizal plants to be symbiotically associated with a N_2-fixing endophyte is spread among eight families whereas this ability is restricted to one family (plus the genus *Parasponia*) for plants in symbiosis with rhizobia; (2) most actinorhizal plants are good and sometimes agressive colonizers, capable of regenerating poor soils or disturbed

sites. Since an excellent review on the tropical actinorhizal plants has been published recently (Becking, 1982), our presentation will be restricted to the family Casuarinaceae which comprises three main genera: *Casuarina*, *Allocasuarina*, *Gymnostoma*.

2. N_2 FIXATION BY NFT

The estimates of N_2 fixation by NFT are still very few and based on methods that are susceptible to error: the acetylene reduction method, the difference method and methods based on N balance studies. These methods have already been appropriately criticized (e.g. Knowles, 1980; LaRue, Patterson, 1981; Herridge, 1981; Vose et al., 1982). However, one should add that whatever method is used, the assessment of N_2 fixation by trees is more difficult than that of annual crops because of the great heterogeneity of forest stands and the variations in the N_2-fixing activity which occur daily and throughout the year (Roskoski et al., 1982) and the whole life of the trees. The presumption, based on nursery and field observations of the nodulation of some Sahelian *Acacia* species, is that the N_2-fixing activity of the trees increases when trees are still young, then decreases with the progressive accumulation of N in the soil below their canopies. Taking into account the imperfections of the methods of evaluation used, one can infer from data already published that the following NFT have a high N_2-fixing potential: *Acacia mearnsii* (Orchard, Darby, 1956); *Leucaena leucocephala* (National Academy of Sciences, 1977; Högberg, Kvarnstrom, 1982), *Parasponia* sp. (Trinick, 1981), *Casuarina equisetifolia* (Dommergues, 1963 and this paper), *Allocasuarina littoralis* (Silvester, 1977), N_2-fixation expressed in kg per ha and per year being respectively 200 (A.m); 500-110 (L.l.);850 (P. sp.); 58-262 (C.e.); 218 (A.l.). If we postulate that the N_2-fixing potential of a system is measured by the amount of N_2 fixed in the absence of any limiting factor, most of these figures, with the exception of *Parasponia*, probably do not reflect the potential of the different NFT. Further studies are needed to quantify these potentialities. Preliminary screenings of the nodulation ability of NFT which are under way (e.g. Basak, Goyal, 1980) will facilitate the choice of the systems which deserve N_2-fixing evaluation, together with other extensive physiological and ecological studies.

3. SPECIFICITY BETWEEN SYMBIONTS AND NEED FOR INOCULATION

3.1. NFT in symbiosis with rhizobia. Taking into account data published in the literature (Bowen, 1956; Lange, 1961; Campelo, Döbereiner, 1969; Trinick, 1980; Vincent, 1980; Dreyfus, Dommergues, 1981) one can classify NFT into three broad groups according to nodulation response patterns with fast- and slow-growing tropical strains of rhizobia: group 1, which nodulates with fast-growing strains (e.g. *Leucaena leucocephala*, *Mimosa caesalpiniaefolia*, *Sesbania sp.*,*Acacia farnesiana*, *A. nilotica*, *A. raddiana*, *A. senegal)*; group 2, which nodulates both with fast- and slow-growing strains (e.g. *Acacia seyal*, *A. cyanophylla*, *Parasponia sp*.);group 3, which nodulates with slow-growing strains (e.g. *Acacia albida*, *A. holosericea*, *A. mearnsii*).

The first group apparently exhibits a symbiotic range narrower than that of the other two groups and probably has a "special requirement for high soil pH, nutrient status and available calcium" (Trinick,1980). The strains of fast-growing rhizobia associated with group 1 NFT are probably related to the "advanced degenerate forms" (Norris,1956) represented by the fast-growing rhizobia of temperate legumes (*R. meliloti, R. trifolii, R. phaseoli, R. leguminosarum*). The practical implication of the specificity of group 1 NFT is that their establishment requires inoculation with the compatible fast-growing strains, which are generally less ubiquitous than the typical slow-growing strains of the cowpea miscellany. This explains the spectacular response to inoculation of *Leucaena leucocephala* (see, for example, Diatloff,1973) or *Mimosa caesalpiniaefolia* (Döbereiner, 1967) in the field whereas inoculation of field-grown *Acacia mearnsii* rarely results in a significant yield increase since most tropical soils harbor the competent rhizobia of the cowpea miscellany.

3.2. Actinorhizal NFT. Presently there is little data available on the *Frankia* requirement of Casuarinaceae so it is not yet possible to classify these plants into groups according to their responses to the different strains of *Frankia*. However, preliminary investigations reported by Coyne at the International *Casuarina* Workshop held at Canberra in 1981, provide evidence of inter-generic specificity in host-endophyte relations within the Casuarinaceae family. Coyne used six species belonging to two genera: *Casuarina (C. cristata, C. cunninghamiana, C. glauca)* and *Allocasuarina (A. verticillata, A. torulosa, A. littoralis)*. Seedlings of the six species were inoculated with crushed nodules of the same species and from each of the other species. According to the response to inoculation, the different species could be classified into two main groups: one comprising the three species of the genus *Casuarina* which behaved similarly, the other group comprising the species of the *Allocasuarina* genus, which could be divided into two subgroups, (1) *A. verticillata* and *A.torulosa* and (2) *A. littoralis*.
Using a pure strain of *Frankia* (ORS 021001) isolated from *Casuarina junghuhniana* to inoculate different species of *Casuarina* and *Allocasuarina*, we found that only the species belonging to the *Casuarina* genus could be nodulated, and none of the species of the *Allocasuarina* genus were nodulated. Experiments under way showed that, inoculated with crushed nodules of *Casuarina equisetifolia* instead of a pure culture of *Frankia*, a low percentage of *Allocasuarina torulosa* (ca 20%) and *A. verticillata* (ca 30%) could form 2-3 nodules per plant. Thus the host spectrum of the inoculum was apparently wider when crushed nodules were used instead of a pure strain. This result might be attributed to the fact that the crushed nodules contain more than one strain of *Frankia* (contaminant strains or mutants) or strains with a host spectrum wider than that of strain ORS 021001. Even within their natural distribution area the different species of native Casuarinaceae are not always regularly nodulated. Thus a survey of 77 sites in Victoria, Australia, revealed that nodulation occurred in less than 30% of these sites (Lawrie, 1982). This may indicate

an absence of endophyte or inhibition of nodulation. Outside their natural distribution area, Casuarinaceae are seldom nodulated. In N-deficient soils, the absence of the competent strain of *Frankia* is the main cause of the failure to establish Casuarinaceae (Torrey, 1982; Gauthier et al., unpublished). In such situations inoculation is required. Up to now, inoculants have been prepared from suspensions of crushed or dried nodules (Torrey, 1982). The first inoculation trial with a pure culture of *Frankia* was carried out at the ORSTOM research station of Bel-Air, Dakar, Senegal, between November 1982 and May 1983. 4.5-month old *Casuarina equisetifolia* seedlings were transplanted in $1m^2$ microplots (4 plants per m^2) containing a typical N deficient sandy soil of the Cap Vert peninsula. The plants were harvested when they were 11 months old. The amount of N_2 fixed during the interval of time between transplantation and harvest (6.5 months) was estimated to be 131 kg/ha by the difference method. Using the fertilizer N equivalent method we calculated that N_2 fixation by the *Casuarina* during 6.5 months was equivalent to the application of 288 kg N/ha (Gauthier et al., in press).
It is puzzling to find nodulated Casuarinaceae outside their natural distribution area since *Frankia* is probably not transmitted with the seeds. When the distance involved is not too great, one can assume that spores of *Frankia* have been circulated by the winds (Torrey, 1981). But such an explanation does not hold in some circumstances. It is tempting to hypothesize that local strains of *Frankia* have acquired by mutation the ability to nodulate newly introduced host-plants.

4. LIMITATIONS

During the last several years, many reviews have been published dealing with the environmental factors controlling the establishment and the functionning of N_2-fixing symbioses (e.g. Vincent, 1980; Gibson et al., 1982; Dommergues, 1982). The present communication concentrates on three limiting environmental factors which are probably among the most important under tropical conditions: soil deficiency in P, moisture deficiency, and combined N.

4.1. Soil deficiency in P. Many tropical soils are deficient in available P, which is known to limit the growth of the host-plant, nodulation and N_2 fixation of the legumes. In such situations endo- or ectomycorhizae may greatly improve P supply of the host-plant by increasing the absorbing capacity of the roots (Mosse, 1977), thus restoring the N_2-fixing capacity of the system. The beneficial effect of inoculation with an endomycorhizal fungus (*Glomus mosseae*) on N_2 fixation by NFT such as *Acacia raddiana* (Cornet et al., 1982) or *Casuarina equisetifolia* (Diem, Gauthier, 1982) has already been reported. However, these results were based on pot experiments, which cannot be safely extrapolated to the field. To the best of our knowledge the only field experiment on the effect of NFT inoculation with an endomycorhizal fungus was carried out recently in Bandia, Senegal. Seedlings of *Acacia holosericea* were grown for ca 3 months in a nursery with the following three treatments: control (C), inoculation with

Rhizobium strain ORS 841 (R), double inoculation with *Rhizobium* and *Glomus mosseae* (RM). When the plants were 2 years old it appeared that treatments R and RM had similarly and slightly increased the height (+ 6 and 7%) and number of surviving plants (+ 13 and 15%) compared to the control. The most striking effect of the treatments was on the coefficients of variation which were 40, 35 and 18% for treatments C, R and RM respectively when the trees were 10 months old; 23, 20, 11% for treatments C, R and RM when the trees were 24 months old. Thus inoculating the seedlings with *Glomus mosseae* clearly decreased the heterogeneity of the plant growth, a result confirming a previous experiment with field-grown soybeans (Ganry et al., 1982).

4.2. Moisture deficiency. This stress is known to seriously reduce nodulation and N_2 fixation in a number of N_2-fixing systems (Sprent,1973; Pate, 1976). However, even during drought spells some perennial deep-rooted legumes (e.g. alfalfa) may still fix N_2 (Johnson et al., 1981). Similarly, drought resistant NFT, such as *Prosopis* spp. (Felker, Clark, 1980) or *Acacia raddiana* (Muthana, Arora, 1980) are probably only slightly affected in their N_2-fixing activity by moisture deficiency in the upper horizons since these plants probably receive adequate moisture from below the nodule zone, as suggested by Hume et al. (1976). By improving P nutrition in P-deficient soils or by means of other mechanisms not yet quite elucidated, mycorhizae probably improve the tolerance of NFT to the moisture stress, a fact that has been shown by different investigators (see review by Gianinazzi-Pearson, Diem, 1982).

4.3. Combined N. The reduction of N_2-fixing activity of the nodule is generally attributed to a decrease in carbohydrate flux to the nodule associated with nitrate assimilation. Two approaches may be suggested to improve N_2 fixation in the presence of combined nitrogen: (1) develop *Rhizobium* or *Frankia* strains insensitive to N or specific partnerships with hosts that are more tolerant of combined N than others or (2) develop plants that can simultaneously absorb soil N and fix atmospheric N_2. The first approach is not merely speculatory, as some N_2-fixing systems moderately sensitive to combined N are known to exist. A second promising approach is based on the discovery that *Sesbania rostrata*, a legume profusely bearing stem nodules all along the stems, has great potential for continued N_2 fixation in soil with high levels of inorganic N (Dreyfus, Dommergues, 1980). Obviously, efforts to transfer the stem nodulation character of *Sesbania rostrata* to *S. grandiflora*, a non-stem nodulated tree, could be very fruitful. One should note that at least one tree is known to bear stem nodules, *Aeschynomene elaphroxylon*. Unfortunately the stem nodules of this tree, which grows in waterlogged soils of the Sahelian zone, are restricted to the lower and immerged section of the stems so that its tolerance of combined N is probably less marked than that of *Sesbania rostrata* (Gibson et al., 1982).

5. CONCLUSION

The potential of some NFT species such as *Leucaena leucocephala* or some species of *Casuarina* is now fully realized and field experiments have been carried out or are under way which provide us with valuable information about the ecology and physiology of NFT so that their use in forestry can now be safely recommended. However these studies deal with only a few of the thousands of potentially significant NFT that thrive in the tropics. The surveys that have been initiated to screen the most promising of them should be further developed; ecological and physiological studies of the selected species should be encouraged. Technologies emerging from the last decade of biological research should facilitate the genetic improvement of both partners of the symbiosis not only by maximizing the N_2-fixing ability but also by obtaining symbiotic associations more tolerant to the environmental constraints, such as nutrient deficiencies that so often impede N_2 fixation in the field. Applying the advances in tree breeding and tissue culture techniques along with good managerial practises should ensure a substantial yield increase of NFT planted in pure stands, mixed with non-NFT,or used as N providers in agroforestry.

REFERENCES

Basak MK and Goyal SK (1980) Plant Soil. 56, 33-37.
Becking JH (1982) In Dommergues YR and Diem HG, eds, Microbiology of Tropical Soils and Plant Productivity, pp. 109-146, Nijhoff/ Junk, The Hague.
Cornet F Diem HG and Dommergues YR (1982) In Les mycorhizes: biologie et utilisation, pp. 287-293, INRA, Paris.
Diatloff A (1973) Qld Agric. J. 99, 642-644.
Diem HG and Gauthier D (1982) C.R. Acad. Sc., Paris. 294, sér 3, 215-218.
Döbereiner J (1967) Pesq Agropec. Brasil. 2, 301-335.
Döbereiner J and Campelo AB (1977) In Hardy RWF and Gibson AH, eds, A treatise on Dinitrogen Fixation, Section 4, pp. 191-220, Wiley, New York.
Dommergues YR (1963) Agrochimica. 105, 179-187.
Dommergues YR (1982) In Graham PH and Harris SC, eds, Biological Nitrogen Fixation Technology for Tropical Agriculture, pp. 395-411 CIAT, Cali.
Dreyfus B and Dommergues YR (1980) C.R. Acad. Sci., Paris 291, sér. D, 767-770.
Dreyfus B and Dommergues YR (1981) Appl. Env. Microbiol. 41,97-99.
Felker P and Clark PR (1980) Plant Soil. 57, 177-186.
Ganry F Diem HG and Dommergues YR (1982) Plant Soil. 68, 321-329.
Gianinazzi-Pearson V and Diem HG (1982) In Dommergues YR and Diem HG, eds, Microbiology of Tropical Soils and Plant Productivity, pp. 209-251, Nijhoff/Junk, The Hague.
Gibson AH Dreyfus BL and Dommergues YR (1982) In Dommergues YR and Diem HG, eds, Microbiology of Tropical Soils and Plant Productivity, pp. 37-73, Nijhoff/Junk, The Hague.
Herridge DF (1982) In Graham PH and Harris SC, eds, Biological Nitrogen Fixation Technology for Tropical Agriculture, pp. 593-608 CIAT, Cali.

Högberg P and Kvarnstrom M (1982) Plant Soil. 66, 21-28.
Hume DJ Criswell JG and Stevenson KR (1976) Can. J. Plant Sci. 56, 811-815.
Johnson DA Rumbaugh MD and Asay KH (1981) Plant Soil. 58,279-303.
Knowles R (1980) In Bergersen FJ, eds, Methods for Evaluating Biological Nitrogen Fixation, pp. 557-582, Wiley, Chichester.
Lange RT (1961) J. Gen. Microbiol. 61, 351-359.
LaRue TA and Patterson TG (1981) Advances in Agronomy. 34, 15-38.
Lawrie AC (1982) Austr. J. Bot. 30, 447-460.
Mosse B (1977) In Vincent JM Whitney AS and Bose J, eds, Exploiting the Legume-*Rhizobium* Symbiosis in Tropical Agriculture. pp. 275-292, University of Hawaii.
Muthana KD and Arora GB (1980) Annals Arid Zone. 19, 110-118.
National Academy of Sciences (1977) *Leucaena* Promising Forage and Tree Crop for the Tropics. NAS, Washington.
National Academy of Sciences (1979) Tropical Legumes: Resources for the future. NAS, Washington.
Norris DO (1956) Empire J. Expt. Agric. 24, 246-270.
Orchard ER and Darby GD (1956) In CR 6ème Congrès International Science Sol, Paris.D, 305-310.
Pate JS (1976) In Nutman PS, ed, Symbiotic Nitrogen Fixation in Plants, pp. 335-360, Cambridge Univ. Press.
Roskoski JP Montano J Van Kessel C and Castilleja G (1982) In Graham PH and Harris SC, eds, Biological Nitrogen Fixation Technology for Tropical Agriculture, pp. 447-454, CIAT, Cali.
Silvester WB (1977) In Hardy RWF and Gibson AH, eds, A Treatise on Dinitrogen Fixation,Section 4, pp. 141-190, Wiley, New York.
Sprent JI (1973) New Phytol. 72, 1005-1022.
Torrey JG (1982) In Graham PH and Harris SC, eds, Biological Nitrogen Fixation Technology for Tropical Agriculture, pp. 427-439 CIAT, Cali.
Trinick MJ (1980) J. Appl. Bacteriol. 49, 39-54.
Trinick MJ (1981) In Gibson AH and Newton WE, eds, Current Perspectives in Nitrogen Fixation, p. 480, Austr. Ac.Sc.,Canberra.
Vincent JM (1980) In Newton WE and Orme-Johnson WH, eds, Nitrogen Fixation, Vol. 2, pp. 103-129, University Park Press, Baltimore.
Vose PB Ruschel AP Victoria RL Saito SMT and Matsui E (1982) In Graham PH and Harris SC, eds, Biological Nitrogen Fixation Technology for Tropical Agriculture, pp. 575-592, CIAT, Cali.

Biological dinitrogen fixation in temperate zone forestry: current use and future potential.

John C. Gordon
School of Forestry and Environmental Studies
205 Prospect Street
Yale University
New Haven, Connecticut 06511

1. INTRODUCTION

Biological dinitrogen fixation (BNF) occurs in most managed temperate forests, and thus is a part of the application of most silvicultural systems. Often, however, the presence of BNF is not explicitly recognized or used by foresters, even though several recent symposia and books indicate that their interest in BNF and its use is increasing, (e.g. Gordon et al. 1980, Dawson, in press, Gordon and Wheeler, 1983.) In the long view, in addition to recognizing the presence of BNF and being sensitive to it when making silvicultural prescriptions, the large-scale use of introduced nitrogen-fixers is a primary target for both basic and applied forestry research. The three primary reasons for this are economic. First, natural gas, the basis for industrially-fixed nitrogen, is now artificially low in cost because of regulation, and Haber-process ammonia produced with underpriced natural gas is even lower because of current world overcapacity. Neither the low gas prices nor the overcapacity can be expected to persist a long time relative to forest production cycles (Beuter 1980). Second, the use of industrially-fixed nitrogen can be expected, under conditions of future scarcity, to be allocated tofood, rather than forest production. Current harvests are increasingly based on the added wood production available because of nitrogen fertilization, as a result of the observation in many places that available nitrogen limits forest growth on most sites. If, however, industrially-fixed nitrogen becomes too scarce or costly, harvests will be reduced immediately on regulated, "sustained yield" forests unless an alternative is available. The obvious alternative is BNF, which even in its current crude state of development is marginally cost competitive with industrially-fixed nitrogen in some temperate forest production systems (Helgerson, in press; Turvey et al 1983, Atkinson et al 1979). Third, and perhaps most important, the reclamation of abused land and perhaps the maintenance of the long-term productive capacity of some sites is currently done most economically using BNF (Tarrant and Miller, 1963, Fessenden, 1980, Beuter, 1980). It is the purpose of this paper to briefly summarize current use of, and research on BNF in temperate-zone forestry, and to speculate on future directions for research and application. To do this, I will rely heavily on the recent syntheses listed earlier.

2. CURRENT USE

2.1. Limitations. Use of BNF is limited everywhere in the temperate zone, except in land reclamation, because of its perceived lower financial rate of return in comparison with the application of chemical fertilizer (Rottink et al, 1980). Also, application in passive systems (Gordon, 1983) is often limited by the perception of naturally-occuring nitrogen fixers as competitors with crop species for light, water and nutrients and possible sources of allelopathic damage to crop species (DeBell, 1980). Large-scale application of active systems, where nitrogen-fixing species are planted as a rotation or murse crop, or form the main crop or a portion of it, are limited by the cost comparison with nitrogen fertilizer cited above. It may be, however, that the primary limit on the active use of BNF is technological; that is, currently, BNF technology is more complex than chemical fertilizer technology and therefore more difficult to apply. In addition, there is much uncertainty about how much nitrogen is actually fixed in most systems, making managers act conservatively and economic calculations questionable (Turvey and Smethurst, 1981). Active systems seem to be most widely-used in tropical or subtropical locations with limited access to forest fuels (Domingo, 1983). Even in the temperate zone, however, several successful examples of active BNF application currently exist.

2.2. Passive System Examples

2.2.1. Snowbrush and Douglas-fir. In the Pacific Northwest of the United States, snowbrush (Ceanothus Velutirus Dougl.), a potentially nodulated actinorhizal plant, often arises in dense stands after old-growth Douglas-fir (Pseudotsuga menziesii (Mirb) Franco) has been clearcut and burned. Foresters disagree on the net effect of the snowbrush on Douglas-fir, the crop species, but there is evidence that well nodulated snowbrush may, on some sites, enhance Douglas-fir growth through nitrogen inputs and the provision of shade (Youngberg et al, 1980), whereas on other sites, snowbrush appears to nodulate poorly, if at all, and to be a particularly efficient competitor.

2.2.2. Alder and Douglas-fir. Red and sitka alder (Alrus rubra Bong and Alrus sinuata (Regel) (Rydb.) often rapidly occupy exposed mineral soil in Western Canada and the Northwestern United States. Thus, when Douglas-fir and other conifer species are planted or regenerated naturally, mixtures with alder often occur. Binkley (1982), in a series of comparisons of such mixtures with pure stands found that sitka alder greatly increased Douglas-fir stem growth, and that red alder enhanced ecosystem productivity on relatively poor sites, but retarded Douglas-fir growth on relatively good sites. The site interaction indicates that sensitive silvicultural prescriptions with the primary objective of wood production will retain and adjust some mixtures on some sites and eliminate them on others. Binkley (1982) concluded that sitka alder merited increased silvicultural attention and derived a preliminary table relating alder stem diameter and numbers per unit area to N fixation rate (Table 1).

TABLE 1. Numbers per hectare and sizes (breast height diameter) of sitka alder required to achieve selected nitrogen fixation rates (after Binkley, 1982).

Diameter, CM	N fixation rate, kg $ha^{-1}a^{-1}$			
	20	40	60	80
	stems ha^{-1}, thousands			
2	20	40	60	80
4	4	8	12	16
6	1.5	3	4.5	6
8	.8	1.5	2.3	3.1

Obviously, large numbers of alder are needed to achieve high fixation rates. The implication is that if sitka alder is to be used in active systems, either large numbers must be planted or fixation per tree enhanced. Similarly, it is apparent that red alder is a vigorous competitor, and that it should be used with care or eliminated on better sites if Douglas-fir wood production is the primary objective.

2.3. Active System Examples

2.3.1. Lupins and Pinus radiata. The Monterey pine (Pinus radiata D. Don) plantations of New Zealand form perhaps the premier example of plantation wood production in the world. On dune (sandy) soils, the introduction of Lupinus arboreus allowed the development of highly productive planted forests on these inherently unproductive soils (Gadgil, 1971). Although initially prominent in the pine stands, lupins are eliminated after 5 to 6 years when the pine canopy closes, only to re-emerge from soil-stored seed after the pines are thinned. This regrowth of lupin restores N-fixation rate to more than 10% of the original, and thus enhances the growth response of the pines to thinning (Silvester et al, 1980). In comparisons of lupin and nitrogen fertilizer effects on pine growth, lupins were demonstrated to produce a growth response equivalent to a fertilizer (urea) application of 112 kg N per hectare and year (Silvester, 1980).

2.3.2. Autumn olive and black walnut. Black walnut (Juglans nigra) is a high-value crop planted at many locations in the temperate zone of eastern North America. Although the aggregate area planted is not large relative to conifers, its high unit value and significance to non-industrial landowners make it an important plantation species. Funk et al (1979) have convincingly demonstrated that autumn olive (Elaeagnus umbellata) positively affects black walnut wood production by providing higher soil nitrogen levels and enhancing the rate at which walnut self-prunes and thus improves wood quality. Early attempts to use BNF concentrated on using alders in mixture with walnut, but this combination failed because the alders outgrew and overtopped the walnut. This illustrates the simple but important principle that growth rates of the nitrogen fixers and the crop species must be accurately known and carefully blended to produce successful active systems.

3. FUTURE POTENTIAL

3.1. Opportunities and limits. The major opportunity related to BNF and temperate forestry is to substitute current photosynthate for fossil fuel as an energy source for the reduction of dinitrogen. Contrary to popular opinion, this may be more feasible in forestry than in annual agriculture, for two simple reasons. First, in contrast to temperate, annual agriculture, forestry, including single-crop species plantation culture (monoculture), invariably includes a high proportion of non-crop species and leaf area, particularly early in the rotation. In other words, even in "intensive" forestry, clean tillage is the exception rather than the rule. The leaf area not devoted to the crop, for reasons of product quality or lack of a suitable control technology, could be allocated to a nitrogen-fixing species. Second, the threat of loss of aerially applied, industrially-fixed, nitrogen is a real prospect in forestry. Food will draw resources, whether through market mechanisms or government allocation, that timber will not. Also, the broadcast aerial application of "chemicals" is coming under steadily-increasing public scrutiny in many parts of the world, and the legal and social turmoil now observable will doubtless increase as an increasingly urban population questions production processes applied to lands they regard as "nature" or "life support systems". Thus, it seems to me to be prudent to develop BNF technology fully even before it is everywhere competitive with current fertilization practise. If we wait until it is needed, we will again be without good alternatives while the research is carried out, and we will again (as with herbicides) be in the wistful and losing mode of "if only we had done what we knew was sensible earlier".

The major limitations on the potential of BNF in temperate forestry are more economic and social than biological. The current cost of nitrogen fertilizer is artificially low, as indicated above. Further, current views of enhanced or "high" technology have mostly to do with the creation of gadgets rather than with production as viewed by some economists (Georgescu-Roegen, 1971) who describe primary production in much the same way as ecologists. Thus, we neglect enhancement of production of low-entropy raw materials that have high potential for increased value through manufacturing and marketing. Even the biological revolution has so far been viewed as relevant to industrial microbiology and annual agriculture, and not to forestry. Nevertheless, 90% of the world's population depends on wood for fuel, or would like to, and most people depend on paper for daily needs and information transfer, and many people live in dwellings that have wood or wood products as major components. Until the desire for knowledge about wood production (not to mention the other vital uses of forested lands) becomes commensurate with our desire to use wood products, we will not progress much.

3.2. Priority research tasks.

Knowledge is the primary technical limitation on use of BNF in temperate forestry. Current research is strong and growing, and all the questions below are being addressed to some degree. All, however, need more attention.

3.2.1. More accurate estimates of fixation rates. Estimates of N fixation and accretion rates in active systems are still sparse and uncertain. As Turvey et al (1983) state: "There is an aura of uncertainty surrounding the data used for economic analysis of nitrogen management in forestry." A large part of this uncertainty stems from the difficulty of estimating the several parameters of such rates over several field seasons, not to mention crop rotation periods of 20 to 100 years. Better methods of estimating nodule biomass and less destructive and faster assays of nitrogenase activity and nitrogen accretion in soils and biomass are needed.

3.2.2. Better predictions of effects on crop trees. Until we can predict quantitative responses to accurately estimated fixation rates, managers will be reluctant to apply BNF. The diversity of sites, nitrogen fixers and crop trees virtually dictate that knowledge of detailed mechanisms will be the only economical aid to prediction. Thus, we must isolate and measure the major variables influencing the transfer of N to crop plants from the time it is fixed until the time it is incorporated into crop plant tissue, with carefully selected ranges of N fixers, crop plants and site conditions.

3.2.3. Development of domesticated forest legumes. Legumes aren't as widely used in temperate forestry as they might be because legumes developed over centuries for pasture use have often been transferred to the forest with poor initial results. Legumes more tolerant of acid soils and more competitive with other understory species would be a good start. Can we domesticate legumes native to forests now? Can we produce superior endophytes for them?

3.2.4. Systematic assessment of non-legumes. Actinorhizal plants are characteristic of temperate forests (Dawson, in press) and still are virtually unstudied relative to legumes. Careful collection into several temperate research sites of full sets of actionorhizal species and a large, careful sample of Frankia isolates will allow their assessment for forestry use both as N fixers and as crop plants. Some host plant species probably remain to be discovered. Frankia variation is poorly understood, and its culture and manipulation is in its infancy.

3.2.5. Transfer of N fixing ability. The actinorhizal hosts present a great array of genetic variation already, with at least 7 families included. Obviously, Frankia is more taxonomically promiscuous than Rhizobium.. If successful transfer of nodulation ability to currently non-nodulating taxa is to occur, the current diversity recognized by actinomycetes argues for work on them rather than Rhizobium. Because most current Frankia-rodulated taxa are woody plants, the relevance to forestry is obvious.

3.2.6. Understand free-living N fixation. The microbiology of forest soils is fascinating but little studied. = The long term of temperate forest crops offers the opportunity for low rates to be significant by cumulation over long times. Can we enhance free-living fixation by inoculation or environmental manipulation? Ifso, perhaps we could cheaply increase available nitrogen in non-disruptive ways over large areas. Certainly, our ability to study and manipulate microorganisms has vastly increased recently. The detritus-based forest soil system is quite different than agricultural ones. Can we use new techniques to explore and exploit these differences?

Many basic and applied research tasks remain unmentioned here. Forests currently cover a major fraction of the temperate zone and satisfy major human needs. Let's give them the attention they deserve, and let those of us interested in biological nitrogen fixation lead the way.

4. REFERENCES

Atkinson WA, Bormann BT and DeBell DS (1979) Botanical Gazette Supplement 140: S134-151.
Beuter JH (1980) in Gordon et al eds. Symbiotic Nitrogen Fixation in the Management of Temperate Forests, pp. 4-13, Forest Research Laboratory, Corvallis OR.
Binkley D (1982) Case Studies of Red Alder and Sitka Alder in Douglas-fir Plantations. Doctoral Thesis, Oregon State University, Corvallis OR.
Dawson JO (In Press) Canadian Journal of Microbiology Special Issue on 2nd Int. Symp. on N_2 Fixation with non-legumes.
DeBell DS (1980) In Gordon et al eds. Symbiotic Nitrogen Fixation in the Management of Temperate Forests, pp. 451-466, Forest Research Laboratory, Corvallis OR.
Domingo IL (1983) In Gordon JC and Wheeler CT. Biological Nitrogen Fixation in Forest Ecosystems, pp. 295-315, Nijhoff/Junk, The Hague.
Fessendon RJ (1980) In Gordon et al eds. Symbiotic Nitrogen Fixation in the Management of Temperate Forests, pp. 403-419. Forest Research Laboratory, Corvallis, OR.
Funk DR, Schlesinger RC and Ponder F (1979) Botanical Gazette Supplement 140: S110-114.
Cadgil RL (1971) Plant and Soil 134: 357-367.
Georgescu-Roegen N (1971) The Entropy Law and the Economic Process. Harvard Univ Press, Cambridge.
Gordon JC (1983) In Gordon JC and Wheeler CT. Biological Nitrogen Fixation in Forest Ecosystems, pp. 2-5, Nijhoff/Junk, The Hague.
Gordon JC and Wheeler CT (1983) Biological Nitrogen Fixation in Forest Ecosystems. Nijhoff/Junk, The Hague.
Gordon JC, Wheeler CT and Perry DA (1981) Symbiotic Nitrogen Fixation in the Management of Temperate Forests. Forest Research Laboratory,

Corvallis, Or.
Helgerson OT (in press) In Akkermans A ed, Proceedings, Intl. Workshop on Frankia Symbioses, Plant and Soil.
Rottink BA, Strand RF and Bentley WR (1980) In Gordon et al eds. Symbiotic Nitrogen Fixation in the Management of Temperate Forests, pp. 14-22, Forest Research Laboratory, Corvallis, OR.
Silvester WB, Carter DA and Sprent JL (1980) In Gordon et al eds. Symbiotic Nitrogen Fixation in the Management of Temperate Forests, pp. 253-265. Forest Research Laboratory, Corvallis, OR.
Tarrant PF and Miller RF (1963) Soil Sci. Soc. Amer. Proc. 27: 231-234.
Turvey ND and Smethurst PJ (1981) In Rummery RA and Hingston FJ eds. Managing Nitrogen Economies of Natural and Man-Made Forest Ecosystems. pp. 124-145, CSIRO Division of Land Resources Management. Perth.
Turvey ND and Smethurst PJ (1983) In Gordon JC and Wheeler CT eds. Nitrogen Fixation in Forest Ecosystems. pp. 233-259, Nijhoff/Junk, The Hague.
Youngberg CT, Wollnm AG and Scott W (1980) In Gordon JC et al eds. Symbiotic Nitrogen Fixation in The Management of Temperate Forests. pp. 224-233, Forest Research Laboratory, Corvallis, OR.

MIRCEN PROGRAMMES AND THE IMPACT ON THE APPLICATION OF BIOLOGICAL NITROGEN FIXATION

J.R. JARDIM FREIRE
IPAGRO-UFRGS/MIRCEN, C.P. 776, 9000 - PORTO ALEGRE, RS - BRASIL

1. INTRODUCTION

In 1974 the UNEP/UNESCO/ICRO Panel of Microbiology coined the concept of the Microbiological Resources Centers (MIRCENs) and a team of specialists outlined its goals : (1) building of a world network incorporating regional and interregional cooperating laboratories (2) efforts for the conservation of microorganisms with emphasis on *Rhizobium* gene pools in developing countries (3) development and promoting of technology for the strenghtening of rural economies (5) training of manpower and diffusion of microbiological knowledge.
With emphasis in the area of biological nitrogen fixation five MIRCENs have already been established. One at the University of Nairobi, Kenya in 1977, another one at Porto Alegre, Brasil, in 1978, at two integrated institutions : The Institute of Agronomic Research of the State Department of Agriculture and The Department of Soils of the University of Rio Grande do Sul. In 1981 two other organizations were recognized as MIRCENs : The NifTAL/University of Hawaii/USAID Project based in Maui, Hawaii, and the Cell Culture and Nitrogen Fixation Laboratory, USDA, Beltsville, USA. In 1982 a new Center was established at the Centre National de Recherches Agronomiques, Bambey, Senegal.
We will try to summarize in this paper the main achievements of these MIRCENs toward the proposed goals.

2. TRAINING

Adequate training has been one of the most important limiting factors in relation to a more intensive and wider application of the *Rhizobium* technology in the developing countries. In all areas from *Rhizobium* microbiology to legume breeding, legume nutrition and soil fertility, legume extension, etc., it is of a high importance to have trained specialists with a perfect awareness of the technology involved and the potentialities of the *Rhizobium*/Legumes symbiosis.
It is worth mentioning that probably in only three countries a high attention was given and there was a wide application of the *Rhizobium* technology in the years before the oil and energy crises. These countries were Australia and Uruguay because of the demand for pasture improvement programs and Brasil because of the expansion of the soybean crop. In industrialized countries the low prices of the nitrogen fertilizers resulted in a reduced interest on biological nitrogen fixation. Now the science is booming again. In the last aproximately five years the increased number of symposiums, workshops, and other types of scientific meetings reflects increased interest. However it is felt that this scientific interest has not been adequately followed by increasing efforts on the application aspects of the *Rhizobium*/Legumes technology. Good rhizobia strains are in general available at national or international laboratories but the use of these strains for the production of inoculants and the application of the informations gathered from the research activities is not significant in most of the developing countries. International organizations are trying in some way to fill the gap but institutional capability at national level

in many cases lack trained personnel in number and/or qualification at the research laboratories up to the extension agencies and the administration level.
Training has certainly been the most important activity of the MIRCENs toward the established objective of development of research capability and diffusion of the *Rhizobium* technology. Through short courses, intern practical training and graduate degree work a large number of rhizobiologists, agronomists, teachers at agriculture schools and legume investigators have been trained at the different Centers. At the MIRCENs of Porto Alegre, Nairobi, Hawaii and Beltsville a total of more than 300 people have already been trained (Table 1) but the number is higher if one takes into account the short courses held at other institutions around the world in the recent past years. There are no statistics for the training at the graduate level but from a rough estimate one can see that the number is low at least for the needs of the developing world.

TABLE 1. Nº of trainees on *Rhizobium* technology at MIRCENs Network.*

Rhizobium MIRCENs	Type of training			Nº of countries
	Short courses	Intern	Graduate	reached
Porto Alegre	94	17	11	15
Nairobi	29	**	4	11
Hawaii	94	36	18	**
Beltsville		5***		

* Up to 1983 ** No data available *** In 1980, 1981

Another aspect of the training either through short courses or long term is the convenience of exposing the trainee not just to the *Rhizobium* technology aspects but also to the problems related to the diffusion and application of the technology there is to induce an extension philosophy in the training which will help the trainee to be useful to his home country and not just a paper producing scientist.
Besides the training conducted at their own headquarters the MIRCENs have organized or co-sponsored, supplied teachers and/or fellowships to a large number of short courses held around the world. The Hawaii NifTAL MIRCEN is the Center that has done more in this aspect. NifTAL has cooperated in short courses organized at Kenya, Porto Alegre, Peru, Mexico, Venezuela, Costa Rica and Thailand. Porto Alegre MIRCEN has cooperate with courses in Argentina, Peru, Rio de Janeiro, Costa Rica and Indonésia.

3. CULTURE COLLECTIONS

Considering the four MIRCENs specialized on *Rhizobium* around 4,000 strains are kept in their collections. Over 5,000 strains are kept in collections in 8 institutions in Latin America. (Table 2) We have no up date information on culture collections in other parts of the world but it is important to mention the large *Rhizobium* collection of the WDC Brisbane MIRCEN. The 1978 IBP World Catalogue of *Rhizobium* Collections (F.A. Skinner, 1973) lists about 3,000 strains from 59 collections in 29 countries. Now the second edition is being prepared by the WDC with jointly support of FAO and UNESCO. The World Directory of Collections of Cultures of Microorganisms lists 61 collections holding *Rhizobium* cultures

(VF McGowan, VBD Skerman, 1982).

TABLE 2. Rhizobia Culture Collections *

Institutions	Country	Nº of strains
MIRCENs		
Porto Alegre	Brasil	650
Nairobi	Kenya	208
Hawaii	USA	2000
Beltsville	USA	938
		3796
Others Centers*	Argentina, Brasil, Chile, Colombia ** Peru, Uruguay	5410

* No data available for large collections of Australia
** CIAT, Colombia - 3000 strains, mainly of pasture tropical legumes

It was an important step to give recognition as MIRCENs to the NifTAL Center at Hawaii and the Cell Culture and Nitrogen Fixation Laboratory (USDA) at Beltsville, U.S.A. Their large Culture collections have in the past and now are going to play an important role by the rhizobia cultures and services provided.
Main function of a culture collection is as a gene bank there is the maintenance and preservation of pure authenticated microorganisms. An important function must also be the preservation of valuable strains. Many valuable strains have been lost in the past because of fire accidents, retirement or death of the curator, etc. A world wide catalogue of valuable strains and preservation at a world recognized laboratory center would be important for the preservation of the *Rhizobium* gene pools. The *Rhizobium* MIRCENs are now starting the preparation of a *Rhizobium* Catalogue which will try to cover their culture collections and of the colaborators laboratories.

4. RHIZOBIA STRAINS DISTRIBUTION

Distribution of *Rhizobium* cultures has a long past of liberal and altruistic work. After the establishment of the MIRCENs, dissemination of efficient strains to other laboratories and inoculant factories had a large increase. From 1978 to 1982 Porto Alegre MIRCEN distributed 943 sub-cultures of strains to 15 countries most of them in Latin America. The Beltsville USDA-MIRCEN serviced 57 requests for 508 cultures. The number of cultures distributed by the Nairobi and Hawaii MIRCENs is also very high and they are giving a high contribution for the dissemination of efficient rhizobia strains throughout the world. Most of the request is for efficient strains for inoculation trials, evaluation a against native isolates or inoculant production. Beltsville MIRCEN in its *Rhizobium* Culture Catalogue has a list of 74 recommended strains for 32 legume species. Porto Alegre MIRCEN put out a special bulletin with a list of 123 strains recommended for inoculationof 49 species. Some of these strains are recommended for inoculant production in Brasil where a federal law requires that commercial inoculants must use only the strains recommended by the national research laboratories. Inoculant factories periodically or upon special request receive sub-cultures of these strains. *Rhizobium* strains have never been patented nor are private property. Even from commercial inoculants strains can be easily isolated and there is no mark or label. The new developments in genetic engineering can however change this situation. The development of short life super-strains might be a possibility for the future which is interesting by a commercial stand point but not for the diffusion of the *Rhizobium* technology among poor farmers areas of the world.

5. INOCULANT PRODUCTION

The production of inoculants for legume trials, demonstrations and farmers at the IPAGRO Laboratory was started in 1950 and was kept at a medium scale (1500 Kg/year) up to 1956 when the first private company went into the business. Since then a production of 400-500 Kg per year is maintained. The Nairobi MIRCEN has also a fairly high production of inoculants (1400 packages in 1982) which attended the request of 753 farmers,according to its 1982 Report. At an early stage of development of the practice in a country or area the ample distribution of a high quality laboratory made inoculant has the important function of the diffusion of the inoculation practice and prevents the possible discredit among farmers, legume investigators and agronomists by the use of bad quality commercial inoculants. These laboratory inoculants may also function as standards for comparison to commercial inoculants either imported or made in the country. National and international organizations should now apply stronger efforts toward the establishment of laboratories for this small scale production of inoculants. It is worth mentioning that equipment for these laboratories can be simple and not too expensive. The Hawaii MIRCEN has recently developed a technique for the dilution of the broth which can multiply enormously the capacity for the production of the inoculants.

6. INOCULANT QUALITY CONTROL

Inoculant quality control is critical to avoid damages to farmers and governmental programs. It is also true that the investments on research on strain selection is useless and the work of the investigators is wasted if the inoculants don't perform well at the field. Since its establishment in 1950 the IPAGRO Laboratory has evaluated the inoculants available to the farmers first on an informal basis and then after 1977 under agreement with the Ministry of Agriculture became responsable for the official control of the inoculants in the market in Brasil. The informal control however played and is playing an important role toward the maintenance and improvement of the quality of the inoculants. Cooperation with the inoculant factories by supplying of strains and informs on the quality of samples has continued along the years and the extension agencies and farmers association are also kept informed on the quality of the inoculants and by this way pressing the factories.
In areas where the inoculation practice is being introduced the quality control is of and even higher importance. It is considered that the laboratories for production of high quality inoculants as recommended before should also be equipped with quality control facilities. There are however examples where the inexistence of an official law for quality control keep rhizobiologists away from running an informal control themselves on the premise that it is not their responsibility. It is

felt that every rhizobiologist should feel himself responsable to keep an eye on the quality of the inoculants in order to avoid discredity to product he is promoting and to his own work.

7. DISSEMINATION OF INFORMATION

Isolation of the researchers, poor library support and deficient dissemination of the research findings are among the most important constraints in developing countries which hampers the development of the research capability and application of the technology. The MIRCENs have contributed to minimize these problems by ample production of Newsletters, bulletins, culture catalogues, etc. The Panel of Microbiology has published already five numbers of the MIRCEN NEWS which summarizes reports from the MIRCENs and other useful informations. The BNF Bulletin from NifTAL MIRCEN with an ample distribution around the world is now taking the place of "RHIZOBIUM NEWSLETTER" unfortunately defunct. From UNESCO a constant flow has supplied the MICRENs with useful bibliography to the training and general information on the scientific staff. It is also important to mention the second edition of "World Director of Collections of Cultures of Microorganisms" from the WDC MIRCEN. It is felt however that the MIRCENs lack (except for NifTAL) especialized personnel and funds for this important job. A better flow of publications within each region would be certainly useful.

The Global Impact of Applied Microbiology Conferences (GIAM) have played an important role for dissemination of Microbiology around the world. These conferences have been organized by the UNEP/UNESCO/ICRO Panel of Microbiology, the first one in Sweden in 1963 and the following ones in Adis Ababa, Bonbay, São Paulo, Bangkok and Lagos aiming to take microbiology to the developing countries.

8. RESEARCH

All the *Rhizobium* MIRCENs have a strong program of applied research however we have no data concerning the newly established Center in Senegal. The results from this work have been promptly spread by the MIRCENs Newsletters and bulletins, etc. and have already been of significance to application of the *Rhizobium* technology in the different regions. Below we summarize the data available on the work under way at the centers :

Hawaii - *R. japonicum* evaluation for tolerance to nitrate and competitiveness in relation to native soil population. Technology of inoculant production by dilution of the broth. Technology of strain identification by immunofluorescence and agglutination using ovendried root nodules. Evaluation of N fixing trees. Evaluation of the *Azolla*-*Anabaena* system.

Nairobi - Photosynthesis vs dinitrogen fixation in legumes intercroped with maise. Methodology of cultures storage. Survey of non-symbiotic N fixers in Kenyan soils. Technology of inoculant production using different carbon sources and carriers. Evaluation of lines of beans and cowpea. Soil stresses tolerance of *Rhizobium* strains. Evaluation of soil factors and techniques of inoculation in beans. Adaptation of the tree legume black wattle to different environments.

Porto Alegre - Evaluation of *R. japonicum* for efficiency, competitiveness and soil colonization. Evaluation of *R. phaseoli* for efficiency, competitiveness and antibiotic tolerance. Efficiency and competitiveness of *Rhizobium spp.* for peas, lens chickpeas, alfalfa, peanuts and clover. Symbiotic characteristics of a tetraploid of *T. riograndensis* and a hybrid of *T. riograndensis/T. repens*. Biochemical and symbiotic characterization of variants of *R. japonicum* and *R. phaseoli*. Technology of alfalfa inoculation. Effect of inoculation and liming on forage tropical legumes. Effect of liming on survival of *R. trifolii* in basaltic soils. Effect of soil moisture, organic residues and N levels on *R. phaseoli*. USDA - Large scale symbiotic evaluation of *Rhizobium spp.* Freeze-drying methods for rhizobia preservation.

9. REGIONAL NETWORK DEVELOPMENT

A guideline of the MIRCENs Program recommended the provision of "infraestruture for the building of a world network which would incorporate regional and interregional cooperating laboratories geared to the manegement, distribution of the microbial gene pool". Considering an assessment made in 1978 on the situation of the research and application of the *Rhizobium*/Legumes technology in Latin America (J.R.J. Freire, 1982) a significant progress was achieved. Research work was started in three countries where there was no activity before and had a strong improvement in several others. The contribution given by the former MIRCEN trainess was certainly of significance for this change. However, it has to be also considered that the activity on rhizobiology has a long past in several countries in Latin America. Since 1964 eleven meetings of the RELAR were held at eight different countries. Since 1968 exists a Latin America Association of Rhizobiologists (ALAR). The RELAR meetings and ALAR have strongly contributed to the development and application of the technology in the region.
It is important to mention the close cooperation FAO/BNF Program and P.Alegre MIRCEN in relation to the FAO projects established in Argentina, Bolivia and Peru which consisted on training of personnel, technical assistence, supply of bibliography, and inoculants.
And also for the strenghtening of the regional network by a colaboration MIRCEN and ALAR led by Carlos Batthyany a project sponsored by UNESCO and FAO is now being implemented for dissemination of information.
The situation was not at an early stage at the starting of the MIRCEN Program in the region at least in relation to the research activities. The majority of the countries had one or more centers of rhizobia investigation but only two or three had really a high level of application of the rhizobia technology (J.R.J. Freire, 1982). The rationale for this varies according to the country. However as it was clearly stated by another author (M. Alexander, unpublished) " Nitrogen fixation in agricultural land is rarely limited by the lack of availability of highly active nitrogen-fixing microorganisms".
Outstanding progress was achieved by the Nairobi MIRCEN towards the fostering of the network within eastern Africa. Eight countries are now integrated by means of cooperative research, training, exchange of staff, literature and materials. Situation in Africa is different from that one in Latin America in the sense that almost all the countries are at an very early stage of development in rhizobiology. Number of researchers is very limited and constitute and important limiting factor along with other constrains common also to other regions related to the stage of development of the countries which restrict the rapid expansion of the

application of the *Rhizobium* technology.
NifTAL Project established in connection with the University of Hawaii under sponsoring by USAID is not restricted to a region but conceived to cover the tropical areas of the world. Its recognition in 1981 as a MIRCEN by the UNEP/UNESCO/ICRO Panel of Microbiology was of a very high significance because of the role already played and the one that can be played by the Center in the diffusion of the BNF technology in the world but also because of the objective of strengthening of inter-MIRCENs network which Hawaii-MIRCEN can successfully activate. Besides their activities in training, distribution of strains and bibliography and research work the NifTAL-MIRCEN group has developed and international network of inoculation trials (INLIT) which now lists over more than seventy colaborators around the world.
The strenghtening of the regional networks have also received a strong support from UNESCO by research grants to colaborators laboratories in Latin America and Africa. In addition, valuable support has been made available from UNEP, USAID and FAO. This last agency has supported with UNESCO research and training programs and provided equipment to the MIRCENs at Nairobi and Porto Alegre and more particularly to the co-labs of these MIRCENs network.

10. REFERENCES

Freire JRJ (1982) Plant Soil 67,227-239.
McGowan VF and Skerman VBD, eds (1982) World DIrectory of Collections of Cultures of Microorganisms, WDC, University of Queensland.
Skinner FA ed (1973) IBP World Catalogue of Rhizobium Collections, International Biological Programme.

POSTER DISCUSSION 1A MEASURING NITROGEN FIXATION

JOHN E. BERINGER, Soil Microbiology Department, Rothamsted Experiental Station, Harpenden, Herts AL5 2JQ, UK

The aim of this session was to discuss the methods commonly used to measure N_2 fixation in the field and to assess their value. This was done partly by relating the discussion to posters, but mainly through debating issues as they arose.

The discussion started with an asessment of the role of acetylene reduction. This simple and accurate method is known to be of limited value because it is a time point assay. We were reminded that acetylene inactivates ethylene oxidase activity (Poster 1A11) and thus that in the absence of accurate controls endogenous ethylene production can have a significant effect on C_2H_2 reduction values obtained with plants fixing low levels of nitrogen. The more recent observation that nodule respiration and nitrogenase activity can be almost halved by exposure to acetylene (Posters 1A 1-3) caused further consternation as this aspect of the assay has also been ignored by many who have used it. It alone can lead to errors of up to 50%. The need to calibrate the C_2H_2 assay was stressed. I saw no concensus of opinion about the C_2H_2 assay. Dr. T.A. LaRue felt that it was the most effective method available, while at the other extreme Dr. J.F. Witty felt that there were so many serious flaws in it that most studies using it for measuring N_2 fixation in the field were meaningless.

Methods utilizing ^{15}N are becoming more popular and the merits of methods based on natural abundance and ^{15}N fertilizer application were discussed. My simplicitic idea that this is the only sure measurement of the total N fixed during the life of a crop was rejected. There are still serious problems in determining which plants make appropriate non-fixing controls, and how to ensure that ^{15}N is equilibrated in the soil. It appears that inclusion of ^{15}N labelled organic matter well in advance of plannned experiments is extremely desirable. It also seems likely that to test a legume in a given environment it will be necessary to calibrate the procedure against a range of controls that will be relevant to the situation. It was felt that it was potentially a very good method for measuring the proportion of N_2 fixed and hence had an important role for screening germplams for plants and rhizobia that would be less demanding of soil N reserves. The allantoin assay method was recommended by Dr. D. Herridge for soybeans though it appeared to be little used by others present. Genotype selection based on differences in growth in soils low in N was favored by Dr. C. Sloger, though he and others pointed out that this gave little indication of the proportion of N derived from N_2 fixation.

I started the discussion thinking that we were not able to measure N_2 fixation in the field. I left it feeling that this was largely true though it was clear that those members of the audience who were aware of the limitation of their methods were obtaining useful data. It was also clear that large amounts of useless data have been, and continue to be, collected.

We need much more information about N_2 fixation rates and effort should continue to go into obtaining it. However, unless great care is taken to calibrate techniques and to interpret results with an intelligent understanding of their limitations our knowledge of N_2 fixation rates with soil-grown plants will remain limited. Nevertheless, people are encouraged to get out into the field and make measurements, continually refining their techniques as the various constants upon their interpretations become apparent.

THE ACETYLENE TO NITROGEN RATIO IN *LUPINUS ANGUSTIFOLIUS* L.

P. R. GIBSON and #A. M. ALSTON
South Australian Department of Agriculture, Box 1671 GPO, Adelaide 5001
South Australia.
Waite Agricultural Research Institute, University of Adelaide,
Glen Osmond 5064, South Australia.

Acetylene reduction (A), hydrogen evolution (H) and nitrogen fixation (N) were monitored throughout the life cycle of narrow-leafed lupin crops (*Lupinus angustifolius* L.) grown in consecutive years in the field. N was monitored by measuring the uptake of ^{15}N- enriched N_2.

The A/N and H/N ratios varied throughout plant ontogeny in a congruous manner. High values of both ratios were associated with low specific rates of N (sN) and *vice versa*. The relationships of the specific rates of A (sA) and H (sH) to sN were curvilinear, and were linearized by plotting sA and sH against ln sN. Figures 1 and 2 depict these relationships for cv. Marri in 1980 and for cv. Illyarrie in 1981 respectively.

Although diurnal variation occurred in sA, sH and sN, no diurnal variation was evident in either the A/N or H/N ratios or in the relative efficiency of nitrogen fixation (defined as 1-H/A).

Measurements of plant and nodule growth rates indicated that low A/N and H/N ratios occurred when the growth of nodules did not keep pace with the demand for nitrogen by the plant.

The results are consistent with the hypothesis that a change occurs in the substrate kinetics of nitrogenase during periods of high sN, pre-disposing the enzyme more to the reduction of N_2 than to the reduction of either H^+ or C_2H_2.

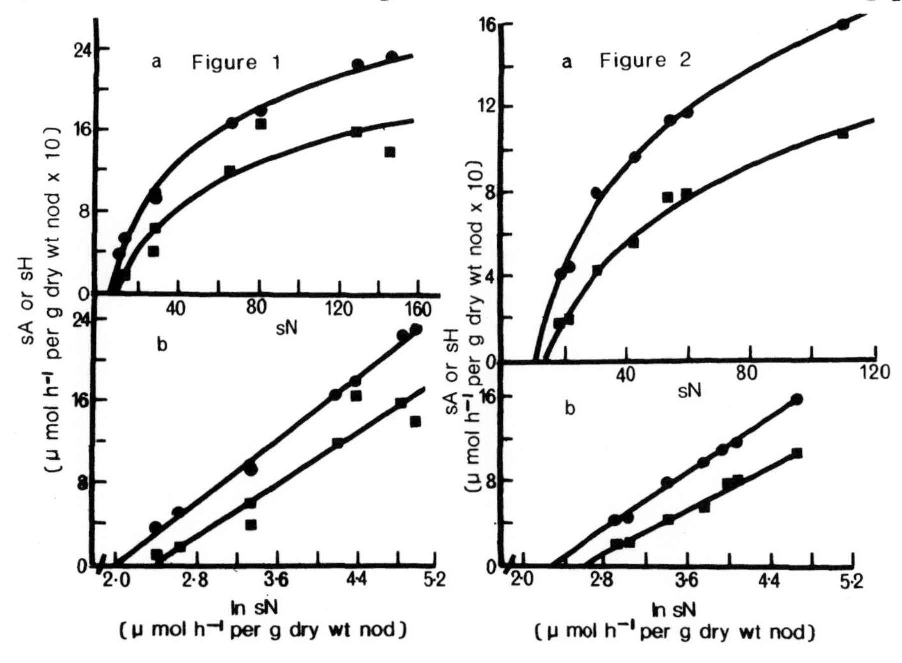

SYMBIOTIC NITROGEN FIXATION BY SOYBEAN VARIETIES AS AFFECTED BY NITROGEN FERTILIZER APPLICATION.

G. Hardarson, F. Zapata and S.K.A. Danso
IAEA Seibersdorf Laboratory, A-2444 Seibersdorf, Austria and Joint FAO/IAEA Division, Wagramerstrasse 5, P.O.Box 100, A-1400 Vienna, Austria.

^{15}N substrate labelling methodology (1,2,3) as well as acetylene reduction and total difference methods were used to assess symbiotic nitrogen fixation of eight soybean (Glycine max L.Merr)varieties grown at 20 and 100kg N/ha levels of nitrogen fertilizer under field conditions.

Large differences were found between soybean varieties in their abilities to support nitrogen fixation as measured by the ^{15}N substrate labelling methodology. In almost all cases, the application of 100kg N/ha resulted in lower N_2 fixation in soybean than at 20kg N/ha in the first year of the study. The mean value of %N derived from fixation for the eight varieties was 16% at the 20kg N/ha level but dropped to 5% at the 100kg N/ha level. Similarly, the amount of N_2 fixed was reduced from 26 to 8kg N/ha by raising the N fertilizer level from 20 to 100kg N/ha. However, N_2 fixation in one variety, Dunadja, was not significantly affected by the higher rate of N fertilizer application, being 18kg N/ha at both N fertilizer levels. These results were supported by measurements of acetylene reduction activity, nodule dry weight and total N difference between fixing and non-nodulating soybean. Further proof of differences in N_2 fixation within soybean varieties was obtained during a second year experiment. The level of N_2 fixation in this experiment was higher than in the first year experiments but the ability of Dunadja to fix N_2 was again less affected by N fertilizer application compared to the other varieties. Dunadja produced larger yield at 100kg N/ha than at 20kg N/ha. This type of variety could be particularly useful in situations where soil N levels are high or where there is need to apply high amounts of N fertilizers. N fertilization could in this case ensure high crop yields without significantly affecting N_2 fixation.

The present study reveals the great variability in soybean varieties to fix N_2 at different inorganic N levels, and the potential that exists in breeding for nitrogen fixation supportive traits. The ^{15}N methodology offers a unique tool to evaluate germplams directly in the field for their N_2 fixation abilities at different N fertilizer levels.

1: Fried M. and H. Broeshart, 1975. An independent measurement of the amount of nitrogen fixed by legume crops. Plant Soil 43:707-711.
2: Freid M. and V. Middelboe, 1977. Measurement of amount of nitrogen fixed by a legume crop. Plant Soil 47:713-715.
3: Hardarson G. S.K.A. Danso and F. Zapata. Biological nitrogen fixation in field crops. Handbook of Plant Science in Agriculture, CRC Press (in press).

FOLIAR NITROGEN CHARACTERISTICS AS INDICATORS OF NITROGEN FIXATION ABILITY OF TROPICAL LEGUME TREES

Peter Högberg, Department of Forest Site Research, Swedish University of Agricultural Sciences, S-901 83 Umeå, Sweden.

In the sub-humid and drier tropics, root nodules of nitrogen-fixing trees are often inaccessible due to deep root systems and compact soils. It is thus not possible to study their nitrogen fixation abilities by, for example, the acetylene reduction method.

The natural abundance of ^{15}N in leaves may give indications on the nitrogen fixation ability of trees. The sampling procedures involved are simple, but the nitrogen isotope analysis is costly. It is of great interest if other chemical characteristics, which are less expensive to analyse, could be used as alternatives.

A pilot study was therefore carried out to test if the foliar ^{15}N abundance, N content and N:P ratio of presumed nitrogen-fixing tree species differed from that of non-nitrogen-fixing species. Three potentially nodulated legume species were compared with three non-nodulated legume species and one non-legume species. All species were deciduous and growing intermingled at a site in Tanzania. The sampling was conducted during the later part of the rainy season. During this period changes in leaf N and P are relatively slow according to previous studies.

The ^{15}N abundance was significantly lower in potentially nodulated than in non-nodulated species. The N content was on average higher in potentially nodulated species as compared to non-nodulated species (difference not statistically significant). Potentially nodulated species also had a significantly higher N:P ratio than non-nodulated species.

Thus, all three nitrogen characteristics used indicated differences between presumed nitrogen-fixing species and non-nitrogen-fixing species.

Future work will involve extended studies of larger numbers of species, confirmation of presence of nitrogenase activity, and comparisons between non-nodulated endo- and ectomycorrhizal species.

NITROGEN FIXATION, ACETYLENE REDUCTION AND HYDROGEN EVOLUTION BY LEGUMES GROWN IN POTS

P. HOPMANS*, P. M. CHALK†, L. A. DOUGLAS†
* FORESTS COMMISSION VICTORIA, GPO BOX 4018, MELBOURNE 3001, AUSTRALIA.
† UNIVERSITY OF MELBOURNE, PARKVILLE 3052, AUSTRALIA.

Calibration of the C_2H_2 reduction assay is necessary in order to obtain accurate measurements of N_2 fixation. The ratio of C_2H_2 reduced : N_2 fixed may be significantly higher than the theoretical ratio of 3:1 due to H_2 evolution (Hudd *et al.* 1980; Saito *et al.* 1980). The aim of this study was to determine the relationships between N_2 fixation, C_2H_2 reduction and H_2 evolution of *Trifolium subterraneum*, *Medicago truncatula* and *Acacia dealbata*.

Materials and Methods

N_2 Fixation: Plants were grown in a N_2 depleted atmosphere ($N_2 < 0.1$ kPa) of $Ar:O_2:CO_2$ (77:20:3) for 100 h in a controlled environment chamber following addition of 8 mg N/pot of KNO_3 at 24.75 atom % ^{15}N excess (non-fixing plants). Another set of plants treated identically was grown in a normal atmosphere (fixing plants).

C_2H_2 reduction: Assays of undisturbed fertilised and unfertilised plants grown in a normal atmosphere were carried out every day as described by Hopmans *et al.* (1982).

H_2 evolution: Plants were carefully removed from the pots, soil was washed from the roots and H_2 evolution was measured.

Results and Discussion

The amount of symbiotically fixed N_2 was calculated from the difference in uptake of fertiliser-N between the fixing and non-fixing plants. The addition of nitrate decreased C_2H_2 reduction of the legumes by 62-69%. The ratios of C_2H_2 reduced : N_2 fixed for *T. subterraneum*, *M. truncatula* and *A. dealbata* were 5.2, 5.8 and 2.8, respectively. In a separate experiment N_2 fixation was measured by exposing *T. subterraneum* to an atmosphere enriched in $^{15}N_2$ and a C_2H_2 : N_2 ratio of 5.1 ± 0.4 was obtained (Hopmans *et al.* 1983). H_2 evolution by the pasture legumes utilised a large proportion (50-60%) of the electron flux available for N_2 fixation compared with *A. dealbata* (2%). The molar ratios corrected for H_2 evolution (C_2H_2 - H_2) : N_2 for *T. subterraneum*, *M. truncatula* and *A. dealbata* were 3.0, 3.1 and 2.8, respectively.

The indirect isotopic technique used in this study is a yield dependent N-difference method for estimating N_2 fixation. This technique uses identical plants as a non-fixing control and can be applied at different stages of growth.

Hopmans P, Chalk PM and Douglas LA (1983) Plant Soil (in press)
Hopmans P, Douglas LA and Chalk PM (1982) Soil Biol. Biochem. 14, 495-500.
Hudd CA, Lloyd-Jones CP and Hill-Cottingham DG (1980) Physiol. Plant. 48, 111-115.
Saito SMT, Matsui E and Salati E (1980) Physiol. Plant. 49, 37-42.

NITROGENASE (C_2H_2) ACTIVITY OF FINNISH FORAGE LEGUMES *IN SITU*

KRISTINA LINDSTRÖM
DEPARTMENT OF MICROBIOLOGY, UNIVERSITY OF HELSINKI
SF-00710 HELSINKI 71, FINLAND

1. INTRODUCTION

In order to give consistant yields despite weather fluctuations, nitrogen fixing agricultural plants should perform well under a wide range of environmental conditions. We have designed a practical system for making acetylene reduction assays in the field, and we have used the *in situ* assay to determine seasonal patterns of nitrogenase activity for three perennial temperate forage legumes.

2. MATERIALS AND METHODS

For acetylene reduction assays *in situ*, fibreglass cylinders (d=15 cm, h=20 cm) were placed in field plots of *Trifolium pratense*, *Galega orientalis* and *Medicago sativa* grown under standard management in 1981-82. Once a week plastic bags were gas-tightly attached to the top ends of the cylinders to enclose the plants, which were then incubated with 10 % acetylene for 4 h, with sampling at 2 and 4 h. Propylene was used as an internal standard. The cylinders could be left in the ground throughout the growing season causing minimal disturbance of the legumes. At the time of the assays irradiance, soil and air temperature, soil humidity and the dry weight of shoot samples representing the assayed plants were also determined.

3. RESULTS AND CONCLUSIONS

Nitrogenase activity was still detected in the field plots in November 1982, when soil temperature was 1.5 °C and air temperature 0.5 °C. The acetylene reduction data were analyzed for correlation with the measured environmental parameters and with plant growth rates, which were determined from growth curves constructed from the dry weight data. The following regression equations were statistically significant:

1981

T. pratense $Y = 7.33 + 5.46\,X_1$ $(r = 0.830^{**})$
G. orientalis $Y = -7.39 + 14.35\,X_1$ $(r = 0.920^{***})$
M. sativa $Y = -9.48 + 22.85\,X_1$ $(r = 0.848^{**})$

1981+1982

T. pratense $Y = 5.72 + 8.84X_1$ $(r = 0.769^{***})$
M. sativa $Y = 0.572 + 3.34_{X1} + 0.682X_2$ $(F_{reg} = 9.1^{**})$

where Y = acetylene reduction activity (umol h-1/cylinder)
X_1 = plant shoot growth rate (g dw/week cylinder)
X_2 = air temperature (°C)

There was a good correlation between nitrogenase activity and plant growth rate. Residual fluctuations in activity were only correlated with environmental factors in one case. The nitrogenase activity of *M. sativa* was dependent on air temperature in addition to growth rate, indicating that photosynthesis or metabolite transport responded to temperature changes, influencing the nitrogenase activity indirectly. Thus, the nitrogen fixing systems of these forage legumes seem to be an integrated part of the plants, being fairly insensitive to short-term environmental changes.

USE OF ^{15}N NATURAL ABUNDANCE FOR QUANTIFICATION OF BNF IN SOYABEAN

KAUSER A. MALIK and Y. ZAFAR
Soil Biology Division, Nuclear Institute for Agriculture and Biology (NIAB)
P.O.Box 128, Faisalabad, Pakistan

Soyabean is an introduced legume in Pakistan. Cross inoculant group (*Rhizobium japonicum*) of this crop is absent in our soils and it is possible to use non-inoculated soyabean as non fixing control plant. *Glycine max* (var. Davis) was grown in plots in summer 1982. Three treatments N_oI_o, N_+I_o and N_oI_+) were used at two fertility levels (F = Farmer level and M = Maximal level) with four replicates.
Comparison of inoculated (N_oI_+) vs. control (N_oI_o) indicates a yield increase of 137 kg/ha ±¿4.99. Isotopic analysis of N-15 in the aerial parts of 77 days crop (pod filling stage) clearly revealed a significantly lower N-15 content in inoculated treatments than in non-inoculated ones. Estimates based on ΔN-15 values indicated that 10-16% of nitrogen is derived from BNF in soyabean.

NITROGEN FIXATION BY WHITE CLOVER (TRIFOLIUM REPENS) IN IMPROVED HILL AND UPLAND PASTURES IN SCOTLAND

CAROL MARRIOTT[+], A. HAYSTEAD[++], P. NEWBOULD[+] and ANNE RANGELEY[+]
\+ HILL FARMING RESEARCH ORGANISATION, BUSH ESTATE, PENICUIK, MIDLOTHIAN EH26 OPY, SCOTLAND, U.K.
++ RUAKURA SOIL AND PLANT RESEARCH STATION, HAMILTON, NEW ZEALAND.

In hill and upland areas of the United Kingdom pasture production is limited by poor climate, low soil fertility, poor quality indigenous species and the practice of traditional management systems. Introduction of more productive grasses and white clover plays a key role in pasture improvement. White clover not only provides high quality herbage but also fixes atmospheric nitrogen, thus contributing to the nitrogen economy of the pasture system. In order to effectively manage improved pastures the response of nitrogen fixation to climatic and management variables must be more fully understood.

Seasonal nitrogen fixation and the response to defoliation in grazed and cut swards was measured at 5 hill and upland sites in Scotland, using the acetylene reduction technique calibrated by cross-reference to ^{15}N isotope dilution experiments.

Seasonal curves of acetylene reduction activity (ARA) were obtained on 2 grazed and 2 cut swards. ARA was negligible over the winter months but rose rapidly in the spring as soil temperature increased, reaching a peak in early summer. There was a close relationship in spring between clover dry matter production and ARA. Summer drought conditions caused a decline in ARA, but a second smaller peak occurred in the autumn on one site (the brown earth hill site). Maximum levels of ARA were 18.7 and 9.7 mmol $C_2H_4 m^{-2} d^{-1}$ respectively on the grazed and cut swards. High levels of soil inorganic nitrogen may have limited ARA on the upland cut pasture.

There was an immediate fall in ARA in response to defoliation, with recovery by about 12 days after harvest. Under grazing conditions the highest level of ARA occurred when the pasture was less frequently and more leniently defoliated; continuous close grazing severely depressed ARA.

Conversion of the acetylene reduction data indicates that between 50 and 140 kg N ha^{-1} is fixed annually, in a seasonal pattern which is affected by climatic variables - especially temperature and rainfall. Although little can be done to overcome climatic limitations, nitrogen fixation can be optimized by manipulating grazing management systems. Work continues to investigate the reasons for interseasonal variations in nitrogen fixing activity.

ERRORS IN THE ACETYLENE REDUCTION ASSAY: DETERMINED USING A FLOW-THROUGH SYSTEM

F.R. MINCHIN*, J.F. WITTY[+] and J.E. SHEEHY*
* THE GRASSLAND RESEARCH INSTITUTE, HURLEY, MAIDENHEAD, BERKS, SL6 5LR, UK.
[+] ROTHAMSTED EXPERIMENTAL STATION, HARPENDEN, HERTS., AL5 2JQ, UK.

Measurements on both attached nodulated roots and detached nodules have revealed that nitrogenase activity in many legume symbioses declines rapidly in the presence of acetylene, with a concurrent reduction in respiration. These observations were made using an open, flow-through system which measures the rate of C_2H_4 production rather than its cumulated concentration (Minchin et al., 1983). The reduction in nitrogenase activity begins within a few minutes of exposure to acetylene and continues for 30-60 min before a new steady-state is attained. This lower value commonly represents 40-60% of the maximum (pre-decline) rate of C_2H_4 production. This maximum rate represents the pre-assay rate of nitrogenase activity as determined by $^{15}N_2$ fixation. When C_2H_2 is removed respiratory activity returns to the pre-exposure level, demonstrating that permanent damage to the nodules is not involved in the decline. A decline and recovery also occurs when N_2 is replaced with argon or helium. This suggests that the decrease in nodule activity is linked to the cessation of ammonia production.

For those symbioses where an C_2H_2-induced decline occurs calculations of nitrogenase activity based on cumulative C_2H_4 production in closed assay vessels will underestimate actual rates. This will produce low apparent values for both N_2 fixation and relative efficiency of H_2 production. Furthermore, the extent of the C_2H_2-induced decline can be influenced by host, strain or environment. Thus, the assay may not be valid even for comparative studies. Indeed, changes in the extent of decline may account for many of the variations in the C_2H_2/N_2 conversion ratios reported in the literature (e.g. Hardy et al., 1973, Sprent, Bradford, 1977).

REFERENCES

Hardy RWF Burns RC and Holsten RD (1973) Soil Biol. Biochem. 5, 47-81.
Minchin FR Witty JF Sheehy JE and Müller M (1983) J. exp. Bot. 34, 641-649.
Sprent JI and Bradford AM (1977) J. Agric. Sci. 88, 303-310.

O_2 DIFFUSION RESISTANCE CONTROL IN LEGUME NODULES: LINKAGE WITH THE C_2H_2-INDUCED DECLINE

M.I. MINGUEZ*, J.F. WITTY[+], F.R. MINCHIN[++] and J.E. SHEEHY[++]
* CATEDRA DE FITOTECNIA I, E.T.S.I. AGRONOMOS, CORDOBA, SPAIN
+ ROTHAMSTED EXPERIMENTAL STATION, HARPENDEN, HERTS., AL5 2JQ, UK.
++ THE GRASSLAND RESEARCH INSTITUTE, HURLEY, MAIDENHEAD, BERKS, SL6 5LR, UK.

The O_2 concentration within the bacteroid zone of a legume nodule is the resultant of inward O_2 diffusion and respiratory O_2 consumption. The rate of O_2 diffusion depends on the concentration gradient across the nodule i.e.:-

$$\text{diffusion rate} = \frac{[O_2]\,\text{external} - [O_2]\,\text{internal}}{\text{resistance}}$$

If it is assumed that the diffusion resistance is a constant then a 50% decrease in respiration, such as that produced by the addition of C_2H_2 (Minchin et al., 1983), would result in an internal O_2 concentration of approximately 10%, which would inactivate nitrogenase. However, the C_2H_2-induced decline does not result in nitrogenase damage (Witty et al., 1983), indicating that a change in O_2 diffusion resistance is involved in the C_2H_2 response.

This hypothesis is further substantiated by several related observations; (a) nodules of C_2H_2-sensitive symbioses (e.g. white clover, lucerne and pea) are tolerant of high (80%) external O_2 levels, despite the saturation of respiratory activity, whilst the nitrogenase of C_2H_2-insensitive symbioses (e.g. sainfoin and soyabean) is denatured at O_2 levels above 40%, (b) even in O_2 tolerant systems the enzyme can be damaged if O_2 concentration is increased to 80% within a 45 sec. period, indicating that an induced change in permeability is involved rather than any permanent structural or biochemical feature, and (c) the magnitude of the C_2H_2-induced decline is substantially reduced when white clover nodules are pre-incubated for 45 min. at high O_2 concentrations (60 or 80%).

It is concluded, therefore, that the C_2H_2-induced decline in nodule activity is due to a decreased O_2 flux, resulting from an increase in the O_2 diffusion resistance. The currently accepted model of nodule function should be expanded to include the concept of a biologically controlled diffusion resistance.

REFERENCES

Minchin FR Witty JF Sheehy JE and Müller M (1983) J. exp. Bot. 34, 641-649.
Witty JF Minchin FR Sheehy JE and Minguez MI (1983) Ann. Bot. (in press).

EVALUATION OF IMMEDIATE ACETYLENE REDUCTION BY FRESHLY COLLECTED ROOTS OF CORN AND SORGHUM

CHARLES SLOGER AND PETER VAN BERKUM
USDA-ARS,BARC-W, Bldg. 011, HH-19, Beltsville, MD 20705, USA

Immediately detectable acetylene reduction by roots of field-grown corn has been reported (van Berkum, Sloger 1979). Our objective was to examine acetylene reduction and endogenous ethylene production as sources of ethylene accumulation during acetylene reduction assays.

Materials and Methods

Corn (Zea mays L. cv. Funk G-4545) and sorghum (*Sorhum bicolor* (L.) Moench. cv. CK-60A) were grown in field plots at Beltsville without N-fertilizer or bacterial treatments. Roots were collected at grain filling stage and prepared for acetylene reduction assay according to van Berkum (1980). Acetylene reduction and ethylene oxidation were measured as described by van Berkum, Sloger (1979) and effects of combined N were studied with 1 h exposures of roots to solutions of KNO_3 or NH_4Cl in 10 mM phosphate buffer pH 7.5 prior to acetylene reduction.

Results

During 4 h acetylene reduction assays ethylene accumulated at accelerating rates for the first hour followed by linear rates of 0.5 and 0.3 nmoles/g dry wt/h for corn and sorghum, respectively. Diurnal variation was not observed and rates in 0.0, 0.2 and 0.8 atm O_2 were similar. Endogenous ethylene production was not detectable with excised roots incubated in air or N_2. Incubating roots at 50°C enhanced ethylene production during acetylene reduction assays. Carbon monoxide (0.1 atm), an inhibitor of nitrogenase activity, had no effect. Ethylene production, in the presence of acetylene, was inhibited by 10 mM KNO_3 and was increased with 100 mM NH_4Cl. The rate of oxidation of 10 ppm ethylene (1.7 nmoles/g dry wt h) was higher than the rate of ethylene accumulation (0.8 nmoles/g dry wt h) during the acetylene reduction assays. Carbon monoxide or chloramphenicol (100mM) inhibited ethylene production at 30°C by 50% when measured with 4 h assays but no inhibition was observed with 1 h determinations.

Discussion

Root associated acetylene reduction activity, endogenous ethylene production, and ethylene oxidase activity occur in corn and sorghum. Ethylene is produced endogenously and by acetylene reduction during assays at 30°C. At 50°C endogenous ethylene production increases while acetylene reduction is absent. Endogenous ethylene production cannot be estimated in air because of ethylene oxidation. Our results indicate that nitrogenase activity is induced at the beginning of short-term acetylene reduction assays with freshly collected roots of field-grown corn and sorghum.

References

van Berkum P (1980) Soil Biol. Biochem. 12, 141-145.
van Berkum P and Sloger C (1979) Plant Physiol. 64, 739-745.

MEASUREMENTS OF THE EFFICIENCY OF N_2 FIXATION: A NEW APPROACH

J.F. WITTY*, F.R. MINCHIN[+] and J.E. SHEEHY[+]
* ROTHAMSTED EXPERIMENTAL STATION, HARPENDEN, HERTS., AL5 2JQ, UK
[+] THE GRASSLAND RESEARCH INSTITUTE, HURLEY, MAIDENHEAD, BERKS, SL6 5LR, UK.

For many legume symbioses there is a coupled decline in nitrogenase activity and respiration of effective nodules which can be induced by C_2H_2 and decreases in O_2 concentration (Minchin et al., 1983, Witty et al., 1983). This observation has given rise to a method which uses an open, flow-through system to evaluate directly the respiratory costs of nitrogenase activity.

Plotting respiration of detached nodules or nodulated roots against the rate of nitrogenase activity for the 30 to 60 min decline period, produces a linear relationship with an $r^2 > 0.95$. Data from both C_2H_2-induced and O_2-induced declines fall on the same regression line. Therefore, a linear relationship can be obtained for both C_2H_2-sensitive symbioses, using the C_2H_2-induced decline supplemented with an O_2-induced decline, and C_2H_2-insensitive symbioses, using the O_2-induced decline only. The gradient of the line directly measures the carbon cost for the transfer of one pair of electrons by nitrogenase (moles CO_2 mole $C_2H_4{}^{-1}$). The intercept, where nitrogenase activity is zero, estimates growth and maintenance respiration of nodules or nodulated roots. Detachment of nodules does not affect the gradient of the regression line, although the intercept value is reduced.

With intact plants this non-destructive method can be used for repeated evaluations of the proportion of root respiration coupled to nitrogenase, as well as the efficiency of carbon use by this enzyme. Detached nodules can be used for the rapid screening of genetically determined variations in efficiency.

Data obtained using this method show a substantial variation in efficiency. This is associated with both legume species (2.0-4.9 moles CO_2/electron pair transferred by nitrogenase) and Rhizobium strain (2.2-4.5 moles CO_2/electron pair, for Pisum sativum).

REFERENCES

Minchin FR Witty JF Sheehy JE and Müller M (1983) J. exp. Bot. 34, 641-649.
Witty JF Minchin FR and Sheehy JE (1983) J. exp. Bot. 34, 951-961.

POSTER DISCUSSION 1B APPLICATION OF BIOLOGICAL NITROGEN FIXATION FOR INCREASING CROP PRODUCTION

PETER DART, Research School of Biological Sciences, Australian National University, Canberra, Australia

BNF can influence crop production either through provision of nitrogen for the crop itself or via a residual effect on subsequent crops. Three plant-bacteria associations are involved - the nodulated legume, the looser associations of bacteria and plant roots, particularly of grasses and cereals, and cyanobacterial associations, such as with Azolla, cycads, bryophytes and lichens. It is difficult, but very important, to quantify the amounts of nitrogen fixed in all these systems and this problem was addressed in section 1A.

Nodulated Legumes. Legumes are involved in a wide variety of farming systems - as grain and fodder crops, as green manure crops where the nitrogen fixed is utilized by subsequent crops, and as trees for timber, fuel and shade. In many situations, it is not known whether the nitrogen supply from BNF is limiting crop yields. This can be determined by examining whether the crop responds to liberal nitrogen fertilization. Such trials indicate the improvement in BNF we should attempt to achieve.

The challenge is to find ways of increasing BNF through inoculation with *Rhizobium* and this faces the problems of overcoming the competition from indigenous populations, to select cultivars or species which fix more nitrogen, or to amend the farming system to maximise BNF. It is very encouraging that consistent responses to inoculation have been obtained in field trials with particular forage legumes grown in acid, infertile oxisols in Colombia (Sylvester-Bradley *et al.*), and with a particular groundnut cultivar - *Rhizobium* strain combination in India (R.T.C. Nambiar *et al.*; Oleagineux), even though the soils contained large populations of *Rhizobium* capable of nodulating these species. Such combinations of plant and *Rhizobium* strain provide a very useful tool for helping to open the black box that competition still remains, and a genetic analysis of them should prove very fruitful. Competition remains the key problem area for all microbe-plant root interactions and hopefully an increased research effort will be made in this area. A major difficulty to overcome is the development of a suitable rapid screen for assessing competitive ability.

Several posters dealt with the identification of *Rhizobium* strains and the success of inoculant strains in forming nodules in the field. Serological methods (FA & ELISA) offer most promise at present and can be adapted to handle the large numbers of nodules required for ecological studies. Other techniques, such as the use of isoenzyme patterns (Young), are being used to study the natural variation in soil populations and the degree of genetic relatedness between 'species' of *Rhizobium* occurring in the same field. Knowing how the composition of the indigenous soil populations change over time and in relation to cropping history will be an invaluable help in interpreting the response to inoculation.

Host cultivar selection for increased BNF may be difficult to achieve. Can a trait for efficiency in BNF *per se* be identified that is independent of plant vigour? Selection for BNF by a crop needs to be done under plant populations which optimise yield rather than on single plants, to reduce the confounding influence of photosynthate supply arising from use of spaced plants. Even so, the problem of how to distinguish differences between cultivars in BNF as opposed to nitrogen uptake remains. Isotope dilution methods are being tested.

The newly identified, host-controlled ineffectiveness or non-nodulation in lucerne, peas and groundnut (e.g. Heichel *et al.*) are useful tools for measuring nitrogen fixation by nodulated cultivars, and for explaining the genetics and physiology of host plant control over nodule development.

Cereal and grass associations. Inoculation of several crops with *Azospirillum* has given statistically significant increases in yield in several countries, although the response can be variable over location and season. The reason for this increase in yield is not clear as BNF by the root associations is very difficult to measure in the field. Development of suitable methods to study the effect of these microbial associations is still a key requirement - is it BNF or increased nutrient uptake and/or larger and more efficient root system, or desease protection or plant hormone production? We know very little about the factors affecting the survival of *Azospirillum* inoculants in the soil. In some experiments in Florida, they declined rapidly (Gaskin *et al.*), whereas in Israel and Egypt, they appear to survive at high populations on the root surface. Some inoculant strains apparently invade root tissue. The development of suitable identification markers for inoculants is an essential next step. The use of well-defined mutants such as *nif*$^-$ (Pedrosa and Yates; Elmerich *et al.*) will help to indicate the nature of the plant response.
$^{15}N_2$ is fixed by the root associated bacteria and within 3 days, the isotope is taken up by sorghum leaves (Gilber *et al.*), rice (Qui, Yuansberg *et al.*), sugar cane, and *Paspalum notatum*. Much careful effort is still required to sort out the role of the host cultivar and bacterial strain genotype on this response. Long term balance studies in pot cultures of rice, field experiments in Israel with corn and in Brazil with *Paspalum notatum* and *Brachiaria* spp., and pot experiments with sorghum in India suggest that host cultivar effects can be large. There is an interaction between cultivars and their soluble plant root exudates and growth and nitrogenase activity of several bacteria (Giller *et al.*; Bazzi, Calupo and Gallori).
New N_2-fixing plant-associated bacteria, such as *Azospirillum amazonense*, continue to be described (Dobereiner *et al.*). Wheat straw incorporated into soil degrades to provide substrate for N_2 fixation and this may help reduce the requirement for mineral N fertiliser of subsequent crops. Incorporation rather than burning has been adopted by some farmers in Australia in continuous wheat rotations (Roper).
Cyanobacterial Associations. *Azolla* is being used in farming systems in China and Vietnam. Incorporation of *Azolla* into paddy rice soils in Sri Lanka (Kulasooriya *et al.*), China and the Philippines provides nitrogen for crop growth and reduces the N fertiliser requirement. Use of *Azolla* as a fertiliser for rice requires an ability to control the paddy water supply and this is difficult under rainfed conditions in which rice is mostly grown. Tolerance of high temperatures, reduced phosphorus requirement and predator resistance are traits that can be selected for. However, the main limitations to improvement of the association are the difficulty in maintaining germplasm collections, the inability to develop variation by sexual means, and to resynthesize the association from the two partners. Recognition factors, such as lectins, may be involved (Ladha and Watanabe). However, Azolla appears to have a place in several rice based farming systems.
Cycad root nodules containg cyanobacteria also fix nitrogen (Grobbelaar), although it is not clear how much nitrogen is contributed to the ecosystem. Bryophyte associations in natural forest communities (Brasell and Davies; Millbank) also fix nitrogen. An elegant apparatus was devised for measuring the amount of biologically fixed nitrogen leached from the lichen and taken up by an associated moss. Results obtained using $^{15}N_2$ suggested that as much as 5 kg N/ha/year could be involved in such a transfer in a favourable environment.
Overall prospects for enhancing the already considerable use of BNF in agriculture seem bright, but much field work to develop appropriate farming systems needs to be done to capitalise on this potential.

EVALUATING PLANT-BACTERIAL INTERACTIONS WHICH AFFECT RHIZOSPHERE NITROGEN FIXATION

S. L. ALBRECHT[+], M. H. GASKINS[+] AND J. R. MILAM[++]
+ USDA-ARS AND DEPARTMENT OF AGRONOMY, UNIVERSITY OF FLORIDA, GAINESVILLE, FL 32611, USA.
++ DEPARTMENT OF MICROBIOLOGY AND CELL SCIENCE, UNIVERSITY OF FLORIDA, GAINESVILLE, FL 32611 USA

The association of N_2-fixing bacteria with roots of crop plants has attracted considerable attention. These systems may be able to reduce nitrogen fertilizer requirements in several agricultural areas. Inoculation of N_2-fixing bacteria into the rhizosphere has been used to augment or produce N_2-fixing systems. The underlying assumption in using such methodology is that the microorganisms will survive, grow and remain metabolically active in the rhizosphere. Large populations of N_2-fixing bacteria are essential to produce nitrogen fixation rates that are significant for agricultural needs. However, the survival of coliform organisms, often similar morphologically and metabolically to N_2-fixing strains, is limited in natural environments. This poster describes a series of experiments that studied populations of inoculated N_2-fixing bacteria and the edaphic factors that affected their survival. Our results indicate that the lack of adequate soil moisture and non-optimal pH can have deleterious effects on some strains of N2-fixing bacteria. These factors are important because they may pose substantial barriers to the survival of inoculated organisms, especially in acidic soils or arid regions with limited facilities for irrigation. Survival of the inoculated microorganisms was enhanced when growing plant roots were present in the rooting medium. This frequent observation suggests that the bacteria are obtaining nutrients from the root. Extended survival was found in media with increased organic matter. It is not clear if the organic matter provides nutrients or a protective habitat or both. The survival of inoculated N_2-fixing bacteria will be facilitated if the organisms are placed in the rhizosphere of growing plants and, particularly in sandy soils, some exogenous organic matter is provided. Survival was enhanced in experiments when native soil microorganisms were reduced, removing potential parasites, predators and competition. We have evidence that *A. brasilense* is prey for protozoa, myxobacteria and myxomycetes. These organisms effectively reduce populations of *A. brasilense* in laboratory experiments.

NITROGEN FIXATION ASSOCIATED WITH COLONISING BRYOPHYTES

H.M. BRASELL
CSIRO, DIV. OF FOREST RESEARCH, HOBART, AUSTRALIA

Wet sclerophyll eucalpyt forest in southern Tasmania are clearfelled, and regenerated by slash burning and aerial seeding. A variable but usually large portion of the nitrogen in the slash and in the surface layers of the soil are volatilised during the intense burn. Following the burn, these sites are very rapidly colonised by a dense mat of bryophytes. There is little regeneration of other vegetation for about thirty months.

For the earliest colonisers - *Marchantia berteroana, Funaria hygrometrica Ceratodon purpureus* - acetylene reduction rates up to 550 nmol C_2H_4 g^{-1} hr^{-1} were found. These rates were much higher than for species that occurred later in the succession or in the undisturbed forest.

The biomass of the bryophytes remained fairly constant at 140-180 g m^2 from about 16 months after the fire until the end of the study (30 months). The bryophytes showed a flush of growth in spring (Sept.), but were all severely dessicated in summer (Feb.).

Epifluorescence microscopy of *Funaria* plants (Scheirer, NE University, Boston) showed that microorganisms were predominantly heterocystous *Nostoc* and *Anabaena* species, occurring as epiphytes on stems, leaves and in the rhizosphere.

Soil cores containing bryophytes were collected monthly from sites burnt in three successive years, and assayed for acetylene reduction activity for a period of 14 months. The ratio to convert acetylene to nitrogen fixation was determined by Brasell, Bergersen & Turner to be 3.53. This ratio was used to provide estimates of nitrogen fixation rates of 43, 99 and 152 mg N m^2 in the second, third and fourth years after the fire.

Field rates of acetylene reduction were not correlated with field temperature or moisture during assay, or with nitrogen, phosphorous or the total organic matter content of the substrate.

However, under controlled laboratory conditions, reduction rates were dependent on temperature and moisture. Maximum rates occurred at about 25°C and from about 80-100% moisture saturation. Activity was very low below 25% moisture saturation and at 5°C. Reduction rates were very low in the dark, but rates changed only slowly when assay conditions changed from light to dark or the reverse. This slow change may be the result of an extended (18 hr) light equilibration period.

Azospirillum amazonense sp. nov., A NEW ROOT ASSOCIATED DIAZOTROPH BACTERIUM

JOHANNA DÖBEREINER[+]/F.M. MAGALHÃES[++]/J.I. BALDANI[+]/S.M. SOUTO[+]
+ PNPBS-EMBRAPA, SEROPÉDICA, 23460 RIO DE JANEIRO, BRAZIL
++ INPA-CNPq, C. POSTAL 478, MANAUS-AM, BRAZIL

INTRODUCTION

New dinitrogen fixing bacteria have been isolated from various Gramineae, Palmaceae and some other plants in various localities of the Amazon region and of Rio de Janeiro State. Highest numbers were found on the root surface (10^6) but the organisms also occur within roots (32% of the samples as compared to 45% for other *Azospirillum* spp.).

METHODS OF ISOLATION

Semi-solid N-free sucrose medium with pH 6.0 (LGI) is inoculated with soil or root dillutions. Nitrogenase positive vials are streaked out on LGI plates and small dense wrinkled colonies repicated into semi-solid LGI vials, where a thick surface pellicle forms after 5 days without acid production.

DESCRIPTION OF THE NEW SPECIES (Magalhães 1983, Magalhães *et al.* 1983)

The organisms fix N_2 under microaerobic conditions and their shape, size spinning motility, and pellicle formation in semi-solid media are consistent with the characteristics of the genus *Azospirillum*. Also malate and other organic acids are used if the pH of the medium stays below 7.0.

However, they differ from the two previously described species of this genus in the following characteristics: (i) a lower pH optimum with pronounced sensitivity to an alkaline reaction of the medium, (ii) a lower oxygen tolerance for N_2 fixation, (iii) poor or no nitrate dissimilation and no denitrification, (iv) an ability to use sucrose, but not fructose or citrate, as a carbon and energy source (v) smaller cell diameter (0.8 nm). Moreover, colonies on potato agar are large, white, and flat with an elevated margin, different from the pinkish raised colonies with curled margin of other azospirilla. The efficiency of N_2 fixation is 30 mg N per g of sucrose in semi-solid medium. Like other azospirilla they are resistant to penicillin and moderately resistant to chloramphenicol and tetracycline. The mol % G + C of the DNA is 67-68 (Tm), only slightly lower than that of other azospirilla. The name *Azospirillum amazonense* sp. nov. is proposed for these organisms and three strains were deposited at ATCC (35119, 35120, 35121).

REFERENCES

Magalhães FMM (1983) Caracterização e distribuição de uma nova espécie de bacteria fixadora de nitrogênio. M.Sc. thesis, INPA/University of Amazon, Manaus, Brazil.
Magalhães FMM, Baldani, JI, Souto SM, Kuykendall JR and Döbereiner J (1983) An. Acad. Brasil. Sci. (in press).

ISOLATION AND CHARACTERIZATION OF A NEW CYANOPHAGE INFECTING A TROPICAL NOSTOC STRAIN

C. FRANCHE
BIOLOGIE DES SOLS, ORSTOM, BP 1386, DAKAR, SÉNÉGAL

Since the discovery of the phage N1 infecting the Nostoc strain PCC 7120(1), several cyanophages of the N group have been reported either in URSS(2) or in United States(3). Until now, no experiments of lysogenisation or transduction have been described.
The screening of 30 samples of water collected in Sénégal permitted the isolation of a cyanophage termed N(S)1. Its host-range appeared to be limited to a Nostoc strain isolated in West Africa; N(S)1 did not infect the strain PCC 7120. The cyanophage N(S)1 had an hexagonal capsid with a very short tail. Physico-chemical experiments demonstrated that N(S)1 was very stable: the pH and temperature ranges of greatest stability were respectively 5-11 and 4-65°C. Characterization of phage growth cycle by one step growth experiment showed that the latent period lasted 20 hrs and that the average burst size was 70 PFU per infected cell.

(1) Adolph KW and Haselkorn R (1971) Virology 46, 200-208.
(2) Koz'Yakov SY (1977) Biol. Inst. 25, 151-175.
(3) Hu NT, Thiel T, Giddings TH and Wolk CP (1981) Virology 114, 236-246.

RESPONSE OF SOME CEREALS TO INOCULATION BY N_2-FIXING BACTERIA

F. GARNIER AND R. BIGAULT
CENTRE DE RECHERCHES DE LA SOCIÉTÉ COMMERCIALE DES POTASSES ET DE L'AZOTE
ASPACH LE BAS, 68700 CERNAY, FRANCE

Many trials were conducted throughout the world to experiment grasses inoculation by N_2-fixing bacteria. We report here some experiments which took place in France, from 1980 to 1982.
On rice, inoculation by Azospirilla increased the grain yield, the straw yield, but decreased the N-content of both. Results were varying with soil and N-fertilizer.
On wheat, inoculation by N_2-fixing bacteria, including Azospirillum, did not have any effect in one experiment, but in another with Bacillus, we only have noticed a decrease in straw yield, varying with soil.
On corn, in a field trial with high N-fertilization, the treatment with six bacteria isolated from corn roots, increased all the parameters measured, but the increase was never significant.
Preliminary experiments, in earlier stages of growth, seem to be necessary before the use of N_2-fixing bacteria for agricultural purposes.

SURVIVAL OF ROOT-ASSOCIATED BACTERIA IN THE RHIZOSPHERE

M. H. GASKINS, S. L. ALBRECHT, and J. R. MILAM - ARS, USDA, AND
UNIVERSITY OF FLORIDA, GAINESVILLE, FLORIDA, U.S.A.

Inoculation studies with root-associated bacteria usually are based on the premise that adding the exotic strain will increase nitrogen fixation in the rhizosphere. This usually does not occur however, except in gnotobiotic conditions, although plant growth often is stimulated by bacterially produced hormones, and perhaps other mechanisms. Inoculation alone does not substantially change numbers of rhizosphere nitrogen fixers, except briefly.

We grew plants in large containers, modified soil conditions to increase microbial activity, and determined whether these treatments improved growth of an *Azospirillum* (JM125). Soil moisture, organic matter, and oxygen levels were varied. The experiment was maintained for two years. In that time two crops of sorghum and one of oats were grown, and the soil was inoculated with the *Azospirillum* strain each time seeds were planted. In separate tests, effects of adding various carbon substrates were determined.

Early growth of seedlings and dry weights of the mature plants were increased in some cases by inoculation. Survival of the inoculant strain was poor in all cases, but several treatments increased populations of all nitrogen fixing organisms. Thirty days or less after inoculation, populations of nitrogen fixing organisms were too low to supply significant amounts of nitrogen to the host plants. Table 1 shows short-term effects (2-3 weeks) of various soil treatments.

Table 1. Effect of Soil Amendments on Survival and Activity of *A. brasilense* in the Rhizosphere of *Sorghum bicolor*

Treatment	Effect of Treatment (as % of control)		
	Bacterial Survival	C_2H_2 Reduction	CO_2 Evolution
Increased soil moisture	110-150	140-200	120-150
Increased organic matter	130-150	110-130	110-130
Increased carbon substrate	100-150	110-120	200-450
Reduced soil oxygen	100-110	100-250	-

This study indicates that the soil environment can be manipulated to increase numbers and activity of nitrogen fixing organisms, but suggests that addition of an exotic strain to the soil affects nitrogen fixation little if at all, except briefly. The responses occurred whether or not the inoculant strain had been added. Indigenous bacterial strains are as efficient as the *Azospirillum*, in responding to soil treatments which increase bacterial activity.

NITROGEN FIXATION ASSOCIATED WITH THE ROOTS OF SORGHUM AND MILLET

K.E. GILLER[+]/ S.P. WANI[++]/ J.A. KIPE-NOLT[++]/ J.M. DAY[+]/ P.J. DART[++]
and U.K. AVALAKKI[++]

[+] ROTHAMSTED EXPERIMENTAL STATION, HARPENDEN, HERTFORDSHIRE, ENGLAND.
[++] ICRISAT, PATANCHERU P.O., ANDHRA PRADESH 502 324, INDIA.

Acetylene reduction assays of sorghum and millet root systems have indicated high, if variable rates of associated N_2-fixation. An apparatus has been designed, in which the growth medium and root systems of several young plants can be incubated simultaneously in soil atmospheres enriched with $^{15}N_2$. Initial results demonstrated that low amounts of nitrogen were fixed, were readily available and were rapidly incorporated into plant tops. Measurable amounts of ^{15}N incorporation were found with all plants examined.

As a means of assessing differences between cultivars and inoculant strains in stimulating N_2-fixation, studies were done on the relationships between root exudation and nitrogen fixation. The amount of root exudation by sorghum seedlings varies between cultivars; and growth and nitrogenase activity of bacterial strains in differentially stimulated by this organic matter. Current research is examining the use of ^{15}N isotope dilution techniques to measure N_2-fixation associated with roots of millet and sorghum in the glass house.

METABOLIC STUDIES ON THE CORALLOID ROOTS OF *ENCEPHALARTOS TRANSVENOSUS* AND ITS ENDOPHYTE.

N. Grobbelaar, J.G.C. Small, J. Marshall & W. Hattingh.
Department of Botany, University of Pretoria, Pretoria, Republic of South Africa

The endophyte of all 28 species of *Encephalartos* as well as of *Stangeria eriopus* which are indigenous to South Africa, was isolated and are being identified taxonomically. All appear to be Nostoc species and at least half of them correspond to the description on *N. commune* Vauch. Other species of Nostoc are, however, also evident. The endophyte occurs intercellularly embedded in mucilage in the central cortex. The adjoining root cells are relatively rich in mitochondria and have convoluted plasmalemmae which suggests that these cells are metabolically very active and specialize in the exchange of substances between the cycad and its endophyte.

In all cases the infected roots as well as the isolated endophyte was shown to reduce acetylene. All the results reported on below was obtained with infected coralloid roots of *E. transvenosus* Stapf & Burtt Davy.

A concentration of 5% acetylene in air gave lower rates of acetylene reduction than acetylene concentrations of 10, 15, 20 or 25%. Although the rates obtained for the latter four concentrations did not differ significantly, 20% acetylene was subsequently used in all experiments.

White light was found to stimulate acetylene reduction of whole infected coralloid roots significantly at all oxygen concentrations from 0 to 30% (Oxygen concentrations up to 70% was tested). The optimal oxygen concentration was found to be 20% both in the light and the dark. Under anaerobic conditions in the dark virtually no acetylene reduction could be detected.

White light from tungsten filament bulbs which is relatively rich in light with wavelengths in the range 625-750 nm had a significantly stronger stimulatory effect on acetylene reduction than white light of the same irradiance from cool white fluorescent tubes. The relative values at 27°C being 411 and 87 mm^3 acetylene reduced per hour by one gram of fresh roots.

The blotting dry of wet roots lowered the concentration of oxygen which is optimal for nitrogenase activity from 50% to 20%. In an atmosphere with 20% oxygen and 20% acetylene, nitrogenase activity was decreased 50% by not blotting the roots dry prior to incubation.

KCN at 1 mmol dm^{-3} completely inhibited the nitrogenase activity of the roots both in the light and the dark. DNP at 0,15 and 0,30 mmol dm^{-3} on the other hand did not have a significant effect on the nitrogenase activity of the roots in either the light or the dark. At a concentration of 0,25 mmol dm^{-3} SHAM tended to stimulate nitrogenase activity in both the light and the dark whereas it tended to have an inhibitory effect in the light but no effect in the dark at a concentration of 1,5 mmol dm^{-3}.

EFFECT OF AZOSPIRILLUM INOCULATION ON ROOT DEVELOPMENT OF WHEAT

Y. KAPULNIK, Y. OKON AND Y. HENIS
DEPARTMENT OF PLANT PATHOLOGY AND MICROBIOLOGY, FACULTY OF AGRICULTURE, THE HEBREW UNIVERSITY OF JERUSALEM, REHOVOT 76100, ISRAEL.

1. INTRODUCTION

Significant increases in field grain yield of wheat, following inoculation with *Azospirillum* have been reported (Reynders, Vlassak, 1982, Kapulnik *et al*. 1983). In these positive experiments, the low acetylene reduction rates measured could not account for the marked increases in N yield obtained above non-inoculated controls (Kapulnik *et al*. 1983).

2. MATERIALS AND METHODS

Wheat seeds (*Triticum aestivum* L. cv Miriam) were germinated in Petri dishes. They were inoculated with one ml either in separate or in mixtures of *Azospirillum brasilense* Cd, Sp 7 and the local isolate Cd-1 and with other types of bacteria. All at a final concentration of 10^2-10^9 colony forming units (CFU) ml^{-1}. Pots with vermiculite were inoculated by mixing the bacteria to a final concentration of 10^2-10^8 CFU g^{-1} vermiculite.

3. RESULTS AND DISCUSSION

Inoculation with 10^5-10^6 CFU caused the largest root elongation and total surface, whereas 10^8-10^9 CFU of *Azospirillum* caused inhibition of root development. Similar effects were obtained in 10 different cultivars of wheat inoculated with *Azospirillum*. Only *Azospirillum* caused a significant effect (10^5-10^6 CFU ml^{-1} in petri dishes or g^{-1} vermiculite) in enhancing wheat root surface when compared to *Azotobacter chroococcum*, *Klebsiella pneumoniae*, *Pseudomonas* sp., and *Bacillus* sp. Scanning electron micrographs of inoculated wheat root segments showed denser and longer root hairs as compared to the control inoculated with dead cells. In inoculated roots bacteria were located mainly on the cell elongation area and on the basis of root hairs, but much less bacterial cells were present on the root cap or adsorbed to root hairs.

The reported enhancement of mineral uptake by roots inoculated with *Azospirillum* (Lin *et al*. 1983) seems to be derived mainly from enhanced development of root hairs and root surface area as shown in this work. Since these effects start early in plant development, they could account for the faster accumulation of dry matter observed in inoculated plants in the field, thus resulting in higher crop yield (Kapulnik *et al*. 1983). The partial contribution of effects of *Azospirillum* on root development and of nitrogen fixation on yield remains to be assessed.

4. REFERENCES

Reynders L and Vlassak K (1982) Plant and Soil 66, 217-223.
Kapulnik Y Sarig S Nur I and Okon Y (1983) Can. J. Microbiol. 29 In press.
Lin W Okon Y and Hardy RWF (1983) Appl. Environ. Microbiol. 45, 1775-1779.

STUDIES ON AZOLLA PINNATA IN SRI LANKA

S.A. KULASOORIYA[1]/W.K. HIRIMBUREGAMA[1]/S.W. ABEYSEKERA[2]/T.M. NAGARAJAH[3] AND I.M. KULASINGHE[3]

[1]DEPARTMENT OF BOTANY, UNIVERSITY OF PERADENIYA, SRI LANKA

[2]RICE RESEARCH STATION, AMBALANTOTA, SRI LANKA

[3]MARGA INSTITUTE, 61, ISSIPATHANA ROAD, COLOMBO 5, SRI LANKA

The ability of *Azolla pinnata* to grow and establish itself in monoculture under field conditions was examined in three locations, falling within different agro-ecological regions of Sri Lanka. Rapid growth was observed in all the localities examined.

In situ nitrogenase activity (acetylene reduction assay) gave values ranging from 2.0 - 2.3 x 10^{-6} mol $C_2H_4.g^{-1}$ (f.w. of *Azolla*)h^{-1}.

Phosphorus applied as concentrated super phosphate (45% P_2O_5) to *Azolla* grown in polythene lined, upland pits in soil-water culture, at the rates of 1.5 g and 3.0 $g.m^{-2}.wk^{-1}$, produced better growth than at the rates of 4.5 g and 6.0 $g.m^{-2}.wk^{-1}$. Under these conditions, the relative growth rate of *Azolla* for 27 days was similar when concentrated super phosphate was applied either as a single, initial dose of 6.0 $g.m^{-2}$ or in split applications of 3.0 $g.m^{-2}$ every two weeks, or 1.5 $g.m^{-2}$ every week.

Field grown *Azolla* survived 2-day periods of desiccation and recovered when water was re-supplied, but could not survive 5-day periods of desiccation.

Azolla grown in dual culture with rice under different planting patterns, produced grain yield increases of 13, 24 and 47% respectively in broadcast seeded, transplanted and avenue planted rice.

Two incorporations of *Azolla* during a single crop cycle resulted in grain yield increases of rice equivalent to field plots that received 55 to 84 kg $N.ha^{-1}$ of chemical nitrogen in the form of urea.

Azolla growth also brought about a 50% suppression of weed growth during dual culture with rice.

These results provide encouraging evidence for the suitability of *Azolla* as a biofertilizer for rice.

SOME ADVANCES ON AZOLLA RESEARCH

LIU CHUNG-CHU, WEI WEN-CHUNG and ZHENG DE-YING
Soil and Fertilizer Institute, Fujian Academy of Agricultural Sciences, Fuzhou, Fujian, People's Republic of China

In China most azolla is cultivated in the winter-spring season because it survives only with difficulty in large scale through the summer due to high temperature, pests, disease and algae. To extend azolla growth into summer, an improved method of cultivating and using the multi-tolerant capacities of azolla strains is needed (1,2). Results from screening showed that *Azolla caroliniana* has multi-tolerant capacities in comparison with other strains (4).

We have conducted recombination experiments of algae-free azolla and *Anabaena azollae* since 1978 (2). Algae-free azolla could be obtained by the *Stem Tip Cultural* Technique. Of the 35 experiments carried out in the recent years, fifteen proved to be successful. The frequency of recombination was about 43%. The evidence for successful recombination was the indication that leaf cavities had *Anabeana azollae*, while algae-free azolla had none, with the N_2-fixing capacity of reconstituted azolla increasing gradually. The success of recombination opens up a new method of studying the relationship between the fern and *Anabaena azollae*, the characteristics of resistance to pests and ecological conditions as well as breeding azolla.

Recently, the Stem Tip Cultural Technique has been developed for azolla preservation (5). This technique involves removing the stem tip of azolla under the microscope after sterilization and transferring it onto agar medium containing mineral nutrients. By using this method, most azolla strains can be kept successfully with only two changes of medium a year.

Besides its N_2-fixing capacity, azolla also has a strong capacity for absorbing K from water (3). The K content of azolla is closely related to the K concentration in external solution. Normally, the K content of azolla is around 0.2-0.3%. The greatest percentage of K removal from solution occurs when the solution contains 0.85 ppm K_2O. It may be regarded as the physiologically critical point of K requirement for azolla, which is 10-times less than that of rice. The K concentration in the irrigation water and rainfall in Fujian is about 1-5 ppm. At this low level, it is difficult for the paddy to absorb and utilize it. However, azolla can rapidly pick up the trace K and enrich the plants with K by its incorporation into the soil, followed by decomposition and release of the K for the use of rice. Hence, Azolla may be a potential potassium source for rice production.

References

1. Liu Chung Chu (1979) Use of azolla in rice production in China. In IRRI (ed.) Nitrogen and Rice. International Rice Research Institute, Laguna, Philippines. p. 375-394.
2. Liu Chung-Chu et al. (1981). The potential of azolla as a nitrogen source for paddy soil. In A.H. Gibson and W.E. Newton (eds.) Current perspectives in nitrogen fixation. p. 501.
3. Liu Chung-Chu (1982). Study on the potassium enriching physiology of azolla. Scienta Agricultura Sinca NO. 4 p. 83-87.
4. Wei Wen-Shung et al. (1982). Studies on resistant capacities of *Azolla caroliniana*. "Tu Fei Jian She" No. 4, p. 108-121. Soil and Fertilizer Institute, Fujian Academy of Agricultural Sciences.
5. Zheng De-Ying (1982). Use of tissue culture technique in preservation of azolla. Fujian Agr. and Tech. No. 4, p. 34.

USE OF ^{15}N ISOTOPE DILUTION TECHNIQUE FOR QUANTIFICATION OF ASSOCIATIVE BNF IN KALLAR GRASS

K.A. MALIK and Y. ZAFAR
Soil Biology Division, Nuclear Institute for Agriculture and Biology, Faisalabad, Pakistan

Biological nitrogen fixation associated with the roots of Kallar grass (Leptochloa fusca), a salt tolerant grass, has been earlier demonstrated by us using acetylene reduction technique with excised roots. Presently we report here results of quantitative estimation of BNF in Kallar grass when grown under controlled conditions in nutrient solution and inoculated with N_2 fixing bacteria, using ^{15}N isotope dilution technique. Three bacteria namely Klebsiella pneumonae (strain NIAB-1) Biejrinckia gumosa (Strain ISO-2) and Azospirillum brasiliense were used. First two strains have been isolated from Kallar grass roots. All the treatments received $(^{15}NH_4)_2SO_4$ 50% a.e. at the rate of 0.5 mg/50 ml nut.solu. The plants were grown in a growth room for 4 weeks with adequate facilities of light and temperature of 28±2°C. After harvest acetylene reduction of roots, total yield, total N and ^{15}N analyses were made. In case of control (uninoculated) no acetylene reduction was observed whereas all other inoculated treatments were able to reduce acetylene which confirmed the presence of microorganisms. Total N in inoculated treatments was 2-3 times higher than in control and so were the fresh and dry weight yields.
The estimates based on isotope dilution indicated that 50-70% N in the plant was derived from BNF in case of inoculated treatment. The results based on N balance gave relatively lower values of 40-60% of total N derived from fixation. This data has indicated that in Kallar grass a substantial amount of plant N is derived from BNF.

ON THE N_2-FIXING EFFICIENCY IN MASS CULTURE OF DIFFERENT AZOLLA SPP.

MARGHERI M.S./SILI C./VINCENZINI M./BARSANTI L./MATERASSI R.
INSTITUTO DI MICROBIOLOGIA AGRARIA E TECNICA DELL'UNIVERSITÀDEGLI STUDI E
CENTRO DI STUDIO DEI MICRORGANISMI AUTOTROFI DEL C.N.R. - FIRENZE (ITALY)

Outdoor mass cultures of Azolla pinnata, A. filiculoides and A. caroliniana were carried out on synthetic medium and on flooded soil. In both the media the three species showed the highest specific growth rate (about 0.35 replications day^{-1}) by keeping the plant density between 20 and 40 g dry weight m^{-2}, while the highest yields (10 g d.w. m^{-2} day^{-1} for A. filiculoides, 11 for A.pinnata and 11.5 for A. caroliniana) were obtained at densities ranging from 40 to 70 g d.w. m^{-2}.

A. pinnata, A. filiculoides and A. caroliniana showed an average nitrogen content of 48.4, 51.5 and 52.3 mg (g d.w.)$^{-1}$ with a N_2-fixing capacity of 0.475, 0.51 and 0.52 nmoles N_2 (g d.w.)$^{-1}$ respectively. The rates of fixation varied markedly over the day as the temperature increased from 18-20°C of the early morning to 34-36°C. At about 30°C all the species showed the highest activity with values higher than 100 nmoles C_2H_4 (mg d.w.)$^{-1}$ $hour^{-1}$. The daily C_2H_2-reducing activities (here including the low dark values) for A. pinnata,A.filiculoides and A. caroliniana were 0.960, 0.915 and 0.780 nmoles C_2H_4 (g d.w.)$^{-1}$ day^{-1} respectively.

Under air the three species did not produce detectable hydrogen gas even after 4 hours of incubation while under Ar-0.3% CO_2 the H_2-producing activity compared to the C_2H_4-formation was 28% in A. pinnata, 9.4% in A. filiculoides and 8.2% in A. caroliniana. These results suggest that under natural conditions of temperature and illumination the uptake hydrogenase activity of the Azolla-Anabaena azollae symbiosis is particularly high. Furthermore it was found for outdoor mass cultures that the average ratios of acetylene to N_2 reduced at a pN_2=0.8 atm. were for A.pinnata, A.filiculoides and A.caroliniana respectively 2.02:1; 1.79:1; 1.5:1 so that the estimates of nitrogen fixation rates based on a ratio of 3:1 need to be corrected. Peters et al. (1977) pointed out that also for the laboratory cultures of A. caroliniana the average ratio of acetylene to N_2-reduced is lower than 3:1. The present results show that the rates of nitrogen fixation are in mass culture as high as in laboratory culture.

In absence of detectable H_2-production under air, the C_2H_2/N_2 ratio may be a valuable way for determining the efficiency of nitrogen fixation and may be that low efficiencies should correspond to high C_2H_2/N_2 ratios and "vice versa".

Among the species assayed A. caroliniana gave higher biomass yield and better N_2-fixing efficiency so that it must be considered the proper species for the italian summer climatic conditions.

IS IT NECESSARY TO IMPROVE NITROGEN FIXATION OF BEAN IN AGRICULTURAL FIELDS IN MEXICO?

ESPERANZA MARTINEZ / RAFAEL PALACIOS
CENTRO DE INVESTIGACION SOBRE FIJACION DE NITROGENO. UNAM. Apdo. Postal 565-A Cuernavaca, Morelos, Mexico.

1. INTRODUCTION

Bean, <u>Phaseolus vulgaris</u>, had its origin and diversified in Mexico. To evaluate nitrogen fixation we collected data of fertilization assays carried out in agricultural areas of Mexico during the last five years by different institutions: National Institute of Agricultural Research, INIA; Postgraduate College of Chapingo, University of Morelos State, UAEM.

2. RESULTS.

2.1 <u>Nitrogen is not limiting bean yield</u>. We found in 70-90% of the works we analyzed, the types of response shown in figure 1. The other works report yield increases as a function of N dose. The lack of response to nitrogen fertilizer may be explained, in some cases, by the low yields obtained due to agronomic problems such as: water unavailability, diseases and insect damage.

2.2 <u>Fertilizer costs do not reduce farmer's profit</u>. We performed an economic analysis to determine the effect of fertilizer costs on farmer's profit using Tarquin and Blanck's equation:
cost = yield x price ($\frac{1}{1+i}$). We considered different yields, the possible net incomes and calculated "i" (% of a ratio benefit/cost) as a function of a broad range of costs. The cost of nitrogen fertilizer is indicated by a bar under the "x" axis in fig. 2. N fertilizer cost represents a small proportion of the total costs, so it can only change "i" within a very limited range. Fertilizer usage reduces profits when very low yields (v.g. 0.2 ton/ha) are obtained.

2.3 <u>Bean is naturally nodulated in Mexico by a great diversity of strains of R. phaseoli</u>. Bean inoculation has not improved yield in Mexico. To evaluate the diversity of <u>R. phaseoli</u> nodulating bean, we have analyzed their plasmid profiles. We found a great diversity of strains and not a single ubiquitous strain.

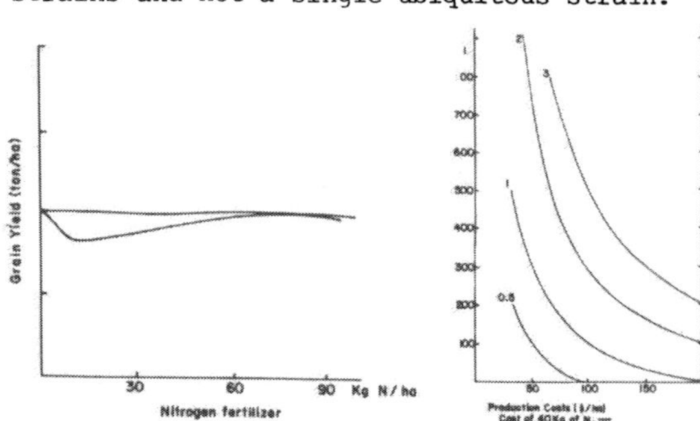

Fig 1 Bean yield response to nitrogen fertilizer.

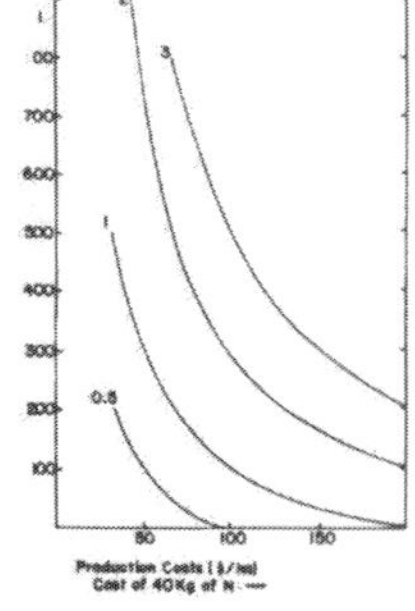

Fig 2 "i" as a function of cost for yields:0.5,1 2,3 ton/ha.

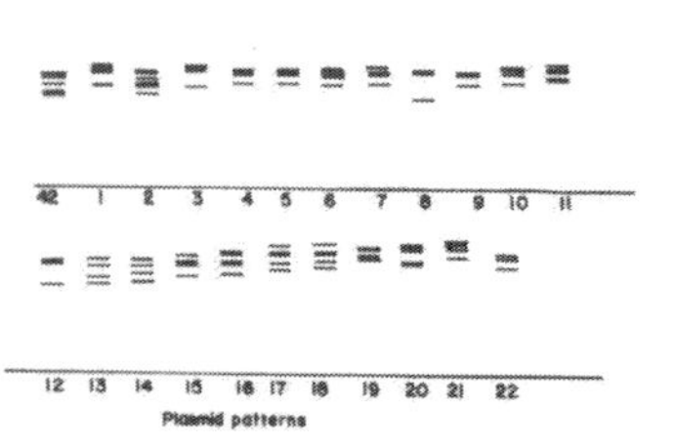

Fig 3 Plasmid profiles of native <u>R. phaseoli</u> isolated in Mexico.

THE USE OF SIMULATED ENVIRONMENTS FOR ASSESSING NITROGEN THROUGHPUT BY CYANOPHYLIC LICHENS

J.W. MILLBANK
DEPT. OF PURE & APPLIED BIOLOGY, IMPERIAL COLLEGE, LONDON, SW7, UK.

The controlled environment chamber for the study of lichens and other thallophytes that has been described previously (Millbank, Olsen, 1981) has been developed and improved. Long term incubations under atmospheres enriched with $15N_2$ can be carried out, with fully automatic monitoring and control of atmospheric CO_2. Night and day air and thallus temperatures, day length and light intensity can be programmed and continuously recorded. Sampling of the atmosphere is possible at any time. Rainfall is fully programmable, with a maximum interval of 24 hrs between falls and a minimum fall equivalent to 1mm of water upon the experimental area. Water runoff collects beneath the experimental material and is automatically pumped out for analysis. Thus a nitrogen balance sheet can be prepared and the nutrient contribution to the ecosystem of the nitrogen fixing lichen assessed. Winter, spring and autumn can be simulated, but summer conditions are not possible as the low relative humidity between showers of rain cannot be achieved with the equipment at present available. Using 15N labelled gas at known enrichment, the rate of N fixed and incorporated in the lichen thallus is easily found. Qualitative demonstration of the transfer of fixed nitrogen from the thallus to associated plant and other substrata is also straightforward, but quantitative estimates require that the 15N enrichment of the eluted compounds be known. This requires lichens uncontaminated by moss to be incubated and the 15N enrichment of the eluate assayed. Thus, duplication of specimens is imposed with consequent reduced precision. The equipment also lends itself to studies of the nature of the substances eluted.

RESULTS

Spring conditions, using Peltigera membranacea
Temperatures: $10^{o}C$ day, $5^{o}C$ night. Day length 12hr.
Light intensity: 120 $\mu E/m^2/sec$. Rainfall: 4.5 mm/day

Nitrogen fixed and retained by lichen in 28 days: 18.8 mg
= 4.0 mgm/gm dry weight lichen.

Nitrogen released and retained by moss substrate in 28 days: 5.8 mg
= 1.12 mgm/gm dry weight moss. (1.23mg/gm lichen)

Nitrogen released and assayed in runoff water in 28 days: 2.14 mg
= 0.45 mgm/gm dry weight lichen.

REFERENCE

Millbank J.W. and Olsen J.D. (1981) New Phytol.89, 657-665.

ACKNOWLEDGEMENT
The author expresses his gratitude to J.D. Olsen for skilled assistance.

WHEAT STRAW : AN ENERGY SOURCE FOR BIOLOGICAL NITROGEN FIXATION

MARGARET M. ROPER
DIVISION OF PLANT INDUSTRY, CSIRO, G.P.O. BOX 1600, CANBERRA, AUSTRALIA

1. INTRODUCTION

Burning of wheat straw residues has been traditional practice in Australia. However, there is an increasing trend towards straw retention and this has the advantages of reduced erosion and improved soil moisture and soil structure. In addition, straw retention and incorporation has resulted in a decline in the requirement for nitrogenous fertilizers in some areas. Cereal straw residues represent a carbon resource and, following decomposition by soil microflora, provide energy that can be used by other bacteria including nitrogen-fixing bacteria.

2. MATERIALS AND METHODS

Nitrogen fixation (nitrogenase activity) was estimated in the field and in the laboratory using the acetylene reduction assay (Turner, Gibson 1980). Straw decomposition was measured indirectly by monitoring evolution of CO_2 using a gas chromatograph fitted with a thermal conductivity detector. In the field, *in situ* assays in soils containing wheat straw were done using open ended cylinders which could be sealed with a lid containing a sampling port for the introduction and removal of gases (Roper 1983). Throughout each experiment, soil moisture content was measured gravimetrically and soil temperature was monitored using a recording soil thermograph. In the laboratory, nitrogenase activity in soils collected from sites with different histories of straw retention was determined in miniaturized systems using glucose as an energy source.

3. RESULTS AND DISCUSSION

Field measurements of nitrogenase activity indicated that there was a response by nitrogen-fixing bacteria to the addition of straw to soil and this increased with increasing amounts of straw. Nitrogenase activity was related to the rate of straw decomposition and to soil temperature. Nitrogenase activity in the field was best at soil moisture contents of about field capacity but diminished rapidly once the soil dried. This relatively high moisture requirement probably was necessary for the development of anaerobic and micro-aerobic microsites needed by many nitrogen-fixing bacteria.

In the laboratory, nitrogenase activity was highest in soils in which straw had been retained for a number of years. This indicates that the soil micro-flora adapt slowly towards enhanced straw decomposition and nitrogen fixation.

The results imply that there is considerable potential for nitrogen fixation in soils amended with wheat straw. This may contribute significantly to the nitrogen status of the soil.

4. REFERENCES

Roper MM (1983) Australian Journal of Agricultural Research 34, in press.
Turner GL and Gibson AH (1980) In Bergersen FJ, ed, Methods for Evaluating Biological Nitrogen Fixation, pp. 111-138, J. Wiley and Sons, New York.

RESPONSES OF TROPICAL FORAGE LEGUMES TO RHIZOBIUM INOCULATION IN UNDISTURBED CORES OF A COLOMBIAN OXISOL

R. SYLVESTER-BRADLEY/ E. BURBANO/ M. A. AYARZA/ C. BALAGUERA

CENTRO INTERNACIONAL DE AGRICULTURA TROPICAL, A. A. 6713, CALI, COLOMBIA

Inoculation with selected *Rhizobium* strains of tropical forage legumes adapted to grow in acid, infertile Oxisols of Latin America could increase their production, nutritional quality and contribution to beef production in these extensive under-developed areas. Selected *Rhizobium* strains must be acid tolerant and able to compete with native soil rhizobia. Screening experiments using unsterilized, undisturbed cores of a Colombian Oxisol showed that in 10 out of 16 trials on 10 promising forage legume species more than 2-fold increases of N in the tops (N yield) due to inoculation with at least one of the inoculated strains occurred. The most effective strains in these 10 successful trials were: CIAT 1780 (*Centrosema macrocarpum*, CIAT germplasm accession no. 5065); CIAT 1670 (*C. macrocarpum* no. 5065); CIAT 2287 (*Centrosema* sp, no. 5112); CIAT 2335 (*Desmodium ovalifolium* no. 3666); CIAT 3418 (*D. ovalifolium* no. 3784); CIAT 1502 (*D. canum* no. 3005); CIAT 2487 (*D. canum* no. 13032); CIAT 2469 (*D. heterophyllum* no. 349); CIAT 2434 (*Pueraria phaseoloides* no. 9900); CIAT 2304 (*Stylosanthes capitata* no. 1019). In the six remaining trials *Centrosema brasilianum* no. 5234, *Centrosema* hybrid no. 5931; *Desmodium ovalifolium* no. 350; *Stylosanthes capitata* nos. 1019 and 1315 and *S. guianensis* "tardío" no. 1283 showed less than 2-fold increases in N yield due to inoculation, although in some cases the increases were significant. Apparently the native soil rhizobia were effective or semi-effective in these cases, or other factors were limiting to N_2 fixation, since in some cases the legumes responded to N fertilization and in other cases not. The ability of the most effective strains to cause responses to inoculation should be investigated further under a range of conditions, in order to determine their potential as commercial inoculants.

SOME PROPERTIES OF THE NITROGEN-FIXING ASSOCIATIVE SYMBIOSIS OF ALCALIGENES FAECALIS A-15 WITH RICE PLANTS

QIU YUANSHENG*/MO XIAOZHEN*/ZHANG YAOLIN*/LI XIN**/YOU CHONGBIAO**
* GUANGDONG MICROBIOLOGY RESEARCH INSTITUTE, GUANGZHOU, CHINA
** INSTITUTE FOR APPLICATION OF ATOMIC ENERGY, CHINESE ACADEMY OF AGRICULTURAL SCIENCES, BEIJING, CHINA

A strain of N_2-fixing bacteria has been isolated from rice roots in China. This strain, A-15, resembles the genus Alcaligenes and has been identified as A. faecalis. This bacterium is widespread in paddy soil in China and has a high nitrogenase activity. However, many questions remain to be answered with regard to the properties of associative system.

1. Association between the bacterium A-15 and rice roots.

A lot of bacteria accumulates on the root surface of a fresh rice root. The capillary assay determining motility of bacteria shows chemotaxis of A-15 in response to rice root extract and exudate and that NH_3 has no effect on the chemotaxis. Only a few A-15 are chemotactic to some organic acids which can be used as a carbon source for their growth (1). When an aseptic rice seedling was inoculated with A-15, many bacteria grew and became attached to the root surface. After incubation for 1-3 days, only a few bacteria enter the root. These results indicate that the association between A-15 and the rice root is intimate and that the main site of association is located on the root surface.

2. Associative N_2-fixing activity of A-15 with rice.

Even though the field assay for determining associative N_2 fixation with rice is unsatifactory, under lab conditions, it was found that the associative N_2-fixing activity (ARA) of A-15 with the rice plant is rather high. ARA of the associative system could reach 3140 nmol C_2H_4/g dry root/hr. ARA was dramatically decreased when the rice was supplemented with NH_3, but MSX could reduce this inhibiting effect of NH_3. The nitrogenase activity of the associative system declined dramatically after washing the rice root with water. The same results were obtained by the use of a tracer assay (2).

3. Nitrogenous compound exuded by the bacterium A-15.

It has been proven by N tracer assays that about 1/3 of the nitrogen fixed by A-15 could translocate rapidly into the roots and the leaves. A-15 can exude a portion of the newly fixed nitrogen during its growth. Exudation of a nitrogenous compound has also been observed in pure cultures of A-15. The amount of nitrogen exuded was about 20-30% of the total amount of nitrogen fixed by A-15. This as yet unidentified nitrogenous compound is different from glutamine, glutamate, asparagine, citrulline or allantoin, but it can be absorbed by the rice plant. Hence, strain A-15 has a character similar to symbiotic N_2-fixing bacteria.

REFERENCES

(1) You CB, Qiu YS (1982) Scientia Agricultura Sinica (6), 1-6.
(2) Zhou SP, Mo XZ, Ye CG, Qiu YS, Song W, Li JW, You CB (1981) Acta Agronomica Sinica 7, 59-62.

CHEMISTRY OF DINITROGEN FIXATION

POSTER DISCUSSION 2

RECENT PROGRESS IN THE CHEMISTRY OF DINITROGEN FIXATION

THE BINDING AND REACTIVITY OF DINITROGEN

G. JEFFERY LEIGH
A.R.C. UNIT OF NITROGEN FIXATION, THE UNIVERSITY OF SUSSEX, BRIGHTON, BN1 9RQ, U.K.

1. INTRODUCTION

The survey given here covers the approximately two years since the last International Nitrogen Fixation Symposium. Information available before mid-1981 will not be discussed in any detail. The discussion will cover new dinitrogen complexes, information on aqueous and non-aqueous systems in which intermediates have not yet been detected by physical techniques, postulated and identified intermediates and mechanisms, and parallels with nitrogenase.

2. DINITROGEN COMPLEXES

There are several new dinitrogen complexes, and some of them exhibit new fashions of bonding for N_2. The complexes $[Zr(cp)_2\{CH(SiMe_3)_2\}N_2]$ (cp = C_5H_5 or C_5H_4Me) remain the only possible examples of side-on N_2, crystal structures having yet to be determined (Jeffery et al., 1979). These complexes yield some hydrazine upon reaction with acid. Another new mode of bonding is exemplified by $[\{WCl(PMe_2Ph)_3(C_5H_5N)(\mu_3\text{-}N_2)(AlCl_2)\}_2]\cdot 2C_6H_6$ which has the basic structure represented below (Takahashi et al., 1983).

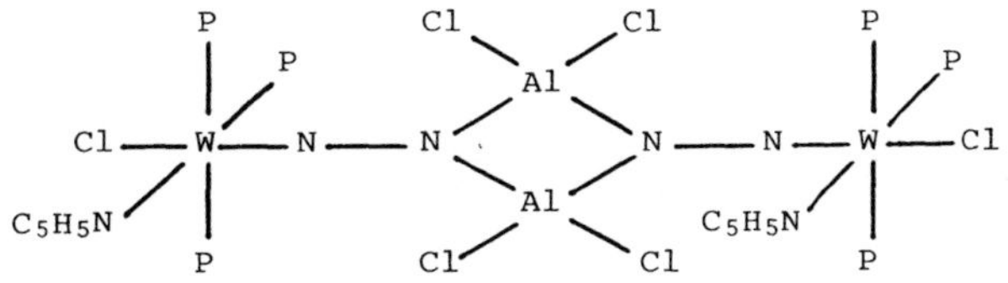

P, P, Cl, W, N, N, Al, Cl, Cl, Al, Cl, Cl, N, N, W, Cl, P, P, C_5H_5N, P, C_5H_5N, P

More complex structures have been identified in titanium systems (Pez et al., 1982). These are only isolated after an extremely careful work-up, and one is shown below. It is a mixed crystal containing two species.

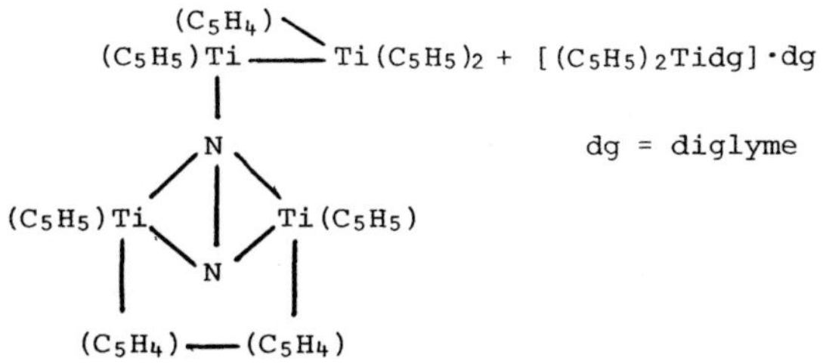

(C_5H_4) — $(C_5H_5)Ti$ — $Ti(C_5H_5)_2$ + $[(C_5H_5)_2Tidg]\cdot dg$

dg = diglyme

N, $(C_5H_5)Ti$, $Ti(C_5H_5)$, N, (C_5H_4) — (C_5H_4)

This complex has a long N-N bond (1.31 Å) and reacts with HCl to give principally NH_3 but also some N_2H_4. However, the most interesting group of complexes isolated (Rocklage, Schrock, 1982; Rocklage et al., 1982; Churchill, Wasserman, 1982a, 1982b) comprises Nb and Ta complexes such as $[\{NbCl_3(tetrahydrofuran)_2\}_2N_2]$ which possess a linear metal-N-N-metal system with long N-N bonds (ca. 1.3 Å) and which have a very low N-N stretching frequency (ca. 850 cm^{-1}). They react with acids to yield hydrazine quantitatively, and are written by the authors as hydrazine derivatives, for example, containing the structure Nb=N-N=Nb. These complexes are of interest because they represent a new class of materials which contain metals in high oxidation states and with hard ligands. They contrast with the bulk of complexes known hitherto, with low-oxidation-state metals with soft ligands.

The other N_2 complexes reported during the period under review are normal end-on materials for the most part. The most interesting are $[\{Ru^{II}(edta)\}_2N_2]^{2-}$ which can be prepared in aqueous solution from $[Ru(edta)(H_2O)]^-$ and N_2 (Diamantis, Dubrawski, 1981), $[Mo(PMe_3)_5N_2]$ (Cloke et al., 1982), $[Mo(N_2)(C_2H_4)_2(PMe_3)_3]$, $[W(PMe_3)_5(N_2)]$ and cis and trans-$[Mo(N_2)_2(PMe_3)_4]$ (Carmona et al., 1982a; Carmona et al., 1982b), $[\{W(N_2)_2(PEt_2Ph)_3\}_2N_2]$ (Anderson et al., 1982), and $[RuH_2(PPh_3)_2(\mu\text{-}H)_4\text{-}Ru(PPh_3)_2(N_2)]$ (N-N = 1.11 Å) (Chaudret et al., 1983).

No dinitrogen complexes with metal sites resembling nitrogenase have been synthesised. It is known that iron-sulphur and iron-molybdenum-sulphur clusters can reduce C_2H_2 to C_2H_4 (Tanaka et al., 1981), C_2D_2 to cis- $C_2H_2D_2$ (Tanaka et al., 1982a), and MeNC to CH_4 and $MeNH_2$, but MeCN is not reduced step-wise. $[Fe_4S_4(SPh)_4]^{4-}$ and $[Fe_6Mo_8S_8(SPh)_9]^{4-}$ are also claimed to reduce N_2 to NH_3 (Tanaka et al., 1982b), but the ligand binding properties of "monocubane" clusters with the Fe_3MoS_4 core do not suggest that they will bind and activate N_2, even though analogous carbonyls with low ν(CO) have been prepared (Mascharak et al., 1983; Palermo, Holm, 1983).

The development of ^{15}N and ^{95}Mo n.m.r. has now allowed the spectral characterisation of N_2 complexes as well as of protonation intermediates such as hydrazide(2-) (Dilworth et al., 1981; Donovan-Mtunzi et al., 1983) and promises to be of value in following chemical reactions.

3. AQUEOUS AND NON-AQUEOUS FIXING SYSTEMS

3.1 Aqueous systems: The heterogeneous system $Cr(OH)_2$-$Mo(OH)_3$ in aqueous methanol reduces N_2 to N_2H_4 and NH_3, and the yield of NH_3 is five times greater than that of N_2H_4, though the chromium reduces hydrons about three times as fast as it does N_2. The evidence suggests that chromium can reduce N_2 to NH_3 directly without the formation of hydrazine (Burbo et al., 1982a). Subsequently it was shown that $Cr(OH)_2$ in the absence of other metal hydroxides can also reduce N_2 to NH_3, albeit in low yield (1.5%, based on Cr). Since Cr(II) is a d^4-system, and not the previously favoured d^3-system, then the earlier interpretation of the significance of the number of d electrons has been modified, although a dinuclear mechanism has not, apparently, been abandoned (Burbo et al., 1982b).

The $V(OH)_2/Mg(OH)_2$ system has properties claimed to be consonant with a diazene (NH=NH) intermediate (Schrauzer et al., 1982), though this interpretation of the mechanism of fixation has been challenged (Nikonova et al., 1980).

The vanadium(II)-catechol system remains unique in its ability to convert N_2 to NH_3 stoichiometrically. A kinetic analysis has been taken to confirm a dinuclear mechanism with V_4^{II} groups bridged by N_2 (Luneva et al., 1980). E.p.r. studies show that there are several species present in V^{II}/catechol mixtures, but the data are generally consistent with the mechanism proposed earlier (Luneva et al., 1982).

3.2 Non-aqueous systems: There has been relatively little work in this area. The reaction of $[Ti(C_5H_5)_2Cl_2]$ with magnesium in the presence of N_2 has been studied by ^1H and ^{13}C n.m.r. spectroscopy (Sobota, Janas, 1983). Nitrides are formed and the cyclopentadienyl rings suffer C-H bond fission. The system is clearly complex. The system $FeCl_3/LiPh/N_2$ has previously been considered to involve dinuclear $Fe-N_2-Fe$ systems. It has now been shown that $FeCl_3$ and naphthyl lithium react to form a structurally characterised tetrahedral species, $[Fe(naphthyl)_4]^{2-}$ (Bagenova et al., 1981). This can be reduced to produce a material which does react with N_2. From the system $FeCl_3$/LiPh a complex $Li_4[FePh_4]\cdot 4Et_2O$ can be isolated. This contains square plane iron(0), $[FePh_4]^{4-}$, and is the material which reacts with N_2 to yield a species which, with acid, produces ammonia and/or hydrazine (Bazhenova et al., 1982; Bazhenova et al., 1983). The joint interaction of iron and lithium with N_2, viz., $Fe-N_2-Li$, is proposed.

4. REACTIONS OF COORDINATED DINITROGEN

4.1 New reaction types: The involvement of metal hydrides in protonation systems has long been postulated. Cluster hydrides such as $[H_2Fe(CO)_4]$ and $[HCo_3Fe(CO)_{12}]$ are sufficiently acid to convert $[W(N_2)_2(PMe_2Ph)_4]$ in methanol to NH_3 and N_2H_4, isolated by base distillation (Nishihara et al., 1982). No products were isolated from reactions involving $[Mo(N_2)_2(PMe_2Ph)_4]$. Under similar conditions, $[W(N_2)_2(dppe)_2]$ yields $[W(OMe)(NNH_2)(dppe)_2]^+$ whereas $[Mo(N_2)_2(dppe)_2]$ gives $[Mo(OMe)_2(dppe)_2]^+$ and $[Mo(OMe)(CO)(dppe)_2]^+$ (dppe = $Ph_2PCH_2CH_2PPh_2$).

Nucleophilic attack on N_2 has also been claimed for the reaction of $(CF_3CO)_2O$ with $[M(N_2)_2(dppe)_2]$ (M = Mo or W), which produces $[M(OOCCF_3)(N_2COCF_3)(dppe)_2]$ as follows (Colquhoun, 1981a). Rather less certain in mechanism are some fascinating

$$W-N{\equiv}N \rightarrow \underset{OOCCF}{\overset{CF_3}{C=O}} \longrightarrow [W-N{=}NCOF_3]^+ + OOCCF_3^-$$

reactions involving hydrazido(2-)-complexes, the key step in which is likely to be the attack of a radical on the species $[WBr(N_2)(dppe)_2]$ (Colquhoun, Henrick, 1981a, 1981b; Colquhoun,

1981b, 1981c). The reaction mechanisms proposed for these transformations have well established precedents.

4.2 Reaction mechanisms: The mechanisms of the initial stages of protonation of terminally coordinated N_2 in molybdenum and tungsten complexes have been established by stopped-flow techniques. They are summarised below. The reactions of cis-$[M(N_2)_2(PMe_2Ph)_4]$ in methanol involve solvent (Henderson, 1981).

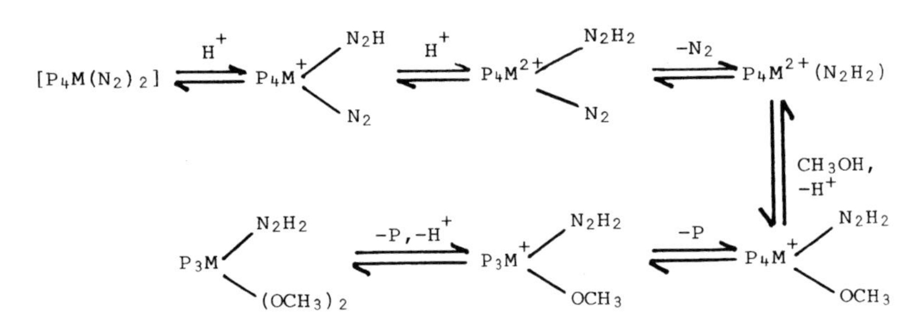

P = PMe_2Ph

Reactions with diphosphine complexes in tetrahydrofuran are more complex (Henderson, 1982; R.A. Henderson, unpublished) (depe = $Et_2PCH_2CH_2PEt_2$).

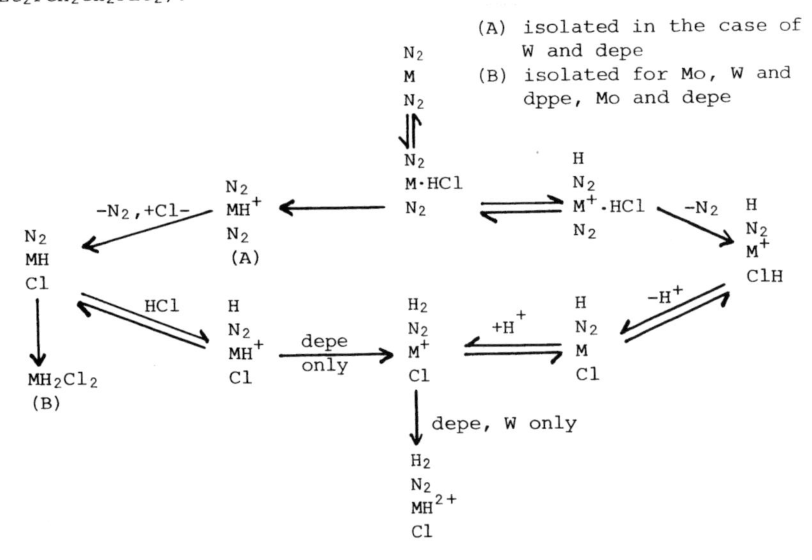

The pathway chosen by a given complex is governed by four different factors: acid strength, solvent, ligand donor power, and metal basicity. Parallel studies on bridging-N_2 are in progress.

Subsequent steps in the protonation process are, as yet, unclear. Calculations on the $[CrN_2(Ph_3)_4]$ (hypothetical) suggest that $NHNH_2$ could be the next stage (Yamabe et al., 1982), but empirical evidence is lacking. Nitrogen monoxide, often used as a model probe for nitrogenase, is known to protonate cleanly to give an imido-complex and water when bridging three manganese atoms (Legzdins et al., 1983).

$$Mn_3(NO) \underset{NEt_3}{\overset{H^+}{\rightleftharpoons}} Mn_3(NOH)^+ \xrightarrow{2H^+} Mn_3(NH)^+ + H_2O$$

It is questionable whether this represents a pattern for N_2 reactions.

Hydrazide(2-) seems a particularly stable ligand. An _iso_-diazene form of this ligand has been identified in a complex (Dilworth et al., 1982). Diazenes themselves rarely bind symmetrically, side-on, though their chemistry is beginning to be explored (Angoletta, Caglio, 1982; Barrientos-Penna et al., 1982; Einstein et al., 1983). Hydrazide(1-), the putative protonation product of hydrazide(2-), can bind end-on, as in $[Mo(pyrazolyl\ borate)(NO)I(NHNH_2)]$ (McCleverty et al., 1983), but more often binds side-on as in $[Ti(C_5H_5)Cl_2NH_2NPh]$ (J.R. Dilworth et al., unpublished) or $[W(C_5H_5)_2NH_2NPh]^+$ (Einstein et al., 1982). The isoelectric group NH_2O^- also prefers to bind side-on (Wieghardt, Quilitzsch, 1981; Wieghardt et al., 1983). The predeliction for this kind of side-on bonding by hydrazide(1-) may well obstruct protonation of end-on hydrazide(2-).

At least two other routes may be available for proceeding from hydrazide(2-) to ammonia. One route is a disproportionation, schematically represented as follows (Bossard et al., 1982; Bossard et al., 1983).

$$\begin{array}{l} Mo^{IV}{=}NNH_2 \\ \downarrow \\ Mo^{IV}{=}NNH_2 \end{array} \xrightarrow{+2H^+} 2Mo^{III} + N_2 + 2NH_3$$

The other route couples reduction and protonation (Hussain et al., 1982) as below (R = alkyl).

$$Mo^{IV}{=}NNR_2 + 2e \longrightarrow Mo^{II}{=}NNR_2 \xrightarrow{H^+} Mo^{IV}{\equiv}N + HNR_2$$

Only the second has been unequivocally demonstrated, the first is as yet unproven.

5. NITROGENASE

The dinitrogen chemistry of nitrogenase still seems, on balance, likely to be terminally-bound-dinitrogen chemistry. Side-on dinitrogen and diazene, however bound, appear improbable intermediates at present but their chemistry is still somewhat limited. Apart from the protonation reaction sequence in its later stages, major problems yet to be explained satisfactorily by the chemistry are the stoichiometry of the nitrogenase reaction, indicated below, and the formation of HD from D_2.

$$N_2 + 8H^+ + 8e \longrightarrow 2NH_3 + H_2$$

Displacement of H_2 from a metal polyhydride by N_2 does not seem a likely cause of H_2 production because it is not a necessary concomitant of N_2 binding; in fact, a bare metal site would bind N_2 just as well. In addition, the evolution of H_2 is apparently specifically related to N_2 reduction and not necessarily to that of other nitrogenase substrates. At least two chemical mechanisms can be adapted to explain it (Luneva et al., 1982; Bossard et al., 1982). The formation of HD is more of a problem and as yet admits of no compelling explanation based on known chemistry. Because HD formation consumes only a small amount of the electron flux in a fixing nitrogenase system, its explanation is probably not at present the most urgent of the chemical tasks.

6. REFERENCES

Anderson SN, Richards RL and Hughes DL (1982) J. Chem. Soc. Chem. Commun. 1291-1292.
Angoletta M and Caglio G (1982) J. Organomet. Chem. 234, 99-105.
Bagenova TA, Shilova AK, Deschamps E, Grunelle M, Levy CR and Tchoubar B (1981) J. Organomet. Chem. 221, C1-C4.
Barrientos-Penna CF, Einstein FWB, Jones T and Sutton D (1982) Inorg. Chem. 21, 2578-2585.
Bazhenova TA, Lobkovskaya RM, Shibaeva RP, Shilova AK, Shilov AE, Gryuzel M, Leni Zh and Chubar B (1982) Kinet. Catal. 23, 210-211.
Bazhenova TA, Lobkovskaya RM, Shibaeva RP, Shilov AE, Shilova AK, Gruselle M, Levy G and Tchoubar B (1983) J. Organomet. Chem. 244, 265-272.
Bossard GE, George TA, Howell DB and Lester RK (1982) Proc. 4th Int. Conf. Chem. Uses Molybdenum, Barry HF and Mitchell PCH eds., Climax Molybdenum Co., Ann Arbor, Michigan (1982), pp. 71-73.
Bossard GE, George TA, Howell DB, Koczan LM and Lester RK (1983) Inorg. Chem. 22, 1968-1970.
Burbo EM, Denisov NT and Kobeleva SI (1981) Kinet. Catal. 22, 1108-1113.
Burbo EM, Denisov NT, Shestakov AF and Shilov AE (1982) Kinet. Catal. 23, 191-193.
Carmona E, Marin JM, Poveda ML, Atwood JL, Rogers RD and Wilkinson G (1982a) Angew. Chem. (Int. Ed. Engl.) 21, 441-442.
Carmona E, Marin JM, Poveda ML, Rogers RD and Atwood JL (1982b) J. Organomet. Chem. 238, C63-C66.
Chaudret B, Revillers J and Poilblanc R (1983) J. Chem. Soc. Chem. Commun. 641-643.
Churchill MR and Wasserman HJ (1982a) Inorg. Chem. 21, 223-226.
Churchill MR and Wasserman HJ (1982b) Inorg. Chem. 21, 218-222.
Cloke GN, Cox KP, Green MLH, Bashkin J and Prout K (1982) J. Chem. Soc. Chem. Commun. 393-394.
Colquhoun HM (1981a) Trans. Met. Chem. 6, 57-59.

Colquhoun HM (1981b) J. Chem. Res. 9, 3416-3437.
Colquhoun HM (1981c) J. Chem. Res. 9, 3401-3408.
Colquhoun HM and Henrick K (1981a) J. Chem. Soc. Chem. Commun. 85-87.
Colquhoun HM and Henrick K (1981b) Inorg. Chem. 20, 4074-4078.
Diamantis AA and Dubrawski JV (1981) Inorg. Chem. 20, 1142-1150.
Dilworth JR, Donovan-Mtunzi S, Kan CT, Richards RL and Mason J (1981) Inorg. Chim. Acta, 53, L161-L162.
Dilworth JR, Zubieta JA and Hyde JR (1982) J. Amer. Chem. Soc. 104, 365-367.
Donovan-Mtunzi S, Hughes M, Leigh GJ, Mohd. Ali H, Richards RL and Mason J (1983) J. Organomet. Chem. 246, C1-C4.
Einstein FWB, Jones T, Hanlan AJL and Sutton DW (1982) Inorg. Chem. 21, 2585-2589.
Einstein FWB, Jones T, Sutton D and Xiaoheng Z (1983) J. Organomet Chem. 244, 87-96.
Henderson RA (1981) J. Organomet. Chem. 208, C51-C54.
Henderson RA (1982) J. Chem. Soc. Dalton Trans. 917-925.
Hussain W, Leigh GJ and Pickett CJ (1982) J. Chem. Soc. Chem. Commun. 747-748.
Jeffery J, Lappert MF and Riley PI (1979) J. Organomet. Chem. 181, 25-36.
Legzdins P, Nurse CR and Rettig SJ (1983) J. Amer. Chem. Soc. 105, 3727-3728.
Luneva LP, Nikonova LA and Shilov AE (1980) Kinet. Catal. 21, 1041-1045.
Luneva NP, Moravsky AP and Shilov AE (1982) Nouv. J. Chim. 6, 245-251.
Mascharak PK, Armstrong WH, Mizobe Y and Holm RH (1983) J. Amer. Chem. Soc. 105, 475-483.
McCleverty JA, Rae AE, Wołochowicz I, Bailey NA and Smith JMA (1983) J. Chem. Soc. Dalton Trans. 71-80.
Nikonova L, Rummel S, Shilov AE and Wahren M (1980) Nouv. J. Chim. 4, 427-430.
Nishihara H, Mori T, Tsurita Y, Nakano K, Saito T and Sasaki Y (1982) J. Amer. Chem. Soc. 104, 4367-4372.
Palermo RE and Holm RH (1983) J. Amer. Chem. Soc. 195, 4310-4318.
Pez GP, Apgar P and Crissey RK (1982) J. Amer. Chem. Soc. 104, 482-490.
Rocklage SM and Schrock RR (1982) J. Amer. Chem. Soc. 104, 3077-3081.
Rocklage SM, Turner HW, Fellmann JD and Schrock RR (1982) Organometallics 1, 703-707.
Schrauzer GN, Strampach N and Hughes LA (1982) Inorg. Chem. 21, 2184-2188.
Sobota P and Janas Z (1983) J. Organomet. Chem. 243, 35-44.
Takahashi T. Kodama T, Watakabe A, Uchida Y and Hidai M (1983) J. Amer. Chem. Soc. 105, 1680-1682.
Tanaka K, Tanaka M and Tanaka T (1981) Chem. Lett. 895-898.
Tanaka K, Imasaka Y, Tanaka M, Honjo M and Tanaka T (1982a) J. Amer. Chem. Soc. 104, 4258-4260.
Tanaka K, Hozumi Y and Tanaka T (1982b) Chem. Lett. 8, 1203-1206.
Wieghardt K and Quilitzch U (1981) Z. Naturforsch. 366, 683-687.
Wieghardt K, Backes-Dahmann G, Swiridoff W and Weiss J (1983) Inorg. Chem. 22, 1221-1224.
Yamabe T, Hori K and Fukui K (1982) Inorg. Chem. 21, 2816-2818.

7. ACKNOWLEDGEMENTS

I acknowledge gratefully the enormous debt I owe to all my colleagues in the Agricultural Research Council, Unit of Nitrogen Fixation.

HYDRAZIDO(1-), 1,2-HYDRAZIDO(2-), and HYDRAZIDO(4-) COMPLEXES OF TUNGSTEN(VI)

RICHARD R. SCHROCK, LAUREN BLUM, AND KLAUS H. THEOPOLD
DEPARTMENT OF CHEMISTRY 6-331, MASSACHUSETTS INSTITUTE OF TECHNOLOGY, CAMBRIDGE, MASSACHUSETTS 02139 U.S.A.

INTRODUCTION

We have found (Rocklage, Schrock, 1982) that niobium or tantalum neopentylidene complexes of the type $M(CHCMe_3)(THF)_2Cl_3$ react with 0.5 eq of an azine such as PhCH=N-N=CHPh to yield cis and trans-Me_3CCH=CHPh and sparingly soluble orange (M = Ta) or plum-colored (M = Nb) complexes of the composition $[MCl_3(THF)_2]_2(\mu\text{-}N_2)$. The THF ligands in $[TaCl_3(THF)_2]_2$-$(\mu\text{-}N_2)$ can be replaced by phosphine ligands such as PMe_2Ph or PEt_3 to give related complexes of the type $[TaCl_3L_2]_2(\mu\text{-}N_2)$. If one equivalent (per Ta) of PBz_3 (Bz = CH_2Ph) is employed a product having the composition $[TaCl_3(THF)(PBz_3)]_2(\mu\text{-}N_2)$ can be isolated. An x-ray structural study (Churchill, Wasserman, 1982, 218) shows it to have the structure in Figure 1. The Ta=N distance (1.796(5)Å) compares favorably with that

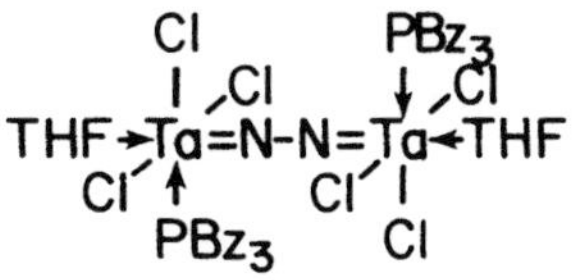

Figure 1. Structure of $[TaCl_3(THF)(PBz_3)]_2(\mu\text{-}N_2)$

(1.765(5)Å) found in the analogous imido complex, $Ta(NPh)Cl_3(THF)(PEt_3)$ (Churchill, Wasserman, 1982, 223). The rather long N-N distance (1.282(6)Å) and essentially linear Ta-N-N arrangement (178.9(4)°) suggest that the $\mu\text{-}N_2$ ligand is essentially a μ-hydrazido(4-) ligand. In keeping with this description this and similar complexes react readily with excess HCl in ether to produce 85-90% of the expected hydrazine dihydrochloride. They also react with acetone to give Me_2C=N-N=CMe_2 in ~50% yield, an unprecedented type of reaction of a $\mu\text{-}N_2$ complex, but one which is analogous to the reaction of related phenylimido complexes with acetone to give Me_2C=NPh in high yield. Treatment of the $[MCl_3(THF)_2]_2(\mu\text{-}N_2)$ complexes with $Mg(CH_2CMe_3)_2$ or $KOCMe_3$ yields neopentyl and t-butoxy complexes of the type $[Ta(CH_2CMe_3)_3(THF)]_2(\mu\text{-}N_2)$ and $[Ta(OCMe_3)_3(THF)]_2(\mu\text{-}N_2)$, respectively, in good yield.

Similar μ-N_2^{4-} complexes have been prepared from molecular nitrogen and reduced (M(III)) complexes (Rocklage et al., 1982). An x-ray structure (Churchill, Wasserman, 1981) of one of them, $[Ta(CHCMe_3)(CH_2CMe_3)(PMe_3)_2]_2(\mu$-$N_2)$, again shows a relatively short Ta=N bond (1.84Å) and long N-N bond (1.298Å). This and related species also yield hydrazine upon hydrolysis and Me_2C=N-N=CMe_2 upon treatment with acetone. An added feature of this chemistry is the ease of labelling complexes with $^{15}N_2$; a sharp medium-strength absorption at 847 cm^{-1} in the IR spectrum of $[Ta(CHCMe_3)Cl(PMe_3)_2]_2(\mu$-$N_2)$ shifts to 820 cm^{-1} in $[Ta(CHCMe_3)Cl(PMe_3)_2]_2(\mu$-$^{15}N_2)$. We proposed that this absorption, the only one observable in the IR spectrum, results from a vibration involving the entire Ta_2N_2 linkage.

The metal in all of the μ-N_2^{4-} complexes has the d^0 configuration (or the d^2 configuration if viewed as uncharged μ-N_2). In either case the metal is in the highest oxidation state ever observed in a dinitrogen complex. One way of rationalizing these results is that early transition metal d^2 complexes are good reducing agents, good enough to reduce molecular N_2 to the hydrazido(4-) level. An obvious question is to what extent do these principles extend to Group VI d^2 metals? We have begun a program aimed at answering this question and here report some of the first results.

RESULTS AND DISCUSSION

Attempts to prepare the known $[TaCl_3(PMe_3)_2]_2(\mu$-$N_2)$ by reducing $TaCl_5(PMe_3)_2$ in the presence of molecular nitrogen led only to $TaCl_3(PMe_3)_3$ or $[TaCl_3(PMe_3)_2]_2$, even in the presence of 100 atm of N_2 (Rocklage et al., 1982). Therefore we felt that Mo(IV) and W(IV) dinitrogen complexes probably could not be prepared readily in this manner. Unfortunately, the few known tungsten neopentylidene complexes (Pedersen, Schrock, 1982; Wengrovius, Schrock, 1982) also do not react with azines (or imines). In our search for alternative sources of N_2^{4-} we turned to hydrazine itself, reasoning that under the proper circumstances it should be deprotonated by (e.g.) alkyl, alkoxide, or amido ligands to yield (ultimately) the desired $M_2(\mu$-$N_2)$ complex. We chose to look for tungsten complexes first as the high oxidation state chemistry of tungsten is more developed than that of molybdenum.

$W(NPh)Me_3Cl$ reacts with LiMe to yield thermally unstable $W(NPh)Me_4$ (L. Blum, unpublished results). Addition of hydrazine to $W(NPh)Me_4$ yields a sparingly soluble yellow powder in high yield. ^1H and ^{13}C NMR spectra suggest that it has the composition shown in equation 1. The

$$W(NPh)Me_4 + N_2H_4 \longrightarrow CH_4 + W(NPh)Me_3(N_2H_3) \qquad (1)$$

three N_2H_3 protons are observed in the ^1H NMR spectrum in C_6D_6 as two singlets at 3.17 and 2.37 ppm in the ratio of 1:2, respectively. The possibility that the imido ligand has been protonated to an amido ligand can be ruled out by the observation that the ^1H NMR spectrum of the product obtained using $^{15}NH_2{}^{15}NH_2$ showed two doublets at 3.17 and 2.37 ppm. To our knowledge this is the first example of an unsubstituted hydrazido(1-) complex. On the basis of the structures of substituted hydrazido(1-) complexes of tungsten in a lower oxidation state (Cowie,

Gautier, 1980, Frisch et al., 1979) one would assume that this complex is a monomer, A (Figure 2). However, its relatively low solubility and the fact that a peak at m/e 688 was observed in an FD mass spectrum suggests that it is, in fact, the dimer, B. (The mass spec peak corresponds to B less a methyl group.) We rationalize structure B on the basis of

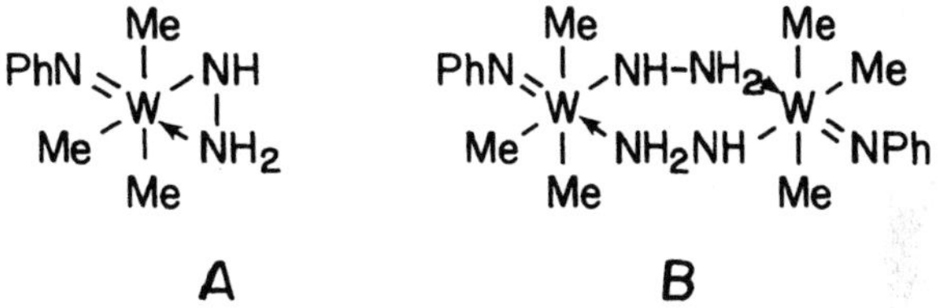

A B

Figure 2. Two possible structures for $W(NPh)Me_3(N_2H_3)$.

the strong donation of the -NH- π-electrons to the metal. The -NH- nitrogen is then forced to be sp^2 hybridized and the NH_2 consequently forced to donate its electron pair to a second tungsten atom. Another way of stating it is that the α-nitrogen in A would have to be sp^3 hybridized and its π-electron pair therefore not available for π-donation; a 16e electron count results (counting both the imido and the N_2H_3 ligands as 3e donors). The electron count in B is 18 for each metal since the N_2H_3 ligand donates a total of 5 electrons.

The ^{15}N NMR spectrum of $W(NPh)Me_3(^{15}NH^{15}NH_2)$ in CD_2Cl_2 is somewhat unusual. The doublet ascribed to the α-^{15}N at 57 ppm (vs. liquid NH_3) overlaps the triplet ascribed to the β-^{15}N at 55 ppm. This result is almost certainly circumstantial as one would have expected the two signals to be separated significantly.

A second type of hydrazido complex, this time 1,2-hydrazido(2-), results when $W(\eta^5\text{-}C_5Me_4Bu^t)(CCMe_3)X_2$ (X = Cl, I) (S.J. Holmes, unpublished results) is treated with hydrazine in the presence of triethylamine (equation 2). An x-ray structure of the iodo derivative

$$2\ W(\eta^5\text{-}C_5Me_4Bu^t)(CCMe_3)X_2 + N_2H_4 \longrightarrow [W(\eta^5\text{-}C_5Me_4Bu^t)(CCMe_3)X]_2(\mu\text{-}N_2H_2) \qquad (2)$$

(Churchill et al., 1983) is shown in Figure 3. The WNNW linkage is planar and the nitrogen atoms therefore sp^2 hybridized. The W-N distance of 1.932Å and N-N distance of 1.410Å we feel force us to conclude that the μ-N_2H_2 ligand is a dianion rather than the alternative diimine. In contrast, the x-ray structure of $[Cr(CO)_5]_2(\mu\text{-}N_2H_2)$ (Huttner et al., 1974) in which Cr-N = 2.076Å and N-N = 1.25Å shows that the μ-N_2H_2 ligand is essentially diimine; "$Cr(CO)_5$" is evidently a much poorer reducing agent than "$W(\eta^5\text{-}C_5Me_4Bu^t)(CCMe_3)I$".

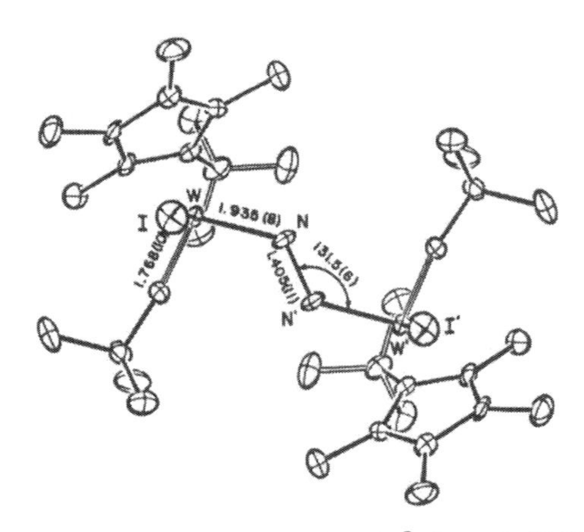

Figure 3. X-ray structure of $[W(\eta^5\text{-}C_5Me_4Bu^t)(CCMe_3)I]_2(\mu\text{-}N_2H_2)$

We then began exploring the chemistry of recently discovered tungsten(IV) acetylene complexes (Greco et al., 1973; Hey et al., 1983; K.H. Theopold, unpublished results) which, in view of (inter alia) the fact that the diphenylacetylene ligand in $W(PhC{\equiv}CPh)(OCMe_2CMe_2O)_2$ can be protonated by t-butanol to give $W(PhC{=}CHPh)(OCMe_3)(OCMe_2CMe_2O)_2$ (K.H. Theopold, unpublished results), can be regarded as tungsten(VI) complexes containing the $PhC{=}CPh^{2-}$ ligand. $W(C_2Ph_2)(OCMe_3)_4$ reacts smoothly and completely with 0.5 eq of hydrazine to produce a single product which can be crystallized with difficulty in the form of dense, smooth, brown nodules (equation 3). The extreme sensitivity of this complex to water

$$2\ W(C_2Ph_2)(OCMe_3)_4 \xrightarrow[-4\ Me_3COH]{N_2H_4} [W(C_2Ph_2)(OCMe_3)_2]_2(\mu\text{-}N_2) \quad (3)$$

we believe is the reason why we have not yet been able to analyze it successfully. (A relative below has been analyzed successfully for C, H, Cl, and N.) However, its ^1H and ^{13}C NMR spectra and its hydrolysis by aqueous HCl to give hydrazine quantitatively clearly point to the formulation shown. The IR spectrum shows an absorption at 1665 cm^{-1} characteristic of these "metallacyclopropene" complexes, but no obvious absorption ascribable to a mode involving the $\mu\text{-}N_2$ ligand. We propose that the dinitrogen is a $\mu\text{-}N_2{}^{4-}$ ligand so that any absorption associated with that linkage should be found at low energy (cf. ~850 cm^{-1} in Ta=N-N=Ta complexes). We believe the structure of $[W(C_2Ph_2)(OCMe_3)_2]_2$-$(\mu\text{-}N_2)$ will be pseudo-tetrahedral, a fact which would account in part for its extreme sensitivity to water.

Addition of 4 eq of HCl to $[W(C_2Ph_2)(OCMe_3)_2]_2(\mu\text{-}N_2)$ in the presence of 1,2-dimethoxyethane yields a purple crystalline complex in high yield. Elemental analysis and ^1H and ^{13}C NMR data suggest it is the complex shown in equation 4. Again, we see a "$\nu_{C\equiv C}$" absorption in its IR

$$[W(C_2Ph_2)(OCMe_3)_2]_2(\mu\text{-}N_2) \xrightarrow[-4\ Me_3COH]{4\ HCl,\ 2\ dme} [W(C_2Ph_2)Cl_2(dme)]_2(\mu\text{-}N_2) \quad (4)$$

spectrum at 1715 cm^{-1}, but no obvious absorption that could be attributed to the μ-N_2 ligand. We expect its structure to be pseudo-octahedral with the μ-N_2^{4-} ligand cis to the acetylene ligand and the oxygen donors trans to each, as shown in Figure 4.

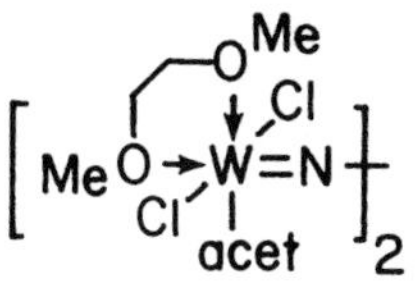

Figure 4. The proposed structure of $[W(C_2Ph_2)Cl_2(dme)]_2(\mu$-$N_2)$.

We have attempted to prepare other hydrazido(4−) complexes by adding hydrazine to $W(C_2Ph_2)(OCMe_2CMe_2O)_2$, $W(C_2Ph_2)(OCMe_3)_2Cl_2$, and $W(C_2Ph_2)Cl_4$(ether). In each case only sparingly soluble adducts containing the μ-hydrazine ligand are formed. We have also found the analogous reaction between W(3-hexyne)$(OCMe_3)_4$ and 0.5 eq of hydrazine to be considerably more complex than that involving $W(C_2Ph_2)(OCMe_3)_4$. Although NMR studies suggest that $[W$(3-hexyne)$(OCMe_3)_2]_2(\mu$-$N_2)$ is present, at least one other product is also, and no solid products have yet been isolated.

Preliminary CV studies on $[W(C_2Ph_2)(OCMe_3)_2]_2(\mu$-$N_2)$ and $[W(C_2Ph_2)Cl_2$-(dme)$]_2(\mu$-$N_2)$ show that they are difficult to reduce ($<$ -1.5V vs. SHE) but fairly easy to oxidize. For $[W(C_2Ph_2)(OCMe_3)_2]_2(\mu$-$N_2)$ in acetonitrile an irreversible oxidation at +0.41V (vs. SHE) was observed. At high scan rates a metastable oxidation product could be reduced and apparently reoxidized reversibly at -0.35V (vs. SHE). For $[W(C_2Ph_2)Cl_2$-(dme)$]_2(\mu$-$N_2)$ in dichloromethane an irreversible oxidation wave was found at +0.93V (vs. SHE) and a reversible wave ascribed to a metastable oxidation product at +0.58V. We are in the process of defining the electrochemical behavior more thoroughly. Our first suspicion is that dinitrogen is lost when either complex is oxidized by one or two electrons. We do know that odixation of $[W(C_2Ph_2)Cl_2(dme)]_2(\mu$-$N_2)$ with chlorine yields what appears to be $W(C_2Ph_2)Cl_3(OCH_2CH_2OMe)$, a type of product which has been observed in studies involving tungsten oxo complexes (Fowles et al., 1978).

CONCLUSIONS

Several features of hydrazido(x−) (x = 1-4) complexes of tungsten(VI) appear to be unique and important. Analogous molybdenum complexes should be preparable. Even though $Mo_2(\mu$-$N_2)$ linkages are thought not to be present in nitrogenase, at least some of the principles surrounding the chemistry of dinitrogen or hydrazido complexes of high oxidation state molybdenum or tungsten should guide us toward more accurate nitrogenase models and ultimately, a better idea of how to reduce molecular nitrogen.

REFERENCES

Churchill MR and Wasserman HJ (1981) Inorg. Chem. 20, 2899.
Churchill MR and Wasserman HJ (1982) Inorg. Chem. 21, 218.
Churchill MR and Wasserman HJ (1982) Inorg. Chem. 21, 223.
Churchill MR, Li J-Y, Blum L, and Schrock RR (1983) Organometallics 2, in press.
Cowie M and Gautier MD (1980) Inorg. Chem. 19, 3142.
Fowles GAW, Rice DA, and Shanton KJ (1978) J. Chem. Soc. Dalton Trans., 1658.
Frisch PD, Hunt MM, Kita WG, McCleverty JA, Rae AE, Sedden D, Swann D and Williams J (1979) J. Chem. Soc. Dalton Trans., 1819.
Greco A, Pirinoli F, and Dall'asta G (1973) J. Organometal. Chem. 60, 115.
Hey E, Weller F, and Dehnicke K (1983) Naturwiss. 70, 41.
Huttner G, Gartzke W, and Allinger K (1974) Angew. Chem. Int. Ed. Engl. 13, 822.
Pedersen SF and Schrock RR (1982) J. Am. Chem. Soc. 104, 7483.
Rocklage SM and Schrock RR (1982) J. Am. Chem. Soc. 104, 3077.
Rocklage SM, Turner HW, Fellmann JD and Schrock RR (1982) Organometallics 1, 703.
Wengrovius JH and Schrock RR (1982) Organometallics 1, 148.

ACKNOWLEDGEMENTS

We thank the National Institutes of Health for supporting this research (GM 31978) and J. Bentsen for his assistance in the preliminary CV studies.

THE SYNTHESIS STRUCTURES AND ELECTRONIC PROPERTIES OF Fe-Mo-S POLYNUCLEAR AGGREGATES. MOLECULES OF ELEMENTARY STRUCTURAL COMPLIANCE WITH THE Fe-Mo-S AGGREGATE OF NITROGENASE.

DIMITRI COUCOUVANIS
Department of Chemistry, University of Michigan, Ann Arbor, Michigan 48109 U.S.A.

INTRODUCTION

The synthesis and study of complexes containing iron, molybdenum and sulfur recently have been subjects of considerable interest to Inorganic chemists (Coucouvanis, 1981; Holm, 1981). This interest derives from recent advances in the chemistry of the nitrogenase enzymes (Eady, Smith, 1979) and the recognition that molybdenum is present in these enzymes in an extractable cofactor that contains iron, molybdenum and sulfur (Shah, Brill, 1977). The iron, molybdenum cofactor (FeMoco) from various nitrogen-fixing organisms seems to contain the same basic unit and extracts of the Fe-Mo component protein from inactive mutant strains of different micro-organisms are activated by the FeMoco. The apparently critical importance of the FeMoco in the function of nitrogenase has stimulated intensive research efforts toward an understanding of its structure and properties. The design and synthesis of Fe-Mo-S clusters as models for the FeMoco must be directed by and confined within the constraints imposed by the analytical, spectroscopic and structural data available for the FeMoco and the Fe-Mo protein component of nitrogenase. These data which in certain instances are subject to considerable uncertainty, when integrated, define a rather diffuse focal point toward which the synthetic studies on model complexes must be aimed.

THE FeMoco COMPOSITION

Originally, the FeMoco was reported to contain iron, molybdenum and sulfide in a 8:1:6 atomic ratio (Shah, Brill, 1977). Since then the Fe:Mo ratio is still open to question with reported ratios of 7:1 (Burgess et al., 1980; Newton et al., 1980) and more recently of 8.2±0.4:1 (Nelson et al., 1983). The sulfur content in the cofactor (which does not contain aminoacids, lipoic acid or coenzyme A (Burgess, Newton, 1983)) also has not been settled. Since the originally reported Mo:S ratio of 1:6 (Shah, Brill, 1977) ratios of 1:4 (Burgess, Newton, 1983) and 1:8 or 9 (Nelson et al., 1983) have been reported. The molecular weight of the FeMoco based on the comparative elution volumes of various inorganic complexes on Sephadex G-100 in N-methyl formamide (NMF) is estimated to be not much greater than 800 (Burgess, Newton, 1983).

SPECTROSCOPIC STUDIES

The unique, complicated structure of the FeMoco which apparently does not contain "common" iron sulfur clusters such as $Fe_2S_2(SR)_4$ or $Fe_4S_4(SR)_4$ (Rawlings et al., 1978) is evident in the results of elegant EPR and Mössbauer spectroscopic studies of the Fe-Mo protein. These studies show six iron atoms each in a distinctive magnetic environment coupled to an overall S = 3/2 spin system (Huyuh et al., 1979; Hoffman et al., 1982). Electron-nuclear double resonance (ENDOR) and EPR studies suggest one molybdenum per spin system and two noninteracting spin systems per Fe-Mo protein (Hoffman et al., 1982; Münck et al., 1975). In the ENDOR studies six distinct

^{57}Fe peaks (doublets) are observed and the iron atoms roughly can be grouped into trios having very similar hyperfine parameters (Hoffman et al., 1982).

EXAFS STUDIES

Extended X-ray absorption fine structure (EXAFS) analyses at the molybdenum K-edge of both the Fe-Mo protein and the FeMoco (Cramer et al., 1978) have shown as major features 3-4 sulfur atoms in the first coordination sphere at 2.35 Å and 2-3 iron atoms further out from the Mo atom at ∿2.7 Å. Two possible models for the immediate Mo environment proposed (Cramer et al., 1978) to be consistent with the EXAFS results are shown in Fig. 1. The iron EXAFS of the FeMoco (Antonio et al., 1982) shows the irons to be surrounded by an average of 3.4±1.6 S(Cl) atoms at 2.25(2) Å, 2.3±0.9 Fe atoms at

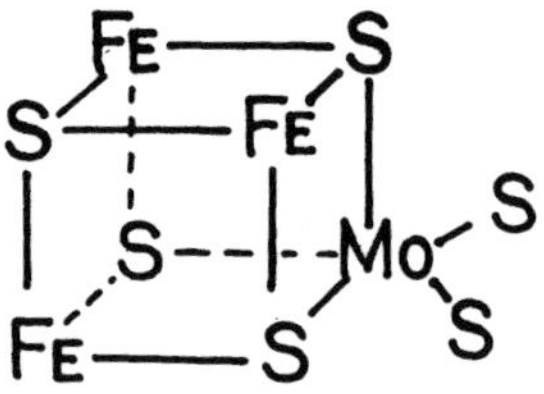

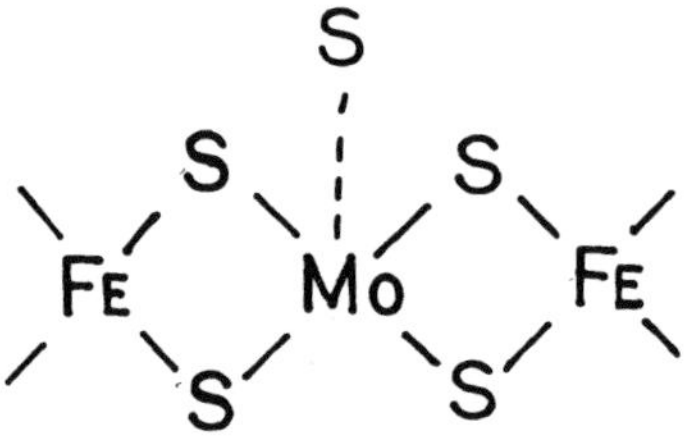

FIGURE 1. A B

2.66(3) Å, 0.4±0.1 Mo atoms at 2.76(3) Å and 1.2±1.0 O(N) atoms at 1.81(7) Å. The Fe and Mo EXAFS results suggest that the S:Fe ratio in the cofactor must be ≥1 if a reasonable structure is to be envisioned and consequently the recently reported (Nelson et al., 1983) S:Fe ratio of ∿9:6 is more compatible with the EXAFS results.

CHEMISTRY OF THE Fe-Mo-S COMPLEXES

From the EXAFS results it appears very likely that a partial structural feature of the Fe-Mo-S cluster in nitrogenase is the FeS_2Mo structural unit. Until 1978 only one Fe-Mo-S compound was known to contain this unit. In this compound, $(Cp)_2Mo(S\text{-}nBu)_2FeCl_2$, the Mo and Fe atoms are bridged by two $n\text{-}BuS^-$ ligands (Cameron, Prout, 1972). Since then, numerous Fe-Mo-S complexes have been obtained in reactions that employ the tetrathio molybdate, MoS_4^{2-}, anion as a reagent. These complexes, which without exception all contain the FeS_2Mo unit, can be obtained by either the use of the MoS_4^{2-} anion as a ligand in $MoS_4^{2-}\text{-}Fe(L)_n$ ligand exchange reactions, or by spontaneous self-assembly reactions (Wolff et al., 1979). In general, oligonuclear Fe-Mo-S complexes are obtained by the former method (Coucouvanis, 1981) and polynuclear clusters containing the Fe_3MoS_4 units are obtained by the latter method (Holm, 1981).

OLIGONUCLEAR Fe-Mo-S COMPLEXES

The synthesis (Fig. 2) structures (Fig. 3; Table I) and electronic properties of the oligonuclear Fe-Mo-S complexes have been described in detail previously (Coucouvanis, 1981, 1983; Müller, Diemann, 1983). In this paper emphasis will be placed on recent studies on the $[Cl_2FeS_2MoS_2FeCl_2]^{2-}$ complex (Coucouvanis et al., 1980) which resembles one of the two proposed models for the Mo coordination environment in nitrogenase (Cramer et al., 1978) (Fig. 1b).

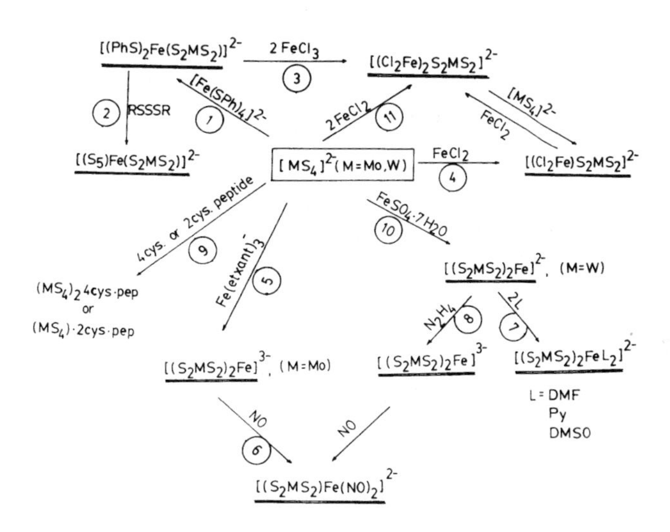

FIGURE 2.

THE $[Cl_2FeS_2MS_2FeCl_2]^{2-}$ COMPLEXES

Stable mixed ligand complexes of the type $[L_2FeS_2MS_2]^{2-}$ (L = PhS^- (Coucouvanis et al., 1979; 1983; Tieckelmann et al., 1980), L = Cl^- (Coucouvanis et al., 1980; Müller et al., 1980; Tieckelmann et al., 1980)) can be obtained by metatheses reactions (eq. 1)

$$[(L)_xFe]^n + [S_2MS_2]^{2-} \rightarrow [L_2FeS_2MS_2]^{2-} + x - 2L \quad (M = Mo, W) \qquad (Eq.\ 1)$$

The $[Cl_2FeS_2MS_2FeCl_2]^{2-}$ trinuclear complexes (Coucouvanis et al., 1980, 1983) can be obtained from the $[L_2FeS_2MS_2]^{2-}$ complexes by either addition of $FeCl_2$ ($L=Cl^-$) or by oxidation with $FeCl_3$ ($L=RS^-$). The interactions of the $FeCl_2$ molecules with the doubly bridging $MS_4{}^{2-}$ centers are rather weak and the trinuclear complexes solvolyze in coordinating solvents. The equilibrium constant Q for M = Mo (eq. 2) at 298 K in DMF solution is 334±10 M^{-1}.

$$[Cl_2FeMS_4]^{2-} + FeCl_2 \overset{Q}{\rightleftharpoons} [Cl_2FeS_2MS_2FeCl_2]^{2-} \qquad (Eq.\ 2)$$

The $[Cl_2FeS_2MS_2FeCl_2]^{2-}$ complexes undergo chemical reduction with R_4N^+ salts of the $BH_4{}^-$ anion, in either CH_3CN or CH_2Cl solution, and the overall stoichiometry of the reaction is:

$$(R_4N)_2(Cl_2FeS_2MoS_2FeCl_2) + R_4NBH_4 \rightarrow (R_4N)_2MoS_4Fe_2Cl_3 + R_4NCl + \tfrac{1}{2}B_2H_6 + \tfrac{1}{2}H_2. \qquad (Eq.\ 3)$$

The release of one mole of Cl^- per mole of $[Cl_2FeS_2MoS_2FeCl_2]^{2-}$ following reduction has been confirmed by elemental analysis of both products in the reaction (eq. 3). In NMF solution the $[MoS_4Fe_2Cl_3]^{2-}$ anion is rapidly converted to the $[Fe(MoS_4)_2]^{3-}$ complex. The latter complex for which the structure has been determined (Coucouvanis et al., 1980) is characterized by a

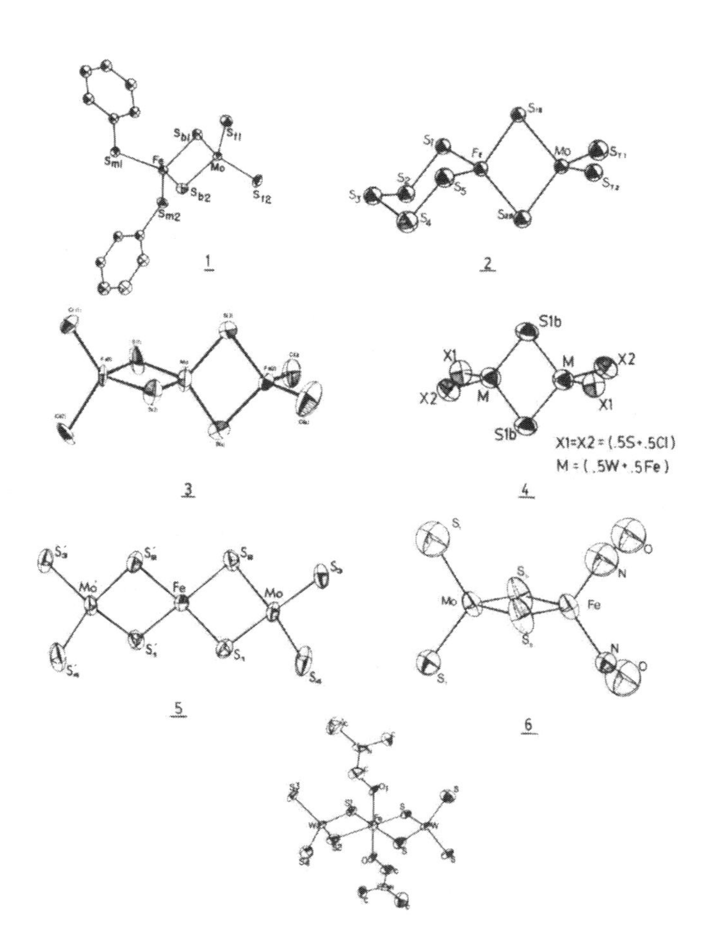

FIGURE 3.

TABLE 1. Selected structural parameters (Å) and Mössbauer parameters mm/sec of the Fe-MoS_4 complexes

	Complex	Mo-S_b [a,b]	Mo-S_4 [e]	Fe-Mo	I.S. [d,f]	Q.S. [e,f]	Ref.
1)	$[(SPh)_2FeMoS_4]^{2-}$	2.246(6)	2.153(6)	2.750(3)	0.44(1)	1.96(1)	g
		2.255(2)	2.153(2)	2.756(1)			h
2)	$[(S_5)FeMoS_4]^{2-}$	2.253(8)	2.145(7)	2.731(3)	0.38(2)	1.18(2)	i,g
3)	$[Cl_2FeS_2MoS_2FeCl_2]^{2-}$	2.204(5)		2.775(6)	0.57(1)	1.98(1)	j
4)	$[Cl_2FeMoS_4]^{2-}$	o	o	2.786(1)	0.59(1)	2.14(1)	j
				2.775	0.480(6)	2.100(5)[n]	k,h
5)	$[Fe(MoS_4)_2]^{3-}$	2.255(5)	2.171(5)	2.740(1)	0.42(3)	1.04(3)	ℓ
6)	$[(NO)_2FeMoS_4]^{2-}$	2.259(8)	2.182(8)	2.835(13)			m

a) For more than one bond, the mean value is reported with the standard deviation of the mean $\sigma = [\sum_o^N (\chi_i - \overline{X})^2/(N-1)]^{\frac{1}{2}}$. b) Bridging sulfur atoms. c) Terminal sulfur atoms. d) Isomer shift. e) Quadrupole splitting. f) Values reported at 77 K. g) Coucouvanis et al., 1979; 1983. h) Tieckelmann et al., 1980. i) Coucouvanis et al., 1980. j) Coucouvanis et al., 1980.

k) Müller et al., 1980. ℓ) Coucouvanis et al., 1980. m) Coucouvanis et al., 1981.

n) T = 295 K. o) Disordered structure.

S = 3/2 ground state (McDonald et al., 1980, 1983). The one electron electrochemical reduction of the $(Cl_2FeS_2MoS_2FeCl_2)^{2-}$ complex in CH_2Cl_2 proceeds with $E_{1/2}$ = -0.56 V. vs SCE. Cyclic voltammetric data show the reduction product to be unstable and to undergo a slow irreversible chemical reaction with a rate constant of 0.5 S^{-1}. Cyclic voltammetric and chronoamperometric measurements (Coucouvanis et al., 1983) demonstrate that the "decomposition" of the $[Cl_2FeS_2MoS_2FeCl_2]^{3-}$ reduction product is a first order and not a coupling reaction, and the following reduction mechanism can be described:

$$[Cl_2FeS_2MoS_2FeCl_2]^{2-} + e^- \rightarrow [Cl_2FeS_2MoS_2FeCl_2]^{3-} \xrightarrow{k=0.5\ S^{-1}} [(MoS_4)Fe_2Cl_3]^{2-} + Cl^-$$

A determination of current functions for the two reduction waves of the $[MoS_4Fe_2Cl_3]^{2-}$ complex (at -0.90 V and -1.21 V vs SCE) and a comparison to the value expected for a one-electron diffusion controlled process support that the wave currents correspond to 0.5 electrons per $[MoS_4Fe_2Cl_3]^{2-}$ unit. It appears therefore that the $[Cl_2FeS_2MoS_2FeCl_2]^{2-}$ reduction product, $[MoS_4Fe_2Cl_3]^{2-}$, is at least a $[(MoS_4Fe_2Cl_3)_2]^{4-}$ dimer. Possible structures for this dimer are shown in Fig. 4.

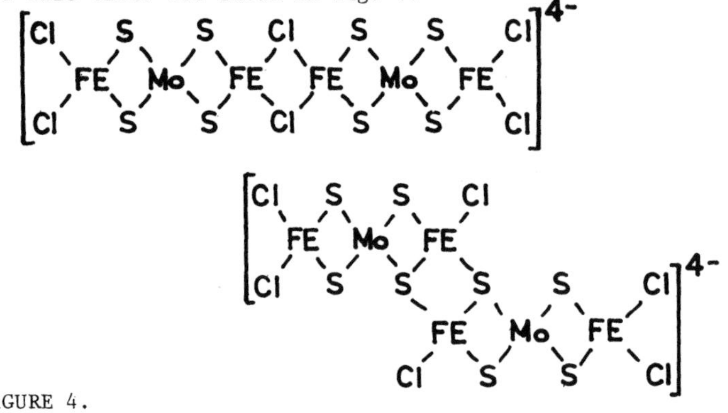

FIGURE 4.

Both of the structures in Fig. 4 contain two different iron sites. The presence of a single quadrupole doublet in the Mössbauer spectrum of the $[MoS_4Fe_2Cl_3]_2^{4-}$ anion, (I.S. = 0.50 mm/sec, Q.S. = 1.12 mm/sec at 77K) suggest that, if either of the structures in Fig. 4 is correct, the electronic differences between the two iron sites are too small to be resolved into two quadrupole doublets.

THE CRYSTAL STRUCTURE AND EXAFS STUDIES OF THE $[Cl_2FeS_2MoS_2FeCl_2]^{2-}$ COMPLEX ANION

The X-ray crystal structure of the $[Cl_2FeS_2MoS_2FeCl_2]^{2-}$ complex shows a tetrahedral MoS_4^{2-} unit bridging two $FeCl_2$ groups (Fig. 3-3), and resembles one of the structures proposed for the immediate coordination environment of the Mo atom in nitrogenases (Fig. 1b). The Mo-Fe distance at 2.775(6) Å is quite similar to the Mo-Fe distance revealed by EXAFS studies (Cramer et al., 1978; Antonio et al., 1982) in nitrogenase and the FeMoco. The Mo-S distance at 2.204(5) Å is considerably shorter than the 2.35 Å determined (by EXAFS analysis) for the Mo-S bonds in nitrogenase.

The Mo and Fe K-edge EXAFS of $(Ph_4P)_2(Cl_2FeS_2MoS_2FeCl_2)$ have been analyzed (Teo et al., 1983). The Mo-S, Mo-Fe and Fe-S(Cl) distances as determined by these analyses agree with the crystallographic values (Table I) to better than 0.5%. Similarly the number of neighboring atoms (around the Mo and Fe) agree to better than 9%. A comparison of the Fourier transforms of the Mo EXAFS of $[(PhS)_2FeS_2MoS_2]^{2-}$, $[Cl_2FeS_2MoS_2FeCl_2]^{-2}$, $[Mo_2Fe_6S_9(SEt)_8]^{3-}$ (vide infra) with that of the Fe-Mo protein of nitrogenase is shown in Fig. 5. It is clear that the Fourier transform of the MoEXAFS of the Fe-Mo protein of nitrogenase (Fig. 5-d) is quite different than the corresponding transforms of the Fe-Mo-S model complexes at least as far as peak intensities are concerned.

FOURIER TRANSFORMS OF Mo EXAFS
IN: A) $((PhS)_2FeS_2MoS_2)^{2-}$
B) $(Cl_2FeS_2MoS_2FeCl_2)^{2-}$
C) $(Mo_2Fe_6S_9(SEt)_8)^{3-}$
D) NITROGENASE

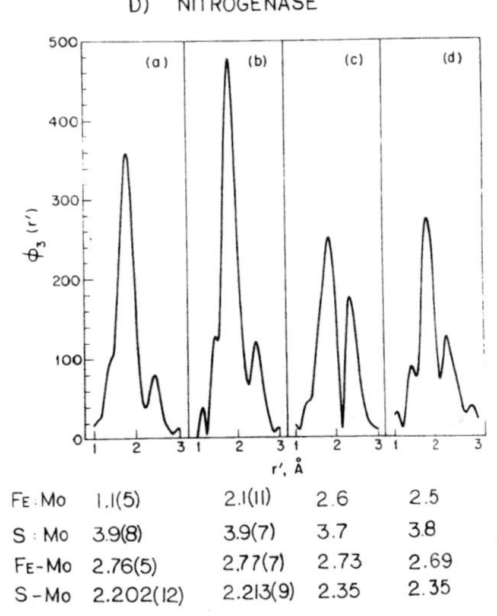

	(a)	(b)	(c)	(d)
Fe:Mo	1.1(5)	2.1(11)	2.6	2.5
S:Mo	3.9(8)	3.9(7)	3.7	3.8
Fe-Mo	2.76(5)	2.77(7)	2.73	2.69
S-Mo	2.202(12)	2.213(9)	2.35	2.35

FIGURE 5.

Qualitatively the Mo-S and Mo-Fe peaks in the Fourier transform of the Fe-Mo protein data resemble the corresponding peaks in the transforms of the EXAFS data for the $[Mo_2Fe_6S_9(SEt)_8]^{3-}$ cluster and the $[Cl_2FeS_2MoS_2FeCl_2]^{2-}$ complex.

POLYNUCLEAR Fe-Mo-S COMPLEXES

An impressive number of polynuclear Fe-Mo-S clusters constitutes the remainder of the Fe-Mo-S complexes known. These polynuclear clusters, that contain the $MoFe_3S_4$ unit, have been obtained in spontaneous assembly reactions of MoS_4^{2-} with iron and mercaptide ligands. A brief account of these complexes is given in the synthetic scheme in Fig. 6 and schematic structures are shown in Fig. 7.

(I) $MS_4^{2-} + 3\text{-}3.5\ FeCl_3 + 9\text{-}12\ NaSR \xrightarrow{MeOH,EtOH}$

a) $[M_2Fe_6S_9(SR)_8]^{3-}$ (M=Mo,W; R=Et) (Wolff et al., 1980)

b) $[M_2Fe_6S_8(SR)_9]^{3-}$ (M=Mo, R=Ph) (Christou et al., 1978)

(M=Mo, R=CH_2CH_2OH) (Christou et al., 1979)

(R=Et) (Wolff et al., 1980)

c) $[M_2Fe_7S_8(SR)_{12}]^{3-}$ (M=Mo,W; R=Et) (Wolff et al., 1980)
d) $[M_2Fe_7S_8(SR)_{12}]^{4-}$ (M=Mo, $R=SCH_2Ph$; M=W, R=Et) (Wolff et al., 1980)

(II) $[M_2Fe_7S_8(SEt)_{12}]^{3-} \xrightarrow[MeCN]{R_2catH_2} [M_2Fe_6S_8(SEt)_6(R_2'cat)_2]^{4-}$ (e)
(Armstrong et al., 1982)

(III) $[Mo_2Fe_6S_8(SR)_6(R_2'cat)_2]^{4-} \xrightarrow{solvent(L)} MoFe_3S_4(SR)(L)(R_2'cat)]^{3-}$
(f)
$L = RS^-, CN^- PhO^-$ (Armstrong et al., 1982).

FIGURE 6.

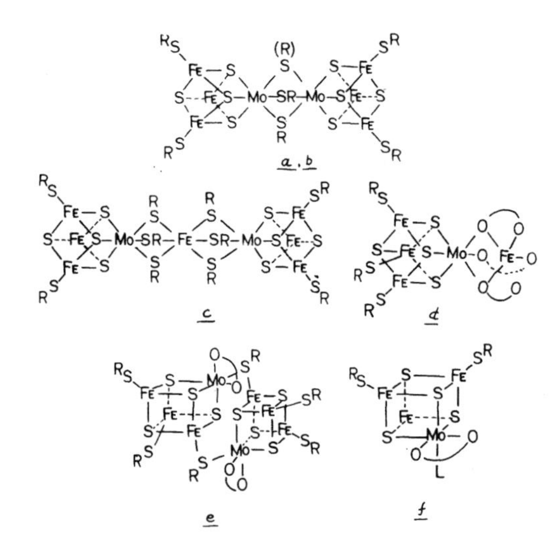

Fig 7

CONCLUSIONS

It is apparent that studies of the coordination chemistry of the Fe-Mo-S complexes, stimulated by the structural information on the nitrogenase active site and the FeMoco have shown considerable progress. In none of the Fe-Mo-S complexes isolated thus far does the Fe/Mo ratio come close to the approximate 6:1 ratio suggested by the Mössbauer and EPR studies and none of them realize as yet the simultaneous presence of the minimal basic characteristics.

A possible model for the Fe-Mo-S site in nitrogenase that seems to satisfy various of the restrictions imposed by the available spectroscopic and structural data on the Fe-Mo cofactor is shown in Fig. 8b. The Fe-S framework in this proposed model already exists in the $[Fe_6S_9(SR)_2]^{4-}$ cluster synthesized and structurally characterized by Holm and coworkers (Christou et al., 1982) (Fig. 8a). Addition of a MoL_3 fragment to either of the two vacant apices in the two Fe_3S_4 cubic units of the $[Fe_6S_9(SR)_2]^{4-}$ cluster would result in the $[L_3MoFe_6S_9(SR)_2]^n$ cluster. Depending on the oxidation state of the Mo atom ((III) or (IV)) and the nature of the ligands L (neutral or anionic) the overall charge of this cluster could vary from 0 to -4.

The RS^- ligands and the L ligands on the molybdenum in this cluster could be wholly or partially replaced by such ligands as citrate, Cl^- or NMF. With the exception of the spin state which is almost impossible

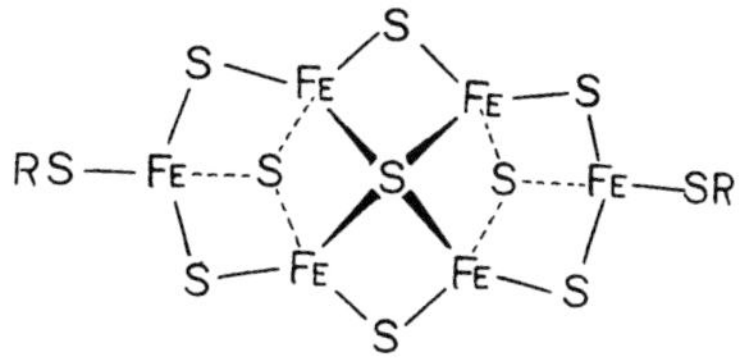

a

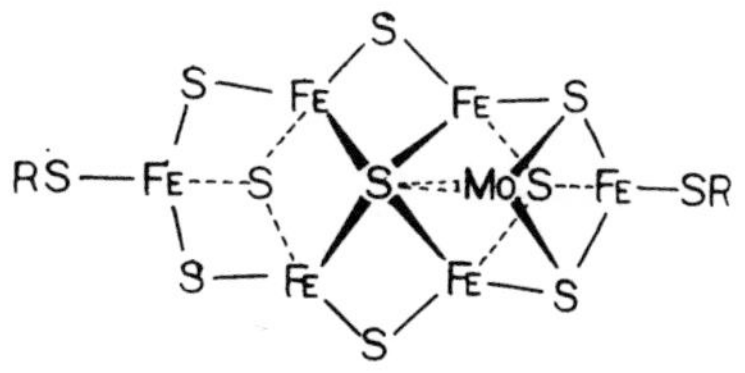

b

FIGURE 8.

to suggest Appriori for the proposed $[L_3MoFe_6S_9(L')_2]^n$ cluster, the stoichiometry and structure roughly are or could be in agreement with a) the Mo EXAFS on the cofactor, b) the Fe EXAFS, c) the most recent Mo:F:S assay and d) the Mössbauer and ENDOR data.

ACKNOWLEDGEMENTS

Support of this work by the National Institute of Health (GM33080-01) and the National Science Foundation (CHE-8306855) is gratefully acknowledged.

REFERENCES

Acott SR, Christou G, Garner CD, King TJ, Mabbs FE and Miller RM (1979) Inorg. Chim. Acta 35, L337.
Antonio MR, Teo BK, Orme-Johnson WH, Nelson MJ, Groh SE, Lindahl PA, Kauzlarich SM and Averill BA (1982) J. Am. Chem. Soc. 104, 4703.
Armstrong WH, Holm RH (1981) J. Am. Chem. Soc. 103, 6246
Armstrong WH, Mascharak PK and Holm RH (1982) J. Am. Chem. Soc. 104, 4373.
Armstrong WH, Mascharak PK and Holm RH (1982) Inorg. Chem. 21, 1700.
Burgess BK, Jacogs DB and Stiefel EI (1980) Biochim. Biophys. Acta 614, 196.
Burgess BK and Newton WE (1983) In Müeller A and Newton WE, eds, Nitrogen Fixation, p. 83, Plenum Press, New York.

Cameron TS and Prout CK (1972) Acta Cryst. B28, 453.
Christou G, Garner CD, Mabbs FE and King TJ (1978) J. Chem. Soc. Chem. Commun., 740.
Christou G, Garner CD, Mabbs FE and Drew MGB (1979) J. Chem. Soc. Chem. Commun., 91.
Christou G, Sabat M, Ibers JA and Holm RH (1982) Inorg. Chem. 21, 3518.
Coucouvanis D, Simhon ED, Swenson D, Baenziger NC (1979) J. Chem. Soc. Chem. Commun., 361.
Coucouvanis D, Stremple P, Simhon ED, Swenson D, Baenziger NC, Draganjac M, Chan LT, Simopoulos A, Papaefthymiou V, Kostikas A and Petrouleas V (1983) Inorg. Chem. 22, 293.
Coucouvanis D, Baenziger NC, Simhon ED, Stremple P, Swenson D, Kostikas A, Simopoulas A, Petrouleas V and Papaefthymiou V (1980) J. Am. Chem. Soc. 102, 1732.
Coucouvanis D, Baenziger NC, Simhon ED, Stremple P, Swenson D, Kostikas A, Simopoulos A, Petrouleas V and Papaefthymiou V (1980) J. Am. Chem. Soc. 102, 1730.
Coucouvanis D, Simhon ED and Baenziger NC (1980) J. Am. Chem. Soc. 102, 6644.
Coucouvanis D, Simhon ED, Stremple P and Baenziger NC (1981) Inorg. Chim. Acta 53, L135.
Coucouvanis D (1981) Acc. Chem. Res. 14, 201.
Coucouvanis D (1983) In Müller A and Newton WE, eds, Nitrogen Fixation, p.211. Plenum Press, New York.
Coucouvanis D, Simhon ED, Stremple P, Ryan M, Swenson D, Baenziger NC, Simopoulos A, Papaefthymiou V, Kostikas A and Petrouleas V (1983) Inorg. Chem., in press.
Cramer SP, Gillum WD, Hodgson KO, Mortenson LE, Stiefel EI, Chisnell JR, Brill WJ and Shah VK (1978) J. Am. Chem. Soc. 100, 4630.
Cramer SP, Hodgson KO, Gillum WD, Mortenson LE (1978) J. Am. Chem. Soc. 100, 3398.
Eady RR and Smith EB (1979) In Hardy RWF, Bottomley F and Burns RC, eds, A Treatise on Dinitrogen Fixation, p. 399, Wiley Interscience, New York.
Hoffman BM, Roberts JE and Orme-Johnson WH (1982) J. Am. Chem. Soc. 104, 860.
Hoffman BM, Venters RA, Roberts JE, Nelson M and Orme-Johnson WH (1982) J. Am. Chem. Soc. 104, 4711.
Holm RH (1981) Chem. Soc. Rev. 10, 455.
Huynh BH, Münck E and Orme-Johnson WH (1979) Biochim. Biophys. Acta 527, 192.
McDonald JW, Friesen GD and Newton WE (1980) Inorg. Chim. Acta 46, L79.
Müller A, Tölle HG and Bögge H (1980) Z. Anorg. Allg. Chem. 471, 115.
Müller A and Diemann E (1983) In Müller A and Newton WE, eds., Nitrogen Fixation, p. 211, Plenum Press, New York.
Münck E, Rhodes H, Orme-Johnson WH, Davis LC, Brill WJ and Shaw VK (1975) Biochim. Biophys. Acta 400, 32.
Nelson MJ, Levy MA and Orme-Johnson WH (1983) Proc. Natl. Acad. Sci. U.S.A. 80, 147.
Newton WE, Burgess BK and Stiefel EI (1980) In Newton WE and Otsuka S, eds., Molybdenum Chemistry of Biological Significance, p. 191, Plenum Press, New York.
Rawlings J, Shah VK, Chisnell JR, Brill WJ, Zimmerman R, Münck E and Orme-Johnson WH (1978) J. Biol. Chem. 253, 1001.
Shah VK and Brill WJ (1977) Proc. Natl. Acad. Sci. U.S.A. 74, 3249.
Teo BK, Antonio MR, Coucouvanis D, Simhon ED and Stremple PP (1983) J. Am. Chem. Soc., in press.
Tieckelmann RH, Silvis HC, Kent TA, Huynh BH, Waszczak JV, Teo BK and Averill BA (1980) J. Am. Chem. Soc. 102, 5550.

Wolff TE, Berg JM, Power PP, Hodgson KO, Holm RH and Frankel RB (1979) J. Am. Chem. Soc. 101, 5454.
Wolff TE, Power PP, Frankel RB, and Holm RH (1980) J. Am. Chem. Soc. 102, 4694.

POSTER DISCUSSION 2 RECENT PROGRESS IN THE CHEMISTRY OF DINITROGEN FIXATION

G.J. LEIGH and A.E. SHILOV
ARC Unit of Nitrogen Fixation, University of Sussex, Brighton BN1 9RQ, England and Institute of Chemical Physics, USSR Academy of Sciences, Chernogolovka, USSR

The posters presented at this Conference represent three areas in which progress has been considerable over the past two years. However, since so few posters were submitted, not all the significant advances have been encompassed.

Those posters which can be grouped under the heading of clusters represent a particularly active area of work. The design of new clusters with specific properties is a problem which admits of no easy solution. The attempts described in 2-7 show that S_2^{2-} and maybe other polysulphur derivatives have to be considered as plausible units in nitrogenase model structures. Poster 2-5 introduces a further variation, attempts to use a solid state crystal structure to provide a specific environment for a model dinitrogen-activating center. Studies of the chemistry of these cluster complexes, such as those reported in poster 2-6, are in their preliminary stages. The recent work of Holm et al. on the synthesis and chemical properties of 'mono-cubane' clusters shows that a systematic investigation of the ligand-binding sites on a specific metal atom of a cluster is now possible. This should lead to many developments in the near future.

The single most important development over the last two years has been the preparation and complete characterization by Schrock et al. of dinitrogen complexes of tantalum (III), mobium (III) and tungsten (IV). These demonstrate unequivocally that dinitrogen can bind to transition metals in high oxidation states, and with hard ligands. Structural data are more consistent with the formation M=N-N=M (for example, as a tantalum (V) hydrazido(4-)-complex) rather than with the formulation as dinitrogen complexes. Some of these complexes can be made from dinitrogen, and they protonate to give hydrazine. They are of a type postulated by Shilov et al. in their fixing systems and they may be of relevance to the function of nitrogenase. It is of interest, however, that the production of hydrazine by the protonation of briding dinitrogen also occurs in more classical cases, such as $[(\mu\text{-}N_2)\{W(N_2)_2(PEt_2Ph)_3\}_2]$ (Anderson et al.).

Developments in the chemistry of the classical dinitrogen complexes have also been considerable. Posters 2-2 and 2-3 demonstrate the way in which generalizations can now be made of reactivity and electron distribution in dinitrogen complexes. However, we should be aware of too facile explanations. The position of attack by nucleophile on the nitrogens of diazenido-ligands can be a function of the particular nucleophile involved, for example, H^- as compared to CH_3^-. In some cases attack can take place concomitantly on the metal, or at other positions in the ligands. The factors governing these observations (Sutton et al.) are so subtle that it will be a long time before they are completely understood. However, the development of multinuclear n.m.r. spectroscopy, with its potential for characterizing the electronic conditions of particular atoms (see poster 2-1), will undoubtedly be a considerable help in solving such problems.

Apart from all these developments, the following also appear to us to be of particular significance. The demonstration (Hussain et al.) of the splitting of a nitrogen-nitrogen bond in a hydrazido(2-)-complex to yield an amine and a nitrido-complex points a way to catalytic pathways to produce amines from dinitrogen. The function of a phospholipid (Shilov et al.) in increasing the reactivity of the molybdenum-sodium amalgam fixing system so that it can reduce dinitrogen catalytically at room temperature and atmospheric pressure opens the way to the design of new catalysts supported on membranes. It is to be expected that significant advances in all the areas cited will be reported at the next Conference in two years time.

SYNTHESIS AND CHARACTERIZATION OF $[Fe_4S_5Cp_4][MoOCl\ (thf)]$: A "CUBANE" TYPE CLUSTER WITH ONE PENTA-COORDINATED Fe SITE.

NADINE DUPRE/HUGO M.J. HENDRIKS/JEANNE JORDANOV
LABORATOIRES DE CHIMIE (LA CNRS n° 321), DÉPARTEMENT DE RECHERCHE FONDAMENTALE
CENTRE D'ETUDES NUCLÉAIRES DE GRENOBLE, 85 X, F.38041 GRENOBLE CEDEX, FRANCE.

Recent findings show that natural Fe-S clusters (1) are not confined to the by now well studied species having planar 2Fe-2S and cubane-type 4Fe-4S core units. This has triggered a series of synthetic efforts to obtain new polynuclear Fe/S geometries, where one of the Fe sites is different from the others in its coordination number and/or type of ligands (2). In this paper we report the preparation and properties of a new structural Fe-S cluster type. The X-rays analysis reveals a S_2^{2-} ligand bonded to three of the Fe sites, and one Fe atom is pentacoordinated. Also, EPR and Moessbauer studies indicate the presence of a $[3Fe^{III}, 1Fe^{II}]$ oxidation state.

Materials and Methods

The preparative reactions were carried out in dry solvents under argon using starting compounds prepared by literature methods. A filtered solution of $Fe_4S_6Cp_4.1/2\ CH_2Cl_2$ (5×10^{-3} mol) in dichloromethane (100 cm3) is added to a solution of $MoOCl_3(thf)_2$ (1×10^{-3} mol) in dichloromethane (20 cm3). The mixture is stirred (15 min) and the brown precipitate (yield 60%) filtered, washed with dichloromethane and dried. The title compound was recrystallized (red-black crystals) from acetonitrile/thf/hexane.

Results and Discussion

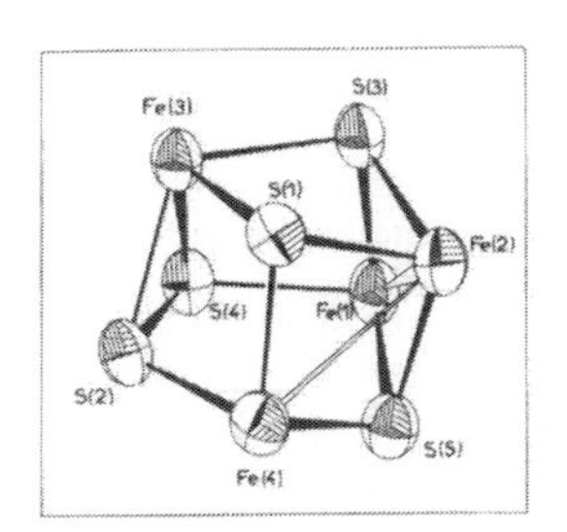

Fe(1)-S(3)	2.189 Å	Fe(4)-S(1)	2.210 Å
Fe(1)-S(4)	2.218	Fe(4)-S(2)	2.194
Fe(1)-S(5)	2.186	Fe(4)-S(5)	2.200
Fe(2)-S(1)	2.240	Fe(3)-S(1)	2.284
Fe(2)-S(3)	2.188	Fe(3)-S(2)	2.276
Fe(2)-S(5)	2.208	Fe(3)-S(3)	2.277
		Fe(3)-S(4)	2.271

In the Fe-S core, 3 different types of Fe are present : (a) Fe(1) and Fe(4) each bonded to 3S and 1Fe ; (b) Fe(2) bonded to 3S and 2Fe ; (c) Fe(3) bonded to 4S. Moreover, this last Fe has expanded its ligation to penta-coordination, while the remaining Fe are tetra-coordinated. Preliminary Moessbauer studies in the solid state indicate the presence of a strongly coupled species with IS : 0.31 mm/s (QS = 1.075 mm/s) at 1.4 K. Epr studies at low concentration ($<10^{-3}$ mol) are consistent with a paramagnetic species (S = 1/2) with g = 2.118, 1.998, 1.968 (4K). The IS value and the $g_{eV} > 2$ imply that the average oxidation state of the Fe atoms is 2.75.

(1) Spiro TG (1982) Iron-Sulfur Proteins. Wiley-Interscience, New York.
(2) Kanatzidis MG. et al. (1983) Inorg. Chem. 22, 179-181.

^{15}N AND ^{95}Mo NMR STUDIES OF METAL COMPLEXES RELEVANT TO NITROGEN FIXATION PROCESSES.

M. HUGHES[++], G.J. LEIGH[+], H. MODH-ALI[+], J. MASON[++] and R.L. RICHARDS[+]
+ ARC UNIT OF NITROGEN FIXATION AND CHEMISTRY DEPT., UNIVERSITY OF SUSSEX, BRIGHTON BN1 9RQ, UK.
++ OPEN UNIVERSITY, MILTON KEYNES, MK7 6AA, UK.

In recent times, availability of high field, multinuclear F.T. N.M.R. spectrometers has opened up the opportunity to observe ^{15}N and ^{95}Mo nucleii in chemical systems and possibly in nitrogenase. Nevertheless, the low sensitivity of these nuclei (and low natural abundance in the case of ^{15}N) still necessitate long accumulation times and isotopic enrichment for successful spectral measurement.

We have determined the NMR parameters of a range of complex compounds of molybdenum (and tungsten) which contain ligands relevant to the possible mechanism of reduction of dinitrogen on molybdenum in nitrogenase. Thus ^{15}N shift ranges have been mapped for the proposed (J.R. Dilworth et al 1981) reduction cycle:

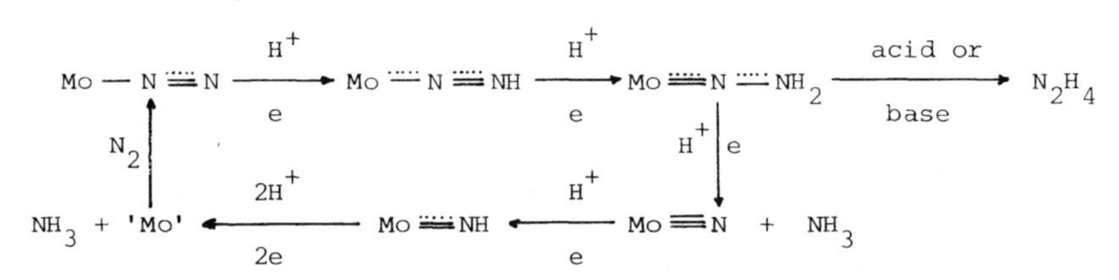

Chemical shifts have also been obtained for $M-N_2$ (M = W, Re, Fe, Ru, Os, Rh) and M - N = N - M' (M = Re or W; M' = Ti, Zr, Hf, Nb, Ta, W, Mo, Al etc.) (J. Mason, R.L. Richards, unpublished).

These parameters can be used to determine the basicity of N_β in MN_2, to follow the dynamics of the stages of reduction of dinitrogen in its complexes and to determine the geometry of N_2R_2, N_2R, NR and NO groups attached to metals in solution (J.R. Dilworth et al 1981 and unpublished work).

Study of ^{95}Mo shifts of d^6 octahedral complexes of N_2 and related ligands has led to an NMR spectrochemical series of ^{95}Mo shielding, which depends on the ligand field splitting: PF_3 $\sim$ phosphite $\sim$ CO > aryl phosphine > alkyl phosphine > MeCN > pyridine $\sim$ piperidine > N_2 > No. $^1J(Mo^{31}P)$ values reflect the σ-electron-donor ability of the ligands, showing e.g. N_2 < CO (S. Donovan-Mtunzi et al 1983).

REFERENCES

Dilworth JR, Donovan-Mtunzi S, Kan CT, Mason J, Richards RL (1981) Inorg. Chim.Acta. 53, L161 and references therein.
Donovan-Mtunzi S, Hughes M, Leigh GJ, Mason J, Mohd-Ali H, Richards RL, (1983) J.Organometallic Chem. 246, C1.

ACKNOWLEDGEMENTS

We thank the SERC, UK for a studentship to M.H. and the Government of Malaysia for a studentship to H.M.A.

PREPARATION AND X-RAY STRUCTURE OF mer-$|Re(\eta^1-S_2PPh_2)(N_2)(CNMe)(PMe_2Ph)_3|$, A MIXED DINITROGEN-ISOCYANIDE COMPLEX STABILIZED BY A SULPHUR LIGAND

ARMANDO J.L. POMBEIRO
CENTRO DE QUÍMICA ESTRUTURAL, COMPLEXO I, INSTITUTO SUPERIOR TÉCNICO
1000 LISBOA, PORTUGAL
PETER B. HITCHCOCK, RAYMOND L. RICHARDS
SCHOOL OF MOLECULAR SCIENCES, UNIVERSITY OF SUSSEX, U.K.

In a dinitrogen complex, if the M-N_2 bond is stabilized by a convenient electron releasing co-ligand (X) in _trans_ position, a displacement reaction may occur with retention of N_2. Hence, the replacement of a tertiary phosphine, bound to a _trans_-$\{XRe(N_2)\}$ metal site (X = Cl or S_2PPh_2), by an isocyanide may occur in preference to N_2 loss, leading to mixed dinitrogen-isocyanide species (_e.g._ equation 1, L = PMe_2Ph) (Pombeiro 1982), where the strong organic π-acceptor ligand coexists with N_2.

$$\textit{trans}\text{-}|ReCl(N_2)L_4| + CNMe \rightarrow \textit{mer}\text{-}|ReCl(N_2)(CNMe)L_3| + L \qquad (1)$$

A related complex, _mer_-$|ReCl(N_2)(CNMe)L'_3|$ (L' = $P(OMe)_3$) (Carvalho _et al._, 1982) may be prepared through a reductive elimination reaction of $|ReCl_2(NNCOPh)L'_3|$ with addition of the isocyanide.
Moreover, the dithiophosphinato ligand, $S_2PPh_2^-$, presents a potential versatile coordinating ability and a bidentade to a monodentate arrangement in the _mer_-$|Re(\eta^2-S_2PPh_2)(N_2)L_3|$ complex allows the coordinating of an isocyanide ligand to afford _mer_-$|Re(\eta^1-S_2PPh_2)(N_2)(CNMe)L_3|$ (reaction 2)whose molecular structure is authenticated by X-rays.

$$|Re(\eta^2-S_2PPh_2)(N_2)L_3| + CNMe \rightarrow \textit{mer}\text{-}|Re(\eta^1-S_2PPh_2)(N_2)(CNMe)L_3| \qquad (2)$$

The Re-N_2 bond stabilizing $S_2PPh_2^-$ ligand is _trans_ to N_2, as expected,and the Re-N and N-N bond lengths |1.83(1) and 1.13(1)Å, respectively| are somewhat shorter and longer, respectively, than those of the related _mer_-$|ReCl(N_2)(CNMe)L'_3|$ complex |corresponding values 1.98(1) and 1.04(2)Å|, thus evidencing the stronger net electron-donor character of the dithiophosphinato/phosphine ligand combination relative to the chloro//phosphite ligands; the former stabilizes to a greater extent the bond from the metal to the dinitrogen π-acceptor ligand and $\nu(N_2)$ for the dithiophosphinato complex (1980 cm^{-1}) occurs at a lower value than for the chloro species (2030 cm^{-1}).
This type of study is being extended to other sulphur-donor and isocyanide ligands in order to study the influence of the electronic and structural properties of metal sites on the metal-dinitrogen bond.

REFERENCES

- Carvalho FNN, Pombeiro AJL, Orama O, Schubert U, Pickett CJ, Richards RL (1982) J. Organometal. Chem., 240, C18
- Pombeiro AJL (1982), XXII ICCC, Budapest, 61

ACKNOMLEDGEMENTS - The support given by the JNICT (contract 216.80.56), the INIC (Portugal, to AJLP) and NATO (research grant 1604 to AJLP and RLR) are gratefully acknowledged.

ISOCYANIDES AS PROBES IN CHEMICAL NITROGEN FIXATION

ARMANDO J.L. POMBEIRO
CENTRO DE QUÍMICA ESTRUTURAL, COMPLEXO I, INSTITUTO SUPERIOR TÉCNICO
1000 LISBOA, PORTUGAL

Isocyanides are isoelectronic with dinitrogen and also reduced by nitrogenase with complete rupture of the triple bond in an overall $6H^+/6e^-$process.

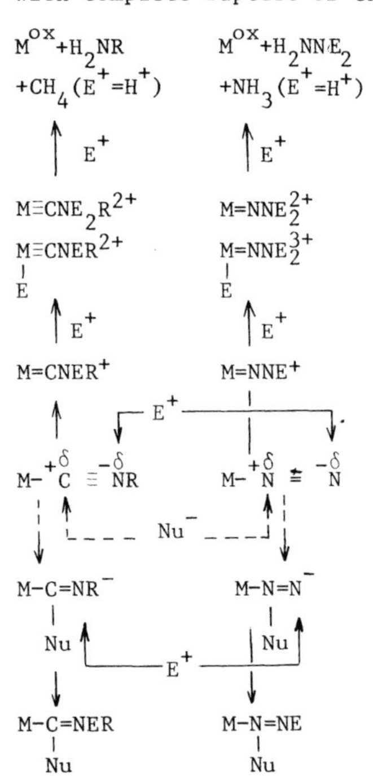

Both CNR and N_2 are activated towards attack by an electrophile(E^+) at the β atom when ligating the high electron rich{ML_4}(M=Mo or W, L=monophosphine or 1/2 diphosphine)centres |where ν(CN)or ν(N_2) occurs at very low wavenumbers| to give carbyne-type($M\equiv CNER^+$)(Pombeiro 80) or diazenido-type ($M=NNE^+$) species (Richards 83);activation towards a nucleophilic (Nu^-) attack at the α atom (which is followed by a ready electrophilic attack at the β position)is known for the low electron-rich {$Mn(\eta^5-C_5H_5)(CO)_2$}type centre where ν(CN) or ν(N_2)occurs at high wavenumbers(Treichel 73; Sellmann 77,78).Further electrophilic attack at the carbyne- or diazenido-type ligands my lead to amine+hydrocarbon (Pombeiro 80) or hydrazine+ammonia (Richards 83), respectively. The isocyanide, however, is more susceptible to protonation than N_2 and only the former undergoes attack by acid when ligating the {$ReCl(dppe)_2$} site (Pombeiro 81) with a weaker electron-rich character than the Mo(O) centre. In accord,$E^{ox}_{1/2}$ of |$ReCl(N_2)(dppe)_2$| lies off (by a cathodic shift) the linear plot followed by $E^{ox}_{1/2}$|$ReCl(L)(dppe)_2$| (L=CNR, CO) vs. $E^{ox}_{1/2}$|$MoL_2(dppe)_2$|, thus evidencing the weaker electronic versatility of accommodation of N_2 to a decrease in the electron releasing character of the metal centre (Pombeiro 81).

This type of combined chemical/electrochemical approach is being extended to other metal centres with different electron releasing prperties in order to study the relative behaviour of N_2 and CNR.

REFERENCES

- Pombeiro AJL and Richards RL (1980) Trans.Metal Chem. 5,55; ibid, 5,281; Pombeiro AJL(1980) In Chatt J et al.,eds, New Trends in the Chemistry of Nitrogen Fixation, Ch. 10, Academic Press
- Pombeiro AJL (1981) Rev. Port. Quím., 23, 179
- Richards RL (1983) In Müller A and Newton WE, eds.,Nitrogen Fixation,the Chemical-Biochemical-Genetic Interface, Plenum Press, 275
- Sellmann D and Weiss (1977)Angew.Chem.Int.Ed.,16,880;(1978),ibid, 17, 269
- Treichel PM (1973), Advan. Organometal. Chem., 11, 21

ACKNOWLEDGEMENTS: The support given by the JNICT (contract 216.80.56) and the INIC (Portugal) is gratefully acknowledged.

CAGE EFFECT IN THE ATTEMPTED SYNTHESIS OF CHEMICAL MODELLING COMPOUNDS FOR THE ACTIVE CENTER OF NITROGENASE--SYNTHESIS AND STRUCTURE OF $[Fe(DMF)_6][(FeCl_2)_2MoS_4]$

LIU QIOUTIAN/HUANG LIANGREN/KANG BEISHENG/LU JIAXI
FUJIAN INSTITUTE OF RESEARCH ON THE STRUCTURE OF MATTER, CHINESE ACADEMY OF SCIENCES, FUZHOU, FUJIAN, CHINA

We have proposed that by combination of a binuclear Mo-Fe-S cluster "fragment" with another bi- or tri-nuclear cluster, it is possible to obtain Fuzhou Model I or II type compounds through "spontaneous self-assembly" reactions. These "string-bag" modelling compounds are believed to be responsible for acetylene and dinitrogen reduction activities.

In the syntheses of these "fragments", we have obtained a Mo-Fe-S cluster $[Fe(DMF)_6][(FeCl_2)_2MoS_4]$ (I) by reacting $FeCl_2.4H_2O$ with $(NH_4)_2MoS_4$ in DMF in molar ratios of 3:1. This compound, obtained in the form of dark red needles, easily dissolves in DMF but decomposes readily in H_2O, CH_2Cl_2, CH_3CN, or C_2H_5OH, possibly due to disruption of the cation by extrusion of the DMF ligands.

IR absorptions at 320 and 340(Fe-Cl),460(Fe-S-Mo), 380 and 420(DMF), and 1645 cm^{-1} (C=O) are similar to those of $(Ph_4P)_2[(FeCl_2)_2MoS_4]$ (II) and incorporated DMF. Uv-vis. can only be obtained in DMF which causes the following equilibrium shift to take place: $[(FeCl_2)_2MoS_4]^{2-} \rightleftharpoons [(FeCl_2)MoS_4]^{2-} + FeCl_2$. The absorption bands (310, 430, 470 nm) are characteristic of $[(FeCl_2)MoS_4]^{2-}$ (III). The magnetic moment of (I) (μ= 8.52 BM at 298.5 K) is higher than that of (II) (6.64 BM at 300 K), and is attributed to the paramagnetism of the cation.

Single crystals of (I) have been obtained from DMF/THF in triclinic space group C_i^1-P$\bar{1}$ with a=12.624(2) Å, b=17.852(3) Å, c=18.867(4) Å, α=78.46(1)^o, β=72.57(2)^o, γ=79.88(1)^o, and Z=4. On the basis of intensity data of 4100 independent reflections ($F_o^2 > 2.5\sigma(F_o)^2$), the crystal structure has been determined and refined to R=7.1%. It is also clear that the cation $[Fe(DMF)_6]^{2+}$ does not form cages as the case of $[Mg(DMF)_6]^{2+}$ possibly due to the larger atomic diameter of Fe^{2+}.

In order to study the cage effect in the 'G' series model compounds, we have been interested in the attempted synthesis of $[Mg(DMF)_6][(FeCl_2)_2MoS_4]$ (IV). By either reacting $FeCl_2$ and $(NH_4)_2MoS_4$ in DMF in the presence of $MgCl_2$ or reacting (II) with $MgCl_2$ in DMF, we have isolated only $[Mg(DMF)_6][(FeCl_2)MoS_4]$ (V). On treating (V) with excess of $FeCl_2$ in DMF, we have been able to recover the starting material after stirring at room temperature for 16 hrs. We have not been able to obtain (IV). Thus we have come to the conclusion that it is characteristic of the 'G' series compounds to enclose dinuclear cluster inside a cationic cage made of Mg^{2+} each surrounded by six DMF. It is the cage effect of $[Mg(DMF)_6]^{2+}$ that degrades any trinuclear cluster formed into binuclear cluster to fit into the cationic cage.

THE CO COMPLEXATION OF $(Et_4N)_3[Fe_4S_4(SPh)_4]$

M.A. WALTERS/W.H. ORME-JOHNSON
Department of Chemistry, Massachusetts Institute of Technology,
77 Massachusetts Avenue, Cambridge, Massachusetts 02139 USA

Carbon monoxide inhibits the reduction of substrates by nitrogenase both in vivo and in vitro. A possible mechanism for this may involve the binding of CO to $Fe_4S_4(SR)_4$ clusters of the molybdenum-iron protein. We have carried out an investigation of CO binding to synthetic $Fe_4S_4(SPh)^{2-,3-}$ clusters in order to understand more fully the nature of CO binding in the MoFe protein of Av1 nitrogenase.

FTIR spectra of the reduced $Fe_4S_4(SPh)_4^{3-}$ cluster in acetonitrile solution indicate that the cluster binds three to four equivalents of carbon monoxide in sequential steps. Binding is complete at ∿600 torr and room temperature. The vibrational frequencies of the bound CO are indicative of exclusively terminal ligation through the carbon (Cotton, F.A., Wilkinson, G., 1972). A concentration dependence is observed in the vibrational spectrum wherein 4 bands appear at ∿600 torr and ∿0.2 mM cluster concentration while 8 bands appear in spectra in the >1.0 mM range. This increase in the number of vibrational bands is presumed to be the result of aggregation products. As the CO pressure above ∿0.2 mM solution is increased, the first bands to appear are those at 2044 and 1911 cm^{-1} respectively, followed by those at 2012 and 1966 cm^{-1}. The latter appear to increase in intensity at approximately the same rate. The number of bands as well as their intensity relationship to increasing CO partial pressure suggest the binding of 3-4 CO ligands. Formally the one electron reduced cluster is 3Fe(II), 1Fe(III). The ferrous and ferric iron sites provide high and low affinity Co binding sites respectively so that the binding constant of a fourth CO might be expected to be small. It is suggested that the vibrational bands at ∿2012 and 1966 cm^{-1} result from conformational isomers resulting from a variable coordination orientation of one CO ligand. In this way three CO ligands could give rise to a total of four vibrational bands.

The EPR spectra of the CO free and CO bound complexes are nearly identical, indicating only a minor perturbation of the cubane-like geometry with CO-binding. Proton NMR reveals a decrease in the thiolate proton isotropic shifts in the CO complex, perhaps due to the electron withdrawing effects of bound CO.

REFERENCES

Cotton, F.A., Wilkinson, G. (1972) Advanced Inorganic Chemistry. Interscience, New York.

ENZYMOLOGY OF NITROGENASE

POSTER DISCUSSION 3

STRUCTURE AND REACTIVITY OF NITROGENASE - AN OVERVIEW

BARBARA K. BURGESS
CHARLES F. KETTERING RESEARCH LABORATORY
YELLOW SPRINGS, OHIO 45387

1. INTRODUCTION

Nitrogenase is composed of two separately purified proteins called the molybdenum-iron protein (MoFe protein) and the iron protein (Fe protein).[1] N_2 fixation and all other reductions catalyzed by nitrogenase require both component proteins, a source of reducing equivalents, MgATP, protons and an anaerobic environment. This overview covers what has been learned recently about the composition, structure and redox properties of the two component proteins and the events that occur during nitrogenase turnover. Fortunately, diffraction quality crystals have recently been obtained for both the MoFe protein (Cp1 and Av1; Weiniger, Mortenson, 1982) and the Fe protein (Av2; Rees, Howard, 1983). Thus, many of the questions raised below may soon have definitive answers.

2. Fe PROTEIN

2.1. Polypeptide

The Fe protein is a dimer composed of two identical subunits. It has a native molecular weight (mw) of ca. 60,000 daltons and migrates as a single band during SDS gel electrophoresis with a subunit mw of ca. 30,000 daltons (e.g. Lowe *et al.*, 1980)[2]. This view is confirmed by the recent complete amino acid sequences. No exeptions to this general rule are known.
The amino acid sequence of the Fe protein has apparently been highly conserved throughout evolution. In recent years, Cp2 (Tanaka *et al.*, 1977), Av2 (Hausinger, Howard, 1982), Kp2 (Sundaresan, Ausubel, 1981), An7120 2 (Mevarech *et al.*, 1980) and Rm2 (Török, Kondorosi, 1981) have been sequenced. The last three were deduced from the DNA sequences of the *nif* H genes. Comparisons (e.g., Török, Kondorosi, 1981; Scott et al., 1981a; Hausinger, Howard, 1980 and 1982) show the strikingly high degree of homology. Regardless of the mechanism for evolution of the *nif* H gene (Postgate, 1974; Ruvkun, Ausubel, 1980), there must have been strong evolutionary pressure to conserve a specific primary structure for the Fe protein, no doubt in response to its functional demands.
The conserved regions of the Fe protein include a large hydrophobic region, which may represent the protein core, and three clusters of acidic amino acids (Hausinger, Howard, 1982). As an iron-sulfur protein, its cysteine residues are of

---

[1] The designations Ac, An, Av, Cp, Ca, Kp, Rm, Rhc and Rc are used for *Azotobacter chroococcum*, *Anabaena*, *Azotobacter vinelandii*, *Clostridium pasteurianum*, *Corynebacterium autotrophicum*, *Klebsiella pneumoniae*, *Rhizobium meliloti*, *Rhodopseudomonas capsulata* and *Rhodospirillum rubrum* respectively. 1 and 2 refer to the MoFe protein and Fe protein respectively. EXAFS, extended X-ray absorption fine structure; ENDOR, electron nuclear double resonance; EPR, electron paramagnetic resonance; CD, circular dichroism; MCD, magnetic circular dichroism; NMR, nuclear magnetic resonance.

[2] Inactive Rr2 migrates as two bands during SDS gel electrophoresis (see Ludden, this volume)

particular interest. Each subunit has from nine for Kp2 to five for Rm2 (with these five conserved in the four other species) located in highly homologous regions (Török, Kondorosi, 1981). The Fe protein shows no sequence homology with other [Fe-S] proteins and its cysteines are not clustered (Mortenson, Thorneley 1979; Hausinger, Howard, 1982). These differences and its large size emphasize the uniqueness of the Fe protein.
The recent sequence information gives subunit MW's of: 29,685 for Cp2; 31,416 for Av2; 31,753 for Kp2; 32,740 for Rm2; and 33,000 for An7120 2, showing variation by at least 6630 daltons.

2.2. Metal and Sulfide Composition

The Fe protein contains 3.5-4.1 g-atoms Fe per mole for homogenous preparations of Ac2, Ca2, Cp2, Kp2 (in Lowe et al., 1980), Av2 (Burgess et al., 1980; Rees, Howard, 1983), Cp2 (Nelson et al., 1983), Rr2 (Ludden, Burris, 1978) and Rhc2 (Hallenbeck et al., 1982), plus ∿4 g-atoms S^{2-} per mole (Lowe et al., 1980; Nelson et al., 1983; Hallenbeck et al., (1982). Core extrusion of Cp2 gives one [4Fe-4S] cluster (Averill et al., 1978; Gillum et al., 1977). In addition, Mössbauer experiments on Kp2 (Smith, Lang, 1974), linear electric field experiments on Cp2 (Orme-Johnson et al., 1977) and CD and MCD studies on Av2 and Kp2 (Stephens et al., 1979, 1983) are all consistent with a single [4Fe-4S] clusters as in bacterial ferredoxins.[3] Hausinger and Howard (1983) show by thiol reactivity that cystines 97 and 132 (referenced to Av2, conserved in all other species) are the probable ligands for the single [4Fe-4S] cluster, which is bound symmetrically between the subunits.

2.3. Redox Properties.

All Fe proteins with excess dithionite exhibit a rhombic, somewhat unusual EPR signal with g values close to those of reduced ferredoxins, strongly indicating the presence of $[Fe_4S_4(Cys)_4]^{3-}$ (e.g., Lowe et al., 1980; Mortenson, Thorneley 1979). The EPR signal should then integrate to one spin per molecule. In fact, values of 0.79, 0.25 and 0.2 for Cp2, 0.45 for Kp2, 0.17 for Ac2 (in Orme-Johnson et al., 1977) and most recently 1 for Av2 (Braaksma et al., 1982) are found. It is very unlikely that the integrations are incorrect (Orme-Johnson et al., 1977) or the Fe proteins were not completely reduced (see below) or Ac2 preparations of specific activity 2000 and spin integration of 0.17 contain 80% inactive Fe protein (Orme-Johnson et al., 1977). Perhaps a more feasible explanation is that the Fe protein contains a second, rapidly relaxing paramagnet, which is spin coupled to the [4Fe-4S] center (Lowe et al., 1980).
The Fe protein can be reversibly oxidized with dyes or with the MoFe protein plus MgATP to an EPR-silent state. Thus reduced Fe protein is viewed as a one-electron donor, whose $[Fe_4S_4(Cys)_4]^{n-}$ cluster functions between the n=3 and n=2 oxidation states (Mortenson, Thorneley, 1979). This view is supported by: (a) potentiometric titrations of the isolated Fe protein, where a one-electron transfer has been observed for Cp2 (E^o = -295 mV; Zumft et al., 1974), Av2 (E^o = -305 mV; G.D. Watt, personal communication) and Ac2 (in Lowe et al., 1980); (b) Mössbauer studies (Smith, Lang, 1974; Münck et al., 1975); and (c) CD and MCD studies (Stephens et al., 1979). But, because one electron develops fully an EPR signal which integrates to only ∿0.2 spin or two electrons (E^o = -397 mV) develop fully a signal, which only integrates to 1 spin, the situation is unclear and demands further experimentation.

[3] Braaksma et al. (1983) report at least 8Fe and $8S^{2-}$ atoms per Av2 in direct conflict with data cited above. The significance of this result must await independent confirmation.

2.4. Binding of MgATP and MgADP to Isolated Fe Protein

MgATP and MgADP both bind to the Fe protein alone and MgATP is not hydrolyzed by the Fe protein alone (e.g. Orme-Johnson *et al.*, 1977). The Fe protein possesses two MgATP-binding sites (Table 2). For reduced *Cp*2, MgADP appears to bind very tightly to one of the MgATP-binding sites with a K_{diss} of 5 μM (Tso, Burris, 1973) and weakly to the second MgATP site (Ljones, Burris, 1978). (For discussion of cooperativity, see Mortenson, Thorneley, 1979). For oxidized *Kp*2 (Stephens, 1983) and *Av*2 (P. Stephens and C. McKenna, personal communication), there are two MgADP-binding sites. *Cp*2 and *Av*2 also have quite different responses to the presence of ATP analogues (Weston *et al.*, 1983).

Table 2.

Species Number	Number MgATP Sites	K_{diss} μM	Redox state	References
*Cp*2	2	300	red	Mortenson *et al.*, 1976
*Cp*2	2	17	red	Tso, Burris, 1973
*Cp*2	2	-	red	Zumft *et al.*, 1973
*Cp*2	2	85	red	Ljones, Burris, 1978
*Cp*2	2	53	red	Emerich *et al.*, 1978
*Av*2	2	430,220	red	Hageman *et al.*, 1980
*Av*2	2	290	ox	Cordewener *et al.*, 1983
*Av*2	1	560	red	Cordewener *et al.*, 1983
*Kp*2	2	400	red	in Orme-Johnson, 1977
*Kp*2	2	-	ox	Stephens *et al.*, 1983

The binding of MgATP to the Fe protein causes: (a) the redox potential to be lowered by 50-100 mV (Lowe *et al.*, 1980; Braaksma *et al.*, 1982); (b) the conformation to be altered, to expose SH groups and the [4Fe-4S] cluster (Mortenson *et al.*, 1976; Ljones, Burris, 1978; Mortenson, Thorneley, 1979); (c) a dramatic change in the CD spectrum (Stephens *et al.*, 1979, 1983) and the EPR signal (in Lowe *et al.*, 1980), although the last has been questioned (Orme-Johnson *et al.*, 1977). The binding of MgADP causes similar changes in the redox potential, although the shape of the titration curve is different (Thorneley, Mortenson, 1979; Braaksma *et al.*, 1982), and in the CD spectrum (Stephens *et al.*, 1983). MgADP appears to bind to the same sites as MgATP (Mortenson *et al.*, 1976). Hausinger and Howard (1983) point out that MgATP can also bind reversibly to apo-*Av*2. Thus, although the effects of MgATP binding to the oxidized and reduced forms of the Fe protein may differ, Table 2 suggests that MgATP binds to both forms.

The two MgATP-binding sites may be either equivalent and interact directly with the [4Fe-4S] cluster (Ljones, Burris, 1978) or remote from the [4Fe-4S] cluster (Mortenson, Thorneley, 1979). Hausinger and Howard (1983) have recently shown that cysteine 85 (reference to *Av*2, conserved in other species) is a likely nucleotide binding site. This residue is not normally an exposed, chemically reactive residue, but one which becomes exposed when the [4Fe-4S] center is destroyed. All recent data show that MgATP binding to the Fe protein dramatically influences the environment around the [4Fe-4S] center, which suggests close interaction.

3. MoFe PROTEIN

3.1. Polypeptide

The MoFe protein is an $\alpha_2\beta_2$ tetramer with a native mw of ca. 220,000 daltons. No known species differences exist with respect to subunit pattern. Mw and amino acid composition data are compiled elsewhere (Lowe et al., 1980). The complete amino acid sequences of the α and β subunits of An7120 1 have been deduced from the DNA sequences of the nif D (Lammers, Haselkorn, 1983) and nif K (Mazur, Chui, 1982) genes, respectively. In addition, amino acid sequences of the N-terminal (Lundell, Howard, 1978) and all cysteinyl peptides (Lundell, Howard, 1981) for the α and β subunits of Av1 have been reported. N-terminal amino acid sequences for the α subunits of Cp1 (Hase et al., 1981), Kp1 (Scott et al., 1981b) and Rm1 (Török, Kondorosi, 1981) are also available, the latter two having been deduced from the DNA sequences of the nif D genes.
For the α subunit (nif D), partial sequences show quite good homology (47-66%) (Lammers, Haselkorn, 1983). It contains nine cysteines for Av1 and eight for An7120 1, five of which are conserved. Only one of the five homologous cysteinyl peptides has a sequence which is also found in bacterial ferredoxins. For the subunit, eight cysteines are found in Av1 and six in An7120 1. Again, five are conserved with one of these showing homology to HiPIP (Mazur, Chui, 1982). Thus, the MoFe protein has, at most, 20 conserved cysteins (Lammers, Haselkorn, 1983) for ligation of metal centers (see section 3.2).
A symmetrical relationship is found with respect to the length of the Fe protein C-terminal region and the length of the N-terminal region of the α subunit of the MoFe proteins for An7120, Rm, Kp, Av and Cp (J.-S. Chen, personal communication). The lengths decrease in the order An7120>Rm>Kp>Av>Cp. The C-terminal region (and/or size) of the Fe protein and the N-terminal region of the α subunit of the MoFe protein may be crucial to the formation of an active nitrogenase complex, as the length differences correlate nicely with the varying ability of MoFe proteins and Fe proteins to form active heterologous complexes (e.g. Emerich, Burris, 1978).

3.2. Metal Composition

The MoFe protein contains two atoms of Mo and 24-32 Fe atoms per molecule (Lowe et al., 1980; Mortenson, Thorneley, 1979). Mössbauer analysis of Av1, Kp1 and Cp1 suggests 33±3 Fe atoms per molecule (for review, see Orme-Johnson, Münck, 1980; Smith, 1983; Zimmerman and Trautwein, 1983) and analytical values are 30±3 (e.g. Orme-Johnson, Münck, 1980; Burgess et al., 1980; Kurtz et al., 1979; Stephens et al., 1981a). Sulfide analysis generally agrees closely with the number of Fe atoms (Mortenson, Thorneley, 1979). There are no known species differences with respect to metal content.
Quantitative extrusion of the non-FeMoco-associated Fe atoms from Av1 and Cp1 gives 4±0.2 (Orme-Johnson, Münck, 1980) and 3.4-4.0 (Kurtz et al., 1979) [4Fe-4S] centers per MoFe protein molecule, although the latter data have been recalculated at 2.9-3.5 centers (Watt et al., 1981). The extrusion studies provides the only direct evidence for the presence of [4Fe-4S] centers in the MoFe protein. Mössbauer studies on Kp1, Av1, and Cp1 (Orme-Johnson, Münck, 1980; Smith, 1983; Zimmerman and Trautwein, 1983) show two classes of Fe atoms, called D (or M5) and Fe^{2+} (or M4) in a 3:1 ratio, which account for about 60% of the total Fe. These 16 Fe atoms are suggested to be in four [4Fe-4S] clusters, each of which contains three D (or M5) and one of Fe^{2+} (or M4). CD and MCD spectra of Av1 and Kp1 are similar in some respects to spectra obtained from simpler [4Fe-4S]-containing proteins (Stephens et al., 1981a). However, their EPR, Mössbauer (in Orme-Johnson, Münck, 1980) and especially MCD (Stephens et al., 1981a;

Johnson et al., 1981) spectra differ markedly, suggesting unusual protein environments and/or unusual oxidation states (Orme-Johnson, Münck, 1980). There are no known species differences concerning the number or type of these centers. Because only 20 conserved cysteines are available per $\alpha 2\beta 2$ (Lammers, Haselkorn, 1983) and four $[Fe_4S_4(Cys)_4]$ would use 16, only four remain for the other metal centers (see below). As each subunit contains only five conserved cysteines, neither is likely to accommodate more than one [4Fe-4S] cluster, even allowing for some non-cysteine ligation. The minimal homology between the cysteinyl peptides of the α and β subunits suggests at least two different environments for [4Fe-4S] clusters in the MoFe protein (Lundell, Howard, 1981) with clusters bridging subunits as a further option.

A second, minor class of Fe atoms has been implicated by Mössbauer analysis and extrusion studies accounting for about 2 Fe atoms per molecule (Orme-Johnson, Münck, 1980; Smith, 1983; Zimmerman and Trautwein, 1983). Other studies (Kurtz et al.,1979; Stephens et al., 1979; 1981a), however, could not confirm these results. The significance of these Fe atoms is not yet clear.

The rest of the metal atoms appear to occur as two identical metal centers per protein molecule, referred to as FeMo cofactor. This center has been studied: (a) in the intact MoFe protein (FeMoco-bound) (e.g. Orme-Johnson, Münck, 1980); and (b) as isolated in NMF solution (FeMoco-free) (Shah, Brill, 1977). FeMoco-free can reconstitute to an active state the cofactor-deficient MoFe proteins from mutant organisms and appears to be identical regardless of bacterial source. This section will briefly summarize and update information on FeMoco reviewed elsewhere (Burgess, Newton, 1983).

FeMoco-bound gives rise to the EPR spectrum of the MoFe protein in the dithionite-reduced state, which is due to two S=3/2 centers per molecule with g values of 4.3, 3.7, and 2.0 (Orme-Johnson, Münck, 1980). FeMoco-free gives rise to a similar, albeit broader, S=3/2 EPR spectrum. Quantitative analysis of these data, correlated with Mössbauer, ENDOR and magnetic susceptibility data, indicate that FeMoco contains one S=3/2 center per Mo and that each center contains about six spectroscopically distinct iron and one Mo atoms in a novel spin-coupled structure (Orme-Johnson, Münck, 1980; Smith, 1983; Mascharak et al., 1982; Hoffman et al., 1982a,b). All spectroscopic studies are consistent with a stoichiometry of $\sim$6Fe atoms per Mo. Analytical data yield values of 7-8 Fe/Mo. Although one report has appeared suggesting these extra Fe atoms are necessary for activity, they could also be adventitious (Burgess, Newton, 1983). Using a ^{35}S-labeling technique, Nelson et al. (1983) have arrived at a minimum stoichiometry of $MoFe_6S_{8-9}$, which is consistent with recent Fe EXAFS studies (Antonio et al., 1982).

Recent Mo EXAFS data on FeMoco free and bound (Newton et al., this volume) show that only a minor perturbation of the cluster occurs when it is extracted from the protein. As only three of the six Fe atoms are seen by the Mo atom in the EXAFS, there must be at least two different environments for the Fe atoms in FeMoco. Three S and three O neighbors are also observed probably in a pseudo-octahedral arrangement. Fe EXAFS on FeMoco-free indicate that, on the average, each Fe atom has as nearest neighbors approximately: 3-4 S (or Cl) at 2.23 Å; 2-3 Fe at 2.64 Å; 0.35 Mo at 2.80 Å and possibly 1-2 O (or N) at 1.84 Å (Antonio et al., 1982). The Mo EXAFS and possibly Fe EXAFS show the presence of atoms other than Mo, Fe and S^{2-}. The Mo atom in both the free and bound forms of FeMoco have O (or N) ligands which might arise from the solvent (e.g. OH^-) and may represent the exchangeable sites for substrate binding. FeMoco-free appears as a small inorganic entity that does not contain recognizable [Fe-S] clusters (Burgess, Newton, 1983).

How is FeMoco held in the protein? FeMoco cannot be extruded like [Fe-S] centers, which may suggest non-cysteine ligation. FeMoco-free does undergoe a number of complexation reactions. In particular, ϕSH sharpens its EPR signal, making it more like FeMoco-bound (in Burgess, Newton, 1983), although it does not bind to the Mo atom (Mascharak et al., 1982; Newton et al., this volume). As the EPR change requires one ϕSH per Mo, ϕSH may bind specifically to one of the ∿6 Fe atoms in FeMoco. One can speculate that this Fe atom is ligated to the protein via a cysteine in FeMoco-bound.

3.3. Redox Properties

Oxidative titrations of the dithionite reduced MoFe protein inevitably show two distinctive phases (Orme-Johnson, Münck 1980; Watt et al., 1981; Stephens et al., 1981b). Orme-Johnson and Münck (1980) state that the first phase of oxidation, by four equivalents of thionine, results in no change in the EPR signal, while the second phase, upon oxidation by two more equivalents, causes loss of the EPR signal. Their Mössbauer titrations are consistent with this view. Watt et al., (1980, 1981) report that oxidation by three equivalents of methylene blue or thionine show no change in the EPR but do show rapid incremental increases in the visible spectral absorptions at A_{425} and A_{700}. Oxidation beyond three equivalents causes the EPR signal to attenuate linearly until complete quenching occurs at five equivalents, during which time A_{700} does not change. Further oxidation by one more electron, to give a total of six electrons removed, then occurs as evidenced by controlled potential coulometry, by reductive titration and optical changes in the 425 nm region. Stephens et al. (1981b; personal communication) observe a sequential, monotonic change in the CD spectrum as 0-to-3 equivalents of oxidant are added during which time the EPR signal does not change. Additional oxidant produces no further changes in the CD, but does cause EPR signal loss, although the endpoint was not specified.

All agree that the first phase corresponds to oxidation of the non-FeMoco, [4Fe-4S]-type clusters and that the second phase corresponds to oxidation of FeMoco. Thus, Watt et al. (1981) divide these six electrons into groups of 3, 2 and 1, with the two FeMoco centers and one other EPR-silent center reacting at the same potential and causing similar changes in the visible spectrum, while Orme-Johnson and Münck (1980) propose a 4- and a 2-electron process. It is possible that the protein contains four [4Fe-4S] clusters that are located very close together such that they might interact under some conditions to accommodate a loss of only three electrons prior to removal of electrons from the FeMoco center (P. Stephens, personal communication). Which, if any, of these oxidized states of the MoFe protein is relevant to turnover is not known (for discussion, see Smith, 1983).

The non-FeMoco, [4Fe-4S]-type centers are diamagnetic in the dithionite-reduced state of the protein and become paramagnetic upon oxidation by one electron per center (Orme-Johnson, Münck, 1980). Magnetic susceptibility studies suggest they are S=5/2 systems in the oxidized protein (Smith et al., 1982), which is consistent with values (S=5/2 or 7/2) obtained from low temperature MCD (Johnson et al., 1981) studies. In spite of their paramagnetism in the oxidized state, only a fleeting EPR (g=1.933, S=1/2) resonance is observed during oxidative titrations (Smith et al., 1983). The above observations have led to the view that the [4Fe-4S]-type centers might be present in the -4 oxidation state (all Fe^{2+}) in the dithionite-reduced protein. Midpoint potentials for the oxidation and re-reduction of the [4Fe-4S]-type clusters are about -340 to -400 mV (Smith, 1983; Johnson et al., 1981; Watt et al., 1981). Midpoint potentials for the one-electron oxidation of FeMoco-bound to an EPR-silent (diamagnetic) state appear to vary greatly (-260 mV to 0 mV) with species and reaction conditions (Smith, 1983). FeMoco-free can also be reversibly oxidized to an EPR-silent state although the redox potential and numbers of electrons required to instigate this change are unknown (Burgess, Newton, 1983).

4. NITROGENASE TURNOVER

The overall steps in nitrogenase turnover have been previously reviewed (Mortenson, Thorneley, 1979; Lowe et al., 1980) and this section will serve as a brief update. There are no known species differences with respect to the overall steps. However, the individual rate constants may vary. The general direction of electron flow is: reductant → Fe protein → MoFe protein → substrate. Although the physiological electron donors in some organisms have been identified, their use has been limited. Most information is derived from studies with dithionite, where $SO_2^{\cdot -}$ is the actual one-electron donor to the Fe protein (Mortenson, Thorneley, 1979).
Electron transfer between $SO_2^{\cdot -}$ and Ac2 alone is extremely rapid, whereas electron transfer between $SO_2^{\cdot -}$ and $Av2_{ox}$ during turnover is much slower. As MgADP, but not MgATP, inhibits the rate of reduction of $Ac2_{ox}$ by $SO_2^{\cdot -}$, the species being reduced is probably Fe $protein_{ox}$ $(MgADP)_2$ (Mortenson, Thorneley, 1979; Thorneley and Lowe, 1983), which with MgADP/MgATP exchange gives equation 1.[4]

$$(1)\quad Fe_{ox}(MgADP)_2 + SO_2^{\cdot -} + 2MgATP \rightarrow Fe_{red}(MgATP)_2 + 2MgADP + HSO_3^-$$

Watt (1977) has provided evidence that MgATP binding precedes Av2 reduction. Complexation with $MoFe^o$, representing the dithionite-reduced state, then occurs. Although Cordewener et al. (1983) suggest that the second MgATP-binding site is generated only after complex formation occurs, Thorneley and Lowe (1983) prefer equation 2.

$$(2)\quad [Fe_{red}(MgATP)_2]_x + MoFe^o \rightleftarrows [Fe_{red}(MgATP)_2]_x MoFe^o$$

The value of x has been extensively investigated (Lowe et al., 1980; Mortenson, Thorneley, 1979; Hageman, Burris, 1978a,b, 1979; Wherland et al., 1981) to show: (a) two binding sites for Fe on MoFe; (b) active 2Fe/MoFe complexes exit; and (c) active 1Fe/MoFe complexes can also exist when Fe is limiting. Thus, the complex has a maximum value of x=2. Electron transfer follows complex formation (e.g. Thorneley, Lowe, 1983; Hageman, Burris, 1978a,b) coupled with MgATP hydrolysis (equation 3).

$$(3)\quad [Fe_{red}(MgATP)_2]_x MoFe^o \rightarrow [Fe_{ox}(MgADP + Pi)_2]_x MoFe_{red}$$

Numerous data suggest a minimum of 2MgATP's hydrolyzed per electron transferred to substrate during turnover (e.g. Lowe et al., 1980; Mortenson, Thorneley, 1979). We have no molecular level understanding of how MgATP hydrolysis is coupled to electron transfer.
The final step in the catalytic cycle is the dissociation of the complex (e.g. Hageman, Burris, 1978a,b, 1979) as shown in equation (4), which returns us to equation (1).

$$(4)\quad [Fe_{ox}(MgADP + 2Pi)]_x MoFe_{red} \rightleftarrows [Fe_{ox}(MgADP)_2]_x + MoFe_{red} + [2Pi]_x$$

Thorneley and Lowe (1983) have recently measured rate constants for all steps and their data indicate that the dissociation of $Kp2_{ox}$ $(MgADP)_2$ from Kp1 (equation 4) is the rate-limiting step in a single catalytic cycle for N_2ase turnover. Presteady state kinetic studies (Hageman, Burris, 1978a; Thorneley, Lowe, 1982)

[4]For the sake of simplicity, Fe represents the Fe protein and MoFe represents the MoFe protein.

indicate that a minimum of two complete catalytic cycles must occur prior to reduction of $2H^+$ to H_2. In situations where two Fe proteins are involved in a single catalytic cycle, $MoFe_{red}$ (equation 4) would contain two electrons. Why then isn't H_2 evolution observed following a single catalytic cycle? One way to rationalize this situation is to view the MoFe protein as two αβ dimers. Each would contain one FeMoco center and two [4Fe-4S]-type centers (Thorneley, Lowe, 1982, 1983; Orme-Johnson, Münck, 1980; Hageman, Burris, 1978a,b, 1979). So even if two reduced Fe proteins transfer a total of two electrons simultaneously to the MoFe protein in a single catalytic cycle, no substrate reduction can occur because the electrons reside on separate halves of the MoFe protein, which do not communicate. The only physical evidence to support this view (Orme-Johnson and Munck, 1980) is that a stable EPR-silent form of the $MoFe_{red}$ (equation 4) has been isolated in which both FeMoco's are one-electron reduced, but which does not reduce substrate.
There is no direct evidence for the reduction of the [4Fe-4S]-type clusters during turnover, and it is probably premature to eliminate the possibility that the two halves of the MoFe protein might communicate via the [4Fe-4S]-type clusters.
The above discussion suggests that, after two catalytic cycles, a free, $2e^-$/FeMoco, reduced MoFe protein exists which cna evolve H_2. Does this mean that the MoFe protein alone is the enzyme and the Fe protein is simply its reductant (Hageman, Burris, 1978a)? No one has definitively demonstrated substrate reduction by the MoFe protein in a system that did not also contain Fe protein and MgATP. For that reason and because of a number of suggested possible second roles for Fe protein and MgATP which remain to be tested (e.g. Lowe et al., 1980; Miller et al., 1980; Thorneley and Lowe, 1982), it would appear premature to adopt the nomenclature proposed by Hageman and Burris (1978a).

5. SUBSTRATE REDUCTION

Details of the myriad of possibilities concerning substrate reduction by N_2ase appear elsewhere (Hardy, 1979; Burris, 1979; Burns, 1979) so that only a few questions of interest will be considered here.
Is total electron flow independent of the substrate being reduced? If, as indicated above, the rate-limiting step for nitrogenase turnover occurs prior to substrate reduction, then the answer to this question should be yes (e.g., Watt, Burns, 1977). Recent kinetic studies indicate, however, that electrn flow is not strictly independent of the substrate. Specifically there are decreases of 35% under 1 atm N_2 (Hageman, Burris, 1980) and 13% under 0.5-1 atm N_2 (Burgess et al., 1981; Wherland et al., 1981). The decrease is even more dramatic with increasing CH_3NC concentration (K_i = 0.68 mM; Rubinson et al., 1983). For the cyanide system, HCN (the substrate) does not affect electron flow, but CN^-, which is also present, is a potent reversible inhibitor of nitrogenase turnover (K_i=0.027 mM; Li et al., 1982). CO appears to stimulate the rate of nitrogenase turnover (Mortenson, Upchurch 1981). It is still unclear how some of these entities influence the rate-limiting step in nitrogenase turnover.
Can H_2 evolution be eliminated during N_2 fixation? Under physiological N_2-fixing conditions, nitrogenase evolves some H_2. Complete elimination of H_2 evolution by N_2ase in vitro under N_2-fixing conditons has never been achieved. Thus, the answer to the question is, probably not. If H_2 evolution can occur at state E_2, which is upstream of the state that interacts with N_2 (Thorneley, Lowe, 1982), then some leakage of electrons to H_2 evolution can be easily rationalized. If both H_2 evolution and at least the first step in N_2 fixation usually occur at E_3 (Thorneley, Lowe, 1982), then the situation could be viewed as a simple competition for protons and electrons at a highly reduced state of the enzyme and

again some leakage of electrons to H_2 evolution could be easily rationalized. Is one H_2 evolved per N_2-fixed? This is an important and very different question from the previous one. If the answer is yes, then it is possible, although not required, that H_2 evolution is an intimate part of the mechanism of N_2 fixation. This suggestion was originally made by Hadfield and Bulen (1969) who found that H_2 evolution could not be eliminated by N_2 fixation. Rivera-Ortiz and Burris (1975) give values of 0.56-to-0.9 H_2 evolved per N_2 fixed at infinite N_2 which could be used to argue either for or against a 1:1 stoichiometry. This study measured ΔH_2 and did not consider the possible inhibition of flux by N_2. At present, there is no direct experimental evidence to prove a minimum of one H_2 must be evolved per N_2 fixed (see Burgess *et al.*, this volume).
What is the chemical mechanism of N_2 reduction? Thorneley *et al.* (1978) have shown that an intermediate is formed during the 6-electron reduction of N_2 to NH_3 which, when quenched with acid or base, produces free N_2H_4. They suggest this intermediate as a hydrazido(2-) ligand (=N-NH_2), which is protonated at pH 0 or 14 to give N_2H_4. The hydrazido(2-) species is proposed based on a comparison of the reactivity patterns of low valent Mo- and W-phosphine dinitrogen and dinitrogen-hydride complexes, which have no structural relationship to MoFeco. The formation of a 2-electron-reduced, diazene (H-N=N-H)-level intermediate (e.g. Burgess *et al.*, 1981) has been proposed based on extensive studies of H_2-inhibition and HD-formation (Guth, Burris, 1983; Li, Burris, 1983). These studies show that both reactions: (a) occur for all N_2 ases; (b) require N_2; (c) require 1e⁻ per HD; (d) are not exchanges with the solvent; and (e) are representatives of the same reaction. Burgess *et al.* (1981) believe these reactions to be an intimate part of the N_2-reduction mechanism and indicative of a bound diazene-level intermediate.
Thorneley and Lowe (1982) and Guth and Burris (1983) propose that H_2 inhibition and HD formation are not directly related to N_2 reduction. They explain HD formation by equation A, with further rounds of either N_2 binding and protonation or protonation alone (Guth and Burris, 1983) giving rise to HD.

$$\text{(A)}\quad E{<}^{H'}_{H}{\scriptstyle -H}\ \underset{N_2\ \ H_2}{\overset{N_2\ \ H_2}{\rightleftharpoons}}\ E{<}^{H'}_{N_2}\ \underset{D_2\ \ N_2}{\overset{D_2\ \ N_2}{\rightleftharpoons}}\ E{<}^{H'}_{D}{\scriptstyle -D}$$

(H' absent from Guth and Burris scheme)

Both suggestions *require* that N_2 fixation and H_2 evolution have a 1:1 minimum stoichiometry. The other suggestion (Burgess *et al.*, 1981; this volume) does not require, but is not inconsistent with, such a stoichiometry. If reaction A occurs, there is no chemical reason to propose that reaction B does not also occur.

$$\text{(B)}\quad E{<}^{H'}_{H}{\scriptstyle -H}\ \underset{D_2\ \ H_2}{\overset{D_2\ \ H_2}{\rightleftharpoons}}\ E{<}^{H'}_{D}{\scriptstyle -D}$$

It does not. Thus, if HD formation (H_2 inhibition) is unrelated to N_2 reduction, why is N_2 required (Burgess *et al.*, this volume)?
Where do substrates bind? Evidence to support the idea that FeMoCo is the site of substrate binding and reduction is reviewed elsewhere (Burgess, Newton, 1983). Physical evidence for the binding of C_2H_2 and CH_3NC to the dithionite-reduced state of FeMoco-bound has been reported (e.g. Smith *et al.*, 1983; Hoffman and

and W.H. Orme-Johnson, personal communication). Both studies suggest that substrates can bind prior to reduction of the MoFe protein by Fe protein/MgATP. But no study suggests to which state of the MoFe protein N_2 binds during turnover. CO can apparently bind to two sites, one is a $[4Fe-4S]^{2-}$-type cluster (Davis et al., 1979) and the other FeMoco-bound (Hawkes et al., 1983).
When FeMoco, isolated from wild type Kp1, is used to reconstitute Kp1 Nif B^-, a wild type N_2-fixing Kp1 results (Hawkes, Smith, 1983a). When FeMoco, isolated from Kp1 Nif V^- (a mutant defective in its ability to fix N_2) is used to reconstitute Kp1 Nif B^-, the resulting Kp1 has the phenotype of Kp1 Nif V^- (Hawkes et al., 1983). This is the first direct indication that FeMoco is the site of N_2 binding and reduction. It is probable that all substrates interact at this center. Further, N_2O (or C_2H_2) stimulates HCN reduction (Li et al., 1983), which suggests they are bound simultaneously. Thus, if all substrates bind to FeMoco, it can probably accommodate more than one substrate simultaneously.

6. REFERENCES

Antonio MR, Teo B-K, Orme-Johnson WH, Nelson, MJ, Groh SE, Lindahl PA, Kauzlarich SM and Averill BA (1982) J. Am. Chem. Soc. 104, 4703-4705.
Averill BA, Bale JR and Orme-Johnson WH (1978) J. Am. Chem. Soc. 100, 3034-3043.
Braaksma A, Haaker H, Grande HJ and Veeger C (1982) Eur. J. Biochem. 121, 483-491
Braaksma A, Haaker H and Veeger C (1983) Eur. J. Biochem. 133, 71-76.
Burgess BK, Jacobs DB and Stiefel EI (1980) Biochim. Biophys. Acta 614, 196-209.
Burgess BK, Wherland S, Newton WE and Stiefel EI (1981) Biochemistry 20, 5140-5146.
Burgess BK and Newton WE (1983) in Muller A and Newton WE, eds., Nitrogen Fixation: The Chemical-Biochemical-Genetic Interface, pp. 1-19, Plenum, New York
Burns RC (1979) in Hardy RWF, Bottomley F and Burns RC, eds., A Treatise on Dinitrogen Fixation, I and II, pp. 491-514, Wiley, New York.
Burris RH (1979) in Hardy RWF, Bottomley F and Burns RC, eds., A Treatise on Dinitrogen Fixation, I and II, pp. 569-604, Wiley, New York.
Cordewener J, Haaker H and Veeger C (1983) Eur. J. Biochem. 132, 47-54.
Davis LC, Henzl MT, Burris, RH and Orme-Johnson WH (1979) Biochemistry 18, 4860-4869.
Dilworth MJ and Thorneley RNF (1981) Biochem. J. 193, 971-983.
Emerich DW and Burris RH (1978) J. Bacteriol. 134, 936-943.
Emerich DW, Ljones T and Burris RH (1978) Biochim. Biophys. Acta 527, 359-369.
Gillum WO, Mortenson LE, Chen J-S and Holm RH (1977) J. Am. Chem. Soc. 99, 584-595.
Guth JH and Burris RH (1983) Biochemistry, in press.
Hadfield KL and Bulen WA (1969) Biochemistry, 8, 5103-5108.
Hageman RV and Burris RH (1978a) Biochemistry 17, 4117-4124.
Hageman RV and Burris RH (1978b) Proc. Natl. Acad. Sci. USA 75, 2699-2702.
Hageman RV and Burris RH (1979) J. Biol. Chem. 254, 11189-11192.
Hageman RV and Burris RH (1980) in Coughlan MP, ed., Molybdenum and Molybdenum-Containing Enzymes, pp. 403-426, Pergamon Press, Oxford.
Hageman RV, Orme-Johnson WH and Burris RH (1980) Biochemistry 19, 2333-2342.
Hallenbeck PC, Meyer CM and Vignais PM (1982) J. Bacteriol. 149, 708-717.
Hardy RWF (1979) in Hardy RWF, Bottomley F and Burns RC, eds., A Treatise on Dinitrogen Fixation, I and II, pp. 515-568, Wiley, New York.
Hase T, Nakano T, Matsubara H and Zumft WG (1981) J. Biochem. 90, 295-298.
Hausinger RP and Howard JB (1983) J. Biol. Chem., in press.
Hausinger RP and Howard JB (1982) J. Biol. Chem. 257, 2483-2490.
Hausinger RP and Howard JB (1980) Proc. Natl. Acad. Sci. USA 77, 3826-3830.
Hawkes TR, McLean PA and Smith BE (1983) Biochem. J., in press.

Hoffman BM, Roberts JE and Orme-Johnson WH (1982a) J. Am. Chem. Soc. 104, 860-862.
Hoffman BM, Venters RA, Roberts JE, Nelson M and Orme-Johnson WH (1982b) J. Am. Chem. Soc. 104, 4711-4712.
Johnson MK, Thomson AJ, Robinson AE and Smith BE (1981) Biochim. Biophys. Acta 671, 61-70.
Kurtz DM, McMillan RS, Burgess BK, Mortenson LE and Holm RH (1979) Proc. Natl. Acad. Sci. USA 76, 4986-4989.
Lammers PJ and Haselkorn R (1983) Proc. Natl. Acad. Sci. USA 80, in press.
Li J-G, Burgess BK and Corbin JL (1982) Biochemistry, 21, 4393-4402.
Li J-L and Burris RH (1983) Biochemistry, in press.
Ljones T and Burris RH (1978) Biochemistry 17, 1866-1872.
Lowe DJ, Smith BE and Eady RR (1980) in Subba Rao NS, ed., Recent Advances in Biological N_2 Fixation, pp. 34-87, Oxford and IBH Publishing Co, New Delhi.
Ludden PW and Burris RH (1978) Biochem. J. 175, 251-259.
Lundell DJ and Howard JB (1981) J. Biol. Chem. 256, 6385-6391.
Lundell DJ and Howard JB (1978) J. Biol. Chem. 253, 3422-3426.
Mascharak PK, Smith MC, Armstrong WH, Burgess BK and Holm RH (1982) Proc. Natl. Acad. Sci. 79, 7056-7060.
Mazur BJ and Chui C-F (1982) Proc. Natl. Acad. Sci. USA 79, 6782-6786.
Mevarech M, Rice D and Haselkorn R (1980) Proc. Natl. Acad. Sci. USA 77, 6476-6480.
Miller RW, Robson RL, Yates MG and Eady RR (1980) Can. J. Biochem. 58, 542-548.
Mortensen LE and Thorneley RNF (1979) Ann. Rev. Biochem. 48, 387-418.
Mortensen LE and Upchurch RG (1981) in Gibson AH and Newton WE, eds., Current Perspectives in Nitrogen Fixation, pp. 75-78, Aust. Acad. Sci., Canberra.
Mortenson LE, Walker MN and Walker GA (1976) in Newton WE and Nyman CJ, eds., 1, pp. 117-149, Wash. State Univ. Press, Pullman, Wash.
Münck E, Rhodes H, Orme-Johnson WH, Davis LC, Brill WJ and Shah VK (1975) Biochim. Biophys. Acta 400, 32-53.
Nelson MJ, Levy MA and Orme-Johnson WH (1983) Proc. Natl. Acad. Sci. USA 80, 147-150.
Orme-Johnson WH, Davis LC, Henzl MT, Averill BA, Orme-Johnson NR, Münck E and Zimmerman R (1977) in Newton WE, Postgate JR and Rodriguez-Barrueco C, eds., Recent Developments in Nitrogen Fixaton, pp. 131-178, Academic Press, London.
Orme-Johnson WH and Münck E (1980) in Coughlin MP, ed., Molybdenum and Molybdenum Containing Enzymes, pp. 427-438, Pergamon Press, Oxford.
Postgate JR (1974) Symp. Soc. Gen. Microbiol. 24, 263-292.
Rees DC and Howard JB (1983) Proc. Natl. Acad. Sci. USA, in press.
Rivera-Ortiz JM and Burris RH (1975) J. Bacteriol. 123, 537-545.
Rubinson JF, Corbin JL and Burgess BK (1983) Biochemistry, in press.
Ruvkun GB and Ausubel FM (1980) Proc. Natl. Acad. Sci. USA 77, 191-195.
Scott KF, Rolfe BG, Shine J, Sundaresan V and Ausubel FM (1981a) in Gibson AH and Newton WE, eds., Current Perspectives in Nitrogen Fixation, pp. 393-395, Aust. Acad. Sci., Canberra.
Scott KF, Rolfe BG and Shine J (1981b) J. Mol. Appl. Genet. 1, 71-81.
Shah VK and Brill WJ (1977) Proc. Natl. Acad. Sci. USA 74, 3249-3253.
Smith BE and Lang G (1974) Biochem. J. 137, 169-180.
Smith BE (1983) in Müller A and Newton WE eds., Nitrogen Fixation: The Chemical-Biochemical-Genetic Interface pp. 23-62, Plenum Press, New York.
Smith JP, Emptage MH and Orme-Johnson WH (1982) J. Biol. Chem. 257, 2310-2313.
Stephens PJ, McKenna CE, McKenna M-C, Nguyen HT and Devlin F (1981a) Biochemistry 20, 2857-2867.
Stephens PJ, McKenna CE, McKenna M-C, Nguyen HT, Morgan TV and Devlin F (1981b) in Gibson AH and Newton WE, eds., Current Perspectives in Nitrogen Fixation, pp. 357, Aust. Acad. Sci., Canberra.

Stephens PJ, McKenna CE, McKenna M-C, Nguyen HT and Lowe DJ (1983) in Ho C, ed., Electron Transport and Oxygen Utilization pp. 405-409, MacMillan Press, New York.

Stephens PJ, McKenna, Smith BE, Nguyen HT, McKenna M-C, Thomson AJ, Devlin F and Jones JB (1979) Proc. Natl. Acad. Sci. USA 76, 2585-2589.

Sundaresan V and Ausubel FM (1981) J. Biol. Chem. 256, 2808-2812.

Tanaka M, Haniu M, Yasunobu KT and Mortenson LE (1977) J. Biol. Chem. 252, 7093-7100.

Thorneley RNF, Eady RR and Lowe DJ (1978) Nature 272, 557-558.

Thorneley RNF and Lowe DJ (1982) Israel J. Botany 31, 1-11.

Thorneley RNF and Lowe DJ (1983) Biochem. J., in press.

Török I and Kondorosi A (1981) Nucleic Acids Res. 9, 5711-5723.

Tso M-YW and Burris RH (1973) Biochim. Biophys. Acta 309, 263-270.

Watt GD (1977) in Newton WE, Postgate JR and Rodriguez-Barrueco C, eds., Recent Developments in Nitrogen Fixation, pp. 179-190, Academic Press, London.

Watt GD and Burns A (1979) Biochemistry 16, 264-270.

Watt GD, Burns A, Lough S and Tennent DL (1980) Biochemistry 19, 4926-4932.

Watt GD, Burns A and Tennent DL (1981) Biochemistry 20, 7272-7277.

Weininger MS and Mortenson LE (1980) Proc. Natl. Acad. Sci. USA 79, 378-380.

Weston MF, Kotake S and Davis LC (1983) Arch. Biochim. Biophys., in press.

Wherland S, Burgess BK, Stiefel EI and Newton WE (1981) Biochemistry 20, 5132-5140.

Zimmerman R and Trautwein AX (1983) In Muller A and Newton WE, eds., Nitrogen Fixation: The Chemical-Biochemical-Genetic Interface, pp. 63-81, Plenum Press, New York.

Zumft WG, Mortenson LE and Palmer G (1974) Eur. J. Biochem. 46, 525-535.

Zumft WG, Palmer G and Mortenson LE (1973) Biochim. Biophys. Acta 292, 413-421.

SUBSTRATE INTERACTIONS WITH NITROGENASE AND ITS FE-MO COFACTOR: CHEMICAL AND SPECTROSCOPIC INVESTIGATIONS

CHARLES E. MCKENNA/PHILIP J. STEPHENS/HARUTYUN ERAN/GUI-MIN LUO/F.X. ZHANG
MATAI DING/HOP T. NGUYEN
DEPARTMENT OF CHEMISTRY, UNIVERSITY OF SOUTHERN CALIFORNIA, LOS ANGELES, CA 90089-1062, USA

1. INTRODUCTION

Substrate interactions with nitrogenase embrace two formally separable phenomena: binding and reduction of N_2, or alternate reducible substrates; and binding and hydrolysis of ATP, which requires dissociation of ADP from the hydrolysis site to continue turnover. Not considered here is the interaction of nitrogenase with its immediate electron donor, which is a metalloprotein _in vivo_ and usually sodium dithionite for assays _in vitro_. Investigations of reductions by nitrogenase have chiefly examined chemical aspects, such as substrate reactivities or product formation kinetics. Alternate substrates are useful as probes of reduction mechanisms, as we have shown by the discovery that certain molecules having a double bond within a three-membered ring are reduced by nitrogenase (McKenna, 1980; McKenna _et al._, 1980a, 1980b). It is generally assumed that the active site for N_2 reduction is located on the nitrogenase Fe-Mo protein, although direct, conclusive evidence to support this assumption is still lacking. Indirect evidence and also intuition supports an important role for Mo in the reduction mechanism. Since both Mo atoms in the FeMo protein are contained in its Fe-Mo-S clusters (FeMo-cofactor, FeMo-co) this cluster has accordingly been regarded as a likely component of the active site of reduction. Prior evidence for nucleotide binding to the Fe protein has come from spectroscopic, equilibrium partitioning and other studies.
In this paper we present a summary of recent work from our laboratory on both categories of substrate interactions, using chemical and spectroscopic approaches.

2. DIAZIRINE INTERACTIONS WITH NITROGENASE

2.1. Diazirine: Properties and Reduction Pathways

The discovery that the cyclic C_3H_4 isomer, cyclopropene, is an effective nitrogenase substrate suggested that the photosensitive cyclic CH_2N_2 isomer, diazirine, might interact with the enzyme (McKenna, 1980). This molecule has several attributes that are pertinent to nitrogenase studies. Unlike the linear CH_2N_2 isomer, diazomethane, diazirine is relatively stable in aqueous solutions, thus allowing use of standard reduction assay conditions. Formally, it is a neutral azo (-N=N-) compound, like the postulated N_2 reduction intermediate diazene (N_2H_2), although its location within a strained, three-membered ring gives its -N=N- bond some -N≡N- character, just as cyclopropene formally contains a -C=C- double bond but in fact has chemical properties intermediate between an alkene and an alkyne. Diazirine is thermodynamically unstable and chemically reactive, in contrast to the inert N_2 molecule.
The cyclic structure of diazirine also increases its level of unsaturation and provides a number of potential reduction pathways. If an initial $2e^-/2H^+$ step involving breaking of the C-N single bond occurs, ring cleavage would result and an intermediate bound methyl diazene (CH_3-N=N-H) might be formed. Methyl diazene cannot disproportionate in the presence of D_2 to give 2 HD + N_2, as is known to occur for free diazene. Decomposition of methyl diazene to CH_4 and N_2 is possible, however. If this occurred at the active site and the N_2 remained

bound, subsequent reduction would give NH_3 as the other alternative product and diazirine-dependent HD formation would be predicted by the diazene-intermediate mechanism. If methyl diazene were stabilized by being bound at the active site, further reduction, possibly via methyl hydrazine, should yield methyl amine and ammonia (cleavage of the -N=N- bond). CH_4 formation (cleavage of the C-N bond) would be less likely; it would require an N_2H_2 intermediate, which according to the diazene hypothesis also ought to give HD formation. Interestingly, this would have to be a $4e^-/4H^+$ turnover process, in contrast to the postulated $N_2 \rightarrow N_2H_2$ reaction which is a $2e^-/2H^+$ turnover process. A second pathway would involve initial reduction of the diazirine -N=N- bond with retention of the ring structure, in a process analogous to the reduction of cyclopropene to cyclopropane. The resultant product, diaziridine, as a hydrazine-like species, is likely to be bound and reduced to methyl hydrazine; this, if bound, would be further reduced to methyl amine and ammonia. An alternative reduction path from diaziridine via cleavage of the NH-NH bond would yield diaminomethane, which ought to be released as an ultimate reduction product. In summary, diazirine presents a complex set of possible reduction pathways, full elucidation of which is expected to provide useful mechanistic insights.

2.2. Diazirine as Nitrogenase Inhibitor

N_2 is a non-competitive inhibitor of C_2H_2 reduction, and in distinction from all other nitrogenase substrates, cannot completely suppress ATP-dependent H_2 evolution. Diazirine is a competitive inhibitor of C_2H_2 reduction (McKenna et al., 1981), with a K_i value of 4×10^{-4} atm, corresponding to an estimated molar K_i of 3×10^{-5} M, a notably low value. We find that diazirine also inhibits ATP-dependent H_2 evolution by nitrogenase (C.E. McKenna, H. Eran and F.X. Zhang, unpublished). A plot of the reciprocal velocity (1/v) of H_2 formation vs diazirine partial pressure extrapolates to a 1/v value of 0.95 ± 0.05, predicting that at a sufficiently high pressure of diazirine, H_2 evolution will be completely surpressed. These two results are self-consistent, since C_2H_2 can also completely abolish H_2 evolution, and reemphasize the special nature of N_2 as a substrate, as does our failure to detect diazirine-dependent HD formation under conditions for which N_2-dependent HD formation is observed (C.E. McKenna, H. Eran and F.X. Zhang, unpublished).

2.3. Diazirine as Nitrogenase Substrate

We have previously reported that diazirine is a substrate of nitrogenase, showing normal assay requirements, and that NH_3 and CH_4 could be detected as products, the former using indophenol or Nessler colorimetric methods, and the latter by flame ionization detection (FID) gas chromatography (GC) (McKenna et al., 1981; McKenna, Eran, 1982). We were unable to detect methyl amine (CH_3NH_2) (cf: Orme-Johnson, 1981) at levels at or greater than those seen for NH_3, using standard colorimetric procedures, but could not rule out the possibility that it might be formed at lower levels and remain undetected due to inadequate analytical sensitivity. We also lacked methods to analyze for several other possible products at the necessary level of sensitivity. Recently, we have succeeded in developing high pressure liquid chromatography (HPLC) analyses for several possible nitrogenase products from diazirine (C.E. McKenna, H. Eran and F.X. Zhang, unpublished). After treatments to remove protein (trichloroacetic acid) and to destroy dithionite (KIO_3), the assay mixture is treated to form an appropriate fluorescent (NH_3, CH_3NH_2) or UV-absorbing (N_2H_4, CH_3N-NH_2) product derivative. These are separated by HPLC on a hydrophobic column and quantitated using a fluorometric or UV absorbance detector as appropriate. The methods are fairly rapid, convenient (no microdistillation is required), quite sensitive (1 nmol) and specific (defined retention time of product derivative is determined; both NH_3 and CH_3NH_2 can be measured simultaneously). Using this HPLC technique, in combination with FID GC analysis (for CH_4) and thermal conductivity (TC) GC analysis (for H_2), we have established that CH_3NH_2 is a third product of diazirine reduction (C.E. McKenna, H. Eran, F.X. Zhang, unpublished). Simultaneous measurements of CH_4, CH_3NH_2, NH_3 and H_2 formation show that the product ratio (Table 1) can be understood by invoking two parallel reduction processes (a and b),

$$\text{a)}\quad CH_2N_2 \;+\; 6e^- \;+\; 6H^+ \;\rightarrow\; CH_3NH_2 \;+\; NH_3$$

$$\text{b)}\quad CH_2N_2 \;+\; 8e^- \;+\; 8H^+ \;\rightarrow\; CH_4 \;+\; 2\;NH_3$$

through which electrons are about evenly partitioned. H_2 evolution occurs concurrently, but the total electron flux summed over all detected products is still somewhat lower than the electron flux to H_2 formation, measured in the absence of diazirine under otherwise identical conditions. If there is a further 'missing' major reduction product, it is not hydrazine or methyl hydrazine, since we were unable to detect either at a level sufficient to restore the overall electron balance. Efforts are currently directed to resolving this problem in terms of either additional, as yet unidentified product(s), a degree of self-inhibition by diazirine such as reportedly exhibited by some other substrates (e.g. aleene), or an undetected futile turnover.

TABLE 1. Rates of Product Formation in Diazirine Reduction by Nitrogenase (A.vinelandii).

Product	Detection Method	Amount Formed/6 min (nmol)
CH_4	GC/FID	98
CH_3NH_2	HPLC	97
NH_3	HPLC	325
H_2	GC/TC	268

TABLE 2. Km Values for Diazirine Reduction Products (A.vinelandii).

Product	Km (atm^{-1})
CH_4	0.0012
CH_3NH_2	0.0014
NH_3	0.0012

The Km values determined for reduction to NH_3, CH_3NH_2 and CH_4 are essentially the same within experimental error, under similar conditions of assay (Table 2). The observed value for the three products is on the order of 1×10^{-3} atm. Based on an approximate (uncorrected for salt effects) estimate of the water solubility of diazirine, this corresponds to molar Km of about 8×10^{-5} M which is comparable to that of N_2 itself, and considerably lower than the Km of other substrates having a similar molecular weight.

2.4. Conclusions

Diazirine is an effective inhibitor of nitrogenase, but its behaviour towards C_2H_2 reduction and H_2 formation indicates that, like all other known substrates, it is kinetically different from N_2. The fact that it is a substrate and is reduced to completely saturated products demonstrates the expected but previously unproved result that nitrogeanse can readily reduce completely a symmetrical -N=N- bond, such as occurs diazene. Diazirine is reduced in $6e^-$ and $8e^-$ processes, in which C-N bond reduction occurs. The absence of observable HD formation suggests that the CH_4 product is not formed via a partial decomposition reaction of bound methyl diazene intermediate.

3. SUBSTRATE REDUCTIONS BY FEMO-CO

3.1. General Considerations and Prior Studies

There is a substantial body of spectroscopic and analytical data comparing various physical properties of the native Fe-Mo-S cluster in FeMo-protein (FeMo-co (nat)) with FeMo-co isolated in NMF solutions from the denatured protein (FeMo-co (nmf)). In light of the intense speculation on the possible function of FeMo-co (nat) in active site reduction, it is therefore surprising that there has so far been little information available to relate catalytic properties of the two entities. FeMo-co (nmf) would not be expected to display the full catalytic capabilities of FeMo-co (nat) with its associated panoply of additional Fe-S clusters, possible protein-derived active site ligands and energy-generating, Fe-protein-dependent reduction-coupled ATP hydrolysis mechanism. Nevertheless, FeMo-co (nmf) may represent a 'stripped down' active site, and thus a valuable model system for nitrogenase reduction (given that assumptions concerning its essential role are in fact correct). It may be able to mimic at least some catalytic function of the intact enzyme, making it possible to gain insight into the role of any other moieties that may comprise the active reduction site. An earlier study (Shah *et al.*, 1978) reported FeMo-co (nmf)-catalyzed, aqueous BH_4^- reduction of C_2H_2 to C_2H_4 and suggested this result as evidence that FeMo-co (nat) is an essential component, if not the key constituent, of the active site. N_2 reduction was not observed, but this could perhaps be attributed to not having available the ATP-hydrolysis process needed to activate the much less reactive N≡N triple bond. We proposed the more stringent reaction criterion of cyclopropene reduction by nitrogenase to propene and cyclopropene, in a specific 2:1 ratio (McKenna *et al.*, 1979). The cyclopropene strained ring is highly activated to be opened (reductively to propene), thus selection of this pathway will be catalyst-dependent but need not be also dependent on a special energy-supplying process, as just argued in the case of N_2. We observed that cyclopropene was reduced only to propene by the aqueous BH_4^--FeMo-co system. On this basis, we noted that if the experiment were valid, then it could be concluded that FeMo-co (nat) alone is not the sole essential part of the active site. However, we also pointed out that results obtained with this system must be regarded as ambiguous, given the known instability of FeMo-co (nmf) when transferred to aqueous solution. Connected with this was the observation that the observed C_2H_2 reduction activity continued in air-exposed systems for hours whereas FeMo-co (nmf) is known to lose its ability to reconstitute an inactive FeMo protein in UW45 'apoprotein' extracts within a minute or two.

3.2. Catalytic Reductions by BH_4^--FeMo-co (nmf)

We have recently studied the properties of BH_4^--FeMo-co (nmf) solutions as a function of their H_2O content, comparing catalytic C_2H_2 reduction with UW45 reconstitution activity as a measure of the integrity of the FeMo-co species (C.E. McKenna, G.M. Luo, C.G. Zhang, unpublished). As shown in Table 3, in aqueous BH_4^- mixtures, FeMo-co (nmf) rapidly loses its activity in the UW45 extract assay, but the rate of decomposition slows as the percent H_2O decreases. In the 'non-aqueous' BH_4^--NMF system (which actually contains about 0.5% to 1% H_2O), reconstitution activity is fully retained by FeMo-co (nmf) assayed directly from the reducing mixture (Table 3). Data for C_2H_2 reduction in systems of varying H_2O content are also presented in Table 3. While the highest apparent activities are obtained in the aqueous systems (possibly due to the higher concentrations of available protons needed for reduction to proceed), these activities continue even after virtually complete decomposition of FeMo-co (nmf) has occurred. In the non-aqueous system, for which FeMo-co (nmf) is demonstrated by the UW45 assay to be intact, a smaller, but nevertheless substantial, reduction of C_2H_2 is observed. Replacement of BH_4^- by dithionite or formamidine sulfinic acid did not support reduction under the conditions used.

TABLE 3. Catalytic (C_2H_2 Reduction) and Reconstitutive (UW45) Activities of FeMo-co (nmf) in H_2O-NMF-$NaBH_4$ systems

Incubation Time (min)	C_2H_4 Formed (nmol)			UW45 Activity (% units FeMo Protein)		
$\%H_2O$:	80	50	0	80	50	0
0	0	0	0	100	100	100
5	649	436	91	0	55	100
10	904	592	115	0	20	100
15	1064	673	161	0	0	100

Significantly, on exposure to air, the catalytic activity falls off rapidly, as does the UW45 reconstitution activity of FeMo-co (nmf) in the reducing mixture. Preliminary experiments with cyclopropene and the non-aqueous system indicate that no propene is formed, as was previously reported for an aqueous system (McKenna *et al.*, 1979).

3.3. Conclusion

Our results re-emphasize the ambiguity of experiments employing aqueous BH_4^- systems to evaluate independent catalytic activity in FeMo-co (nmf). To remedy this, we have deivsed a non-aqueous BH_4^- system in which FeMo-co (nmf) is demonstrated to retain all of its definitive UW45 reconstitution activity over the period during which its catalytic activity is measured. The observation that demonstrably intact FeMo-co (nmf) is an active catalyst for C_2H_2 reduction unequivocally proves that the isolated cofactor can reduce this alternate nitrogenase substrate. This result strengthens the case for FeMo-co (nat) as an essential part of the active site for substrate reduction. It also provides a useful model system for nitrogenase reductions, having necessary structural relevance. Confirmation that propene formation from cyclopropene cannot be reproduced by this system indicates that other components, such as protein-derived non-constitutive ligands, chemically active protein amino acid side chain residues, or other metal centers are also important for full expression of the catalytic activity of the enzyme. The observation that dithionite or formamidine sulfinic acid cannot replace BH_4^- as electron donors in the FeMo-co (nmf) system is consistent with the existence of a hydride or a hydride-like reducing species at the active site which BH_4 can replace for C_2H_2 reduction.

4. ATP/ADP Fe PROTEIN BINDING STUDIES USING CD SPECTROSCOPY

4.1. Background: Chiroptical Spectroscopy of Nitrogenase

A variety of spectroscopic techniques have been used in attempts to better characterize structure and function in the metal centers of nitrogenase. These have principally included EPR, EXAFS, Mossbauer and, UV-visible spectroscopy. Each technique has particular advantages and attendent, inherent disadvantages. For example, diamagnetic metal cluster states, such as the oxidized state of the Fe protein Fe_4S_4 cluster (Fe (ox)) do not have EPR. An advantage of CD and MCD is that the spectra may be obtained in solutions at room temperature (in contrast to EPR and Mossbauer, which require cryogenic samples). A further advantage is that both CD and MCD can be detected in diamagnetic, as well as paramagnetic cluster states. We have reported near ir-visible-uv circular dichroism (CD) and magnetic circular dichroism (MCD) spectra for the FeMo (Av1) and Fe (Av2) nitrogenase proteins from A.vinelandii (Stephens et al., 1979, 1982). Structured, redox state-dependent spectra were observed in both proteins. Stephens et al. (1978) demonstrated that in general, MCD of Fe-S proteins is characteristic of cluster type and relatively insensitive to protein conformation, while the CD varies with protein and should be sensitive to protein conformation. In this respect, the two techniques are complementary. From the point of view of substrate interactions, the CD of Fe protein is of particular importance, since it can be used to differentiate the two mechanistically significant redox states of the Fe-S cluster, and also to detect changes in cluster environment arising from nucleotide binding, as discussed below.

4.2. CD Studies of ADP/ATP-Nitrogenase Interaction

We previously observed that distinct, structured and characteristic CD spectra are obtained for Fe (red) and Fe (ox). The added effect of ADP or ATP on the CD of Fe(red) is negligible. However, when a fixing system containing excess Av2 and limiting dithionite is allowed to exhaust the reductant, a third, different spectrum is obtained which we assigned to ATP(ADP)-bound Fe(ox). Subsequently, we carried out CD studies in which oxidized K.pneumoniae Fe protein (Kp2(ox)) prepared by dye oxidation was titrated with ADP or ATP (Stephens et al., 1982). Both nucleotides gave equivalent final spectra, showing that either produces the same effect on the CD chromophore. Tight binding was indicated by the observed linear plot of ΔCD vs [ATP] or [ADP] added, which gave a binding stoichiometry (uncorrected for any inactive Kp2 present in the sample) of greater than one. It was concluded that there are two ATP/ADP binding sites on Fe(ox), which are indistinguishable by CD. In these studies, no data were given on the amount of inactive Fe present in the sample used. In parallel studies using the Fe protein from A.vinelandii (Av2), we have also found that the CD of Fe(red) is insensitive to added ATP or ADP. Given other evidence that ATP and ADP bind to this state of the protein, the results confirm a fundamental difference in the binding interaction for the reduced and oxidized Fe protein species, insofar as it affects the CD chromophore. When Av2(ox) is titrated with excess ATP or ADP, the characteristic change in the CD to the ATP/ADP-bound form is observed. A plot of ΔCD vs added nucleotide gives an almost linear plot virtually to the saturation 'end point', again demonstrating strong binding (Fig. 1). When ΔCD is plotted against the ratio of added ADP: Av2(ox), this end point corresponds to a nonunitary value of 1.2 - 1.3 ADP equivalents/Av2(ox). However, analysis for active Fe-centers in the Av2 sample using the ATP-sensitive bathophenanthroline Fe chelation assay of Ljones and Burris (1978) shows that only 2.3 ATP-sensitive Fe atoms were present per protein molecule, implying the presence of inactive protein in the sample (specific activity for C_2H_2 of 1750). If a correction is made on the assumption that 4.0 active Fe atoms should be present in fully active Fe protein, the titration plot yields an ADP/Av2 ratio value close to 2.0. This result, that there are two ADP sites on Av2, is also consistent with the linear plot obtained. The assumption of only a single ADP site does not give a reasonable fit to the titration data.

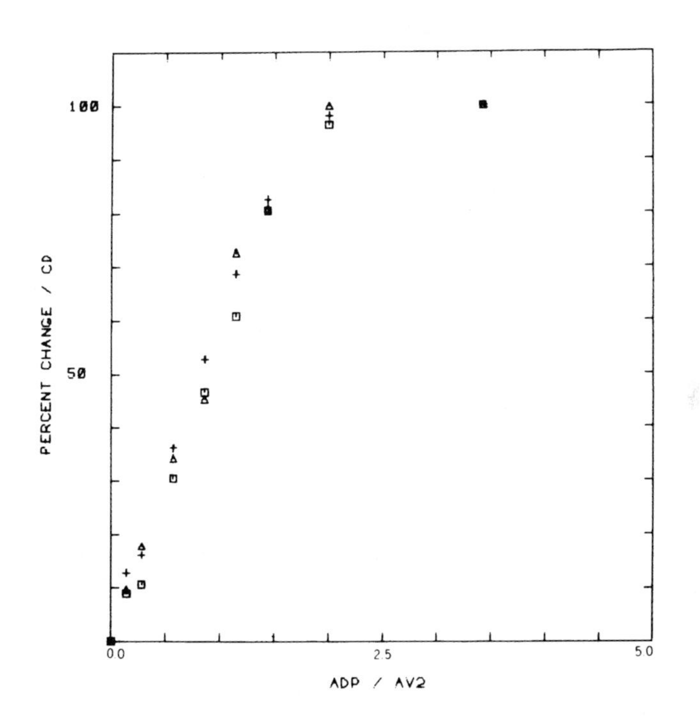

FIGURE 1. CD change (as % of total change) for indigo carmine oxidized Av2 at 260(x), 410(Δ) and 470 (□) nm as a function of added ADP. Av2 is 0.2 mM in 25 mM HEPES, pH 7.4 containing 2 mM Mg^{2+} and 0.1 mg/ml dithiothreitol. The Av2 concentration is corrected for inactive Fe using the ATP-sensitive bathophenanthroline Fe chelation assay.

4.3. Conclusion

The Fe-S chromophore CD of nitrogenase Fe protein in the oxidized state can be used to detect ATP or ADP binding. The spectral changes resulting from binding of these nucleotides are specific, readily observed, and characteristic. Plots of the change at several wavelengths in the CD of Kp2(ox) or Av2(ox) resulting from added increments of ATP or ADP indicate that tight binding occurs. Correction of the data for Av2(ox) to account for inactive Fe yields a binding stoichiometry of 2 ADP sites.

5. REFERENCES

Ljones, T and Burris RH (1978) Biochemistry 17, 1866-1872.
McKenna CE, Jones, JB, Eran, H and Huang CW (1979) Nature 280, 5723, 611-612.
McKenna CE, Huang CW, Jones JB, McKenna MC, Nakajima T and Nguyen HT (1980a) In Newton WE and Orme-Johnson WH, eds., Nitrogen Fixation, Volume 1, pp. 223-235, University Park Press, Baltimore.
McKenna CE, Nakajima T, Jones JB, Huang CW, McKenna MC, Eran H and Osumi A (1980b) In Newton WE and Otsuka S, eds., Molybdenum Chemistry of Biological Significance, pp. 39-57, Plenum Publishing Corporation, New York.
McKenna CE (1980) In Coughlan MP, ed., Molybdenum and Molybdenum-Containing Enzymes, pp. 441-461, Pergamon Press, Oxford.
McKenna CE, Eran H, Nakajima T and Osumi A (1981) In Gibson AH and Newton WE, eds., Current Perspectives in Nitrogen Fixation, pp. 358, Australian Academy of Science, Canberra.
McKenna CE and Eran H (1982) Fed. Proc., 41, 891.
Orme-Johnson WH (1981) In Gibson AH and Newton WE, eds., Current Perspectives in Nitrogen Fixation, p. 53, Australian Academy of Science, Canberra.
Shah VK, Chisnell JR and Brill WH (1978) Biochem. Biophys. Res. Commun. 81, 232-236.
Stephens PJ, Thomson AJ, Keiderling TA, Rao KK and Holl DO (1978) Proc. Nat. Acad. Sci. USA, 75, 5273.
Stephens PJ, McKenna CE, Smith BE, Nguyen HT, McKenna MC, Thomson AJ, Devlin F and Jones JB (1979) Proc. Nat. Acad. Sci. USA, 78, 2585.
Stephens PJ, McKenna CE, McKenna MC, Nguyen HT and Devlin F (1981) Biochemistry, 20, 2857.
Stephens PJ, McKenna CE, McKenna MC, Nguyen HT and Lowe DJ (1982) In Ho C, ed., Electron Transport and Oxygen Utilization, pp. 405-409, MacMillan Press, New York.

6. ACKNOWLEDGEMENTS

We wish to thank NIH (CE McK), the Frasch Foundation (CE McK), USDA (CE McK), NSF (PJS, CE McK) and UNESCO (MD, FXZ) for generous support of this research.

IRON-SULFIDE CONTENT AND ATP BINDING PROPERTIES OF NITROGENASE COMPONENT II FROM AZOTOBACTER VINELANDII

HUUB HAAKER/ARNOLD BRAAKSMA/JAN CORDEWENER/JAN KLUGKIST/HANS WASSINK/
HANS GRANDE/ROBERT EADY* and CEES VEEGER
DEPARTMENT OF BIOCHEMISTRY, AGRICULTURAL UNIVERSITY, DE DREIJEN 11, 6703 BC WAGENINGEN, THE NETHERLANDS
*ARC UNIT OF NITROGEN FIXATION, UNIVERSITY OF SUSSEX, BRIGHTON, BN1 9RQ, ENGLAND

INTRODUCTION

Nitrogenase is the enzyme system that catalyses the reduction of N_2 to ammonia. Nitrogenase consists of two separable proteins. A tetrameric MoFe-protein and a dimeric Fe-protein (Mortenson, Thorneley, 1979). Electrons are donated to the Fe-protein and pass to the MoFe protein ATP-dependently (Mortenson, Thorneley, 1979; Hageman, Burris, 1980). It is generally accepted that the Fe-protein of any nitrogenase enzyme complex has one [4Fe-4S] cluster (Mortenson, Thorneley, 1979; Eady, 1980; Orme-Johnson, Munck, 1980). This is based upon iron and sulfide determinations of various Fe-proteins of nitrogenase isolated from different bacteria (Eady, 1980). Cluster extrusion experiments indicate that the iron and sulfide in the Cp2 preparations are quantitatively recovered as a similar product as found with ferredoxin, i.e. a [4Fe-4S] cluster (Gillum *et al.*, 1977). The Fe-protein from *C.pasteurianum* behaves like a one electron donor/acceptor upon redox titrations in EPR experiments (Zumft *et al.*, 1974) and in colorimetric measurements (Ljones, Burris, 1978). However, there are reports that the Fe-protein from Azotobacter can accommodate two electrons (Thorneley *et al.*, 1976, Braaksma *et al.*, 1982). Lowe (1978) proposed that in view of the anisotropic linewidth and the low integrated EPR signal intensities, a second rapidly relaxing paramagnetic center is present in the protein. In this article we present data that highly-active Fe protein can be isolated from *A.vinelandii* that contains 8 iron and 8 sulfide atoms per dimer of 63 kDa.
The reduction of N_2 and H^+ to NH_3 and H_2 is accompanied by the hydrolysis of ATP to ADP and orthophosphate. The specific role of ATP in nitrogenase catalysis has not been resolved although several hypotheses have been proposed (Eady *et al.*, 1980; Hageman *et al.*, 1980). It has been demonstrated that ATP hydrolysis is coupled to electron transfer between the nitrogenase components (Eady *et al.*, 1978; Hageman *et al.*, 1980). Binding of ATP to the Fe-protein has been reported (Yates, 1972; Tso, Burris, 1973; Zumft *et al.*, 1973). Binding of ATP to the MoFe- protein has only been found for the MoFe protein isolated from *K.pneumoniae* (Miller *et al.*, 1980). In the earlier experiments, complications that may arise during binding experiments like oxidation and inactivation of the Fe-protein under fully anaerobic conditions have not been recognized. In this article the binding properties of MgATP and MgADP to reduced and oxidized Av2 is described.
The implications of these binding properties and the redox properties of Av2 with respect to a model for the nitrogenase reaction will also be discussed.

MATERIALS AND METHODS

Azotobacter vinelandii ATCC 478 was grown in a batch culture of 2500 l. Harvesting of the cells started just before the mid-logarithmic growth. During harvesting of the cells, the culture was aerated and growth continued. Cells collected at different growth stages were collected separately. For physiological studies, cells were grown in a chemostat as described earlier (Haaker, Veeger, 1976). Isolation of the nitrogenase proteins and standard nitrogenase activity assays were run as described earlier (Braaksma *et al.*, 1982). Av2 purified by the described method has a specific activity between 1500-1800 nmoles C_2H_4 formed.min^{-1}.

mg $Av2^{-1}$. To produce more active Av2 an additional concentration step is necessary. The isolated Av2, stored in liquid nitrogen, was thawed in 25 mM Hepes-KOH (pH = 7.5) and 2 mM $Na_2S_2O_4$ and concentrated on a small DEAE-cellulose (DE-32 Whatman) column (1 x 5 cm) and eluted with 25 mM Hepes-KOH (pH = 7.5), 2 mM $Na_2S_2O_4$ and 90 mM $MgCl_2$. This gives routinely an increase in specific activity provided that Av2 was isolated from cells harvested before the mid-logarithmic phase. Desalting procedures, protein estimations and analytical methods were performed as described earlier (Braaksma _et al._, 1983).
ATP and ADP binding to Av1 and Av2 was studied with the flow dialysis technique. This technique was originally developed by Colowick and Womack (1969) and slightly modified for the oxygen labile nitrogenase proteins (Cordewener _et al._, 1983). The whole cell nitrogenase activity was measured in a closed gas flow system as described earlier (Haaker _et al._, 1974).
The concentrations of Av1, Av2, flavodoxin (fld) and Fe/S protein II (Fe/SII, or nitrogenase protective protein) were measured by means of Western blotting. Poly acrylamide gel electrophoresis in the presence of dodecylsulphate was carried out essentially according to Laemmli (1970). Details of the Western blotting procedure have been given elsewhere (Bowen _et al._, 1980). Antisera against the purified proteins were elicited in New Zealand white rabbits using the procedures described earlier (Voordouw _et al._, 1982). The immunologically reactive polypeptides on the nitrocellulose paper were identified using the antisera prepared against the purified proteins. The reacted immunoglobins were detected with ^{125}I-labeled protein A (1-40 μCi) in 50-150 ml RIA buffer. Spots containing radioactivity, were cut out and counted.

RESULTS AND DISCUSSION

From studies of the redox dependence of the nitrogenase reaction catalyzed by nitrogenase from _A.vinelandii_, it was found that nitrogenase activity is independent of the redox potential up to -440 mV. At higher potentials the nitrogenase activity declines and no significant activity was detectable at potentials above -350 mV (Braaksma _et al._, 1982). It was ascertained in these experiments that there was always redox equilibrium between free Av2 and the redox mediator (electron donor). This was done by varying the Av2 concentration or the electron donor concentration. When the same redox dependence of the nitrogenase reaction was observed, it was assumed that the decrease in nitrogenase activity measured at higher potentials was caused by the increased concentration of oxidized Av2. In these experiments it was observed that at a redox potentials of around -390 mV with dithionite, methyl viologen or _Megasphaera elsdenii_ flavodoxin as electron donors, the nitrogenase activity was about 50% of the maximal activity. When under the same conditions the redox properties of Av2 were measured, the results as shown in Fig. 1, were found.
In the three cases, the height of the g_y peak was used as a measure of the redox state of Av2. It was controlled that the shape of the EPR signal did not change during the redox titration by the addition of high concentrations of Na_2SO_3. Three important conclusions can be drawn from these results. By comparing the redox dependence of the nitrogenase reaction with the redox state of free Av2 (both are half maximal at -390 mV), it appears that the redox state of free Av2- thus without adenine nucleotide bound-determines the nitrogenase activity when the redox potential is limiting the nitrogenase activity. This observation has consequences for the nitrogenase mechanism. It requires a sequential mechanism. First a complex must be formed between Av2 and Av1. After the complex is formed ATP can bind to Av2 and coupled to ATP hydrolysis, electron transfer occurs from Av2 to Av1. This implicates also that the association constant for the Av2 (red). Av1 (ox) complex must be an order of magnitude higher than that of the Av2 (red). MgATP. Av1 (ox) complex. The second important

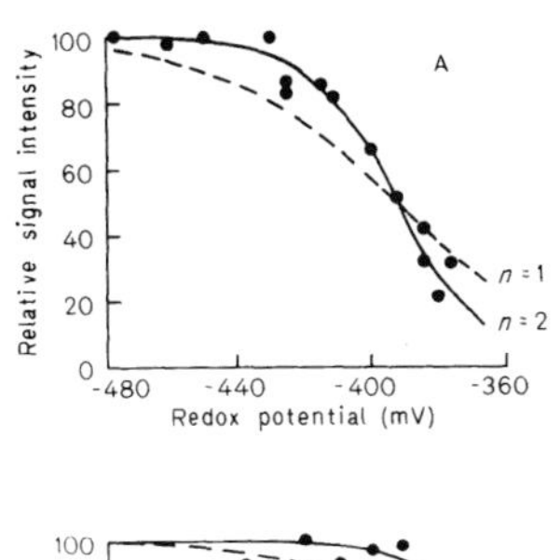

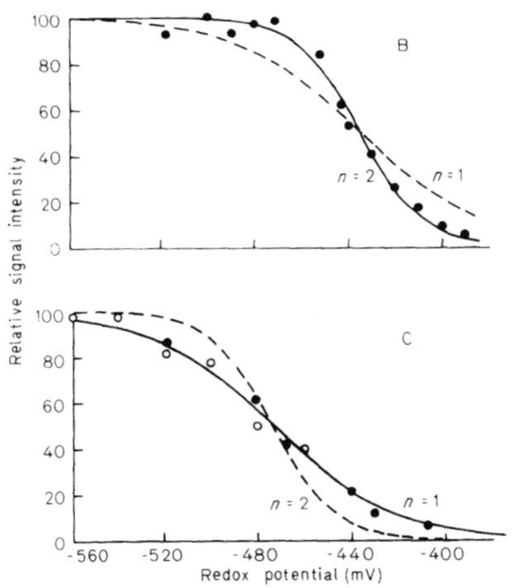

Figure 1. Dependence of the electron paramagnetic resonance signal of Av2 in the presence or absence of Mg^{2+}-adenine nucleotides on the redox potential. The redox potential was set with $Na_2S_2O_4$ and Na_2SO_3. (A) (●-●) 56 μM Av2, 5 mM $Na_2S_2O_4$, 7.5 mM $MgCl_2$, 112 Hepes/KOH and varying amounts of Na_2SO_3 to obtain the redox potential indicated, final pH 7.0. The g = 1.941 signal is plotted against the redox potential. The dotted line is the theoretical curve for a one-electron transition, the solid line for a two-electron transition with an E_m = -393 mV. (B) (●-●) 31.5 μM Av2, 5 mM $Na_2S_2O_4$, 7.5 mM $MgCl_2$, 5 mM ATP, 112 mM Hepes/KOH and varying amounts of Na_2SO_3 to obtain the redox potential indicated, final pH 7.0. The g = 1.940 signal is plotted against the redox potential. The dotted line is the theoretical curve for a one-electron transition the solid line for a two-electron transition with E_m = -435 mV. (C) (●-●) 56 μM Av2, 5 mM $Na_2S_2O_4$, 7.5 mM $MgCl_2$, 5 mM ADP, 112 mM Hepes/KOH, and varying amounts of Na_2SO_3 to obtain the redox potential indicated, final pH 7.0. The g = 1.947 signal is plotted against the redox potential. The normalized maximum signal intensity at -600 mV at pH 7.8 was taken as 100%. (O-O) 21.5 μM Av2, 5 mM $Na_2S_2O_4$, 7.5 mM $MgCl_2$, 5 mM ADP, 112 mM Hepes/KOH and varying amounts of Na_2SO_3 to obtain the redox potential indicated, final pH 7.8. The solid line is the theoretical curve for a one-electron transition, the dotted line for a two-electron transition with an E_m = -473 mV. (From Braaksma _et al._, 1982, permission Eur. J. Biochem.).

observation is that Av2, considered to be containing one [4Fe-4S] center protein, shows redox properties characteristic for a two electron transferring protein. This redox behaviour of free Av2 or Av2-MgATP, can be explained by assuming that the EPR signal of the protein is caused by two coupled Fe-S centers or by one Fe-S center coupled to a fast relaxing paramagnetic center (EPR invisible).

The third important observation is that Av2-MgADP bound behaves as a one-electron donor/acceptor and operates at much lower potentials than Av2 or as Av2-MgATP. These properties might explain the strong inhibition of nitrogenase by MgADP. This will be discussed later.

From the redox titrations, described in Fig. 1, it is clear that Av2 must contain at least two redox centers. We used Proton Induced X-ray Emission Spectroscopy to detect the transition metal content of Av2. Only unexpected high levels of iron were detected. These high levels were also found with two other analytical methods namely colorimetric chelation with bathophenanthroline disulfonate and with atomic absorption spectroscopy. The three different methods gave the same results. The results of a large number of Av2 isolations from different batches of bacteria are shown in Fig. 2. There is a tendency that above a specific activity of 1000, the increase in specific activity is associated with an increase in iron-content of the protein, but there is a large scatter around the visual fit. For instance, it is possible to obtain Av2 with a specific activity of 1600 with an iron content of 4 or with an iron content of 8. The relation between iron content and sulfide content is reasonably good.

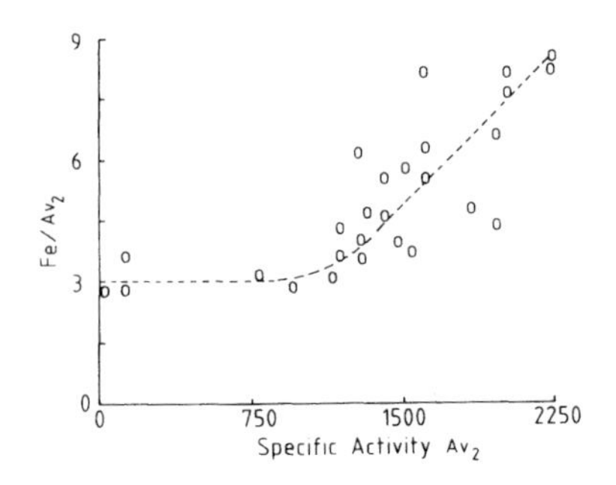

Figure 2. Relation between the number of iron atoms per molecule Av2 and the specific activity of different Av2 preparations prepared as described in Materials and Methods. Iron was determined colorimetrically; specific activity is expressed as a nmol C_2H_4 produced.min^{-1}.mg $Av2^{-1}$. All data points represent different preparations. The dotted line is a visual fit through the data points. (From Braaksma et al., 1983, permission Eur. J. Biochem.)

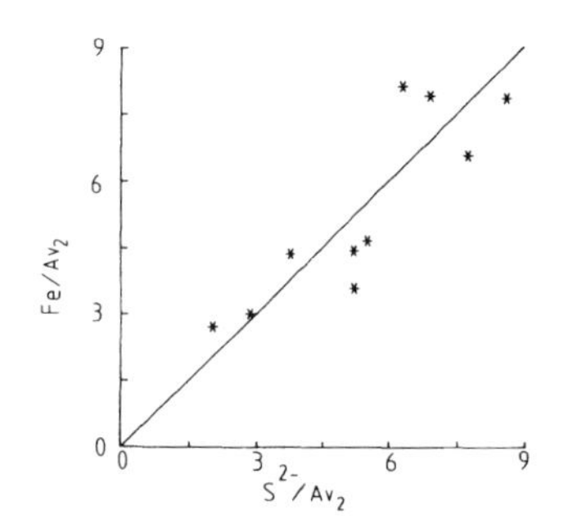

Figure 3. Relation between the number of iron and sulfide atoms per molecule Av2 of several different preparations with different specific activities. Each plotted point is the result of an iron and sulfide determinations of one Av2 preparation. The plotted line gives the expected ratio Fe/S = 1. (From Braaksma et al., 1983, permission Eur. J. Biochem.).

shown in Fig. 3 indicates that a higher iron content of the protein is associated with a higher sulfide content indicative for [Fe-S] centers. Since [Fe-S] centers absorb light at 430 nm one expects an increase in molar absorbance with increasing [Fe-S] center content of the protein. As can be seen in Fig. 4., there is a direct relation between molar absorbance and specific activity of Av2.
It is clear that from A.vinelandii, Fe-protein can be isolated that probably contains 2[4Fe-4S] centers. The redox centers are partly coupled so that the integral of the EPR signals integrates below 1. By uncoupling the two clusters one expects an increase in spin intensity. As shown in Table 1, this can be done by addition of ethylene glycol. When the concentration of ethylene glycol increases, an increase in signal intensity of both Av2 and Kp2 is observed. The addition of MgATP diminishes this effect but MgADP uncouples the two centers further as indicated by a spin integration of 1.38 spins/molecule Av2 . This indicates that the EPR signal of Av2 and probably also of Kp2 is caused by two coupled paramagnetic centers. As shown in Fig. 2, it is possible to isolate Av2

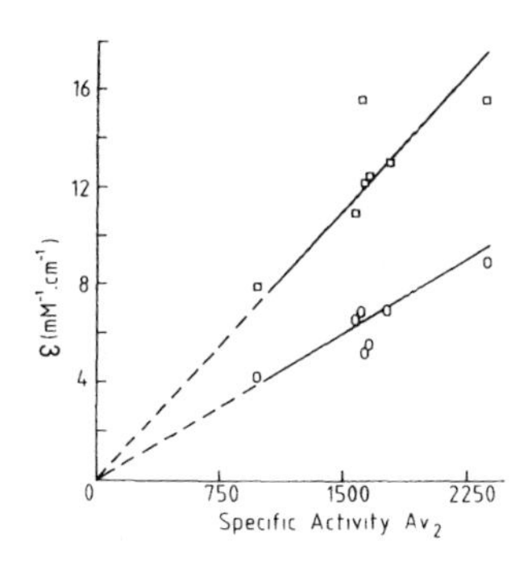

Figure 4. Relation between molar absorbance of oxidized (□) and reduced (O) Av2 at 430 nm of several preparations with different specific activities prepared as described in Materials and Methods. Specific activity of Av2 is expressed as nmol C_2H_4 produced.min^{-1}.mg $Av2^{-1}$. Molar absorbance, ε, was measured at 430 nm. (From Braaksma *et al.*, 1983, permission Eur. J. Biochem.)

with a variable Fe content and different specific activities. One can argue that the different specific activities are caused by a variable degree of inactivation during isolation. But this does not explain the variability of Fe numbers found. We therefore measured the specific activity of Av2 *in vivo* in cells grown under different conditions (Table 2).

Table 1. The effect of ethylene glycol on the integration of the EPR signal of Av2 and Kp2.

addition of ethylene glycol (v/v)	integrated intensity (spins/mole) Av2	Kp2
0	0.51	0.39
10	0.69	-
50	0.96	0.65
50 + 5 mM MgATP	0.49	-
50 + 5 mM MgADP	1.38	1.03

Table 2. The effect of the growth conditions on the specific activity of Av2 *in vivo* and *in vitro*.

Dilution rate (h^{-1})	nitrogenase activity (nmoles C_2H_4 formed min^{-1}.mg total protein^{-1}) whole cell act.	extract act.	μg Av2/ mg total cell protein	specific activity of Av2 (nmoles C_2H_4 formed. min^{-1}.mg $Av2^{-1}$) whole cell act.	extract act
0.1	44	44	20	2200	2200
0.2	124	97	32	3875	3031
0.1 + O_2 shock	210	60	30	7000	2000

Cells were grown in a chemostat at different dilution rates and with a constant

oxygen input rate. Cells were harvested from the bulk of the growth medium in the chemostat, centrifuged and the whole cell nitrogenase activity was measured as the maximum activity at different oxygen input rates as described in Materials and Methods. Cells were grown oxygen-limited, except in the case of oxygen stress. These cells were exposed to an increase in oxygen input. Growth stopped and after approximately 2 hours the respiration rate of the cells was adapted and growth continued. As shown in Table 2 whole cell activity and cell free extract nitrogenase activity are similar of cells grown oxygen-limited and the specific activity of Av2 in these cells is about 2200, which is a good value for isolated enzyme. But when the cells were forced to grow faster and growth switched from oxygen-limited to nitrogen-limited, the activities of the whole cell and cell free extract increased and highly-active Av2 was detected in both whole cells and cell free extracts. This Av2 could be the highly-active Av2 with a high iron and sulfide content (see Fig. 2). When the cells were exposed to an oxygen shock, whole cell nitrogenase activity increased and also the whole cell specific activity of Av2, but this highly-active Av2 could not be detected in a cell free extract. The reason for this failure is not known at the moment.
The results clearly show that the specific activity of Av2 *in vivo* is not constant. This may explain why protein with different specific activities and different Fe/S cluster content has been isolated. There are also indications (R. Robson, this volume) that Azotobacter contains two genes for the Fe protein. It might be possible that the two genes code for two slightly different proteins with modified specific activities which is related to the accommodation of one or two [4Fe-4S] centers in the protein. This possibility will be tested soon.
To gain more insight in the rôle of MgATP and MgADP in nitrogenase function, the interaction of both nucleotides with the individual proteins was investigated. The binding was studied with the flow dialysis technique. This is a relative fas technique which was found to be important for such binding studies (Cordewener *et al.*, 1983). At protein concentrations around 5-10 mg/ml, which are necessary for such experiments, dithionite is rapidly oxidized. When no special attention is paid to the dithionite concentration during storage of the protein and during the binding experiments, the Fe-protein could become oxidized and partly inactivated. This would explain the differences published by us and others with respec to binding to Fe-protein (Tso, Burris, 1973). Table 3 shows the binding properties of Av2 in the different redox states.

Table 3. Binding of MgATP and MgADP to the Fe-protein and the MoFe-protein isolated from A.vinelandii

	Number of binding sites (mol/mol)		Dissociation constant Kd (mM)	
	MgATP	Kd (mM)	MgADP	Kd (mM)
Av2 (reduced)	1	0.33	1	0.15
Av2 (oxidized)	2	0.14	1	0.14
Av1 (reduced)	0	0	0	0
Av1 + Av2 (reduced)	-	-	1	0.15

As can be seen from Table 3, binding of MgATP to Av2 depends on the redox state of the protein. Binding of MgADP is independent upon the redox state of the protein and the binding properties are not different in the complex of Av2 with

Av1. No binding of MgATP or MgADP to Av1 could be detected with the flow dialysis method or by equilibrium dialysis (not shown).
Inhibition studies of ATP and ADP binding to Av2 were performed to establish the nature of the adenine nucleotide binding sites. As can be seen from Fig. 5 binding of MgATP to reduced Av2 is much more inhibited by MgADP than the binding of MgADP is inhibited by MgATP. This difference in inhibition pattern cannot be explained by the difference in binding constants for MgATP and MgADP (Table 3). As shown in Fig. 6 a similar pattern was found for phenazine methosulfate-oxidized Av2. In this case oxidized Av2 binds two molecules of ATP and binding of these two molecules can be inhibited by only one molecule MgADP. Therefore we conclude that the binding sites of MgATP and MgADP are independent sites and suggest that the ATP binding site(s) is(are) a catalytic site(s) and the ADP binding site is a regulatory site.
In Fig. 7 a working hypothesis of electron transfer between highly active Av2 and Av1 is shown. This model includes the most important findings presented in this paper: isolated highly active Av2 contains two redox centers; Av2 behaves as a two electron redox enzyme; reduced Av2 binds one MgATP and oxidized Av2 two molecules of MgATP. Only in the presence of MgATP electrons are transferred from Av2 to Av1. Because free Av2 with or without MgATP bound behaves as a n=2 redox protein, this means that in the Av2-Av1 complex, Av2 could donate one electron above its observed midpoint potential of catalysis and that the second electron must be donated in that case to a redox center below this midpoint potential. The two possible accepting redox centers in Av1 are the FeMo-cofactor and the so-called P-centers. These centers have redox potentials around -290 mV and -480 mV respectively (Watt et al., 1980).

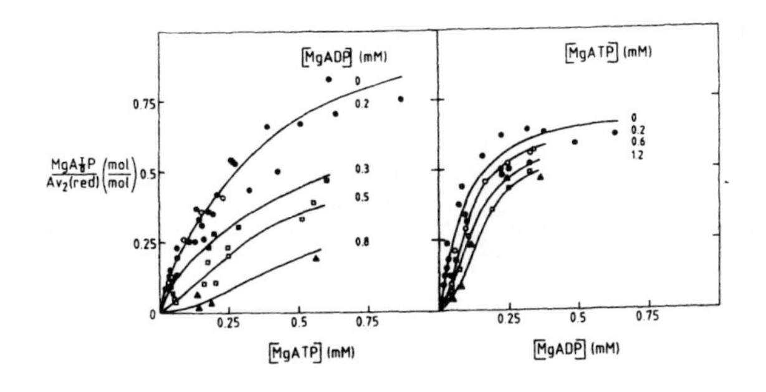

Figure 5. Inhibition of the MgATP or MgADP binding to reduced Av2 as determined by flow dialysis. Protein concentrations varied over 100-160 μM. Experimental conditions are described in Cordewener et al., 1983.

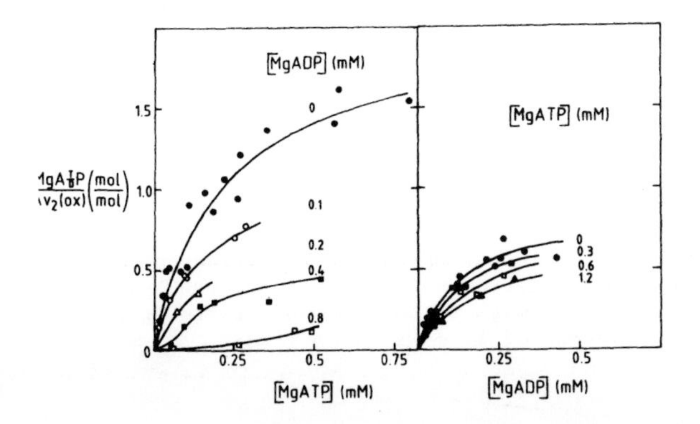

Figure 6. Inhibition of MgATP or MgADP binding to oxidized Av2 as determined by flow dialysis. Other conditions as described in legend to Fig. 5.

Back flow of electrons in the complex from Av1 to Av2 is inhibited by a ATP hydrolysis driven change in conformation of Av2 - complexed to Av1 - to its oxidized protein conformation. In this conformation Av2 cannot be reduced - possibly due to the presence of ADP bound to this form and a change in redox properties (from n=2 to n=1); thus the back flow of electrons is inhibited. In the oxidized conformation Av2 has two binding sites for MgATP. MgATP can be bound to the protein complex - replacing bound ADP - and can be hydrolyzed. The energy obtained by hydrolysis, can be used for electron transfer within Av1 to substrates or by supplying by hydrolysis H^+ to the catalytic site at Av1. The binding and/or hydrolysis of a fourth MgATP could be used to change the conformation of oxidized Av2 complexed to Av1, to allow reduction of oxidized Av2 still bound to Av1. With this working hypothesis an explanation can be given for the presence of two redox centers in highly active Av2 and the presence of only one MgATP binding site on reduced Av2 and two MgATP binding sites on oxidized Av2. ADP inhibition of the nitrogenase reaction can be explained by the inhibition of MgATP binding to reduced as well as oxidized Fe-protein and also by the inability of Av2 with MgADP bound to behave as a n=2 electron donor.

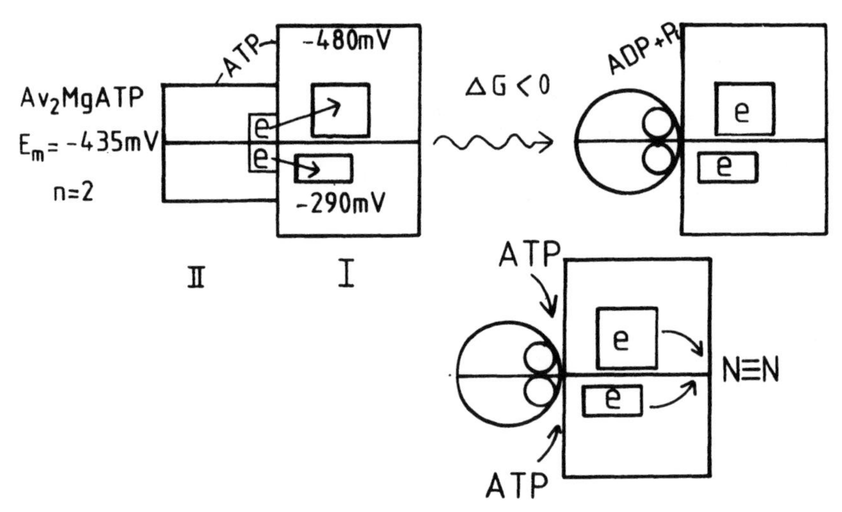

Figure 7. Working hypothesis for electron transfer between Av2 and Av1.

ABBREVIATIONS

The MoFe- and Fe-proteins of the nitrogenase of Azotobacter vinelandii and Klebsiella pneumoniae are referred to as Av1, Av2, Kp1 and Kp2 respectively. EPR, electron paramagnetic resonance.

ACKNWOLEDGEMENTS

We thank Dr. J. van Leeuwen for the useful discussions and Mrs. J. Toppenberg-Fang for typing the manuscript. This investigation was supported by the Netherlands Foundation for Chemical Research (SON) with financial support from the Netherlands Organization for the Advancement of Pure Research (ZWO).

REFERENCES

Bowen B, Steinberg J, Laemmli UK and Weintraub H (1980) Nucleic Acids Res. 8, 1-20.
Braaksma A, Haaker H, Grande HJ and Veeger C (1982) Eur. J. Biochem. 121, 483-491.
Braaksma A, Haaker H and Veeger C (1983) Eur. J. Biochem. 133, 71-76.
Colowick SP and Womack FC (1969) J. Biol. Chem.244, 744-777.
Cordewener J, Haaker H and Veeger C (1983) Eur. J. Biochem., 132, 47-54.
Eady RR, Lowe DJ and Thorneley RNF (1978) FEBS Letters 95, 211-213.
Eady RR (1980) Methods Enzymol. 69, 753-792.
Eady RR, Iman S, Lowe DJ, Miller RW, Smith BE and Thorneley RNF (1980) In Stewart WDP and Gallon JR eds., Nitrogen Fixation, pp. 19-35, Academic Press, London.
Gillum WO, Mortenson LE, Chen J-S and Holm RH (1977) J. Am. Chem. Soc. 99, 584-595.
Haaker H, De Kok A and Veeger C (1974) Biochim. Biophys. Acta 357, 344-357.
Haaker H and Veeger C (1976) Eur. J. Biochem. 63, 499-507.
Hageman RV, Orme-Johnson WH and Burris RH (1980) Biochemistry 19, 2333-2342.
Hageman RV and Burris RH (1980) Curr. Top. Bioenerg. 10, 279-291.
Laemmli UK (1970) Nature (London) 227, 680-685.
Ljones T and Burris RH (1978) Biochem. Biophys. Res. Commun. 80, 22-25.
Miller RW, Robson RL, Yates MG and Eady RR (1980) Can. J. Biochem. 58, 542-548.
Mortenson LE and Thorneley RNF (1979) Annu. Rev. Biochem. 48, 387-418.
Orme-Johnson WH and Munck E (1980) in Coughlan MP, ed., Molybdenum and Molybdenum-Containing Enzymes, Pergamon Press, Oxford.
Thorneley RNF, Yates MG and Lowe DJ (1976) Biochem. J. 155, 137-144.
Tso, M-Y and Burris RH (1973) Biochim. Biophys. Acta 309, 263-270.
Voordouw G, De Haard H, Timmermans JAM, Veeger C and Zabel P (1982) Eur. J. Biochem. 127, 267-274.
Watt GD, Burns A, Lough S and Tennent DL (1980) Biochemistry 19, 4926-4932.
Yates MG (1972) Eur. J. Biochem. 29, 386-392.
Zumft WG, Palmer C and Mortenson LE (1973) Biochim. Biophys. Acta 292, 413-421.
Zumft WG, Mortenson LE and Palmer G (1974) Eur. J. Biochem. 46, 525-535.

THE MECHANISM OF SUBSTRATE REDUCTION BY NITROGENASE

D.J. LOWE, R.N.F. THORNELEY and J.R. POSTGATE
ARC UNIT OF NITROGEN FIXATION, UNIVERSITY OF SUSSEX, BRIGHTON BN1 9RQ, UK.

A computer model for nitrogenase action has been developed from the scheme first presented at the Fourth Symposium (Canberra) in this series (Thorneley, Lowe 1981). The model simulates both steady state and pre-steady state kinetic data for H_2 evolution, N_2 reduction and H_2 inhibition of N_2 reduction. It also accurately simulates the unusual damped oscillatory time course for hydrazine that is obtained from an enzyme bound intermediate in N_2 reduction (Thorneley et al, 1978). The model predicts HD formation rates both in the presence and absence of N_2 at 23°C with Kp nitrogenase that are consistent with those reported by Burgess et al (1981). An important feature of the model that has been verified by stopped-flow kinetic spectrophotometry (Thorneley, Lowe 1983), is that the rate-limiting step for nitrogenase is the dissociation (k_{-3}) of oxidised Kp2 from reduced Kpl after MgATP-induced electron transfer has occurred between these proteins.

The Fe protein oxidation-reduction cycle

Scheme 1 shows a cycle in which one electron is transferred from Kp2 to $Kpl^{\dagger}$ with the concomitant hydrolysis of 2MgATP to 2MgADP + $2P_i$ (Eady et al 1978; Hageman et al 1980).

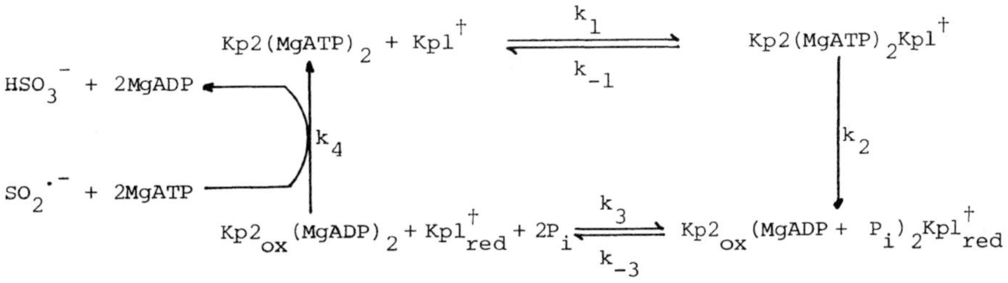

$$Kp2(MgATP)_2 + Kpl^{\dagger} \underset{k_{-1}}{\overset{k_1}{\rightleftharpoons}} Kp2(MgATP)_2Kpl^{\dagger}$$

$$Kp2(MgATP)_2Kpl^{\dagger} \xrightarrow{k_2} Kp2_{ox}(MgADP + P_i)_2Kpl^{\dagger}_{red}$$

$$Kp2_{ox}(MgADP)_2 + Kpl^{\dagger}_{red} + 2P_i \underset{k_{-3}}{\overset{k_3}{\rightleftharpoons}} Kp2_{ox}(MgADP + P_i)_2Kpl^{\dagger}_{red}$$

$$Kp2_{ox}(MgADP)_2 \xrightarrow[SO_2^{\cdot -} + 2MgATP \;\rightarrow\; HSO_3^- + 2MgADP]{k_4} Kp2(MgATP)_2$$

Scheme 1

$Kpl^{\dagger}$ represents one of two independently functioning halves of the tetrameric ($\alpha_2\beta_2$ structure) Kpl. Each $Kpl^{\dagger}$ is assumed to contain one Mo substrate-binding site and one Kp2 binding site.

Table 1 contains the values of all the rate constants in Scheme 1.

The value of the specific activity for Kpl calculated using k_{-3} = 6.4 s^{-1} as the rate-limiting step (1140±150 nmol min^{-1}(mg Kpl)$^{-1}$ at 23°) agrees with that measured using standard steady state assays for H_2 production (1160±50 nmol min^{-1} (mg Kpl)$^{-1}$). The latter procedure uses an excess of Kp2 over Kpl with a high concentration of dithionite (30 mM) which minimises the inhibition due to the back reaction of $Kp2_{ox}$ with Kpl (k_3).

TABLE 1. Rate constants used in the simulation of the kinetics of nitrogenase

Rate constant	Value	Comment	Reference
k_1	[1] $4.4 \times 10^6 M^{-1} s^{-1}$	Responsible for dilution effect	Lowe, Thorneley unpublished.
k_{-1}	[1] $6.4\ s^{-1}$		
k_2	$200\ s^{-1}$	Electron transfer coupled to MgATP hydrolysis	Thorneley (1975)
k_3	[1] $4.4 \times 10^6 M^{-1} s^{-1}$	Responsible for inhibition at high protein concentrations	Thorneley, Lowe (1983)
k_{-3}	[1] $6.4\ s^{-1}$	Rate limiting when Kp2 and substrates are saturating	"
k_4	$3.0 \times 10^6 M^{-1} s^{-1}$	Rate of reduction of $Kp2_{ox}(MgADP)_2$ by $SO_2{}^{\cdot -}$	"
k_5	[1] $4.4 \times 10^6 M^{-1} s^{-1}$	Responsible for inhibition of H_2 evolution	"
k_{-5}	[1] $6.4\ s^{-1}$	when MgATP but not reductants is limiting	"
k_6	$1.2 \times 10^9 M^{-1} s^{-1}$	$S_2O_4{}^{2-} \underset{k_{-6}}{\overset{k_6}{\rightleftharpoons}} 2SO_2{}^{\cdot -}$	"
k_{-6}	$1.7\ s^{-1}$		
k_7	[2,3] $230\ s^{-1}$	Responsible for enhanced H_2 evolution at low e^- flux	Lowe, Thorneley unpublished.
k_8	[2,3] $0.8\ s^{-1}$	Slow in order to maximise N_2 binding to E_3	"
k_9	[2,3] $600\ s^{-1}$	Rapid H_2 evolution from the most reduced hydridic species	"
k_{10}	[3] $5 \times 10^4 M^{-1} s^{-1}$	Determined from $K_m{}^{N_2}$ at low e^- flux	"
k_{-10}	[3] $2 \times 10^4 M^{-1} s^{-1}$	Determined from $K_i{}^{H_2}$ at low e^- flux	"
k_{11}	[3] $8.9 \times 10^6 M^{-1} s^{-1}$	Determined from $K_m{}^{N_2}$ at high e^- flux	"
k_{-11}	[3] $3.0 \times 10^6 M^{-1} s^{-1}$	Determined from $K_i{}^{H_2}$ at high e^- flux	"
k_{12}	[3] $6.0 \times 10^4 M^{-1} s^{-1}$	Responsible for HD formation in absence of N_2	"

[1] Kp1-Kp2 association-dissociation rates assumed to be independent of oxidation level. k_3, k_{-3} experimentally determined. See also poster abstract Thorneley, Lowe in this volume.

[2] H_2 evolution rates. These depend on small differences between large numbers and are subject to errors of factors of about two.

[3] Since these rate constants determine K_Ms and K_Is only their ratios are absolute values.

The assumption that Kp2 interacts independently with two sites on Kp1 is supported by two types of experimental evidence. Firstly, Ac2 (Thorneley et al 1976), Cp2 (Ljones, Burris 1978) and Av2 (Hageman et al 1980) are reported to be one electron donor proteins. Secondly, the pre-steady state kinetics of H_2 and C_2H_4 formation require two slow steps (k_{-3}) before the two electrons reduced product is detected (Thorneley, Lowe 1981, 1982; Lowe et al 1983). Kp2 cannot interact with only one site on Kp1, since otherwise the steady state rate of H_2 or C_2H_4 formation would be half of that observed. The protein dissociation reaction (k_{-3}) determines the maximum rate of substrate reduction that can be obtained per Mo centre when all other reactants, including Fe protein, are saturating. The use of limiting dithionite ion or another reductant can also cause a change in rate-limiting step. The values of k_3, k_{-3} and the linear dependence of k_4 on $[S_2O_4^{2-}]^{\frac{1}{2}}$ provide an explanation for the dependence of the apparent K_m for $S_2O_4^{2-}$ on the ratio of Fe:MoFe proteins described by Hageman and Burris (1978). $k_4 = k_{-3} = 6.4 \pm 0.8\ s^{-1}$ when $[S_2O_4^{2-}] = 2$ mM. Thus 2 mM is an upper limit for the apparent K_m for $S_2O_4^{2-}$ at protein concentrations when inhibition due to the back reaction k_3 is negligible (i.e. standard assay conditions). This apparent K_m will decrease as the ratio of Fe:MoFe protein increases since at high ratios, a lower concentration of $S_2O_4^{2-}$ will be required to maintain an optimum steady state concentration of reduced Fe protein.

The MoFe protein or substrate reduction cycle.

The cycle shown in Scheme 2 comprises eight consecutive repetitions of the Fe protein cycle (Scheme 1). Thus one cycle of Scheme 2 involves eight electron transfers which are used to effect the reactions shown in equation 1.

$$N_2 + 8H^+ + 8e^- \longrightarrow 2NH_3 + H_2 \qquad (1)$$

The values of the rate constants used to simulate time courses for intermediates and products shown in Scheme 2 are given in Table 1 together with comments on their determination and mechanistic significance. The poster abstract of Lowe, Thorneley (1983) in this volume provides additional

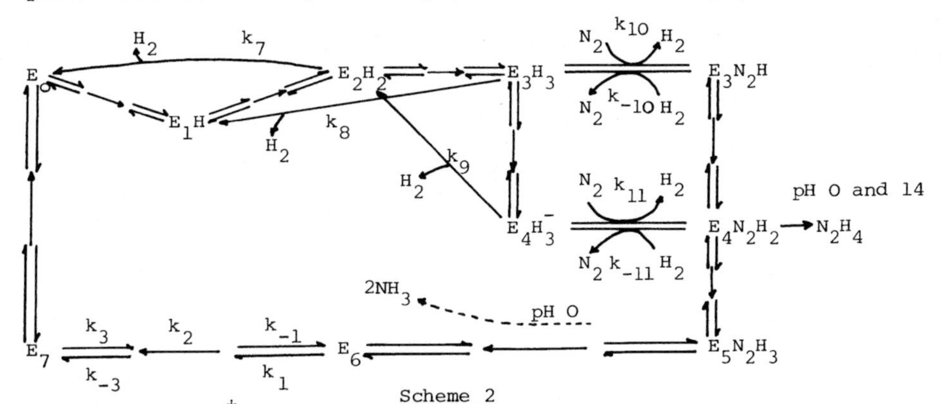

Scheme 2

E_n represents Kp1† (Scheme 1) that has been reduced by n electron equivalents.

details of Scheme 2. A feature of this scheme is its ability to accurately simulate pre-steady state time courses for H_2, NH_3 and N_2H_4. The latter is of particular note since it is derived from an enzyme bound intermediate (Thorneley et al 1978) and exhibits an unusual damped oscillation before a constant steady state concentration is achieved (Fig. 1). H_2 evolution under Ar, H_2 evolution concomitant with N_2

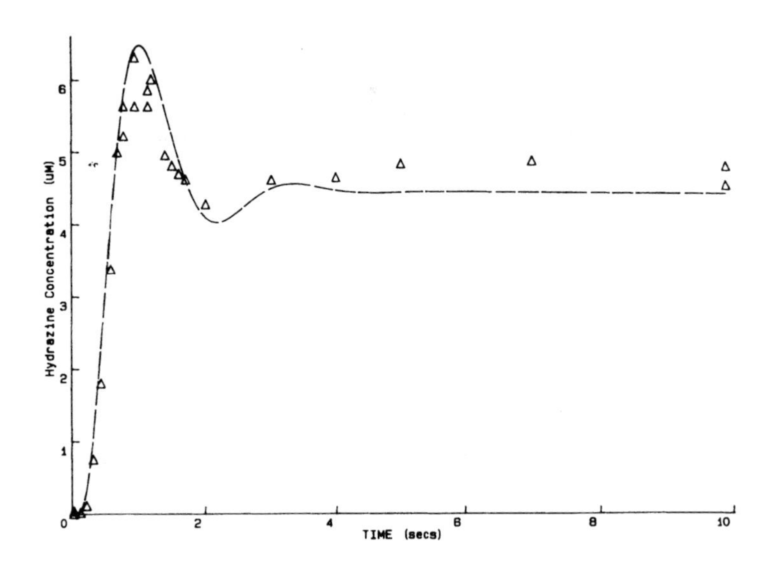

Fig. 1. Pre-steady-state time course of N_2H_4 formed by nitrogenase quenched in acid. The curve is a simulation based on Scheme 2 using the rate constants in Table 1. Protein concentrations: Kp1 - 37 μM, Kp2 - 125 μM.

reduction and NH_3 formation have also been simulated (data not shown). The curve in Fig. 1 is based on the assumption that only one Kp1 intermediate (and its complexes with Kp2) gives N_2H_4 on quenching and that this species is associated with one Mo centre. An active site involving two Mo atoms would give half the amplitude obtained. We conclude that the two Mo centres in Kp1 function independently and that bridging of N_2 across 2 Mo atoms does not occur. The extent of the lag phase (ca. 200 ms) and the time at which the maximum amplitude occurs (ca. 1100 ms) means that it is species E_4 that gives rise to N_2H_4 on quenching. This species involves bound N_2 (two electrons have been used to form H_2 that is displaced when N_2 binds) and E_4 is reduced by two electron equivalents relative to the oxidation level of Kp1 as isolated in the presence of dithionite (E_o). In view of the stability of the hydrazido(2-)-species in Mo complexes, which also give rise to N_2H_4 on treatment with acid or base (Chatt et al 1977), the intermediate bound to E_4 is also considered to be a hydrazido(2-)-species. When quenched with acid, the protein is oxidised by two equivalents and N_2H_4 is released.

The species $E_4H_3^-$ has been introduced into Scheme 2 in order to explain the steady state kinetics of N_2 reduction as a function of electron flux through Kp1. Silverstein, Bulen (1970) first reported that as the ratio of Fe to MoFe protein is increased, a higher percentage of the total electron flux is used to reduce N_2 and proportionately less H_2 is evolved. We have confirmed this for K. pneumoniae nitrogenase, and by determining the $K_M^{N_2}$ at a series of Kp2:Kp1 ratios, have shown that when [Kp2]>>[Kp1], an extrapolation of reciprocal plots to infinite N_2 concentration gives 73±5% of the total electron flux being used to reduce N_2. Thus the stoichiometry of equation 1 has been confirmed. In addition to V_{max} for N_2 reduction increasing with increasing Kp2:Kp1, surprisingly $K_M^{N_2}$ also increases. Both these effects are simulated by Scheme 2. High electron flux through Kp1 is necessary to reduce E_2 to species E_3 (the N_2 binding species). At low flux, species E_2H_2 evolves H_2 to give E_o at a rate that competes effectively with the conversion of E_2 to E_3. At high electron flux, reduction of E_3H_3 to $E_4H_3^-$ competes with the N_2 binding reaction (E_3H_3 to E_3N_2H) and it is this competition that causes the apparent $K_M^{N_2}$ to increase with increasing Kp2:Kp1. Since the apparent $K_M^{N_2}$ over the whole range of Kp2:Kp1 ratios is always <0.1 atm, this effect is not of physiological consequence providing that in vivo the enzyme is in equilibrium with more N_2 than this and that the kinetic parameters of Scheme 2 are not significantly different.

We have confirmed that H_2 is a competitive inhibitor of N_2 reduction, as is predicted by Scheme 2. The Kettering model for H_2 inhibition of N_2 reduction (Burgess et al 1981), which assumes that H_2 interacts with a diazene level intermediate (Scheme 3) seems to us however to predict non-competitive inhibition.

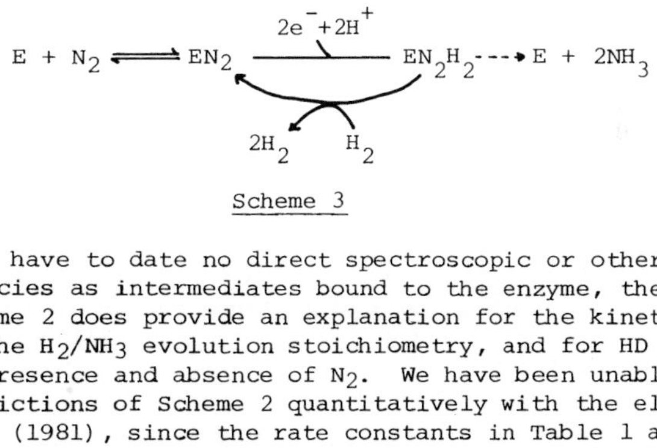

$$E + N_2 \rightleftharpoons EN_2 \xrightarrow{2e^- + 2H^+} EN_2H_2 \dashrightarrow E + 2NH_3$$

$$EN_2H_2 + H_2 \rightarrow EN_2 + 2H_2$$

Scheme 3

Although we have to date no direct spectroscopic or other evidence for hydridic species as intermediates bound to the enzyme, the mechanism outlined in Scheme 2 does provide an explanation for the kinetics of H_2 evolution, for the H_2/NH_3 evolution stoichiometry, and for HD formation from D_2 in the presence and absence of N_2. We have been unable to compare the predictions of Scheme 2 quantitatively with the elegant data of Burgess et al (1981), since the rate constants in Table 1 are at 23° for K. pneumoniae nitrogenase and the results of Burgess et al (1981) were obtained at 30° with A. vinelandii nitrogenase. However, Scheme 2 qualitatively predicts the observed dependences on protein component ratio and, in contrast to the mechanism of Burgess et al (1981), it also explains N_2-independent H_2 formation. Scheme 2 predicts, under our conditions, that only 5% of the total HT formed, when T_2 is used as an inhibitor, would appear in solution as T^+. Burgess et al (1981) gave a limit for T^+ of 2±1% and since we have ignored isotope effects we consider Scheme 2 to be consistent with the somewhat lower limit reported by them.

CONCLUSIONS:

1) The analysis based on Schemes 1 and 2 explains the kinetic data on K. pneumoniae nitrogenase quantitatively and published data on nitrogenase from other organisms qualitatively.

2) The simulations require that solvent protons, together with N_2 and H_2, interact only with the free MoFe protein. Thus the Fe protein protects the substrate reducing site from solvent. This rôle of the Fe protein is additional to that of a specific reductase of the MoFe protein.
3) At high enzyme concentrations (∿100 μM), with limiting reductant or MgATP, the ratio of $NH_3:H_2$ evolved is higher than at low enzyme concentration at a similar electron flux through the MoFe protein. Under these conditions the concentration of free MoFe protein is lowered by reaction k_3.
4) Only models which predict competitive inhibition of N_2 reduction by H_2 are valid.
5) The MoFe protein contains two independent active sites for substrate reduction, the concentration of which is proportional to the Mo content of the protein.

REFERENCES

Burgess BK, Wherland S, Newton WE, and Stiefel EI (1981) Biochemistry 20, 5140-5146.
Chatt J, Pearman AJ, and Richards RL (1977) J.Chem.Soc.Dalton 1852-1860.
Eady RR, Lowe DJ, and Thorneley RNF (1978) FEBS Lett, 95, 211-213.
Hageman RV and Burris RH (1978) Biochemistry 17, 4117-4124.
Hageman RV, Orme-Johnson WH and Burris RH (1980) Biochemistry 19, 2333-2342.
Ljones T and Burris RH (1978) Biochemistry 17, 1866-1872.
Lowe DJ, Thorneley RNF and Smith BE, In Harrison PM, ed, Metalloproteins 1, Macmillan, London, in the press.
Silverstein R and Bulen WA (1970) Biochemistry 9, 3809-3815.
Thorneley RNF (1975) Biochem.J. 145, 391-396.
Thorneley RNF, Eady RR and Lowe DJ (1978) Nature 272, 557-558.
Thorneley RNF and Lowe DJ (1981) In Gibson AH and Newton WE, eds, Current Perspectives in Nitrogen Fixation, pp.360.
Thorneley RNF and Lowe DJ (1983) Biochem.J. in the press.
Thorneley RNF, Yates MG and Lowe DJ (1976) Biochem.J. 155, 137-144.

Acknowledgements

We thank Mr. K. Baker and Miss L. Sones for growing *K. pneumoniae* cells and Mrs. G.A. Ashby and Mr. K. Fisher for skilled technical assistance.

NITROGENASE FROM nifV MUTANTS OF KLEBSIELLA PNEUMONIAE

B.E. SMITH, R.A. DIXON, T.R. HAWKES, LIANG, Y.-C., P.A. McLEAN+,
J.R. POSTGATE
ARC UNIT OF NITROGEN FIXATION, UNIVERSITY OF SUSSEX, BRIGHTON BN1 9RQ, UK.
+ 18-408 MASSACHUSSETTS INSTITUTE OF TECHNOLOGY, CAMBRIDGE MA 02139, USA.

1. INTRODUCTION

NifV mutants of Klebsiella pneumoniae are unable to fix N_2 but can reduce C_2H_2 to C_2H_4. The Fe protein (Kp2) of their nitrogenase is normal but the MoFe protein (NifV⁻Kp1) in combination with wild-type Kp2 (to form NifV⁻ nitrogenase) exhibits the mutant phenotype, but can form a poor nitrogenase with a large excess of Kp2. The H_2-evolution activity of NifV⁻ nitrogenase, unlike that of the wild-type enzyme, is partially inhibited by CO (McLean, Dixon 1981). In this paper we describe our recent investigations into the differences between wild-type Kp1 and NifV⁻Kp1.

2. THE IRON-MOLYBDENUM COFACTOR

An iron-molybdenum cofactor (FeMoco) can be extracted from precipitated, denatured, MoFe protein into N-methylformamide and can activate the inactive MoFe protein from mutants unable to synthesise FeMoco (Shah, Brill 1977). NifB mutants of K. pneumoniae have this phenotype and the MoFe protein (NifB⁻Kp1) from such mutants has been purified and characterized (Hawkes, Smith 1983) and shown to contain the 'P' clusters (probably unusual 4Fe-4S clusters: Zimmermann et al 1978)) in their normal state (Hawkes 1981).

When the FeMoco (NifV⁻FeMoco) was extracted from NifV⁻Kp1 and combined with NifB⁻Kp1 the resultant MoFe protein (with excess Kp2) exhibited the NifV⁻ phenotype (Table 1) i.e. its H_2 evolution activity was partially inhibited by CO and was only 30% inhibited by N_2. The same experiment using FeMoco from wild-type Kp1 yielded MoFe protein with the wild-type phenotype (Table 1).

TABLE 1. H_2 evolution activity of nitrogenase containing MoFe protein prepared by reacting NifB⁻Kp1 with wild-type or NifV⁻FeMoco.

FeMoco from	% H_2 evolution activity under		
	Argon	CO	N_2
Wild-type Kp1	100	100	30
NifV⁻Kp1	100	27	70

These data provide strong evidence that FeMoco contains the enzyme's site for binding N_2 and CO (Hawkes et al 1983). We conclude that the nifV gene product modifies NifV⁻FeMoco in order to generate a more effective nitrogenase. We cannot say, as yet, whether this modification occurs before or after insertion of the cofactor into the MoFe protein but it probably occurs towards the end of FeMoco synthesis.

We have been unable, as yet, to distinguish any physical differences between wild-type and NifV⁻FeMoco. The electron paramagnetic resonance spectra of the extracted cofactors are indistinguishable and the metal contents of samples prepared by the acid treatment (Shah, Brill 1977) or the dimethyl sulphoxide precipitation methods (Smith, 1980) are essentially identical with Mo:Fe ratios of 1:7±1 (Hawkes et al 1983).

3. SUBSTRATE REDUCTION PROPERTIES

The evidence presented above strongly implies that FeMoco is the site of substrate binding to nitrogenase. We have therefore attempted to probe the differences between the wild-type and NifV$^-$ nitrogenases by comparing their ability to reduce a number of substrates.

3.1 H_2 evolution

CO is not reduced by NifV$^-$ nitrogenase but inhibits total electron flow through the enzyme. It is a partial (73% at pH 7.3) inhibitor of H_2 evolution. Inhibition is half-maximal at 0.031 matm CO at pH 7.3. However MgATP-hydrolysis by the enzyme is not inhibited and thus the ATP/2e ratio increases in the presence of CO. This is presumably due to futile cycling of electrons back from the MoFe protein to the Fe protein whence they are returned to the MoFe protein with concomitant hydrolysis of MgATP (Eady et al 1978). The inhibition is pH-dependent being less at low pH and reaching a maximum of about 80% near pH 8.5. The pH optimum of the uninhibited enzyme is near 7.5 whereas that of the CO-inhibited enzyme is near 6.6 (McLean et al 1983).

3.2 N_2-reduction

The poor N_2-reduction activity of NifV$^-$ nitrogenase is reflected in its apparent K_m which at 0.6 atm.N_2 is about 10 fold higher than that of the wild-type enzyme (0.05 atm.N_2) (Hawkes et al 1983). The relative inability of NifV$^-$ nitrogenase to reduce N_2 compared with C_2H_2 was demonstrated by comparing assays performed under an atmosphere of 15% C_2H_2/85% N_2 with those under an atmosphere of 15% C_2H_2/85% Ar. With a 30 fold molar excess of Kp2 over MoFe protein, wild-type Kp1 C_2H_2-reducing activity was inhibited by 33% under N_2 relative to Ar whereas NifV$^-$Kp1 C_2H_2-reducing activity was not affected.

3.3 C_2H_2-reduction

NifV$^-$ nitrogenase has an apparent K_m for C_2H_2 that is indistinguishable from that of the wild-type enzyme (≃0.17 mM). However the activity/ng atom Mo of the wild-type enzyme is 275±27 nmol C_2H_2 reduced/min/ng atom Mo whereas that of NifV$^-$ nitrogenase is only 181±10 nmol C_2H_2 reduced/min/ng atom Mo. These data provide further evidence that the substrate-reducing power of the enzyme is related to its molybdenum content and the environment of its molybdenum atoms (Hawkes et al 1983).

3.4 HCN-reduction

Li et al (1982) have investigated cyanide reduction by *Azotobacter vinelandii* nitrogenase in considerable detail. They found that HCN, the substrate, was reduced in a 6-electron reaction to CH_4 and NH_3 or in a 4-electron reaction to CH_3NH_2. Some excess (over CH_4) NH_3 was detected which was attributed to a 2-electron reduction yielding NH_3 and HCHO (undetected).

CN^- was a potent inhibitor of total electron flow. Since $NifV^-$ nitrogenase was ineffective in the 6 (or 8)-electron reduction of N_2 we decided to investigate its ability to reduce HCN.

The apparent K_m (CH_4 production) for HCN reduction by $NifV^-$ nitrogenase (0.20±0.07 mM) is lower than that of the wild-type enzyme (0.75±0.22 mM). Furthermore HCN is reduced by a greater proportion of the total electron flow through $NifV^-$ nitrogenase than that observed with the wild-type enzyme (Table 2).

TABLE 2. Distribution of electrons to products during HCN reduction by wild-type and $NifV^-$ nitrogenases from K. pneumoniae (Fe:MoFe protein ratio = 15).

	% electrons to				% electrons
	H_2	CH_4	CH_3NH_2	(NH_3-CH_4)	to HCN
Wild-type nitrogenase	75.5	18	4.1	2.3	24.4
$NifV^-$ nitrogenase	30.5	33.3	27.2	9.0	69.6

These data show that $NifV^-$ nitrogenase is more effective at reducing HCN than is wild-type nitrogenase.

In common with Li et al (1982) we found that as the component protein ratio was increased from a Fe:MoFe ratio = 1 to 30, evolution of CH_4 increased more rapidly (maximal at a ratio of 5) than did H_2 evolution which continued to increase throughout the titration. These data suggest that HCN is reduced by a higher oxidation level of the enzyme than is H^+. (The above data are described in more detail in the poster paper by Liang, Smith in this book.)

4. DISCUSSION

We have earlier (McLean et al 1983) discussed the partial inhibition by CO of H_2 evolution by $NifV^-$ nitrogenase in terms of three models. Two of these models were classical biochemical models for partial inhibition and postulated respectively two CO-binding sites with one H_2-evolution site, and one CO-binding site with two H_2-evolution sites. The third model required only one site each for CO-binding and H_2-evolution and involved consideration of the redox potential of the enzyme's active site. We have provided very strong evidence that this site is the FeMoco metal cluster. The binding of CO to such a cluster would be expected to raise its potential (make it more positive). If the functional redox potential of $NifV^-$-nitrogenase is higher than that in the wild-type enzyme then CO-binding could raise the potential to slightly higher than that of the H_2-electrode (-400 mV) and result in less efficient H_2-evolution.

This third explanation of the H_2 evolution data would also explain our N_2- and HCN-reduction data. On the model of Lowe et al (previous paper) N_2-binding only occurs at relatively low potentials after the enzyme has been reduced by at least three electrons (i.e. to E_3 Scheme 1).

SCHEME 1. Model for nitrogenase action (Lowe et al)
E_n = the MoFe protein reduced by n electrons by the Fe protein with concomitant hydrolysis of MgATP.

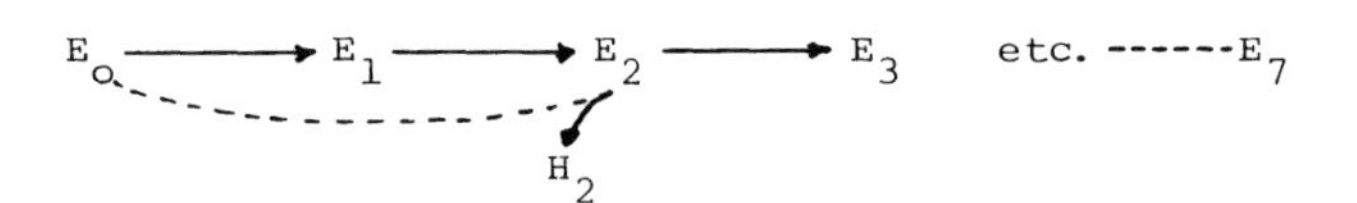

On our hypothesis the redox potentials of E_1, E_2 etc. are higher in $NifV^-$ nitrogenase than in the wild-type enzyme. Therefore the $NifV^-$ enzyme finds it more difficult to reach a potential equivalent to that of E_3 in the wild-type enzyme before H_2-evolution occurs. Consequently the $NifV^-$ enzyme is a poor nitrogenase.

Conversely both our data and that of Li et al (1982) indicate that HCN binds to higher oxidation levels of the enzyme than those required for H_2 evolution. If the potentials of the various E_n species in $NifV^-$ nitrogenase are higher than those of the wild-type enzyme then HCN will be more competitive with H_2 evolution in the mutant enzyme as observed.

ACKNOWLEDGEMENTS

We thank Mrs C. Gormal for expert technical assistance and Mr K. Baker and Miss L. Sones for growth of cells. T.R.H. and P.A.McL. were recipients of studentships from the Science and Engineering Research Council and the Agricultural Research Council respectively.

REFERENCES

Eady RR, Lowe DJ and Thorneley RNF (1978) FEBS Letts, 95, 211-213.
Hawkes TR (1981) Ph.D. Thesis, University of Sussex.
Hawkes TR and Smith BE (1983) Biochem.J. 209, 43-50.
Hawkes TR, McLean PA and Smith BE (1983) Biochem.J. submitted.
Li J-g, Burgess BKand Corbin JL (1982) Biochemistry, 21, 4393-4402.
McLean PA and Dixon RA (1981) Nature, 292, 655-656.
McLean PA, Smith BE and Dixon RA (1982) Biochem.J. 211, 589-597.
Shah VK and Brill WJ (1977) Proc.Natl.Acad.Sci.USA, 74, 3249-3253.
Smith BE (1980) In Newman W and Otsuka S, eds, Molybdenum Chemistry of Biological Significance, pp. 179-190, Plenum Press, New York.
Zimmermann R, Münck E, Brill WJ, Shah VK, Henzl MT, Rawlings J and Orme-Johnson WH (1978) Biochim.Biophys.Acta, 623, 124-138.

POSTER DISCUSSION 3 ENZYMOLOGY OF NITROGENASE

R.H. BURRIS, Department of Biochemistry, University of Wisconsin-Madison, WI 53706, USA

Although homogeneous nitrogenase components have been available for some years, there still is a lively controversy about their exact nature and composition. Arguments about the Fe-protein (dinitrogenase reductase) were concentrated on whether there are 4 or 8 Fe atoms per active unit. It generally has been accepted that there are 2 equivalent subunits of about 30,000 molecular weight and a shared 4Fe-4S center in a cubical array. This idea was challenged by Braaksma et al who maintain that there are molecules with a 3Fe and 3S composition up to a specific activity of 1000 nmol C_2H_4 produced/(min.mg Av2). They reported a linear increase in Fe and specific activity from 1000 to 2250 with a maximum of 8.8 Fe and 8 S per Av2 molecule, a value that suggests two 4Fe-4S centers/molecule. (Nobody would claim to match the specific activity of 7,000 as reported for the Fe-protein by the Wageningen group).

Howard, Anderson and Hausinger offered a different explanation, i.e. that there are 4Fe and 4S atoms per Av2 molecule, but that two of these Fe can be removed immediately and the other half slowly. The removal was reflected in change in both absorbance and EPR signals. The Mössbauer spectra indicated only four Fe, but by partial removal of Fe, one 4Fe center could appear to act like two centers. The altered center appeared different from the spinach ferredoxin center a [2Fe-2S] center in S and cysteine concentration, E_o', spin quantitation and spectral response. It is apparent that this difference between groups must be resolved.

There also is no meeting-of-the-minds yet on the report of Bishop et al, that there is an alternative N_2-fixing system elaborated by A.vinelandii on Mo-deficient media. Tarzaghi et al tested for growth of wild type and nif mutants on a Mo-deficient medium and found that NH_4^+ derepressed mutants grew if NH_4^+ was supplied and that they produced apparently normal MoFe protein. Bishop and coworkers have extended their earlier work on an alternative N_2-fixation system and now report a dinitrogenase-reductase-like protein from mutant cells grown on a Mo-deficient medium. Its formation was repressed by Mo or W. The mobility of the protein was similar to that of normal Fe-protein (dinitrogenase reductase).

Thorneley and Lowe reported that data from stopped-flow-spectrophotometry indicated that the rate-limiting step in the nitrogenase system is the dissociation of Kp2 from Kp1 and not substrate reduction. The Sussex group also indicated that when Kp1 was oxidized by dyes, it gave an EPR signal characteristic of a [4Fe-4S] center. These centers (P centers, not FeMoco) apparently are at the zero oxidation level with dithonite present.

Moulis et al reported comparison of resonance Raman spectra from ferredoxin and rubredoxin, plus ferredoxin from C.pasteurianum in which Se was substituted for S. These studies will be extended to nitrogenase components.

Zumft reported for his group on the sequencing of the α and β subunits of Cp1. The 530 residues of the α unit are defined and about 70% of those in the β unit. The cysteines have been located, and it is suggested that the active centers of the MoFe protein are ligated in the N terminal regions of the subunit. Newton et al have continued EXAFS and other studies on FeMoCo and have concluded that there are two or more environments for Fe in the Mo coordination sphere. Electrochemical investigations have suggested electron-transfer paths in Av2. Trinchant and Rigaud reported that NO_2^- inhibited C_2H_2 reduction but not H_2

evolution by nitrogenase, whereas NO inhibited both. They concluded NO_2^- inhibited directly rather then by conversion to NO. Three laboratories (in Yellow Springs, Peking and Perth) collaborated to determine that CH_3NC is an inhibitor of electron-flow through nitrogenase and that it uncouples MgATP hydrolysis. Guth and Burris found that $D_2(H_2)$ inhibits by binding first before N_2 in an ordered sequential reaction that yields HD and N_2 with no measurable change in the N_2. N_2 is reduced only if it is bound before $D_2(H_2)$. Shilov indicated that such a scheme was incompatible with his model system that operates via a hydrazido intermediate.

Wassink and Haaker limited electron flux through nitrogenase and predictably found reduction of H^+ favored over reduction of N_2. Despite this and contrary to generally held opinion, they concluded that ADP *in vivo* controls electron allocation to H^+ and N_2 rather than electron flux through MoFe-protein (dinitrogenase). The group also concluded that reduced Fe protein binds one MgATP and one MgATP, whereas the oxidized Fe protein binds two MgATP and one MgADP.

Studies of the effect of covalent modification of the Fe-protein dominated investigations of control in *R.rubrum* and *Rhodopseudomonas capsulata*. Ludden and coworkers have purified the modifying and inactivating group, have found that it can be removed by heating, and have verified that it contains phosphate, pentose and an adenine-like group, that is not AMP. They have prepared the activating enzyme in high specific activity. Studies of 'switch-off' indicated that NH_4^+, darkness, O_2 or uncouplers modified the Fe protein and its activity. Nordlund reported on how divalent cations function in the activation of *R.rubrum* Fe-protein. *R.capsulata* exhibits a similar control mechanism, and Vignais reported that her group had demonstrated that inactive Fe-protein exhibited two different subunits of different molecular weights. Release or attachment of a modifying group served as an indicator of activation or inactivation.

EVIDENCE FOR TWO (4Fe-4S) CLUSTERS IN Av2

A. BRAAKSMA, H. HAAKER, H.J. GRANDE, C. VEEGER and R.R. EADY*
Department of Biochemistry, Agricultural University, De Dreijen 11, 6703 BC Wageningen, The Netherlands
*ARC Unit of Nitrogen Fixation, University of Sussex, Brighton, England.

The Fe-protein of the Azotobacter vinelandii nitrogenase enzyme complex contains a variable iron and sulphide content. In fig. 1 the dependency of the iron content upon the specific activity is shown. The scattering in these determinations is considerable, as can be seen.
The iron content is associated with the sulphide content as can be seen in fig. 2. Each point represents a sulphide and iron analysis done on the same sample. The maximum values in this figure are resp. 8.8 and 8.6 atoms per molecule for the iron and sulphide contents. Occasionally we found in preparations of specific activity 3000 nmoles. min^{-1} mg^{-1} (which are very rare and labile) values of resp. more than 10 and 11 for the iron and sulphide contents.

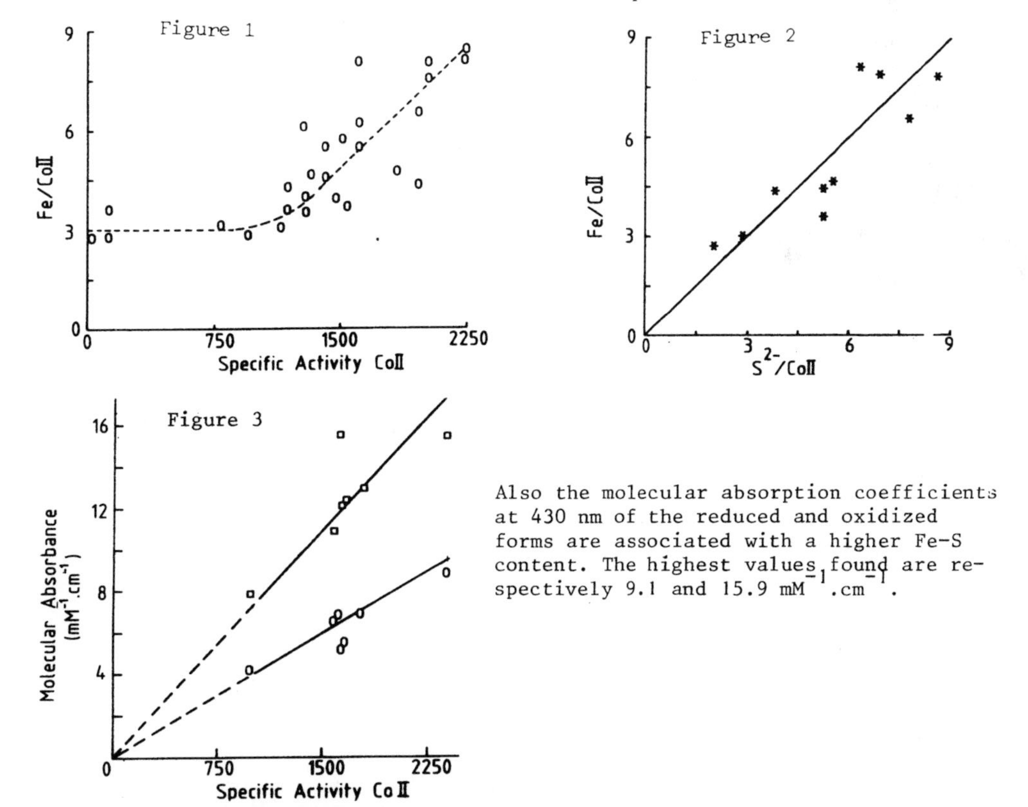

Also the molecular absorption coefficients at 430 nm of the reduced and oxidized forms are associated with a higher Fe-S content. The highest values found are respectively 9.1 and 15.9 $mM^{-1}.cm^{-1}$.

In EPR experiments evidence is also found for two clusters: the integral of Av2 in the presence of MgADP and 50% (v/v) ethylene glycol increased to 1.4 electron/Av2. Redox titrations Av2, with or without MgATP, have shown a two-electron behaviour. Preliminary results of Mössbauer experiments indicate that only clusters of the [4Fe-4S]-type are present.

NITROGENASE REACTIVITY

B.K. BURGESS/J.L. CORBIN/J.F. RUBINSON/J.-g. Li[a]/ M.J. DILWORTH[b] and W.E. NEWTON
C.F. Kettering Research Laboratory, Yellow Springs, OH 45387, USA
a) Institute of Botany, Peking, PRC. b) Murdoch University, Western Australia.

Electron Allocation. At low Fe protein/MoFe protein ratios (limited electron flux), reduction of 2e- substrates (H^+, C_2H_2) appears favored over the 6e$^-$ substrate, N_2. However, for the 6e$^-$ reduction of either HCN or CH_3NC, H_2 evolution is favored at high Av2/Av1. Thus, HCN and CH_3NC are reduced at an enzyme redox state more oxidized than for H_2 evolution or N_2 reduction. N_2 also differs from these other 6e$^-$ substrates in its inability to eliminate H_2 evolution at ∞ concentration. Dilworth and Thorneley (Biochem.J. 193, 971, 1981) suggest H_2 evolution could be eliminated by ∞ azide at high pH. Our data, measuring all products not just ΔH_2, suggest otherwise. Like them, we find at low pH (6.7) ∿75% e$^-$ going to azide reduction at ∞ azide. Although for N_2, 75%/25% to N_2/H_2 would be stoichiometric, for azide, which has 2, 6 or 8e$^-$ pathways, it represents 0.43 H_2 per azide reduced.

N_2-Dependent HD formation. H_2 specifically inhibits N_2 reduction and under D_2/N_2, N_2ase redirects e$^-$ from N_2 fixation to HD formation (1e$^-$/HD) (Burgess et al., Biochem., 20, 5140, 1981). Involvement of a N_2H_2-level intermediate, susceptible to decomposition by $H_2(D_2)$, is suggested. To distinguish between our mechanism (eq. 1) and the trihydride of Thorneley and Lowe (Israel J.Bot., 31, 1, 1982; eq. 2), we have used a N_2/HD atmosphere (see Table).

$$(1)\quad E + N_2 + 2H^+ + 2e^- \longrightarrow E\text{-}N_2\!\!<^{H}_{H} + \begin{matrix} D \\ | \\ D \end{matrix} \begin{bmatrix} H \\ | \\ D \end{bmatrix} \longrightarrow \begin{matrix} E + N_2 + 2HD \\ [E + N_2 + H_2 + HD] \end{matrix}$$

$$(2)\quad E\!\!<^{H}_{H}\!\!-H \underset{N_2\ \ H_2}{\overset{N_2\ \ H_2}{\rightleftharpoons}} E\!\!<^{H}_{N_2} \underset{HD\ \ N_2}{\overset{HD\ \ N_2}{\rightleftharpoons}} E\!\!<^{H}_{D}\!\!-H \underset{N_2\ \ H_2}{\overset{N_2\ \ H_2}{\rightleftharpoons}} E\!\!<^{D}_{N_2} \underset{D_2\ \ N_2}{\overset{HD\ \ N_2}{\rightleftharpoons}} E\!\!<^{H}_{D}\!\!-D$$

TABLE. Product Distribution[a] under HD/N_2 and D_2/N_2 Atmospheres.

	H_2	HD	D_2	NH_3	Total 2e$^-$
50% HD/40% N_2/10% Ar	448	b	0	79	567
50% D_2/40% N_2/10% Ar	357	202	b	91	595
40% N_2/60% Ar	343	0	0	152	571

a) In nmoles min^{-1}mg^{-1}; Av2/Av1 = 8
b) Not measurable under 0.5 atmospheres of same gas.

Mechanism (1) does not allow D_2 formation, while mechanism (2) does by scrambling. No D_2 was detected.

ACKNOWLEDGEMENT. Supported by the USDA / CRGO (to BKB and JLC).

ON THE BINDING OF MGATP AND MGADP TO THE NITROGENASE PROTEINS FROM *AZOTOBACTER VINELANDII*

J. CORDEWENER/H. HAAKER/C. VEEGER
DEPARTMENT OF BIOCHEMISTRY, AGRICULTURAL UNIVERSITY, DE DREIJEN 11, 6703 BC WAGENINGEN, THE NETHERLANDS

To gain insight on the role of MgATP and MgADP in nitrogenase function and regulation, we investigated the interaction of both nucleotides with the nitrogenase proteins from *A.vinelandii*. The method used for these binding studies was the flow dialysis technique.

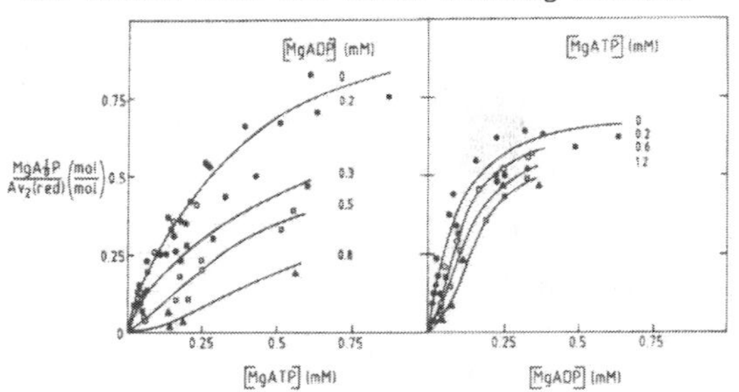

Fig. 1 shows that reduced Av_2 binds one molecule of MgATP (K_d = 0.33 mM) and one molecule of MgADP (K_d = 0.15 mM). Inhibition studies were performed to establish whether the binding site for MgATP on Av_2 is the same as the one for MgADP. Up to concentrations of 0.2 mM, MgADP does not have any effect on the binding of MgATP to reduced Av_2. At higher concentrations of MgADP there is an increasing inhibition of the MgATP binding. A completely different inhibition pattern is obtained when MgADP binding is measured in the presence of MgATP. By increasing the concentration of MgATP, the hyperbolic binding curve of MgADP becomes sigmoidal. These results indicate that there are different binding sites for MgATP and MgADP present on the reduced protein, but that these two binding sites have interaction with each other. Fig. 2 shows that oxidized Av_2 binds two molecules of MgATP (K_d = 0.14 mM) and one molecule of MgADP (K_d= 0.14 mM). The binding of MgATP is strongly inhibited by MgADP. MgATP, on the other hand, has only a small effect on the binding of MgADP. No binding of MgATP and MgADP to Av_1 could be demonstrated.

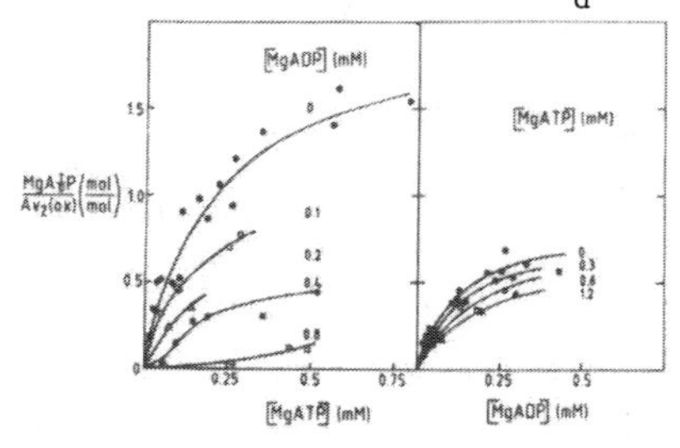

Tso and Burris (1) found two MgATP binding sites and one MgADP binding site for (reduced) Cp_2. It was suggested (2) that these two molecules of MgATP are hydrolyzed during the one electron transfer from the Fe protein to the MoFe protein. Braaksma *et al*. (3) found that Av_2 is a two electron donor. Together with our binding results, this would mean that only one molecule of MgATP is hydrolyzed for each pair of electrons transferred. Since a minimum of 4 ATP per 2 electrons is found for the overall nitrogenase reaction, there should be MgATP hydrolysis after electron transfer.

REFERENCES

1. Tso MYW and Burris RH (1973) Biochim.Biophys.Acta, 309, 263-270
2. Hageman RV Orme-Johnson WH and Burris RH (1980) Biochemistry, 19, 2333-2343.
3. Braaksma A Haaker H Grande HJ and Veeger C (1982) Eur.J.Biochem. 121, 483-491

INHIBITION OF N_2 FIXATION BY H_2.

JOSEPH H. GUTH AND ROBERT H. BURRIS
DEPARTMENT OF BIOCHEMISTRY, UNIVERSITY OF WISCONSIN, MADISON, WI 53706 USA

Li and Burris (Biochemistry, in press) demonstrated the formation of HD from D_2 with nitrogenase preparations from *Azotobacter vinelandii*, *Clostridium pasteurianum*, *Klebsiella pneumoniae* and *Azospirillum* sp. From these observations it is concluded that the ability to form HD is a general property of nitrogenases. However, there are differences among the nitrogenases in their K_i values for D_2 (N_2 fixation) and in the rates at which they are capable of catalyzing HD formation. Among the nitrogenases examined, *C. pasteurianum* supported the slowest formation of HD. We have investigated the inhibition by H_2 (D_2) of NH_3 formation by nitrogenase from *Klebsiella pneumoniae* and have confirmed that the inhibition is competitive vs N_2. D_2 inhibits NH_3 formation by diverting nitrogenase from production of NH_3 to production of HD (1 electron/HD). By careful exclusion of N_2 from the reaction mixture, we have been able to place an upper limit on N_2-independent HD formation by nitrogenase, under 1 atm of D_2, at 1% of the total electron flux. Formation of NH_3 and of HD were inhibited identically by CO. We observed that as the ratio of dinitrogenase reductase is increased, the ratio of HD formed to NH_3 formed rises, and D_2 becomes a stronger inhibitor of N_2 reduction. This may be caused in part by an accompanying increase that is observed in the K_m of nitrogenase for N_2. Examination of the kinetics of HD formation indicates that it arises via an ordered sequential reaction with H_2(D_2) binding first. We propose a model for D_2 inhibition of NH_3 formation in which D_2 and N_2 compete for the same form of nitrogenase. According to our proposal (see Figure), when N_2 reacts with nitrogenase, N_2 reduction either proceeds to completion (yielding NH_3) if H_2 (D_2) is absent, or if D_2 already is bound to nitrogenase, N_2 reduction is aborted (N_2 and enzyme are released) and 2 molecules of HD are produced at the net expense of 1 electron per HD. Key consequences of the model are that it predicts that H_2 (D_2) is a competitive inhibitor of NH_3 formation, and that the apparent K_m (N_2) for formation of HD and NH_3 may differ. It also predicts that at infinite pN_2, no HD should be produced.

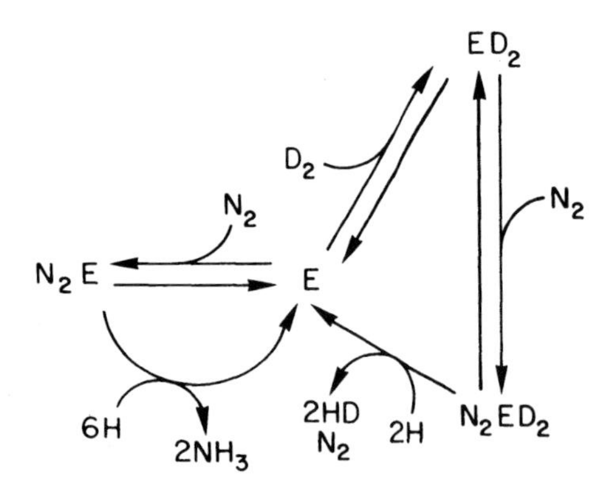

PROGRESS IN THE STUDY OF THE PRIMARY STRUCTURES OF THE α- AND β-SUBUNITS OF CLOSTRIDIAL MoFe PROTEIN

T. Hase, S. Wakabayashi, T. Nakano, W.G. Zumft* and H. Matsubara
Department of Biology, Faculty of Science, Osaka University, Toyonaka, Osaka 560, Japan and *Lehrstuhl für Mikrobiologie, Universität Karlsruhe, Kaiserstrasse 12, D-7500 Karlsruhe, West Germany

We have been sequencing the α- and β-subunits of MoFe protein from Clostridium pasteurianum by conventional method (1). The complete sequence of the α-subunit and about 70% of the total sequence of the β-subunit are now elucidated. The sequence of 529 amino acid residues shown below is the first complete sequence of the α-subunit of a MoFe protein. The molecular weight (58,774) and amino acid composition deduced from the sequence agree well those of the original protein. One of our interests focusses on the distribution of cysteine residues as ligands to the FeS and MoFeS clusters. In the α-subunit, 5 out of a total of 9 cysteine residues at positions 52, 78, 144, 173 and 261, are invariant compared to MoFe proteins from other species, and the sequences surrounding these cysteines are homologous to more than 70%. Thus, some or all of these residues are expected to be indispensable for cluster ligation. Based on the criteria mentioned above, 3 cystein residues in the N-terminal region of the β-subunit appear to have an essential function. Therefore, we suppose that active centers in MoFe protein are ligated predominantly by the N-terminal domains of both subunits.

From a comparison of the complete sequence of the clostridial α-subunit to those predicted from the DNA sequences of nifD genes of Anabaena (2) and Parasonia (3), the overall homology is calculated. The clostridial sequence is 40 and 41% identical with that of Parasponia and Anabaena, respectively, whereas the latter two are 68% identical. The clostridial α-subunit also shows relatively low homology to the α-subunits from other sources whose partial sequences are available. This agrees with the fact that the clostridial MoFe protein cross-reacts relatively little with almost all Fe proteins from other organisms.

The complete sequence of the α-subunit of MoFe protein from Clostridium pasteurianum

SENLKDEILEKYIPKTKKTRSGHIVIKTEETPNPEIVANTRTVPGIITARGCAYAGCKGV(60)
VMGPIKDMVHITHGPIGCSFYTWGGRRFKSKPEDGTGLNFNEYVFSTDMQESDIVFGGVN(120)
KLKDAIHEAYEMFHPAAIGVYATCPVGLIGDDILAVAATASKEIGIPVHAFSCEGYKGVS(180)
QSAGHHIANNTVMTDIIGKGNKEEKKYSINVLGEYNIGGDAWEMDRVLEKIGYHVNATLT(240)
GDATYEKVQNADKADLNLVQCHRSINYIAEMMETKYGIPWIKCNFIGVNGIVETLRDMAK(300)
CFDDPELTKRTEEVIAEEIAAIQDDLDYFKEKLQGKTACLYVGGSRSHTYMLKSFGVDSL(360)
VAGFEFAHRDDYEGREVIPTIKIDADSKNIPEITVTPDEQKYRVVIPEDKVEELKKAGVP(420)
LSSYGGMMKEMHDGTILIDDMNHMEVVLEKLKPDMFFAGIKEKFVIQKGGVLSKQLHSYD(480)
YNGPYAGFRGVVNFGHELVNGIYTPAWKMITPPWKKASSESKVVVGGEA(529)

References: 1. Hase, T., Nakano, T., Matsubara, H. and Zumft, W.G. (1981) J. Biochem.(Tokyo) 90, 295-298. 2. Lammers, P.J. and Haselkorn, R.(1983) Proc. Natl. Acad. Sci. USA in press. 3. Fellow, F., Scott, K.F., Weinman, J. and Shine, J. unpublished data

Fe-PROTEIN Fe:S LIGANDS

Robert P. Hausinger and James Bryant Howard
4-225 Millard Hall, University of Minnesota, Minneapolis, MN 55455 (USA)

We report functional roles for some Av2 (Azotobacter vinelandii Fe-protein) cysteinyl (Cys) residues using the chemical reactivity of the Cys with iodoacetic acid (IAA) as a probe. Cysteinyl residues were modified with $[^{14}C]$-IAA under a variety of conditions. The percent carboxymethylation for each of the 7 residues (Cys-5, 38, 85, 97, 132, 151 and 184) was determined by ion exchange peptide maps and their correlation to the known amino acid sequence (Hausinger, R.P., Howard, J.B.). We found that Av2 has no disulfides, surface thiols or hyperreactive thiols. For example, <30% of any residue was labeled in 20 min. If Av2 is incubated with IAA, MgATP, and α,α'-dipyridyl, the Fe:S center is rapidly chelated (2 min) with subsequent modification of Cys-97 and 132 (results are given in Table). Greater than 80% alkylation of Cys-97 and 132 occurs in 10-15 min with <20% of other residues modified. The rate for Cys-97 and 132 is similar to an exposed residue. When MgADP, IAA, and α,α'-dipyridyl are included with Av2, Cys-85, and, to a lesser degree, Cys-38, were labeled in addition to Cys-97 and 132 (see Table). The simplest interpretation of our results is that Cys-97 and 132 are the 4 thiol ligands (2 from each subunit) for a single 4Fe:4s center shared symmetrically between subunits of Av2. Cys-85 appears to be associated with the nucleotide binding site. Our results support the hypothesis of a major conformational change upon binding MgATP. If Av2 is denatured, Cys-5 and 184 are rapidly and completely alkylated (<5 min) while Cys-38, 85, and 151 rapidly exchange with the Fe:S ligands (Cys-97 and 132). Cys-38, 85, and 151 may be near the Fe:s center in the teriary structure which allows for exchange in the denatured protein or during catalysis. The probable ligands are common to all nitrogenase Fe-proteins so far sequenced. Our results are summarized in the Figure below.

Carboxymethylation of Av2

Residue		5	38	85	97	132	151	184
		(Mole of cysteine modified/mole protein)						
+MgATP	(10 min)	0.15	0.14	0.18	0.61	0.71	0.17	0.13
	(30 min)	0.17	0.21	0.37	0.80	0.89	0.19	0.18
+MgADP	(10 min)	0.14	0.17	0.23	0.61	0.59	0.17	0.14
	(30 min)	0.10	0.38	0.64	0.83	0.88	0.16	0.17

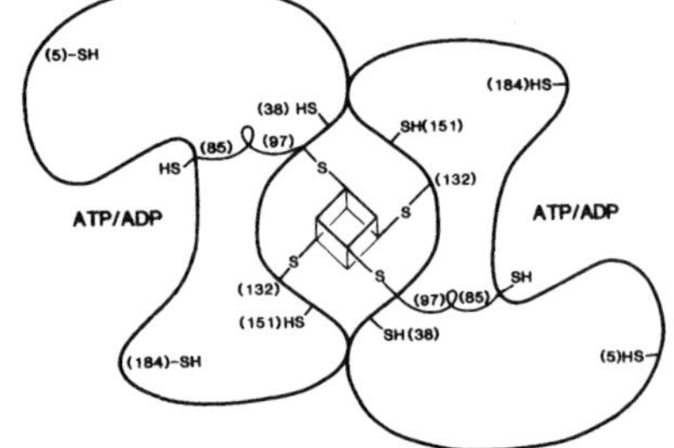

Proposed model for thiol distribution in Av2.

Reference: Hausinger, R.P. and Howard, J.B. (1982) J. Biol. Chem. 257:2483-90
Research supported by USDA Grant GRCR-1-1119.

STRUCTURAL INTERACTION BETWEEN NITROGENASE MoFe PROTEIN AND MEMBRANE IN AZOTOBACTER VINELANDII AND RHODOSPIRILLUM RUBRUM

KAREN S. HOWARD[+*]/BRIAN J. HALES[++]/M. D. SOCOLOFSKY[+]
[+]DEPARTMENTS OF MICROBIOLOGY AND [++]CHEMISTRY, LOUISIANA STATE UNIVERSITY, BATON ROUGE, LA, USA 70803
[*]PRESENT ADDRESS: CHEMISTRY 18-404, M.I.T., CAMBRIDGE, MA, USA 02139

Nitrogenase MoFe protein is routinely isolated from the cytoplasmic fraction of disrupted cells and, clearly, is not an intrinsic membrane protein. However, evidence does exist for extrinsic structural and functional interactions between MoFe protein and membrane in several organisms. Nitrogenase activity and, specifically, MoFe protein can be found in the pellets of osmotically lysed *A. vinelandii* and chromatophore preparations of osmotically lysed *R. rubrum*, respectively; MoFe protein has been localized in the vicinity of the cell membrane in *A. vinelandii* by immunoferritin labeling; and *in situ* nitrogenase activity has been shown to be enhanced by the presence of an intact energized membrane. We have used electron spin resonance (esr) spectroscopy and spatially oriented whole bacterial systems to obtain evidence for a structural association between MoFe protein and membrane in *A. vinelandii*, *R. rubrum*, and to a lesser extent, in *Rhodopseudomonas palustris*.

Spatially oriented cytoplasmic and intracytoplasmic membranes were prepared in *A. vinelandii* and *R. rubrum* cells by gently collapsing them into whole cell multilayers (WCM) on a flat quartz surface under anaerobic conditions. Two-dimensional arrangement of membranes was demonstrated using the orientation-dependent properties of 5-doxyl stearate spin label in *A. vinelandii* and bacteriochlorophyll *a* dimer triplet in *R. rubrum*. *R. palustris* was oriented and frozen in place under anaerobic conditions in a 23k Gauss magnetic field.

The *g*4.3 and *g*3.6 MoFe protein signals showed orientation-dependent characteristics in all three organisms. As the normal to the membrane plane was rotated from perpendicular to parallel with the esr magnetic field, the amplitude of the *g*3.6 signal decreased from maximum to 35% of maximum in *A. vinelandii* and to 88% of maximum in *R. rubrum*. The angular dependence of the *g*4.3 peak varied in *A. vinelandii*, but decreased from maximum to 63% of maximum in *R. rubrum*. Supporting evidence for a similar MoFe protein-membrane relationship was obtained in *R. palustris*.

Variations in the angular dependent properties of the signals may relate to 1.) different spatial and functional relationships between MoFe protein and membrane or membrane-bound molecules, 2.) the possibility of overlapping signals generated by two paramagnetic sites per MoFe protein, and 3.) the possibility of a greater freedom of motion of the MoFe protein with respect to the membrane surface due to an extrinsic rather than intrinsic association between the two. These properties suggest that the MoFe protein of nitrogenase was oriented in response to the physical orientation of cellular membranes and that a structural association exists between this nitrogenase component and membrane in these organisms.

MOLECULAR ASPECTS OF THE REGULATION OF NITROGENASE ACTIVITY IN *RHODOPSEUDOMONAS CAPSULATA*

Y. JOUANNEAU, C. MEYER and P.M. VIGNAIS
Laboratoire de Biochimie (CNRS/ER 235, INSERM/U.191), Département de Recherche Fondamentale, Centre d'Etudes Nucléaires, 85X, 38041 Grenoble cedex, France

Nitrogenase activity in *R. capsulata* is regulated in response to the supply of fixed nitrogen through a quick and reversible inhibitory system. Inhibition of nitrogenase involves a covalent modification of the Fe protein component. We have purified the Fe protein in the active form from N-starved cells and in the inactive form from glutamate-grown cells. On the basis of the electrophoretic mobility in SDS polyacrylamide gel, the active form has two equal subunits (MW 33,500) whereas the inactive form is composed of two subunits of apparent molecular weights 33,500 and 38,000 resp. Immunological experiments performed with antibodies against the active Fe protein, attest that the active and inactive forms have structural homologies. When *in vivo* nitrogenase activity was switched off by ammonia addition, Fe protein molecule changed from a single type of subunit (active form) to two different subunits (inactive form). Further incubation of the cells in a N free medium allowed restoration of nitrogenase activity and, at the same time, Fe protein shifted back to the active form. The heavier subunit of the inactive Fe protein appeared to be selectively labelled in $[^{32}P]$ phosphate containing culture grown on glutamate. The radioactive labelling was removed upon *in vitro* activation of nitrogenase. These results indicate that regulation of nitrogenase activity in *R. capsulata* is mediated by interconversion of the Fe protein. Interconversion could be monitored by attachment (or release) of a phosphate containing group (still unidentified) to one of the subunit of the Fe protein.

FACTORS AFFECTING THE IN VIVO MODIFICATION OF Fe PROTEIN OF NITROGENASE IN RHODOSPIRILLUM RUBRUM

Roy H. Kanemoto, Theresa E. Dowling and Paul W. Ludden*
from the Dept of Biochemistry and the Center for the Study of Nitrogen Fixation at the University of Wisconsin, Madison, WI, USA 53706

The switch-off of nitrogenase activity observed in R. rubrum has been correlated with the modification of the Fe protein. The modified, inactive form of the protein can be distinguished from the active form on polyacrylamide gels containing SDS. The inactive form exhibits two subunits and the active form a single subunit; the upper subunit of the inactive form contains the covalently attached modifiying group.
A method was developed to rapidly extract small samples of cells and analyze the extract for the subunit composition of the Fe protein. The method involves the collection of cells on a glass fiber filter, freeze quenching to stop metabolism and extracting the cells by grinding with carborundum in a small volume. The Fe protein is then immunoprecipitated. The Fe protein-antibody complex is resuspended in SDS gel cocktail and electrophoresed. The gel is analyzed for subuit composition and, in some cases, labelling of the upper subuint with ^{32}P.

Using this method, it has been demonstrated that merely placing the cells in the dark results in loss of whole cell activity and the modification of the Fe protein. In addition, uncouplers such as CCCP, oxidizing dyes such as phenazine methosulfate and methylene blue result in loss of activity and Fe protein modification. Exposure of cells to oxygen results in modification. The glutamine synthetase inhibitor methionine sulfoximine blocks modification due to darkness as well as ammonia. The Ferredoxin antagonist metronidizole inhibits whole cell activity, but does not result in modification.

ATP concentration, energy charge or ATP/ADP ratio do not correlate with Fe protein modification.

The modification/demodification of Fe protein can be followed for several cycles (at least three). Because there are two identical subunits in the Fe protein, there are presumably two sites for modification available (although only on seems to be modified at any time) thus the entire modifying group seems to be removed during activiation. An experiment in which subunit composition (% inactive Rr2) and whole cell nitrogenase activity are followed through three cycles of light/dark is shown below.

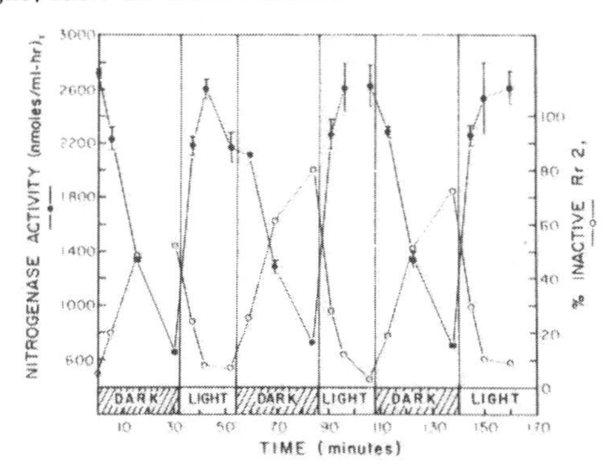

ON THE SPECIFIC ACTIVITY OF AV2 *IN VIVO*

J. KLUGKIST/H. WASSINK/H. HAAKER/C. VEEGER
DEPARTMENT OF BIOCHEMISTRY, AGRICULTURAL UNIVERSITY, DE DREIJEN 11, 6703 BC WAGENINGEN, THE NETHERLANDS

The concentrations of the two nitrogenase proteins (Av1, Av2), flavodoxin (Fld), Fe-S protein II (FeS II) in intact *Azotobacter vinelandii* cells were measured (Table 1). This was done by means of Western blotting as described by Zabel et al. (1982).

Table 1.

Medium ± NH_4Cl	Dilution rate(hr^{-1})	Density culture (mg protein/ml)	Concentration (μM) Av1	Av2	Fld	Fes II
-	0.05	0.60	45	50	180	110
-	0.1	0.45	52	75	195	64
-	0.2	0.35	60	88	200	57
-	0.25	0.20	75	115	220	45
+	0.10	0.65	0	0	85	120

Table 2.

Medium ± NH_4Cl	Dilution rate(hr^{-1})	Nitrogenase activity (nmoles C_2H_4/min.mg) intact cells	Av1	Av2
-	0.05	70	1200	3800
-	0.1	80	1400	3400
-	0.2	155	2400	5600
-	0.25	170	2100	4700
+	0.1	0	0	0

Because the nitronase activity of the cells was measured (Table 2), the activity of Av1 and Av2 in intact cells could be calculated (Table 2) by using the data of Table 1.

Conclusions

- In intact cells Av1 is not fully active.
- The specific activity of purified Av2 is about 2000 $nmoles.min^{-1}.mg^{-1}$. In intact cells the activity of Av2 is much higher (up to 5600 nmoles/min.mg).
  Explanation: -Av2 loses activity during purification.
  -The electron donating system to Av2 *in vivo* is superior to the system used *in vitro*.
- *In vitro* a ratio Av1:Av2:FeS II of 1:2:1 gives an oxygen stable complex. *In vivo* this ratio is variable. We wonder whether nitrogenase *in vivo* is always completely protected against oxygen by Fe-S protein II.
- The concentration of flavodoxin in N_2 fixing cells is higher, than in cells grown on NH_4Cl. This may be an indication for a role of flavodoxin as physiological electron donor to nitrogenase.

Reference: Zabel, P. et al. (1982) J.Virol. 41, 1083-1088.

HCN REDUCTION BY NITROGENASE FROM NIFV MUTANTS AND WILD-TYPE *KLEBSIELLA PNEUMONIAE*

LIANG YIN-CHU/BARRY E. SMITH
ARC UNIT OF NITROGEN FIXATION, UNIVERSITY OF SUSSEX, BRIGHTON BN1 9RQ, UK.

The MoFe protein ($NifV^-$Kp1) of nitrogenase from *nifV* mutants of *K. pneumoniae* is defective. When it is combined with an excess of wild-type Fe protein (Kp2) the resultant enzyme ($NifV^-$ nitrogenase) is a poor N_2-fixer, but can catalyse the six electron reduction of cyanide as well as the two electron reduction of C_2H_2 to C_2H_4 (McLean, Dixon, 1981). Li et al (1982) studied the reduction of cyanide and found that HCN was the substrate and CN^- was a potent inhibitor of total electron flow. We have used cyanide reduction to probe the differences between the active sites of wild-type and $NifV^-$ nitrogenase from *K. pneumoniae*.

In common with *A. vinelandii* nitrogenase (Li et al 1982) both wild-type and $NifV^-$ nitrogenase reduced HCN to CH_4, NH_3 and CH_3NH_2. NH_3 production was higher than CH_4 production for both enzymes. The apparent K_m (CH_4 production) was lower (0.20 0.07 mM) for $NifV^-$ nitrogenase than for the wild-type enzyme (0.75 0.22 mM). Furthermore the percentage of electrons going to HCN reduction rather than H_2 evolution was much higher with $NifV^-$ nitrogenase than with the wild-type enzyme (Table 1).

TABLE 1. Component ratio effect on product formation by $NifV^-$ and wild-type nitrogenase from *K. pneumoniae*.

Nitrogenase	Fe:MoFe ratio	Product (nmoles/assay) H_2 (2e)	CH_4 (6e)	CH_3NH_2 (4e)	(NH_3-CH_4) (2e)	% electrons to HCN
Wild-type	0.5:1	268	126	31.6	-	62.2
	8:1	2294	223	60.9	47	26.8
	30:1	2987	211	60.9	91.7	22.1
$NifV^-$	0.5:1	116	113	18.1	-	76.4
	8:1	528	290	271	189	75.2
	30:1	895	259	325	297	65.8

With increasing component protein ratio CH_4 formation by both enzymes rises more rapidly than concomitant H_2-evolution and reaches a plateau at about 5:1. H_2 evolution activity continues to rise as the component protein ratio increases (Table 1). These observations indicate that HCN binds to and is reduced by nitrogenase at higher oxidation levels than those responsible for H_2 evolution. Since $NifV^-$ nitrogenase is more effective at reducing HCN than the wild-type enzyme it probably operates at higher potentials. This would be consistent with earlier data (McLean et al, 1983) on the partial inhibition by CO of H_2 evolution which was interpreted in terms of a higher functional potential for $NifV^-$ nitrogenase (see Smith et al, this volume).

REFERENCES

Li J-g, Burgess BK and Corbin JL (1982) Biochemistry, 21, 4393-4402.
McLean PA and Dixon RA (1981) Nature, 292, 655-656.
McLean PA, Smith BE and Dixon RA (1983) Biochem.J. 211, 589-597.

STUDIES ON NITROGENASE OF BLUE-GREEN ALGAE

LIN HUIMIN/HE ZHENRONG/DU DAIZIAN/DAI LINGFERS/XING WUSEN & LI SHANGHAO
INSTITUTE OF HYDROBIOLOGY, ACADEMIA SINICA, WUHAN, HUBEI, P.R. CHINA

Seven species of nitrogen-fixing blue-green algae, different in their morphology, ecology and physiology, were comparatively studied with reference to their acetylene reduction activity and hydrogen release activity in intact cell. Some species of blue-green algae, for example *Anabaena azollae* and *Anabaena azotica* (HB 686), have higher growth rate and higher acetylene reduction activity in intact cell than *Anabaena cylindrica*, but the acetylene reduction activity of their crude free-cell extract on the contrary was lower. So we use *Anabaena cylindrica* for studying free-cell nitrogenase.

With DEAE-cellulose DE52 columns (3 times) and anaerobic preparative gel electrophoresis, the crude extract from *A.cylindrica* was separated and purified into two homogenous components ---- Mo-Fe protein and Fe protein. The procedure used by us gave pure a Mo-Fe protein or Fe protein from paste of algae within 48 hours. The purity of these components was corroborated by finding only a single band upon anaerobic polyacrylamide gel electrophoresis.

Some properties of nitrogenase components from *A.cylindrica* have been studied. Amino acid composition of purified Mo-Fe protein and Fe protein was determined according to J.P. Thornber and J.M. Olson. Samples were analyzed on 835 Hitachi amino acid analyzer. The number of amino acid residues/molecular of MoFe protein and Fe protein is 2284 and 556 respectively. So their molecular weight was 250000 and 62000 respectively. But the molecular weight of MoFe protein was estimated by molecular sieve and SDS gel electrophoresis to be 360000. Only one type of subunit was observed. Metal content of Mo-Fe protein, analyzed on Atomic absorption spectrometer (Perking Elmer 503, Mop-E), was Mo-1.2 and Fe-18 in moles of metal per mole of MoFe protein. The Km for acetylene of the partially purified nitrogenase was determined from a standard double reciprocal plot. It was found to be 3.33×10^{-3} atm. This was within the range previously reported. A Specord UV/vis spectrophotometer was used to measure absorption spectrum of Mo-Fe protein and Fe protein. The absorption spectrum was similar to result previously obtained with bacterial nitrogenase. The effect of NaCl KCl and $MgCl_2$ to partially purified nitrogenase was studies. High levels of K^+ Na^+ and Mg^{2+} markedly inhibited nitrogenase activity. The inhibitory order was $Na^+ > K^+ > Mg^{++}$. They inhibited nitrogenase activity of bacteria, too. But the inhibition was weaker. The complementary functioning of cross components of nitrogenase of the blue-green algae *Anabaena cylindrica* and of the strictly aerobic bacterium *Azotobacter vinelandii* was studied. The results showed that Mo-Fe protein of *A.cylindrica* crossed with Fe protein of *A.vinelandii* gives a positive nitrogenase reaction. The reciprocal cross also gives activity, but much lower.

INTRAMOLECULAR ELECTRON TRANSFER IN THE MoFe PROTEIN OF NITROGENASE

D.J. LOWE, B.E. SMITH, G-X. CHEN, M.J. O'DONNELL, T.R. HAWKES
ARC UNIT OF NITROGEN FIXATION, UNIVERSITY OF SUSSEX, BRIGHTON BN1 9RQ, UK.

Oxidation of the MoFe protein of nitrogenase from Klebsiella pneumoniae by ferricyanide or dyes gave a transient epr signal with g-values 2.05, 1.95 and 1.81. These g-values are typical of [4Fe-4S] centres at the +1 oxidation level. Selective isotopic substitution experiments showed that this signal arises from the non-FeMo-cofactor iron, the 'P'-centres, and provide additional evidence that these centres are at the zero oxidation level in the dithionite-reduced protein.

When the concentration of oxidant was from 4 to 10 times that of the protein, the rate constant for the disappearance of the transient signal was $4.1 \pm 0.8\ s^{-1}$ and was independent of the concentration of the MoFe protein, ferricyanide or ferrocyanide. It was the same as that for the disappearance of the g = 4.3, 3.7, 2.01 epr signal from FeMo-cofactors. These data indicate that both signals decayed by an intramolecular process.

We have interpreted these results in terms of Scheme 1 in which an electron is transferred from a FeMo-cofactor to a P centre in the +2 oxidation level, followed by a spin state change of the P^{+1} centres from an S = 1/2 to an S = 5/2 (epr invisible) state. This spin state change is rapid when there are two P^{+1} centres per half Kp1 molecule.

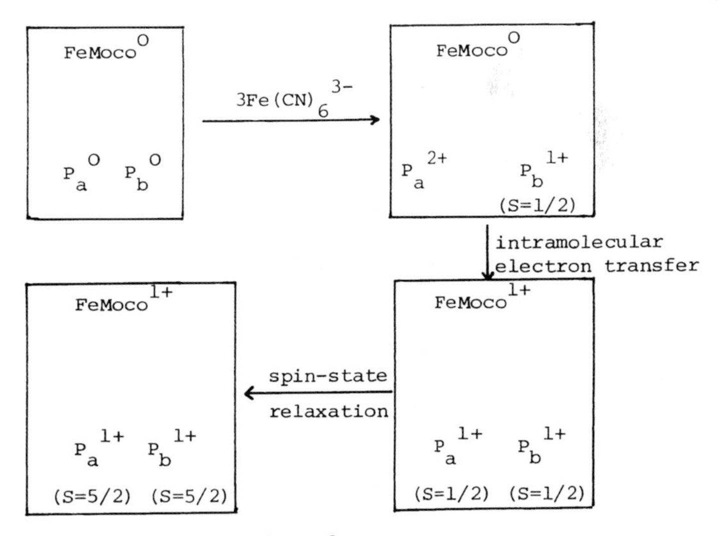

Scheme 1

The rate of this intramolecular electron transfer is essentially the same as that of the rate-limiting step in enzyme turnover and we suggest that both processes are related to the same conformational change in Kp1 which triggers or is triggered by the electron transfer.

REFERENCE

Smith BE, Lowe DJ, Chen G-X, O'Donnell MJ and Hawkes TR (1983) Biochem.J. 209, 207-213.

PRE-STEADY STATE KINETICS OF THE FORMATION OF PRODUCTS AND INTERMEDIATES IN DINITROGEN REDUCTION BY NITROGENASE

D.J. LOWE and R.N.F. THORNELEY
ARC UNIT OF NITROGEN FIXATION, UNIVERSITY OF SUSSEX, BRIGHTON BN1 9RQ, UK.

Scheme 1 simulates the pre-steady state kinetics of H_2, N_2H_4 (derived from an enzyme bound intermediate after quenching) and NH_3 formation and steady state ($K_m{}^{N_2}$, $K_i{}^{H_2}$ and V_{max}) data for K. pneumoniae nitrogenase at 23°C pH 7.4. This mechanism also explains HD formation from D_2 in the presence and absence of N_2.

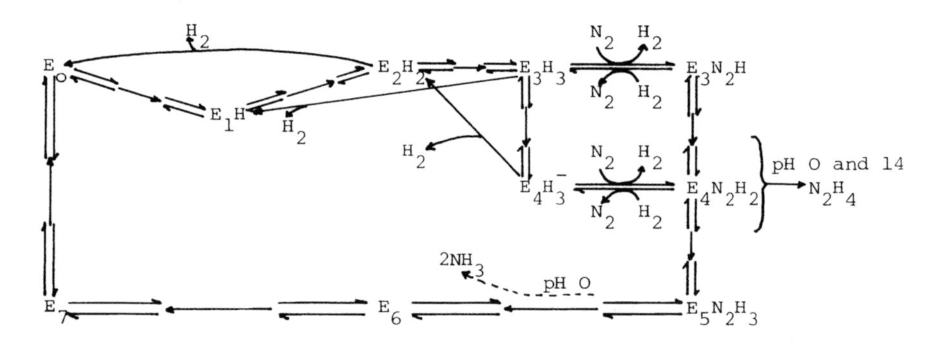

E_n represents free MoFe protein (Kp1) that has had 'n' electrons transferred to it from the Fe protein (Kp2)

Scheme 1

1) Each step from E_n to E_{n+1} is divided into 3 partial reactions, e.g.

$$E_o \underset{Kp2_{red}(MgATP)_2}{\overset{Kp2_{red}(MgATP)_2}{\rightleftharpoons}} Kp2_{red}(MgATP)_2E_o \rightarrow Kp2_{ox}(MgADP)_2E_1 \underset{Kp2_{ox}(MgADP)_2}{\overset{Kp2_{ox}(MgADP)_2}{\rightleftharpoons}} E_1$$

2) Reactions with D_2 and HD not shown; but E_3H_3 can exchange H_2/HD/D_2 e.g.

$$E_3H_3 \underset{D_2 \;\; k_{12} \;\; H_2}{\overset{D_2 \;\; k_{12} \;\; H_2}{\rightleftharpoons}} E_3HD_3$$

and hence E_1,E_2,E_3 and E_4 can have any combination of H and/or D bound.

3) Only free Kp1 can release H_2 or react with N_2 or H_2 i.e. Kp2 restricts access to the active site of Kp1. This mechanism is discussed in detail in the full paper in this book by the same authors.

4) The structure of intermediates of the type $E_nN_xH_y$ is not specified although $E_4N_2H_2$ is most likely a hydrazido(2-) species (=N-NH_2).

RESONANCE RAMAN SPECTROSCOPY OF IRON-SULFUR PROTEINS : APPLICATION TO THE Se/S SUBSTITUTION IN [4Fe-4S] CENTERS

J-M. MOULIS[+], M. LUTZ[++], and J. MEYER[+]
+ DRF-BIOCHIMIE, CEN-Grenoble, 85 X, 38041 GRENOBLE Cédex, FRANCE
++ DB-BIOPHYSIQUE, CEN-SACLAY, BP 2, 91190 GIF-sur-YVETTE, FRANCE

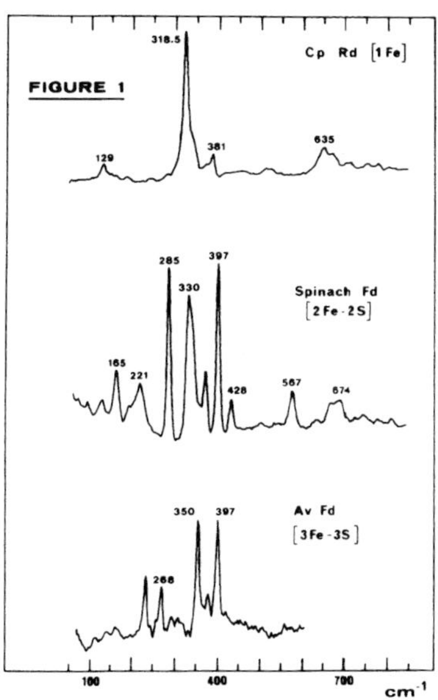

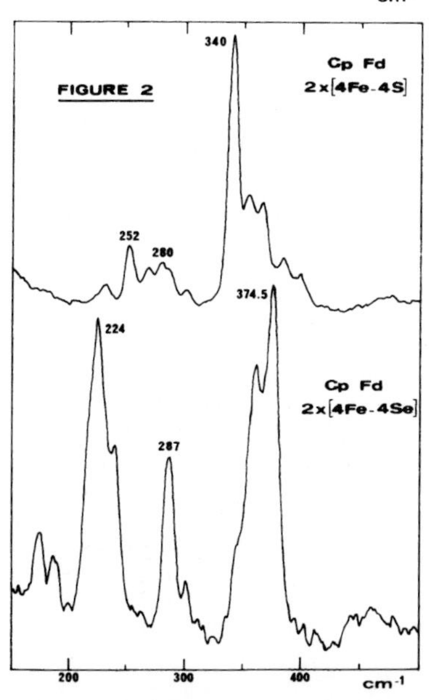

The molecular structures of chromophores can be investigated in detail by Resonance Raman (RR) spectroscopy.

Low temperature (≃ 20°K) RR spectra of C. pasteurianum rubredoxin ($\lambda_{exc.}$ = 488 nm), of spinach ferredoxin (Fd), A. vinelandii FdI and C. pasteurianum Fd ($\lambda_{exc.}$ = 457.9 nm) constitute the reference data for each type of well characterized iron-sulfur clusters (Fig.1 and 2). The RR spectrum of the [3Fe-3S] cluster of Av FdI (Fig.1) was obtained by substracting the contribution of the [4Fe-4S] cluster (Fig.2) from the experimental spectrum of Av FdI.

The substitution of S by Se in the [4Fe-4S] centers of Cp Fd (Moulis and Meyer, 1982) brings useful information on the structure of this 4Fe-type cluster : its vibrational behaviour can be described in D_{2d} local symmetry using isotopic substitutions on core chalcogenide atoms and polarization studies. The terminal Fe-Scys stretching modes centered near 360 cm^{-1} for both ferredoxins are distinct from the 8 bridging Fe-X (X = S, Se) modes which extend from 250 to 400 cm^{-1} and 170 to 315 cm^{-1} for sulfur and selenium containing ferredoxins, respectively.

These studies will be extended to the metal centers of larger proteins such as the components of nitrogenase.

REFERENCE

Moulis, J-M and Meyer, J (1982) Biochemistry 21, 4762-4771.

STRUCTURAL ASPECTS AND REACTIVITY OF THE IRON-MOLYBDENUM COFACTOR FROM NITROGENASE

W.E. NEWTON/B.K. BURGESS/S.C. CUMMINGS[a]/S. LOUGH/J.W. McDONALD/
J.F. RUBINSON/S.D. CONRADSON[b] and K.O. HODGSON[b]
C.F. Kettering Research Laboratory, Yellow Springs, OH 45387, USA
a) Wright State University, Dayton, OH 45431, USA b) Stanford University, Stanford, CA 94305, USA

We have recently investigated the structure and composition of FeMoco by both Mo Kα X-ray absorption spectroscopy (XAS) and O_2 degradation and its reactivity with $[Fe_4S_4(SEt)_4]^{3-}$, a synthetic analogue of reduced ferredoxin. FeMoco used had Mo:Fe ratios of 1:6-7, the typical S=3/2 EPR signal, <20% easily complexed Fe and Azotobacter vinelandii UW45 reconstitution activities of 249 ±22 nmoles C_2H_2 reduced min^{-1} (ng-atom Mo)$^{-1}$. XAS samples were 0.6 to 1.3 mM in Mo with less than 10% activity variation after the 8-12 hour runs at -20°C.

MoEdge, XANES and EXAFS Spectra. The spectral shape close to the Mo edge (20,000 - 20,030 eV) reflects the general chemical and structural nature of its first coordiation shell (XANES). The spectra of FeMoco and the MoFe protein closely resemble those of Mo compounds where the first coordination shell is a pseudo-octahedron with O and S ligands rather than either a tetrahedron or a pseudo-octahedron with only S ligands. EXAFS data for three states of FeMoco and for the MoFe protein have been analyzed by curve fitting, using empirical parameters derived from 16 synthetic Mo compounds of known structure.

Sample	FeMoco			MoFe protein			FeMoco/ØXH		
Atom	O	S	Fe	O	S	Fe	O	S	Fe
Number	3.0	3.0	2.8	1.9	4.5	3.5	∿2.7	∿4.0	∿3.5
Distance	2.09	2.36	2.68	2.12	2.37	2.67	2.12	2.36	2.70

Therefore, removal of FeMoco from the protein causes a decrease in the apparent number of S and Fe neighbors of Mo and an increase in O. Addition of thiol or selenol (ØXH; X=S,Se) to FeMoco produces a Mo site more similar to the protein. No Se shell in FeMoco + ØSeH shows that ØSH and ØSeH do not bind to Mo.

Elicitation of Thiomolybdates. Controlled O_2 degradation releases all of FeMoco's Fe while changing its virutally featureless visible spectrum to show initially a shoulder at 500-600 nm (a MoS_2Fe intermediate?) and finally major absorption bands at 465 and 395 nm due to tetrathio- and trithio-molybdate. Preliminary spectral simulation yields a 3:1 mixture of $MoOS_3^{2-}$ and MoS_4^{2-} representing ≥70% of the total Mo. Comparative reactions with model compounds, $[MoS_4FeCl_2]^{2-}$ and $[Mo_2Fe_6S_8(SØ)_9]^{3-}$, suggest a MoS_3 core in FeMoco with Mo in a formal oxidation state less than VI.

Reactivity and Redox Properties. As $[Fe_4S_4]$ units appear to be involved in electron transfer to FeMoco in N_2ase, we have investigated mixtures of $[Fe_4S_4(SEt)_4]^{3-}$ in NMF solution with $Na_2S_2O_4$-reduced FeMoco (S=3/2 EPR signal) and its oxidized form (produced by methylene blue or 1:1 DMF dilution) which is EPR silent. Electron transfer between the cluster and FeMoco can be monitored by EPR-spectral changes. With EPR-active FeMoco, a simple superposition of the two individual EPR spectra results indicating no electron transfer. However, EPR-silent oxidized FeMoco is rapidly reduced. Thus, synthetic analogue and FeMoco communicate, even though both are anions. Thus, FeMoco requires a potential more negative than -1.3 V (E^o for cluster) for reduction to the substrate-reducing level.

REGULATION OF NITROGENASE ACTIVITY IN RHODOSPIRILLUM RUBRUM: METAL DEPENDENCE OF THE ACTIVATION OF INACTIVE Fe-PROTEIN BY THE ACTIVATING ENZYME

STEFAN NORDLUND and AGNETA NORÉN
DEPARTMENT OF BIOCHEMISTRY, ARRHENIUS LABORATORY, UNIVERSITY OF STOCKHOLM, S-106 91 STOCKHOLM, SWEDEN

INTRODUCTION

The Fe-protein (Rr2) of nitrogenase from *R. rubrum* grown with N_2 or glutamate as N-source is inactive as isolated (Ludden, Burris, 1976, 1978, Nordlund et al 1977, 1978). This inactive form can be converted to the active in a reaction catalyzed by the activating enzyme, which can be isolated from chromatophores (Ludden, Burris, 1976, Nordlund et al 1977, Zumft, Nordlund 1981). The activation also is dependent on ATP and a divalent cation.

MATERIALS AND METHODS

The MoFe-protein, the Fe-protein and the activating enzyme were purified according to our published methods. The activation mixture contained inactive Rr2, activating enzyme, 50 mM Hepes pH 7.6, 5 mM ATP, 3 mM dithionite and the metal ions studied. The activation was run for 15 min and then an aliquote was transferred to the reaction mixture which contained MoFe-protein, 50 mM Hepes pH 7.3, 5 mM ATP, an ATP-generating system, 10 mM $MgCl_2$ and 6 mM dithionite. The carry over from the activation mixture was compensated for. The reaction was run for 7 min and the ethylene produced was analyzed by gas chromatography.

RESULTS AND DISCUSSION

When the concentration of free Mn^{2+} was varied in the presence and in the absence of Mg^{2+} it was demonstrated that Mn^{2+} alone will support maximal activation. However increasing the Mg^{2+}-concentration leads to decreasing requirement for Mn^{2+}. Both free Mn^{2+} and MnATP are needed for activation and in the presence of 1 mM free Mn^{2+}, maximal rate is obtained with 1 mM MnATP. It has previously been reported, by us and others, that a high concentration of Mg^{2+} is needed for high nitrogenase activity in the absence of added Mn^{2+}. However, when the activation is run separately, with no Mn^{2+} added, only low rates are obtained with Mg^{2+} alone. At 40 mM total Mg^{2+} the activation is only 25% of that under optimal conditions with Mn^{2+} added. The rather high rates that have been reported could possibly be due to Mn^{2+} or Fe^{2+} being present in the enzyme preparations used. Under such conditions the high Mg^{2+}-concentration would prevent complex formation between ATP and the cations.

We have previously reported that Fe^{2+} will substitute for Mn^{2+} in the activation and we have now shown that maximal activation is obtained at 7.5 mM total Fe^{2+} in the presence of 5 mM ATP but no other cations. Under the same conditions 6 mM total Mn^{2+} is needed for maximal rate. When investigating the effect of other cations we found that Ba^{2+} is a potent inhibitor of the activation with I_{50}=0.05 mM free Ba^{2+} in the presence of 1 mM free Mn^{2+}. The inhibition is competitive with respect to Mn^{2+} and there is no effect on the nitrogenase reaction by the concentrations used.

REFERENCES

Ludden, P.W. and Burris, R.H. (1976) Science 194, 424-426.
Ludden, P.W. and Burris, R.H. (1978) Biochem. J. 175, 251-259.
Nordlund, S., Eriksson, U. and Baltscheffsky, H. (1977) Biochim. Biophys. Acta 462, 187-195.
Nordlund, S., Eriksson, U. and Baltscheffsky, H. (1978) Biochim. Biophys. Acta 504, 243-254.
Zumft, W.G. and Nordlund, S. (1981) FEBS Lett. 127, 79-82.

EVIDENCE FOR TWO DINITROGENASE REDUCTASES UNDER REGULATORY CONTROL BY MOLYBDENUM IN AZOTOBACTER VINELANDII

R. PREMAKUMAR[+]/ELIANA M. LEMOS[++]/PAUL E. BISHOP[+++]
+ DEPT OF MICROBIOLOGY, NC STATE UNIVERSITY, RALEIGH, NC 27650 USA
++ FACULDADE DE CIENCIAS AGRARIAS E VETERINANIAS, UNESP JABOTICABAL (SP) BRASIL
+++U.S. DEPT OF AGRICULTURE, AGRICULTURAL RESEARCH SERVICE, DEPT OF MICROBIOLOGY, NC STATE UNIVERSITY, RALEIGH, NC 27650 USA

The conventional nitrogenase system in Azotobacter vinelandii that is responsible for the reduction of N_2 to ammonia consists of dinitrogenase, a molybdenum-iron protein, and dinitrogenase reductase, an iron-protein.

Evidence for an alternative nitrogen fixation system which is expressed under conditions of molybdenum (Mo) deficiency has been reported in Azotobacter vinelandii (Bishop et al. Proc. Natl. Acad. Sci. U.S.A. (1980) 77, 7342-7346). It is hypothesized that Mo derepresses the conventional nitrogenase system and represses the alternative system. In view of the proposed opposing effects of Mo on the regulation of the two nitrogen fixation systems, we decided to study the role of Mo on the expression of dinitrogenase reductase using strains which lack (UW1, UW3) or have inactive (UW91) dinitrogenase reductase under Mo-sufficient conditions. Nitrogenase activities were determined by the acetylene reduction method in a complementation assay where cell extracts of UW91 derepressed in the presence of 1μM Na_2MoO_4 were used as the source of dinitrogenase and cell extracts of the mutant strains, to be tested, as the source of dinitrogenase reductase. When incubated alone, cell extracts of strains UW1, UW3 and UW91, derepressed for nitrogenase in the presence or absence of Mo, had only a trace of nitrogenase activity. In the complementation assay, extracts obtained from cells derepressed in the presence of 1μM Na_2MoO_4 or 10μM Na_2WO_4, did not show any enhancement of nitrogenase activity. However, cell extracts of the same Nif^- strains, derepressed under Mo-deficient conditions in the presence or absence of 2μM V_2O_5 increased the acetylene reduction activity between 9 and 35 fold over that of the strain UW91 (+Mo) extract alone. Extracts from strain UW3 were also found to complement each other when derived from cells derepressed under Mo-sufficient and Mo-deficient conditions. This complementation resulted in an 11-fold increase in activity over that of the strain UW3 extract alone. Two-dimensional gel electrophoretic analysis showed that extracts from Mo-starved cells contain a protein which had a molecular weight and pI similar to dinitrogenase reductase, though this protein was slightly more basic than dinitrogenase reductase. In conclusion, the results with the Nif^- strains provide evidence for the presence of two reductases, one of which is expressed in the presence of Mo (dinitrogenase reductase) and the other in the absence of Mo (alternative reductase).

GROWTH AND PROTEINS PRODUCED BY WILD-TYPE, NIF^-, AND AMMONIUM-DEREPRESSED *AZOTOBACTER VINELANDII* IN THE PRESENCE AND ABSENCE OF MOLYBDATE AND AMMONIUM.

B.E. TERZAGHI/B.D. SHAW/A.D. PATERSON
PLANT PHYSIOLOGY DIVISION, DSIR, PALMERSTON NORTH, NEW ZEALAND.

1. INTRODUCTION

Bishop *et al.* (1980) postulated an alternative nitrogen fixing system in *Azotobacter vinelandii* to explain the growth of wild-type and some Nif^- mutant strains in nitrogen-free medium which is also deficient in molybdenum. Several papers (Bishop *et al.* 1982; Page, Collinson 1982; Riddle *et al.* 1982) subsequently presented data in support of the model. We have tested ammonium-derepressed mutants of *A. vinelandii* (which reduce acetylene in the presence of NH_4^+) for their ability to grow in Mo-deficient media, and to see what nitrogenase polypeptides are made under these conditions.

2. PROCEDURE

Strains were grown in Burk's medium, with and without ammonium acetate and sodium molybdate, to mid-log phase. The cultures were then enumerated and tested for acetylene reduction activity, centrifuged, lysed, and the proteins subjected to two-dimensional analysis (isoelectric focusing and SDS-polyacrylamide gel electrophoresis) as described by O'Farrell (1975) and Bishop *et al.* (1980). Purified component I from wild-type strain UW and a collection of standard proteins were used as MW markers. Strains tested included UW, Nif^- mutant PPD 82 and ammonium-derepressed mutants PPD 80 and PPD 85, both derived from PPD 82.

3. RESULTS

In contrast to other results (Bishop *et al.* 1980, 1982; Page, Collinson 1982), neither our wild-type strain nor our Nif^- mutant grew at all in Mo-deficient N-free medium; they grew well, however, in N-containing Mo-deficient medium. The ammonium-derepressed strains grew in all conditions and reduced acetylene.

Proteins corresponding in size and pI to normal nitrogenase polypeptides were found in the ammonium-derepressed strains in all four growth conditions. However, differences in total protein profiles were observed between some strains and within a strain, depending on the growth conditions, and despite the use of markers, the MoFe protein was sometimes difficult to identify. Thus, it has not been necessary to postulate an alternative nitrogen fixation system to explain our results.

4. REFERENCES

Bishop, P.E., Jarlenski, D.M.L. and Hetherington, D.R. (1980) Proc. Natl. Acad. Sci. USA 77, 7342-7346.
Bishop, P.E., Jarlenski, D.M.L. and Hetherington, D.R. (1982) J. Bacteriol. 150, 1244-1251.
O'Farrell, P.M. (1975) J. Biol. Chem. 250, 4007-4021.
Page, W.J. and Collinson, S.K. (1982) Can. J. Microbiol. 28, 1173-1180.
Riddle, G.D., Simonson, J.G., Hales, B.J. and Braymer, H.D. (1982) J. Bacteriol. 152, 72-80.

A STOPPED-FLOW STUDY OF THE KINETICS OF THE DISSOCIATION OF THE MoFe AND Fe PROTEINS OF NITROGENASE

R.N.F. Thorneley and D.J. Lowe
ARC Unit of Nitrogen Fixation, University of Sussex, Brighton, BN1 9RQ, U.K.

Stopped-flow spectrophotometry and epr spectroscopy have been used to study the kinetics of reduction by dithionite of the oxidised Fe protein of nitrogenase from Klebsiella pneumoniae ($Kp2_{ox}$) in the presence of MgADP at 23°, pH 7.4 (Thorneley, Lowe 1983). The active reductant, $SO_2{}^{\cdot -}$, produced by the pre-dissociation of $S_2O_4{}^{2-} \rightleftharpoons 2SO_2{}^{\cdot -}$, reacts with $Kp2_{ox}(MgADP)_2$ with $k_4 = 3.0 \pm 0.4 \times 10^6 M^{-1}s^{-1}$. The inhibition of this reaction by the MoFe protein (Kp1) has enabled the rate of dissociation of $Kp2_{ox}(MgADP)_2$ from $Kp1^{\dagger}$ (the Kp2 binding site on Kp1) to be measured ($k_{-3} = 6.4 \pm 0.8\ s^{-1}$). Comparison with the steady state rate of substrate reduction shows that the dissociation (k_{-3}) of the complex $Kp2_{ox}(MgADP)_2Kp1^{\dagger}$, which is formed after MgATP induced electron transfer from Kp2 to $Kp1^{\dagger}$, is the rate limiting step in the catalytic cycle for substrate reduction.

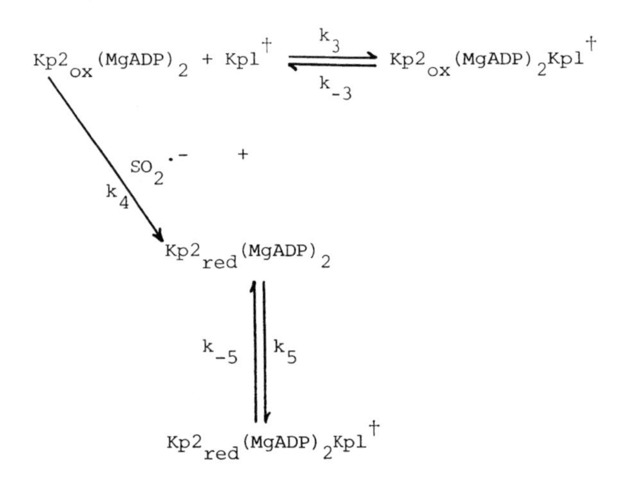

Rate constant	Value
k_3	$4.4 \pm 0.5 \times 10^6 M^{-1} s^{-1}$
k_{-3}	$6.4 \pm 0.8\ s^{-1}$
k_4	$3.0 \pm 0.4 \times 10^6 M^{-1} s^{-1}$
k_5	$>2 \times 10^6 M^{-1} s^{-1}$
k_{-5}/k_5	$2.2 \pm 1.0\ \mu M$

REFERENCES

Thorneley RNF and Lowe DJ (1983) Biochem.J. in the press.

EFFECT OF NITRITE AND NITRIC OXIDE ON NITROGENASE EXTRACTED FROM SOYBEAN BACTEROIDS.

J.C. TRINCHANT/J. RIGAUD
LABORATOIRE DE BIOLOGIE VEGETALE, FACULTE DES SCIENCES, 06034 NICE CEDEX, FRANCE.

Nitrite, the usual reduction product of nitrate, can be generated in nodules of Legumes both by bacteroids (Rigaud et al. 1973) and plant cytosol (Streeter 1982). It appeared as a strong inhibitor for C_2H_2 reduction by purified nitrogenase (Trinchant, Rigaud 1980) but in this paper we reported its inability to affect H_2 evolution. In contrast, NO inhibited both C_2H_2 reduction and H_2 evolution and nitrogenase was more sensitive to NO than to NO_2^- during C_2H_2 reduction (K_I values were respectively 0.056 mM and 5.2 mM). The possibility to reduce NO_2^- to NO in nitrogenase assays was reported by Meyer (1981). In the absence of preincubation of NO_2^- and dithionite, the generation of NO was low and the characteristics of the respective inhibitions of nitrogenase ruled out the suggestion concerning an inhibition by NO_2^- through NO. These results confirmed the possibility for NO_2^- to inhibit directly nitrogenase under natural conditions.

Meyer J (1981) Arch. Biochem. Biophys. 210, 246-256.

Rigaud J, Bergersen FJ, Turner GL and Daniel RM (1973) J. Gen. Microbiol. 77, 137-144.

Streeter JG (1982) Plant Physiol. 69, 1429-1434.

Trinchant JC and Rigaud J.(1980) Arch. Microbiol. 124, 49-54.

H_2-UPTAKE ACTIVITY OF THE MO-FE PROTEIN

Z. WANG AND G. WATT
C.F. KETTERING RESEARCH LABORATORY, 150 EAST SOUTH COLLEGE STREET, YELLOW SPRINGS, OHIO, USA 45387

Nitrogenase, consisting of the MoFe and Fe proteins, catalyzes the evolution of H_2 when supplied with a low potential reductant and MgATP. During enzymatic reduction of dinitrogen, but not during reduction of other substrates, molecular dihydrogen is a competitive inhibitor. Also, during dinitrogen reduction, HD is produced by a reductive process from gaseous O_2 and solvent protons. The evolution of and inhibition by dihydrogen is clearly an enzymatic process but the mechanism of these reactions is still unclear.

We report four studies which suggest that the MoFe protein alone in the absence of reductant and MgATP acts as a hydrogen uptake protein. The first of these studies monitors the decrease in added oxidant (methylene blue, $Fe(CN)_6^{3-}$, O_2 etc) concentration in the presence of H_2 and the MoFe protein. All oxidants added were completely reduced by H_2 when in the presence of MoFe protein. The second study measured $^3H_2$ transfer from the gaseous phase to the aqueous phase only when the MoFe protein was present. The third study demonstrated that the EPR signal was present continually during the oxidant reduction demonstrating that only a partially oxidized form of the MoFe protein is formed in the presence of oxidants which normally oxidize the EPR centers. Study four demonstrated the absence of a contaminating hydrogenase which could possibly explain the observed results.

From these studies we conclude that the MoFe protein has H_2-uptake activity.

ELECTRON ALLOCATION TO H^+ AND N_2 BY NITROGENASE OF RHIZOBIUM LEGUMINOSARUM BACTEROIDS

Hans Wassink and Huub Haaker
Department of Biochemistry, Agricultural University, De Dreijen 11, 6703 BC Wageningen, The Netherlands

Electron allocation to H^+ and N_2 by nitrogenase has been studied in intact Rhizobium leguminosarum bacteroids and in cells made permeable for small molecules. The flux of electrons through nitrogenase in intact cells was modified by three methods. 1) Nitrogenase activity was inhibited by depolarizing the membrane potential. This inhibits electron transport to nitrogenase. Under these inhibited conditions no effect on electron allocation to H^+ and N_2 by nitrogenase was observed (fig. 1). 2) Nitrogenase activity was inhibited by O_2-limitation (fig. 2). 3) Nitrogenase activity was inhibited by H^+-conducting ionophores (not shown). In both cases, the flux of electrons through nitrogenase was inhibited by an inhibition of electron transport to nitrogenase and by decrease in ATP concentration and an increase in ADP concentration. Under these conditions allocation to H^+ was favoured above allocation to N_2.

figure 1 figure 2

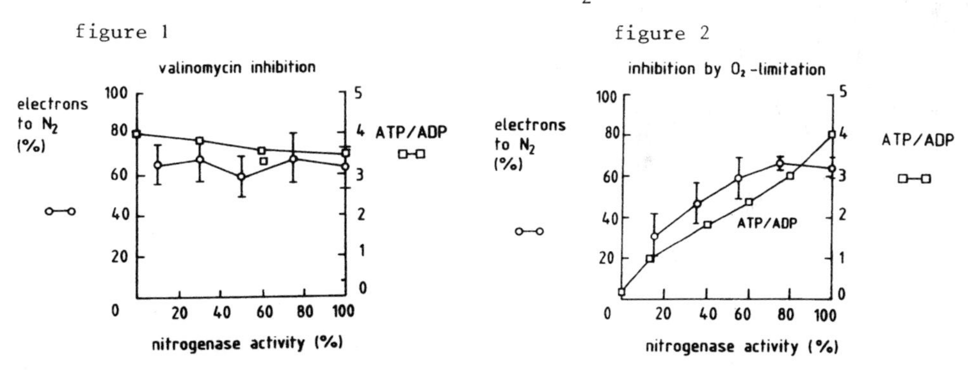

By treating bacteroids with cetylhexadecyl trimethylammoniumbromide cells were made permeable for small molecules. In these cells the flux of electrons through nitrogenase was inhibited by limiting reductant $Na_2S_2O_4$. As in intact cells, the allocation of electrons to H^+ and N_2 by nitrogenase was hardly affected. Only by inhibition with MgADP, electron allocation to H^+ was favoured above allocation to N_2 (fig 3). From these experiments it is concluded that in vivo only ADP influences electron allocation to H^+ and N_2 by nitrogenase and not the flux of electrons through nitrogen.

figure 3

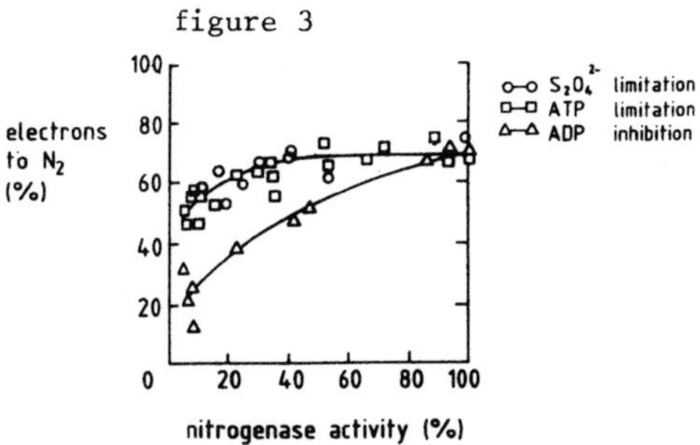

INTERACTION OF MGATP AND MGADP WITH COMPONENT PROTEINS OF NITROGENASE FROM *AZOTOBACTER VINELANDII*

YOU CHONGBIAO/WANG HUIXIAN/GAO MENGSHENG/PING SHUZHEN
INSTITUTE FOR APPLICATION OF ATOMIC ENERGY, CHINESE ACADEMY OF AGRICULTURAL SCIENCES, P.O. BOX 5109, BEIJING, CHINA

Nitrogenase *in vitro* requires a sizeable free energy input in order to reduce N_2 to NH_3. One of this energy requirement is met by MgATP hydrolysis. However, MgADP is a potent inhibitor of nitrogenase activity. To gain additional insight into the role of ATP and ADP in nitrogenase function, we investigated the interaction of ATP and ADP with nitrogenase protein by fluorescence probe and calorimetric assay.

With the aid of fluorescence probes fluorecein mercuric acetate (FMA) and fluorescamine, it has been shown that Fe protein (Av2) bind MgATP and MgADP, while no binding of MgATP and MgADP to the MoFe protein (Av1) could be demonstrated. There are two binding sites for ATP as described previously (1-2) and at least two for ADP. The binding of MgATP and MgADP was at sulfhydryl groups. Addition of MgATP and MgADP is accompanied by an increase of the number of amino groups reactive with fluorescamine in Av2 and the binding of ATP and ADP apparently causes the same conformational change in this protein.

The results of the quenching reaction of FMA and the fluorescence intensity of the product formed by the reaction of fluorescamine with Av1-Av2 complex suggest that Av1 can react with Av2 in the absence of MgATP and that the binding of Av1 to Av2 is via the sulfhydryl groups and causes a change of reactive amino groups with fluorescamine.

The binding of MgATP and MgADP to Av2 has been studied by batch microcalorimetric assays and a ΔH^o value of -4.78 kcal/mole for ATP and a ΔH^o value of -5.45 kcal+mol for ADP have been measured. The $-\Delta H^o$ of Av2-MgATP increased with increasing concentration of MgATP. It reaches saturation at the concentration of ATP>60 μM, when Av2 is 38 μM. These small enthalpy values, combined with a ΔG^o value of -6.66 kcal/mol found for ATP and -6.85 kcal/mol for ADP, give a ΔS^o values for +6.31 and +4.82 entropy units, respectively. These data suggest that the strong binding energy resulting from ATP and ADP interacting with Av2 is conversed in the entropy change as $T\ \Delta S^o$ energy amounting to +1.88 kcal/mol and +1.44 kcal/mol, respectively. The positive sign of the entropy change indicates a losening of the Av2 structure with the energy somehow stored in the Av2-MgATP complex.

REFERENCES

1. You CB, Li JW, Song W and Li X (1978) Acta Phytophysiologia Sinica 4, 123-131.
2. You CB, Li JW, Song W and Li X (1979) Acta Phytophysiologia Sinica 5, 215-224.

PHYSIOLOGY OF DIAZOTROPHS

POSTER DISCUSSION 4A

PHYSIOLOGY OF FREE-LIVING DIAZOTROPHS INCLUDING CYANOBACTERIA AND *AZOSPIRILLUM* WITH SPECIAL ATTENTION TO: NH_4 METABOLISM, H_2 METABOLISM, O_2 PROTECTION

POSTER DISCUSSION 4B

METABOLISM OF *RHIZOBIUM* AND *AZOTOBACTER*

OXYGEN AND THE PHYSIOLOGY OF DIAZOTROPHIC MICROORGANISMS

F.J. BERGERSEN
CSIRO DIVISION OF PLANT INDUSTRY, CANBERRA, AUSTRALIA

1. INTRODUCTION

This synthesis paper discusses some aspects of O_2-related problems using representative examples from recent literature seen by the author up to June 1983. A large proportion of diazotrophic microorganisms require O_2 as the terminal electron acceptor for energetic pathways; these are the most energetically efficient. However, some are facultative aerobes/anaerobes which are generally believed to fix N_2 only anaerobically and a few are strict anaerobes. It is well-known that nitrogenase proteins are easily inactivated in the presence of O_2: a progressive series of changes involves metal-S centres. Therefore aerobic diazotrophes (and perhaps also some facultative anaerobes) have developed strategies by means of which the deleterious effects of O_2 are avoided or minimised, whilst at the same time preserving the energetic advantages of aerobic metabolism. Robson, Postgate (1980) distinguished between aerobic and microaerobic organisms. However, in terms of efficient N_2-fixation, most O_2-tolerant diazotrophs are microaerobes. Pearson et al. (1982) recently described nitrogenase activity (confirmed by $^{15}N_2$ incorporation) in 6 strains of Mycoplana species and type cultures of M. dimorpha and M. bullata, which are obligate aerobic, gram-negative, branched bacteria. Nitrogenase was produced microaerobically but best activity was anaerobic. McClung, Patriquin (1980) described a N_2-fixing, microaerophilic, root-associated strain of a species of Campylobacter (Spirillaceae). Malik et al. (1981) described N_2 fixation by 3 strains of the aerobic H_2-oxidizing Alciligenes latus. In these organisms also active nitrogenase was confined to microaerobic conditions. These examples emphasize the importance of microaerobic conditions for N_2 fixation in many microbial habitats.

The requirements for ATP and low potential reductant for activity suggest that nitrogenase may be located in or adjacent to the bacterial membrane where many electro-chemical reactions are located (e.g. Veeger et al. 1981). Although nitrogenase is not bound in the membrane (Haaker et al. 1977), recent electron microscopic studies by D.J. Goodchild (private communication) have shown nitrogenase (the FeMo-protein) to be located just beneath the membrane in bacteroids in soybean nodules. This location may increase the hazard of exposure to O_2. Apart from nitrogenase itself, other components of metabolic pathways in N_2-fixing microorganisms are sensitive to O_2, e.g. the nitrogenase reductase activating enzyme of Rhodospirillum rubrum (Zumft, Nordlund 1981), hydrogenase (e.g., Arp, Burris 1982; Bowien, Schlegel 1981), the electron donors ferredoxin and flavodoxin (review: Robson, Postgate 1980). Oxygen damage sometimes involves the free radicle $O_2^{\bullet -}$, especially when O_2 concentration is high (e.g. Fridovich 1975; Kulakova et al. 1982) and this needs to be taken into account. A light-dependent adaptation to supra-atmospheric concentrations of O_2 has been described in Anabaena cylindrica, where protection of nitrogenase involved superoxide dismutase and catalase (Mackey, Smith 1983).

2. STRATEGIES TO MINIMISE DAMAGE CAUSED BY O_2 (O_2-PROTECTION).

2.1. Regulation of synthesis of nitrogenase

Repression of nitrogenase synthesis in the presence of O_2 has been observed in representatives of all diazotrophic groups. In *Klebsiella pneumoniae*, this involves the product of the *nif* L gene, which acts as a negative effector (Kennedy *et al*. 1981; Hill *et al*. 1981; Merrick *et al*. 1982). Recently, a modified model of the role of the *nif* LA operon in regulation of nitrogenase synthesis and the relationships of this operon with *gln* regulation and other genes concerned with N-assimilation has been presented (Ow, Ausubel 1983). This model accommodates observations of O_2 regulating these other pathways (e.g. Goldberg, Hanau 1980). Bergersen *et al*. (1982) found that derepression of *nif* H was inhibited by 50% at a dissolved O_2 concentration of 0.1 µM, very close to the apparent K_s of the principal terminal oxidase. The mode of action of the *nif* L gene product and of the interaction with O_2 has yet to be determined. The net effect is to prevent wastage of cellular resources on nitrogenase synthesis when available N is present or when O_2 poses a threat of inactivation.

Other diazotrophs have analagous control systems regulating nitrogenase synthesis (cf. Ow, Ausubel 1983; Kennedy, Robson 1983). Sometimes O_2 interacts with control of other related metabolic pathways (e.g. with hydrogenase in *Rhodospseudomonas capsulata*; Colbeau *et al*. 1980). In *Rhizobium* sp. strain CB756, growing in O_2-limited continuous culture, steady-state levels of nitrogenase activity in the presence of NH_4^+ (4 mM), were greatest when O_2 supply rates resulted in 1.1 nmoles ATP mg^{-1} (dry wt) (adenylate energy charge = 0.65). Low (giving <0.8 nmoles ATP mg^{-1}), or high rates of O_2 supply (giving >1.8 nmoles ATP mg^{-1}) resulted in repression of nitrogenase (Ching *et al*. 1981). This narrow "energy window" for nitrogenase synthesis may be a common feature in diazotrophs (cf. Upchurch and Mortenson 1980). Concentration of O_2 alone may not be the critical factor. For example, for different *Rhizobium* strains the limiting concentrations of dissolved O_2 for nitrogenase synthesis vary from 0.2 µM (Ching *et al*. 1981) to 12 µM (Gebhardt *et al*. 1983). Shaw (1983) showed that synthesis of the two nitrogenase proteins by bacteroids from lupin nodules was differentially controlled by O_2; this contrasts with other reports in which synthesis of the two proteins was coordinately controlled.

2.2. Regulation of activity of nitrogenase

Rapid regulation (switch off-switch on) and conformational protection in *Azotobacter* spp. has been reviewed by Robson, Postgate (1980). It may not be necessarily associated with protection from O_2, but may involve diversion of electrons from nitrogenase by rapid autoxidation of flavodoxin hydroquinone. Conformational protection requires Mg^{2+} and involves equimolar associations between nitrogenase and protective proteins containing Fe-S centres. Pienkos *et al*. (1983) recently interpreted experiments with *Anabaena* spp. as indicating a possible conformational protection mechanism and studies of *Azotobacter* spp. continue (e.g. Xu, Li 1981; Post *et al*. 1983).

2.3. Protection through O_2 demand

In any microbial environment the O_2 concentration is the resultant of rate of O_2 entry, rate of O_2 consumption and the affinity for O_2 of the consuming systems. An equilibrium concentration near ambient is reached when the entry rate is not limiting. The concentration will be near the K_s of the consuming system when O_2 entry rate is lower than the maximum consumption rate (O_2 limited). It follows that for a system with a fixed O_2 permeation rate, an increase in O_2 demand will cause a fall in O_2 concentration. If the fall is sufficient, nitrogenase is protected. There have been many recent papers about these effects (e.g. Hochman, Burris 1981; Peterson, La Rue 1981). In some diazotrophs the presence of an uptake hydrogenase may increase O_2 demand, conferring protection (e.g. Nelson, Salminen 1982; Volpon et al., 1981). However, H_2-dependent respiration may also be O_2-sensitive (e.g. Pedrosa et al. 1982). The synthesis of hydrogenase may itself be negatively regulated by O_2 (e.g. Eisbrenner et al. 1981; Maier, Merberg 1982; Mutaftschiev et al. 1983). Some O_2 scavenging systems are associated with membranes (e.g. Wang et al. 1981,1982).

Some diazotrophic microorganisms have an additional capacity to scavenge O_2 by means of energetically poorly coupled but enhanced respiration. The best known of these is in Azotobacter spp. (reviewed by Robson, Postgate 1980). It involves an O_2-adaptable, divided terminal respiratory pathway in which the dominant terminal branch bypasses energy conservation site III and there is a loss of energy coupling at site I. If growth is O_2-limited, electron transport is fully coupled. In addition to this, there are complex interactions with other metabolic pathways. Some aspects of O_2 protection at high levels of dissolved O_2 in continuous cultures of A. vinelindii are being re-examined (Post et al. 1983).

2.4. Specific biochemical protection

The possibility exists that there are specific biochemical mechanisms which protect nitrogenase in vivo. These might be expected to occur in organisms whose environment may contain wide limits of O_2 concentration. Hamadi, Gallon (1981) described a Ca^{2+} dependent system in the cyanobacterium Gleocapsa sp., which was specifically inactivated by EDTA. They concluded that the mechanism specifically protected nitrogenase from inactivation by O_2 when Ca^{2+} was present in adequate amounts in the cells.

2.5. "Behavioural" and growth adaptations

Associations with aerobic heterotrophs probably serve to protect the nitrogenase of diazotrophs in environments with a high C/N ratio (Robson, Postgate 1980). Similarly, clustering growth habit would tend to limit O_2 tension within the cluster, as has been reported for certain cyanobacteria (Carpenter, Price 1979). We have noted that Azotobacter vinelandii, grown in continuous culture at limiting O_2 levels of 0.2-0.6 μM, reacted by switching off nitrogenase and clumping when exposed to 5% O_2 in well-shaken vessels. Following development of the clumps, nitrogenase activity was restored. At 0.4% O_2 there was no switch-off of nitrogenase and little clumping, the data indicating that O_2 was limiting throughout (Bergersen, Turner 1980). Aggregations of cells may provide an intensity of O_2 consumption which exceeds the rate at which O_2 can enter from the environment. The result is probably that only a proportion of the cells experience suitable O_2 concentrations for nitrogenase activity. Similar reasoning applies to conditions within colonies on solid culture media.

Barak et al. (1982) have described an aerotactic response in the highly motile diazotroph Azospirillum brasilense. It is well-known that this organism forms a growth band beneath the air/medium surface of soft agar, the depth at which the band forms being greater when air is the gas phase than when a diminished pO_2 is used (Okon et al., 1980). Barak et al. (1982) have quantified these effects, which are rapidly initiated, and have shown that the bacteria congregate to form an O_2 consuming mass which then migrates to a suitable location in an O_2 concentration gradient and then migrates with the gradient as O_2 is consumed. Migration rates of 100-500 µm min^{-1} were observed, depending upon intensity of O_2 demand and O_2 concentration. This phenomenon is related to both protection against O_2 damage and the need for O_2 to supply metabolic energy. It is a special case of bacterial aggregation.

2.6. Morphological Adaptations
Oxygenic photosynthetic cyanobacteria have a special problem in protecting their nitrogenase from O_2. The separation of oxygenic photosynthetic reactions from the N_2-fixing, thick-walled heterocysts is the main solution and has been well described elsewhere (e.g. Stewart 1980). Vesicle clusters within host cells are the site of nitrogenase in most actinorrhizal nodules produced by *Frankia* spp. The isolation of some of these endophytes in N_2-fixing culture has shown that vesicles are also O_2-protective adaptations (Tjepkema *et al*. 1980), since these cultures fix N_2 in air (Torrey *et al*. 1981).

2.7. Symbiotic Structures
Probably the ultimate strategy for protection is the variety of symbiotic structures found in many plants and some animals, within which diazotrophic microorganisms function. The best known are the root nodules of legumes and a smaller number of non-leguminous angiosperms, the coralloid roots of some *Cycadaceae*, the leaf base glands of *Gunnera* spp. and the associations found in Lichens, Bryophytes and Pteridophytes.

Root nodules are of two types with respect to O_2. Those of legumes have low concentrations of O_2 within the air-spaces of the central tissue, whilst non-leguminous nodules do not (Tjepkema 1979). [An exception is found in the nodules of *Parasponia* spp. The gross morphology of these nodules is typical of non-leguminous angiosperms, but the endophyte is a *Rhizobium* sp. In this case the internal pO_2 is also low, as in legume nodules, (Tjepkema, Cartica 1982)]. In legumes and in *Parasponia* spp., the inner cortex provides a barrier to the inward diffusion of O_2, generating a zone in the central tissue of the nodule depleted in O_2 because of consumption of O_2 by the masses of endophyte. Sinclair & Goudrian (1981) modelled various aspects of O_2 and solute transport in legume nodules. In the non-legumes such as *Myrica* spp. and *Alnus* spp, all nodule tissues are penetrated by gas-filled intercellular spaces (Tjepkema 1979). In the endophyte-filled cells the vesicle clusters appear to provide sufficent protection against O_2. It will be interesting to observe the nature of O_2 restriction in nodules of *Casuarina* spp. where vesicle clusters are not present.

Detached soybean nodules are less active than when on the roots of intact plants. Further, they respond with increased N_2 fixation when exposed to pO_2 up to 0.5 atm., whereas attached nodules usually have greatest activity near 0.2 atm O_2. Damage to the spongy outer cortex, including the lenticel structures, during detachment of the nodules from the roots, or due to diminished turgor in the spongy tissues (Ralston, Imsande 1982) may be responsible for these effects, by altering the normal pathway of access of O_2 into the nodules.

3. DEALING WITH THE PROBLEMS OF LIMITED SUPPLY OF O_2

The benefits of restricted O_2 concentration for preservation of nitrogenase are accompanied by the increased probability that the resulting diminished diffusive fluxes of O_2 will limit the availability of ATP and reductant for nitrogenase activity of aerobic or microaerobic diazotrophic microorganisms.

3.1. Aggregates of bacteria

In natural environments N_2 fixation probably occurs frequently in aggregates of diazotrophs or in aggregates containing diazotrophs and other microorganisms. I have calculated diffusive fluxes of O_2 for some hypothetical situations and applied the results to some known examples. Aerobic diazotrophs in aggregates face two immediate problems if the surrounding environment is depleted in O_2: (a) the demand quickly exhausts O_2 within the aggregate and (b) the space-filling of the bacteria themselves limits access of O_2. Extracellular deposits consisting of highly hydrated polysaccharides within which solutes were free to diffuse, may increase the separation between the organisms and promote O_2 permeation. This would be especially so if there were structural capsules of H_2O-containing gels, which produced regular 'porosity' in the aggregates. Their presence increased the calculated O_2 consumption of an aggregate of 1000 *A. vinelandii*, enabling more of the bacteria to experience conditions in which nitrogenase could function (Fig. 1).

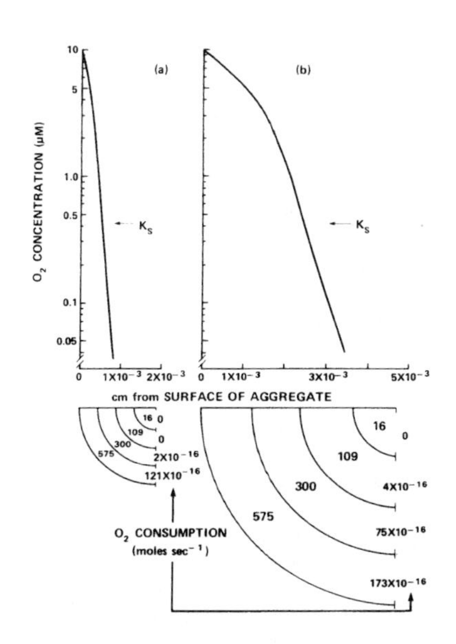

FIGURE 1. The calculated distribution of O_2 in aggregates of 1000 naked (a) or capsulated (b) cells of *Azotobacter vinelandii* suspended in a solution containing 10 µM O_2. In each case the upper diagram shows the profile of O_2 concentration and the lower quadrants show the corresponding distributions of cell numbers within zones of the spherical aggregates, each zone having a thickness of r/4. The rates of consumption of O_2 in each zone are also illustrated. The non-capsulate cells were considered to occupy 50% of the volume of the aggregate and the capsules were considered to be equal in thickness to the thickness of the cells. The capsulate cells then occupied only about 3% of the aggregate space. The increased spacing due to the presence of capsules resulted in deeper penetration of O_2 and the respiration of the total aggregate was doubled.

Recently, Hill *et al.* (1983) showed that the activity of nitrogenase of *Klebsiella pneumoniae* was stimulated relative to anaerobic rates, when O_2 was supplied at a steady concentration of about 30 nM. Considering aggregates of 1000 naked or capsulate cells it can be shown that, when ambient O_2 is limiting and most of the aggregate is anaerobic, the presence of capsules allows penetration of O_2, so that a significant proportion of the aggregate could benefit from the increased efficiency associated with microaerobic N_2 fixation. In the same conditions an aggregate of the same number of naked cells would be almost entirely anaerobic (Bergersen unpublished).

3.2 Adaptations to low concentrations of O_2

3.2.1. High affinity O_2 consumption. In O_2-limited conditions, diazotrophs develop the capacity to utilize O_2 at very low concentrations. For example, Bergersen, Turner (1980) found apparent K_s values of <1 μM for terminal oxidase systems of bacteria from O_2-limited continuous cultures of *Azotobacter vinelandii*, *Azospirillum brasilense*, *Klebsiella pneumoniae* and *Rhizobium* spp. For cultured *A. brasilense* and the *Rhizobium* spp., and for bacteroids from soybean and cowpea nodules the values were extremely low (2-6 nM) and the kinetics sometimes complex. Similar high affinity systems have been implicated in the oxidation of H_2 by soybean bacteroids (Emerich *et al*. 1980a), the pathways involving cytochromes *b* and *c*, non-haem-Fe-proteins and ubiquinone (Emerich *et al*. 1980b; Eisbrenner, Evans 1982a,b; Eisbrenner *et al*. 1982; cf. also O'Brian, Maier, 1982 for free-living *R. japonicum*) and are at least in part membrane-bound (Mutaftschiev *et al*. 1983). Keister *et al*. (1983) described the cytochromes of bacteroids of strain 61A76, differing from an earlier description of a different strain of *R. japonicum* (Appleby 1969); therefore the details of adaptation to low concentrations of O_2 may vary. There may also be differences in affinity for O_2 between oxidase systems of bacteroids from different legumes (cf. soybeans, peas and French beans in Bergersen 1980, Uheda, Syōno 1982, Trinchant *et al*. 1981, 1983). Utilization of various carbon substrates may depend upon the O_2 concentration to which bacteroids are exposed. This has been suggested by Trinchant *et al*. (1981, 1983) using soybean and French bean nodules. Production of free haem by *R. japonicum* bacteroids appears to be a specific adaptation to the near-anaerobic conditions within nodule cells (Keithly, Nadler 1983).

3.2.2. Morphological Adaptations. Post *et al*. (1982) have recently re-examined controversial earlier observations of internal membrane systems in N_2-fixing Azotobacter. In this new work, utilizing N_2 or NH_4^+-grown carbon-limited cells from continuous cultures, the N_2 fixing cells grown at low concentrations of O_2 had the greatest proportion of intracytoplasmic membrane surface. Possible artifactual origin of the intracytoplasmic vesicles was eliminated by rapid cryofixation. The roles of these structure in adaptation to low concentrations of O_2 remain to be determined.

3.2.3. Adaptations in the symbiotic host. The low concentrations of free dissolved O_2 in cells of the legume root nodule (Tjepkema 1979) result in many adaptations. For example, assimilation of fixed N_2, must also function at low concentrations of O_2. Rainbird, Atkins (1981) found that urate oxidase, an important component of assimilatory pathways in cowpea nodules, had a K_m (O_2) of 29 μM. This concentration suggests that the likely location of this enzyme is in uninfected interstitial cells of these nodules rather than in the infected cells where observed *in vivo* oxygenation of leghaemoglobin shows that cytoplasmic concentrations of free O_2 would be at least two orders of magnitude lower. The near-anaerobic conditions in infected cells of nodules results in the production of acetaldehyde and ethanol by cytosol enzymes. These products can be used by bacteroids (DeVries *et al*. 1980; Petersen, La Rue 1981, 1982; Tajima, La Rue 1983).

3.3. Haemoglobins

3.3.1. In vivo functions. No consideration of O_2 in relation to N_2 fixation would be complete without reference to the functions of these haemoproteins which are primarily concerned with the distribution of O_2 within bacteroid-filled cells of the central tissue of legume nodules and with delivery of O_2 to the bacteroids at 'safe' concentrations and adequate flux.

Recent results (Bergersen, Appleby 1981) showed that, in soybean nodules, leghaemoglobin was present between the bacteroids and the surrounding host membrane at a concentration of 0.3 mM and in host cytoplasm between the envelopes (3 mM). With data on structural dimensions of soybean nodule cells, cell spaces and bacteroids, and of the relationships between bacteroid respiration rates and O_2 concentration (Bergersen, Turner 1980), a model of the delivery of O_2 to the bacteroids has been developed (Bergersen 1982). The *in vivo* concentration of free O_2 throughout the envelope space is maintained just above the K_s of the bacteroid oxidases. Analysis of the distribution of O_2 in nodule cells is more difficult because of the interposition of the various structural elements in the cytoplasm (Stokes unpublished). However an indication is obtained by considering the soybean nodule cell as an aggregate of 30,000 soybean bacteroids, spaced evenly within the water-filled volume of a host cell, and calculating diffusive fluxes of O_2. Without leghaemoglobin, an external supply concentration of at least 10 μM O_2 would be required so that O_2 could penetrate the entire mass at respirable concentrations (i.e. more than 10 nM O_2 at the centre). Without leghaemoglobin but at the supply concentrations indicated by the observed *in vivo* oxygenation of leghaemoglobin (about 15 nM free O_2), virtually all the O_2 would be consumed by the outermost layer of bacteria. However, if a solution of only 1 mM ferrous leghaemoglobin was present throughout the interbacterial spaces, O_2 would be distributed throughout the mass at a concentration of 10-15 nM, a concentration sufficient to support O_2-limited but efficient respiration of all of the bacteria (Fig. 2). This is brought about by a diffusive flux of oxyleghaemoglobin (present at 600 μM) which greatly augments the very small diffusive flux of free dissolved O_2, which is present at only about 10 nM.

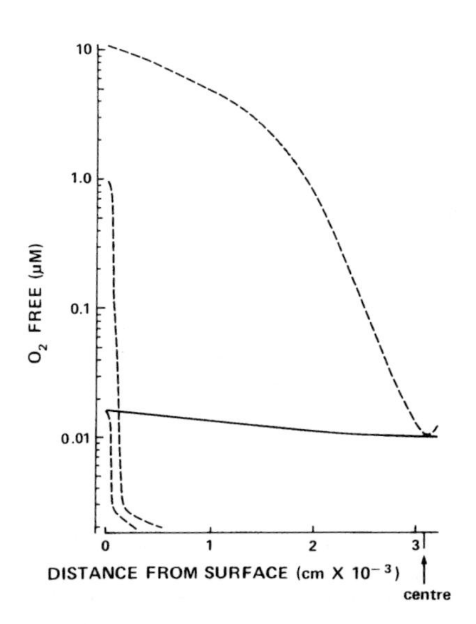

FIGURE 2. The distribution of O_2 in bacteroid-filled soybean nodule cells. The cell was considered as a sphere of radius 3.06 x 10^{-3} cm containing 30,000 bacteroids, which occupied 30% of the cell volume (Bergersen 1982). The O_2 profiles were calculated using data for soybean bacteroids (Bergersen, Turner 1980) and for diffusion of O_2 and leghaemoglobin. The dashed lines are profiles of O_2 concentration with no leghaemoglobin present. Observed *in vivo* oxygenation of leghaemoglobin in soybean nodule tissue, indicates an average of 0.015 μM O_2 (free). The solid line is the profile of O_2 concentration when 1 mM leghaemoglobin is present in solution in the cytoplasm.

For these functions, leghaemoglobin must be maintained in the Fe^{2+} form (Saari, Klucas 1981; Puppo *et al.* 1980,1981,1982). Studies of the structure of the O_2-binding site in leghaemoglobins have continued to explain features associated with the high affinity for O_2 (e.g. Irwin *et al.* 1981; Ollis *et al.* 1981; Appleby *et al.* 1983a). Until recently, O_2-binding kinetics were only established for soybean leghaemoglobins. The extremely high affinity of these (k_2/k_1 = 0.04 x 10^{-6} M for soybean leghaemoglobin *a*) may be an extreme case. Uheda & Syōno (1982) reported lower affinity for O_2 in pea leghaemoglobins (K_{eq} = 0.15-0.21 x 10^{-6} M). This could mean that the prevailing concentrations of dissolved O_2 could differ between the tissues of nodules of different legumes.

3.2.2. Physiological experiments using soybean leghaemoglobin. Experimental systems using soybean leghaemoglobin (Bergersen, Turner 1979) have been used to study the oxyhydrogen reaction in bacteroids (Emerich *et al.* 1980a), derepression of nitrogenase in *K. pneumoniae* (Bergersen *et al.* 1982; Hill *et al.* 1983) and are now being adapted for energetic studies at low concentrations of free O_2, including measurements of steady state values of energy charge, ΔpH and $\Delta\Psi$ (Bergersen unpublished).

3.2.3. Haemoglobins in non-legume root nodules. One of the most interesting developments of the period being reviewed has been the discovery that haemoglobins occur in some non-leguminous nodules. Appleby *et al.* (1983b) described the detection and preliminary characterization of a haemoglobin from nodules of *Parasponia rigida*, containing a *Rhizobium* as endophyte. Tjepkema (1983) described the detection of haemoglobins in slices of non-leguminous actinorrhizal nodules. The study of the properties and roles of these O_2-carrying proteins in nodule function will provide subject matter for future synthesis papers about O_2 in symbiotic tissues.

4. REFERENCES

Appleby CA (1969) Biochim. Pbiophys. Acta 172, 71-87.
Appleby CA Bradbury JH Morris RJ Wittenberg BA Wittenberg JB and Wright PE (1983a) J. biol. Chem. 258, 2254-2259.
Appleby CA Tjepkema JD and Trinick MJ (1983b) Science 220, 951-953.
Arp DJ and Burris RH (1981) Biochem. 20, 2234-2240.
Barak R Nur I Okon Y and Henis Y (1982) J. Bacteriol. 152. 643-649.
Bergersen FJ (1982) Root Nodules of Legumes: Structure and Functions. Research Studies Press/Wiley, Chichester.
Bergersen FJ and Appleby CA (1981) Planta 152, 534-543.
Bergersen FJ and Turner GL (1979) Analyt. Biochem. 96, 165-174.
Bergersen FJ and Turner GL (1980) J. gen. Microbiol. 118, 235-252.
Bergersen FJ Kennedy C and Hill S (1982) J. gen. Microbiol. 128, 909-915.
Bowien B and Schlegel HG (1981) Ann. Rev. Microbiol. 35, 405-452.
Carpenter EJ and Price CC (1979) Science 191, 1278-1280.
Ching T-M Bergersen FJ and Turner GL (1981) Biochim. Biophys. Acta 636, 82-90.
Colbeau A Kelly BC and Vignais PM (1980) J. Bacteriol. 144, 141-148.
De Vries GE InT'Veld P and Kijne JW (1980) Pl. Sci. Lett. 20, 115-123.
Eisbrenner G and Evans HJ (1982a) J. Bacteriol. 149, 1005-1012.
Eisbrenner G and Evans HJ (1982b) Pl. Physiol. 70, 1667-1672.
Eisbrenner G Hickock RE and Evans HJ (1982) Arch. Microbiol. 132, 230-235.
Eisbrenner G Roos P and Bothe H (1981) J. gen. Microbiol. 125, 383-390.
Emerich DW Albrecht SL Russel SA Ching T-M and Evans HJ (1980a) Pl. Physiol. 65, 605-609.

Emerich DW Ruiz-Argüeso T Russell SA and Evans HJ (1980b) Pl. Physiol. 66, 1061-1066.
Fridovich I (1975) Amer. Sci. 63, 54-59.
Gebhardt C Turner GL Dreyfus B and Bergersen FJ (1983) J. gen. Microbiol. (submitted).
Goldberg RB and Hanau R (1980) J. Bacteriol. 141, 745-750.
Haaker H Scherings G and Veeger C (1977) In Newton WE Postgate JR and Rodrigues-Barrueco C eds. Recent Developments in Nitrogen Fixation pp 271-285, Academic Press London.
Hamadi AF and Gallon JR (1981) J. gen. Microbiol. 125, 391-398.
Hills S Kennedy C Kavanagh E Goldberg RB and Hanau R (1981) Nature 290, 424-426.
Hill S Turner GL and Bergersen FJ (1983) J. gen. Microbiol. (submitted).
Hochman A and Burris RH (1981) J. Bacteriol. 147, 492-499.
Irwin MJ Armstrong RS and Wright PE (1981) FEBS Letts. 133, 239-242.
Keister DL Marsh SS El Mokadem NT (1983) Pl. Physiol. 71, 194-196.
Keithly JH and Nadler KD (1983) J. Bacteriol. 154, 838-845.
Kennedy C and Robson RL (1983) Nature 301, 626-628.
Kennedy C Cannon F Cannon M Dixon R Hill S Jensen J Kumar S McLean P Merrick M Robson R and Postgate JR In Gibson AH and Newton WE eds. Current Perspectives in Nitrogen Fixation pp 146-156, Austr. Ac. Sci., Canberra.
Kulakova SM Yakunin AF and Gogotov IN (1982) Prikl. Biokhim. Microbiol. 18, 324-330.
Mackie EJ and Smith GD (1983) FEBS Lett. 156, 108-112.
McClung CR and Patriquin DG (1980) Canad. J. Microbiol. 26, 881-886.
Maier RJ and Merberg DM (1982) J. Bacteriol. 150, 161-167.
Malik KA Jung C Claus D and Schlegel HG (1981) Arch. Microbiol. 129, 254-256.
Merrick M Hill S Hennecke H Hahn M Dixon R and Kennedy C (1982) Mol. gen. Genet. 185, 75-81.
Mutaftschiev S O'Brian MR and Maier RJ (1983) Biochim. Biophys. Acta 722, 372-380.
Nelson LM and Salminen SO (1982) J. Bacteriol. 151, 989-995.
O'Brian MR and Maier RJ (1982) J. Bacteriol. 152, 422-430.
Okon Y Cabmakci L Nur I and Chet I (1980) Microb. Ecol. 6, 277-280.
Ollis DL Wright PE Pope JM and Appleby CA (1981) Biochem. 20, 587-594.
Ow DW and Ausubel FM (1983) Nature 301, 307-313.
Pearson HW Housley R and Williams ST (1982) J. gen. Micribiol. 128, 2073-2080.
Pedrosa FO Stephen M Döbereiner J and Yates MG (1982) J. gen. Microbiol. 128, 161-166.
Peterson JB and LaRue TA (1981) Pl. Physiol. 68, 489-493.
Peterson JB and LaRue TA (1982) J. Bacteriol. 151, 1473-1484.
Pienkos PT Bodmer S and Tabita FR (1983) J. Bacteriol. 153, 182-190.
Post E Golecki JR and Oelze J (1982) Arch. Microbiol. 133, 75-82.
Post E Kleiner D and Oelze J (1983) Arch. Microbiol. 134, 68-72.
Puppo A Rigaud J and Job D (1980) Pl. Sci. Lett. 20, 1-6.
Puppo A Rigaud J and Job D (1981) Pl. Sci. Lett. 22, 353-360.
Puppo A Dimitrijevic L and Rigaud J (1982) Planta 156, 374-379.
Rainbird RM and Atkins CA (1981) Biochim. Biophys. Acta 659, 132-140.
Ralston EJ and Imsande J (1982) J. Exp. Bot. 33, 208-214.
Robson RL and Postgate JR (1980) Ann. Rev. Microbiol. 34, 183-207.
Saari LL and Klucas RV (1981) Abstr. 8th Nth. Amer. Rhizobium Conf. Winnepeg p.56.
Shaw BD (1983) J. gen. Microbiol. 129, 849-857.
Sinclair TR and Goudriaan J (1981) Pl. Physiol. 67, 143-145.
Stewart WDP (1980) Ann. Rev. Microbiol. 34, 497-536.
Tajima S and LaRue TA (1983) Pl. Physiol. (in press).

Tjepkema JD (1979) In Gordon JC Wheeler CT and Perry DA eds, Symbiotic Nitrogen Fixation in The Management of Temperate Forests, pp 175-186, Oregon State
Tjepkema JD (1983) Canad. J. Bot. (in press).
Tjepkema JD and Cartica RJ (1982) Pl. Physiol. 69, 728-733.
Tjepkema JD Ormerod W and Torrey JG (1980) Nature 287, 633-635.
Torrey JG Tjepkema JD Turner GL Bergersen FJ and Gibson AH (1981) Pl. Physiol. 68, 983-984.
Trinchant JC Birot AM and Rigaud J (1981) J. gen. Microbiol. 125, 159-165.
Trinchant JC Birot AM Dennis M and Rigaud J (1983) Arch. Microbiol. (in press).
Uheda E and Syōno K (1982) Pl. Cell. Physiol. (Jap) 23, 85-90.
Upchurch RG and Mortenson LE (1980) J. Bacteriol. 143, 274-284.
Veeger C Haaker H and Laane C (1981) In Gibson AH and Newton WE eds, Current Perspectives in Nitrogen Fixation, pp 101-104, Austr. Ac. Sci., Canberra.
Volpon AGT DePolli H and Döbereiner J (1981) Arch. Microbiol. 128, 371-375.
Wang Y-q He J-w Dai L-f and Li S-h (1981) Acta Bot. Sin. 23, 288-296.
Wang Y-q He J-w and Li S-h (1982) Acta Bot. Sin. 24, 231-240.
Xu J and Li J-g (1981) Microbiology (China), 8, 1-3.
Zumft WC and Nordlund S (1981) FEBS Letts 127, 79-82.

REGULATION OF NITROGEN FIXATION IN PHOTOSYNTHETIC BACTERIA

PAUL W. LUDDEN, SCOTT A. MURRELL, MARK POPE, ROY KANEMOTO, T. E. DOWLING, L. L. SAARI and E. TRIPLETT.

DEPARTMENT OF BIOCHEMISTRY AND THE CENTER FOR THE STUDY OF NITROGEN FIXATION, COLLEGE OF AGRICULTURAL AND LIFE SCIENCES, UNIVERSITY OF WISCONSIN-MADISON, MADISON, WISCONSIN 53706.

INTRODUCTION

Nitrogen fixation by the photosynthetic bacteria was first noted by Kamen and Gest (1949). These workers also observed an inhibition of N_2 uptake when ammonia was added to the cells (Kamen and Gest, 1949; Gest et al., 1950). Nitrogenase activity in crude extracts of R. rubrum exhibits unusual properties when compared to the properties of nitrogenase in extracts of other organisms: high magnesium concentration is required for optimal activity (Burns and Bulen, 1966) and low levels of manganese greatly stimulate activity (Ludden and Burris, 1976); the time course of activity in extracts is nonlinear (Munson and Burris, 1969); the membrane fragments (chromatophores) were found to inhibit nitrogenase activity (Burns and Bulen, 1966). These observations were explained when it was discovered that the Fe protein of nitrogenase from Rhodospirillum rubrum is isolated in inactive form (Ludden and Burris, 1976) and that an oxygen labile, membrane bound activating enzyme was present in extracts (Ludden and Burris, 1976; Nordlund et al., 1977). The activating enzyme requires MgATP and free divalent metal (Guth and Burris, 1983). A covalently bound modifying group (MG) consisting of phosphate, pentose and adenine was found attached to the inactive form of the Fe protein (Ludden and Burris, 1978).

The activating enzyme removes at least the adenine compound during activation (Ludden and Burris, 1979). And Gotto and Yoch (1983) have reported that the entire modifying group is removed. Although the Fe protein is a dimer of identical subunits (Ludden, et al., 1982), a single modifying group unit is attached to the protein. The modified subunit can be separated from the unmodified subunit on SDS gels (Ludden et al., 1982) and both the phosphate and the adenine can be demonstrated to be on the upper subunit by radioactive labeling (Ludden, et al., 1982; Nordlund and Ludden, 1982).

The Fe protein is obtained in the inactive form when grown with glutamate or N_2 as the N source. If cells are grown with limiting ammonia as the nitrogen source, the Fe protein is obtained in the active, unmodified form (Carithers et al., 1979). This active form of the Fe protein exhibits a single subunit on SDS gels. The reason for this difference is not known. While much of the work on switch-off has been done with R. rubrum, similar systems are found in Rhodopseudomonas capsulata (Hallenbeck et al., 1982; Michalski et al., 1983) and Rhodopseudomonas palustras (Zumft and Castillo, 1973).

Activation of the native, reduced Fe protein requires MgATP, free divalent metal ion and activating enzyme. If the Fe protein is oxidized before activation, MgATP is no longer required. A non-physiological method of activating the protein is to heat the protein at 50°C or more at pH 8.0 or above. This results in removal of the modifying group as a unit (Dowling et al., 1982). A model for activation/inactivation is shown in Fig. 1.

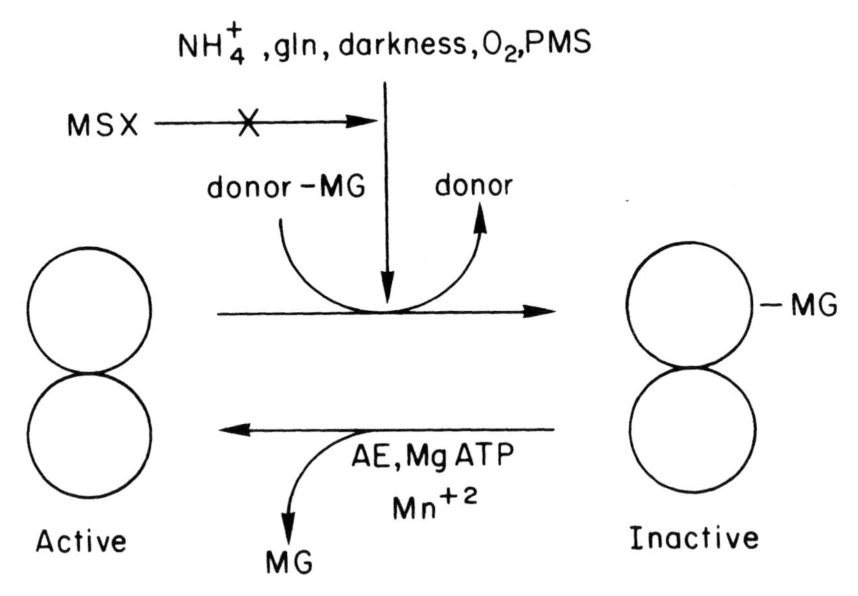

Purification of the Modifying Group

A major goal of our research has been to identify the modifying group attached to the inactive Fe protein. The heat method of modifying group removal was chosen because it requires no additions to the purified Fe protein; the conditions of the activation can be reproducibly set and the components of the modifying group released as a unit. In most cases, ^{32}P or 8-[^3H]-adenine labeled Fe protein was mixed with a large amount of unlabeled, purified protein. The protein was desalted on a G-25 column to remove any contaminating small molecules before heating. The desalted protein was concentrated to a concentration of at least 15 mg/ml, protein concentrations for heat activation were usually 50 mg/ml or more. The protein was heated for 3 hours at 60°C under N_2 at pH 8.5. The thermally released modifying group (MG) was separated from protein on a G-25 column or on a boronate affinity column. The boronate column binds molecules with cis hydroxyls. The isolated small molecule fraction was further purified by ion exchange high pressure liquid chromatography (HPLC) (Fig. 2). Some variation in elution profile was observed but a major peak was always seen. This fraction (peak 3) was considered to be MG. Peak 3 contained both ^{32}P and ^3H label and, under some conditions, peak 3 would break down the yield the other peaks observed in the HPLC elution profile. Peak 3 was concentrated on the boronate affinity column for further study.

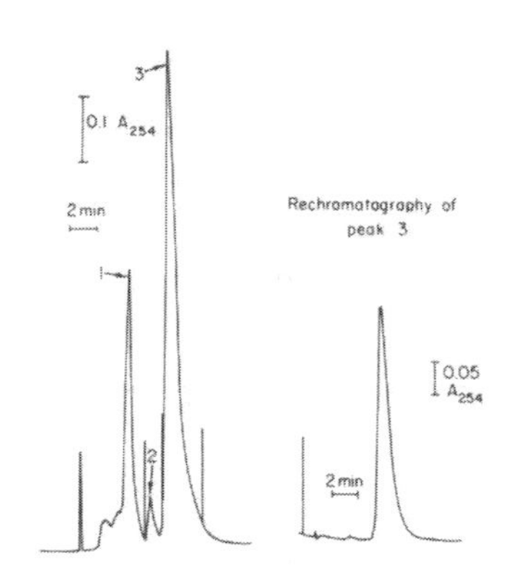

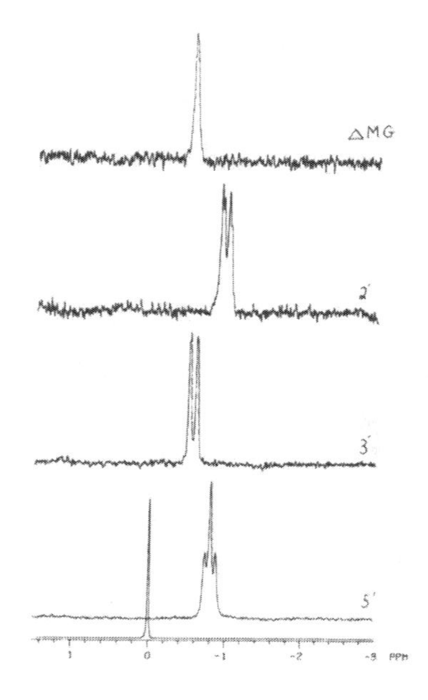

FIGURE 2. (left) Elution profile of heat released MG on ion exchange HPLC column.

FIGURE 4. (right) ^{31}P NMR of isolated modifying group and 2', 3' and 5' AMP standards.

Peak 3 is clearly not AMP as indicated by gel filtration chromatography, HPLC on an AX 300 ion exchange column and gas chromotography of after silation. In addition NMR and mass spectra will be discussed below which indicate that the molecule is not AMP. The unusual chromatographic properties do not appear to be the result of heating as it is not possible to generate peak 3 by heating AMP under the conditions used to remove MG from the protein.

Characterization of MG

The modifying group consists of phosphate, pentose and adenine; chemical assays show these to be present in a 1:1:1 ratio with respect to Fe protein (Ludden and Burris, 1978; Ludden et al., 1982). The original linkage model proposed a protein-pentose-phosphate-adenine linkage based on susceptibility to AE, diesterases and monoesterases (Ludden and Burris, 1979); the adenine has always been referred to as adenine like because the UV spectrum of the group on the protein showed a λ_{max} at 268 nm.

Some linkage information comes from labeling studies. Both 2-[^{3}H]-adenine and 8-[^{3}H]-adenine can be fed to cells and incorporated into the MG _in vivo_ (Nordlund and Ludden, 1982). Thus there are no links to these positions on the adenine ring structure or the tritions would have been lost. Retention of the 2 position trition absolutely rules out any guanosine derivative as the modifying group. Positions 4, 5 and 6 of the adenine ring have no replaceable protons and the 6 amino must be free in order for Yuki's fluorescence assay for adenine to work (Yuki et al., 1973). The linkages to the adenine must be through the ring nitrogens.

Further information about the modifying group is obtained by nuclear magnetic resonance (NMR studies). The phosphorus of the modifying group on the protein exhibits a single peak at -12 ppm (PO_4 as standard with apparent signal splitting of 10 hertz (Hz) (Fig 3). The signal is observed both with the native, reduced protein and with the oxygen denatured protein. This indicates that the protein conformation does not affect the ^{31}P NMR signal. Because the signal is observed from the native reduced protein, it can be argued that the iron sulfur center and the MG are not close (less than 10 Å), otherwise the paramagnetic iron sulfur center would result in broadening of the ^{31}P signal. The position of the signal is unusual for a single phosphorus. There is a minor signal at approximately -11 ppm which may be the MG in a different conformation or a rearrangement of the MG molecule.

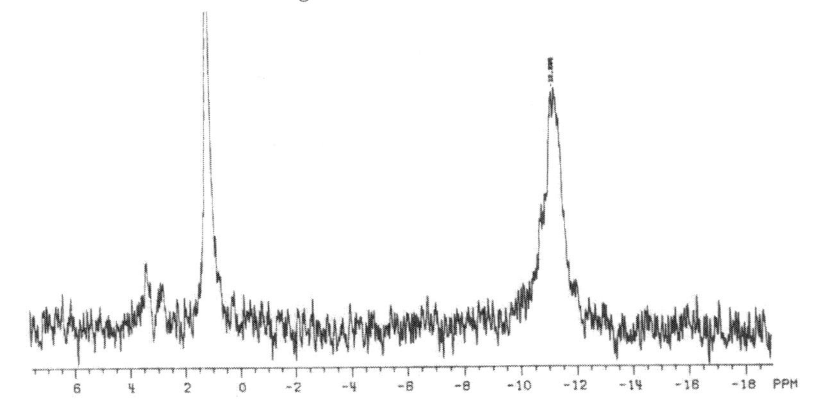

FIGURE 3. ^{31}P NMR of Fe protein. The signal at 1.0 ppm is due to KH_2PO_4 added to the sample as a standard. The signal at -12.2 ppm is due to the P of the modifying group.

The ^{31}P NMR spectrum of the isolated MG is shown in figure 4. After removal from the protein and purification, the MG signal is moved downfield to -0.5 ppm, a region characteristic of phosphomonoesters. The dramatic shift in MG signal position on removel from the protein indicates that the linkage of the phosphate group has changed; this argues for a linkage between phosphate group and the protein in contrast to our previous model. In addition, the splitting of the phosphorous signal is lost on removal of MG from the protein, indicating that the proton or other atom responsible for the splitting is on the protein.

The very minor signal splitting in the ^{31}P NMR spectrum of the isolated MG is important when compared to the ^{31}P NMR of 2', 3' or 5'-AMP. In each of these molecules, the phosphate signal is split by 6 Hz or more. If the phosphorus on the modifying group were on the 5' position, splitting by the 5' protons would be expected. Once again, the ^{31}P NMR of the isolated modifying group rules out any 5' compound which has 5' protons. These include 5' AMP, NAD, NADP and ADPR. Several of these compounds are also ruled out by the presence of a single phosphate species on the molecule.

The proton spectrum of the isolated modifying group in D_2O is shown in figure 5. Peaks characteristic of the 2 and 8 protons of the adenine ring are seen at 8.4 and 8.5 ppm respectively. A peak characteristic of the anomeric proton is seen at 6.1 ppm. The presence of peaks due to 2 and 8 protons confirm the results obtained with ^3H-labeled adenine. The presence of the anomeric proton demonstrates that the modifying group is a single unit: the presence of adenine is confirmed by labeling and NMR; the presence of phosphate is confirmed by labeling and NMR; the presence of pentose is now confirmed by NMR. All the components are also now confirmed by chemical analysis of the isolated modifying group. The linkage of all modifying group components has always been an assumption and is now confirmed. The isolated MG is not soluble in DMSO, so all spectra have been taken in D_2O; thus it is not possible to see the amino or hydroxyl protons, which exchange in D_2O. However, the carbon bound protons of the carbohydrate can be seen. While not definitive, peaks characteristic of 2', 3', 4' and 5' protons can be observed in the region from 3.5 to 4.5 ppm. In addition, an unusual peak is seen at 1.4 ppm and this signal is coupled to another signal, indicating that it is a part of a signal due to the modifying group molecule. Further decoupling and 2-dimensional NMR experiments should confirm all assignments.

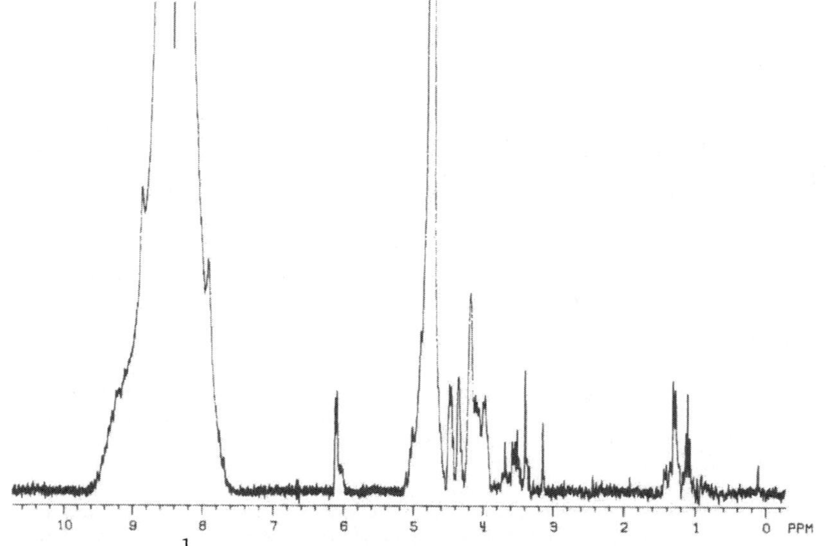

FIGURE 5. The ^1H NMR of isolated MG in D_2O. The large peak at 8.5 is due to residual formate used in elution of MG from the borate column; the large peak at 4.9 ppm is due to HDO.

Regulation of Nitrogenase Activity *in vivo*

The feedback regulation of nitrogenase activity in whole cells of *R. rubrum* by fixed nitrogen that was first observed by Kamen and Gest (1949) has been the subject of numerous investigations. Both nitrogen nutrition (Neilson et al., 1975; Carithers et al., 1979; Sweet and Burris, 1980; Alef et al., 1983) and light intensity (Yoch and Gotto, 1983) are known to have an effect on the NH_4^+-dependent switch-off (the term given to this effect by Zumft).

A number of other factors have been found to affect switch-off in *R. rubrum*. We have developed a method for rapidly extracting small amounts of cells and analyzing the Fe protein in the extract for its subunit composition and incorporation of ^{32}P or ^{3}H-adenine label into the upper subunit. Because the appearance of upper subunit and incorporation of label are correlated with inactivation of Fe protein, this method allows assessment of the activity state of the Fe protein *in vivo*. Using this method, it has been found that darkness results in loss of nitrogenase activity and Fe protein modification even in the absence of added N. When cells are reilluminated, whole cell activity increases rapidly and upper subunit is lost. Because light/dark cycles are easily imposed on the cells with no residual effect (unlike addition of NH_4^+ etc.), this observation allows repeated cycles of activation/inactivation to be imposed on the cells. Because at least three cycles of Fe protein modification/demodification can be observed in much less than the turnover time of the Fe protein, we conclude that the entire modifying group is removed during activation *in vivo*, either in a single step or in a sequence of steps. This conclusion is drawn because the Fe protein has two identical subunits, presumably with only two available sites for modification.

It might be expected that darkness would lead to a decrease in ATP pool or energy charge. Substantial changes in ATP concentration or energy charge are not observed, even though whole cell activity decreases immediately in the dark. Further evidence comes from the treatment of cells with phenazine methosulfate (PMS) which does not affect rates of photophosphorylation in *R. rubrum*. (Baltsheffsky, 1978). PMS causes immediate loss of whole cell nitrogenase activity and Fe protein is modified in the same time scale as it is during dark or NH_4^+-induced switch-off. The ATP pool does not appear to be the decisive factor in regulation of switch-off *in vivo*. Figure 1 shows a model of events observed *in vivo*. As seen in the model, fixed nitrogen sources such as NH_4^+ or glutamine lead to switch-off as do the treatment of the cells with oxygen or PMS and lack of illumination. The glutamine synthetase inhibitor, methionine sulfoximine, has been shown to inhibit switch-off nitrogenase *in vivo* and it also inhibits switch-off by NH_4^+ or PMS in our experiments. (More data on the switch-off effect *in vivo* can be found in the poster abstract of Kanemoto et al. in these proceedings.) Further experiments on pool sizes of different metabolites may lead to an understanding of the events leading to modification *in vivo*.

REFERENCES

Alef K, Arp DJ and Zumft WG (1981) Arch Microbiol. 130, 138-142.
Baltscheffsky M (1978) In Clayton RK and Sistrom WR, eds, The Photosynthetic Bacteria, pp. 595-614, Plenum Press, New York.
Burns RC and Bulen WA (1966) Arch. Biochem. Biophys. 113, 461-463.
Carithers RP, Yoch DC and Arnon DI (1979) J. Bacteriol. 137, 779-789.
Dowling TE, Preston GG and Ludden PW (1982) J. Biol. Chem. 257, 13987-13992.
Gest H, Kamen MD and Bregoff HM (1950) J. Biol. Chem. 182, 153-170.
Gotto JW and Yoch DC (1982) J. Biol. Chem. 257, 2868-2873.
Guth JH and Burris RH (1983) Biochem. J. 213, 741-749.
Hallenbeck PC, Meyer CM and Vignais PM (1982) J. Bacteriol. 149, 708-717.
Kamen MD and Gest H (1949) Science 109, 560.
Munson TO and Burris RH (1969) J. Bacteriol. 97, 1093-1098.
Ludden PW and Burris RH (1976) Science 194, 424-426.
Ludden PW and Burris RH (1978) Biochem. J. 175, 251-259.
Ludden PW and Burris RH (1979) Proc. Natl. Acad. Sci. USA 76, 6201-6205.
Ludden PW, Preston GG and Dowling TE (1982) Biochem. J. 203, 663-668.
Michalski WP, Nicholas DJD and Vignais PM (1983) Biochim. Biophys. Acta 743, 136-148.
Neilson AH and Nordlund S (1975) J. Gen. Microbiol. 91, 53-62.
Nordlund S, Eriksson and Baltscheffsky H (1977) Biochim. Biophys. Acta 462, 187-195.
Nordlund S and Ludden P (1983) Biochem. J. 209, 881-884.
Sweet WJ and Burris RH (1982) Biochim. Biophys. Acta 680, 17-21.
Yoch DC and Gotto JW (1982) J. Bacteriol. 151, 800-806.
Zumft WG and Castillo F (1978) Arch. Microbiol. 117, 53-60.

ACKNOWLEDGEMENTS

This investigation was supported by the College of Agricultural and Life Sciences at the University of Wisconsin-Madison and by USDA-SEA grant #81-CRCR-1-0703 and NSF grant PCM 8302630.

*E. Triplett present address: Department of Plant Pathology, University of California-Riverside, Riverside, California 92521.

ENERGY GENERATION AND HYDROGEN METABOLISM IN RHIZOBIUM

A.H. STOUTHAMER
Biological Laboratory, Vrije Universiteit, de Boelelaan 1087, Amsterdam

1. INTRODUCTION

Biological dinitrogen fixation is ATP- and reductant-dependent; between 12 and 30 mol of ATP are required per mol of dinitrogen reduced (Hill, 1976). Rhizobia are aerobic organisms in which ATP and reductant are generated during the oxidation of a carbon source. In the Rhizobium-legume symbiosis the supply of photosynthetate is one of the major factors that limits dinitrogen fixation (Hardy and Havelka, 1976). A high efficiency of energy coupling within bacteroids would minimise the amount of photosynthetate to be oxidized to supply the ATP and reductant necessary for dinitrogen fixation. For these reasons knowledge of energy metabolism in rhizobia is of great importance. Energy generation has been studied intensively in many bacteria. The methods for the characterization of the efficiency of energy generation are the study of the complexity of the respiratory chain, the stoichiometry of proton translocation and the measurement of molar growth yields (Stouthamer, 1977a, Jones, 1977). Only few studies have been devoted to rhizobia. These studies will be discussed and their results will be compared with those of more detailed studies in some other bacteria.

2. Composition of the respiratory chain and proton translocation

In all Rhizobia studied b-, c- and a-type cytochromes are present and the respiratory chain is branched in the direction of oxygen. As terminal oxidases cytochrome aa_3 and cytochrome o are present. The branching point is located at cytochrome b. In R. trifolii the scheme shown in Fig. 1 was proposed for the respiratory chain of cells grown in the chemostat under carbon-limited conditions (de Hollander and Stouthamer, 1980; de Hollander, 1981). In this scheme b-type cytochromes with absorption maxima at 555 and 561 nm are distinguished. The latter b-type cytochrome should be identical with the CO-reactive cytochrome o (de Hollander et al., 1979). The kinetics of oxygen uptake was consistent with the presence of two oxidases. These two oxidases have K_m values for oxygen which are largely different. In the presence of 45µM cyanide, which inhibits respiration by about 80%, respiration can perfectly be described by a one-component system. The system with the high K_m is completely inhibited by cyanide. Difference spectra of electron transport particles reduced by dithionite measured against aerated particles in the presence of NADH were recorded at 77K. Cyanide inhibition caused a strong decrease in the peak heights of cytochromes a and c, indicating a fairly complete reduction of these cytochromes. In this

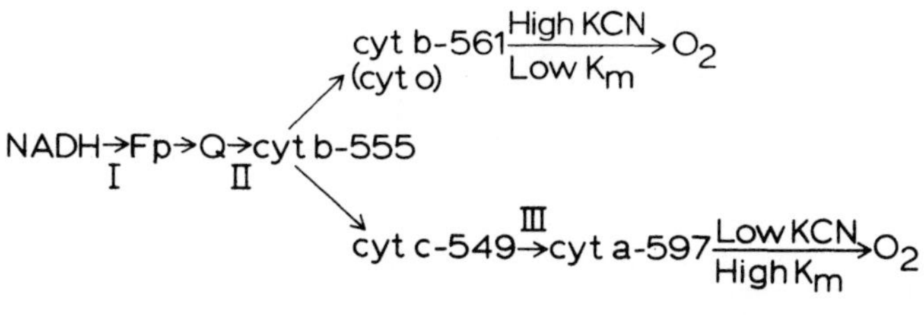

Figure 1. Electron transport chain of R. trifolii. Fp, flavoprotein; Q, ubiquinone. The classical coupling sites are indicated by Roman numerals. Data from de Hollander and Stouthamer (1980).

way the oxidase with the high K_m can be identified as cytochrome aa_3. The stoichiometries of proton translocation and the influence of cyanide on these stoichiometries are shown in Table 1. For endogenous substrates high $\rightarrow H^+/O$ ratios are found indicating a high efficiency of energy conservation. When respiration via cytochrome aa_3 is blocked by cyanide the $\rightarrow H^+/O$ ratios for endogenous substrates and for succinate are about 2 less than in the absence of cyanide. This is an indication that cytochrome aa_3 in R. trifolii is proton-translocating as found before in P. denitrificans (van Verseveld et al., 1981). These data indicate that the cyanide-sensitive branch of the respiratory chain is more efficient in energy generation.

Table 1. H^+/O ratios for cells of Rhizobium trifolii grown in batch culture.

Substrate	KCN absent	KCN present (0.5 mM)
Endogenous	8.6	6.6
Succinate, rotenone added	6.6	4.2

Since the overall stoichiometries of proton translocation in R. trifolii and P. denitrificans are the same it is very likely that the scheme for proton translocation derived for P. denitrificans (Fig. 2) is also valid for R. trifolii. In the scheme for proton translocation in P. denitrificans 3, 4 and 2 protons are translocated by site 1, 2 and 3 respectively. The number of charges separated across the membrane during electron transport are 3, 2 and 4 for the various segments of the respiratory chain. Since the charge separation is proportional to the efficiency of oxidative phosphorylation, this implies that the three sites have a different $P/2e^-$ ratio as reported before for mitochondria (Wikström and Krab, 1980). P. denitrificans is also able to denitrify and to utilize hydrogen. The pathways for electron transfer to nitrate, nitrite and nitrous oxide and for hydrogen oxidation are included in Fig. 2 (Boogerd et al., 1981, 1983). The charge separation for electron transfer from NADH, succinate and H_2 to oxygen and nitrogenous oxides are included in Table 2. The data indicate that electron transfer from NADH or H_2 to nitrogenous oxides has a charge separation which is 55-71% resp. 33-50% from that to oxygen, depending on the oxidase used. These data have been confirmed by measurements of molar growth yields (Stouthamer et al., 1982; Boogerd et al., 1983).

Nitrate reduction is possible in a number of Rhizobia (Zablotowicz et al., 1978; Daniel et al., 1980). Most of the slow-growing rhizobia and R. meliloti strains were found to denitrify (Daniel et al., 1982). Hydrogen oxidation is present in about 20% of the R. japonicum strains (Lim, 1978; Carter et al., 1978; Maier et al., 1979ab) and in very few R. leguminosarum strains (Ruiz-Argueso et al., 1979; Dixon, 1972; Nelson and Child, 1981). A number of R. japonicum strains that are able to oxidize hydrogen can grow chemolithotrophically (Hanus et al., 1979). On basis of the available evidence it is expected that the scheme given in Fig. 2 for P. denitrificans will be general applicable. The electron transport chain for hydrogen oxidation in free-living cells and in bacteroids of R. japonicum is in accordance with this scheme (Eisbrenner and Evans, 1982a; Eisbrenner et al., 1982; O'Brian and Maier, 1982). In the schemes of Fig. 1 and 2 one b-type cytochrome is given. In reality more b- and c-type cytochromes are present. The number and the properties of b- and c-type cytochromes was studied by a combined spectral and potentiometric analysis on cells of R. trifolii grown under carbon-limited conditions in the chemostat by a method developed by van Wielink et al. (1982). In this way at least three c-type cytochromes and four b-type cytochromes are found in R. trifolii (de Hollander, 1981). The midpoint potentials, the peak maxima and the amounts of these components are given in Fig. 3. A number of these b-type cytochromes may function in the site II region of the respiratory chain as found before for mitochondria (Wikström and Krab, 1980; van Wielink et al., 1982). Some other

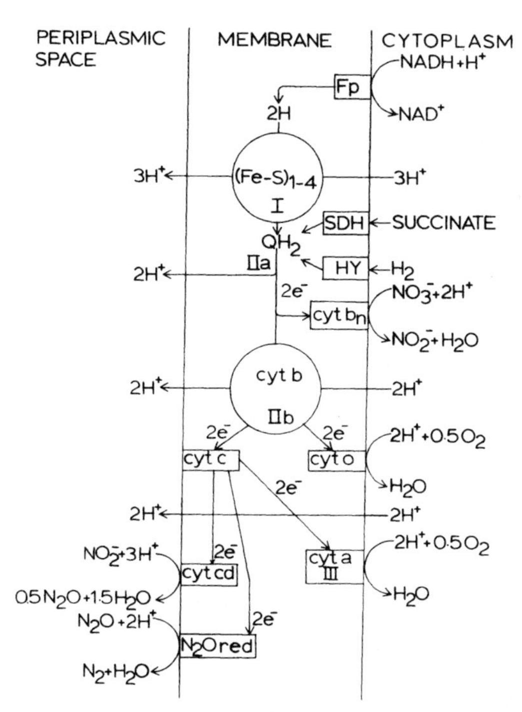

FIGURE 2. Simplified scheme for aerobic electron transport and proton translocation in P. denitrificans. Fp, flavoprotein; QH_2, ubiquinol; Hy, hydrogenase; SDH, succinate dehydrogenase.

TABLE 2. Charge separation for electron transfer from NADH, succinate and hydrogen to oxygen and various nitrogenous oxides.

Hydrogen acceptor	Charge separation/$2e^-$	
	NADH	H_2 or succinate
O_2 (cytochrome aa_3)	9	6
O_2 (cytochrome o)	7	4
Nitrate	5	2
Nitrite	5	2
Nitrous oxide	5	2

may have special functions, e.g. a special b-type cytochrome is involved in nitrate reduction (Stouthamer et al., 1980). Recently a special b-type cytochrome (designated as component 559-H_2) has been identified which is involved in hydrogen oxidation in bacteroids of R. japonicum (Eisbrenner and Evans, 1982ab, Eisbrenner et al., 1982). Consequently, the schemes given in Fig. 1 and 2 are simplifications. The composition of the respiratory chain is strongly influenced by the conditions under which the organism is grown. In R. trifolii oxygen-limitation in the chemostat leads to a decrease in the intensity of the absorption peak of a-type cytochromes (Fig. 4). Consequently respiration by oxygen-limited cells. Therefore respiration is more resistant to KCN and the efficiency of energy generation is less in oxygen-limited cells than in carbon-limited cells (Fig. 1). In accordance with this the $\rightarrow H^+/O$ ratio for oxidation of endogenous substrate is lower in oxygen-limited cells (5.8) than

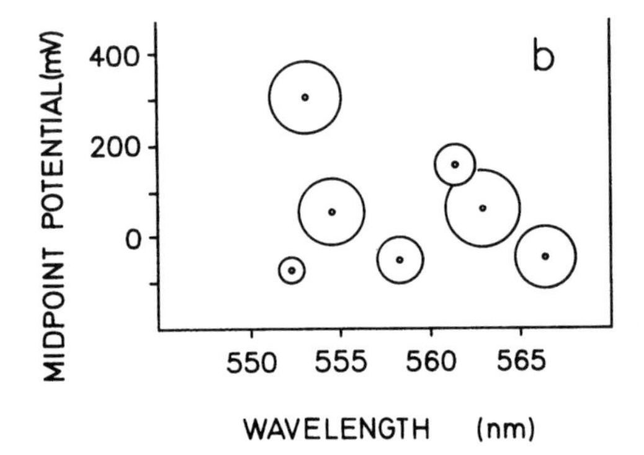

FIGURE 3. Diagrammetical representation of the results of the combined analysis of spectral data for membranes from R. trifolii grown under carbon-limitation. The center of the circle indicates the position of the peak maximum and the midpoint potential. The area of the circles is in proportion to the ralative abundance of the component in the spectrum. Data from de Hollander (1981).

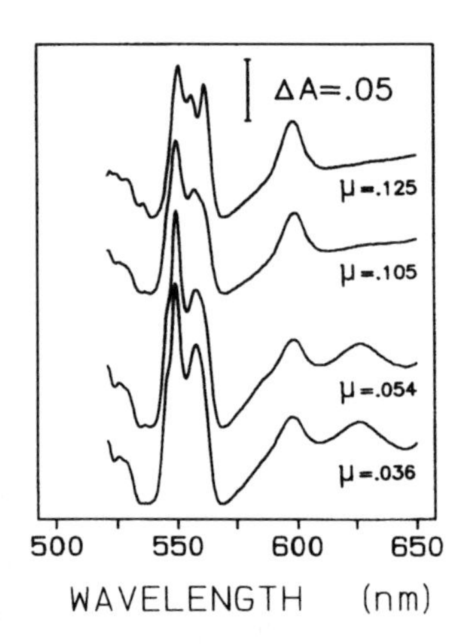

FIGURE 4. Influence of the specific growth rate (μ) on dithionite reduced minus oxidized difference spectra of R. trifolii cells grown under oxygen limitation. Spectra were recorded at 77K.

in carbon-limited cells (de Hollander, 1981). A decrease in the concentration of cytochrome a by limited availability of oxygen has also been reported for R. japonicum and R. leguminosarum (Tuzimura and Watanabe, 1964; Kretovich et al., 1973; Avissar and Nadler, 1978; Appleby, 1969ab). In bacteroids cytochrome aa_3 is generally not present. This means that growth under oxygen-limitation in the chemostat can mimick part of the changes occurring during the differentiation of bacteria to bacteroids. In R. trifolii cells grown under oxygen-limited conditions in the chemostat the composition of the respiratory is influenced by the specific growth rate. In cells grown at low dilution rate cytochrome d is present (Fig. 4) (de Hollander, 1981). This cytochrome was detected before in R. leguminosarum grown under semi-anaerobic conditions (Kretovich et al., 1973; Ratcliffe et al., 1983a) but not in bacteroids.

3. Molar growth yields

Molar growth yields have been measured in chemostat cultures of R. trifolii (de Hollander et al., 1979; de Hollander, 1981) and R. leguminosarum (Ratcliffe et al., 1983ab). The results are presented in Table 3. In the experiments with R. trifolii asparagine was also present in the medium. The interpretation of the results was complicated by polysaccharide formation. The equation derived for growth was: $1C_6H_{14}O_6$ (mannitol) + $bC_4H_8O_3N_2$ (asparagine) + $cO_2 \rightarrow dC_6H_{11.3}O_{3.21}N_{1.24}$ (cell material) + $eC_6H_{9.01}O_{5.09}$ (polysaccharide) + fH_2O + gCO_2 + hNH_3. It was shown that the value of h was dependent on the asparagin concentration and on μ. At high asparagin concentrations a positive value of h is found, indicating that ammonia is produced. At lower asparagine concentrations a negative value of h is found, indicating that ammonia is incorporated into

TABLE 3. Molar growth yields for substrate and oxygen after correction for maintenance energy (Y(sub)MAX resp. $Y(O_2)MAX$) for chemostat cultures of R. trifolii and R. leguminosarum.

Organism	Substrate	Limiting factor	Y(sub)MAX)	$Y(O_2)MAX$	Reference
R. trifolii	mannitol	mannitol	67.6	31.7	De Hollander et al., 1979
	mannitol	oxygen	52.8	22.8	De Hollander, 1981
R. legunino-sarum	glucose	glucose	80.3	25.7	Ratcliffe et al., 1983a
	glucose	glutamine	95.7	30.2	Ratcliffe et al., 1983b

cell material. These data indicate that asparagine functions as nitrogen source and as carbon source (de Hollander et al., 1979). The growth yields indicate that the efficiency of energy conversion in R. trifolii is high: a P/O ratio between 2.0 and 2.5 and a Y(ATP)MAX of about 8 g/mole were found. Ratcliffe et al. (1983a) conclude from their data that the efficiency of energy conservation in R. leguminosarum is low. However it must be realized that Y(sub)MAX and $Y(O_2)MAX$ are dependent on both Y(ATP)MAX and the P/O ratio (Stouthamer, 1977a). For R. leguminosarum a Y(ATP)MAX of about 15 was assumed (Ratcliffe et al., 1983a). However the real values for aerobic bacteria are much lower (de Hollander et al., 1979; Stouthamer et al., 1982). On basis of the growth yields, the $\rightarrow H^+/O$ ratios and the complexity of the respiratory chain I conclude that the efficiency of energy conservation in R. leguminosarum is also high. This conclusion is fortified by the observation of a $Y(O_2)$ value of about 40 for R. leguminosarum 128C30 (H. Stam, unpublished results). Under oxygen-limited conditions the Y(sub)MAX and $Y(O_2)MAX$ values were lower than for carbon-limited cultures. Similarly in R. leguminosarum (Ratcliffe et al., 1983a) and in the cowpea Rhizobium the $Y(O_2)$ values for oxygen-limited cultures were lower than for carbon-limited cultures. These (Ching et al., 1981) observations are in accordance with the earlier conclusion that energy generation in carbon-limited cells is more efficient than in oxygen-limited cells.

R. leguminosarum was also grown in a fed batch culture, which is a batch culture to which a continuous stream of nutrients is fed. In a fed batch culture μ steadily declines and rapidly very low μ values are reached. This system is a good model for the development of bacteria in the root nodule. Three phases could be distinguished with constant Y'(mannitol) values of 80, 47 and 30 respectively (Stam et al., 1983). Generally it is assumed that Y(mannitol) is dependent on μ; the relation between Y and μ is supposed to be given by the equation of Pirt (1965). The transition from phase 1 to phase 2 takes place at a μ value of 0.053 h^{-1}, that of phase 2 to phase 3 at 0.012 h^{-1} (Stam et al., 1983). The occurrence of 3 phases of growth keyed by the nutrient supply has also been found for E. coli (Chesbro et al., 1979; Arbige and Chesbro, 1982a; van Verseveld et al., 1983), Bacillus polymyxa (Arbige and Chesbro, 1982b) and for P. denitrificans (van Verseveld et al., 1983). In the second phase the concentration of cyclic AMP reaches a maximum and there is an increase in the concentration of guanosine-tetraphosphate (ppGpp), which concentration reaches the maximal value in the third phase. The accumulation of ppGpp is responsible for the transition of the second to the third phase. These nucleotides are known to be involved in the regulation of enzyme synthesis. Since nitrogen fixation in rhizobia occurs at very low μ values these nucelotides might be involved in the expression of the nitrogenase genes. The involvement of ppGpp in the expression of the nitrogenase genes has been suggested before for other nitrogen-fixing bacteria (Kleiner and Phillips, 1981; Riesenberg et al., 1982; Zumft and Neumann, 1983).

4. Nitrogen fixation in chemostat cultures of Rhizobium

Nitrogenase activity has been demonstrated in glutamine-limited chemostat cultures of the cowpea Rhizobium strain 32H1 (Bergersen et al., 1976; Bergersen 1977; Bergersen and Turner, 1976, 1978; Ching et al., 1981). The amount of nitrogen fixed was very low however (Bergersen, 1977). In a glutamine-limited chemostat culture with a dilution rate of 0.05 h^{-1} the glutamine supply was suddenly interrupted. This led to a washout of the culture and the authors concluded that no growth with N_2 was possible (Ching et al., 1981). However this conclusion is not valid. Assuming a nitrogen content of the cells of A% and a nitrogenase activity of B (mmol nitrogen fixed per hour per gram dry weight) the maximum specific growth rate can be calculated according to the equation: $\mu = 2.8$ B/A. Generally nitrogenase activities or rhizobia are low (Ching et al. 1981; Keister and Ranga Rao, 1977; Stam et al., 1983). The μ_{max} for growth with N_2 for Rhizobia is therefore lower than 0.02 h^{-1} R. leguminosarum was grown in a nitrogen-limited chemostat culture in which the oxygen-limited chemostat culture in which the oxygen tension was regulated at about 1 μMO_2 (Stam et al., 1983) at a dilution rate of 0.02 h^{-1}. When the supply of nitrogen was switched off the culture washed out but slower than predicted from the dilution rate. When the medium supply was interrupted the culture showed batch growth with N_2. The Ymannitol values decreased from 33 to 23 after the start of growth with N_2 (Stam et al., 1983). This indicates that the energy cost of nitrogen fixation is high. The strain used (R. leguminosarum 128C30) is hydrogenase-positive and in a number of nitrogen-fixing cultures hydrogenase was found to be induced. Recently it was demonstrated that R. sesbania can grow very easily in free-living culture with N_2 (Dreyfuss et al., 1983).

5. Hydrogen oxidation and energy generation

Nitrogen fixation is always linked to proton reduction and hydrogen evolution (Robson and Postgate, 1980). A minimal value of 1 mol H_2 produced per mol N_2 fixed has been reported (Robson and Postgate, 1980). In bacteroids H_2 evolution by nitrogenase may consume 40-60% of the energy supplied to nitrogenase, implicating H_2/N_2 ratios of 2 to 5 (Schuberth and Evans, 1976). Hydrogen evolution by nitrogenase represents a loss of energy and reduction equivalents. A number of rhizobia have the capacity to reoxidize hydrogen. This oxidation is associated with ATP formation (Dixon, 1972; Emerich et al., 1979; Nelson and Salminen, 1982). In this way some of the energy expended during H_2 evolution may be recaptured. Furthermore hydrogen oxidation will consume oxygen and this may aid in the protection of nitrogenase from oxygen damage (Dixon, 1972; Emerich et al., 1979). In bacteroids of R. leguminosarum ATP formation could not be demonstrated for all strains oxidizing hydrogen (Nelson and Salminen, 1982) and these authors concluded that not in all strains hydrogen oxidation was coupled to ATP formation. However this conclusion must be regarded with caution, since a) only in the strains with the lowest hydrogenase activities no ATP formation could be detected, b) the hydrogenase activity is very low in bacteroids of R. leguminosarum and therefore only a very small part of the respiratory capacity is used during hydrogen oxidation. Therefore it seems more likely that in all rhizobia hydrogen oxidation is associated with ATP formation. Nitrogen-fixing chemostat cultures of R. sesbania growing with succinate effectively oxidize hydrogen. For growth with ammonia Y(succ) and $Y(O_2)$ were respectively 43.2 and 24.5. For growth with N_2 these values were 27.0 resp. 12.1. In accordance with the theoretical calculations (Stouthamer, 1977b) especially the $Y(O_2)$ for growth with nitrogen is much lower than for growth with ammonia. It can be calculated that when no hydrogen oxidation would occur these parameters would be 25.7 resp. 12.7. So the difference is very small. However the difference in CO_2/O_2 ratio is very significant: 1.30 when hydrogen oxidation occurs and 1.48 when no hydrogen oxidation would occur. If additional hydrogen is added this is also oxidized and the energy gained from hydrogen oxidation is used to assimilate a larger part of the succinate. By oxidation of 3.5 mol hydrogen per mol succinate Y(succ) is increased to 34.8 and $Y(O_2)$ decreases to 10.0. It can be calculated that the energy gained from the oxidation of about 11 mol H_2 is equal to that from the oxidation of 1 mol succinate. The experiments indicate an ATP/N_2 ratio of about 30 and the operation of 2 sites of oxidative phosphorylation in R. sesbania. The results with R. sesbania will be described in more detail in the abstract of Stam et al. in this volume.

About 10-15% of the photosynthetate is used in the nodules of peas for nitrogen fixation (Minchin and Pate, 1973). In lupine about 30% of the photosynthetate entering the nodule is used for hydrogen formation (Layzell et al., 1979). From these data it can be calculated that 3-4% of the photosynthetate is used for hydrogen formation. This seems only a very small amount, but it can easily be understood that this will have a dramatic influence on crop yield. In greenhouse and field experiments soybean plants inoculated with hydrogenase-positive strains of R. japonicum were reported to have an increased dry weight and a higher percentage of N in their tissues and seeds than plants inoculated with groups of hydrogenase-negative strains (Albrecht et al., 1979; Schubert et al., 1978; Zablotowicz et al., 1980). Hydrogenase-negative mutants have been isolated from hydrogen-oxidizing R. japonicum (Maier et al. 1978; Maier, 1981; Lepo et al., 1981). This enabled growth experiments with isogenic hydrogenase-positive and negative strains, in which also better plant growth was observed for the hydrogenase positive strains (Lepo et al., 1981). In similar comparisons it was found that higher yields were obtained for mungbean (Pahwa and Dogra, 1981) with hydrogen uptake positive Rhizobium strains. The experiments of Layzell et al. (1979) have shown that the amount of photosynthetate required for nitrogen fixation in lupin (no hydrogen oxidation) was much larger than in cowpea. Pea plants inoculated with hydrogenase-positive R. legunimosarum isolates did not exhibit higher dry weight or N content than those inoculated with hydrogenase-negative isolates (Nelson, 1983). In most cases the hydrogenase activity was insufficient to recycle all the hydrogen formed and consequently in most cases hydrogen was evolved. In one of those isolates the genetic determinants for hydrogenase activity resided on a non-transmissible plasmid (Brewin et al., 1980), which could be mobilized (Brewin et al., 1982). With a number of strains, which had received a recombinant plasmid containing the hydrogenase genes, plant dry weight and N-content were higher than with the parent strains (De Jong et al., 1982). With R. leguminosarum it seems that other traits than hydrogenase are of importance in the symbiotic effectiveness, but a positive effect of hydrogenase can be found in a suitable genetic background. Since in a number of cases the capacity to recycle hydrogen has a positive influence on plant yield and N-content there is a vivid interest in trials to transfer the genetic determinants for hydrogenase to strains which cannot form this enzyme. The cloning of a hydrogenase gene of R. japonicum has been achieved by Cantrell et al. (1983). The improvement of symbiotic properties in R. leguminosarum by plasmid transfer was described by De Jong et al. (1982).

REFERENCES

Albrecht SL, Maier RJ, Hanus FJ, Russel SA, Emerich DW and Evans HJ (1979) Science 203, 1255-1257.
Appleby CA (1969a) Bioch. Biophys. Acta 172, 71-87.
Appleby CA (1969b) Bioch. Biophys. Acta 172, 88-105.

Arbige M and Chesbro WR (1982a) J. gen. Microbiol. 128, 693-703.
Arbige M and Chesbro WR (1982b) Arch. Microbiol. 132, 338-344.
Avissar YJ and Nadler KD (1978) J. Bacteriol. 135, 782-789.
Bergersen FJ and Turner GL (1976) Bioch. Biophys. Acta 173, 524-531.
Bergersen FJ, Turner GL, Gibson AH and Dudman WE (1976) Bioch. Biophys. Acta 444, 164-174.
Bergersen FJ (1977) In Newton W, Postgate JR and Rodriquez-Barrueco C eds. Recent developments in nitrogen fixation, pp. 308-321. Academic Press Inc., London.
Bergersen FJ and Turner GL (1978) Bioch. Biophys. Acta 538, 406-416.
Boogerd FC, van Verseveld HW and Stouthamer AH (1981) Bioch. Biophys. Acta 638, 181-191.
Boogerd FC, van Verseveld HW and Stouthamer AH (1983) Bioch. Biophys. Acta 723, 415-427.
Brewin NJ, De Jong TM, Phillips DA and Johnston AWB (1980) Nature 288, 77-79.
Brewin NJ, Wood EA, Johnston AWB, Dibb NJ and Hombrecher G (1982) J. gen. Microbiol. 128, 1817-1827.
Cantrell MA, Haugland RA and Evans HJ (1983) Proc. Nat. Acad. Sci. 80, 181-185.
Carter KR, Jennings NT, Hanus FJ and Evans HJ (1978) Can. J. Microbiol. 24, 307-311.
Chesbro WR, Evans T and Eifert R (1979) J. Bacteriol. 139, 625-638.
Ching TM, Bergersen FJ and Turner GL (1981) Bioch. Biophys. Acta 636, 82-90.
Daniel RM, Limmer AW, Steele KW and Smith IM (1982) J. gen. Microbiol. 128, 1811-1815.
Dixon ROD (1972) Arch. Microbiol. 85, 193-201.
Dreyfuss BL, Elmerich C and Dommergues YR (1983) Appl. Env. Microbiol. 45, 711-713.
Eisbrenner G and Evans HJ (1982a) J. Bacteriol. 149, 1005-1012.
Eisbrenner G and Evans HJ (1982b) Plant Physiol. 70, 1667-1672.
Eisbrenner G, Hickok RE and Evans HJ (1982) Arch. Microbiol. 132, 230-235.
Emerich DW, Ruiz-Argüeso T, Ching TM and Evans HJ (1979) J. Bacteriol. 137, 153-160.
Hanus FJ, Maier RJ and Evans HJ (1979) Proc. Nat. Acad. Sci. 76, 1788-1792.
Hardy RWF and Havelka UD (1976) In: Nutman PS ed. Sybiotic nitrogen fixation in plants, pp. 421-439. Cambridge University Press, Cambridge, United Kingdom.
Hill S (1976) J. gen. Microbiol. 95, 297-312.
De Hollander JA, Bettenhaussen CW and Stouthamer AH (1979) Ant. van Leeuwenhoek 45, 401-415.
De Hollander JA and Stouthamer AH (1980) Eur. J. Biochem. 111, 473-478.
De Hollander JA (1981) Ph.D. Thesis, Vrije Universiteit, Amsterdam.
De Jong TM, Brewin NJ, Johnston AWB and Phillips DA. J. gen. Microbiol. 128, 1829-1838.
Jones CW (1977) Symp. Soc. Gen. Microbiol. 27, 23-59.
Keister DL and Ranga Rao V (1977) In: Newton W, Postgate JR and Rodriguez-Barrueco eds. Recent developments in nitrogen fixation, pp. 419-431. Academic Press, London.
Kleiner D and Phillips G (1981) Arch. Microbiol. 128, 341-342.
Kretovich WL, Romanov VI and Korolyov AV (1973) Plant Soil 39, 619-634.
Layzell DB, Rainbird RM, Atkins CA and Pate JS (1979) Plant Physiol. 64, 888-891.
Lepo JE, Hickok RE, Cantrell MA, Russel SA and Evans HJ (1981) J. Bacteriol. 146, 614-620.
Lim ST (1978) Plant Physiol. 62, 609-611.
Maier RJ (1981) J. Bacteriol. 145, 533-540.
Maier RJ, Postgate JR and Evans HJ (1978) Nature 276, 494-495.
Maier RJ, Campbell NER, Hanus FJ, Simpson FB, Russel SA and Evans HJ (1979a) Proc. Nat. Acad. Sci. 75, 3258-3262.

Maier RJ, Hanus FJ and Evans HJ (1979b) J. Bacteriol. 137, 824-829.
Minchin FR and Pate JS (1973) J. Exp. Bot. 24, 259-271.
Nelson LM (1983) Appl. Env. Microbiol. 45, 856-861.
Nelson LM and Child JJ (1981) Can. J. Microbiol. 27, 1028-1034.
Nelson LM and Salminen SO (1982) Can. J. Microbiol. 151, 989-995.
O'Brian MR and Maier RJ (1982) J. Bacteriol. 152, 422-430.
Pahwa K and Dogra RC (1981) Arch. Microbiol. 129, 380-383.
Pirt SJ (1965) Proc. Roy. Soc. B. 163, 224-231.
Ratcliffe RD, Drozd JW and Bull AT (1983a) J. gen. Microbiol. 129, 1679-1706.
Ratcliffe RD, Drozd JW and Bull AT (1983b) J. gen. Microbiol. 129, 1707-1712.
Riesenberg D, Erdei S, Kondorosi E, and Kari C (1982) Mol. Gen. Genet. 185, 198-204.
Robson, RL and Postgate JR (1980) Ann. Rev. Microbiol. 34, 183-207.
Ruiz-Argüeso T, Hanus FJ and Evans HJ (1978) Arch. Microbiol. 116, 113-118.
Schuberth KR and Evans HJ (1976) Proc. Nat. Acad. Sci. 73, 1207-1211.
Schuberth KR, Jennings NT and Evans HJ (1978) Plant Physiol. 61, 398-401.
Stam H, van Verseveld HW and Stouthamer AH (1983) Arch. Microbiol. 135, 199-204
Stouthamer AH (1977a) Symp. Soc. Gen. Microbiol. 27, 285-315.
Stouthamer AH (1977b) Ant. van Leeuwenhoek 43, 351-367.
Stouthamer AH, van 't Riet J and Oltmann LF (1980) In Knowles CJ ed. Diversity in bacterial respiratory systems, Vol II, pp. 19-48. CRC Press Inc., Boca Raton, Florida, USA.
Stouthamer AH, Boogerd FC and van Verseveld HW (1982) Ant. van Leeuwenhoek 48, 545-553.
Tuzimura K and Watanabe I (1964) Plant Cell Physiol. 5, 157-170.
Van Verseveld HW, Krab K and Stouthamer AH (1981) Bioch. Bioph. Acta 635, 525-534.
Van Verseveld HW, Chesbro WR, Braster M and Stouthamer AH (1983) Arch. Microbiol., in press.
Van Wielink JE, Oltmann LF , Leeuwerik FJ, De Hollander JA and Stouthamer AH (1982) Biochem. Biophys. Acta 681, 177-190.
Wikström M and Krab K (1980) Curr. Top. Bioenerg. 10, 51-101.
Zablotowicz RM, Eskew DL and Focht DD (1978) Can. J. Microbiol. 24, 757-760.
Zablotowicz RM, Russel SA and Evans HJ (1980) Agron. J. 72, 555-559.
Zumft WG and Neumann S (1983) FEBS Lett. 154, 121-126.

PHYSIOLOGY AND BIOCHEMISTRY OF N_2-FIXATION BY CYANOBACTERIA

H. Bothe, H. Nelles, K.-P. Häger, H. Papen, G. Neuer
Botanisches Institut, Universität Köln, Gyrhofstr, 15,
D-5000 Köln 41, FRG

Aspects of the subject have recently been reviewed by several authors (Stanier, Cohen-Bazire 1977; Haselkorn 1978; Stewart 1980; Hawkesford et al. 1983; Wolk 1982; Bothe et al. 1980; Bothe 1982; Neuer et al. 1983). Therefore newer findings, published in the last two years, will preferentially be discussed in the present communication.

1. NITROGEN FIXING CYANOBACTERIA AND THEIR MECHANISMS TO PROTECT NITROGENASE FROM DAMAGE BY OXYGEN

The physiological differences allow a subdivision of N_2-fixing cyanobacteria into three groups: a) nonheterocystous species performing N_2-fixation under anaerobic or microaerobic conditions, b) nonheterocystous forms which express nitrogenase activity in air, c) filamentous, heterocystous species. As regard to group a), Rippka and Stanier (1978) demonstrated that more than 50% of the filamentous, nonheterocystous forms have the capability to synthesize nitrogenase. Some express nitrogenase activity only under strictly anaerobic conditions, whereas others tolerate moderate levels of O_2. In all cases, effective devices to protect nitrogenase from damage by O_2 have apparently not been developed. The best known example of this group is Plectonema boryanum which performs N_2-fixation under microaerobic conditions as shown by Stewart and Lex (1970).

Group b) consists of only few cyanobacteria expressing nitrogenase activity in air. Examples are the unicellular Gloeocapsa (Wyatt, Silvey 1969) now classified Gloeothece (Rippka et al. 1979), Aphanothece (Singh 1973) now Synechococcus (Rippka et al. 1979) and the filamentous Microcoleus chtonoplastes (Pearson et al. 1979). The Synechococcus strain and Microcoleus are now available as pure cultures where they retain the capability to perform N_2-fixation aerobically (Kallas et al. 1983). Stal and Krumbein (1981) recently observed N_2-fixation by an axenic culture of a marine Oscillatoria. The list of this group of organisms is presumably not complete at present. These aerobic cyanobacteria must protect their nitrogenase against the O_2 present in air and that formed by photosynthesis. The extra O_2 produced photosynthetically appears to cause special problems to nitrogenase in these organisms. The enzyme, though readily solubilized upon disintegration of the cells, is probably associated with the thylakoid membranes. This has never been shown directly but follows from the observation that N_2-fixation is drastically enhanced by light. The reaction requires reduced ATP and ferredoxin which are mainly supplied by cyclic photophosphorylation and photosystem I dependent reductions, respectively (see section 2). In aerobic, nonheterocystous species, the thylakoids must also perform O_2-production, and the cellular organisation may not allow a spatial separation of photosystem II and O_2-production. The experiments of Gallon and coworkers (see Gallon 1981) showed that Gloeocapsa copes with the problem by separating N_2-fixation and photosynthetic O_2-production temporarily. Upon constant illumination, batch cultures show maximal nitrogenase activity 4 - 8 d and maximal photosynthetic O_2-formation 12 - 14 d after the inoculation (Gallon et al. 1975). When grown under alternating light and dark cycles, Gloeocapsa performs virtually all N_2-fixation during darkness and very little when O_2 is formed in the light (Mullineaux et al. 1981).

The organisms of the group c), the heterocystous species, solve this problem by separating photosynthetic O_2-evolution and N_2-fixation spatially. The former is performed in the vegetative cells and the latter in the heterocysts which do not possess photosystem II activity and ribulose-1,5 bisphosphate carboxylase. Heterocysts have unambiguously been shown to be the site of nitrogenase under aerobic growth conditions. Under anaerobiosis, vegetative cells probably also express nitrogenase activity (for the experimental evidence see Wolk 1982 or Bothe 1982).

All N_2-fixing cyanobacteria, including Gloeocapsa and heterocystous species, are faced to the problem to protect their nitrogenase against the O_2 from air. Informations in this field are scanty but are now coming. Nitrogenase, when isolated from cyanobacteria, does not show special features. In particular, both subunit proteins of nitrogenase are as sensitive to exposure to O_2 as those from other organisms (Pienkos et al. 1983). The special Fe-protein (Shethna protein) which protects against damage by O_2 by complexing with the nitrogenase of Azotobacter (Haaker, Veeger 1977) has not been found in cyanobacteria. There is no evidence for conformational protection to occur in cyanobacteria as in Azotobacter. Slime does not play a special protective role, because a slimeless mutant of Gloeocapsa is as tolerant to O_2 as the wild strain (Kallas et al. 1983). The cell wall of heterocysts, particularly special compounds in it (glycolipids, see Haury, Wolk 1978), unlikely forms a barrier against the diffusion of O_2, since heterocysts respire indicating that O_2 reaches the inside of heterocysts.

There is positive evidence for protection mechanisms to occur in cyanobacteria. A high turnover of nitrogenase peptides may help to permit aerobic nitrogenase activity (Bone 1972). Heterocysts respire with higher rates than vegetative cells, and respiratory O_2-consumption probably scavenges O_2 from the nitrogenase site. Another related device is the oxyhydrogen reaction (= Knallgas reaction). Whereas all isolated nitrogenases produce significant amounts of H_2 concomitantly with the reduction of N_2 to ammonia, very little H_2 is produced in intact cyanobacteria under aerobic conditions (see Bothe 1982b). H_2 evolved by the nitrogenase reaction is apparently immediately reconsumed by the whole cells. H_2 was shown to be utilized via a membrane-bound uptake hydrogenase and the respiratory chain resulting in a removal of O_2. When H_2 was present in the assay vessels, C_2H_2-reduction was, indeed, less sensitive to exposure to O_2 in experiments performed with intact *Anabaena variabilis* (Bothe et al. 1978), isolated heterocysts (Wolk 1979) and more recently with heterocysts from the marine *Anabaena CA* (Kumar et al. 1982).

Exposure of *Anabaena* to 100% O_2 results in a rapid decline of nitrogenase activity. Such a high O_2-concentration directly acts on both nitrogenase proteins and does not affect other metabolic processes. Restoration of activity requires synthesis of new nitrogenase, as expected (Lambert, Smith 1980; Pienkos et al. 1983). Pienkos et al. (1983) made the remarkable observation that the nitrogenase of such recovered cells of *Anabaena CA* is less sensitive when exposed once more to 100% O_2. Prolonged exposure of *Anabaena CA* resulted in a partial recovery of nitrogenase activity even in 100% O_2. *Anabaena CA* is apparently able to synthesize a nitrogenase which is then better protected against damage by O_2 by some unknown mechanism.

2. THE GENERATION OF REDUCING EQUIVALENTS FOR NITROGEN FIXATION

As mentioned earlier, nitrogenase does not show special features. It is, therefore, only mentioned here that the nif K gene of *Anabaena* 7120, which codes for the ß-subunit of the molybdenum-iron part of nitrogenase, has been sequenced (Mazur, Chui 1982). Janaki and Wolk (1982) showed that isolated heterocysts synthesize nitrogenase proteins. The elucidation of the pathways which provide the reducing equivalents for N_2-fixation in heterocysts has arisen considerable interest. Ferredoxin is most likely the electron carrier which transfers electrons to the nitrogenase complex. The reducing equivalents for the reduction of ferredoxin mainly come from the hexosemonophosphate shunt/oxidative pentose phosphate cycle (for references see Bothe 1982). The activity levels of the key enzymes of the hexosemonophosphate shunt, glucose-6-P dehydrogenase and 6-phosphogluconate dehydrogenase, are considerably higher in heterocysts than in vegetative cells. NADPH generated by these enzyme reactions is used to reduce ferredoxin in a reaction catalyzed by NADPH:ferredoxin oxidoreductase. Ferredoxin (Tel-Or, Stewart 1977) and NADPH:ferredoxin oxidoreductase (Rowell et al. 1981) are present in heterocysts. NADPH:ferredoxin oxidoreductase exists only as a single form in heterocysts whereas multiple forms occur in vegetative cells (Rowell et al. 1981). The physiological function of the NADPH:ferredoxin oxidoreductase in heterocysts is to catalyze the reduction of ferredoxin by NADPH, because heterocysts do not generate NADPH photosynthetically at the expense of H_2O as the electron donor. The reduction of ferredoxin is thermodynamically unfavourable and, therefore, requires a large excess of NADPH. Both glucose-6-P dehydrogenase and NADPH: ferredoxin oxidoreductase are subject to complex regulations. Surprisingly, the activity of glucose-6-P dehydrogenase is negatively affected by NADPH and that of NADPH:ferredoxin oxidoreductase by $NADP^+$. Glucose-6-P dehydrogenase shows greatest activity at reduction charges <0.3 and ratios >0.3 are required for maximal activities of NADPH:ferredoxin oxidoreductase (Apte et al. 1978). Recent experiments (Papen et al. 1983) showed that $NADP^+$-dependent isocitrate dehydrogenase of heterocysts is negatively regulated by the reduction charge (= $NADPH/NADP^+$ ratio) similar as glucose-6-P dehydrogenase. These findings suggest that heterocysts also utilize electron donors and enzymes other than NADPH and NADPH:ferredoxin oxidoreductase for the reduction of ferredoxin. Alternatively, the transfer of electrons from glucose-6-P or isocitrate to ferredoxin via the dehydrogenases, NADPH and NADPH:ferredoxin oxidoreductase may be regulated by the energetization of the thylakoid membranes (Hawkesford et al. 1983; see below).

Other electron donors for the generation of reducing equivalents for N_2-fixation are, indeed, available. Crude extracts from *Anabaena* were shown to catalyze a pyruvate clastic reaction by pyruvate:ferredoxin oxidoreductase (Leach, Carr 1971; Bothe et al. 1974). The enzyme has now also been demonstrated in heterocysts (Neuer, Bothe 1982). These authors also showed that pyruvate supports C_2H_2-reduction by isolated heterocysts. This activity requires the addition of coenzyme A to the reaction mixture. Such observation is not surprising because the pyruvate:ferredoxin oxidoreductase probably follows a ping-pong mechanism of enzyme catalysis for pyruvate and coenzyme A. Pyruvate-dependent C_2H_2-reduction though clearly demonstrable, was low, because coenzyme A and other components of the reaction mixture may not readily penetrate into heterocysts. Pyruvate:ferredoxin oxidoreductase was O_2-labil and is typically occurring in *Clostridium* and other anaerobes. Heterocysts obviously possess anaerobic compartments where they can effectively protect O_2-labile enzymes as nitrogenase, pyruvate:ferredoxin oxidoreductase and glutamate synthase (Häger et al. 1983) against damage by O_2. Glyoxylate also stimulates N_2-fixation (Bergman 1981). Glyoxylate metabolism in heterocysts still awaits elucidation.

All authors agree that H_2 is the best electron donor to support C_2H_2-reduction by isolated heterocysts (Peterson, Wolk 1978; Eisbrenner, Bothe 1979; Kumar et al. 1982; Houchins, Hind 1982; Privalle, Burris 1983; Ernst et al. 1983). This reaction is light-dependent (Bothe et al. 1977; Peterson, Wolk 1978; Houchins, Hind 1982). In the dark, isolated heterocysts perform very little H_2-dependent C_2H_2-reduction by reversed electron flow (Houchins, Hind 1982). The different groups also agree that H_2-consumption catalyzed by uptake hydrogenase proceeds by two different pathways (for a review see Bothe 1982; for a divergent view see Böhme, Almon 1983). In the one pathway H_2 and uptake hydrogenase donate electrons to photosystem I and in the other to the respiratory chain. The electron acceptor by both pathways is probably plastoquinone or a component close to plastoquinone in the electron transport chain. Both pathways of H_2-utilization probably do not operate independently of each other. Evidence is accumulating that photosynthesis and respiration and thus H_2-utilization share a common segment of electron carriers consisting of plastoquinone, cytochrome c_{553} and probably the cytochrome b/f complex on the thylakoid membranes (Eisbrenner, Bothe 1979; Lockau 1981; Houchins, Hind 1982; Peschek 1982). In addition, part of the respiratory activity may be located on the plasmalemma membrane which may be free of photosynthetic activity in some cyanobacteria (Peschek et al. 1982).

Although H_2 is the best electron donor supporting C_2H_2-reduction by isolated heterocysts, it is, of course, not the only one, because aerobic N_2-fixing cyanobacteria are only exposed to the H_2 evolved by the nitrogenase reaction. Thus, the observation that H_2 is the best electron donor, unlikely reflects the situation in the intact filaments. Despite of other claims, all the heterocyst preparations used so far may partially have lost ADP, P_i, coenzyme A, $NADP^+$ and other low molecular weight compounds and may not be so intact as assumed. Upon isolation of heterocysts, the plasma bridges between heterocysts and vegetative cells must break which may not seal afterwards. After the isolation of heterocysts, compounds like $Na_2S_2O_4$, ATP, ATP generating systems and coenzyme A not at all or poorly penetrate into isolated heterocysts (Peterson, Wolk 1978; Eisbrenner, Bothe 1979; Kumar et al. 1982; Houchins, Hind 1982; Ernst et al. 1983). The inability of $Na_2S_2O_4$ and ATP to stimulate C_2H_2-reduction may not be a firm criterium for the intactness of heterocyst preparations as several authors suggest (Peterson, Wolk 1978; Ernst et al. 1983). In this context, it may be mentioned that $Na_2S_2O_4$ and methylviologen support H_2-evolution by intact filaments of Anabaena variabilis (Daday et al. 1979). This observation was taken as a proof for the existence of the reversible, soluble hydrogenase in cyanobacteria which others believe is an artifact of cell-free preparations (Eisbrenner et al. 1981). In conclusion, intact heterocysts which show exactly the same activities as in the whole filament can possibly not be isolated at all (see also Janaki, Wolk 1982). Moreover, it appears often to be difficult to resolve artifacts from physiological conditions in this field.

Houchins and Hind (1982) showed that NADH is oxidized by the same pathway as H_2 in heterocyst preparations from Anabaena 7120. NADH donates electrons both to the respiratory and photosynthetic electron transport chains. In contrast, NADPH could only be oxidized by NADPH:ferredoxin oxidoreductase in the presence of various electron acceptors, and these reactions are not stimulated by light. This latter finding is in sharp contrast to the experiments of Biggins (1969) where site I of the respiratory chain preferentially utilized NADPH. Clearly, site I of respiration of cyanobacteria needs to be characterized in more detail. Houchins and Hind (1982) have also presented a useful scheme for the topography of the thylakoid membranes and for the interaction of H_2 and NADH with the respiratory and photosynthetic electron transport chains in Anabaena heterocysts. The generation of NADH and the electron transfer from NADH to the electron transport chains remained to be elucidated. The data of table 1 indicate that NADH oxidation by a membrane preparation from heterocysts and vegetative cells of Nostoc muscorum is largely stimulated by light and methylviologen (G. Neuer, H. Nelles and H. Bothe, unpublished). The activity observed in the dark is inhibited by KCN to about 50% and therefore partly due to unspecific NADH oxidases. The light-dependent NADH

TABLE 1: Light stimulated NADH oxidation by membrane preparations from heterocysts and vegetative cells of Nostoc muscorum

	heterocysts		vegetative cells	
	light	dark	light	dark
1. without addition	12.7	8.7	8.3	3.6
2. + methylviologen	19.4	8.3	13.1	3.3

Heterocysts and vegetative cells were prepared as described by Papen et al. (1983). Membrane preparations were obtained by breaking heterocysts in a French press at high pressure and by centrifuging at 100 000 x g for 1 h. The washed pellet was used as the membrane preparation from heterocysts. For obtaining membrane preparations from vegetative cells, the filaments were broken at low pressure in the French press which did not affect heterocysts. Particles from this extract sedimenting between 1500 and 100 000 x g were suspended, washed and assayed. The tests contained in a final volume of 3 ml: protein, 0.05 - 0.1 mg and the following in mM: NADH, 0.12; $MgCl_2$, 5; KH_2PO_4/KOH buffer, pH 7.0, 50; methylviologen, 1. The reaction was followed by the NADH oxidation in air at 25 °C and 35 000 lux or in the dark, respectively. Rates are given in nmol NADH oxidized/min x mg protein.

oxidation activity is about 1.5 fold higher in heterocysts than in vegetative cells. A membrane-bound NADH:plastoquinone oxidoreductase can be isolated from Nostoc muscorum in essentially the same way as described for the enzyme from Chlamydomonas (Godde 1982). The main isolation steps are solubilisation of the enzyme from the thylakoid membranes by treatment with Triton X-100 and enrichment by $(NH_4)_2SO_4$ precipitation and chromatography on DE-cellulose. The absorption spectra of the oxidized and reduced forms are given in figure 1. The preparation obviously still contains some chlorophyll proteins. The difference spectrum (Fig. 1) as well as the fluorescence excitation and emission spectra (data not given) indicate the presence of flavin in this preparation. Thus cyanobacteria similar as Chlamydomonas (Godde, Trebst 1980; Godde 1982) and Rhodopseudomonas capsulata (Ohshima, Drews 1981) and unlike chloroplasts oxidize NADH by a membrane bound NADH:plastoquinone oxidoreductase. Details of these experiments will be published elsewhere (G. Neuer, H. Nelles and H. Bothe, in preparation).

All the above mentioned data indicate that heterocysts utilize multiple electron donors in N_2-fixation (NADPH, pyruvate and glyoxylate in the dark, H_2 and NADH in the light). Clearly, electron transport to nitrogenase requires control. Regulation of nitrogenase activity in heterocysts is not understood. The experiments of Bottomley and Stewart (1976, 1977) indicated that cyclic photophosphorylation in the light or respiration in the dark supply adequate amounts of ATP and that the generation of reductant is more critical for nitrogenase activity than the supply with ATP. In a

FIGURE1. Absorption spectrum of the oxidized and reduced form of the NADH:plastoquinone oxidoreductase from membranes of Nostoc muscorum

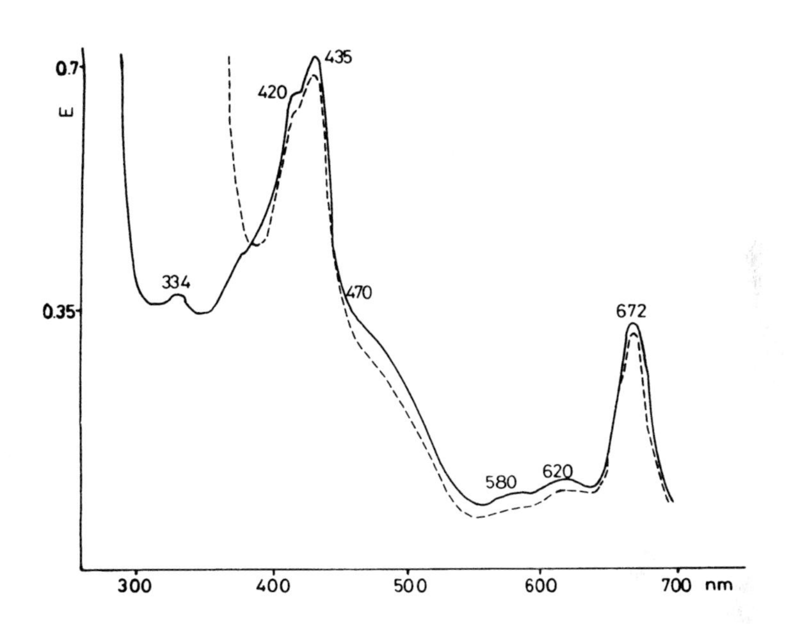

The enzyme was solubilized from a membrane preparation of both heterocysts and vegetative cells by treatment with Triton X-100 and enriched by $(NH_4)_2SO_4$ precipitation and chromatography on DE-52 cellulose. Symbols: straight line: oxidized form, dotted line: reduced form (reduction by dithionite); protein concentration: 0.5 mg/ml

recent paper, Privalle and Burris (1983) redetermined the adenine nucleotide levels in whole filaments and heterocysts of Anabaena 7120 and confirmed that adenine nucleotides do not reflect the energetic expense of N_2-fixation in this organism. On the other hand, when ammonia is added to N_2-fixing Anabaena cylindrica, the ATP pool in the filaments rapidly decreases and parallel to it nitrogenase activity declines (Ohmori, Hattori 1978). These authors suggested that the competition for energy between ammonia assimilation and N_2-fixation causes the inhibition of nitrogenase activity. More recently, Ernst et al. (1983) found a positive correlation between energy charge and nitrogenase activity in heterocysts of Anabaena variabilis. They also reported evidence for a regulatory role of myokinase in the energy balance of heterocysts. At present we have no explanation for the large divergencies in the data of the different groups, e.g. those of Privalle and Burris (1983) and Ernst et al. (1983).

The Dundee group put forward the hypothesis that components of the proton motive force regulate N_2-fixation by cyanobacteria

(Hawkesford et al. 1981;1982;1983). They found a positive correlation between the membrane potential component ($\Delta\Psi$) and nitrogenase activity in both Anabaena variabilis and Plectonema boryanum and suggested that the reverse electron flow from NADPH to nitrogenase is regulated by $\Delta\Psi$. It has been mentioned earlier that glucose-6-P dehydrogenase and NADPH:ferredoxin oxidoreductase are inversely regulated by the reduction charge. A regulation of the electron flow from NADPH to nitrogenase would, indeed, circumvent this difficulty. The regulation of N_2-fixation by $\Delta\Psi$ is not debated. However, other reactions than the electron transfer from glucose-6-P to ferredoxin may be controlled by $\Delta\Psi$ (e.g. exchange of metabolites from heterocysts to vegetative cells and vice versa, transfer of electrons from the Fe-protein to the Mo-Fe-protein of nitrogenase, binding of Mg^{++} to the membranes in order to avoid inhibition by $MgADP^-$). As pointed out above, NADPH generated from glucose-6-P and glucose-6-P dehydrogenase may not be the only electron donor to nitrogenase in heterocysts. It is also noteworthy that Houchins and Hind (1982) found rapid glucose-6-P and $NADP^+$-dependent C_2H_2-reductions when the experiments were performed in the dark and in the presence of nigericin or valinomycin plus K^+. Because either the proton gradient or the membrane potential is abolished under these conditions, Houchins and Hind (1982) did not find evidence for the involvement of $\Delta\Psi$ in the generation of reducing equivalents from glucose-6-P to nitrogenase via glucose-6-P, glucose-6-P dehydrogenase, $NADP^+$, NADPH:ferredoxin oxidoreductase and ferredoxin.

3. ASPECTS OF THE BIOCHEMISTRY OF HETEROCYSTS

Since heterocysts do not possess a Calvin cycle, they must be supplied with carbohydrates from vegetative cells. The exact identity of the carbon compound which moves is unknown. It is possibly a disaccharide which is degraded to glucose-6-P in the heterocysts. As said above, glucose-6-P is mainly degraded by the hexosemonophosphate shunt/oxidative pentose-P cycle to provide NADPH as the reductant for N_2-fixation. Ammonia formed by N_2-fixation is converted to glutamine in the presence of glutamate, ATP and glutamine synthetase. Glutamine is exported to vegetative cells. As an enzyme catalyzing the formation of glutamate (glutamate dehydrogenase, glutamate synthase) appeared to be lacking in heterocysts, glutamate was said to be transported from vegetative cells to heterocysts. Heterocysts were believed not to degrade hexoses via glycolysis and the tricarboxylic acid cycle (see Thomas et al. 1977; for a review Bothe et al. 1980).

We recently examined the enzyme pattern in heterocysts and vegetative cells of cyanobacteria and found all the enzymes that catalyze the conversion of glucose-6-P to glutamate (Neuer, Bothe 1982; Häger et al. 1983; see Fig. 2). Glutamate synthase of heterocysts is dependent on ferredoxin which cannot be substituted by methylviologen (Häger et al. 1983). This observation may explain the failure of other authors to observe this enzyme in heterocysts (Thomas et al. 1977). Heterocysts do not possess an enzyme catalyzing the cleavage of oxoglutarate in heterocysts and vegetative cells. This confirms that the tricarboxylic acid cycle is incomplete in cyanobacteria and that heterocysts cannot regenerate oxaloacetate by this pathway. Therefore they must possess anaplerotic reactions to form oxaloacetate or malate. This is also evident from the observation that the cyanophycin granule polymer consisting of aspartate and arginine is synthesized and degraded with high activity in heterocysts (Gupta, Carr 1981a, b). The synthesis of cyanophycin requires a continuous supply with metabolites of the tricarboxylic acid cycle. Neuer and Bothe (1983) now find the anaplerotic enzymes phosphoenolpyruvate carboxylase and $NADP^+$-dependent malic enzyme in heterocysts and vegetative cells of Anabaena cylindrica. Heterocysts also possess a rather active NAD^+-dependent malate dehydrogenase (Fig. 2). NADH can, therefore, be formed by the malate dehydrogenase or the glyceraldehyde-3-P dehydrogenase reaction and then be reoxidized by photosystem I. Thus the occurrence of a membrane bound NADH:plastoquinone oxidoreductase in heterocysts makes sense physiologically.

Currently we have only demonstrated the presence of the enzymes catalyzing the conversion of glucose-6-P to glutamate in heterocysts. Specific activities in crude extracts do not tell much about the enzyme turnover in the intact organisms. Experiments with labeled tracers have, therefore, to show to which extent glucose-6-P is degraded by glycolysis and the tricarboxylic acid cycle. The occurrence of the enzymes for the formation of glutamate, however, suggests that heterocysts may not be dependent on a supply with glutamate as believed hitherto.

An unexpected outcome of our investigations was that isocitrate dehydrogenase is only poorly active in vegetative cells of A. cylindrica and N. muscorum. The enzyme from heterocysts and vegetative cells of these two cyanobacteria and from cells of Anacystis nidulans was characterized in detail and was found to be regulated in a complex way (Papen et al. 1983; Friga, Farkas 1981). It is activated by thioredoxin and deactivated by oxidized glutathione as $NADP^+$-dependent malate dehydrogenase but unlike glucose-6-P dehydrogenase or NAD^+-dependent malate dehydrogenase. New data (H. Papen, unpublished) indicate that glutamine synthetase from A. cylindrica can also be activated by thioredoxin as described for the enzyme from Chlorella (Tischner, Schmidt 1982). Thioredoxin may therefore function as a coarse activator of the enzymes involved in glutamine biosynthesis. This is particularly noteworthy because glutamine is believed to suppress heterocyst differentiation (see Neuer et al. 1983). Thioredoxin and oxidized glutathione possibly play a key role in the differentiation of vegetative cells to heterocysts in cyanobacteria.

4. REFERENCES

Apte SK, Rowell P and Stewart WDP (1978) Proc. R. Soc. Lond. B 200, 1-25
Bergman B (1981) Planta 152, 302-306
Biggins J (1969) J. Bacteriol. 99, 570-575
Böhme H, Almon H (1983) Biochim. Biophys. Acta 722, 401-407
Bone DH (1972) Arch. Microbiol 86, 13-24
Bothe H, Distler E, Eisbrenner G (1978) Biochimie 60, 277-289

FIGURE 2. Disaccharide metabolism in N_2-fixing heterocysts

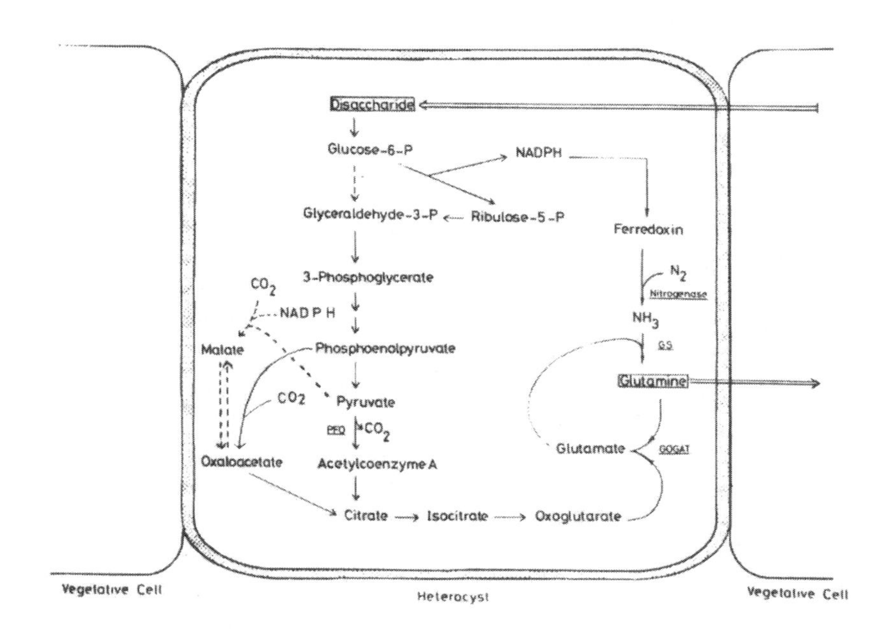

Bothe H, Neuer G, Kalbe I and Eisbrenner G (1980) In Stewart WDP and Gallon JR, eds, Nitrogen Fixation, pp. 83-112, Academic Press, London

Bothe H, Falkenberg B and Nolteernsting U (1974) Arch. Microbiol. 96, 291-304

Bothe H (1982) In Carr NG and Whitton BA, eds, The Biology of Cyanobacteria, pp. 87-104, Blackwell, Oxford

Bothe H (1982b) Experientia 38, 59-64

Bothe H, Tennigkeit J and Eisbrenner G (1977) Arch. Microbiol. 114, 43-49

Bottomley PJ and Stewart WDP (1976) Arch. Microbiol. 108, 249-258

Bottomley PJ and Stewart WDP (1977) New Phytol. 79, 625-638

Daday A, Lambert GR and Smith GD (1979) Biochem. J. 177, 139-144

Eisbrenner G and Bothe H (1979) Arch. Microbiol. 123, 37-45

Eisbrenner G, Roos P and Bothe H (1981) J. Gen. Microbiol, 125, 383-390

Ernst A, Böhme H and Böger P (1983) Biochim. Biophys. Acta 723, 83-90

Friga GM and Farkas G (1981) Arch. Microbiol. 129, 331-334

Gallon JR, Kurz WGW and LaRue TA (1975) In Stewart WDP, ed, Nitrogen Fixation by Free-Living Microorganisms, pp. 159-173, Cambridge University Press, Cambridge

Gallon JR (1981) TIBS 6, 19-23

Godde D and Trebst A (1980) Arch. Microbiol. 127, 245-252

Godde D (1982) Arch. Microbiol. 131, 197-202

Gupta M and Carr NG (1981a) FEMS Microbiol. Lett. 12, 179-181

Gupta M and Carr NG (1981b) J. Gen. Microbiol. 125, 17-23

Haaker H and Veeger C (1977) Eur. J. Biochem. 77, 1-10
Haury JF and Wolk CP (1978) J. Bacteriol. 136, 688-692
Häger, KP, Danneberg G and Bothe H (1983) FEMS Microbiol. Lett. 17, 179-183
Haselkorn R (1978) Ann. Rev. Plant Physiol. 29, 319-344
Hawkesford MJ, Reed RH, Rowell P and Stewart WDP (1981) Eur. J. Biochem. 115, 519-523
Hawkesford MJ, Rowell P and Stewart WDP (1983) In Papageorgiou GC and Packer L, eds, Photosynthetic Prokaryotes: Cell Differentiation and Function, pp. 199-218, Elsevier Biomedical, New York
Hawkesford MJ, Reed RH, Rowell P and Stewart WDP (1982) Eur. J. Biochem. 127, 63-66
Houchins JP and Hind G (1982) Biochim. Biophys. Acta 682, 86-96
Janaki S and Wolk CP (1982) Biochim. Biophys Acta 696, 187-192
Kallas T, Rippka R, Coursin T, Rebiere MC, Tandeau de Marsac N and Cohen-Bazire G (1983) In Papageorgiou GC and Packer L, eds, Photosynthetic Prokaryotes: Cell Differentiation and Function, pp. 281-302, Elsevier Biomedical, New York
Kumar A, Tabita FR and van Baalen C (1982) Arch. Microbiol. 133, 103-109
Lambert GR and Smith GD (1980) Arch. Biochem. Biophys. 205, 36-50
Leach CK and Carr NG (1971) Biochim. Biophys. Acta 245, 165-174
Lockau W (1981) Arch. Microbiol. 128, 336-340
Mazur BJ and Chui CF (1982) Proc. Natl. Acad. Sci. USA 79, 6782-6786
Mullineaux PM, Gallon JR and Chaplin AE (1981) FEMS Microbiol. Lett. 10, 245-247
Neuer G and Bothe H (1982) Biochim. Biophys. Acta 716, 358-365
Neuer G, Papen H and Bothe H (1983) In Papageorgiou GC and Packer L, eds, Photosynthetic Prokaryotes: Cell Differentiation and Function, pp. 219-242, Elsevier Biomedical, New York
Neuer G and Bothe H (1983) FEBS Lett. 158, 79-83
Ohmori M and Hattori A (1978) Arch. Microbiol. 117, 17-20
Ohshima T and Drews G (1981) Z. Naturforsch. 36c, 400-406
Papen H, Neuer G, Refaian M and Bothe H (1983) Arch. Microbiol. 134, 73-79
Pearson HW, Howsley R, Kjeldsen CK and Walsby AE (1979) FEMS Microbiol. Lett. 5, 163-167
Peschek GA, Muchl R, Kienzl PF and Schmetterer G (1982) Biochim. Biophys. Acta 679, 35-43
Peschek GA (1982) Naturwissenschaften 69, 599-600
Peterson RB and Wolk CP (1978) Proc. Natl. Acad. Sci USA 75, 6271-6275
Pienkos PT, Bodmer S and Tabita FR (1983) J. Bacteriol. 153, 182-190
Privalle LS and Burris RH (1983) J. Bacteriol. 154, 351-355
Rippka R and Stanier RY (1978) J. Gen. Microbiol. 105, 83-94
Rippka R, Deruelles J, Waterbury JB, Herdman M and Stanier RY (1979) J. Gen. Microbiol. 111, 1-61
Rowell P, Diez J, Apte SK and Stewart WDP (1981) Biochim. Biophys. Acta 657, 507-516
Singh PK (1973) Arch. Microbiol. 92, 59-62
Stal LJ and Krumbein WE (1981) FEMS Microbiol Lett. 11, 295-299
Stanier RY and Cohen-Bazire G (1977) Ann. Rev. Microbiol. 31, 225-274
Stewart WDP (1980) Ann. Rev. Microbiol. 34, 497-536

Stewart WDP and Lex M (1970) Arch. Microbiol. 73, 250-260
Tel-Or E and Stewart WDP (1977) Proc. R. Soc. Lond. B. 198, 61-86
Thomas J, Meeks JC, Wolk CP, Shaffer PW, Austin SM and Chien WS (1977) J. Bacteriol. 129, 1545-1555
Tischner R and Schmidt A (1982) Plant Physiol. 70, 113-116
Wolk CP (1979) In Nichols JM, ed, Abstract Third Internat. Symp. Photosyn. Prokaryotes, Oxford, p. D1, Univ. Liverpool, England
Wolk CP (1982) In Carr NG and Whitton BA, eds, The Biology of Cyanobacteria, pp. 359-386, Blackwell, Oxford
Wyatt JT and Silvey JKG (1969) Science 165, 908-909

5. ACKNOWLEDGEMENTS

The authors are indebted to Deutsche Forschungsgemeinschaft for financial support of their own experiments cited in the text.

POSTER DISCUSSION 4A PHYSIOLOGY OF FREE-LIVING DIAZOTROPHS INCLUDING CYANOBACTERIA AND AZOSPIRILLUM WITH SPECIAL ATTENTION TO: NH_4^+ METABOLISM, H_2 METABOLISM AND O_2 PROTECTION

WALTER G. ZUMFT, Lehrstuhl für Mikrobiologie, Universität Karlsruhe, Kaiserstrasse 12, D-7500 Karlsruhe, FRG

A large number of contributions was presented in the Physiology Section of free-living diazotrophs, requiring a split into two sessions under specialized topics. Despite considerable organizational effort to convene under coherent groupings some heterogeneity remained among the nearly 40 contributions from 13 countries. Certainly, one perceived this as an indication of the diversity and ramification of the research activities in this area. The following intends to offer some guidance through the majority of contributions, emphasizing points of broader interest or problematic topics.

Protection of nitrogenase from O_2 damage

Oxygen may cause one or more of the following responses in diazotrophic bacteria: reversible inhibition of nitrogenase, oxidative destruction of the enzyme, and repression of nitrogenase synthesis. That the latter has not to be taken as a statement of fact was shown in work with Klebsiella, for which, at very low O_2-concentrations, enhanced nitrogenase expression as well as higher nitrogenase activity have now been reported. Further insights were gained into the nature of the interaction of the Shethna FeS II protein with nitrogenase from Azotobacter. The complex of this [2Fe-2S]cluster-containing protein and nitrogenase, rendering the latter more oxygen-stable, requires the complex to be in its oxidized form. Besides this well-investigated form of stabilization of nitrogenase against the deleterious action of oxygen, different mechanisms were shown to be active in Frankia and in members of phototrophic bacteria. Respiratory protection was invoked in Rhodopseudomonas capsulata, and in O_2-scavenging by glutathione reductase or hydrogenase in the cyanobacteria. Kinetic data in the Frankia-Alnus symbiosis are believed to indicate a structural basis for O_2-protection by formation of a gas diffusion barrier. It is doubtlessly astonishing that a full spectrum of different modalities, rather than a single and uniform mechanism, has been developed for the protection of nitrogenase in different diazotrophic bacteria, each of which adjusts perfectly to the overall cellular physiology into which N_2-fixation is embedded.

Hydrogen physiology

A relationship between H_2-metabolism and N_2-fixation has been recognized for many years. The continuing interest in this area centers on the dual role of hydrogenase to bring about a higher efficiency of nitrogen fixation by recycling part of the reducing power, and indirectly by providing an oxygen sink to exert thus a protective function (see above). Contributions within this area dealt with Klebsiella, Azotobacter, Rhizobium, Gloethece, Rhodopseudomonas, and the sulfate-reducing bacteria.

Ammonium, its transport and regulatory role in nitrogenase activity

An Ammonium transport system is widely distributed among the prokaryotic and eukaryotic kingdom, and diazotrophic bacteria are no exception to that observation. Methylamine was found previously to be a useful model compound to study this transport and was applied to free-living and symbiotic cyanobacteria and to Azospirillum. However, as a study with Rhizobium indicated, caution appears to be necessary in certain systems, and methylamine uptake cannot be taken for granted as model for ammonium transport.

Besides its universal role in nitrogenase repression, ammonium leads to short-term inactivation of nitrogenase by covalent modification of the Fe-protein in specialized cases, notably the Rhodospirillaceae. The mechanism was extended here to the denitrifying strain of Rhodopseudomonas sphaeroides, and its existence was further supported in Azospirillum. The conditions for nitrogenase inactivation were also studied in Azotobacter, a continuing controversial subject, which aroused considerable interest because preliminary evidence was pointing to a possible extension of this regulatory mechanism beyond the non-sulfur purple bacteria and Azospirillum. Although the cyanobacteria appear to show a dual response towards ammonia, no evidence yet indicates there a short-term inhibition akin to the non-sulfur purple bacteria.

N- and C-metabolism in Azospirillum and the cyanobacteria

Interest in Azospirillum is mainly due to its association with roots of grasses and the search to apply this to agriculture. Towards this end, contributions dealt with the effects of fixed N on nitrogenase activity and its stimulation by root exudates. The metabolism of glucose, fructose and sucrose was studied in this bacterium and found to be of different value for three species. Difficulties in growing large batches of N_2-fixing Azospirillum have been overcome now by growth on glutamate, which allows expression of nitrogenase. For growth on malate, a closely controlled chemostat for microaerophilic conditions again allowing nitrogenase expression, was developed. The versatility of Azospirillum with respect to its inorganic nitrogen metabolism was underscored by a demonstration of its ability to denitrify, although N_2-fixation and denitrification are mutually exclusive. NADPH-dependent glutamate synthase as part of the ammonium assimilatory system of Azospirillum was shown to be an iron-sulfur and flavoprotein.

The heterocysts of N_2-fixing cyanobacteria derive an important part of their energy from the oxidative pentose pathway. Their morphogenesis was reported to be influenced by a yet to be identified factor in peptones. Recent results also indicate the presence of other catabolic activities in the heterocysts, such as the enzymes of glycolysis and part of the tricarboxylic acid cycle. Thioredoxin was obtained from Anabaena, and was shown to be the mediator molecule in the light-dependent enzyme regulation. The unicellular cyanobacteria, Gloethece, fixes nitrogen predominantly in the dark period of an alternating 12 hr. light-dark rhythm. The dark fixation is supported by respiration; however, photosynthesis is required for de novo enzyme synthesis. This system was studied in more detail with Ca-alginate-immobilized cells.

A COMPARISON OF H_2-UPTAKE HYDROGENASE NEGATIVE (Hup^-) MUTANTS AND WILD-TYPE (Hup^+) AZOTOBACTER CHROOCOCCUM UNDER DIFFERENT GROWTH CONDITIONS

O. MARIO AGUILAR[+], M.G. YATES[++], J.R. POSTGATE[++]
+ FAKULTAT FÜR BIOLOGIE, UNIVERSITÄT BIELEFELD, POSTFACH 8640, 4800 BIELEFELD 1, WEST GERMANY.
++ ARC UNIT OF NITROGEN FIXATION, UNIVERSITY OF SUSSEX, BRIGHTON BN1 9RQ UK.

H_2 uptake hydrogenase (Hup) activity can support ATP production, nitrogenase activity and respiratory protection of nitrogenase in the presence of exogenous H_2. However, recycling endogenous H_2 produced by nitrogenase activity recovers only about 2.5% of total electron flow to O_2 under optimum conditions. We have attempted to quantitate Hup effects by comparing growth of Hup^+ and Hup^- strains of *Azotobacter chroococcum* (MCD1) in batch and chemostat cultures.

MATERIALS AND METHODS

Three NTG-induced Hup^- mutants of MCD1 with different secondary phenotypes (fast growing or acid-producing) were selected for comparison with MCD1. Adenine nucleotides were measured by the luciferase assay.

RESULTS

Hup^- mutants produced H_2 in N_2- but not NH_4^+-grown conditions. This H_2 production was greater under $Ar:O_2$ (4:1) than under air. Optimum N_2 fixation occurred only during exponential growth in batch culture: electrons diverted largely to H_2 production during early and late growth. Hup^- mutants grew as rapidly as the wild-type (equivalent μ values) but had longer lag phases under N_2-fixing conditions with low carbon substrates and dilute inocula.

N_2-fixing sucrose-limited continuous cultures of mutants and wild-type MCD1 produced similar yields at optimum D values ($0.1\ h^{-1}$). However, at higher D values the mutant yields decreased and they washed out above $D = 0.17\ h^{-1}$ whereas the wild-type gave optimum yields up to $D = 0.22\ h^{-1}$ and washed out at $D = 0.29\ h^{-1}$. $H_2:N_2$ molar ratios in the mutants were 1 at optimum D values but increased significantly (>2) at D values prior to washout. In a mixed N_2-fixing, sucrose-limiting culture of MCD1 and the mutant most stable to high D values, the proportion of mutant in the culture decreased progressively from 78% to 2% between D values of 0.1 and $0.26\ h^{-1}$ and remained at 2% on returning to $D = 0.12\ h^{-1}$. Similar, but less pronounced behavioural differences were observed in SO_4-limited but not in O_2-limited cultures. On the other hand, carbon-limited NH_4^+-grown cultures of both mutants and wild-type gave similar yields and washed out at the same D value ($0.37\ h^{-1}$). ATP:ADP ratios and energy charge (EC) values were similar in the mutants and wild-type. EC values ranged from 0.82 to 0.92; they increased with D values under O_2- or SO_4^--limitation but showed no similar trend under C-limitation.

CONCLUSIONS

Possession of Hup is beneficial to N_2-fixing Azotobacter only at high D values under carbon limitation or during growth initiation with dilute inocula. These observations may help to explain the conflicting results reported with Hup^+ and Hup^- strains of rhizobia associating with legumes (1).

REFERENCE

(1) Yates MG (1983) in Temperate Legumes, physiology, genetics and regulation (ed. Jones DG, Davies DR) Pitman Press, in press.

ROOT EXUDATES STIMULATE NITROGEN FIXATION IN AZOSPIRILLUM

MARCO BAZZICALUPO and ENZO GALLORI

Institute of Genetics, University of Florence, Via Romana 17, 50125 FLORENCE, ITALY

The nitrogen fixing activity in the rhizosphere is probably supported by substances excreted by roots. *Azospirillum* is one of the most common rhizosphere bacteria able to fix nitrogen in association with roots of grasses. Different cereal species and cultivars exhibit different ability in promoting the association with *Azospirillum* (J.Dobereiner and R.Boddey, 1981); therefore it could be useful to develop a test *in vitro* for the preliminary selection of the most effective plant and bacterial genotypes for the association.

In the present communication the ability to stimulate nitrogen fixation of *A.brasilense* strain Sp6 (D.Bani "et al.", 1980) and *A.lipoferum* strain USA 5a by root exudates is reported. Nitrogenase activity was assayed in semisolid nitrogen-free minimal medium with limiting carbon sources, to which concentrated root exudates of the fllowing species and cultivars were added: *Triticum durum* cv. Valgerardo, Trinakria and Creso; *Triticum aestivum* cv. Cheyenne, Chinese spring, Cappelle and Capeti; *Festuca arundinacea* cv. Manade; *Dactylis glomerata* cv. Dora and *Lycopersicum aesculentum*.Higher levels of stimulation were obtained with exudates of *T.durum* cv; Valgerardo and Trinakria and *T.aestivum* cv. Chinese spring; while *T.durum* cv. Creso, *T.aestivum* cv. Cappelle and Capeti and *D.glomerata* showed a lower stimulation.

The same kind of results were obtained with *A.brasilense* and *A.lipoferum*. Preliminary results indicated that the stimulatory substances have small molecular weight (were lost during dialysis) and are thermoresistant. Experiments on fractionation of root exudates by ion exchange chromatography suggested that the stimulating substances were mainly in the anionic and to a less extent in the neutral fractions.

Bani D, Barberio C, Bazzicalupo M, Favilli F, Gallori E, and Polsinelli M (1980) J. Gen. Microbiol. 119, 239-244.

Dobereiner J and Boddey R M (1981) In Gibson AH and Newton WE, eds, Current Perspectives in Nitrogen Fixation, pp 305-312, Austr. Ac. Sc.,Camberra.

Acknowledgments: This work was supported by C.N.R.,P.F. IPRA.

PHOTORESPIRATORY RELEASE OF NH_4-N BY ANABAENA CYLINDRICA IN THE PRESENCE OF METHIONINE SULFOXIMINE (MSX).

BIRGITTA BERGMAN, INSTITUTE OF PLANT PHYSIOLOGY, UNIVERSITY OF UPPSALA, BOX 540 S-751 21 UPPSALA, SWEDEN.

The release of NH_4-N by non-nitrogen-fixing cultures of *Anabaena cylindrica* (grown on NH_4Cl) in the presence of MSX and the absence of external nitrogen was examined. The release was maximal at 0.2 mM MSX, a concentration which did not affect net CO_2 fixation nor glycollate excretion, but inhibited the uptake of ammonium and the glutamine synthetase activity. The release was negligible in absence of MSX while it showed linearity for several hours in its presence.

It is suggested that the major source of the NH_4-N released is the photorespiratory conversion of glycine to serine as

- the release was stimulated by increase in light intensity.
- high CO_2 lowered the release, if not given as a longer pretreatment when a stimulation was observed.
- glutamate had no effect or slightly stimulated the release particularly under N-deficient conditions.
- the inhibitors INH (isonicotinicacid hydrazide), AOA (aminooxyacetate) and AAN (aminoacetonitrile) caused a reduced release.

Furthermore 30-50 % of the NH_4-N being released by N_2-fixing cultures may originate from photorespiration.

The data thus suggest that glyoxylate, besides being metabolized via tartronic semialdehyde (tsa) (Codd & Stewart, Arch. Microbiol. 1973, 94:11-28), may be metabolized via glycine to serine as shown in the tentative metabolic scheme below. The NH_4-N produced is likely further converted to glutamine and then glutamate by the GS-GOGAT enzyme complex.

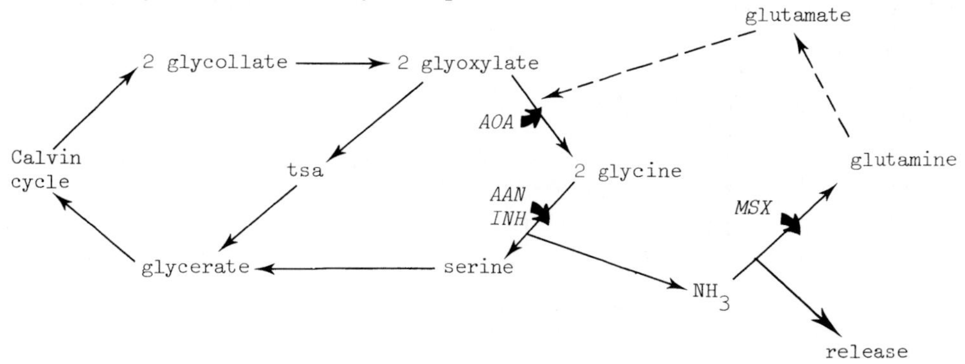

HYDROGENASE ACTIVITIES IN KLEBSIELLA PENUMONIAE GROWN UNDER N_2-FIXING OR NON FIXING CONDITIONS.

Y.BERLIER, G. FAUQUE AND P.A. LESPINAT.
ARBS/CNRS-ECE, CEN CADARACHE. P.O. BOX 1, F 13115 ST PAUL LEZ DURANCE.

The aims of this work with intact cells of Klebsiella pneumoniae were to determine the contribution of hydrogenase and nitrogenase to H_2 production and evaluate the significance of the $H^+/^2H_2$ exchange reaction as a measurement of the hydrogenase activity.

MATERIAL AND METHODS. K. pneumoniae M5a1 (UNF, Brighton) was grown anaerobically in a chemostat on glucose under N_2-fixing or non-fixing conditions. The experiments were performed either on growing cells or on resting cells in glucose-free medium. The gas changes (H_2 or ^2HH evolution, $^2H_2$ uptake and C_2H_2 reduction to $C_2H_2H_2$) were monitored in the liquid phase by mass-spectrometry.

RESULTS AND DISCUSSION. Based on the inhibiting effect of C_2H_2 and CO plus the C_2H_2 reducing activity, the contribution of nitrogenase under N_2-fixing conditions was estimated to approximately 14 % of the total H_2 evolved.

In the presence of $^2H_2$, hydrogenase catalyzed an exchange reaction leading to the disappearance of $^2H_2$ (table, col.2) and the appearance of ^2HH (col.3) and H_2 (col.4). In the absence of net H_2 utilization or production both rates were equal. In contrast, a net H_2 production resulted in a higher value for ($\Delta H_2 + \Delta\ ^2HH$) than for ($\Delta\ ^2H_2$).

Cells growing on glucose catalyzed a net production of H_2 (table, line 1). In the exchange reaction with $^2H_2$ the calculated value for net production ($\Delta H_2 + \Delta HH - \Delta\ ^2H_2$) was only half the observed value under glucose alone. This was probably due to the inhibition of H_2 production by $^2H_2$.

Resting cells from a N_2-fixing culture did not evolve significant H_2 without glucose and then ($\Delta H_2 + \Delta\ ^2HH$) was equal to ($\Delta\ ^2H$)(line 2). Glucose addition resulted in a net H_2 production which was also reflected in the exchange reaction but, as in the growing cells, with a calculated amount only half of that with glucose alone (line 3).

Resting cells from a non N_2-fixing culture did not evolve significant H_2 even from glucose which is confirmed by the nil balance of the exchange reaction (line 4, 5). The reason for the failure of these cells to metabolize glucose with H_2 formation has not been ascertained (nor whether it was specific of non-fixing cells). It is interesting that ^2HH appearance was unmodified by glucose and that the $H_2/^2HH$ ratio varied according to the physiological conditions of the bacteria. This may reflect different hydrogenase complements and electron-carriers formed under the different nutritional conditions.

These results clearly indicate that H_2 metabolism can be evaluated under any circumstances through the parameters of the exchange reaction.

Table. Relationship of H_2 production to the exchange reaction in growing and resting cells of K. pneumoniae (see text for lines and columns correspondance).

	Net H_2 production	Exchange reaction (nmol/min/mg protein)					
	1	2	3	4	5	6	7
		$\Delta\ ^2H_2$	$\Delta\ ^2HH$	ΔH_2	$\Delta(^2HH+H_2)$	$\Delta(^2HH+H_2-^2H_2)$	$H_2/^2HH$
1 G.C.+gluc.	213	323	123	297	420	97	2.4
2 R.C.-gluc.	12	240	81	157	238	- 2	2.0
3 R.C.+gluc.	158	240	84	229	313	72	2.7
4 R.C.-gluc.	4.5	646	296	372	668	23	1.2
5 R.C.+gluc.	14	608	258	362	620	12	1.4

ATP, CARBON METABOLISM AND N_2 FIXATION BY GLOEOTHECE

ALAN E. CHAPLIN/TARIK M.A. RAJAB/JOHN R. GALLON
DEPARTMENT OF BIOCHEMISTRY, UNIVERSITY COLLEGE OF SWANSEA, SWANSEA SA28PP, U.K.

The unicellular cyanobacterium, *Gloeothece*, is unusual in that it is able to fix N_2 aerobically in the light and the dark and anaerobically in the dark. When grown on a 12 h light/12 h dark cycle, *Gloeothece* fixes N_2 predominantly in the dark period. Nitrogenase activity is low at the start of the dark period, rises to a maximum as a result of nitrogenase synthesis about 8 h into the dark period and thereafter declines, reaching a minimum shortly before or soon after the start of the next light period (Mullineaux *et al.*, 1981). Dark metabolism must therefore be capable of generating sufficient ATP to provide for both nitrogenase activity and synthesis.

In order to assess the effect of ATP on the activity of nitrogenase, ADP/ATP ratios were measured under the various environmental conditions known to influence the rate of N_2-fixation. When cultures of *Gloeothece* were grown under continuous light and transferred to dark anaerobic conditions, 2 h after transfer the rate of acetylene reduction was low (2 pmol min^{-1} 10^6 $cells^{-1}$) and the ADP/ATP ratio was 0.7. On transfer to dark aerobic conditions, the rate of acetylene reduction increased (7 pmol min^{-1} 10^6 $cells^{-1}$) and the ADP/ATP ratio fell to 0.4. Transfer to light aerobic conditions resulted in a further increase in the rate of acetylene reduction (20 pmol min^{-1} 10^6 $cells^{-1}$) and a fall in the ADP/ATP ratio to 0.25. These data suggest that under these conditions, the rate of acetylene reduction is related to the ADP/ATP ratio. This may be due either to the availability of ATP for N_2-fixation or to the inhibitory effect of ADP. However, this relationship between the ADP/ATP ratio and the rate of acetylene reduction does not hold true under all conditions. For example, in the light phase of a light/dark cycle, when the rate of acetylene reduction is minimal, the ADP/ATP ratio is low (0.35).

When grown on a light/dark cycle, the sharp decrease in the rate of acetylene reduction after 8 h in the dark coincided with a dramatic increase in the ADP/ATP ratio, values as high as 4.0 being observed. It therefore appears that high ADP/ATP ratios are incompatible with high rates of acetylene reduction.

In the dark phase of a light/dark cycle, ATP must be provided by oxidative phosphorylation. Environmental conditions that affect the rate of acetylene reduction similarly affect the rate of respiration. For example, in the dark phase of a light/dark cycle, $^{14}CO_2$ evolution by *Gloeothece* increased coincidentally with the increase in the rate of acetylene reduction.

When cultures were transferred to an atmosphere of pure O_2 at the start of the dark period, the rates of $^{14}CO_2$ evolution and N_2-fixation were low. However, transfer to air resulted in an increase in both the evolution of $^{14}CO_2$ and the rate of acetylene reduction. It therefore appears that N_2-fixation increases the demand for ATP which is met by increased respiration. These data also suggest that respiratory protection of nitrogenase from O_2 is not a significant feature in *Gloeothece* during prolonged exposure to O_2.

References

Mullineaux, P.M., Gallon, J.R. and Chaplin, A.E. (1981) FEMS Microbiol.Lett. 10, 245-247.

We thank the Royal Society, The Nuffield Foundation and SERC for financial help.

ROLE OF THIOREDOXIN IN REGULATION OF GLUCOSE-6-PHOSPHATE DEHYDROGENASE OF A N_2-FIXING HETEROCYSTOUS CYANOBACTERIUM

J.D. COSSAR/S.M. IP/P. ROWELL/A. AITKEN[+]/W.D.P. STEWART
A.R.C. RESEARCH GROUP ON CYANOBACTERIA AND DEPARTMENT OF BIOLOGICAL SCIENCES,
[+]DEPARTMENT OF BIOCHEMISTRY,
UNIVERSITY OF DUNDEE, DUNDEE DD1 4HN, U.K.

In heterocysts of N_2-fixing cyanobacteria an important and a major route of carbon dissimilation is the oxidative pentose phosphate pathway, with glucose-6-phosphate dehydrogenase (G6PDH) and 6-phosphogluconate dehydrogenase producing NADPH (Smith, 1982). The modulation of G6PDH, the first enzyme in the pathway, is thus important in relation to reductant supply to nitrogenase. There is evidence that in vegetative cells the enzyme is inactive in the light, whereas in heterocysts it may be active both in the light and in the dark (see Apte et al. 1978). Light-deactivation has previously been attributed to inhibition by ribulose-1,5-bisphosphate, ATP and NADPH which are products of photosynthesis (Schaeffer,Stanier, 1978).

We have purified G6PDH from *Anabaena variabilis* to homogeneity and shown that it is modulated by thioredoxin. In cell-free extracts, G6PDH is deactivated by dithiothreitol. During purification the enzyme becomes insensitive to dithiothreitol and sensitivity is restored by adding purified *Anabaena cylindrica* thioredoxin, or a thioredoxin-like moiety from *A. variabilis*. The *A. cylindrica* thioredoxin (molecular weight of ~ 11,800) in addition to deactivating G6PDH, activates other enzymes (fructose-1,6-bisphosphatase of *A. cylindrica* and malate dehydrogenase, and glutamine synthetase of *Scenedesmus obliquus*). The N-terminal amino acid sequence of 39 residues shows extensive homology with the sequences of *Escherichia coli* and *Corynebacterium nephridii* thioredoxins (Holmgren, 1968; Meng,Hogenkamp, 1981). Deactivation of G6PDH by reduced thioredoxin (see also Udvardy et al 1983) can be prevented by glucose-6-phosphate or the amino acids glutamine, alanine and glycine. Other amino acids tested had no effect or promoted deactivation. In vegetative cells of cyanobacteria thioredoxin may play a key role in the light-dependent deactivation of the oxidative pentose phosphate pathway. Activation in the dark may involve the absence of photosynthetically produced reductant or,glucose-6-phosphate derived from fixed carbon reserves. In heterocysts G6PDH is active in the light and dark There may be insufficient reduced thioredoxin in heterocysts to deactivate G6PDH. However, the effects of glucose-6-phosphate, glutamine, alanine, glycine and reduced thioredoxin on G6PDH cannot be viewed in isolation. It is possible that, in heterocysts, glutamine, the primary product of NH_4^+ assimilation accumulates and, together with glucose-6-phosphate derived from fixed carbon transferred from vegetative cells, serves to override the deactivating effect of reduced thioredoxin.

REFERENCES

Smith A (1982) In Carr NG and Whitton BA, eds, The Cyanobacteria, pp. 47-85, Blackwell Scientific, London.
Apte SK, Rowell P and Stewart WDP (1978) Proc. Roy. Soc. Lond. B. 200, 1-25.
Schaeffer F and Stanier RY (1978) Arch. Microbiol. 116, 9-19.
Holmgren A (1968) Eur. J. Biochem. 6, 475-487.
Meng M and Hogenkamp HPC (1981) J. Biol. Chem. 256, 9174-9182.
Udvardy J, Juhaz A and Farkas GL (1983) FEBS Letts. 12, 97-100.

GLUTAMATE SYNTHASE FROM AZOSPIRILLUM BRASILENSE: PARTIAL PURIFICATION AND CHARACTERIZATION

B.Curti, S.Ratti, G.Zanetti and (^) E.Galli
Department of General Physiology and Biochemistry, (^) Department of Biology, Genetics and Microbiology Section, University of Milano, Via Celoria 26, 20133 Milano, Italy.

Glutamate synthase (E.C.1.4.1.13) has been detected in Azospirillum Brasilense Sp6. The enzyme activity was tested under different growth conditions, the specific activity of the enzyme being higher when the microorganism was grown under limiting concentrations of ammonia. Azospirillum Brasilense Sp6 exhibits also glutamate dehydrogenase activity (E.C.1.4.1.2): the maximum specific activity was found with an excess of ammonia in the growth medium. Thus, as shown for other nitrogen fixing microorganisms, the levels of the two enzymes are inversely correlated, their physiological role in the route of nitrogen metabolism strictly depending on the availability of ammonia in the growth culture.
The Azospirillum glutamate synthase uses preferentially NADPH as electron donor, whereas glutamate dehydrogenase depends on NADH.
Glutamate synthase was partially purified starting from Azospirillum Brasilense Sp6 grown on a medium MSP under limiting concentrations of ammonia. Purification steps include heat and pH treatments, ammonium sulphate precipitation, DEAE chromatography, gel filtration and affinity chromatography on 2'-5' ADP Sepharose. The partially purified enzyme shows a specific activity of $\simeq$ 17 units mg^{-1} (μmol NADPH oxidized min^{-1}. mg^{-1}) with a purification factor of $\simeq$300 fold in comparison to the crude extract. The enzyme shows a typical iron-sulfur flavoprotein spectrum with peaks at 440, 376 and 278 nm. It contains FAD and FMN and it is composed of two dissimilar subunits with MW of 135,000 and 50,000, as determined by SDS gel electrophoresis.
Mutants of Azospirillum Brasilense Sp6, strains SPF101 and SPF103 show less than 10% of glutamate synthase activity when compared with the wild type. Azospirillum Brasilense Sp6 has been found to harbour a plasmid whose molecular size, as determined by restriction endonuclease digestion and agarose electrophoresis, corresponds to 265 Kb (177 Md); however, both Azospirillum Brasilense Sp6 and its mutants show no difference in the plasmid fragmentation pattern. The relationship between the plasmid and the phenotypic properties of the microorganism is under investigation.

ACKNOWLEDGEMENTS

This work was supported by CNR, IPRA project, Section 1.6.6.

CARBOHYDRATE METABOLISM IN AZOSPIRILLUM SPP.

MADDALENA DEL GALLO/G. MARTINEZ-DRETS/E.M. GOEBEL/R.H. BURRIS/N.R. KRIEG
DEPARTMENT OF BIOCHEMISTRY, UNIVERSITY OF WISCONSIN, MADISON, WI 53706 U.S.A.

The metabolic preferences of azospirilla for organic acids is well documented (JJ Tarrand et al. 1978). In addition, some carbohydrates (D-fructose, D-gluconate, D-galactose and L-arabinose) support N_2-ase activity in A. brasilense sp7, Cd. A. lipoferum can also grow and fix N_2 on glucose, mannose, sorbose and α-ketoglutarate. Azospirilla have been widely studied, but their pathways of energy-yielding catabolism are unknown with the exception of catabolism of L-arabinose (NJ Novick and ME Tyler, 1982).
Our work deals with the enzymatic mechanisms that operate in the catabolism of carbohydrates and organic acids of azospirilla. These mechanisms are inducible. The bacterial strains used were A. brasilense sp7 and Cd, A. lipoferum sp59b and Col5. The bacteria were grown at 33°C in a mineral medium plus NH_4^+, aerobically or microaerobically in N_2-fixation conditions plus 0.1 g/l of yeast extract. Cell-free extracts were obtained by French pressure cell and after centrifugation, the supernatant was used for the enzymatic assays (spectrophotometrically at 340 nm by following the rate of appearance or disappearance of NAD(P)H).
For growth in various carbon sources, the cells were grown in 21 ml serum bottles containing 5 ml of medium plus 25 mM carbon source for 25 h, and C_2H_4 formed was measured periodically after adding 10% v/v of C_2H_2. The uptake of C-fructose, glucose, sucrose and α-ketoglutarate was determined with a scintillation counter as described (A Gardiol et al. 1980). Our data suggest that both species have all of the enzymes of the Embden-Meyerhof-Parnas pathway and of the TCA cycle and strain sp59b also possesses all the enzymes of the Entner-Doudoroff pathway. Gluconate, but not fructose or organic acids, induces the enzymes of this pathway in sp7 also. Fructose is transported and phosphoyrlated to form F-1-P via PEP-PTS in both species; moreover, sp59b forms F-6-P via hexokinase - an enzyme that is lacking in sp7 - and this sugar might be utilized via F-6-P and G-6-P as mediated by an abundant P-G-isomerase present. The absence of 6-PGD in both species of azospirilla suggests that no oxidative Hexose Monophosphates pathway is present. The absence of growth of A. brasilense in glucose and α-ketoglutarate apparently resulted from the lack of uptake of either substrate. Added NaF (a specific inhibitor of the glycolytic pathway) in Fructose MM inhibits sp7; the EMP pathway appears to operate under conditions suitable for N_2-fixation. The lack of inhibition of malate-grown azospirilla makes it clear that inhibition of C_2H_2 reduction operates only through fructose catabolism.

REFERENCES

Gardiol A, Arias A, Cervenansky C, Gaggero G, Martinez-Drets G (1980) J. Bacteriol. 144, 12-16.
Novick NJ, Tyler ME (1982) J. Bacteriol. 149, 364-367.
Tarrand JJ, Krieg NR, Döbereiner J (1978) Can. J. Microbiol. 24, 967-980.

IS METHYLAMINE TRANSPORT A VALID MODEL FOR AMMONIA MOVEMENT IN *RHIZOBIUM LEGUMINOSARUM*?

M.J. DILWORTH AND A.R. GLENN
NITROGEN FIXATION RESEARCH GROUP, SCHOOL OF ENVIRONMENTAL AND LIFE SCIENCES, MURDOCH UNIVERSITY, MURDOCH, WESTERN AUSTRALIA 6150.

Uptake of radioactive methylamine has been studied in a range of organisms with a view to its use as an index of ammonia transport. While apparently successful in eukaryotes, in prokaryotes it appears unsatisfactory where methylamine is actually used for growth.

In *Rhizobium*, active methylamine uptake has been shown for *R. leguminosarum* (Dilworth & Glenn, 1982), *R. meliloti* (Osburne, 1982; Wiegel & Kleiner, 1982) and cowpea *Rhizobium* 32H1 (Gober & Kashket, 1983). Laane *et al.* (1980) used the unloading of methylamine from pea bacteroids as an index of ammonia movement.

Several fast- and slow-growing rhizobia will grow using methylamine or ethylamine as their sole source of N, but not as the source of both C and N. Cells of *R. leguminosarum* MNF3841 transferred from methylamine or nitrate to NH_4^+ grow immediately, but cells transferred from NH_4^+ to methylamine lag for 4-5 h before growing (generation time 10 h). The system for methylamine utilization is thus inducible, while the system for ammonia is constitutive.

The uptake system for [^{14}C] methylamine in MNF3841 is induced by methylamine or ethylamine, while cells grown on NH_4^+, glutamate or nitrate show 7%, 3% and 1%, respectively, of the fully-induced activity. Uptake is sensitive to 2,4-dinitro -phenol, azide and CCCP. On addition of 1 mM unlabelled methylamine, ethylamine or propylamine, accumulation of radioactivity from [^{14}C] methylamine falls by 98%, 66% and 25% respectively. Addition of 1 mM NH_4^+ only inhibits by about 40%.

The uptake system has an apparent K_m of 40-45 μM methylamine, with a V_{max} of 10-12 nmol.min^{-1}.(mg protein)$^{-1}$. The apparent K_i for ammonia inhibition of [^{14}C]-methylamine uptake is 2 mM. The small but measurable uptake of methylamine in cells grown on NH_4^+ shows the same K_m and K_1 as in methylamine-grown cells; it is unlikely to be an NH_4^+ transport system functioning with methylamine.

The [^{14}C] methylamine taken up by cells is rapidly metabolized, since it does not efflux when excess methylamine is added, nor with CCCP or azide. Ammonia, however, is rapidly leached from cells by washing. After [^{14}C] methylamine uptake, 90% of cell-associated label is TCA-precipitable.

Bacteroids from pea nodules also actively take up [^{14}C] methylamine at the low rate typical of NH_4^+-grown cells (0.2 nmol.min^{-1}.(mg protein)$^{-1}$.

R. leguminosarum MNF3841 resembles methylamine-utilizing *Pseudomonas* spp. (Bellion & Weyland, 1982) in having an inducible methylamine transport system relatively insensitive to NH_4^+, and therefore unlikely to be an NH_4^+ transport system. Methylamine transport in this strain can therefore not be used to study NH_4^+ transport.

Bellion, E. & Weyland, L. (1982). J. Bacteriol., 149, 395-37.
Dilworth, M.J. & Glenn, A.R. (1982). J. gen. Microbiol., 128, 29-37.
Gober, J.W. & Kashket, E.R. (1983). J. Bacteriol., 153, 1196-1201.
Laane, C., Krone, W., Konings, W. & Veeger, C. (1980). Europ. J. Biochem., 103, 39-46.
Osburne, M.S. (1982). J. Bacteriol., 151, 1633-1636.
Wiegel, J. & Kleiner, D. (1982). FEMS Microbiol. Letters, 15, 61-63.

IN VIVO NITROGENASE REGULATION BY AMMONIUM AND METHYLAMINE AND THE EFFECT OF MSX ON AMMONIUM TRANSPORT IN ANABAENA FLOS-AQUAE

SCOTT A. EDIE, DAVID H. TURPIN, AND DAVID T. CANVIN
DEPT. OF BIOLOGY, QUEEN'S UNIVERSITY, KINGSTON, CANADA K7L 3N6

Ammonia inhibits nitrogenase activity of blue-green algal cultures. L-methionine-DL-sulphoximine (MSX) alleviates this effect (2). Since MSX is an inhibitor of glutamine synthetase (GS), Stewart and Rowell (2) suggested that rather than ammonium *per se*, GS or a product of GS was responsible for the inhibition of nitrogenase action. Recently, it has been demonstrated that GS activity is not essential for the MSX alleviation (1) and these authors suggest that MSX may inhibit ammonium transport. We report the effects of MSX on ammonium transport and present data to suggest that methylamine may serve as an ammonium analogue in *in vivo* nitrogenase regulation.

Material and Methods

Anabaena flos-aquae 1444 was grown at 30°C without combined N at pH 9.3, under constant illumination of 40 μE m^{-2} s^{-1}. Experiments were carried out midlog phase. Culture pH was adjusted prior to experiments. Acetylene reduction and oxygen evolution were measured at 300 μE m^{-2} s^{-1}. Ammonium uptake, ^{14}C-methylamine uptake and glutamine synthetase transferase activities were measured under culture conditions.

Results and Discussion

Methylamine and ammonium additions inhibited nitrogenase activity over 24h at all pH values tested. MSX alleviated inhibition by both ammonium and methylamine almost completely at pH 7.1 and by 80% at pH 8.1. Inhibition at pH 9.3 and 10.2 was complete and unaffected by MSX.

Uptake of ^{14}C-methylamine was linear over 30 minutes at pH 7.1. Uptake was completely inhibited by NH_4Cl and unaffected by 100 μM ethylamine suggesting that methylamine is transported by an ammonium permease. At pH 7.1, pretreatment with MSX inhibited methylamine uptake non-competitively. The time-course of pretreatment suggested MSX must enter the cells for inhibition to occur. Methylamine transport at pH 9.3 was approximately 4-fold higher than at pH 7.1 and was insensitive to MSX. These data suggest that protonated methylamine enters the cells at lower pH via an ammonium permease that is inhibited by MSX. At high pH (9.3), methylamine and ammonia can enter the cells passively in the unprotonated form and so the ammonium permease is unimportant. At pH 7.1, amine treatments did not inhibit photosynthesis in the short (1h) or long (24h) term. At pH 9.3, some inhibition by MA was observed. Glutamine synthetase activity inhibited 60% after 24h by 10μM MSX whereas 1 mM MSX inhibited GS activity completely within 1h of addition. MSX alleviation of methylamine and ammonium inhibition was observed at both concentrations of MSX.

Our data suggest that the alleviation of ammonium induced nitrogenase suppression by MSX is due to inhibition of ammonium transport and not inhibition of glutamine synthetase. Furthermore methylamine may act as an ammonium analogue in the regulation in nitrogenase activity. We suggest that ammonium *per se* may be an important regulator of *in vivo* nitrogenase activity in blue-green algae.

(1) SINGH, H.N., U.N. RAI, V.V. RAU, AND S.N. BAGCHI (1983) Biochem. Biophys. Res. Comm. 111, 180-187.
(2) STEWARD, W.D.P. AND P. ROWELL (1975) Biochem. Biophys. Res. Comm. 65, 846.856.

BIOENERGETICS OF NITROGENASE IN BLUE-GREEN ALGAE: III..RATE LIMITATION BY PHOTOSYNTHETIC ELECTRON TRANSPORT OR PHOSPHORYLATION*

A. ERNST/H. ALMON/H. BÖHME AND P. BÖGER
LEHRSTUHL FÜR PHYSIOLOGIE UND BIOCHEMIE DER PFLANZEN, UNIVERSITÄT KONSTANZ, D-7750 KONSTANZ, GERMANY

Adenylate-pool composition and nitrogenase activity were examined in heterocysts isolated from Anabaena variabilis (ATCC 29413). The cellular ATP concentration did not change when nitrogenase activity was varied from 100% to 10% by inhibiting photosynthetic electron flow either with increasing concentrations of 2,5-dibromothymoquinone (DBMIB), by decreasing light intensity, or by decreasing hydrogen concentration, which was used as substrate. Metronidazole, acting as alternative electron acceptor to nitrogenase from reduced ferredoxin, increased the cellular ATP concentration to a certain extent (20%), while inhibiting nitrogenase activity.
On the other hand, uncoupling agents like gramicidin, nigericin or carbonylcyanide-p-trifluoromethoxy phenylhydrazone (FCCP) decreased both nitrogenase activity and cellular ATP concentration. These data are partly shown in Fig.1. Plotting nitrogenase versus the amount of extracted ATP, two distinct relationships are observed. A constant ATP concentration in the heterocysts is established, if electron flow from hydrogen through photosystem I to nitrogenase is varied from 100 to 10%, indicative of a stoichiometrically coupled ATP formation in the light and ATP consumption by nitrogenase (curve A). Under these conditions, electron supply to nigrogenase by hydrogen is rate-limiting, whereas the ATP concentration, hence the energy charge, has no regulatory influence.
Uncouplers, however, decrease the amount of ATP synthesized per electron transported through the photosynthetic chain. This results in an increasing depletion of the ATP pool by nitrogenase activity, since electron supply is no longer rate-limiting. Under this condition, the cellular ATP concentration determines the nitrogenase activity. Comparison of the data of curves A and B leads to the conclusion, that in the light electron supply by H_2, but not ATP, is rate limiting. No membrane potential is effective (see "Bioenergetics" part II).

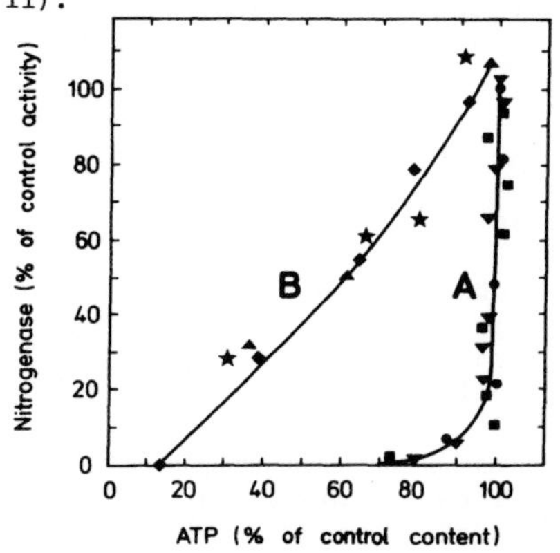

Fig.1. Heterocysts were isolated according to (2). Nitrogenase activity was assayed as acetylene reduction with H_2 as donor, as described in (1); ATP was extracted and determined by bioluminescence according to (1). Activity of preparations (100%) was 60 to 120 µmol C_2H_2 reduced/mg chlorophyll x h. ATP content (100%) varied between 160 and 220 nmol ATP/mg Chl a. Experimental variations were caused by varying light intensity by neutral density filters (3-300 µEinstein/m^2 x s) -▼-; hydrogen (0.1-20%, v/v) -■-; DBMIB (10-200 µM) -●-; nigericin (0.1-10 µM) -▲-; gramicidin (0.1-10 µg/ml) -✱-; FCCP (10-200 µM) -◆-.

(1) Ernst, A., Böhme H., Böger, P.: Biochim. Biophys. Acta 723, 83-90 (1983)
(2) Peterson, R.B., Wolk, C.P.: Proc. Natl. Acad. Sci. USA 75, 6271-6275 (1978)

*Parts I and II were presented at the 6th Int.Congr.Photosynth. 1983, Brussels.

STUDIES ON THE MECHANISM WHEREBY THE PATTERN OF ILLUMINATION AFFECTS N_2 FIXATION BY *GLOEOTHECE*

JOHN R. GALLON/JOÃO S. YUNES/ALAN E. CHAPLIN
DEPARTMENT OF BIOCHEMISTRY, UNIVERSITY COLLEGE OF SWANSEA, SWANSEA SA2 8PP, UK

Cells of the unicellular cyanobacterium *Gloeothece* sp. ATCC 27152, immobilized in 5 mm diameter spheres of calcium alginate (Musgrave *et al.*, 1982) and grown in medium free of combined nitrogen, reduced acetylene almost exclusively in the dark phase of an alternating cycle of 12 h light and 12 h darkness. In this respect they resembled free-living cultures (Mullineaux *et al.*, 1981).

In immobilized cultures of *Gloeothece*, incubated under continuous illumination, this pattern of acetylene reduction could be mimicked by addition of 10 μM 3-(3,4-dichlorophenyl)-1,1-dimethyl urea for 12 h out of each 24 h period. This suggests that the pattern of acetylene reduction may be imposed by the effect of alternating light and darkness on photosynthetic activity.

To examine this further, cultures grown under an alternating cycle of 12 h light and 12 h darkness were transferred at the beginning of the dark period to light of various wavelengths. Under white light, acetylene reduction increased for at least 12 h whilst in the dark, activity initially increased but after 8 h declined. The onset of darkness, which probably imposes the pattern of acetylene reduction on *Gloeothece* (Mullineaux *et al.*, 1981), seems therefore to cause a decrease in nitrogenase activity after 8 h rather than to cause the preceding increase in activity.

The pattern of acetylene reduction under red (615-850 nm) light resembled that under white light whilst under green (500-530 nm) or blue (420-475 nm) light the pattern found was identical to that in the dark. This is consistent with an involvement of photosynthesis in the pattern of acetylene reduction.

Addition of 0.2 mM NH_4Cl to immobilized cultures of *Gloeothece* for 6 h out of each 24 h period also resulted in a pattern of acetylene reduction similar to that under alternating 12 h light and 12 h darkness. This supports earlier suggestions (Mullineaux *et al.*, 1983) that regulation of nitrogenase synthesis causes the pattern of acetylene reduction observed under alternating light and darkness. However, whilst simultaneous addition of 2 mM L-methionine-DL-sulphoximine markedly altered the pattern found with NH_4Cl it did not affect the pattern of acetylene reduction under a cycle of 12 h light:12 h dark.

In conclusion therefore, photosynthesis may provide a link between light and darkness and their effect on N_2 fixation. A possible mechanism is as follows. When cultures of *Gloeothece* enter the dark phase, photosynthesis ceases. This, in turn, triggers a series of events (probably the accumulation or disappearance of a metabolite) that, about 8 h later, inhibits nitrogenase synthesis by a mechanism different from that of NH_4Cl. Inhibition of nitrogenase synthesis results in loss of nitrogenase activity, probably largely the result of O_2-inactivation. Activity returns following synthesis of fresh nitrogenase but this synthesis commences only after a period of illumination.

REFERENCES

Mullineaux PM, Gallon JR and Chaplin AE (1981) FEMS Microbiol.Lett. 10,245-247.
Mullineaux PM, Chaplin AR and Gallon JR (1983) J.Gen.Microbiol. 129, 1689-1696.
Musgrave SC, Kerby NM, Codd GA and Stewart WDP (1982) Biotechnol.Lett. 4,647-652.

ACKNOWLEDGEMENTS
We thank both the Royal Society and CNPq(Brazil) for financial support.

EFFECTS OF NITROGEN SOURCES ON NITROGENASE ACTIVITY IN AZOSPIRILLUM BRASILENSE

ENZO GALLORI and MARCO BAZZICALUPO
Institute of Genetics, University of Florence, Via Romana 17,
50125 FLORENCE, ITALY.

It is known that nitrogen fixation is regulated by different nitrogen compounds. Few data are available on the regulation of nitrogenase in the bacterium *Azospirillum brasilense*, which has stimulated considerable interest owing to its ability to fix nitrogen in association with roots of several grasses (Y.Okon,1982). This communication will attempt to answer three questions concerning the *in vivo* regulation of nitrogenase activity using *A.brasilense* strain Sp6 (D.Bani "et al.",1980):

- Which nitrogen sources inhibit nitrogenase activity of *A.brasilense*?
- Do different nitrogen sources inhibit nitrogenase activity in the same way?
- Is the inhibition mediated by protein synthesis?

The effects of the addition of different compounds on nitrogenase activity were studied on cells of Sp6 grown under microaerophilic conditions with nitrogenase fully derepressed. In Table 1 are reported data which indicate that some nitrogen sources completely block nitrogenase activity while others do not.

Table 1. Effect of nitrogen sources on nitrogenase activity

Compounds inhibiting nitrogenase ♀		% of control activity	Compounds non ≠ inhibiting nitrogenase		% of control activity
NH_4Cl	0.3 mM	0	Aspartate	1.0 mM	106
Glutamine	0.3 "	0	Histidine	1.0 "	104
KNO_3	1.0 "	0	Adenine	1.0 "	101
KNO_2	0.1 "	0			

♀The minimal concentrations which completely inhibit nitrogenase activity are reported. ≠The higher concentrations used are reported.

The same experiments were carried out adding methionine-sulphoximine (MSX) to the cultures. The presence of 10 mM MSX prevents any inhibition by subsequent addition of NH_4^+ or glutamine while it does not prevent the effect of NO_3^- and NO_2^-. This result suggests that the compounds tested affect nitrogenase in different ways. Preliminary experiments on the effect of chloroamphenicol and rifampicin showed that the inhibition of nitrogenase activity by NH_4^+ is not affected by these drugs suggesting that protein and mRNA synthesis are not involved in the mechanism of inhibition.

Bani D.Barberio C,Bazzicalupo M,Favilli F,Gallori E, and Polsinelli M (1980) J. Gen. Microbiol. 119, 239-244.
Okon Y (1982) Israel J. Botany 31, 214-220.

COMPETITION FOR CARBON SCELETONS BETWEEN THE NITROGEN FIXING SYSTEM AND CO_2 FIXATION IN *ANABAENA CYLINDRICA*.

LARS HÄLLBOM

INST. PHYS. BOT. BOX 540, S-751 21 UPPSALA, SWEDEN.

INTRODUCTION

No *in vivo* regulatory system for the activity of cyanobacterial nitrogenase has so far been demonstrated. Three substrates for the nitrogen fixing system have been suggested to be limiting for nitrogenase activity under certain physiological conditions; ATP, reducing power and carbon sceletons. These substrates are produced in the photosynthetic reactions and are common substrates for the Calvin cycle and the nitrogen fixing system. In this paper, I present experiments concerning the role of carbon sceletons in the regulation of nitrogenase activity and photosynthetic CO_2 fixation in *Anabaena cylindrica* (CCAP 1403/2a).

METHODS

All methods used and all culture conditions were described by Bergman (1980).

RESULTS AND DISCUSSION

3 mol m^{-3} NH_4Cl inhibited net photosynthesis of *A. cylindrica* by 80 % within 24 h. This inhibition was reverted by simultaneous addition of 1 mol m^{-3} α-ketoglutarate, Na-citrate or Na-malate. C_2H_2 reduction was inhibited completely within 24 h after addition of 3 mol m^{-3} NH_4Cl. The inhibition was reverted by simultaneous addition of 1 mol m^{-3} α-ketoglutarate and to some extent also by 1 mol m^{-3} Na citrate or Na-malate.
It is proposed that availability of carbon sceletons is an important regulatory factor for the nitrogenase activity in *A. cylindrica* and that competition for carbon sceletons between the N_2 fixing system and the Calvin cycle is governing the NH_4^+-inhibition of photosynthesis.

REFERENCES

Bergman, B. (1980). Physiol. Plant. 49, 398-404.

AMMONIUM UPTAKE AND RELEASE BY *AZOSPIRILLUM*

A. HARTMANN, D. KLEINER[+] AND W. KLINGMÜLLER
LEHRSTUHL FÜR GENETIK UND LEHRSTUHL FÜR MIKROBIOLOGIE[+] DER UNIVERSITÄT BAYREUTH, POSTFACH 3008, D-8580 BAYREUTH, FRG

1. INTRODUCTION

Azospirillum can absorb extracellular ammonium via a repressible specific transport system with the expenditure of energy (Hartmann, Kleiner 1982). The uptake of ^{14}C-methylammonium (MA) gives a valid measure of the ammonium uptake activity, because the label can be chased by non radioactive ammonium and only small amounts are assimilated by the glutamine synthetase (GS) reaction. Here we report on experiments concerning the regulation of ammonium uptake.

2. RESULTS AND DISCUSSION

Glutamine synthetase mutants of *A. brasilense* Sp 7, isolated by Gauthier and Elmerich (1977), which fixed nitrogen constitutively, did not take up ^{14}C-MA. Nitrogen fixing cultures of these mutants, supplemented with 70 µM glutamine, excreted up to 1 mM ammonium. Glutamine independent revertants showed a partial restored ammonium uptake, normal regulation of nitrogen fixation and no ammonium excretion. Therefore, ammonium uptake of *Azospirillum* appeared to be under genetic control of GS, similar as amino acid transport systems of *Salmonella typhimurium* (Kustu et al. 1979) and the ammonium uptake in *Klebsiella pneumoniae* (Kleiner 1982). Glutamate synthase mutants of *A. brasilense* Sp 6, isolated by Bani et al. (1980), were also ammonium uptake negative and excreted ammonium. For asm-mutants of *K. pneumoniae* ammonium excretion have been described by Shanmugam and Valentine (1975).
The antimetabolite of glutamine, methionine sulfoximine (MSX), effectively blocked ammonium uptake (50 % inhibition at 1 µM). In addition, MSX caused ammonium excretion. Glutamine inhibited ammonium uptake, when added above 1 mM to cell suspensions, while glutamate had no effect. Thus, like in *K. pneumoniae* (Kleiner, Castorph 1982), there is evidence for a metabolic control of ammonium uptake by glutamine.
Long term cultures of *A. brasilense* and *A. lipoferum* wild type strains excreted ammonium up to 0.5 mM, when the energy sources malate or glucose were exhausted. A release of amino acids and proteins due to cell lysis was not observed. Ammonium excreting cells did not show ^{14}C-MA-uptake and harboured an adenylylated, inactive GS. After the readdition of the energy source, ammonium was rapidly taken up again. *A. brasilense* Sp 7 did not release ammonium like other strains under the same conditions. Differences in the property to release ammonium may be of relevance for practical application of *Azospirillum*.

3. REFERENCES

Bani D et al. (1980) J. Gen. Microbiol. 119, 239-244.
Gauthier D and Elmerich C (1977) FEMS Lett. 2, 101-104.
Hartmann A and Kleiner D (1982) FEMS Lett. 15, 65-67.
Kleiner D (1982) Biochim. Biophys. Acta 688, 702-708.
Kleiner D and Castorph H (1982) FEBS Lett. 146, 201-203.
Kustu SG et al. (1979) J. Bacteriol. 138, 218-234.
Shanmugam KT and Valentine RC (1975) Proc. Natl. Acad. Sci. USA 68, 1174-1177.

4. ACKNOWLEDGEMENTS

This work was supported by the Deutsche Forschungsgemeinschaft. We thank Mrs. M. Ohlraun for skilful technical assistance.

CAN R. CAPSULATA PROTECT ITS NITROGENASE AGAINST OXYGEN DAMAGE?

AYALA HOCHMAN AND VARDA NADLER, Dept. of Biochemistry, The George S. Wise Faculty of Life Sciences, Tel-Aviv University, Tel Aviv, 69978, Israel.

1. INTRODUCTION

One of the most intriguing characteristics of nitrogen fixing organisms is that nitrogenase has remarkably similar properties irrespective of its source. Among these properties is the rapid, irreversible destruction of purified nitrogenase by exposure to oxygen. Both proteins of nitrogenase lose activity upon exposure to oxygen.

The toxicity of oxygen is well documented, as is the relative ease with which oxygen is reduced to O_2^- and O_2^{-2}, reportedly providing at least part of the basis for the observed oxygen toxicity. However, despite the O_2 lability of nitrogenase in vitro, the capacity to fix N_2 is found in obligate aerobes and oxygenic photosynthetic prokaryotes, as well as in obligate and facultative anaerobes. This implies that cells have protective mechanisms.

It is suggested that nitrogen fixing organisms protect their nitrogenase from oxygen in two general ways: A) control the access of oxygen, which can be achieved, for example, by respiratory protection, conformational protection, microaerophile, etc.; and B) synthesis of enzymes which eliminate the damaging products of oxygen reduction - superoxide dismutase (SOD) to destroy superoxide, and catalase and/or peroxidase to take care of the peroxides.

R. capsulata is a purple, non-sulfur bacterium which can grow anaerobically in the light or under aerobic conditions in the dark. It can grow with N_2 as the sole nitrogen source. The aim of our work was to study oxygen-protection-mechanisms in this bacterium in relation to the mechanisms mentioned above.

2. RESULTS AND DISCUSSION

R. capsulata cells can fix N_2 in the light under anaerobic conditions or in the presence of up to 20.5 uM dissolved O_2. Respiration rates of cultures grown with N_2 in the presence of oxygen were not higher than in anaerobic cultures. Crude extracts from R. capsulata were shown to have activities of catalase peroxidase and SOD. The cellular activities of these enzymes were higher in cells grown in the presence of oxygen relative to anaerobic cultures. However, only the increase in catalase was significantly higher in N_2-grown cells than in NH_4^+ on glutamate cultues. Nitrogenase of cells with higher catalase activity (due to either growth in the presence of O_2 or mutation) is less sensitive to inhibition by H_2O_2. However, higher catalase activity, even up to a 21-fold increase in the mutant, did not provide the cells with better protection of nitrogenase against inhibition by oxygen.

Although R. capsulata shows the switch-off-switch-on phenomenon when O_2 is added to the cell suspension, we could not find anything in crude extracts similar to the Shethma-protein reportedly providing conformational protection to azotobacter nitrogenase.

Our findings suggest that, except for catalase activity, R. capsulata cannot develop any of the known oxygen-protection-mechanisms for nitrogenase.

H_2- UPTAKE ACTIVITY IN AEROBIC NITROGEN FIXING NONHETEROCYSTOUS CYANOBACTERIA

TOIVO KALLAS, ROSMARIE RIPPKA AND GERMAINE COHEN-BAZIRE
UNITE PHYSIOLOGIE MICROBIENNE, DEPARTEMENT DE BIOCHIMIE ET GENETIQUE MOLECULAIRE, INSTITUT PASTEUR, 28 RUE DU DOCTEUR ROUX, 75724 PARIS, 15 FRANCE.

Certain nonheterocystous cyanobacteria, such as those belonging to the unicellular genus Gloeothece, have evolved strategies that allow them to fix nitrogen both in the presence of air and photosynthetically generated oxygen. Many aerobic and microaerobic azotrophs including heterocystous cyanobacteria (Bothe 1978) possess an uptake hydrogenase by which they recapture the reductant generated as H_2 gas by nitrogenase. Here we have asked: (1) Do aerobic nonheterocystous nitrogen-fixing cyanobacteria possess an uptake hydrogenase? (2) If they do, does the hydrogenase play a role in protecting nitrogenase from oxygen?

We found uptake hydrogenase activity in Gloeothece strains PCC 6909, 6909-1 (a sheathless mutant), 6501, 7109, 8302 and 73107; Cyanothece 7424; and Microcoleus 8002. However, we could detect no H_2-uptake activity in Cyanothece 7822, a unicellular strain capable of slow aerobic growth on N_2. Thus hydrogenase is clearly not an obligate requirement for aerobic N_2-fixation. We found no evidence of H_2-uptake in Plectonema 73110, a strain incapable of aerobic growth on N_2, which confirms a similar observation by Bothe (1978).

In all of the hydrogen uptake positive (Hup^+) strains tested, H_2 uptake rates in air were no more than 50% of the microaerobic rates. In some experiments, 20% H_2 stimulated aerobic C_2H_2 reduction but never by more than about 30%. However in an experiment where N_2-fixing cultures were shifted to low (1%) O_2 and limiting reductant (DCMU was added to block electron flow from photosystem II), Hup^+ strains 6909 and 7424 showed almost two fold higher initial rates and final yields of C_2H_2-reduced when incubated with 1% H_2. Strain 7822 (Hup^-) showed no difference under these conditions. This experiment suggested that hydrogenase can improve the efficiency of nitrogen fixation either: (1) by supplying reductant to a terminal oxidase and thus scavenging oxygen or (2) by providing reductant to nitrogenase. In the former case hydrogenase would contribute directly to the protection of nitrogenase against oxygen as well as supply ATP. Such an hydrogenase has been found in all aerobic azotrophs thus far examined (Robson and Postgate 1980). In the latter case hydrogenase would act only as an indirect protection mechanism. By providing electrons to nitrogenase, it would make other endogenous sources of reductant available for respiratory scavenging of oxygen.

In experiments with strain 6909, we found that H_2 uptake can occur anaerobically in the presence of DCMU and light. Upon addition of 1% O_2 under these conditions, H_2uptake activity fell to about 50% of the anaerobic rate. These data (and those above) suggest that the hydrogenase is sensitive to oxygen and not linked to a terminal oxidase. In all strains tested, hydrogenase and nitrogenase activities occurred concomitantly and both were repressed by NH_4^+. Thus both functions may be under the same genetic control.

We have also attempted to clone hup genes from Gloeothece by using a cloned Rhizobium japonicum hup gene (gift of H. Evans; see Cantrell et al. 1983) as a heterologous hybridization probe. However, we have not yet succeeded in hybridizing this probe to Southern blots of total Gloeothece DNA. Thus the hup genes of these organisms may share insufficient homology to permit cloning by this approach.

Bothe H, Distler E and Eisbrenner G (1978) Biochimie 60, 277-289.
Cantrell MA, Haugland RA and Evans HJ (1983) Proc Natl Acad Sci USA 80, 181-185.
Robson RL and Postgate JR (1980) Ann Rev Microbiol 34, 183-207.

THE PHYSIOLOGICAL BASIS FOR THE SCAVENGING OF OXYGEN AND ITS RADICALS IN HETEROCYSTS OF THE CYANOBACTERIUM *Nostoc muscorum*

LEAH KARNI, STEPHEN, J. MOSS AND ELISHA TEL-OR
DEPARTMENT OF AGRICULTURAL BOTANY, THE HEBREW UNIVERSITY OF JERUSALEM, REHOVOT 76100, ISRAEL.

The process of nitrogen fixation is dependent on the provision of reducing power for the nitrogenase and for the protection mechanism removing oxygen and its derivatives from the N_2-fixing cell.
Nitrogenase in cyanobacteria is reduced by ferredoxin which is reduced by photosystem I activity, or by NADPH, provided mainly by isocitrate dehydrogenase.
Isocitrate dehydrogenase activity is enriched in heterocysts as compared with the vegetative cells (Karni & Tel-Or, 1983). The activity of isocitrate dehydrogenase continues also at high NADPH/$NADP^+$ ratio, and may support nitrogenase more effeciently than glucose-6-phosphate dehydrogenase.
Few potential mechanisms were tested in order to elucidate their involvement in the removal of oxygen in the cyanobacterium *Nostoc muscorum*:

i) The respiratory rate observed in both cell types was relatively low, 4.5 and 1.5 nmoles O_2 x min^{-1} x mg protein^{-1}, for heterocysts and vegetative cells, respectively. Such rates may not be sufficient for the removal of oxygen.
ii) The content of glutathione in the filament was found to be 0.6-0.7 mM, suggesting its potential role in the removal of oxygen radicals.
iii) The activity of the NADPH dependent glutathione reductase was chartacterized to test the efficiency of regeneration of reduced glutathione. The activity of the enzyme varied between 50-150 nmoles GSH x min^{-1} x mg protein^{-1} and the Km for NADPH was 0.125 mM and 0.2 mM for heterocysts and vegetative cells, respectively. The Km for GSSG was up to 0.330 mM, which corresponds with the intracellular concentration of glutathione observed. The enzyme was found to be sensitive to Zn^{+2} ions, and a preincubation with GSSG protects the enzyme from the inhibition by Zn^{+2}.

The employment of the isocitrate dehydrogenase activity to generate NADPH for ammonia production and regenerate reduced glutathione, was tested.
The following scheme presents the interaction between the reducing power and the mechanisms of N_2-fixation and O_2 removal in the heterocysts.

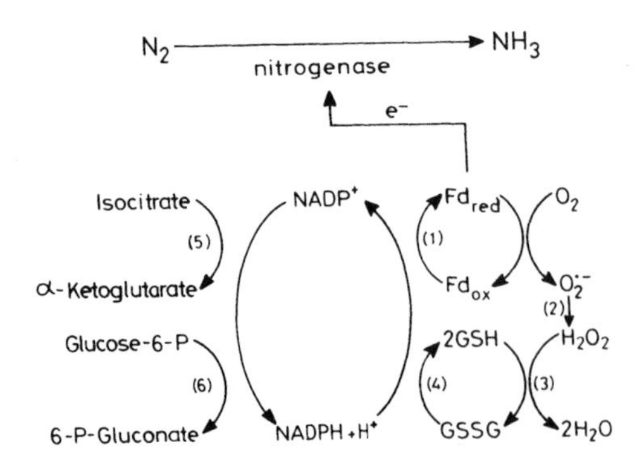

MALATE LIMITED GROWTH OF AZOSPIRILLUM BRASILENSE Sp 7 IN A CHEMOSTAT

M. KLOSS/K.-H. IWANNEK AND I. FENDRIK
INSTITUT FÜR BIOPHYSIK, UNIVERSITÄT HANNOVER, HERRENHAUSER STR. 2, D-3000 HANNOVER 21, FRG

Azospirillum brasilense is living in the rhizosphere of several tropical grasses. Under low pO_2 and supply of organic acids as carbon source it is capable of fixing molecular nitrogen. A chemostat possess certain properties which reflect ecologically important characteristics of the rhizosphere. The growth of the microorganisms is restricted by the availability of one selectable substrate. Some physiological properties of A. brasilense Sp 7 grown in a malate limited continous culture will be shown in this paper.

A chemostat for growth of microorganisms under microaerobic conditions was set up. In this chemostat Azospirillum brasilense Sp 7 was grown under nitrogen fixing conditions. The concentration of dissolved oxygen was kept constant; the concentration of malate was varied by the dilution rate (D). Several steady states could be established. Maximum growth rate was found to be 0,146 (h^{-1}). Malate limitation occured at dilution rates less or equal than 0,05 (h^{-1}). Under malate limitation the metabolism of A. brasilense changed distinctly. The content of poly-ß-hydroxybutyrate was decreasing from 18,5% to about 2% and protein was detectable in the culture filtrate. The biomass concentration, protein content and biomass productivity were found to be optimum functions with a maximum at D = 0,025 (h^{-1}) or D = 0,05 (h^{-1}) respectivily. The decrease of these parameters at low dilution rates might be due to a secondary limitation caused by the malate deficiency. Increasing the dilution rate to values higher than 0,05 (h^{-1}) malate limitation didn't occur. Under this conditions the physiological activity of A. brasilense was influenced by the strong wall growth of these organisms. The total and specific N_2-ase activity were determined by the Acetylene-reduction-assay and by the protein productivity. The calculation of N_2-ase activity in a malate limited chemostat culture was shown to be only possible by deriving from the amount of protein being formed.

EFFECT OF AMMONIA ON THE NITROGENASE ACTIVITY OF AZOTOBACTER VINELANDII

J. KLUGKIST/H. HAAKER/C. VEEGER
DEPARTMENT OF BIOCHEMISTRY, AGRICULTURAL UNIVERSITY, DE DREIJEN 11, 6703 BC WAGENINGEN, THE NETHERLANDS

In *A.vinelandii* nitrogenase activity is inhibited within minutes after addition of NH_4Cl to whole cells. The extent of this inhibition reported by different workers varies between 15 and 100%. Our results show, that this variation can be ascribed to differences in growth and test conditions

Assay conditions: The nitrogenase activity of intact cells depends on the amount of oxygen present during the assay (figure 1). At a low or a high oxygen input rate, the inhibition by added NH_4^+ is strong (up to 95%). At the oxygen input rate (2.0), where nitrogenase activity is maximum, NH_4Cl is only slightly inhibitory (7%).

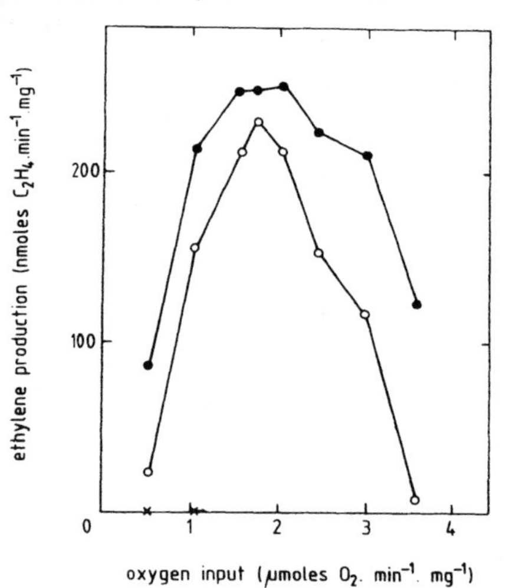

FIGURE 1.

●—● activity without NH_4Cl.
○—○ activity after addition of NH_4Cl.

Two other factors, that are important for the extent of the inhibition of nitrogenase activity by added NH_4Cl are: a) the pH of the incubation mixture (Table 1), b) the stage of growth at which cells are harvested (Table 2).

Table 1

pH	Nitrogenase activity (nmoles.min^{-1}.mg^{-1})	Inhibition by NH_4Cl
6.5	78	78%
7.0	83	67%
7.5	88	43%
8.0	78	24%

Table 2

Time of growth	Nitrogenase activity (nmoles.min^{-1}.mg^{-1})	Inhibition by NH_4Cl
3 hrs	300	21%
6 hrs	230	12%
10 hrs	180	39%
24 hrs	50	71%

NITROGENASE AND HYDROGENASE ACTIVITIES IN SULFATE-REDUCING BACTERIA.

P.A. LESPINAT, Y. BERLIER, G. FAUQUE, R. TOCI, G. DENARIAZ AND J. LE GALL,
ARBS/CNRS-ECE, CEN CADARACHE, P.O. BOX 1, F 13115 ST PAUL LEZ DURANCE.

The present work relates to nitrogenase and hydrogenase relationships in three strains of sulfate-reducing bacteria, *Desulfovibrio (D.) gigas*(NCIB 9332), *D. desulfuricans* (NCIB 8388) and *Desulfotomaculum (Dt) orientis* (ATCC 19365).

MATERIAL AND METHODS. The bacteria were either subcultured daily in batch culture or grown several weeks in chemostat, *D.* strains on Starkey's medium without ammonia and with only 0.01% yeast extract, *Dt. orientis* on Liu et al medium with 0.05% $Na_4P_2O_7$. $^{15}N_2$ reduction and $H^+/^2H_2$ exchange were followed by mass spectrometry (2), H_2 changes and C_2H_2 reduction by gas chromatography.

RESULTS AND DISCUSSION. Under N_2-fixing conditions, all three species grew more slowly than on ammonia and *Dt.orientis* only in the presence of pyrophosphate. The highest nitrogenase activities recorded were 220 nmol N_2 and 500 nmol C_2H_2 reduced per hour per mg protein for *D. desulfuricans* and only 7.3 nmol C_2H_2 for *Dt.orientis*. Although higher than reported earlier (1,4,5,6) these values are below those found for other anaerobes such as *Clostridium pasteurianum*.

Using the $H^+/^2H_2$ exchange reaction, a reversible hydrogenase activity was shown in all three species. In the *D.* strains both exchange components were insensitive to C_2H_2 but were inhibited by CO. With *D.desulfuricans*, the prominence of either H_2 uptake or H_2 production depended on the nutritional status and particularly on the electron donor or acceptor. For instance with pyruvate instead of lactate, an H_2 evolution was sometimes observed even with ammonia and sulfate. Without a carbon source but with sulfate, H_2 uptake was enhanced indicating that H_2 served as an electon donor in sulfate reduction. Under diazotrophic conditions H_2 evolution was observed as long as nitrogenase was not active. When it was, H_2 uptake occurred concomitantly with N_2 fixation but not with C_2H_2 reduction (fig.1). Indeed, H_2 uptake appeared to depend, at least partly, on the nitrogen nutrition of the cells, whether ammonia was provided, or N_2 was actively being fixed. This was probably in connection with increased needs for electron donors in nitrogen metabolism.

A recycling process of H_2 coupled with ATP production was recently reported (3) in sulfate-reducing bacteria. The present results show that such a process does exist when *D. desulfuricans* is fixing N_2 but not, or to a lesser extent, when it is reducing C_2H_2. Hence, in contrast to the data for other diazotrophs, the higher figures found for nitrogenase activity with the $^{15}N_2$ than with the C_2H_2 technique.

(1) Le Gall J et al (1959) Ann. Inst. Pasteur 96, 223-230.
(2) Lespinat PA et al (1978) Biochimie 60,339-341.
(3) Odom JM, Peck HD Jr (1981) FEMS Lett.12, 47-50.
(4) Postgate JR (1970) J. Gen. Microbiol.63, 137-139.
(5) Riederer-Henderson MA, Wilson PW (1970) J. Gen. Microbiol. 61, 27-31.
(6) Sekiguchi I et al (1977) Canad.J. Microbiol.23, 567-572.

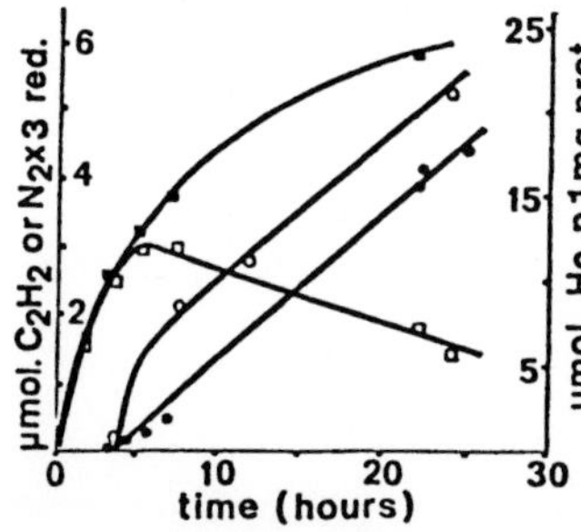

Fig.1 : Relationships between the reduction of N_2 (○) or C_2H_2 (●) and the respective H_2 evolution or uptake (□ under N_2, ■ under C_2H_2).

STIMULATION OF PREFORMED NITROGENASE IN GLOEOTHECE 6909 BY C_2H_2 IS O_2-DEPENDENT

P.S. MARYAN/R.R. EADY/J.R. GALLON*/ AND A.G. CHAPLIN*
AGRICULTURAL RESEARCH COUNCIL, UNIT OF NITROGEN FIXATION, UNIVERSITY OF SUSSEX, BRIGHTON BN1 9RQ, U.K.
*DEPARTMENT OF BIOCHEMISTRY, UNIVERSITY COLLEGE OF SWANSEA, SINGLETON PARK, SWANSEA SA2 8PP, U.K.

In *Azotobacter vinelandii* and in various filamentous cyanobacteria C_2H_2 gradually stimulates the activity of pre-formed nitrogenase (David *et al*, 1978). This observation has been interpreted as being a result of stimulated nitrogenase synthesis caused by N-starvation (David and Fay, 1977) or of a conformational change in the structure of nitrogenase (Apte *et al*, 1975). It has also been shown that under certain conditions, C_2H_2 and its reduction product C_2H_4 bind to the Mo-Fe protein of *Klebsiella pneumoniae* and stimulate electron flux through nitrogenase *in vitro* (Thorneley and Eady, 1977).

In cultures of *Gloeothece*, grown aerobically under constant illumination at 2500 lx addition of C_2H_2 to 10% (v/v) stimulated the rate of nitrogenase activity during the following 30 min, but subsequently the rate declined to a constant level after about 2 h.

The extent of the transient stimulation of nitrogenase by C_2H_2, was dependent on O_2 concentration. For example, no stimulation was observed at 5% dissolved O_2 tension. (DOT) a 1-fold stimulation was found at 10% DOT. 2-fold stimulation was obtained at 15 and 20% DOT. The stimulatory effect was rapid but the final rates of reduction after 2 hours were independent of O_2 concentration. Addition of 0.1 mg/ml of tetracycline or chloramphenicol had no effect on the transient stimulation of nitrogenase activity, though activity subsequently declined with a half-life of 2.5 h.

It therefore appears that rapid stimulation and subsequent decrease in nitrogenase activity to a constant level is not a result of nitrogen starvation caused by the inhibition of nitrogen fixation by C_2H_2. Furthermore it is unlikely that O_2 directly stimulates nitrogenase, because high rates of C_2H_2 reduction are observed under anaerobic conditions. Protein synthesis is not involved as stimulation still occurs when protein synthesis is inhibited. Nevertheless, due account of this effect must be taken when assaying *Gloeothece* for C_2H_2 reduction activity in air.

REFERENCES

1. Apte, S.K., David, K.A.V. and Thomas, J. (1978). Biochem.Biophys.Res.Commun., 83, (3), 1157-1163.
2. David, K.A.V., Apte, S.K. and Thomas, J. (1978). Biochem.Biophys.Res.Commun., 82 (3), 39-45.
3. David, K.A.V., and Fay, P. (1977). Appl.Env.Microbiol., 34, 640-653.
4. Thorneley, R.N.F., and Eady, R.R. (1977). Biochem.J., 167 457-461.

REGULATION OF NITROGEN FIXATION BY GLUTAMINE IN *Rhodopseudomonas sphaeroides* forma sp. *denitrificans*

W.P. MICHALSKI/D.J.D. NICHOLAS
Department of Agricultural Biochemistry,
Waite Agricultural Research Institute,
University of Adelaide,
Glen Osmond, 5064, South Australia.

Purple bacteria utilize N_2 or ammonium salts as a nitrogen source for photosynthetic growth. Nitrogenase (N_2ase), glutamine synthetase, GS (EC 6.3.1.2) and glutamate synthase, GOGAT (EC 1.4.7.1) are key enzymes for the assimilation of inorganic nitrogen in these bacteria.

Addition of either ammonium ions or glutamine to a culture of photosynthetic bacteria fixing N_2 produces a rapid inactivation of N_2ase ("switch on/off" effect) (Zumft, Castillo, 1978). Glutamine and/or GS rather than free ammonia have been implicated in this regulatory mechanism. Recently we have shown that either ammonia or glutamine shock treatment of toluene-permeabilized cells of *R. capsulata* preincubated with [^{14}C] ATP resulted in a labelling of both Fe-protein of N_2ase and GS, resulting in their inactivation (Michalski et al. 1982). Now, we present evidence that glutamine is involved in regulating nitrogenase activity in *R. sphaeroides* forma sp. *denitrificans*.

N_2ase and GS were purified and characterized from *R. sphaeroides* forma sp. *denitrificans*. The molecular and regulatory properties of N_2ase and GS purified from this bacterium were similar to those reported for other photosynthetic bacteria (*R. capsulata, R. palustris*).

An ammonia shock treatment (15 mM NH_4Cl) of nitrogen-starved cells of *R. sphaeroides* forma sp. *denitrificans* preincubated for 20 min with [^{14}C] glutamate in light resulted in a rapid accumulation (within 5 min) of [^{14}C] glutamine, as well as adenylylation of GS. N_2ase was also completely inhibited. However, after incubation for one hour, both [^{14}C] glutamine concentration and the adenylylation state of GS reverted to the original levels before ammonia treatment. The preincubation of washed cells in light for 30 min with 2 mM azaserine (an inhibitor of GOGAT) followed by further incubation with either [^{14}C] glutamate or $^{15}NH_4Cl$ also resulted in an accumulation of ^{14}C- or ^{15}N-labelled glutamine. The concentration of glutamine produced was approx. 15 μmol per mg dry weight calculated from either the ^{14}C or ^{15}N experiments. This increase in glutamine correlated well with increased adenylylation of GS (3 to 60%). Under these conditions N_2ase was also inhibited as determined by both $^{15}N_2$ fixation and C_2H_2 reduction.

Our results indicate that glutamine is an effector compound for the regulation of N_2ase and that the main role of GS in this regulation is to produce glutamine.

Michalski WP, Nicholas DJD and Vignais PM (1982) Biochim. Biophys. Acta 743, 136-148.
Zumft WG and Costillo F (1978) Arch. Microbiol. 117, 53-60.

O_2 PROTECTION OF NITROGENASE IN *FRANKIA* ISOLATE HFPArI3.

M. A. Murry, M. S. Fontaine and J. G. Torrey

Cabot Foundation, Harvard University, Petersham, MA USA. 01366.

O_2 protection of nitrogenase in a cultured *Frankia* isolate from *Alnus rubra* (ArI3) was studied *in vivo*. Acetylene reduction (nitrogenase activity) was correlated with vesicle differentiation following removal of NH_4^+. Vesicles were broken off from the point of attachment to vegetative filaments by brief sonication. Loss of acetylene reduction was correlated with the increase in detached vesicles. Respiratory protection was suggested by the observation that O_2 consumption rates in actively-fixing vesicle-containing cells were 5 times greater than rates in ammonia-grown undifferentiated cells. Furthermore, carbon limitation resulted in a decrease in the optimum O_2 tension for nitrogenase activity. In actively-fixing (10 nMol C_2H_4/min/mg protein), aerobically induced cells, maximum acetylene reduction occurred near 20k PaO_2. Acetylene reduction and O_2 uptake decreased more than 10-fold in cells that were washed and incubated aerobically without a carbon source. Residual acetylene reduction activity was optimal at 5k PaO_2 and nearly eliminated at 20 k PaO_2. Evidence for a passive gas diffusion barrier in the vesicles was obtained by kinetic analysis of *in vivo* enzyme rates. O_2 uptake rates of NH_4^+-grown cells showed an apparent KmO_2 of approximately 1.0 um O_2. In N_2-fixing cultures, a KmO_2 of approximately 160 um O_2 was observed. Thus, respiration remains unsaturated by O_2 at air saturation levels. *In vivo* the apparent Km for acetylene was 10-fold greater than reported *in vitro* values. These data show diffusion-limited kinetics and are interpreted as evidence for a gas diffusion barrier in the vesicles but not vegetative filaments of ArI3.

NITROGEN FIXATION AND DENITRIFICATION BY AZOSPIRILLUM

G. Neuer, A. Kronenberg, M.P. Stephan, W. Zimmer and H. Bothe
Botanisches Institut, Universität Köln, Gyrhofstr. 15, D-5 Köln 41

Azospirillum species participate in all steps of the nitrogen cycle except nitrification. They can fix molecular nitrogen and perform assimilatory nitrate reduction and nitrate respiration. Recent experiments showed that *Azospirillum brasilense* Sp. 7 grows under anaerobic conditions utilizing nitrate (Bothe et al. 1981), nitrite (Stephan et al. 1983) and nitrous oxide (Stephan et al. 1983) as respiratory electron acceptor. Under anaerobic conditions, nitrate respiration can also supply the energy for nitrogen fixation without supporting growth (Bothe et al. 1981). Nitrate-dependent nitrogenase activity lasts only 3 - 4 h until the enzymes of assimilatory nitrate reduction are formed (Bothe et al. 1981; 1983).
An association between wheat plants and *Azospirillum brasilense* Sp. 7 has recently been investigated for N_2-fixation (C_2H_2-reduction and denitrification (N_2O-formation) activity. 25 germinated wheat seeds and an *Azospirillum* culture with about 1.3×10^{7} cells were inoculated onto the surface of a semisolid agar/mineral salt medium in 1 l flasks and incubated for a week in a growth chamber under light/dark cycles (12.5 h at 33 °C in the light and 11.5 h at 23 °C in the dark). C_2H_2-reduction and N_2O-formation activities were then assayed for a day where the gas phase consisted of argon supplemented with 0.5% C_2H_2 and 1 - 2% O_2.
C_2H_2-reductions and N_2O-formations were strictly dependent on both *Azospirillum* and wheat plants in the assays. Both activities commenced 3 - 5 h after removal of air out off the flasks. C_2H_2-reduction occurred when the concentration of nitrate in the semisolid agar medium was under 1 mM. Above 1 mM KNO_3 in the medium, C_2H_2-reduction stopped and the association evolved N_2O by denitrification. Both C_2H_2-reduction and N_2O-formation were strictly dependent on the presence of C_2H_2 indicating that the reaction was not due to abiotic effects. The addition of carbohydrates to the medium did not enhance the activities indicating that the bacteria must have lived from the carbon sources of the plants. C_2H_2-reduction and N_2O-formation were maximal at a temperature comparable to those of the tropical zone and marginal at temperatures prevailing in the temperate regions. Some C_2H_2-reduction and no N_2O-formation were observed when the assays were performed in air.
Thus depending on the O_2 and NO_3^- amount available, the association may either perform nitrogen fixation or denitrification.
Details of the experiments will be published in Bothe, H., Kronenberg, A., Stephan, M.P., Ismailcelebioglu, Y.N. and Neuer, G. In Klingmüller, W. (ed) *Azospirillum* workshop, Experientia Suppl., Birkhäuser, Basel.

REFERENCES

Bothe H, Barbosa G and Döbereiner J (1983) Z. Naturforsch. in press
Bothe H, Klein B, Stephan MP and Döbereiner J (1981) Arch. Microbiol. 130, 96-100
Stephan MP, Zimmer W and Bothe H (1983) manuscript in preparation

This work was kindly supported by a grant from the BMFT (PTA 8424)

MUTANTS OF AZOSPIRILLUM BRASILENSE DEFECTIVE IN NITROGEN FIXATION

F.O. PEDROSA+ and M.G. YATES++

+ DEPT. OF BIOCHEMISTRY, UNIVERSIDADE FEDERAL DO PARANA, C. POSTAL 939, 80,000 CURITIBA, PR, BRAZIL.

++ ARC UNIT OF NITROGEN FIXATION, UNIVERSITY OF SUSSEX, BRIGHTON BN1 9RQ UK.

We have isolated and characterised Nif^- mutants of Azospirillum brasilense Sp7 $Nal^{15}Sm^{200}$ (FP2) following NTG mutagenesis. One mutant lacked active MoFe protein of nitrogenase and three others were complemented by nifA from Klebsiella pneumoniae.

MATERIALS AND METHODS

FP2 was mutagenised (20 µg/ml NTG) in NFb HP.NH_4Cl medium (50% survivors), outgrown (30 subcultures) for 10 generations and selected for slow growth on N-free agar under 1% O_2:N_2. Nitrogenase MoFe component (Ab1) Fe component (Ab2) and Ab2 activating factor were isolated from glutamate-grown FP2 (glutamate allowed Nif expression under air in liquid medium). Nitrogenase activity was measured by C_2H_2 reduction.

RESULTS

Nif^- mutants: Small colonies comprised 3.7% of the mutagenised FP2; 27% of these showed <10% of wild-type acetylene-reducing activity in N-free semi-solid medium at 37° but only 8 isolates (1%) were inactive in NFb HP glutamate medium at 30°. Of these FP3, 8, 9 and 10 had zero, FP5, 6 and 7 <5% and FP <10% of wild-type activity.

Growth on N sources: All strains grew well with 20 mM NH_4Cl. FP5 failed or grew poorly on all other N sources (1 mM NH_4Cl, His, Pro, Arg, NO_3^-). FP4, 8 and 9 failed to grow on NO_3^-; FP3, 4, 5 and 8 grew poorly on 1 mM NH_4Cl and 3 and 4 grew poorly on histidine. FP8 and 9 had low GS (30%) and FP8 no GOGAT.

Genetical complementation: Plasmid pCK3 (K. pneumoniae $nifA^c$ Tc^r) was mobilised by pRK2013 into FP2 and all the Nif^- mutants at $\sim 10^{-3}$/recipient, confirmed by agarose gel electrophoresis. Purified transconjugants of FP8, 9 and 10 showed 30 to 50% of wild-type nitrogenase activity in semi-solid N-free medium at 30° but not at 37°. FP2, 8, 9 and 10 transconjugants derepressed nif in NFb HP.NH_4Cl medium; NH_4^+ 'switched off' nitrogenase activity in glutamate-grown FP2 and FP10 transconjugants but not in FP7 and FP8 transconjugants. FP8 and 9 transconjugants grew well in NFb HP.NO_3^- medium. pCK3-dependent nif expression was eliminated by introducing plasmid R68.45.

Biochemical complementation: All the mutants had Ab2 activating factor; FP3, 8, 9 and 10 had no Ab1 or Ab2. FP6 had very low Ab1 activity and produced high levels of Ab2. FP4, 5 and 7 had low (<10%) levels of both Ab1 and Ab2.

CONCLUSIONS: FP8, 9 and 10 were complemented by K. pneumoniae nifA and are therefore regulatory mutants. FP10 is a $nifA^-$ type mutant but FP8 and 9 may be $ntrC^-$ (nifA complements ntrC mutations[1]) since pCK3 transconjugants grow on NO_3^-. FP6 is a structural mutation affecting either one of the subunits or the FeMo cofactor of Ab1. These results suggest that nif expression in A. brasilense is under nif specific (nifA like) and ntr type regulation analogous to that in K. pneumoniae.

REFERENCES AND ACKNOWLEDGEMENTS

1. Merrick MM (1983) EMBO Journal 2 39-44.

We thank Dr. C. Kennedy for pCK3 and FOP the British Council and CNPq.

METHYLAMINE TRANSPORT IN N_2-FIXING CYANOBACTERIA

AMAR N. RAI[+]/PETER ROWELL[++]/WILLIAM D.P. STEWART[++]
[+]SCHOOL OF LIFE SCIENCES, UNIVERSITY OF HYDERABAD, HYDERABAD 500134, A.P.; INDIA.
[++]DEPARTMENT OF BIOLOGICAL SCIENCES, UNIVERSITY OF DUNDEE, DUNDEE DD1 4HN, U.K.

Whereas nitrogenase in free-living organisms is inhibited by ammonium, this inhibition is reduced, or does not occur, when the free-living organism develops into symbiosis (Stewart, Rowell 1977; Stewart et al. 1980; Laane et al. 1980; Rai et al. 1980). Laane et al. (1980) have suggested that ammonium does not inhibit nitrogenase activity of bacteroids because they lack an ammonium transport system. Kleiner et al. (1981) hypothesised that the ammonium transport system may also be affected in symbiotic cyanobacteria. We have examined both the free-living heterocystous cyanobacterium Anabaena variabilis and the symbiotic packets of Anabaema azollae isolated from the water fern Azolla caroliniana for an ammonium transport system using methylamine, an analogue of ammonium.
The free-living cyanobacterium A. variabilis showed a biphasic pattern of methylamine uptake. Initial accumulation (upto 60s) was independent of methylamine metabolism, but long term uptake was dependent on methylamine metabolism via glutamine synthetase (GS). The methylamine was converted into methylglutamine which was not further metabolised. The addition of MSX, to inhibit GS, inhibited methylamine metabolism, but did not affect the methylamine transport system. Ammonium, when added after the addition of methylamine, caused the efflux of free methylamine; when added before methylamine, ammonium inhibited methylamine uptake indicating that both ammonium and methylamine share a common transport system. Carbonylcyanide m-chlorophenylhydrazone (CCCP) and triphenylmethylphosphonium ($TPMP^+$) both inhibited methylamine accumulation indicating that the transport system was $\Delta\psi$-dependent. At pH7 and at an external methylamine concentration of 30 μ mol dm^{-3}, A. variabilis showed a 40-fold intracellular accumulation of methylamine (internal concentration 1.4 mmol dm^{-3}). Packets of the symbiotic cyanobacterium A. azollae also showed a $\Delta\psi$-dependent ammonium transport system suggesting that the reduced inhibitory effect of ammonia on nitrogenase cannot be attributed to the absence of an ammonium transport system but is probably related to the reduced GS activity of the cyanobiont.

Kleiner D, Phillips S and Fitzke E (1981) In Bothe H and Trebst A, eds, Biology of Inorganic Nitrogen and Sulphur, pp.131-140, Springer-Verlag, Berlin.
Laane C, Krone W, Konings W, Haaker H and Veeger C (1980) Eur. J. Biochem. 103, 39-46.
Rai AN, Rowell P and Stewart WDP (1980) New Phytol. 85, 545-555.
Stewart WDP and Rowell P (1977) Nature 265, 371-372.
Stewart WDP, Rowell P and Rai AN (1980) In Stewart WDP and Gallon JR, eds, Nitrogen Fixation, pp.239-277, Academic Press, London New York.

ON THE FORMATION OF AN OXYGEN-TOLERANT THREE COMPONENT NITROGENASE COMPLEX FROM AZOTOBACTER VINELANDII

G. SCHERINGS/H. HAAKER/J.H. WASSINK
DEPARTMENT OF BIOCHEMISTRY, AGRICULTURAL UNIVERSITY, DE DREIJEN 11, 6703 BC WAGENINGEN, THE NETHERLANDS

Complex formation between FeMo-protein, Fe protein and the so called Fe/S protein II (Fe/S II) was studied by gel chromatography (see Table 1).

TABLE 1. Presence of Av1, Av2 and Fe/SII in high molecular mass fractions obtained by gel filtration of mixtures of the three proteins in the presence and absence of $MgCl_2$ and $Na_2S_2O_4$ and after oxidation with phenazine methosulfate.

Conditions	Av1	Av2	Fe/SII	nitrogenase activity (nmoles C_2H_4 formed min^{-1}. mg total protein^{-1})
reducing, $MgCl_2$	+	-	-	0
reducing	+	±	-	10
oxidizing	+	±	±	0
oxidizing, $MgCl_2$	+	+	+	250

As can be seen from Table 1 only in the presence of Mg^{2+} and with oxidized proteins a stable three component complex can be isolated. By starting with different ratios of the three proteins in the mixture applied to the molecular sieve column the maximal values for Av2 and Fe/SII in the complex with Av1 could be estimated. The values are Av1:Av2:Fe/SII = 1:2.4±0.5:1.1±0.2. The Fe/S protein II contains 2[2Fe-2S] centers per dimer of 26 kDa. When protein mixtures at thes ratios are mixed and oxidized, either anaerobically by PMS or by low concentrations of O_2, a reasonably oxygen-stable complex is obtained (see Table 2).

TABLE 2. Inactivation by O_2 (250 μM) of phenazine methosulfate or O_2 (1 μM) oxidized mixtures of Av1, Av2 and Fe/SII. Fe/SII contains 1.7 [2Fe-2S] centers per dimer of 26 kDa

ratio of Av1, Av2 and Fe/SII	Oxidation method	nitrogenase activity before O_2 exposure	2 min. after O_2 exposure
1/2.4/1.8	-	460	115
1/2.4/1.8	PMS	460	435
1/2.4/1.2	O_2	490	434
1/2.5/1.4 + 2 mM MgATP	PMS	436	436
1/2.5/1.4 + 2 mM MgADP	PMS	436	412

Analysis in the ultracentrifuge showed that the major fraction of the reconstituted complex sediments with S-values centered around 34S. This suggests an average mass for the oxygen stable nitrogenase complex of 1.5 x 10^6 Da. Taking into account the determined stoichiometry of the individual proteins, the molecular composition of the complex is 4 molecules of Av1, 8-12 molecules of Av2 and 4-6 molecules of Fe/SII containing two [2Fe-2S] clusters per dimer of 26 kDa.

ALTERATION OF HETEROCYST PATTERN AND AKINETE FORMATION IN ANABAENA CYLINDRICA PRODUCED BY NEOPEPTONE.

PAWAN SHARMA
INSTITUTE OF PHYSIOLOGICAL BOTANY, BOX 540, S-751 21 UPPSALA, SWEDEN.

Anabaena cylindrica does not form heterocysts on N-supplemented medium, but on transfer to N-free medium, proheterocysts appear at regular intervals in filaments after 6 hrs of incubation. These proheterocysts, develop into mature heterocysts after about 14 hrs of incubation. Akinete formation occurs during the late exponential phase of both N-free and N-supplemented cultures and akinetes are always formed adjacent to the heterocyst in N-free cultures. On adding Neo-peptone B_{119} (Difco) at concentration 0.4 g/l to lag phase cultures of A. cylindrica, following effects were observed after 40 hrs incubation: (i) increased heterocyst frequency with altered heterocyst spacing and presence of double and multiple heterocysts in cultures grown on N-free medium, (ii) induction of regular pattern of heterocysts in ammonia-grown cultures, (iii) induction of akinete formation both adjacent and non adjacent to heterocysts in both N-free and ammonia-grown cultures. Chloramphenicol (5µM) negated these effects when given at 0, 6, and 12 hrs but not after 24 hrs of growth with peptone. Bacto-peptone B_{118} (Difco) and Peptone-Bacteriological L_{37} (oxoid) at 0.4 g/l or higher, caused no change in pattern of heterocyst and akinete formation in both N-free and ammonia-grown cultures.

Gel-filtration of Neo-peptone through Sephadex G-50 (superfine) showed the presence of active factor(s) with molecular weight between 15,000-25,000. Autoclaving the active fraction had no effect on its activity while acid hydrolysis or digestion with trypsin (1 ml fraction with 1 mg trypsin for 60 min), resulted in complete loss of activity indicating that active factor(s) may be polypeptide(s). The active fraction caused lysis of A. cylindrica at higher concentrations, but on dilution, it showed the same effects as observed with Neo-peptone in both ammonia-grown and N-free cultures. In cultures grown in N-free medium, most of the vegetative cells are enlarged 2-3 times after 12 hrs incubation with the active fraction of Neo-peptone. After 18 hrs of incubation, these enlarged cells develop into proheterocysts which may divide to give rise to pair of proheterocysts and eventually develop into mature heterocysts after about 24 hrs of incubation. Some of the vegetative cells do not divide but keep on elongating and develop into akinetes after about 30 hrs. In ammonia-grown cultures, proheterocysts and pro-akinetes are seen after 24 hrs of incubation with active-fraction, which eventually develop into mature heterocysts and akinetes.

These results clearly indicate the involvement of some polypeptide(s) in regulation of heterocyst and akinete formation. The nature of such polypeptide(s) and whether it is the same polypeptide(s) which regulate heterocyst and akinete formation respectively, is being investigated. Under normal condition, such polypeptide(s) may be produced by vegetative cells undergoing N-starvation, so that they may form a gradient against N-gradient in the cells. A cell may require a particular optimum level of such polypeptide(s) for its complete development into heterocyst. Adjacent cells which may have levels lower than this optimum may develop into proheterocyst, which, in turn, may be reverted back completely by the products of the mature heterocyst, to vegetative cells or may form pro-akinetes, depending upon the culture conditions. In the cyanobacteria with akinetes non-adjacent to heterocysts e.g. Nostoc the cells adjacent to heterocyst may have more capacity to revert back to vegetative cells. While during akinete formation, the cells in centre of interheterocyst interval may have levels of polypeptide(s) lower than optimum for heterocyst formation, resulting in their conversion into akinetes.

CHANGE IN NITROGENASE ACTIVITY OF ANAEROBIC BACTERIA, GENUS CLOSTRIDIUM DEPENDING ON THE CONDITIONS OF GROWTH AND PHASE OF DEVELOPMENT

SHERAJUL ISLAM SHELLEY/L.K. NITZSE/ V.T.EMTSEV
DEPARTMENT OF MICROBIOLOGY, TIMIRYAZEV AGRICULTURAL ACADEMY, MOSCOW, USSR.

Heterotrophic nitrogen-fixing anaerobic bacteria genus Clostridium can use a wide range of simple carbon-bearing compounds and energy sources. The effectiveness of usage of such sources by the anaerobic nitrogen-fixing bacteria depends on the chemical origin of these compounds.

The model tests with the usage of Cl.pasteurianum and Cl.saccharobutyricum growing on nutrient substrate with mixture of carbon-bearing compounds (glucose + mannit or glucose + starch) showed two peaks of nitrogenous activity ($C_2H_2 \rightarrow C_2H_4$). The first peak corresponds to an easily accessible source of carbon and energy and when its concentration in the substrates decreases the nitrogenous activity goes down. The rate of decrease of the nitrogenous activity depends on the rate of changing over of the bacteria culture to the usage of the second component of the substrate. The mechanism of this phenomena - diaxin - represents a regulation of catabolic reactions, type of catabolic repression. If there are two organic substrates, the organism synthesizes the ferments necessary for assimilation of a more accessible compound and in this case a repression of synthesis of ferments required for assimilation of less accessible compounds takes place.

The most important conclusion of the model is that the rate of development of anaerobic nitrogen-fixing bacteria and their physiological activity are limited by the complex of the carbon-bearing compounds.

The factors making for the process of spore formation of anaerobic nitrogen-fixing bacteria genus Clostridium at the same time stimulate intensification of the nitrogenous activity since the maximum value of this process with Cl.pasteurianum and Cl.saccharobutyricum is observed during the phase of spore formation.

REDOX-DEPENDENCE OF THE SHETHNA IRON-SULFUR PROTEIN II - NITROGENASE INTERACTION AND THE EFFECT OF HIGH pN_2 ON H_2 EVOLUTION BY NITROGENASE

FRANK B. SIMPSON AND R. H. BURRIS
DEPARTMENT OF BIOCHEMISTRY, UNIVERSITY OF WISCONSIN, MADISON, WI 53706 USA

The Shethna iron-sulfur protein II (FeS II) is believed to protect the nitrogenase of Azotobacter vinelandii from inactivation by O_2 by forming a complex with the nitrogenase components. From the "switch-off - switch on" response of Azotobacter it is reasonable to predict that formation of this O_2-stable complex is redox-dependent. SDS gel electrophoresis shows that FeS II is composed of two 15,000 MW subunits. Anaerobic chromatography on Sepharose CL-6B in the absence of dithionite followed by resolution of the column fractions on SDS gels has shown that FeS II forms a high MW complex with the thionin-oxidized, mixed nitrogenase components; mixed proteins reduced with dithionite form no complex. Thus complex formation is redox-dependent. A high MW complex forms between the oxidized FeS II and Fe proteins but none forms between the thionin-oxidized MoFe and FeS II proteins. The stability of nitrogenase activity toward O_2 in the presence of FeS II depends upon the redox state of the proteins immediately prior to exposure to O_2. When the redox potential was adjusted with a potentiostat to measured values between -425 mV and +15 mV (vs. H_2) prior to exposure to O_2, protection due to FeS II began at -300 mV and increased to a maximum at +15 mV at which potential the half-life of nitrogenase activity was 25 min as compared to 2.5 min for the minus FeS II control. FeS II protected the isolated Fe protein slightly only at potentials above -300 mV but did not protect isolated MoFe protein. That complex formation and O_2 stability mediated by FeS II are redox-dependent is compatible with its midpoint potential of -225 mV and with the suggested redox-dependent role of FeS II in conformational protection.

Results of previous studies of nitrogenase performed at low pN_2 when extrapolated to high pN_2 predicted that N_2 cannot completely block H_2 evolution and that 13% to 23% of the electron flux is directed toward H_2 production at infinite pN_2. To test the effect of high pN_2 on H_2 evolution we have conducted experiments with purified nitrogenase at 50 atm N_2. The Fe and MoFe proteins used had specific activities of 2100 and 1400 nmoles electron pairs · min^{-1} · mg^{-1} protein, respectively, and protein concentrations were selected to saturate the MoFe protein for electron flow. The following results were obtained.

Experiment #	µmoles N_2 reduced	µmoles H_2 produced	% of total electron flux to H_2	H_2:N_2 ratio
1	1.42	1.40	24.7	0.985
2	7.85	5.69	19.5	0.725

We conclude that H_2 production is not blocked, rather 19% to 25% of the electron flux is allocated to H_2 evolution under 50 atm N_2. That the H_2 evolved : N_2 fixed ratio is close to 1:1 implies that H_2 evolution may be obligatory in the N_2 fixation reaction of nitrogenase.

NITROGENASE-MEDIATED H_2 PRODUCTION BY *RHODOPSEUDOMONAS CAPSULATA*. ROLE OF UPTAKE HYDROGENASE IN THE EFFICIENCY OF H_2 PRODUCTION

J.C. WILLISON, D. MADERN, A. COLBEAU, Y. JOUANNEAU and P.M. VIGNAIS
Biochimie (CNRS/ER 235, INSERM U.191), Département de Recherche Fondamentale, Centre d'Etudes Nucléaires, 85X, 38041 Grenoble cedex, France

As in other N_2 fixers, the hydrogenase of *R. capsulata* functions as an uptake hydrogenase. It is an intrinsic membrane protein with a high affinity for H_2 (K_M^{H2} = 0.2 µM) (Colbeau, Vignais, 1981) and a true component of the respiratory chain. It has recently been shown, with the use of hydrogenase antibodies, to be oriented towards the cytoplasmic compartment (Colbeau et al., 1983) and to contain N_i (Colbeau, Vignais, 1983). The role of uptake hydrogenase in the efficiency of H_2 production by *Rhodopseudomonas capsulata* has been studies as follows : 1. Isolation and biochemical characterization of hydrogen uptake-deficient (Hup^-) mutants. 2. Comparison of the efficiencies of H_2 production by the Hup^- mutants and wild type strain, B10. 3. Analysis of the relationship between hydrogenase activity and nitrogenase activity in continuous culture. Two Hup^- mutants, IR2 and IR4, were isolated, which contained hydrogenase activities corresponding to about 10% and 5%, respectively, of the activity found in strain B10. When L-lactate was given as electron donor, both mutants differed little from strain B10 in the rate and efficiency of H_2 production. However, with DL-malate as electron donor, strain IR4 produced approx. 50% more H_2 (per mole of malate consumed) than did strain B10. Data are presented, which suggest that factors other than the loss of hydrogenase activity were responsible for the enhanced efficiency of H_2 production in strain IR4.

Colbeau A and Vignais PM (1981) Biochim. Biophys. Acta 662, 271-284.
Colbeau A, Chabert J and Vignais PM (1983) Biochim. Biophys. Acta (in press).
Colbeau A and Vignais PM (1983) Biochim. Biophys. Acta (in press).

REGULATION OF SYNTHESIS OF H_2-UPTAKE HYDROGENASE IN *AZOTOBACTER CHROOCOCCUM* CORRELATION WITH THE NAD:NADH RATIO.

M.G. YATES, C.D.P. PARTRIDGE
ARC UNIT OF NITROGEN FIXATION, UNIVERSITY OF SUSSEX, BRIGHTON BN1 9RQ UK.

O_2, H_2 NH_4^+ ions, NO_3^- ions and carbon substrates affect the *in vivo* activity of H_2 uptake hydrogenase (Hup) in *Azotobacter*. These substrates could affect the redox state of the cell and, hence, the intracellular NAD:NADH ratios.

MATERIALS AND METHODS

A. chroococcum was grown in N_2-fixing SO_4^{2-}-limited chemostat cultures in Burk's medium. Nitriloacetate (NTA) was present to inhibit synthesis of active Hup. Washed cells resuspended in NTA-free medium were used to follow Hup derepression at optimum O_2. Alternatively, Hup levels were determined on chemostat samples. One unit of activity is one μmole H_2 absorbed/mg protein/min.

NAD:NADH ratios were determined on samples withdrawn into acid (NAD) or alkali (NADH) for assay by NADH specific FMN reduction coupled to light production catalysed by bacterial luciferase.

RESULTS

Hup synthesis first increased, then declined with increasing O_2; the optimum O_2 level for derepression increased with the carbon substrate concentration.

Carbon substrates: Hup levels and NAD:NADH ratios decreased progressively (2.5 to 0.5 units and 8.4 to 1.7 respectively) with increasing dilution rates (0.03 to 0.25 h^{-1}) under sucrose limitation. Hup levels and NAD:NADH ratios were high in SO_4^{2-}-limited cultures supported by organic acids (lactate, acetate, gluconate, malate, succinate; >0.8 units and >5 respectively) compared with sugars (sucrose, glucose, mannitol: <0.4 units and <2.5).

Other substrates: NO_3^- had two opposite effects on Hup synthesis: both synthesis and the NAD:NADH ratio were increased during short term exposure. During this time NO_3^- did not repress nitrogenase. Adaptation to growth on NO_3^- required 26 days in a chemostat. Subsequently both Hup and nitrogenase were repressed. NH_4^+ repressed Hup but did not affect the NAD:NADH ratio. H_2 (4%) enhanced Hup synthesis by only 30-50% in NH_4^+-grown SO_4^{2-}-limited cultures.

Catabolite repression studies: There was no apparent relationship between cAMP and Hup levels: (a) Hup activities were 5 fold higher in sucrose than SO_4^{2-}-limited cultures whereas the cAMP levels were identical. (b) The glucose analogue α-methyl glycoside (10 mM) did not repress Hup. (c) Addition of 0.5 to 1 mM dibutyryl cAMP failed to stimulate Hup synthesis.

CONCLUSIONS

Hup synthesis is repressed in Azotobacter by glucose-related sugars more so than by carboxylic acids. This effect and the enhancement of synthesis by short-term exposure to NO_3^- may be due to the effect of these substrates on a redox couple affecting the NAD:NADH ratio. Alternatively, sugars may effect catabolite repression through a regulator other than cAMP. Repression of Hup synthesis by NH_4^+ is, apparently, by a different, unknown, mechanism.

POSTER DISCUSSION 4B METABOLISM OF RHIZOBIUM AND AZOTOBACTER

M.J. DILWORTH, School of Environmental and Life Sciences, Murdoch University, Western Australia.

The convenor apologized to poster presenters whose work did not get fully discussed, but pointed out that he hoped the topics chosen for discussion were those where the most people would be able to contribute to debate.

The discussion covered four areas of physiology of nitrogen-fixing bacteria - the physiology of *Rhizobium* in relation to carbon metabolism in laboratory culture and in the nodule, molybdenum metabolism in relation to nitrogenase, electron transport pathways and energy yield in *Rhizobium*, and the occurrence of nitrogenase peptides in ammonia-grown *Azotobacter*.

A number of problems surfaced in relation to *Rhizobium* carbon metabolism. The view was expressed that parallels to rhizobial physiology should not be sought in the behaviour of *Escherichia coli*, but that comparison to *Pseudomonas* was likely to be more useful. The problem of dealing with the physiology of a very diverse group of organisms was also emphasized, and the general agreement was that it was too early to be able to form a generalized idea of rhizobial physiology, and that too much generalizing was done from particular individual organisms.

The use of chemostats for establishing principles in rhizobial physiology was discussed, with the caution that a number of events in physiology were very much limited to particular stages of growth curve in batch culture. The problem of pH control was also raised - with the general difficulties of acidification of media in sugar- or polyol-grown cultures, and alkalinization in organic acid-based media.

Some discussion focussed on the occurrance of catabolite repression in *R.meliloti*. While one study had emphasized the occurrence of classical catabolite repression, another experience showed that other strains of *R.meliloti* did not show diauxic growth but simultaneous utilization of substrates.

Another area discussed was the effect of high concentrations of sugars on the respiration rate of isolated bacteroids. It appeared that actual utilization of the sugar had not really been shown - a stimulation of endogenous respiration was also a possible explanation.

Other discussion concerned the status of bacteroids in legume nodules - whether they were essentially carbon- or oxygen-limited. A change in attitude was evident from plant physiology in regarding the nodule as more oxygen-limited than carbon-limited. From the microbiologists viewpoint, bacteroids with extensive reserves of poly-β-hydroxybutyrate were signalling a carbon-rich situation with limitation imposed by some other nutrient, probably oxygen. This discussion was suspended to allow the nodule physiology discussion the following day to explore that area. The presence of other carbohydrates in soybean bacteroids - sucrose and trehalose - was discussed.

On the question of alterations in rhizobial electron transport pathways, it was pointed out that chemostat cultures differed in their cytochrome patterns in response to the limitation imposed - particularly carbon or oxygen. It was also emphasized that, while in a number of bacteroids the cytochrome aa_3 complex was decreased or eliminated, this was not generally true. The question of which rhizobial oxidases were responsible for oxygen consumption remained an area requiring further understanding. Knowledge of energetic coupling points in the respiratory chain of bacteroids was clearly still slight.

Another area of discussion brought up from a poster was the occurence or otherwise of nitrogenase peptides in ammonia-grown A.vinelandii. Other evidence presented showed that antibodies prepared agains purified MoFe protein were cross-reactive with a considerable range of proteins transferred to nitrocellulose by Western blotting. However, no identification of genuine nitrogenase proteins could be made from ammonia-grown cells.

There was considerable discussion of the effect of molybdenum deficiency on production of nitrogenase activities in Azotobacter species. One poster reported the underestimates of N_2 fixation by acetylene reduction assays with Mo-deficient cultures. The basic question of whether non-classical nitrogenases are produced in response to Mo deficiency or whether the activities reported were altered properties of classical nitrogenase was discussed. It was emphasized that this was a basic question of genetics, with the possibility of other genes specifying non-classical nitrogenases. It appeared that in A.chrooccum some areas of the chromosome besides the classical nitrogenase region showed hybridization with nitrogenase DNA, but that this was not apparent in A.vinelandii. It was pointed out that no other non-classical nitrogenase-appeared in K.pneumoniae in response to Mo-deficiency. Further work was obviously necessary before the occurrence of multiple nitrogenases in Azotobacter species was resolved.

LOCALIZATION OF NITROGENASE USING MONOCLONAL ANTIBODIES

DWIGHT BAKER/CRAIG LENDING/DENNIS DEAN, C.F. KETTERING RESEARCH LABORATORY
150 E. SOUTH COLLEGE ST., YELLOW SPRINGS, OHIO 45387 USA

1. INTRODUCTION At present there exist two theories on the location of nitrogenase in the bacterial cell. The first, reported by Reed *et al.* (1974), suggested that nitrogenase is associated with a vesicular membrane system called an "azotophore". More recent research by Haaker and Veeger (1977) suggested no membrane involvement by nitrogenase based on rates of sedimentation. Our approach was to physically localize nitrogenase in the bacterial cell by using monoclonal antibodies and a protein A-gold EM labelling procedure.

2. PROCEDURE Nitrogenase was isolated and purified from *Azotobacter vinelandii* using the procedure of Burgess *et al.* (1980). Monoclonal antibodies were prepared against the nitrogenase and characterized by ELISA and immunoblotting procedures. Selected monoclonals were used as an ultrastructural stain on sections of *Azotobacter vinelandii, Azosprillum brazilense* and *Enterobacter cloacae* and visualized using an indirect protein A-gold label. The location of nitrogenase in active nitrogen-fixing bacterial cells was determined from the number and site of the precipitated gold particles.

3. RESULTS Characterization of our monoclonals by ELISA demonstrated that two, A_2 and C_2, had specific activity for native nitrogenase from both *Azotobacter* and *Clostridium*. Further characterization using Western blots and immunochemical staining indicated that A_2 retained its high specific activity even for denatured protein but that C_2 did not. The binding of A_2 and C_2 was used to localize nitrogenase within *Azotobacter* (Fig 1), *Azospirillum* and *Enterobacter*. No association of the label with membranes was observed in any of the bacteria, *fix*- and non-immune serum controls gave no specific labelling (Fig 2). From these results it was concluded that nitrogenase is freely distributed in the bacterial cytoplasm; confirming the reports of Haaker and Veeger (1977). The use of monoclonal anti-nitrogenase and the protein A-gold labelling procedure will be a useful tool in studying regulation and physiology of nitrogenase.

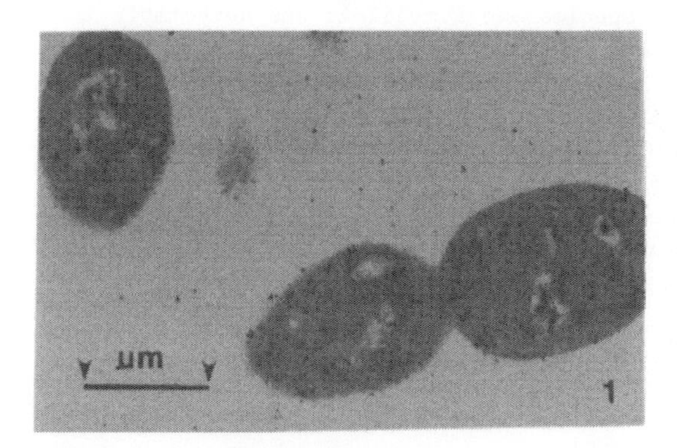

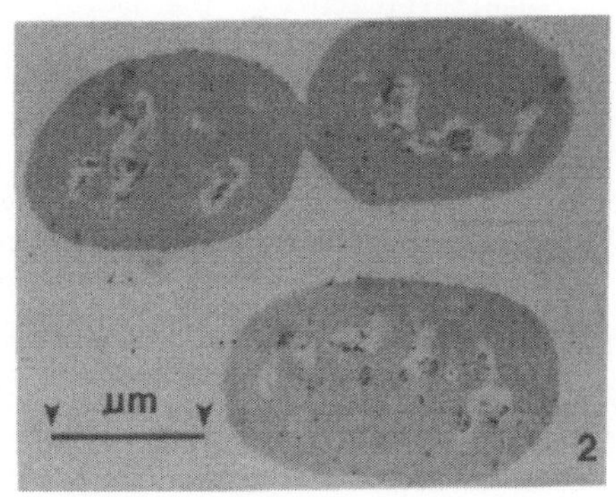

Figure 1. *Azotobacter* cells stained with anti-nitrogenase and protein A-gold.
Figure 2. *Azotobacter* cells stained with non-immune serum and protein A-gold.

Burgess, B, Jacobs, D and Stiefel, E (1980) BBA 614: 196-209.
Haaker, H and Veeger, C (1977) Eur. J. Biochem 77: 1-10.
Reed,D,Toia, R Jr and Raveed, DL (1974) BBRC 58: 20-26.

DENITRIFICATION BY FAST-GROWING RHIZOBIA

CASELLA G.*, LEPORINI C.*, NUTI M. P.**

* Istituto di Microbiologia agraria, Università di Pisa
**Istituto di Chimica e Industrie agrarie, Università di Padova, Italy

1. INTRODUCTION

Nitrate has been shown to serve as an electron acceptor during chemoorganotrophic growth by different rhizobia, particularly R.japonicum and strains of the cow-pea group, while according to literature data fast growing rhizobia do not reduce nitrate by dissimilatory means, at least in rich media. Here we describe conditions suitable to detect nitrous oxide formation by fast-growing rhizobia both in planta and ex planta.

2. MATERIAL AND METHODS

Strains of R.trifolii, R.leguminosarum, and R. "hedysarum" (Casella et al. 1983a, submitted) were grown under anoxyc conditions in a chemically defined medium (Casella et al. 1983b, submitted) or in rich media such as yeast-mannitol-broth. N_2O formation was monitored by gas-chromatography. for in planta studies, acetylene-reduction (ARA) and N_2O formation were assessed with intact plants and excised nodules. Symbiotic properties were scored by conventional methods.

3. RESULTS AND DISCUSSION

Gaseous nitrogen (N_2O) was formed during anoxyc growth of R.trifolii, R.leguminosarum, and R. "hedysarum"; after 4 days at 25-26°C in minimal medium, but not in rich media, about 0.2 μM of N_2O were formed x mg^{-1} cells d.w. in the presence of 1 to 12 mM nitrate, nitrite, and ammonium nitrate. The N_2O production is enhanced by vaste availability of organic matter.

Denitrification of the above three species and R.meliloti was detectable also in planta by using N_2O-reductase inhibition test as well as ^{15}N measurement. N-N_2O evolved in the gas phase increases from 10 to 45 μM/mg nodules d.w. almost linearly with increasing concentrations of N-NO_3 supplied to the plants (1 to 12 mM).

Since the above results indicate that gaseous nitrogen losses and nitrogen fixation (ARA) may occur contemporarily in the nodules, a better knowledge of the reciprocal rates of these activities is needed, as well as more insights of their regulatory mechanisms.

4. ACKOWLEDGEMENTS

This investigation was supported by CNR, P. F. IPRA.

NITROGENASE ACTIVITY IN MO-LIMITED CHEMOSTAT CULTURES OF *AZOTOBACTER VINELANDII* HAS AN ALTERED SUBSTRATE SPECIFICITY

R.R. EADY and R.L. ROBSON
Unit of Nitrogen Fixation, University of Sussex, Brighton, BN1 9RQ, UK.

Chemostat cultures of *A. vinelandii* UW growing on N_2 in purified Burk's medium without added Mo were established in order to characterise Mo-limited growth and nitrogenase activity.

Steady-state cultures growing at 30°C were established in an all-glass chemostat (Baker, 1968). The medium was substantially freed (< 10^{-9} M) of trace Mo by extraction with 8 hydroxyquinoline. All other components of Burk's medium were present in excess.

We showed that growth was dependent on N_2 by:
1) changing the gas supply from air to Ar/N_2 (80:20% $^v/v$) whereupon the population declined to < 1% closely following the theoretical washout rate.
2) $^{15}N_2$ was incorporated to a 3% atom percent excess.

Mo-limitation was indicated by:
1) addition of Mo at 3nM to the culture medium resulted in an increase in culture density. No other element tested limited growth of the population (see below).
2) Increasing the dilution rate (over the range from 0.089 to 0.26) decreased the population density and N content per ml of culture but the rate of N_2 fixation per ml of culture was relatively constant. This is consistent with an intrinsic N-limitation consequent on limitation of available Mo.

These cultures showed substantial rates of H_2 evolution *in situ* (80 nmol min^{-1} mg of $protein^{-1}$), both in the presence and absence of C_2H_2 (0.1 atm.). Comparison of the rate of N_2 fixation with the rate of C_2H_2 reduction showed that C_2H_2 measurements considerably underestimated (3-4 fold at D = 0.206) the actual rate of N_2 fixation. Analysis of polypeptides present in steady state cultures showed low levels of both nitrogenase components and suggested a high ratio of Fe protein to Mo-Fe protein.

Samples removed from the chemostat were used as a source of Mo-starved organisms for batch growth experiments. No additional trace metals were required for good growth with NH_4^+ as N source. Unsupplemented chemostat samples continued to grow with a doubling time of 14h on N_2 with linear, not exponential growth kinetics, consistent with a rate-limiting step associated with Mo limitation. Diazotrophic growth was stimulated by Mo (2.5-10 nM) provided that sufficient Fe was present. Fe^{3+} alone (30μM); Mn^{2+} (0.3μM); Ni^{2+} (0.66μM); Cu^{2+} (0.4μM); Zn^{2+} (7.3μM); ReO_4^- (0.03μM); BO_3^{3-} (6.6μM) and WO_4^{2-} (6.6μM) were without effect. The growth response to added Mo (10nM) was biphasic. Initially a rapid (within 30 min) small increase in growth rate was observed, followed 9h later by a further marked increase in growth. Pulse-labelling with $^{35}SO_4^{2-}$ showed that there was no rapid (within 5h) effect on the rate of synthesis of either component of nitrogenase.

Tungstate up to 6.6mM did not antagonise the stimulation in growth by Mo added at 10 nM or 5μM. It is apparent that organisms derived from Mo-limited continuous culture are insensitive to WO_4^{2-} inhibition of growth. Vanadate when added to Mo deficient cultures stimulated growth at 50 nM, a low level which makes it unlikely to be due to contamination with Mo.

REFERENCE: Baker (1968) Lab Practice 17: 817-824.

HEAVY METAL TOLERANCE OF FAST GROWING RHIZOBIUM ISOLATES

A. FYSON/R.P. WHITE/R.B. WADEY
ROTHAMSTED EXPERIMENTAL STATION, HARPENDEN, HERTFORDSHIRE. AL5 2JQ, ENGLAND

Tolerance of Rhizobium isolates to heavy metals was screened by multi-point inoculation of defined medium agar plates each with a metal ion at a particular concentration.
Forty four Rhizobium isolates from a wide range of environments were exposed to nine heavy metal ion (Ag^{2+}, Cd^{2+}, CrO_4^{2-}, Cu^{2+}, Hg^{2+}, Ni^{2+}, Pb^{2+}, Zn^{2+}) and Al^{3+} over a wide range of concentrations at pH 6.7. Presence or absence of colonies was noted after 96 h. Ag^{2+} was the most toxic ion (no isolates able to grow at 1.2 μM) followed by Hg^{2+}, CrO_4^{2-}, CU^{2+}, Cd^{2+}, Co^{2+}, Ni^{2+}, Al^{3+}, Pb^{2+} and Zn^{2+} which was tolerated by some isolates at 3 mM in these conditions. There was much variation between isolates and between cross inoculation groups. A multivariate analysis (principal coordinate analysis) of the data separated R.meliloti, R.leguminosarum and R.trifolii as three distinct groups on the basis of heavy metal tolerance.
The effects of Cd^{2+}, Cu^{2+} and Ni^{2+} on nodulation of Trifolium repens L. ∿S184 by two R.trifolii isolates on Fahråeus slopes were examined. One isolate (tri 10) was able to form nodules and fix nitrogen (acetylene reduction) at higher metal ion concentrations than the other (tri 6). This isolate was also found to be more tolerant to these ions in the plate tests.

The plate test described provides a simple and efficient method of screening rhizobia for heavy metal tolerance.

METABOLISM OF ORGANIC ACIDS AS RELATED TO BACTEROID DIFFERENTIATION AND FUNCTION IN THE *RHIZOBIUM MELILOTI*-ALFALFA SYMBIOSIS

A. E. GARDIOL, G. L. TRUCHET, AND F. B. DAZZO
DEPARTMENT OF MICROBIOLOGY, MICHIGAN STATE UNIVERSITY, EAST LANSING, MICHIGAN, 48824 U. S. A.

Succinate and malate are abundant organic acids within legume root nodules, and *Rhizobium* mutant strains defective in their uptake and metabolism are ineffective in nitrogen-fixing symbiosis with the host plant (Ronson et al. 1981; Gardiol et al. 1982). The role of succinate and malate in bacteroid differentiation and function was investigated with a wild type strain *R. meliloti* L5-30, a succinate-dehydrogenase mutant strain UR6, and a spontaneous revertant strain UR7.

PROCEDURES Minimal media and growth conditions were as previously described (Gardiol et al. 1982). Methods to study the electron microscopy of alfalfa nodule ultrastructure (Truchet et al. 1980), bacterial morphology (Mutaftschiev et al. 1982), ^{14}C-mannitol uptake (Gardiol et al. 1980), protein determination and mannitol dehydrogenase activity (Drets et al. 1980) were all published. Oxygen consumption by cell suspensions was measured with an NBS oxygen electrode.

RESULTS AND DISCUSSION Meristematic nodules induced by strains L5-30 and UR7 were effective with typical histologies and ultrastructures. Nodules induced by UR6 were ineffective. 3 weeks after inoculation, very few transformed bacteroids were observed, and premature degeneration and lysis occurred in the central zone. 6 weeks after inoculation, the nodules were fully senescent and the bacteria were lysed in all zones. L5-30 and UR7 bacteroids were typically degenerated only in the normal senescent zone.

In vitro growth studies showed that succinate or malate affected cell growth in 27 mM mannitol minimal medium. At 20 mM succinate or malate, L5-30 showed growth inhibition, and repression of mannitol uptake and mannitol dehydrogenase activity. 15% of the cells displayed elongation and pleomorphism, similar to results with *R. trifolii* 0403 (Urban, Dazzo 1982). At 10 mM mannitol plus 5 mM succinate or malate, no inhibition of growth was observed. UR6 grew on fumarate or malate but not on succinate. 20 mM succinate did not inhibit growth of UR6 in mannitol, nor induce cell pleomorphism or repress mannitol uptake and mannitol dehydrogenase activity. 20 mM malate induced "Y" morphology in 3% of UR6 cells, but no cellular elongation or growth inhibition. L5-30 consumed oxygen at a greater rate in 27 mM mannitol plus 20 mM succinate than did UR6. These results suggest that metabolism of organic acids through a complete TCA cycle is necessary for normal differentiation and maintenance of functional alfalfa bacteroids.

REFERENCES Gardiol A, Arias A, Cervenansky C, Gaggero C and Martinez-Drets G (1980) J. Bacteriol. 144,12-16.
Gardiol A, Arias A, Cervenansky C and Martinez-Drets G (1982) J. Bacteriol. 151,1621-1623.
Martinez-Drets G and Arias A (1970) J. Bacteriol. 103,97-103.
Mutaftschiev S, Vasse J and Truchet G (1982) FEMS Microbiol. Lett. 13,171-175
Ronson CW, Lyttleton P and Robertson JG (1981) Proc. Natl. Acad. Sci. U.S.A. 78,4284-4288.
Truchet G, Michel M and Denarie J (1980) Differentiation 16,163-172.
Urban J, Dazzo FB (1982) Appl. Env. Microbiol. 44, 219-225.

CONTINUOUS CULTURE OF A RHIZOBIUM STRAIN (ORS 571) FIXING NITROGEN IN CULTURE WITH NEW AND UNUSUAL CHARACTERISTICS.

Christiane Gebhardt (1) and Fraser J. Bergersen
CSIRO, Division of Plant Industry, P.O. Box 1600, Canberra 2601, Australia
(1) Present address: Biochemistry Department, Rothamsted Experimental Station, Harpenden, Herts. AL5 2JQ, England.

INTRODUCTION

ORS 571 is a fast growing Rhizobium strain which induces stem as well as root nodules on its host plant Sesbania rostrata, a tropical legume. It is the first Rhizobium strain to show nitrogen fixing growth on agar plates and in batch cultures (Dreyfus, Dommergues 1981, Elmerich et al. 1982). In contrast, other strains which fix nitrogen in culture (e.g. 32H1, CB756) are slow growers and require a reduced nitrogen source for growth. Our aim was to find out whether ORS 571 can grow continuously with N_2 as the sole nitrogen source.

METHODS

ORS 571 was cultured in a chemostat. Dissolved O_2 and O_2-supply were adjusted and controlled by the head gas composition and the stirring rate feedback-controlled by an O_2-electrode. Nitrogen fixation was measured by the total nitrogen balance of the culture (Kjeldahl-analysis) and by acetylene reduction. The inflowing medium did not contain nitrogen except for the vitamins Ca-pantothenate and nicotinate (20 mg/l each = 3.4 mg N/1 or 2 mg/l each = 0.34 mg/l).

RESULTS

1. We were able to induce and to maintain continuous cultures of ORS 571 for up to 37 days irrespective of whether the medium contained high or low concentrations of the vitamins. Taking into account the nitrogen content of the vitamins and small impurities in the medium we estimated that at least 80-90% of the total nitrogen was provided by N_2-fixation.
2. The maximum values for the amount of nitrogen fixed and for nitrogenase activity were 380 nmoles N_2 mg^{-1} dry wt. h^{-1} and 1800 nmoles C_2H_4 mg^{-1} dry wt. h^{-1} respectively. These are the highest values yet recorded for Rhizobium in culture.
3. A continuous N_2-fixing culture could be maintained in up to 12 μM dissolved O_2. Therefore the nitrogenase activity of ORS 571 was much less O_2-sensitive than of any other known strain. However nitrogenase activity was sensitive to the O_2-supply, N_2-fixing growth could be maintained at greater than 40 nmoles O_2 mg^{-1} dry wt. min^{-1} but not at 30 nmoles O_2 mg^{-1} dry wt. min^{-1}.

REFERENCES

Dreyfus BL and Dommergues YR (1981) In Gibson AH and Newton WE, eds, Current Perspectives in Nitrogen Fixation, p.471, Austr. Ac. Sc., Canberra.
Elmerich C, Dreyfus BL, Reysset G, Aubert JP (1982) The EMBO J. 1, 499-503.

ACKNOWLEDGEMENTS

The authors thank BL Dreyfus for providing the strain and AH Gibson and GL Turner for their help. This work was supported by the Deutsche Forschungsgemeinschaft.

CARBON METABOLISM IN THE *RHIZOBIUM*-LEGUME SYMBIOSIS

A.R. GLENN, I. McKAY, R. ARWAS AND M.J. DILWORTH,
NITROGEN FIXATION RESEARCH GROUP, SCHOOL OF ENVIRONMENTAL AND LIFE SCIENCES,
MURDOCH UNIVERSITY, MURDOCH, WESTERN AUSTRALIA, 6150.

This paper presents data on sugar and organic metabolism in *Rhizobium leguminosarum* MNF 3841.

R. leguminosarum MNF 3841 metabolises glucose, fructose and sucrose via the Entner-Doudoroff and pentose phosphate pathways and not via glycolysis since it lacks phosphofructokinase. The enzymes of C_6 and C_{12} sugar catabolism are essentially constitutive, though the "Entner-Doudoroff" enzymes are lower in Pi-limited chemostat cells grown on fumarate (18 nmol min^{-1}(mg protein)$^{-1}$) than in similar cells grown on glucose (66 nmol min^{-1}(mg protein)$^{-1}$). Isolated pea bacteroids possess a complete C_6 sugar catabolic system at specific activities comparable to organic acid grown free-living cells. Phosphate-limited chemostat cells grown on sucrose have only low activities of fructose bis phosphate aldolase (7 mU.mg $protein^{-1}$) and PEP carboxykinase (0 mU.mg $protein^{-1}$) whereas cells grown on fumarate contain 30 mU and 128 mU mg $protein^{-1}$ for the aldolase and PEPCK respectively. Isolated bacteroids contain significant amounts of these two enzymes, suggesting that the bacteroids are gluconeogenetic. Free-living and bacteroid forms of strain MNF 3841 possess a full complement of enzymes of the TCA cycle: malate dehydrogenase is present at particularly high levels in the bacteroid.

In Pi-limited chemostat cultures containing fumarate and glucose in differing molar ratios, strain MNF 3841 utilizes both simultaneously, but exhibits a preference for the dicarboxylic acid. Even when glucose is present at twice the molar concentration of fumarate, MNF 3841 utilizes 3 times more fumarate than glucose.

Tn5-induced mutants unable to use various carbon compounds for growth were all able to nodulate, but varied in their capacity to fix N_2 (Glenn *et al*, 1983a). All the sugar-negative mutants were effective, though the fructose mutants were delayed in the onset of nodulation. The fructose mutants showed "fructose-inhibition" i.e. growth on a second carbon source was retarded by fructose (Glenn *et al.* 1983b). A pyruvate dehydrogenase mutant unable to utilize pyruvate, or sugars, was similarly able to nodulate and was effective. Inability to utilize sugars appears, therefore, to have no significant effect on either nodulation or fixation. Two succinate negative strains nodulated peas but were ineffective. Two mutants unable to utilize C_4 dicarboxylic acids, or a range of other compounds as sole carbon source, were able to nodulate; one was effective and one ineffective. This suggests that the capacity to utilize C_4 dicarboxylic acids is important in the establishment of an effective symbiosis.

Taken together these results suggest a picture of an N_2-fixing bacteroid with gluconeogenic capacity consuming organic acids in preference to sugars.

References

Glenn, A.R., McKay, I., Arwas, R. & Dilworth, M.J. (1983a). Sugar metabolism and the symbiotic properties of carbohydrate mutants of *Rhizobium leguminosarum*. Journal of General Microbiology (in press).

Glenn, A.R., Arwas, R., McKay, I. & Dilworth, M.J. (1983b). Fructose metabolism in wild type, fructokinase negative and revertant strains of *Rhizobium leguminosarum*. Journal of General Microbiology (in press).

CONTROL OF HEME BIOSYNTHESIS IN *RHIZOBIUM SP.*

R. GOLLOP AND Y.J. AVISSAR
DEPARTMENT OF BIOLOGY, BEN GURION UNIVERSITY OF THE NEGEV
P.O. BOX 653 BEER SHEVA 84105 ISRAEL

Heme synthesis in Rhizobium is regulated at the level of 5-aminolevulinate synthase (ALAS) activity (1). Free-living cells produce small amounts of heme and have correspondingly low levels of ALAS activity, while symbiotic rhizobia produce larger amounts of heme and exhibit high ALAS levels (3). The extra heme produced by bacteroids is exported into the surrounding cytoplasm and serves as a precursor of leghemoglobin (Lb) production. The globin moiety of Lb is produced by the host plant (2). Lb is essential for nitrogen fixation in the nodules of legumes. The study of the regulation of heme synthesis in Rhizobium is a first step towards understanding the regulation of Lb biosynthesis.

Rhizobium sp. 2206 is a slow growing microsymbiont of peanuts. Cultures were grown on yeast-mannitol (YM) medium at 30°C on rotary shakers. ALAS activity was assayed *in vitro* (4). Growth was followed by turbidity measurements using a Klett Summerson colorimeter with a blue (#42) filter.

Rhizobium cultures grown on YM exhibit a short log phase (generation time 5.5 h), followed by a prolonged intermediate phase and eventually reach stationary phase. In the intermediate phase, about 100 K.U. turbidity, the activity of ALAS was about 3 times as high as either in the log or the stationary phase. This transitory increase in ALAS activity was prevented by the presence of 0.1 mM hemin in the growth medium. However, the effect of hemin was prevented by the addition of 1 mg/ml bovine serum albumin in the growth medium. Neither hemin nor bovine serum albumin in the *in vitro* assay mixture affected ALAS activity.

ALAS activity is apparently modulated by the concentration of heme (or hemin) in the surrounding medium. A basal level of ALAS activity, probably that required for intracellular syntheses, is maintained in the presence of high levels of free heme, but the synthesis of large amounts of ALA and (heme) is depressed. This mechanism could be used for the coordination of heme synthesis with globin synthesis in the legume root nodule.

REFERENCES

1. Avissar Y.J. & Nadler K.D. (1978) 135:782-789.
2. Cutting J.A. & Schulman H.M. (1971) Biochim. Biophys. Acta 192:486-493.
3. Nadler K.D. and Avissar Y.J. (1977) Plant Physiol. 60:433-436.
4. Roessler P.G. and Nadler K.D. (1982) J. Bacteriol. 149:1021-1026.

This research was funded in part by the Israel Academy of Science and in part by BARD (project no. I-298-80).

REGULATION OF NITROGENASE IN PARASPONIA-RHIZOBIUM STRAIN ANU289

PETER M. GRESSHOFF[†], G.L. BENDER[††], S. HOWITT[†], R. SANDEMAN[†], B.G. ROLFE[††] and S.S. MOHAPATRA[†]
[†]Botany Department, [††]Genetics Department (RSBS),
AUSTRALIAN NATIONAL UNIVERSITY, CANBERRA, AUSTRALIA

1. INTRODUCTION:

Rhizobium strain ANU289, capable of effective nodulation and symbiotic nitrogen fixation in several tropical legumes (such as siratro) and the tropical non-legume Parasponia was found to derepress nitrogenase under defined in vitro conditions. This permitted an analysis of the physiological and genetic parameters of nitrogenase control (Mohapatra et al, 1983, Mohapatra, Gresshoff, 1983). Findings from the in vitro studies were further tested in isolated bacteroids from siratro and Parasponia rigida nodules. The study represents part of a general attempt to describe the non-legume symbiosis involving Parasponia at the bacterial and plant level (see other contributions in this book).

2. PROCEDURES:

Strain ANU289 was isolated as a streptomycin, rifampicin resistant mutant from strain CP283 (Trinick, Galbraith, 1980). It is non-mucoid on mannitol containing media. Derepression of nitrogenase activity was possible on agar cultures, in stationary liquid and agitated liquid culture. The derepression conditions are as follows: preculture cells in mannitol containing medium (such as RGM30M, Mohapatra et al, 1983) to a density of 2×10^9 cells/ml. Dilute cells 1:4 (after washing) into derepression medium J50GM (containing N-free B5 salts, 50mM succinate, 10mM mannitol, 3mM glutamate, pH 6.8). Place 1.0ml of cell mix into a scintillation vial, seal and gas to 91% argon, 5% oxygen and 4% acetylene. Incubate at 27°C, stationary. For agitated culture, cells were resuspended at a density of 1.1×10^9 cells/ml in a medium similar to J50GM, except that mannitol was deleted and 50mM HEPES was added. Flat bottom flasks (400ml volume with 10ml culture) were agitated at 170 rpm at 27°C. Oxygen was maintained at 0.2-0.3% throughout the experiment.

3. RESULTS AND DISCUSSION:

Using the agitated liquid culture nitrogenase activities of up to 77 nmoles C_2H_4/hr/mg protein were determined at 48-60h after derepression. Succinate and fumarate were optimal carboxylic acids. A reduced nitrogen source (3mM glutamate) was needed for derepression. Glutamine and ammonium (above 5mM) severely affected nitrogenase activity. The ammonia effect was reversable by the addition of MSO. Mucoidy (exopolysacharide) inhibited nitrogenase activity. Bacteroids isolated from siratro nodules in the absence of hemoglobin showed a low oxygen optima in the absence of succinate. No ammonia or glutamine effects were demonstrable in ANU289 bacteroids. Ammonia assimilation of strain ANU289 occurs via the GS-GOGAT pathway, as both enzymes were measured in vegetative cells. GDH was not detectable. The conclusion gained from the analysis of in vitro control of nitrogenase is that (a) the Parasponia strain has no special adaptation to handle higher oxygen levels, (b) glutamine exerts a very pronounced inhibition of in vitro but not bacteroid nitrogenase and (c) the in vitro derepressed state is only a marginal representation of the symbiotic state within the legume or non-legume nodule.

4. REFERENCES:

Mohapatra, SS et al (1983) Arch. Microbiol. 134, 12-16
Mohapatra, SS and Gresshoff PM, (1983) Curr. Sci. 52, 352-357
Trinick MJ and Galbraith J (1980) New Phytol. 86, 17-26

EFFECT OF MANGANESE ON AMMONIA EXCRETION AND AMMONIA ASSIMILATORY ENZYMES IN AZOTOBACTER CHROOCOCCUM

K.G. GUPTA AND NARULA NEERU
DEPARTMENT OF MICROBIOLOGY, PANJAB UNIVERSITY, CHANDIGARH, INDIA

Some of the *Azotobacter chroococcum* strains have been shown to release ammonia in liquid culture medium. This observation could be of interest in agriculture and industrial economy. Studies have shown that a number of strains of *Azotobacter chroococcum* showing ammonia release can be isolated and modified by mutation with UV irradiation. It is also shown that ammonia electrode is more sensitive and useful than indophenol method of ammonia estimation. Among the cations studies 0.1 μM Mn^{++} increased ammonia excretion by 230% in culture medium, 120% in sterilized soil and 190% in unsterilized soil over the control. Whereas cations like Co^{++}, Cu^{++} and Zn^{++} were inhibitory to ammonia excretion. It was observed that at 20 days of incubation (in presence of Mn^{++}) where maximum ammonia was released there was increase in GS (from 9.22 units) and GOGAT (from 2700 to 11400 units) activities. No GD activity was observed at 20 days incubation. Absence of GDH suggest that there is inhibition of ammonia assimilation. These results indicate that Mn^{++} might be useful in increasing the ammonia excretion by Azotobacter. It is also possible that optimum concentration of Mn^{++} in soils might help in increasing the yield of a crop by providing more ammonia via excretion by Azotobacter.

REVERSAL OF THE TETRAMETHYL THIURAM DISULFIDE (TMTD) EFFECT ON SYMBIOTIC NITROGEN FIXATION AND SOIL ENZYMES BY *PSEUDOMONAS AERUGINOSA*

K.G. GUPTA AND C.K. SHIRKOT
DEPARTMENT OF MICROBIOLOGY, PANJAB UNIVERSITY, CHANDIGARH, INDIA

Pesticides are often found to act against nontarget organisms like Rhizobia, etc. The pesticides are also known to be degraded by microorganisms. The results showed that the effect of tetramethylthiuram disulphide (TMTD) on symbiotic nitrogen fixation could be reversed by using TMTD degrading strains of *Pseudomonas aeruginosa*. It was observed that this strain of Pseudomonas could degrade 45% and 40% of TMTD in salt medium and soil, respectively.
Adverse effect of TMTD on growth of Rhizobia and nodulation was reversed up to 100 ppm TMTD concentration. TMTD effect on dry weight and total nitrogen was reversed to a significant level when soil was inoculated with *P. aeruginosa*. The adverse effect of TMTD on soil enzyme activity was also found to be reversed in the presence of *P. aeruginosa.* It is suggested that this strain of *P. aeruginosa* could be used as inoculant along with rhizobia to protect rhizobia against the harmful effects of seed dresser like TMTD.

SYNTHESIS AND ACTIVITY OF NITROGENASE IN *KLEBSIELLA PNEUMONIAE* EXPOSED TO LOW CONCENTRATIONS OF OXYGEN

SUSAN HILL[+], G.L. TURNER, F.J. BERGERSEN
CSIRO, DIVISION OF PLANT INDUSTRY, CANBERRA, A.C.T. 2601, AUSTRALIA
[+]ARC UNIT OF NITROGEN FIXATION, UNIVERSITY OF SUSSEX, BRIGHTON, BN1 9RQ, UK.

1. INTRODUCTION

O_2 rapidly inhibits both activity and synthesis of nitrogenase in *K. pneumoniae* (Eady et al 1978). Nevertheless limiting O_2 enhances N_2 fixation in chemostats (Hill 1976). Bergersen et al (1982) exploited leghaemoglobin to maintain low defined dissolved O_2 and established, in a strain carrying a chromosomal *nifH::lac* fusion (Dixon et al 1980) with the Nif^+ plasmid pRD1, 50% repression of *nifH::lac* by 100 nM O_2 (close to the apparent K_s of the principal terminal oxidase). Here we report that lower dissolved O_2 accelerates derepression, that nitrogenase activity is no more sensitive to O_2 than expression of *nifH::lac* and that nitrogenase activity is enhanced microaerobically.

2. METHODS

As for Hill (1976) for shaken flask assays and as for Bergersen et al (1982) and Bergersen, Turner (1979) for flow chamber reactions except that C_2H_2 (20% v/v) was included in gas mixture for O_2 and anaerobic treatments in order to measure nitrogenase activity *in situ*, and anaerobic derepressed populations were obtained by overnight growth with limiting NH_4^+.

3. RESULTS AND DISCUSSION

3.1 *Acceleration of nitrogenase derepression by O_2.* Derepression of C_2H_2 reduction by sucrose-grown cells, supported by sucrose [in flasks] was unaffected by an initial 0.5 kPa O_2. However, when glucose-grown cells were used, 0.5 kPa O_2 accelerated derepression with sucrose. Anaerobic derepression of such cells was delayed, this probably arose because fermentation was restrained, as judged by a QH_2. O_2 caused a 3-fold expansion of the ATP pool suggesting that nitrogenase derepression was correlated with the level of ATP, consistent with Jensen and Kennedy's (1982) report of selective curtailment of anaerobic nitrogenase synthesis by low ATP.

3.2 *Influence of O_2 concentration on nitrogenase derepression.* Dissolved O_2, maintained at steady concentrations within the range 8 to 36 nM, accelerated derepression. Near 80 nM, O_2 delayed and partially inhibited derepression of *nifH::lac*, but C_2H_2 reduction was still detected. Thus derepression of C_2H_2-reducing activity is not markedly more sensitive to O_2 than is derepression of *nifH::lac*. Hence synthetic processes required to produce active nitrogenase, ATP and reducing power are also tolerant of this micro-aerobic condition.

3.3 *Influence of O_2 concentration on activity of nitrogenase.* When populations were derepressed anaerobically, with sucrose, nitrogenase activity was observed with O_2 maintained at steady values from 9 to 550 nM. With 20 to 40 nM O_2, activity was about twice the anaerobic rate. This is consistent with Hill's (1976) evidence that nitrogenase activity in *K. pneumoniae* can be supported in part by aerobic processes.

4. REFERENCES

Bergersen FJ and Turner, GL (1979) Anal.Biochem. 96, 165-174.
Bergersen FJ et al. (1982) J.gen.Microbiol. 128, 909-915.
Dixon R et al. (1980) Nature 286, 128-132.
Eady RR et al. (1978) J.gen.Microbiol. 104, 277-285.
Hill S (1976) J.gen.Microbiol. 93, 335-345.
Jensen JS and Kennedy C (1982) EMBO J. 1, 197-204.

THE STIMULATION OF THE RESPIRATION RATE OF ISOLATED RHIZOBIUM LEGUMINOSARUM BACTEROIDS BY GLUCOSE AND FRUCTOSE TRANSPORTED INTO THE BACTEROIDS BY A DIFFUSION PROCESS

J.J.M. HOOYMANS /G.J.J. LOGMAN
DEPARTMENT OF PLANT MOLECULAR BIOLOGY, UNIVERSITY OF LEIDEN, NONNENSTEEG 3, 2311 VJ LEIDEN, THE NETHERLANDS

It has been shown that isolated Rh.leguminosarum bacteroids can accumulate the organic acids fumarate, malate and succinate. Uptake of glucose was not demonstrable. Repression of the glucose uptake system or destroying of the uptake mechanism during the isolation of the bacteroids may underlie this phenomenon (Glenn et al. 1981, De Vries et al. 1981). To investigate whether isolated bacteroids were able to catabolize carbohydrates, the respiration rate of bacteroids in the presence of high concentrations of carbohydrates was determined.

Material and methods

Bacteroids were isolated from root nodules of Pisum sativum inoculated with Rh.leguminosarum strain RPL 1003. The isolation procedure was as described by Bergersen (1981). Mannitol was used as osmoticum in the isolation and experimental media. Respiration rates were measured polarographycally.

Results

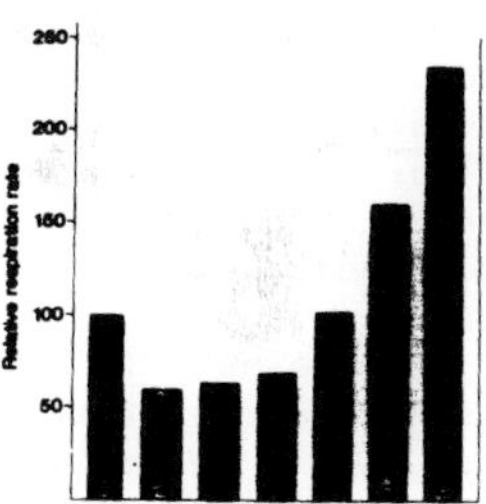

Fig. 1. Respiration rate of bacteroids with rhamnose, sorbose, sucrose, arabinose, glucose or fructose replacing mannitol in the isolation and experimental media. Respiration rates were plotted as percentage of the respiration rate of bacteroids with mannitol in the isolation and experimental media (19 n moles $O_2.min^{-1}.mg\ prot^{-1}$). Arabinose, glucose and fructose as well as mannitol increased the respiration rate.

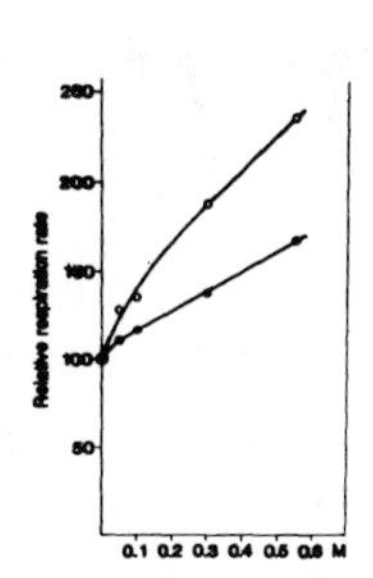

Fig. 2. Respiration rate of bacteroids in experimental medium in which 0, 0.05, 0.10, 0.30 or 0.54 M mannitol was replaced by glucose (●) or fructose (o). Respiration rates were plotted as percentage of the respiration rate of bacteroids with mannitol in the experimental medium (21 n moles $O_2.min^{-1}.mg\ prot^{-1}$). The respiration rate increased with increasing concentrations of glucose or fructose, suggesting an uptake by a diffusion process.

In the presence of 0.5 M fructose the respiration rate of the bacteroids equalled those in the presence of 1 mM malate, suggesting the respiration rate of the bacteroids was nearly maximal in the presence of 0.5 M fructose.

Conclusion

The obtained data suggest that bacteroids are able to catabolize glucose and fructose transported into the bacteroids by a diffusion process.

References

Bergersen, FJ et al. (1981) Planta 152, 534-543.
Glenn, AR et al. (1981) J. of Gen. Microbiol. 126, 243-247.
De Vries, GE et al. (1981) J. of Gen. Microbiol. 149, 873-879.

DIAUXIC GROWTH AND CATABOLITE REPRESSION IN *RHIZOBIUM MELILOTI*

HORNEZ J.P./THEODOROPOULOS A.P./COURTOIS B./DERIEUX J.C.
LABORATOIRE DE MICROBIOLOGIE, UNIVERSITE DES SCIENCES ET TECHNIQUES DE LILLE, 59655 VILLENEUVE D'ASCQ CEDEX, FRANCE

Tricarboxylic acid cycle intermediates, like succinate, can support N_2 fixation in *Rhizobium*-legume symbiosis (F.J. Bergersen, L.G. Turner, 1967). Moreover it has been shown that dicarboxylic acids repress carbohydrate transport (A. Arias et al., 1982 ; G.E. De Vries et al., 1982) and other enzymatic activities (S.T. Lim, K.T. Shanmugan, 1979 ; D.S. Ucker, E.R. Signer, 1978). Nevertheless, catabolite repression has not been observed in *Rhizobium meliloti*. We present the effect of sodium succinate complemented with glucose or fructose on growth of free-living forms of *Rhizobium meliloti* M_5N_1.

Precultures were grown in minimal medium RHB_1 with glucose or fructose as sole carbon source. Growth was observed either in fermentor vessels or in a biophotometer.
- Cells cultured in RHB_1 with glucose plus succinate showed diauxic growth. During the first phase of growth, glucose concentration remained unchanged and the glucose uptake activity was very low comparatively to the second phase where high levels of glucose uptake were observed.
- Bacteria grown in RHB_1 with fructose plus succinate showed diauxic growth. Surprisingly fructose consumption and fructose uptake activity were observed during both phases of growth. These results suggest that there are likely several fructose uptake systems.
- In both cases with cAMP (10 mM) diauxic was abolished and the growth rate was increased.

Regulatory processes in carbon metabolism in rhizobia have been little explored and results describing catabolite repression in these organisms were differently discussed. Indeed catabolite repression have been reported by S.T. Lim, K.T. Shanmugan (1979) and C.W. Ronson, S.B. Primrose (1979) in *Rhizobium japonicum* and *Rhizobium trifolii*, respectively. On the other hand, D.S. Ucker, E.R. Signer (1978) reported in *Rhizobium meliloti* a repression termed 'catabolite repression like phenomenon'. Our results suggest that in *Rhizobium meliloti* the glucose and at least one of the fructose uptake systems are submitted to catabolite repression by succinate.

REFERENCES

Arias A, Gardiol A and Martinez-Drets G (1982) J. Bacteriol. 151-3, 1069-1072.
Bergersen FJ and Turner GL (1967) Biochim. Biophys. Acta 141, 507-515.
De Vries GE, Van Brussel AAN and Quispel A (1982) J. Bacteriol. 149-3, 872-979.
Lim ST and Shanmugan KT (1979) Biochim. Biophys. Acta 584, 479-492.
Ronson CW and Primrose SB (1979) J. Bacteriol. 139, 1075-1078.
Ucker DS and Signer ER (1978) J. Bacteriol. 136, 1197-1200.

BIOCHEMICAL AND GENETIC STUDIES OF SALT TOLERANT *RHIZOBIUM*

S.-S. T. Hua, A. Guirao, W. Zing and M. Durant, WRRC, USDA/ARS, 800 Buchanan Street, Berkeley, CA, 94710, U.S.A.

INTRODUCTION AND RESULTS

Soil salinity and pH limit the survival of rhizobia in many areas. The viable number of rhizobia was greatly reduced in peat inocula containing NaCl (Steinborn J, Roughley RJ, 1974). A recent report indicated that cowpea inoculated with salt-tolerant *Rhizobium* could recover nitrogen fixation activity after drought (Zablatowicz RM, Focht DD, 1981). Thus salt tolerance may be an important trait to maximize N_2-fixation potential in semi-arid regions. The investigation of osmotically modulated enzymes, genes, and membrane functions is an emerging field in rhizobial biology.

Current studies from several laboratories indicate that intracellular glutamate increases markedly in several species of *Rhizobium* under osmotic stress (Hua et al. 1972). Thus the control system for modulating glutamate biosynthesis may play a vital role in osmoregulation. Biochemical properties of glutamine synthetase and glutamate synthase extracted from WR1001 were determined. Two glutamine synthetases, GSI and GS II were identified. GSI was very stable in high salt. 500 mM NaCl or KCL inhibited the enzyme activity by 10%. A new NADPH-GOGAT was detected in strain WR1001. The enzyme has a pH optimum at 8.2-8.3 and its activity is stimulated by glutamate *in vitro*. Because GDH is not present in *Rhizobium*, GOGAT may play a key role in adaptation to high salt environment. High isocitrate dehydrogenase activity was found in WR1001. Procedures for enzyme preparation and assay were described by Hua et al. (1982).

Adaptation of living cells to salinity probably involves a sequence of steps. Elucidation of the mechanisms requires application of genetic and biochemical techniques and is facilitated by the availability of specific mutants blocked at each step. Recently, transposons have been used to induce mutations in a wide range of bacterial species. We have tried to transfer transposon Tn5 from *E. coli* 1830 (PJB4J1) and *E. coli* SM10 (1011) to *Rhizobium* SP WR1001 for this purpose (Beringer JE et al. 1978). Since WR1001 is resistant to high concentrations of kanomycin, other transposons may have to be used. Although plasmids were identified in both salt-tolerant *Rhizobium* WR1001 and WR1002 by the method of Eckhard (1978), the correlation of plasmids to salt tolerance needs to be defined.

REFERENCES

Beringer JE, Beynon JL, Buchanan-Wollaston AV and Johnston AWB (1978) Nature (London) 276, 63634.
Eckhard T (1978) Plasmid 1, 584-588.
Hua, SST, Tasi VY, Lichens G M and Noma AT (1982) Appl. Environ. Microb. 44, 135-140.
Steinborn J and Roughley RJ (1974 J. Appl. Bacterol. 37, 93-99.
Zablatowicz RM and Focht DD (1981) Appl. Environ. Microb. 41, 679-685.

PHENOTYPIC CHARACTERISTICS OF NITROGENASE-POSITIVE STRAINS OF RHIZOBIUM JAPONICUM

D.L. KEISTER/T.A. HUBER/S.S. MARSH and A.K. AGARWAL
C.F. KETTERING RESEARCH LABORATORY, YELLOW SPRINGS, OH 45387

Knowledge of the physiology and genetics of nitrogenase synthesis has advanced considerably during the past eight years. However, one gap in our knowledge is that we do not know the physiological controls of nitrogenase derepression in the fast-growing rhizobia and in many slow-growing strains.

We have just completed a survey of about forty wild-type *Rhizobium japonicum* strains for their ability to derepress nitrogenase in culture under microaerobic conditions. The results revealed that only one half of the strains could reduce more than trace amounts of C_2H_2. During these studies, we noted that the *R. japonicum* strains which developed high levels of N_2ase in culture had certain phenotypic characteristics which distinguished them as a group from non-derepressible strains. The most obvious characteristic was that strains which developed good N_2ase activity were less slimy on agar and produced markedly less extracellular polysaccharide (EPS) under both aerobic and microaerobic conditions (Agarwal and Keister, 1983).

Hollis *et al.* (1981) recently demonstrated that the DNA homology among *R. japonicum* strains is insufficient, based on DNA:DNA hybridization studies, to be consistent with a single species. We have determined the carbohydrate composition of the major acidic EPS of twenty strains. Twelve of the strains have an EPS composed of mannose, glucose, galactose, 4-O-methylgalactose and galacturonic acid which is consistent with a structure recently determined by Mort and Bauer (1982). Seven strains have an EPS containing rhamnose and 4-O-methylglucuronic acid consistent with a structure determined by Dudman (1978). All of these latter strains fell within a group designated "DNA homology group II" and includes the strains giving high rates of *ex planta* nitrogenase activity. All of the other strains were in group I. Thus the type of EPS appears to be a phenotypic expression of a genotype.

Preliminary studies on the cytochrome composition of bacteroids from several strains have revealed another possible phenotype associated with the DNA homology group. Appleby (1969) reported that bacteroids from strain 505 did not contain the major aerobic oxidases, cytochromes aa_3 or *o*, which were present in cultured cells. We reported that bacteroids of strain 61A76 retained these cytochromes. We have recently analyzed the cytochromes from several strains of bacteroids and have found that strains from DNA homology group II (USDA 83, 86, 117 and 61A76) retain significant amounts of cytochrome aa_3 whereas bacteroids of six strains from DNA homology group I retain only trace amounts of this terminal oxidase.

In summary, some phenotypic characteristics of DNA homology group II as contrasted with group I are: (a) High *ex planta* nitrogenase activity; (b) low EPS production; (c) small colony morphology on agar; (d) EPS composed of rhamnose and 4-O-methylglucuronic acid; (e) lower O_2 consumption under microaerobiosis; and (f) retention of cytochrome aa_3 in the bacteroid state.

Agarwal AK and Keister DL (1983) Appl. Environ. Microbiol. 45(5), 1592-1601.
Appleby CA (1969) Biochem. Biophys. Acta 172, 71-87.
Dudman WF (1978) Carbohydrate Res. 66, 9-23.
Hollis AB, Kloos WE and Elkan GE (1981) J. Gen. Microbiol. 123, 215-222.
Keister DL, Marsh SS and El Mokadem MT (1983) Plant Physil. 71, 194-196.
Mort AJ and Bauer WD (1982) J. Biol. Chem. 257, 1850-1857.

These studies were supported by grants PFR77-27269 from the National Science Foundation, and 82-CRCR-1-1042 and 59-2394-1-1-653-0 from the USDA.

DOES AZOTOBACTER SYNTHESIZE NITROGENASE IN THE PRESENCE OF NH_4^+ ?

DORIS J. MARIS, LI-WEN WANG, P.B. NEW and Y.T. TCHAN
DEPARTMENT OF MICROBIOLOGY, UNIVERSITY OF SYDNEY, N.S.W. 2006, AUSTRALIA.

It is known that NH_4^+ inhibits nitrogenase activity in Azotobacter vinelandii, and it is generally accepted that synthesis of the enzyme is inhibited under these conditions. Previous serological studies using Ouchterlony plates or Preer tubes gave conflicting reports as to the presence of nitrogenase in NH_4^+ grown cells. We believe that these serological techniques are not sufficiently specific to positively identify nitrogenase.

We found that rocket, crossed and line immunoelectrophoresis provided greater resolution, but the nitrogenase precipitation band could not be identified due to the numerous antibodies present even in antiserum obtained by injecting biochemically homogeneous nitrogenase (chemically undetectable trace impurities can still induce production of irrelevant antibodies).

Sucrose-density-gradient-ultracentrifugation (SDG) of extracts of N_2 fixing cells removed contaminating antigens. Fused rocket immunoelectrophoresis revealed that only three antigen-antibody precipitation peaks were confined to the SDG region containing nitrogenase activity. These three peaks were identified as nitrogenase peaks because:

(1) The antiserum contains specific antibodies which inhibit the nitrogenase activity of broken N_2 fixing cells (using preinoculation serum as the control).

(2) The three peaks are located in the region of the SDG expected to contain nitrogenase.

(3) There is a relationship between nitrogenase activity and the height of the three peaks.

(4) The immunoelectrophoresis peaks were analysed using an electron probe microanalyser. The three peaks contain the following nitrogenase associated metals: peak 1 contains Mo; peaks 2 and 3 contain Mo and Fe.

Immunoelectrophoresis of SDG fractions prepared from NH_4^+ grown cells revealed three bands in the region equivalent to the active region of the SDG prepared from N_2 fixing cells. Fused rocket immunoelectrophoresis of equivalent fractions demonstrated that these three peaks are immunologically indistinguishable from those in the N_2 fixing cells.

Our results show that contrary to general belief, proteins immunologically indistinguishable from nitrogenase are synthesized by A. vinelandii in the presence of NH_4^+, even though N_2 fixing activity is absent.

FUNCTIONAL DIFFERENTIATION OF *R. MELILOTI* BACTEROIDS

R.W. Miller and P.A. Tremblay, Chemistry and Biology Research Institute, Agriculture Canada, Ottawa, Canada K1A 0C6

INTRODUCTION - A complex process of differentiation of *R. meliloti* bacteroids within the peribacteroidal membrane of alfalfa nodule cells is associated with the expression of the nitrogenase enzyme system. Functional differentiation of the bacteroid respiratory membrane presumably allows regulation of the nitrogenase reaction primarily through energy transduction and provision of a sufficient transmembrane electrical potential (Veeger et al 1981).

METHODS AND MATERIALS - Alfalfa plants were inoculated and grown for 6 weeks from seed at 25°C as previously described (Miller, Sirois 1983). Cells of strain 102F70 (Nitragin Co.) were grown aerobically on yeast extract-mannitol media for comparison with isolated bacteroids. Nodules were harvested from 60 plants, flushed with O_2-free N_2 and disrupted by blending in an ascorbate-sucrose-albumin medium buffered at pH 6.8 with sodium TES (80 mM). Bacteroids were isolated by differential centrifugation and average intracellular volumes were determined with permeating (tritiated water) and non-permeating (^{14}C-dextran) probes. Membrane potentials developed on aeration of cells were determined from the uptake of $^{86}Rb^+$ in static or flow dialysis experiments (Feldman 1978). Intracellular pH was determined by NMR from the chemical shift of phosphate ion relative to an external standard. Protein was determined by the Lowry method. Optical absorbance difference spectra of bacteroids and free living rhizobia were recorded with a Perkin Elmer 356 spectrophotometer. The binding of 9-aminoacridine to the bacterial cytoplasmic membranes was monitored with a Hitachi MPF 2A spectrofluorimeter.

RESULTS AND DISCUSSION - Bacteroids had an internal volume 6-7 times greater and a protein per cell 3-4 times greater than log phase air grown *R. meliloti*. Aeration of bacteroids with 200 mM O_2 caused a 25% greater accumulation of $^{86}Rb^+$ on a protein basis by bacteroids than by air grown cells. Either calcium or magnesium ions (10 mM) or a combination of both cations increased the differential accumulation of the membrane potential probe by bacteroids by 50% while respiratory inhibitors (CCCP, CN^-, azide) eliminated any net accumulation. Air grown cells (but not bacteroids) anomalously expelled organic cation probes on energization. Measurement of the magnitude of the bacteroid membrane potential by flow dialysis with $^{86}Rb^+$ indicated that bacteroids developed a maximum potential of 140 mv (negative inside) while free living cell membrane potentials were proportionally lower. Isolated bacteroids maintained membrane potentials for less than 2 min after energization with O_2 while the potentials of air grown cells were longer lived. Free living rhizobia maintained a pH gradient of up to 1 pH unit (alkaline inside) while isolated bacteroids did not develop a significant pH gradient on energization. Calcium ion caused the release of the fluorescent probe 9-aminoacridine from bacteroid membranes but had no effect on probe binding by air grown cells. The rate of oxygen uptake by bacteroids was increased 30% by calcium ion as was the fluorescence differential between energized and deenergized bacteroid membranes. In addition to calcium sensitivity and increased cellular volume and protein content, *R. meliloti* bacteroid cytoplasmic membranes lacked cytochromes a and a_3 which were replaced with an alternate terminal oxidase capable of rapid turnover at low oxygen tension. The observed alterations in bacteroid membrane function were also associated with an altered lipid composition and fluidity change as determined from the ESR spectra of intercalated spin probes (Miller, Tremblay 1983).

REFERENCES - Miller RW and Sirois JC (1983) Physiolog. Plantarum (in press).
Miller RW and Tremblay PA (1983) Can. J. Biochem. and Cell Biol. (in press).
Feldman K (1978) Anal. Biochem. 88: 225-235.
Veeger C, Haaker H, and Laane C (1981) in Current Perspectives in Nit. Fix., AH Gibson & WE Newton, Eds. Aust. Acad. Sci., Canberra, p. 101-104.

ECOLOGICAL ASPECTS OF NITROGEN-FIXING ACTIVITY OF ANAEROBIC BACTERIA GENUS CLOSTRIDIUM

L.K. NITZSE/V.T. EMTSEV/SHERAJUL ISLAM SHELLEY
DEPARTMENT OF MICROBIOLOGY, TIMIRYAZEV AGRICULTURAL ACADEMY, MOSCOW, USSR

In 1884 S.N. Vinogradsky isolated Cl.pastorianum (later on Cl.pasteurianum) from soil. He was the first to determine non-symbiotic fixation of molecular nitrogen. But despite a long history of study of Clostridium, many aspects of ecology and ecologo-biochemical changeability of these bacteria have been developed insufficiently. Therefore we studied ecology and nitrogen-fixing activity of anaerobic bacteria genus Clostridium taken at different lattitudes and soil zones: in Arctic soil (Spitsbergen island), alluvial-marshel soil (Egypt), alluvial-calcareous soil (Bangladesh) and red ferralitic soils (Cuba).

Our investigation showed that groups of anaerobic microorganisms possessing saccharolytic properties are rather spread in all types of the soils under study. Certain regularities were determined in distribution of anaerobes in the soils of a number of the countries of the world (Sherajul Islam Shelley et al., 1983). The difference in biological fixation of atmospheric nitrogen ($C_2H_2 \rightarrow C_2H_4$) by the anaerobes isolated from soils of various soil and climatic zones was established. Cl.pasteurianum and Cl.tyrobutyricum possess a high nitrogenous activity (75 to 110 nmol C_2H_4). Cl.saccharobutyricum isolated from soils of northern latitudes (podzolic and meadow-marshel soils of the USSR) possess this property in a lesser degree. Cl.acetobutylicum and Cl.butyricum are characterized by a low nitrogenous activity.
Anaerobes (Cl.acetobuytlicum, Cl.saccharobutyricum, Cl.beijerinsky) isolated from the soils of south latitudes (alluvial-marshel from Egypt, alluvial-calcareous from Bangladesh) possess the highest nitrogenous activity (250 to 280 nmol C_2H_4) per hour per 100 ml of medium). Cl.beijerinsky and Cl.kluyvery are active nitrogen-fixing bacteria in ferrolytic soils of Cuba.

The biochemical properties of anaerobic nitrogen-fixing bacteria depend not only on the soil and climatic conditions but also on the specific antrhopogenous factors influencing the energy and trophic substrates on which they develop. The degree of trophic and energy potential inside the ecological system creating favourable conditions for the development of certain species play an important role in ecologo-geographical changeability of anaerobic nitrogen-fixing bacteria.

EVALUATION OF RHIZOBIA GENETICALLY ENGINEERED FOR PESTICIDE RESISTENCE

M. P. NUTI, S. CASELLA*, M. PASTI**

Istituto di Chimica agraria e Industrie agrarie, Università di Padova
*Istituto di Microbiologia agraria dell'Università di Pisa, Italy

1. INTRODUCTION

There is evidence that many herbicides, applied as seed dressing or as a pre-emergence soil treatment, may interfere with nodulation of legumes by rhizobia. An approach for the relief of such a limitation in Rhizobium-legume symbiosis could involve the use of strains specifically constructed for pesticide resistence. As a model system, we evaluated laboratory-constructed strains of *R. trifolii* and *R.leguminosarum* resistant to the widely used herbicide 2,4-dichlorophenoxyacetic acid (2,4-D).

2. MATERIAL AND METHODS

Alcaligenes eutrophus JMP 134 and *E.coli* JMP 501 (courtesyof Dr. J.M. Pemberton) were used as donor bacteria in crosses with several *R.trifolii* and *R.leguminosarum* strains. Conjugal transfer of plasmid determinants specifying pesticide degradation was performed on solid media in the presence of the appropriate selective agent. Plasmid profiles and symbiotic properties were assessed by routine methods.

3. RESULTS AND DISCUSSION

In *A.eutrophus* the ability to utilize 2,4-D has been correlated with the presence of a specific 52 x 10^6 plasmid, p JP4 (Don, Pemberton 1981). While no conjugational system could be established between JMP 134 and rhizobia, by using JMP 501 genetic determinants originated in p JP4 could be transferred into *R. trifolii* and *R.leguminosarum* at frequencies 7 x 10^{-4} to 1 x 10^{-5}; recipient strains were rifamycin resistant and selection was carried out on media supplemented with rifampin and mercuric chloride. All rhizobia tested sofar, from different crosses, exhibited resistence to 2,4-D,100 mg x 1^{-1}, but none of them was able, like the donor, to use the chemical as a sole carbon source. This phenotypic change, which possibly involves physiological shifts at cell permeability level, awaits further studies.

DNA profiles of 2,4-D resistant rhizobia showed the absence of p JP4 or plasmid rearrangements. Hybridization studies between labelled p JP4 and Southern blots of the above DNAs are in progress for a better localization of "resistence" genes. Symbiotically defective strains apparently retained their phenotype,while 2% of Nod^+Fix^+ strains had unaltered symbiotic properties.

4. REFERENCES

Don RH and Pemberton JM (1981) J. Bacteriol. 145, 681-686.

5. ACKNOWLEDGEMENTS

This work was supported by CNR, P. F. IPRA.

** Present address: FAO/IAEA Research Labs., Seibersdorf, Vienna (Austria)

NITROGENASE ACTIVITY AND WHOLE CELL FLUORESCENCE STUDIES IN AZOTOBACTER VINELANDII

J.B. Peterson[+] and T.A. LaRue[++]
+Botany Dept., Iowa State Univ., Ames, IA 50011. USA
++Boyce Thompson Inst., Ithaca, NY 14853. USA

We used a fluorescence method for observing flavin and NAD(P)H oxidation/reduction during shifts in physiological conditions in intact *Azotobacter vinelandii* cells. The shifts were designed to provide information on the metabolism of compounds which influence nitrogen fixation. Fluorescence shifts due to flavin oxidation/reduction were of particular interest, since flavodoxin is thought to be the proximal electron donor to nitrogenase.

The fluorescence of flavins (flavoproteins) and nicotinamide nucleotides (NAD(P)H) is dependent on their redox state. The excitation and emission spectra of flavins do not overlap significantly with the corresponding spectra of NAD(P)H and the oxidation/reduction of these compounds can be monitored separately. When used on intact systems, specific redox reactions or their effect on whole cell redox state can be observed.

Azotobacter vinelandii cells were grown with mannitol as carbon source and atmospheric N_2 as nitrogen source. Nitrogenase activity of a suspension of washed cells was determined in serum-stoppered vials. The gas phase was 10% C_2H_2 in Ar with various levels of O_2. Vials were rotated continuously to equilibrate gasses with the cell suspension. Ethylene production was measured by gas chromatography. Oxygen uptake was measured separately with a Clark-type electrode. The fluorescence experiments were performed in a 4 ml rubber-stoppered fluorescence cuvette. The cuvette contained 2 ml cell suspension and 2 ml gas phase; both were equilibrated with Ar. Additions of carbon substrates or inhibitors or gasses were made with syringes. Gasses (O_2, H_2 and C_2H_2) were mixed by rapidly inverting the cuvette. A recording spectrofluorimeter was used to monitor fluorescence over time.

Acetate, acetaldehyde, ethanol and mannitol supported O_2-dependent nitrogenase activity and promoted O_2-uptake in *A. vinelandii* cells. Nitrogenase activity was inhibited by the electron acceptors metronidazole and menadione bisulfite but not by dicoumarol. Metronidazole had no effect on O_2-uptake; menadione bisulfite and dicoumarol promoted O_2-uptake under some conditions.

Fluorescence studies indicated that three flavins were oxidized by addition of various oxidants to an anaerobic cell suspension. One flavin and NAD(P)H were oxidized by O_2. The duration of the oxidation was decreased by the addition of carbon substrates or H_2. Metronidazole, menadione bisulfite, and especially dicoumarol also shortened the duration of the flavin oxidation. The effects of the inhibitors on NAD(P)H redox could not be studied due to spectral interference. The oxidation of this flavin and NAD(P)H are apparently associated with respiration.

Metronidazole and menadione bisulfite, but not dicoumarol, oxidized a second flavin. Acetate or H_2 reduced this oxidized flavin if O_2 was present. The flavin was reduced in the absence of O_2 by addition of acetaldehyde, ethanol, or mannitol. In separate experiments, however, acetaldehyde added alone oxidized NAD(P)H whereas ethanol and mannitol kept NAD(P)H reduced. This second flavin might be involved in electron transport to nitrogenase, and its reduction seems not to require NAD(P)H.

The third flavin and NAD(P)H were oxidized by addition of acetaldehyde. Both were reduced, anaerobically, by addition of mannitol or ethanol but not acetate or H_2. This flavin apparently serves as a source of reductant for alcohol dehydrogenase under anaerobic conditions.

THE F_1 ATPASE FROM FREE LIVING AND BACTEROID FORMS OF RHIZOBIUM LUPINI

J.G. ROBERTSON, J.E. THOMAS, L.R. GOWING AND M.J. BOLAND

Applied Biochemistry Division, Department of Scientific and Industrial Research, Palmerston North, New Zealand.

The F_1 ATPase from free-living and bacteroid forms of *Rhizobium lupini* strain NZP2257 was purified from Triton X-100 extracts of cell envelopes, using gel filtration on Ultrogel AcA-22 (LKB) followed by agarose gel electrophoresis.

The enzyme from free-living *R. lupini* appears typical of Mg^{++} (Ca^{++}) F_1 type ATPases from bacteria: Mg^{++} was necessary for activity at pH 7.0, although high concentrations caused noncompetitive inhibition (K_I = 1.2 mM). Maximum activity with ATP occurred over a broad pH range, from 6.0 to 10.5. ATP, GTP and UTP were hydrolysed by the enzyme, little activity was obtained with CTP and ADP and none with glucose-6-P. The K_M for ATP at pH 7.0 was 0.67 mM. ATPase activity was inhibited by ADP, competitive with ATP (K_I = 0.18 mM). Azide also caused inhibition, but fluoride and DCCD had no effect. Comparison of the enzymes from free-living and bacteroid forms of *R. lupini* using native and SDS gel electrophoresis revealed no obvious differences and indicated, from estimates of the molecular weight of the subunits, that the purified enzymes contained both F_1 and F_0 proteins of the ATPase complex.

NITROGEN FIXATION AND HYDROGEN OXIDATION IN CHEMOSTAT CULTURES OF RHIZOBIUM SESBANIAE.

H. STAM, W. de VRIES, H.W. van VERSEVELD, A.H. STOUTHAMER. BIOLOGICAL LABORATORY VRIJE UNIVERSITEIT, P.O. Box 7161, 1007 MC AMSTERDAM, THE NETHERLANDS.

Rhizobium sesbaniae is able to grow in batch cultures with N_2 as the sole nitrogen source (Dreyfus et al., 1983). We cultivated *R. sesbaniae* with N_2 as the sole nitrogen source in continuous cultures (dilution rate 0.1 h^{-1}) with succinate as the growth-limiting substrate. The dissolved oxygen tension (d.o.t.) was kept automatically at 2 μM. Growth yields with N_2 were much lower than those with NH_4Cl. H_2 was needed for the induction of hydrogenase (Table 1). The increase of Y_{succ} with H_2 indicates that H_2 is used as an additional energy source; the decrease of Y_{O_2} shows that H_2 oxidation yields less ATP than NADH oxidation. 11 mol H_2 yields as much energy as 1 mol succinate.

Table 1. Hydrogenase activities and growth yields of *R. sesbaniae*.

N-source	H_2 added	Hydrogenase[1]	Y_{succ}[2]	$Y_{O_2}^{exp}$[2]	$Y_{O_2}^{theor}$[2,3]	H_2/succ[4]
N_2	-	633	27.0	11.8	12.1	0
N_2	+	1420	30.6	11.3		1.3
N_2	+	2200	34.8	10.0		3.2
NH_4Cl	-	0	43.2	27.0	24.5	
NH_4Cl	+	1517	57	-		

[1]nmol H_2/mg protein.min. [2]g dry wt/mol. [3]calculated from Y_{succ} and the composition of the biomass ($C_3H_{5.42}N_{0.7}O_{1.53}$; m.w.= 76). [4]mol H_2 used/mol succinate: calculated from Y_{succ}, $Y_{O_2}^{exp}$ and the biomass composition.

Cytochrome spectra show the presence of cytochromes b and c and cytochrome o in cells grown at low d.o.t. (2 μM). Cytochrome aa_3 was absent. The $\rightarrow H^+/O$ value (with endogeneous substrate) was 6.3. These results indicate the functioning of site I and site II of oxidative phosphorylation and the absence of site III. With an H^+/ATP value of 3 (Kashket, 1982), the ATP/2e values for site I and site II are 1.0 and 1.33, respectively. This should imply that the oxidation of NADH, FADH and H_2 yields 2.33, 1.33 and 1.33 mol ATP per 2e, respectively, Hydrogenase enables *R. sesbaniae* to reoxidize H_2 formed at N_2 fixation. Using the ATP/2e values mentioned above Y_{ATP} values for different H_2/N_2 ratio's (mol H_2 formed/mol N_2 fixed) can be calculated. From the Y_{ATP} value calculated for growth with NH_4Cl and those found with N_2, ATP/N_2 ratio's (mol ATP used/mol N_2 fixed) can be calculated (Stouthamer, 1977). Table 2 shows that the H_2/N_2 ratio has only a limited effect on the Y_{ATP} value (about 3) and the ATP/N_2 ratio (between 31 and 37). The H_2/N_2 ratio is about 5 (Table 2).

Table 2. Y_{ATP} and ATP/N_2 values of *R. sesbaniae* at different H_2/N_2 ratio's.

N-source	Y_{succ}	H_2/N_2	Y_{ATP}	ATP/N_2[1]	ATP/N_2[2]
N_2	27	0	2.86	37.1	12
		3.0	2.98	34.1	24
		4.5	3.05	32.4	30
		6.0	3.11	31.3	36
NH_4Cl	43		5.93		

[1]calculated from Y_{succ}.
[2]calculated from ATP/2e = 4.

REFERENCES
Dreyfus BL, Elmerich C and Dommergues YR (1983) Appl.Env.Microbiol.45, 711-713.
Kashket ER (1982) Biochemistry 21, 5534-5538.
Stouthamer AH (1977) Ant. van Leeuwenhoek 43, 351-367.

CYTOCHROME-c REDUCTION STATE OF ISOLATED FRENCH-BEAN BACTEROIDS DURING NITROGEN FIXATION (C_2H_2 REDUCTION).

J.C. TRINCHANT/J. RIGAUD
LABORATOIRE DE BIOLOGIE VEGETALE, FACULTE DES SCIENCES, 06034 NICE CEDEX, FRANCE.

A rapid spectrometry method, providing 400 successive spectra with a scanning speed of about 80 spectra per s (Denis, Ducet 1975), was used to determine the reduction level of cytochrome-c in whole bacteroids. Optimal C_2H_2 reduction by French-bean bacteroids was correlated with a high cytochrome-c reduction state with glucose and succinate as energy-yielding substrates. However, these conditions were achieved at low oxygen tensions with glucose and in a large range of O_2 concentrations with succinate. The strong decline in C_2H_2 reduction occurring in the presence of glucose and raising O_2 tensions was associated with a low reduction level of cytochrome-c. Since significant ATP level was generated under these conditions, a progressive oxidation of cytochrome-c, in altering the electron allocation to nitrogenase, was responsible for the decline in C_2H_2 reduction activity of bacteroids.

Denis M and Ducet G (1975) Physiol. vég. 13, 709-720.

MOLYBDENUM METABOLISM AND NITROGEN FIXATION

R. A. UGALDE, J. IMPERIAL, V. K. SHAH, W. J. BRILL
DEPT. BACTERIOLOGY, UNIVERSITY OF WISCONSIN, MADISON, WI 53706, U.S.A.

Molybdenum is an essential element for nitrogen fixation. No nitrogen fixation can occur in its absence in *Klebsiella pneumoniae* and no other metal can replace it. However, this is difficult to prove under normal experimental conditions due to the high affinity of nitrogen-fixing cells for molybdenum. Traces present in media or glassware are usually enough to support nitrogen fixation. We devised a method to remove these traces from the medium taking advantage of the high affinity Mo-binding protein of *Azotobacter vinelandii*. The addition of 10 nM MoO_4^{2-} to *A. vinelandii*-treated medium is sufficient for maximum nitrogenase activity in *K. pneumoniae*. Other sources of Mo, either Mo(VI), Mo(V), or Mo(IV), are also efficiently utilized. In particular, MoS_2, the insoluble major natural reservoir, could satisfy the requirements for both Mo and S.

Molybdate is transported into the cells by a very specific, two-component system which is constitutive. The high affinity component exhibits binding kinetics (maximum uptake in ca. 30 seconds). In nitrogenase-repressed cells, the internal Mo pool is totally exchangeable by extracellular MoO_4^{2-}. In derepressed cells, uptake is followed by the transformation of MoO_4^{2-} into non-exchangeable forms. The transformation depends on the functions codified by at least five genes: *mol*, *nifB*, *nifE*, *nifN*, and *nifQ*. Mutants in these genes show defective Mo accumulation. The *mol* (functionally and genetically equivalent to *E. coli chlD*) and *nifQ* gene products are required for the accumulation of Mo in a non-exchangeable form at low extracellular molybdate concentrations. Increasing this concentration restores Mo accumulation, Mo presence in nitrogenase component I and nitrogenase activity to wild-type levels. MoO_4^{2-} concentration does not affect Mo accumulation or nitrogenase activity in $NifB^-$, $NifE^-$ or $NifN^-$ mutants. Mo accumulation in these mutants is <10% of the wild-type. Mo was not detected in nitrogenase component I from these mutants.

RHIZOBIUM SP. 127E15 MUTANTS WHICH ARE NO LONGER SALT TOLERANT FORM INEFFECTIVE OR NO NODULES ON LEGUMES

PETER P. WONG
DIVISION OF BIOLOGY, KANSAS STATE UNIVERSITY, MANHATTAN, KANSAS 66506, USA

Rhizobium sp. 127E15 induces effective nodules on lima bean, cowpea, mung bean, lupine, and partially effective nodules on pole bean. *R.* sp. 127E15 was mutagenized. Individual colonies were screened for altered ability to nodulate lima bean. Ten mutants were isolated after screening 1,300 colonies. All ten mutants could still nodulate lima bean, cowpea, and pole bean, but the nodules were ineffective. Three of the ten mutants could no longer nodulate lupine, and five could not nodulate mungbean. Biochemical characterizations revealed that eight of the ten mutants could no longer grow in a chemically defined medium with 55 mM α-ketoglutarate, succinate, fumarate, or malate as carbon source. The eight mutants grew as well as the wild-type when the concentration of the organic acids was 20 mM. Further studies showed that the inhibitory effect of high organic acid concentration is due to salinity of the medium. The eight mutants could not grow in 20 mM organic acid if 100 mM KCl or NaCl was added to the medium. Six of the ten mutants were revertible. When legumes were inoculated with more than a billion cells of a mutant, few pink and acetylene-reducing nodules could be harvested.
Revertants were also isolated by plating the six mutants on agar containing a defined medium and 100 mM KCl. The reversion frequency ranged from 1.2×10^{-6} to 8.0×10^{-8}. Revertants isolated by nodulation or by plating on salt medium formed partially effective nodules on pole bean and effective nodules on all other legumes tested. They also regained their salt tolerance.

BACTEROID "REPRODUCTION" - FACT OR FICTION

J.C. ZHOU, Y.T. TCHAN, J.M. VINCENT
DEPARTMENT OF MICROBIOLOGY, UNIVERSITY OF SYDNEY, N.S.W. 2006, AUSTRALIA.

The controversy on reproduction of bacteroids has been revived recently. In this report the term "bacteroid" is restricted to bacteria within cells of legume root nodules.

1. MATERIALS AND METHODS

The reproductivity of bacteroids of white clover and soybean was tested by protoplast isolation, microchamber incubation with video recording, and comparison of viable and total counts.

2. RESULTS AND CONCLUSIONS

2.1 White clover with *Rhizobium trifolii* T1

(i) Only 3.2% of the bacteria released from protoplasts of 5-week old nodules formed colonies even with osmoprotection.

(ii) Small protoplasts, with high colony-forming ability (CFA), contained only small rods. Medium, with low CFA, contained immature bacteroids and some rods. Large, with very low CFA, comprised almost entirely mature bacteroids.

(iii) No morphologically differentiated bacteroid showed any sign of growth, even with osmoprotection. Any growth came from small rods, found freely in crushed nodules or, in some cases, within protoplasts.

(iv) Generation time for the small rods from nodules was progressively reduced and became identical with the cultured rhizobium after 6 divisions.

2.2 Soybean with *Bradyrhizobium japonicum* CB 1809

(i) Immature bacteroids (2 weeks after inoculation) are long and thin; and mature (5-week old nodules) are slightly larger, straight or curved with granular structure.

(ii) Soybean bacteroids were able to reproduce; their CFA increased with the age of nodule (from < 0.01% at 2 weeks to 60% at 13 weeks).

(iii) A heat-labile extract from soybean nodules or roots increased the recovery of bacteroids.

(iv) Mannitol in growth medium (3 g/l) increased CFA, but an osmoprotective concentration (either 37 g/l in the medium or 0.5 M/l in protoplast dilution buffer) was not beneficial and even detrimental.

(v) pH 5-8 in the dilution medium was a suitable range for the growth of bacteroids; less growth was obtained from pH 8-9.

(vi) Generation time of soybean bacteroids was progressively reduced and became identical with the cultured bradhryizobium after 4 divisions.

3. DISCUSSION

These results give no support to the proposition that mature, morphologically differentiated, bacteroids formed by *R. trifolii* in white clover are capable of growth *in vitro*. Those of soybean (with *B. japonicum*), which are morphologically less differentiated, are capable of growth to a degree affected by nodule age and supplementation of the growth medium with extract of soybean nodules or roots. Osmoprotection during the isolation procedure and in the growth medium had no beneficial effect.

MICROBIOLOGY AND PHYSIOLOGICAL ECOLOGY OF DIAZOTROPHS

POSTER DISCUSSION 5A

STEM-NODULATED LEGUMES AND NITROGEN-FIXING TREES

POSTER DISCUSSION 5B

FRANKIA SYMBIOSIS

ECOLOGY AND MICROBIOLOGY OF SYMBIOTIC DIAZOTROPHS

W.J. BROUGHTON/U. SAMREY/C.E. PANKHURST/G.M. SCHNEIDER/C.P. VANCE
MAX-PLANCK INSTITUT FUER ZUECHTUNGSFOERSCHUNG, D-5000 KOELN 30, FEDERAL REPUBLIC OF GERMANY

1. INTRODUCTION

Nodulation of legumes by rhizobia occurs in a series of steps which include:
(1) rhizobial growth in the plant rhizosphere;
(2) attachment of the rhizobia to roothairs of the legume;
(3) rhizobial induced root-hair curling;
(4) infection thread formation within the root hairs
(5) growth of the infection thread containing the rhizobia towards the root pericycle;
(6) branching of the infection thread so that many cells of the root cortex become infected;
(7) release of rhizobia from infection threads into the plant cytoplasm; and
(8) development of rhizobia into bacteroids coupled with the production of leghaemoglobin and the complete apparatus for nitrogen fixation (Dart 1977; Broughton, 1978; Schmidt, 1978; Vincent, 1980; Meijer, Broughton 1982). In so far as the genes for leghaemoglobin are borne by the plant while the structural genes for nitrogense are in rhizobia, this interaction is truly symbiotic. Classically, rhizobia have been divided into two groups - the slow growing and fast-growing strains (Trinick, 1982). Bacteriologically, these two groups are so different that suggestions have been made to group them in two separate genera (1982). Nevertheless, data accumulating since the mid-1950's pointed to the existence of a group of rhizobia, intermediate between the classical types that combine fast growth rates with broad host-range (Trinick, 1982; Broughton et al 1983a; Dommergues, 1984). Since normal slow-growing rhizobia have proven difficult to analyse genetically, this new group of organisms is proving a good substitute in research aimed at studying nodulation in plants normally infected by slow-growing strains.
Accordingly, this review will deal with novel developments in infection as well as with this new group of fast-growing rhizobia.

2. COMPETITION FOR NODULATION

Many investigations of competition amongst rhizobia for nodulation of a certain legume have been reported. Occasionally, a dominant organism selected on the basis of ability to fix nitrogen has been found, but generally "competitive ability" depends on inoculum levels, environmental influences, and other factors difficult to control in the field. Competition amongst strains also requires co-operation from the plant and live bacteria (Broughton et al. 1982). Since it seemed likely that selecting inoculants on the basis of their ability to fix nitrogen may overlook genes involved in competitives, we performed experiments the other way around. Re-constituted earthfill soils from a former open-cut coal mine were sown with Medicago sativa and the resulting nodules typed using intrinsic antibiotic resistance (IAR) (Josey et al, 1979). These same isolates were also typed serologically and in terms of plasmid content.
Furthermore, a strain from the "dominant" IAR group was used to inoculate M.sativa plants sown into the same field in the following year.
As the inoculant strain could not be detected in the nodules in the second year, the data suggest that "dominance" is not better than symbiotic nitrogen

fixation as a measure of competitives (Broughton *et al*. (1984).
Additionally, groupings based on IAR did not correlate with those based on plasmid content which were again different from the serological groups. And whilst there is some reason to doubt the stability of IAR patterns, another possibility seemed worth investigating.

3. EXCHANGE OF GENETIC INFORMATION IN SOIL

One way of explaining lack of competition amongst inoculant rhizobia is to assume that exchange of genetic information between introduced and resident rhizobia readily occurs. In this sense, strains would not be single, unchanging entitics, but rather a constantly changing pool of genes responding to environmental pressures. We tested this hypothesis using a Nod$^-$ *R.meliloti* (IA113) that could be complimented by a Nod$^+$ *R.leguminosarum* (T83K3) which contains the transmissible *sym* plasmid pJB5J1 (Banfalvi *et al*. 1981). Since *R.leguminosarum* cannot nodulate *M.sativa*, any nodules formed on lucerne as a result of mixed inoculation with IA113 and T83K3 most probably result from a conjugation event in soil and/or rhizosphere.
Nodules developed in soil in pots, the time of formation and number depending on the inoculation sequence. IA113 tranconjugents containing PJB5JI were found in 0 to 37% of all nodules. The rest of the nodules contained a mixed population of IA113 and T83K3. Proof that the extra plasmid-band seen in "Eckhardt" gels was pJB5JI was provided by conjugating pJB5JI into *Agrobacterium tumefaciens* cured of the Ti-plasmid (strain C58CI), and re-isolating the introduced plasmid on caesium chloride gradients. Then the plasmid was "nick" translated and hybridized to the transconjugant plasmids separated on "Eckhardt" gels and transferred to nitrocellulose filters (Fig 1).

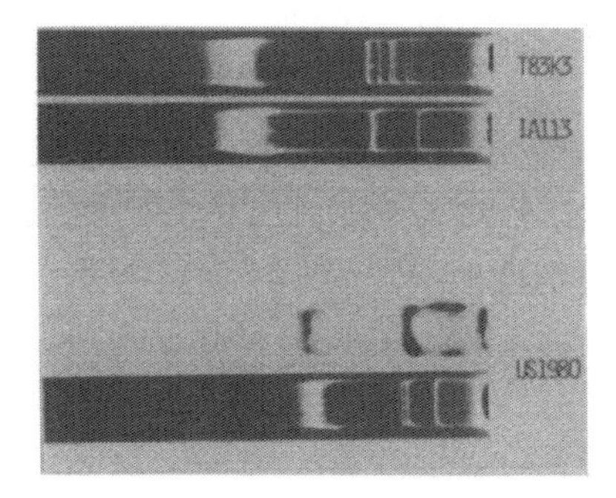

Fig. 1. *R.meliloti* transconjugant (US1980) isolated from nodules of *M.sativa* after soil inoculation first with Nod$^-$ *R.meliloti* strain IA113 and 48 h later with Nod$^+$ *R.leguminosarum* strain T83K3. Hybridisation was to "nick"-translated pJB5JI.

Whilst the fact that a large number of nodules only contained transconjugants strongly suggests that all nodules developed as a result of conjugation events, it was still possible that T83K3 "helped" Nod$^-$ IA113 into the nodules. To test if a "helper" effect was operative, two further experiments were performed The first was simply to see if conjugation *per se* could occur in unamended soil. Again the same strains were used and transconjugants were found.
The "helper" effect was also measured by replacing T83K3 with Nod$^+$ *R.leguminosarum* strain JIIN897 in which the *sym* plasmid is not self-transmissible (frequency < 10^{-9} - G. Hombrecher, per. communication). Since in this case conjugation could not occur, nodules should only develop as a result of one strain "helping" the other into nodules.

Even though the results of this experiment were not complete at the time of writing, it seems that genetic information is readily transferred (at least amongst some strains) between rhizobia in the soil/rhizosphere. As there is ample evidence of instability in plasmids introduced into other strains/species, the data suggest that strains would continually evolve both as a result of acquiring new genes as well as deletion of already existing ones (Broughton et al, 1983b).

4. DO RHIZOBIA TRANSFORM LEGUMES?

Since it was shown that Agrobacterium, the only other genus in the bacterial family Rhizobiaceae, infects plants by transferring a piece of its infective plasmid to the host chromosome, people have speculated that the infection process in rhizobia may be similarly controlled. We tested this hypothesis in two ways. First, we used either plant conditioned ineffective mutants of M.sativa (Vance and Johnson, 1983; Vance and Heichel, 1984) or auxotrophic derivatives of R.meliloti (Truchet et al., 1980) both of which produced "pseudo" nodules on M.sativa. These "pseudo" nodules are characterized by a block in infection thread development which prevents release of bacteroids, yet nodule-like structures develop (Fig. 2).

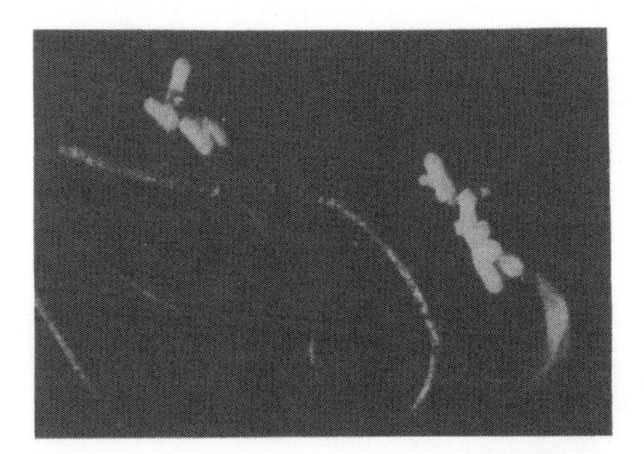

Fig. 2. Pseudo-nodules formed on M.sativa c. Cardinal by R.meliloti strain B13 (Leu^-).

Using appropriate techniques, bacterial-free callus can be isolated from these nodules. Sufficient quantities of callus can be obtained to isolate the DNA, which was digested with appropriate restriction enzymes, and transferred to nitro-cellulose filters. Hybridization of these filters with PRme41b (the megaplasmid of R.meliloti strain AK631 isolated from preparative "Eckhardt" gels) shows an extra hybridizing band in callus derived from "pseudo" nodules (Fig. 3).

Encouraged by this result we tried to localise the hybridizing region on pRme41b by using the R-primes generated by insertion of pJB4JI into AK631 (Forrai et al. 1983). Unfortunately, none of these hybridised with the total DNA preparations shown in Fig. 3.

For this reason, another way of approaching this problem was tested. If one assumes that rhizobial DNA was stabily integrated into plant DNA sometime during nodule development, then it seems reasonable that expression from the integrated sequence would come under control of the plant. In other words, transcription would follow the eukaryotic pattern, and any mRNA would be expected to have a poly A-tail. Accordingly we prepared total RNA preparations from nodules, labelled them with ^{32}P using reverse transcriptase, and hybri-

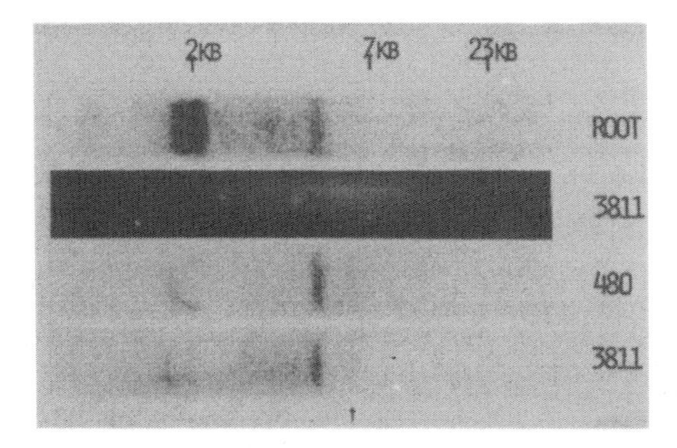

Fig. 3. Hybridization of pRme 41b DNA (labelled with ^{32}P by T4-DNApolymerase to a specific activity of 5 x 10^8 dpm. µg DNA^{-1}) to EcoRI digested total DNA preparations of root callus ("control") and "pseudo" nodule callus of two plant conditioned, ineffective mutants of M.sativa (No. 480 and 3811). Control "Root" digests contained twice as much DNA as in lanes containing "pseudo" nodule DNA. The dark lane (No. 3811) is a photograph of a total DNA digest, the arrow shows the extra hybridising band.

dised them to restriction-digested plasmid preparations transferred to nitrocellulose filters. This technique did not work with nodules on M.sativa induced by AK631 nor with those from Psophocarpus tetragonotodus induced by N6R234. Perhaps fortuitously, we were able to obtain hybridisation of a poly-A RNA containing fraction isolated from nodules of Pisum sativa (induced by T83K3) to plasmid DNA of T83K3 (Fig. 4), however.

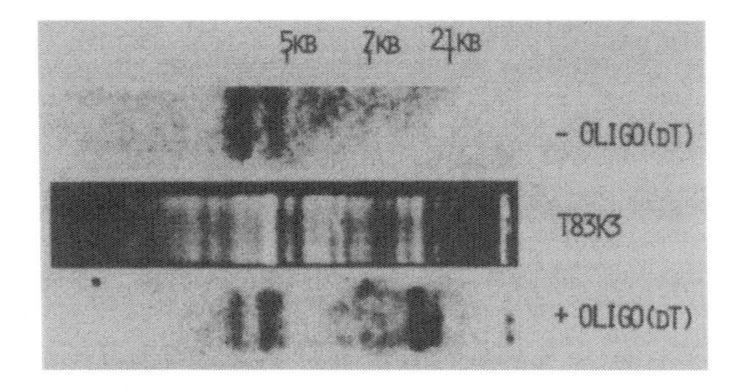

Fig. 4. Hybridization of poly-A containing RNA isolated from P.sativum nodules induced by T83K3 to EcoRI digested plasmid DNA preparations of T83K3. Nodule RNA was isolated by extraction in guanidinium isothiocyanate and centrifugation though caesium chloride (Willmitzer et al., 1981) and labelled with ^{32}P using reverse transcriptase.

Of course, it remains to be seen whether this hybridising band really represents bacterial DNA incorporated into the plant genome, or is an artefact.
Taken together the data neither suggest that transformation occurs nor deny it: they merely suggest that other experiments are necessary to obtain a clear answer.

5. TROPICAL FAST-GROWING RHIZOBIA

Rhizobia isolated from the nodules of some tropical and sub-tropical legumes including Acacia, Desmodium, Glycine, Lablab, Leucaena, Mimosa and Sesbania have (a) generation times of 3-4 h, (b) sub-polar flagellation, (c) the ability to catabolise a wide-range of carbohydrates, (d) produce an acidic reaction on yeast extract mannitol agar, and (e) nodulate large-seeded legumes (Keyser et al., 1982; Trinick, 1982; Broughton et al., 1983; Dommergues et al., 1984). As points (a), (c) and (d) are characteristic of fast-growing rhizobia while (b) and (e) apply more to slow-growing strains, this group appears to be intermediate.
Some of these fast-growing isolates are able to inoculate economically important legumes like Arachis, Glycine, Psophocarpus and Vigna. Since it has been shown that nif genes are plasmid borne in some Glycine isolates (Masterson et al., 1982) and that nif and "nod" genes are linked together on plasmids in isolates from many species (Broughton et al., 1983a) these strains provide a way of genetically analysing associations in which the micro-symbiont is normally slow-growing.
Furthermore, the recent demonstration that the sym plasmid of a least one of these strains can be mobilised using several techniques into other Rhizobium species and into Agrobacteria (Broughton et al., 1983a; Morrison et al., 1983) suggests that they will be important in studying host-range in rhizobia. Cloning of the sym plasmid of one of these strains (the Lablab isolate NGR234) followed by mapping of the "nod"-nif region suggests that the general organisation is similar to that described for R.leguminosarum and R.meliloti (Fig. 5).

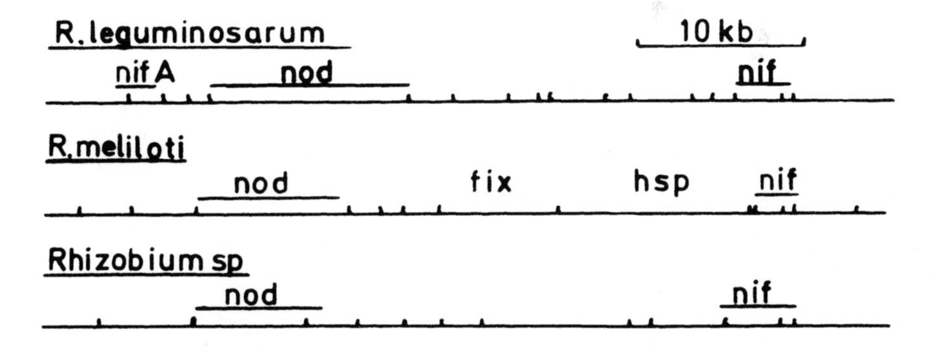

Fig. 5. Comparison of the cloned "nod"-nif regions of R.leguminosarum (Downie et al., 1983), R.meliloti (Kondorosi et al., 1983) and Rhizobium sp (Broughton et al., 1983a).

An interesting development from these studies as well as the work on complementing "nod" functions between different species (Banfalvi, et al., 1981; Long et al., 1982) is the finding that "nod"-genes have been relatively well conserved. This has allowed use of the "nod" region as a hybridisation probe and the identification of sym plasmids purely by homology (Broughton et al., 1983a). Since the "nod" region of pNGR234a (pMPIK3030a) shows broader homology than that of R.meliloti (Broughton et al., 1983b) it is a particularly valuable probe, even hybridising with sequences in R.japonicum strain USDA110.

6. LOCATION OF "NOD"-*NIF* GENES IN SLOW-GROWING RHIZOBIA

Plasmids are present in some slow-growing strains of rhizobia (Gross *et al.*, 1979; Broughton *et al.*, 1983b) but there is no evidence that these plasmids contain *nif* genes. Similarly, two fast-growing *R.meliloti* strains (NZP2037 and NZP2215) contain plasmids that do not hybridize with *nif* DNA of *R.meliloti*. As hybridisation was noted with total DNA preparations of these strains however, it seemed that here too, *nif* genes may not be plasmid borne (Pankhurst *et al.*, 1983). Corroboration (but not proof) that these were not *sym* plasmids was obtained by (a) mobilising pR102037 into NZP2213 without changing the host range (b) heat-curing NZP2037 of pR102037 without effecting its ability to nodulate *Lotus peduncolatus* (C.E. Pankhurst, unpublished observations), and (c) mobilising pCB376a of slow-growing *Rhizobium sp* CB376 (isolated for *Lotononis bainisii*) into Nod⁻ *R.leguminosarum*, Nod⁻ *R.meliloti* and *A.tumefaciens* cured of the Ti-plasmid without changing the host range (U. Samrey, unpublished observations). Proof that symbiotic functions are carried by the chromosome in slow-growing as well as in some fast-growing rhizobia awaits data demonstrating linkage of *nif* and "nod" to accepted chromosomal genes.

7. CONCLUSIONS

Our understanding of the complicated series of interactions between bacteria and plant resulting in nodule development has been augmented by:
(a) evidence that "dominance" (as it applies to a population of rhizobia in the field) is not a better criterion for competitivenes than ability to fix nitrogen symbiotically,
(b) demonstration of genetic exchange amongst rhizobia in soil. This suggests that strains are in a constant state of flux most probably changing their genome as a result of environmental influences,
(c) that a group of rhizobia, intermediate between classical fast- and slow-growing types, exist that are able to infect e.g. peanuts, soyabeans, winged beans and cow-peas. Since *nif* and "nod" are linked on plasmids in these strains, they are a good substitute for slow-growing rhizobia in studying the control of nodule development in these plants, and
(d) that *nif* and "nod" genes are most probably chromosomally located in most slow-growing and some fast-growing strains.

8. ACKNOWLEDGEMENTS

We wish to thank A. and E. Kondorosi, A. Pühler, J.S. Schell and G. Truchet for their help with many aspects of his work as well as H. Meyer z.A. and U. Wieneke for their excellent technical assistance.

9. REFERENCES

Banfalvi Z, Sakanyan V, Koncz C, Kiss A, Dusha I and Kondorosi A (1981) Mol. gen. Genet. 184, 334-339.
Broughton WJ (1978) J. appl. Bact. 45, 165-194.
Broughton WJ, Samrey U and Bohlool BB (1982) Can. J. Microbiol. 28, 162-168.
Broughton WJ, Heycke N, Meyer z.A. H and Pankurst CE (1983a) Proc. Natl. Acad. Sci. USA, submitted.
Broughton WJ, Bohlool BB, Shaw CH, Pankhurst CE, Bohnert HJ, Wieneke U and Schell J (1983b) Mol. gen. Genet., submitted.

Broughton WJ, Heycke N, Schneider GM, Priefer U, Wieneke U, Meyer z.A. H and Pühler A (1984) Can. J. Microbiol., submitted.
Dart PJ (1977) In Hardy RWF and Silver WS, eds, A Treatise on Dinitrogen Fixation, pp. 367-472, Wiley, New York.
Dommergues YR, Diem HG, Gauthier DL, Dreyfus BL and Cornet F (1984) These proceedings.
Downie JA Ma QS, Knight CD, Hombrecher G and Johnston AWB (1983) EMBO Journal 2, 947-952.
Forrai T, Vincze E, Banfalvi Z, Kiss GB, Randhawa GS and Kondorosi A (1983) J. Bact. 153, 635-643.
Josey DP, Beynon JL, Johnston AWB and Beringer JE (1979) J. appl. Bact. 46, 343-350.
Keyser HH, Bohlool BB, Hu TS and Weber DF (1982) Science 215, 1631-1632.
Kondorosi A, Kondorosi E, Banfalvi Z, Broughton WJ, Pankhurst CE, Randhawa GS, Wong CH and Schell J (1983) In Pühler A, ed, Molecular Genetics of the Plant-Bacteria Interaction, Springer-Verlag, Berlin, in press.
Long SR, Buikema WJB, and Ausubel FM (1982) Nature 298, 485-488.
Masterson RV, Russel PR and Atherly AG (1982) J. Bact. 152, 928-931.
Meijer EGM and Broughton WJ (1982) In Kahl G and Schell J, eds, Molecular Biology of Plant Tumors, pp. 107-129, Academic Press, New York.
Morrison NA, Hau CY, Trinick MJ, Shine J and Rolfe BG (1983) J. Bact. 153, 527-531.
Pankhurst CE, Broughton WJ and Wieneke U (1983) J. gen. Microbiol. 129, 2535-2543.
Schmidt EL (1978) In Dommergues YR and Krups SV, eds, Interactions between Non-pathogenic Soil Microorganisms and Plants, pp. 269-303, Elsevier, Amsterdam.
Trinick MJ (1982) In Broughton WJ, ed, Nitrogen Fixation II RHizobium pp. 76-146, Oxford, Oxford.
Truchet G, Michel M and Denarie J (1980) Differentiation 16, 163-172.
Vance CP and Johnson LEB (1983) Can. J. Bot. 61, 93-106.
Vincent JM (1980) In Newton WE and Orme-Johnson WH, eds, Nitrogen Fixation, pp. 103-129, Univ. Park Press, Baltimore.
Willmitzer L, Otten L, Simons G, Schmalenbach W, Schröder J, Schröder G, Van Montagu M, de Vos G and Schell J (1981) Mol. gen. Genet. 182, 255-262.

THE ECOLOGY OF *RHIZOBIUM* IN THE RHIZOSPHERE: SURVIVAL, GROWTH AND COMPETITION

B. BEN BOHLOOL/RENEE KOSSLAK/ROBERT WOOLFENDEN
UNIVERSITY OF HAWAII, DEPT. OF MICROBIOLOGY, HONOLULU, HI 96822, USA.

1. INTRODUCTION

The excitement over prospects for genetic improvement of *Rhizobium* has, unfortunately, overshadowed the need for an understanding of its ecology. It must be kept in mind, however, that in order for any "superbug" engineered in the laboratory to have an economic impact, it must be able to withstand the stresses of the soil, survive and grow in the rhizosphere, and compete successfully with well-adapted indigenous rhizobia for nodulation of the appropriate host.

The neglect for the study of ecology of *Rhizobium* stems mainly from lack of suitable means for distinguishing specific strains directly in the complex environment of the soil. Immunofluorescence (IF) provides the potential for such studies, for it facilitates simultaneous detection and identification of desired strains of *Rhizobium* directly in soil, rhizosphere and nodules (for review see Bohlool, Schmidt 1980).

In this study, we illustrate the usefulness of immunofluorescence for quantitative studies of the ecology of *Rhizobium*. Strain-specific fluorescent antibodies (FA) are used to enumerate inoculum strains in the rhizosphere, and determine their competitiveness (nodule occupancy) on the host plant.

2. Population dynamics of *Rhizobium* in the rhizosphere of host and nonhost plants

2.1. Background: Prior to infection, the rhizobia increase their numbers in the root zone and, somehow, recognize infectible foci on the roots of the appropriate host. Although the concept of "specific stimulation" has received support in the past (Nutman 1965), most available data indicate that rhizosphere stimulation might be nonspecific (for example see Tuzimura, Watanabe 1962; Peters, Alexander 1966). Van Egeraat's studies (1972) on homoserine in pea root exudates and its selective effect on root nodule bacteria are probably the only evidence suggestive of a possible specific stimulation mechanism in peas. The only published results on direct examination of population dynamics of rhizobia in developing rhizosphere soil come from the work of Schmidt and coworkers. They have used strain-specific fluorescent antibodies to monitor populations of *R. japonicum* (Reyes, Schmidt 1979, 1981) and *R. phaseoli* (Robert, Schmidt 1983) in the developing rhizosphere of host and nonhost plants. Their findings with both systems indicate that, although rhizobia are generally stimulated in the rhizosphere, the stimulation is not specific for the particular legume/*Rhizobium* systems tested.

In our studies, we followed the growth and final number of two strains of *R. japonicum* and two strains of *R. leguminosarum* in the rhizosphere of soybeans, peas and corn, growing in a nonsterile soil. We also determined % nodule occupancy of each strain on the appropriate host to discern any relationship between numbers in rhizosphere and nodule occupancy.

2.2. Materials and Methods

2.2.1. Soil: An Inceptisol (typic Eutradept, pH6.5); Kula loam soil was used here. The soil did not contain indigenous *R. japonicum*, but, when used to inoculate peas, it caused the formation of a few ineffective nodules. Test of the recovery of rhizobia added to this soil showed that 70-95% of the cells could be recovered for quantitative immunofluorescence.

2.2.2. *Rhizobium* strain: *R. japonicum* strains USDA 110 and CB1809 (USDA 136b) and *R. leguminosarum* strains Nitragin 92A3 and Hawaii 5-0 (May, Bohlool 1983) were used in all experiments. Cultures were maintained on yeast-extract mannitol agar (Bohlool, Schmidt 1970).

2.2.3. Soil inoculation and plant growth: Strains were first grown in autoclaved Kula soil amended with 1% mannitol before they were thoroughly mixed into the nonsterile Kula soil to a level of about 0.5×10^6 cells of each strain per gram of soil. Peas (cv Wisconsin Perfection), soybean (cv Davis), and corn (cv Hawaiian Supersweet #9) were planted in separate pots and grown in the glasshouse for the specified times. All treatments were done in triplicates. Three pots were left as fallow controls and used as "nonrhizosphere" soil. A randomized-complete block design was used.

2.2.4. Rhizosphere sampling and immunofluorescence staining: At designated times after planting, the content of each pot was carefully removed, and the loosely adherring soil gently shaken off the root system. Each set of roots was placed in a screw-capped bottle containing 100 ml of partially-hydrolyzed gelatin (Kingsley, Bohlool 1982) with 4 drops of Tween 80, and shaken for 30 minutes on a wrist action shaker. The suspension was centrifuged gently (700 xg, 5 minutes) and aliquots of the supernatant filtered through Irgalan Black treated nucleopore filters. Immunofluorescence counts were determined as described previously (Kingsley, Bohlool 1981).
Nodules were typed by the method of Schmidt, *et al.* (1968), using gelatin-rhodamine conjugants (Bohlool, Schmidt 1968) to suppress background fluorescence.

2.3. Results

2.3.1. Growth in the rhizosphere. The increase in the population of each strain in the rhizosphere soil of soybean and peas in the first 3 days of plant growth are shown in Figure 1.
In a separate experiment, numbers of each strain in the rhizosphere of soybean, peas, and corn were monitored over a longer period of time. The results in Table 1 show that the numbers of all the strains were at their highest values at 3 days and decreased thereafter.

2.3.2. Competition among strain for nodulation of their host. The patterns of competition (% nodule occupancy) of the strains on the appropriate host are indicated in Table 1. The % values are averages of data obtained from 5 consecutive harvests (25 days each) of plants grown in the same pots. The incidence of double infection (nodules containing both strains) is also shown on Table 1.

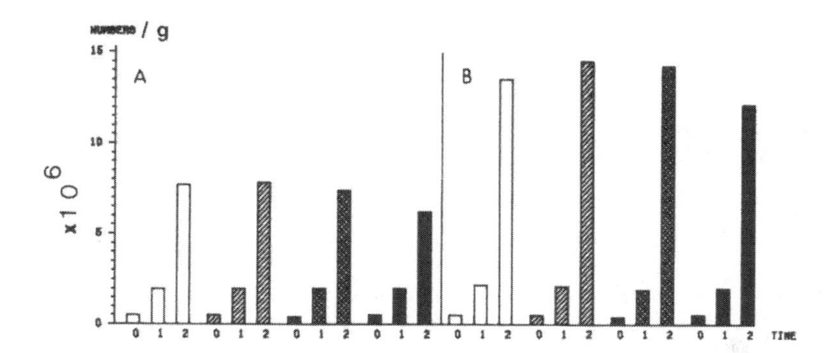

Figure 1. Growth of 4 strains of *Rhizobium* in the rhizosphere of (A) soybeans and (B) peas. *R. japonicum*, □=USDA 110, ▨=CB1809; *R. leguminosarum*, ▩=Nitragin 92A3, ■=Hawaii 5-0.

Table 1. Populations of strains of *R. japonicum* and *R. leguminosarum* in the rhizosphere of different plants as determined by immunofluorescence.

Host/Rhizobium[a]		Rhizosphere Populations (No. x 10^6/g dry soil) at Day:[b] 3	5	7	9	35	Nodule Occupancy of Strain as % of Total[c]
Soybean/	1	21.0	10.0	9.2	7.5	16.7	52 (7,0)
	2	22.1	10.0	9.3	7.7	14.5	41
	3	18.7	9.6	8.7	6.8	16.5	
	4	16.8	8.4	8.4	5.5	7.4	
Peas/	1	14.9	7.2	5.5	9.8	7.7	
	2	12.2	8.0	6.8	9.9	7.8	
	3	11.8	7.4	4.7	7.4	16.8	46 (8,25)
	4	8.9	7.0	4.5	7.3	4.9	21
Corn/	1	8.6	2.9	2.1	1.8	3.8	
	2	9.1	3.5	2.2	2.1	4.1	
	3	8.7	2.5	1.6	1.8	3.9	
	4	7.8	2.1	1.4	1.5	3.2	
None/	1	2.7	2.0	1.8	1.7	1.4	
	2	2.6	2.0	1.9	1.8	2.3	
	3	2.2	1.4	1.6	1.5	1.3	
	4	1.8	1.5	1.4	1.4	1.2	

[a]*Rhizobium* strains: *R. japonicum* - 1=USDA 110, 2=CB1809; *R. leguminosarum* - 3=Nitragin 92A3, 4=Hawaii 5-0.

[b]Values for 35 day sampling was obtained in a separate experiment conducted at a different time.

[c]Values in parenthesis (A,B) refer to: A=% of nodules containing both strains, and, B=% of nodules not reacting with the fluorescent antibodies used.

2.4. Discussion

In the first experiment (Figure 1) all strains were found to be stimulated to approximately the same extent in each rhizosphere in the early stages of plant growth. The numbers in the pea rhizosphere were about two times higher for all strains than those in the soybean rhizosphere. However, in later samplings, Table 1, soybean rhizosphere was generally more stimulating. This was also found to be true for total bacteria enumerated by epifluorescence using acridine orange (data not shown). The data in Figure 1 and Table 1 indicate that rhizosphere stimulation does not seem to be the function of the specific host/*Rhizobium* combination. Rather, it is the rhizosphere competence of a particular strain that determines its population in a particular root zone.

The nodule occupancy data (Table 1) shows that USDA 110 and CB 1809 are about equally competitive on soybeans, while Nitragin 92A3 outcompetes Hawaii 5-0 on peas.

3. Host involvement during the early stages of the infection process

3.1. Background: An important aspect of the legume/*Rhizobium* symbiosis is the differential competitiveness of strains for nodulation of their host. Although there is considerable evidence that both the plant and the bacteria are involved in the selection of the "more competitive" strain, the mechanisms which confer competitive advantage to a strain remain poorly understood (for reviews see Ham 1980; Schmidt, Bohlool 1980; Graham 1981).

Some indications concerning the role of the host in the process of infection and nodulation come from the work of Nutman (1963), Lim (1963), Munns (1968), Bhuvaneswari *et al.* (1980), and Singleton and Stockinger (1983). In this study, using a delayed inoculation design, we found that preexposure of the soybean roots to a poorly competitive strain of *R. japonicum* resulted in a significant increase in the nodule occupancy of that strain. We then used a split-root system to separate strain/strain interactions from those involving the host in the early stages of infection and nodule development.

3.2. Materials and Methods

3.2.1. Delayed inoculation experiments: Two-day old seedlings were placed in Erlenmeyer flasks containing sterile vermiculite, moistened with 1:4 strength Hoagland's nitrogen-free solution (Hoagland 1938). One ml of 4-day old YEM broth cultures was injected into the plant chambers at the designated time. Three different treatments were used. At 0-time: 1) one group received both strains; 2) another, only USDA 110; 3) and the last, only USDA 138. At selected times (6, 48 and 168 hours) thereafter, 3 flasks from treatments 2 and 3 received one of the other strains as the secondary inoculum. Controls consisted of uninoculated and single-strain inoculants at each time period. Four weeks after the secondary inoculation, the plants were harvested and the nodule occupants identified by immunofluorescence as before (2.2.4). A similar experiment was done in a nonsterile midwestern soil, (Waseca) in which 2-day old seedlings were first pre-exposed to USDA 110 and then transplanted into pots containing the soil.

3.2.2. Split root experiments: A modified version of the plant growth vessel of Singleton (1983) was used. Split roots were allowed to extend into large (38 x 250 mm) Bellco test tubes containing sterile vermiculite. At 0-time (one week after planting) one side of the split-root systems received the primary inoculum. At different times thereafter, the other side was inoculated with the secondary inoculum. Plants were harvested 3 weeks after the secondary inoculum and nodule occupancy was determined by immunofluorescence.

3.3. Results

3.3.1. Delayed inoculation (axenic conditions): Results in Table 2 show that in sterilized vermiculite, when the roots are exposed to a less competitive strain, USDA 138, prior to the addition of another, USDA 110, the pattern of nodule occupancy changes in favor of the primary strain. When inoculated at the same time as USDA 110, USDA 138 will cause the formation of only 5% of single-strain nodules (13% including 2-strain nodules). If, however, it is introduced 6, 48, or 168 hours ahead of USDA 110, its occupancy in single-strain nodules increases to 28%, 70% and 82% respectively (58%, 72% and 98% respectively if 2-strain nodules are included). The results with 110 as primary inoculum are even more dramatic, for it dominates in nodules from 0-time on.

Table 2. Effect of pre-exposure of roots of soybean (cv Peking) to one inoculum strain of *R. japonicum* on establishment of a secondary inoculum strain in nodules.[a]

Primary Inoculum (0-time) Strain	Secondary Inoculum Strain	At time (hours)	% of Total Nodules[b] USDA 110	USDA 138
I. USDA 110	None	0	100	0
II. USDA 138	None	0	0	100
III. USDA 110 + USDA 138	None	0	95	13
IV. USDA 110	USDA 138	6	80	29
		48	100	0
		168	97	3
V. USDA 138	USDA 110	6	72	58
		48	30	72
		168	18	98

[a]Plants were grown in sterilized vermiculite and nodules harvested 4 weeks after secondary inoculum was applied.
[b]Percentages include nodules containing both strains, thus values exceeding 100%. Standard error of the mean = ±16.5.

3.3.2. Delayed inoculation (soil conditions): In a soil from the soybean-growing region of the U.S., the vast majority of nodules on uninoculated soybean (cv Chippewa) plants were found to contain USDA 123. Addition of high numbers of USDA 110, did not alter this pattern dramatically. The results in Table 3, however, show that when the seedlings were pre-exposed to USDA 110 and transplanted into this soil, the percentage of nodules containing USDA 110 increased significantly.

3.3.3. Split-root design: In split-root systems, early inoculation of one half-side of the split-root of soybeans (cv Lee) resulted in significant suppression of nodulation on the opposite side (Table 4).

Total nodule numbers and mass for the whole plant, however, did not differ significantly between treatments.

Table 3. Effect of pre-exposure to USDA 110 on competition pattern of *R. japonicum* strains in a soil containing native *R. japonicum* (Waseca).[a]

Inoculation	Pre-exposure Time	Nod. No. per plant	% of Total Nodules[b] 110	123	110+123	Other
None	0	127	2[a]	70[d]	2[h]	26[k]
USDA 110	0	127	15[b]	60[de]	9[h]	16[k]
"	2 hours	163	26[b]	46[e]	4[h]	24[k]
"	48 hours	120	51[c]	26[f]	6[h]	17[k]
"	72 hours	75	55[c]	27[f]	7[h]	11[k]

[a]The native *R. japonicum* include USDA 110, USDA 123 and several other serogroups in approximately equal numbers (E.L. Schmidt, personal communication).
[b]Treatments followed by the same letter do not differ (P=0.05) significantly within a given column.

Table 4. Suppression of nodulation of half-root systems as a result of delayed inoculation.[1]

Side A: Strain (0-time)	Side A: Nod. Mass (mg)/ side	Side A: Nod. No. (mg)/ side	Side B: Strain (days)	Side B: Nod. Mass (mg)/ side	Side B: Nod. No. per side	Total: Nod. Mass per plant	Total: Nod. No. per plant
138	58±18	61±15	Uninoc	0[c]	0[c]	58±18[a]	61±15[a]
138	39±14	48±14	110/138 (0)	32±5[a]	32±8[a]	71±11[a]	80±18[a]
138	49±15	52±21	110/138 (4)	15±6[b]	24±7[b]	64±16[a]	77±21[a]
138	71± 4	77±18	110/138 (10)	0[c]	0[c]	71± 4[a]	77±18[a]

[1]4 to 5 replicates used per treatment; all treatments harvested 3 weeks after inoculation of Side B. Results in Mean±S.E. Numbers in the same column flanked by the same letter are not significantly different (P=0.05) as determined by the Mann-Whitney U Test.

3.4. Discussion

Skrdleta (1970) and Winarno and Lie (1972) have shown that prior inoculation with one strain can alter the pattern of nodulation of another strain introduced later. In our studies we find that the pattern of competition between two strains can be affected by pre-exposing the host to either strain. *R. japonicum* strain, USDA 110 was a better competitor than USDA 138 on soybeans (cv Peking). However, in the midwestern soils of the U.S., strains belonging to the USDA 123 serogroup form the majority of the nodules (Ham 1980), and outcompete USDA 110 (Table 2). The pattern of competition between either USDA 110 and 138 in vermiculite or USDA 110 and 123 in soil could be altered by introducing the less competitive strain into the root zone prior to the more competitive strain.

The role of the host in determining the outcome of competition between strains was highlighted by the observation that young seedlings with only a small radicle at the time of pre-exposure to one strain became predisposed to that strain for further nodulation. The role of the host in controlling the overall process of nodulation was also illustrated in the split-root study, where early inoculation of one half of the split root system resulted in suppression of nodulation on the opposite side. Total nodule numbers and mass per plant, however, remained approximately equal, suggesting that the host can compensate for a lack of nodules on one side by increasing nodule number and mass on the other side.

4. Summary

Quantitative ecology of desired strains of *Rhizobium* in soil, rhizosphere and nodules can be studied directly using strain-specific fluorescent antibodies. Competition among strains for nodulation of the appropriate host is a complex process involving strain/strain, as well as, strain/host interactions. This paper discusses the rhizosphere competence of different strains of *R. japonicum* and *R. leguminosarum* in the rhizosphere of soybeans, peas and corn in relation to nodule occupancies of each strain on the host. Evidence is presented on how the host plant may orchestrate the selection of one strain over another in the early stages of the infection process.

References

Bhuvaneswari, T.V., Turgeon, B.G. and Bauer, W.D. (1980) Plant Physiol. 66, 1027-1031.
Bohlool, B.B. and Schmidt, E.L. (1968) Science 162, 1012-1014.
Bohlool, B.B. and Schmidt, E.L. (1970) Soil Sci. 110, 229-236.
Bohlool, B.B. and Schmidt, E.L. (1980) In Alexander, M., ed., Advances in Microbial Ecology, pp. 203-241, Plenum Publishing Co.
Graham, T.L. (1981) In Giles, K.L. and Artherly, A.G., eds., Biology of the *Rhizobaceae*, Suppl 13, pp. 127-148, Academic Press.
Ham, G.E. (1980) In Newton, W.E. and Orme Johnson, W.H., eds., Nitrogen Fixation, Vol. II, pp. 131-138, University Park Press.
Hoagland, D.R. and Arrnon, D.I. (1938) Calif. Ag. Exp. Std. Circular 347.
Kingsley, M.T. and Bohlool, B.B. (1981) Appl. Environ. Microbiol. 42, 241-248.
Lim, G. (1963) Ann. Bot., N.S. 27, 61-67.
May, S.N. and Bohlool, B.B. (1983) Appl. Environ. Microbiol. 45, 960-965.
Munns, D.N. (1968) Plant and Soil 28, 246-257.
Nutman, P.S. (1965) In Baker, K.F., Snyder, W.C., eds., Ecology of Soil-Borne Plant Pathogens, pp. 231-247, Univ. California Press, Berkeley.
Peters, R.T. and Alexander, M. (1979) Soil Sci. 102, 380-387.
Reyes, V.G. and Schmidt, E.L. (1979) Appl. Environ. Microbiol. 37, 854-858.
Reyes, V.G. and Schmidt, E.L. (1981) Plant Soil 61, 71-80.
Robert, R.M. and Schmidt, E.L. (1983) Appl. Environ. Microbiol. 45, 550-556.
Schmidt, E.L. and Bohlool, B.B. (1980) In Tanner, W. and Loewns, F.A., eds., Encyclopedia of Plant Physiology, sec. IV, Vol. 13B, pp. 658-677, Springer-Verlag.
Schmidt, E.L., Bankole, R.O. and Bohlool, B.B. (1968) J. Bacteriol. 95, 1987-1992.
Singleton, P.W. and Stockinger, K.R. (1983) Crop Sci. 23, 69-72.
Singleton, P.W. (1983) Crop Sci. 23, 259-263.
Skrdleta, V. (1970) Soil Biol. Biochem. 2, 167-171.
Tuzimura, K. and Watanabe, I. (1962) Soil Sci. and Plant Nutrition 8, 13-24.
van Egeraat, A.W.S.M. (1972) Meded. Landbouwhogesch. Wageningen 72-27, H. Veenman en Zonen N.W., Wageningen, 90 pp.
Winarno, R. and Lie, T.A. (1979) Plant and Soil 51, 135-142.

EFFECTS OF DROUGHT AND SALINITY ON HETEROTROPHIC NITROGEN FIXING BACTERIA AND ON INFECTION OF LEGUMES BY RHIZOBIA.

J. I. SPRENT
DEPARTMENT OF BIOLOGICAL SCIENCES, UNIVERSITY OF DUNDEE, DD1 4HN SCOTLAND

INTRODUCTION

Both drought and salinity involve changes in water potential. The situation may be further complicated by the type of salinity (ionic species) and whether it is coupled to pH changes.

Total water potential, measured in Pascals, has a number of components. Although in a higher plant there is usually a temperature difference between roots and leaves, we shall assume a constant temperature, since the discussion centres on water and soil, but not the aerial environment. Similarly, except in tall shrubs and in trees, the gravitational component is negligible. The remaining components are,

$$\psi = \psi_p + \psi_m + \psi_s$$

Total = pressure + matric + solute components

The pressure component (wall or turgor pressure) obtains only in cells with a rigid cell wall exerting pressure on the cell contents. The other components apply to each of the following; environment (water, soil), microorganisms (prokaryotic, in the present case) and cells of potentially nodulated plants.

THE ENVIRONMENT

TABLE 1. Magnitude of ψ_s and ψ_m in different environments.

	WATER		DRYING SOIL	
	FRESH	SALINE	NON-SALINE	SALINE
ψ_s	high	low	high -------------->	low
ψ_m	high	high	low	low

Aquatic

Fresh water is included in Table 1 only for completeness, since neither drought nor salinity can occur in it. Saline water may be inland (salt lakes) or marine: it occurs cyclically in estuaries. In saline water the matric potential is high and the solute potential low. This means, effectively, that polymeric materials such as cell walls remain fully hydrated whereas the normal rules of osmosis apply between components separated by a semi-permeable membrane. Nodulated plants are virtually absent from aquatic saline environments, so we need only consider bacteria. These may be free in water, attached to sediments (particularly in estuarine regions where they may thus be protected from displacement by tides: frequently these sites are anaerobic) or associated with plant roots. Free heterotrophic bacteria in such regions tend to be carbon limited and fix little nitrogen (see Sprent, 1979). Some may survive in the saline conditions without growth; others tolerate salinity and still others are halophytic. Genera with nitrogen fixing species which have been found in saline environments are listed in Table 2.

TABLE 2. Nitrogen fixing genera in saline environments.

GENUS	ENVIRONMENT	COMMENT	REFERENCE
Azotobacter *Klebsiella* *Enterobacter*	estuarine	washed down from fresh water: survive, but do not grow or fix nitrogen	Herbert,1975
Clostridium	estuarine	facultative halophyte	Herbert,1975
Desulfovibrio	estuarine, marine	halophyte	Herbert,1975
Campylobacter	associated with *Spartina*	obligate halophyte	McClung, Patriquin,1980

How the halotolerant species adjust osmotically to their environment is not known. Related species produce glutamate and/or proline (Sleytr,1975). In that the responses can be simulated with glucose, they are truly osmotic, and the sequence of events for Gram negative species may be (a) water is removed osmotically, (b) [K$^+$] rises in consequence, (c) glutamate dehydrogenase (synthetic) is activated, (d) [glutamate] rises; the charge is balanced by K$^+$. Gram positive species normally have high internal [glutamate], balanced by K$^+$. A rise in [K$^+$] would be toxic, therefore glutamate formed in response to salinity is further processed into electroneutral species such as proline.

SOIL AND SOIL BACTERIA

Here there are both matric and osmotic components which vary considerably in magnitude (Table 1). For excellent accounts of the general responses of bacteria to drought and salinity, see Griffin (1981) and Brown (1976). The present account applies especially to *Rhizobium* and *Frankia*. Figure 1 illustrates the major areas of interaction between the environment and bacterial cells. Compared with a "typical" vacuolate plant cell, we should note the following; (a) that bacterial cells are essentially non-vacuolate and do not plasmolyse. (b) that a comparatively high proportion of their cell water is bound matrically, a correspondingly smaller amount is available as solvent water, (c) that turgor pressure as conceived by higher plant scientists may not exist as a factor driving cell expansion. Indeed cells may grow when their contents are hypo- iso-

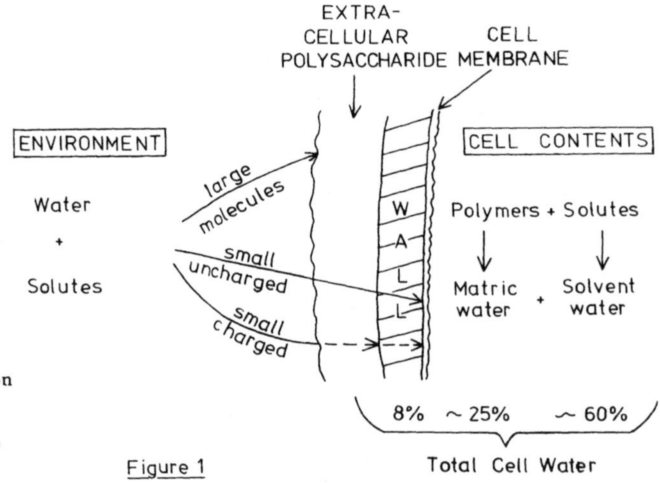

Figure 1

or hypertonic to their surroundings (Sleytr,1975; see also Griffin,1981). This may be why rhizobia can grow in an infection thread, apparently against cell turgor (see Bauer,1982). Thus, in considering symbiotic systems it must be appreciated that the for structural as well as life cycle reasons (see Singleton et al,1982), the two partners are likely to respond quite differently to stress. In many bacteria (Frankia sp. for example; Shipton, Burggraaf,1982) matric stress appears to be more harmful than osmotic stress: in others (some rhizobial strains, Singleton et al,1982) the two are additive. Under natural conditions it is not easy to separate the two since, as soils dry out, the concentration of solutes must rise.

Matric effects.

These vary greatly with soil type, including factors such as pore size and compaction. Published results show great variation, both amongst rhizobia and research workers. Bushby and Marshall (1977) concluded that fast growing rhizobia (R.leguminosarum, R.trifolii) are more sensitive to desiccation than slow growing ones (R.japonicum, R.lupini) because they retain more water at a given stress. Slow growing species dry out sufficiently to reduce metabolism to a low enough level to prevent growth but maintain viability (compare freeze drying of cultures). This difference arises from a greater water adsorptive capacity of the surface of fast growing cells. In certain soils, protection may be afforded to fast growing cells when they become coated with a layer of particles such as montmorillonite clay which have a stronger adsorption for water than the cells. Such particles will remove water from the cells so that they fall below the critical level for survival.

The fast growing species used by Bushby and Marshall (1977) correspond to those which nodulate principally temperate legumes (Sprent,1980). Recently, fast growth has been found in other rhizobia, including a R.japonicum isolate (Yelton et al 1983) and a strain nodulating Leucaena (Trinick,1980). This may be why Singleton et al (1982) did not find a correlation between fast growth and desiccation intolerance:- their fast-growing desiccation tolerant strain was isolated from Leucaena and was probably unrelated to the species used by Bushby and Marshall (1977); the water adsorptive properties of its walls may have thus been quite different. Bushby (1982) observed that the types of clay which afforded protection from desiccation to fast growing rhizobia were those characterised by constant surface charge (rather than constant surface potential), which tend to occur at high latitudes (i.e. temperate regions) which is true also of the protected rhizobia.

Mahler and Wollum (1981) have also found variation in reponse to desiccation with rhizobial strain and soil type. They used a variety of soils each at a range of water potentials and looked at the survival of two isolates each from R.leguminosarum and R.japonicum. In general extremes of soil texture (sand and clay-loam) supported fewest rhizobia. With respect to bacterial growth rate, in the short term (1 week) fast growers were the more sensitive to desiccation (as found by Bushby and Marshall,1977). However, over a 7 week incubation there were marked changes in population and one of the fast growing strains survived as well as the two slow growing strains. A possible explanation is that, over a longer time, changes in the type of carbon substrates available in the soil could have taken place, favouring the selection of certain strains. It is instructive to consider the features of bacterial metabolism which could enable bacteria to take advantage of such changes and also to vary generally in their survival under stress. Table 3 lists some of the possibilities.

It is clear that populations of rhizobia will vary during the imposition of and recovery from stress. How these interact with populations of other soil microorganisms is poorly understood.

TABLE 3. Features of "traditional" fast and slow growing rhizobia which could be related to stress resistance.

FEATURE	POSSIBLE EFFECT
Growth rate	Building of population after stress relieved, using C from dead roots (Zablotowicz, Focht, 1981).
Carbon nutrition	Different carbohydrates used by fast- and slow- growing strains (Elkan, Kuykendall, 1982); different enzymes of carbohydrate breakdown (Martinez de Drets, Arias, 1972).
Ion uptake	Fast growers take up ions (including nitrate and ammonium) more rapidly (Tan, Broughton, 1982).
Internal osmolalities	Much greater in slow- than fast- growing strains (Bushby, Marshall, 1977)
Surface polysaccharides	Vary in composition and amount and with soil. May protect against salt and desiccation (Carlson,1982; Bushby,1982; Upchurch and Elkan,1982).
Adsorption of clays	Constant surface charge minerals (high latitude soils) bound more easily than constant surface potential (low latitude) clays. This may protect against desiccation (Bushby, 1982).

SALINITY

The problem of matric stress is essentially one of survival until water supplies become adequate for growth. Salinity in soils tends to be prolonged and it is desirable that organisms can not only survive but grow.

Over the years numerous experiments have been conducted on the effects of salinity on growth and survival of Rhizobium strains in soil. It is very difficult to draw any general conclusions from these efforts for a variety of reasons, including the confounding of salt and pH effects, and the expression of concentrations in percentages regardless of ionic species and molecular weight.

Most workers agree on two points (a) that there is considerable strain variability in survival and growth under saline conditions and (b) that rhizobia are generally more able to cope with salinity than their host legumes. Singleton et al (1982) point out that this follows from the fact that rhizobia have no dormant phase corresponding to the seed, with which to survive adverse conditions. What has not been remarked is that in the symbiotic state rhizobia live in a cell of approximately 300 - 400 mol m^{-3} solute concentration - much higher than most of the concentrations used in experiments on free-living forms. This suggests that toxicities observed in soil are ionic rather than purely osmotic. Further, the bacterial wall may give protection against solutions of low concentration, since Sutton et al (1977) found that, in order to isolate viable bacteria from lupin nodules, osmoticum must be added.

Growth in a saline environment can only be achieved with osmotic adjustment. Figure 2 shows that, as the external water potential is reduced, a greater proportion of the cells total energy resources is used for osmotic adjustment until a point is reached when no growth is possible (although the cells may still remain viable). The exact point at which growth stops will be genotypically, as well as environmentally (e.g. by substrate availability) controlled. Some of the variables listed in Table 3 could be operative.

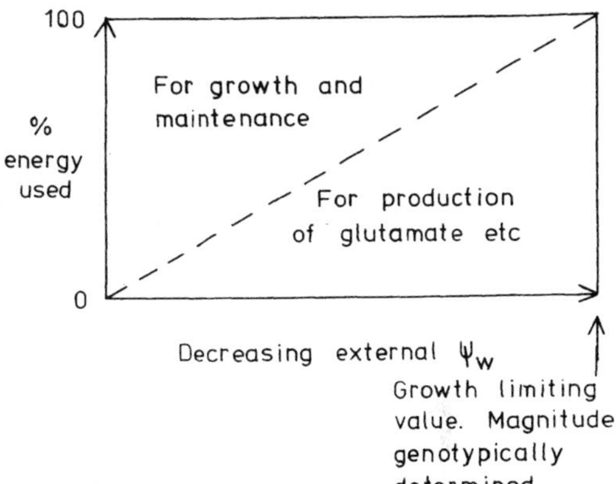

Figure 2

Osmotic adjustment in two strains of rhizobia, a fast-growing isolate of R.japonicum (Yelton et al 1983) and Rhizobium sp WR 1001 (Hua et al 1982) has been shown to be by production of glutamate. In the latter case, increased activity of glutamate dehydrogenase was not found (see earlier rationale for glutamate production): the Rhizobium concerned was isolated from the extremely drought and salinity tolerant legume Prosopis (Felker, Clark, 1982).

Once in their host, neither Frankia nor Rhizobium is likely to be subjected to low matric potentials, which may be why their walls can be greatly modified or even lost without detriment. Although they may be subjected to low solute potentials, these are unlikely to involve particles with a net charge, since the overwhelming evidence is that, even in salt accumulating halophytic higher plants, the NaCl passes into the vacuole and the cytoplasmic compartment (which houses the endophyte) adjusts osmotically by producing enzyme-compatible solutes such as betaines (Flowers et al, 1977). Thus only in the external environment are high inorganic salt concentrations likely to be encountered. There is much evidence that certain ions, particularly HCO_3^- are more toxic than others, but there is insufficient work separating ionic and pH effects to draw any definite conclusions.

INFECTION PROCESSES

All the published evidence for the effects of stress on infection are from species where the endophyte gains access via root hairs. For a recent general account of root hair infection, see Bauer, 1982. Infectibility of hairs appears to be a transient phenomenon (Bhuvaneswari et al, 1981) and infection confined to areas of the hair which are actively growing. Thus any factors which affect the duration of the infectible stage and/or the growth of hairs, is likely to affect infection. But, as Bhuvaneswari et al (1981) point

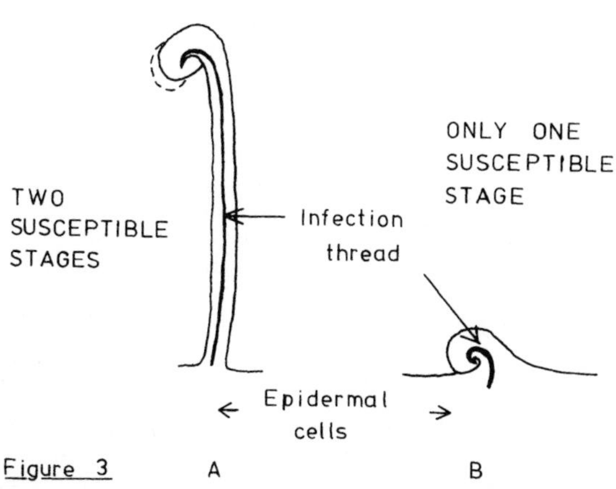

Figure 3

out, not all legume- Rhizobium systems are the same. In particular, hairs of plants like soyabean (Fig.3B) appear to have only one infectible period, whereas others, such as clovers (Fig.3A), have two, one early and one late in hair development. In Trifolium subterraneum growth may be stopped by drought (Worrall, Roughley, 1976), but on rewatering hairs may resume growth and are receptive to rhizobial infection. It seems unlikely that this could occur in soybean, whose hair growth has been shown to be modified by salinity (Tu, 1981). In T. repens calcium has been shown to be associated with root hair infection (Sethi, Reporter; 1981) and this element is generally associated with tip growth, as found in root hairs and pollen tubes (Sievers, Schnepf, 1981). One of the known effects of salinity is to displace calcium in certain soils (Epstein,1983) and addition of calcium may offset some of the adverse effects of salinity on nodulation (e.g. in Vicia faba, Yousef, 1982). However calcium may enhance the effects of salinity in other plants (e.g.soyabean, Nukaya et al, 1981). This could relate to differences in infection processes.

Reduced hair growth is not always found under drought: in Trifolium repens it may be stimulated (Reid, Bowen, 1979). The difference in the response of the two clover species may relate to the type of soil used. Soil texture can affect root growth and mucigel production and the latter may be also be stimulated by drought (Sprent, 1975). How mucigel is related to the ability of rhizobia to grow around and infect root hairs is not known, but it could well be a factor affecting response to stress.

Infection other than via root hairs

Some legumes, such as Phaseolus lunatus can be infected through epidermal cells as well as through root hairs (Bal, Wong, 1982). It would be interesting to see if this alternative route is any different in its susceptibility to stress.

TABLE 4. Effect of 50-55 mol m^{-3} NaCl on nodulation of various legumes. Data expressed as percentage of nodules per control plant at age specified in brackets (days).

SPECIES	cv	% (days)	REFERENCE
Glycine max	Amsoy	100 (30)	Tu, 1981
Glycine wightii	Cooper	63[1] (14[2])	Wilson, 1970
Vigna aureus	Hybrid 45	72 (25)	Balasubramanian, Sinha, 1976a
Vigna unguiculata	C 152	88 (25)	Balasubramanian, Sinha, 1976a
Cicer arietinum	G 62.404	21 (84)	Balasubramanian, Sinha, 1976b
Medicago sativa	not given	40 (12)	Lakshmi-Kumari et al, 1974
Pisum sativum	Bonneville	65 (?[3])	Kumar, Garg, 1980
Trifolium alexandrinum	Juslen	80[1] (45)	Bajpai et al, 1974
Vicia faba	Maris Bead	40 (14[2])	Yousef, Sprent, (1983)
Arachis hypogaea	J 11	114 (21)	Sprent, McInroy, unpub.

[1] by interpolation; [2] days after beginning of salt treatment, nodule number before treatment subtracted; [3] age not given; the value is % of nodules at about 30 mol m^{-3} NaCl, pH 7.0.

Plants such as Arachis hypogaea (groundnut, peanut) where rhizobia enter not via root hairs but at lateral root junctions (Chandler,1978) may be affected less by stresses acting principally via root hair growth. Preliminary experiments (Sprent and McInroy, unpub.) have shown that 50 mol m^{-3} NaCl has no effect on nodulation of groundnut, whilst slightly reducing growth. Double this concentration depressed both growth and nodulation, but, surprisingly, results in the formation of root hairs (normally absent). The significance of these remains to be determined, but the hairs certainly help in adherence to soil particles.

Species variation

The ability of species and cultivars of legumes to nodulate under stress varies widely. Table 4 lists the percentage reduction in number of nodules on plants subjected to 50-55 mol m^{-3} NaCl - the concentration found in irrigation waters of "moderate" quality (Epstein, 1983). Apart from Arachis those species least affected are in the genera Vigna and Glycine having the type of infection shown in figure 3B. Glycine is known to have considerable genetic variation in tolerance to NaCl, but even tolerant lines have growth depressed above 50 mol m^{-3} (Lauchli, Wieneke, 1979). The remaining species belong to temperate legume tribes and conform to the general infection pattern of figure 3A. Above 50 mol m^{-3} nodulation in all widely grown agricultural species that have been tested is reduced.

PROGNOSIS

It has been suggested (Singleton et al, 1982) that, since no agriculturally important legume is truly halotolerant, whereas many rhizobia show some salt tolerance, the host is the factor most limiting symbiotic nitrogen fixation under saline conditions: current evidence supports this. With the increasing occurrence of saline soils and the use of saline irrigation water, it is time to breed salinity tolerance into agricultural legumes, as is being done in other crops, for example, tomato (Epstein, 1983). Variation in tolerance of nitrogen fixing legumes to stress is already known. The Epstein philosophy of breeding "better crops for the soils we have", rather than improving soils to suit the crops should be applied to symbiotic nitrogen fixing systems. The germ plasm exists. Recent work on Prosopis spp. shows that extreme drought and salinity can be withstood. A number of the species listed in the excellent NAS publication (Anon 1979) have drought and salinity tolerance. Some, such as Phaseolus acutifolius appear to be able to yield a metric tonne of fixed N per ha per year in the dry bean and may have considerable potential for breeding as well. Development of such plants may generate a need for more highly drought and salinity tolerant rhizobia. The search should continue with these specific aims in mind.

REFERENCES

Anon (1979) Tropical legumes: resources for the future. National Academy of Sciences, Washington DC.
Bajpai PD Gupta BR and Singh C (1974) J. Ind. Soc. Soil Sci. 22, 375-376.
Bal AK and Wong PP (1982) Can. J. Microbiol. 28, 890-896.
Balasubramanian V and Sinha SK (1976a) Physiol. Plant. 36, 197-200.
Balasubramanian V and Sinha SK (1976b) J. agric. Sci. Camb. 87, 465-466.
Bauer WD (1982) In Asada Y Bushnell WR Ouchi S and Vance CP, eds, Plant Infection: The physiological and biochemical basis, pp. 67-78, Japan Sci. Soc. Press, Tokyo/Springer Verlag, Berlin.
Bhuvaneswari TV Bhagwat AA and Bauer WD (1981) Plant Physiol. 68, 1144-1149.

Brown AD (1976) Bact. Rev. 40, 803-846.
Bushby HVA (1982) In Broughton, WJ ed, Nitrogen Fixation 2, pp. 35-75, Oxford University Press.
Bushby HVA and Marshall KC (1977) J. gen. Microbiol. 99, 19-27.
Chandler MR (1978) J. exp. Bot. 29, 749-755.
Carlson RW (1982) In Broughton WJ ed, Nitrogen Fixation 2 pp. 199-234, Oxford University Press.
Elkan GH and Kuykendall LD (1982) In Broughton, WJ ed, Nitrogen Fixation 2 pp. 147-166, Oxford University Press.
Epstein E (1983) In Better Crops for Food. Ciba Foudation symposium 97 pp.61-76, Pitman London.
Felker P and Clark PR (1982) Pl. Soil 64, 297-305.
Flowers TJ Troke PF Yeo AR (1977) Ann. Rev. Plant Physiol. 28, 89-121.
Griffin,DM (1981) Adv. Microbial Ecol. 5, 91-136.
Herbert RA (1975(J.exp. mar. Biol. Ecol. 18, 215-225.
Hua S-ST Tsai VY Lichens GM and Noma AT (1982) Appl. environ. Microbiol. 44, 135-140.
Kumar S and Garg OP (1980) Ind. J. Pl. Physiol. 23, 55-60.
Lakshmi-Kumari M Singh CS and Subba Rao NS (1974) Pl. Soil 40, 261-268.
Lauchli A and Wieneke J (1979) Z. Pflanzenernaehr. Bodenkd. 142, 3-13.
McClung CF and Patriquin, DG (1980) Can. J. Microbiol. 26, 881-886.
Mahler RL and Wollum AG II (1981) Soil Sci. Soc. Am. J. 45, 761-766.
Martinez de Drets G and Arias A (1972) J. Bact. 109, 467-470
Measures, J.C. (1975) Nature 257, 398-400.
Nukaya, A Masui M and Ishida A (1981) J. Japan. Soc. Hort. Sci. 50, 326-331.
Reid CP and Bowen GD (1979) In Harley JL and Russell RS eds, The Soil-Root Interface, pp. 211-219, Academic Press, London.
Shipton WA and Burggraaf AJP (1982) Pl. Soil 69, 293-297.
Sievers A and Schnepf E (1981) In Kiermeyer,O ed, Cytomorphogenesis in Plants pp.265-299, Springer-Verlag Wien/New York.
Sethi RS and Reporter M (1981) Protoplasma 105, 321-325.
Singleton PW Swaify SA and Bohlool BB (1982) Appl. environ. Micrbiol. 44, 884-890.
Sprent JI (1975) New Phytol. 74, 461-463.
Sprent JI (1979) The Biology Of Nitrogen Fixing Organisms, McGraw-Hill, Maidenhead.
Sprent JI (1980) Plant Cell Environ. 3, 35-43.
Sutton WD Jepson NM and Shaw BD (1977) Plant Physiol. 59, 741-744.
Tan IKP and Broughton WJ (1982) Soil Biol. Biochem. 14, 295-299.
Trinick MJ (1980) J. appl. Bact. 49, 35-53.
Tu JC (1981) Can. J. Pl. Sci. 61, 231-239.
Upchurch RG and Elkan GH (1977) Can. J. Microbiol. 23, 1118-1122.
Wilson JR (1970) Aust. J. agric. Res. 21, 571-582.
Worrall VS and Roughley RJ (1976) J. exp. Bot. 27, 1233-1241.
Yelton MM Yang SS Edie SA and Lim ST (1983) J. gen. Microbiol. 129, 1537-1547.
Yousef AN (1982) PhD thesis, University of Dundee.
Yousef AN and Sprent JI (1983) J. exp. Bot. 34, 941-950.
Zablotowicz RM and Focht DD (1981) Appl. environ. Microbiol. 41, 679-685.

ACKNOWLEDGEMENTS

I should like to thank the numerous people with whom I have discussed this work, but in particular my colleague Dr R.A.Herbert who has attempted to keep my microbiology on sound lines. I apologise to those whose work I have omitted from the necessarily short literature survey. Miss S. McInroy has provided invaluable technical assistance.

RESPONSE OF CEREAL AND FORAGE GRASSES TO INOCULATION WITH N_2-FIXING BACTERIA

YAACOV OKON

DEPARTMENT OF PLANT PATHOLOGY AND MICROBIOLOGY, FACULTY OF AGRICULTURE, THE HEBREW UNIVERSITY OF JERUSALEM, REHOVOT 76100, ISRAEL.

INTRODUCTION

During the past ten years there has been extensive research efforts in order to increase and exploit biological N_2-fixation (BNF) by grasses-N_2 fixing bacterial associations (associative symbioses, diazotrophic rhizocoenoses).

The knowledge and achievements accumulated have been recently reviewed (van Berkum, Bohlool, 1980; Döbereiner, De-Polli, 1980; Boddey, Döbereiner, 1982; Patriquin, 1982; Okon, 1982; Brown, 1982).

In this paper I will try to evaluate recent developments on N_2-fixing bacterial-plant associations, with special reference to response of plants to inoculation with N_2-fixing bacteria. No reference will be made to associations with rice or with salt marsh grass *Spartina alterniflora*.

One possibility for promoting BNF is by selection or breeding of plant genotypes that will respond to and promote activity of nitrogen fixers in the rhizosphere. Some examples such as *Azotobacter paspali* - *Paspalum notatum* cv *batatais* (Boddey, Döbereiner, 1982), disomic chromosome substitution lines of spring wheat and N_2-fixing *Bacillus polymyxa* (Rennie, Larson, 1979) and some initial efforts to select maize that enhance BNF (Ela et al. 1982), have shown that there is a potential in the above approach, but so far reported responses to inoculation have been very small and not been tested in the field.

In order to obtain a positive response in plant development and yield, from inoculation, the organism that is introduced must be in sufficient numbers and capable of colonizing the roots. The developing population must reach significant size in terms of bacterial biomass, only then can one expect marked effects.

The potential use of *Azotobacter* as a biofertilizer has been reviewed by Brown (1982) who concluded that inoculation with *A. chroococcum* occasionally promoted yield probably by mechanisms other than BNF. *A. chroococcum* is not a typical rhizosphere organism and *Azotobacter* numbers counted in the rhizosphere were relatively low.

Only few attempts to inoculate plants with *Klebsiella pneumoniae* have been reported (Klucas, Pedersen, 1980; Wood et al. 1981).

There have been reports on positive responses to inoculation with non-nitrogen-fixing root colonizing bacteria *Pseudomonas fluorescens* and *P. putida*. These bacteria which are now called "rhizobacteria" or plant growth-promoting rhizobacteria have caused significant increases in the yield of potatoes and sugar beet (Suslow, 1982). The main mechanism of action proposed is by competition and antagonism in the rhizosphere by the introduced organism against pathogens that inhibit root development (Suslow, 1982).

Since their initial isolation and characterization, nitrogen-fixing bacteria of the genus Azospirillum were proposed as bacteria with promising potential for BNF and for increasing yields of economically important cereal and forage graminae (Döbereiner, Day, 1976; Tarrand et al. 1978). The use of Azospirillum inoculants has given positive results in increasing the yield of corn, sorghum, wheat and forage grasses (Boddey, Döbereiner, 1982; Kapulnik et al. 1981b, 1983; Reynders, Vlassak, 1982).

Physiological properties of the bacteria in relation to their capability to colonize, proliferate and survive on and in the roots.

The first step for aiding colonization of roots by bacteria, probably resides in their capability to reach colonization sites by chemotaxis and/or by aerotaxis. According to Schmidt, 1979, chemotaxis to root exudates may play an important role in this process. Rhizobium was found to be attracted to root exudates (Currier, Strobel, 1976; Gitte et al. 1978). However, the significance of this process in the specific infection of legumes by Rhizobium is not known. Azospirilla are very motile cells with a long polar flagellum occasionally possessing peritrichous flagella. Azospirillum brasilense responded chemotactically to amino acids, sugars and organic acids. Varying the concentration of attractants affected the chemotatic response. By using an open channeled chamber or in semi-solid agar plates with attractants, chemotaxis was obtained under conditions that prevented aerotaxis (Barak et al. 1983).

Several strains of Azospirillum responded to self created gradients and preformed oxygen gradients by forming (within a few seconds) aerotactic bands in capillary tubes and actively moving toward a specific zone with low dissolved oxygen. High O_2 concentration repulsed the bacteria. Under anaerobic conditions, bacteria remained motile but there was no band formation (Barak et al. 1982).

The capability of Azospirillum of rapidly seeking preferred low dissolved oxygen tension (d.o.t.) by aerotaxis, when fixing nitrogen and also in the presence of combined nitrogen, make Azospirillum a more likely candidate to colonize roots. To the best of my knowledge, no comparable information is available on chemotactic and aerotactic properties of Azotobacter chroococcum or Klebsiella pneumoniae.

Azospirillum preferentially adsorbed to pearl millet and guinea grass roots (Umali-Garcia et al. 1980) when compared to Azotobacter, Klebsiella and other bacteria. It cannot however, be concluded from that work whether the cells reached the root surface by active attraction or in a passive way. The importance of adsorption and the capacity of rhizosphere nitrogen-fixing bacteria to bind to roots of grasses has not yet been assessed.

Other microaerobic properties of Azospirillum that suggest adaptability to the rhizosphere of plants, where O_2 gradients are constantly changing, are in the fact that A. brasilense utilized its energy and carbon source more efficiently for growth with N_2 (Döbereiner, Day 1976) or NH_3 (Nur et al. 1982) under low d.o.t.: it also produced poly -β-hydroxybutyrate (PHB) in large quantities. PHB may serve as further energy and carbon source for survival under limiting nutritional conditions.

At intermediate d.o.t. levels some strains of A. brasilense produced red carotenoids which are apparently capable of protecting the cell against oxidative damage (Nur et al. 1981, 1982). In chemostat studies, it was found that Azospirillum, readily adapted to high d.o.t. (only with combined nitrogen) mainly by increasing the protein content of cells and the specific activities of succinate oxidase and superoxide dismutase (Nur et al. 1982). A. brasilense cells grown under low or high d.o.t.'s has a branched electron transport system with oxidases possessing high and low affinities to O_2 (I. Nur, unpublished), similarly that of Rhizobium (Bergersen, Turner, 1980).

Other properties that may give Azospirillum an advantage in the rhizosphere are in their tendency to form cell aggregates (Nur et al. 1981); weak pectinolytic activity (Umali-Garcia et al. 1980); adaptability to utilize N_2, NH_3, NO_3^- amino acids as N sources and NO_3^- as terminal electron acceptor and the utilization of a wide variety of organic compounds as carbon and energy sources (Tarrand et al. 1978).

Microscopic studies of inoculated roots revealing sites of Azospirillum-grass association have been recently reviewed (Patriguin, 1982). The bacteria have been found mainly embedded in the mucigel layer, but were also present in intercellular spaces of living cells in the cortex, endodermis, xylem and stele. Some degree of specificity, between different Azospirillum strains and C_3 (wheat) and C_4 (maize) has been proposed (Baldani, Döbereiner, 1980). However, this assumption has been made only from percentages of bacterial type isolated after surface sterilization of roots.

In the above mentioned studies on colonization site and specificity, no estimates of bacterial numbers of biomass were made. Their relationship to nitrogen fixation or other activities has not been established. Therefore, they may indicate potential capabilities, but not proof, of Azospirillum or Azospirillum types to colonize roots. Nevertheless, relatively large populations of Azospirillum (1-10% of the total rhizosphere population) have been consistently estimated by the most probable number technique in inoculated plants in the greenhouse and in the field (Okon, 1982). In Setaria italica the highest numbers of Azospirillum, associated with the roots were obtained at the booting stage when the highest acetylene reduction activities could be detected (Okon et al. 1983). Since Azospirilla are larger than other gram negative rhizosphere bacteria and since they appear in aggregates and microcolonies, their contribution to root microbial biomass is more significant.

Fewer reports are available on Azotobacter and Klebsiella colonization and association with roots. Methods for estimating bacterial biomass of the introduced organism on the rhizosphere and inside the root need to be developed

Effect of inoculation on root development and mineral uptake.

It was observed by Tien et al. (1979) that Azospirillum inoculated seedlings of Pennisetum developed more roots and root hairs. Subsequently we observed a marked development of roots of Setaria italica (Kapulnik et al. 1981a) and in wheat growing in pots, hydroponic systems and by sampling roots at different times during growth from field experiments carried out in Israel (Y. Kapulnik, unpublished). In recent work (Y. Kapulnik, unpublished) it was observed that there was almost no effecton root development when wheat was inoculated with 10^2-10^4 cells per gram of soil (4 week old seedlings in pots with vermiculite) whereas, in plants inoculated with 10^5-10^6 cells, root elongation, branching and root surface area increased significantly above non inoculated controls. In plants inoculated with 10^8-10^9 cells there was a significant inhibition in root development. Moreover, these effects were obtained only with Azospirillum but not with other bacteria. Inoculation with Azotobacter chroococcum (Y. Kapulnik, unpublished) enhanced only root elongation as described by Harper, Lynch (1979) in barley.

The inoculum doses effect suggest a similarity of diseases caused by phytopathogenic bacteria (Goodman, 1976). There are sites in the roots where colonization is initiated and there is an optimal inoculum level that may favour colinization. The reaction of roots being positive in the case of Azospirillum.

The production of plant growth promotors by the introduced Azospirillum (Tien et al. 1979) and Azotobacter (Brown, 1982) has been proposed as the main mechanism for enhancement of root development. However, it has not been directly demonstrated to take place during bacterial colonization and association with roots in the soil. Other possibilities are that the plants react to bacterial colonization by producing by themselves plant-growth promoting factors such as when plants are infected with phytopathogenic bacteria (Goodman, 1976).

Inoculation of corn seeds with Azospirillum brasilense significantly enhanced (30-50% over controls) the uptake of NO_3^-, K^+ and $H_2PO_4^-$ into 3-4-day-old or 2-week old root segments (Lin et al. 1983). Corn and sorghum plants grown to maturity on limiting nutrients in the greenhouse showed improved growth from inoculation, approaching that of plants grown on normal nutrient concentrations (Lin et al. 1983). Furthermore in field experiments in Israel, it was clearly observed that inoculated sorghum (Sarig et al. 1984) and wheat (Y. Kapulnik, unpublished) accumulated during the season dry matter, N, P and K at faster rates than the controls, thus resulting in a significantly higher grain yield. We have recently found that the water status of plants as measured by leaf water potential was favoured by Azospirillum inoculation in the field. Significantly less pressure was needed to pull out water from inoculated leaves as compared to controls (S. Sarig, Y. Okon, unpublished).

BNF in inoculated plants

Numerous experiments have demonstrated that in plants such as Setaria, corn and sorghum, inoculated with Azospirillum, vigorous, immediate, linear acetylene reduction (1000-3000 nmoles of ethylene produced per h, per gram dry weight of roots) can be measured in intact plant-soil systems (Kapulnik et al. 1981a; van Berkum, Bohlool, 1980, Döbereiner, De-Polli, 1980).

Acetylene reduction activities in inoculated plants are favoured by higher soil moisture, light intensities and temperatures. They take place mainly during flowering (Kapulnik et al. 1981a, Okon et al. 1983). Inoculated plants in the field have consistently shown higher acetylene reduction activities than non-inoculated controls (Kapulnik et al. 1981, Albrecht et al. 1981). Inoculated wheat has shown activities of only up to 100-200 nmoles at 20-25°C (Y. Kapulnik, unpublished; Lethbridge et al. 1982). But no activities whatsoever are detected below 15°C (Y. Kapulnik, unpublished). We have not found enhanced acetylene reduction in Setaria inoculated with Azotobacter chroococcum (E. Yahalom, Y. Okon, unpublished). Klebsiella pneumoniae enhanced acetylene reduction in Poa pratensis (Wood et al. 1981).

The major question still remains as to what extent the nitrogen that has been fixed by the inoculated roots is incorporated and contributes to the yield. BNF in associative symbiosis has been confirmed in the greenhouse by exposing roots to an environment enriched with $^{15}N_2$ (Eskew et al. 1981; De Polli et al. 1977, Okon et al. 1983). However, only very small amounts of fixed nitrogen have been incorporated into plant parts, suggesting that there was no direct bacteria-to-plant transport of fixed nitrogen, but rather a slow transfer suggesting the gradual death of bacteria and subsequent mineralization of their nitrogen.

By using the ^{15}N isotope dilution technique to quantify N_2 fixation in wheat (Rennie et al. 1983) it was found in some bacteria-cultivar combinations that the amounts of plant N derived from N_2 fixation (% Ndfa) varied from 0% to 32.0%. However, inoculation of wheat had no significant effect on the total N yield of the shoots (Rennie et al. 1983). In a previous work with maize up to 38% of the plant N was derived from associated N_2 fixation by *Azospirillum brasilense* (Rennie, 1980). Many more replications of this technique are needed to assess contribution of BNF to the plants.

Nitrogen balances carried out with Kjeldahl N analysis have not demonstrated clearly gains of nitrogen by the plants above those present in the soil (Okon et al. 1983). Nevertheless, under field conditions with much larger root systems, and with longer growth periods than in the greenhouse there is a better possibility that BNF may contribute to the nitrogen yield in plants inoculated with *Azospirillum*. The contribution of *Azospirillum*-grass association to the nitrogen content of the soil for the next crop season is more feasible, but this remains to be demonstrated.

Yield increases caused by inoculation with Azospirillum in field experiments

Extensive experimentation in the field over the last 8 years, carried out under diverse environmental and soil conditions in Belgium (Reynders, Vlassak, 1982), Brazil (Boddey, Döbereiner, 1982) Egypt (Hegazi et al. 1981), India (Subba-Rao, 1981), Israel (Kapulnik et al. 1981b, 1983) and USA (Smith et al. 1976, Albrecht, et al. 1981), has demonstrated that significant increases (5-30%) over controls in yields of sorghum, maize, wheat and several forage grasses, could be obtained following inoculation with *Azospirillum*.

The benefit of *Azospirillum* to plants was mainly derived from increases in dry weight and nitrogen content of shoots, and in the average numbers of ears per plant in sorghum, corn and wheat (Kapulnik et al. 1981b, 1983; Sarig et al. 1984, Reynders, Vlassak, 1982).

The most consistent positive results have been obtained when the inoculum has been applied in wet soil as peat inoculant containing 10^8-10^9 cells per gram of peat and when poured over the row at sowing or soon after (Kapulnik et al. 1981b, 1983).

Although positive results have been reported when the inoculum was given in cell suspensions (Vlassak, Reynders, 1982, Smith et al. 1976, Boddey, Döbereiner, 1982). In my opinion colonization of roots is less likely to succeed and this may explain occasional failures and inconsistencies.

In general, the highest yield increases in inoculated plants as compared to controls were obtained in soils fertilized with 30-50% less N-fertilizer than that recommended for the particular field where the experiment was carried out.

CONCLUSIONS

We seem to have an organism, *Azospirillum* sp., with a good potential for being used as a biofertilizer for economically important cereal and forage grass crops. Following inoculation, Azospirillum colonizes and proliferates "on and in" the roots, and it causes an enhanced root development, and mineral and water uptake by the roots (by mechanisms not yet demonstrated). This results in faster accumulation of dry matter in the plant and in many instances produces higher crop yield. As a result of inoculation BNF apparently contributes small amounts of combined N to the plant, mainly after flowering; whereas its contribution to the soil might be more significant.

Much remains to be done in order to elucidate the colonization dynamics, the mechanism of action, the methods of application and the optimal bacteria-plant-soil type combinations.

REFERENCES

Albrecht SL Okon Y Lonnquist J and Burris RH (1981) Crop Sci. 21, 301-306.
Baldani VLD and Döbereiner J (1980) Soil Biol. Biochem. 12, 433-439.
Barak R Nur I Okon Y and Henis Y (1982) J. Bacteriol. 152, 643-649.
Barak R Nur I and Okon Y (1983) J. Appl. Bacteriol. 54, 399-403.
Bergersen FJ and Turner GL (1980) J. Gen. Microbiol. 118, 235-252.
van Berkum P and Bohlool BB (1980) Microbiol. Rev. 44, 491-517.
Boddey RM and Döbereiner J (1982) 12th International Congress of Soil Science New Delhi India 8-16 February 1982. Symposia Papers I pp. 28-47.
Brown ME (1982) In Rhodes-Roberts ME and Skinner FA, eds, Bacteria and Plants, pp. 25-41, Academic Press, London.
Currier WW and Strobel GA (1976) Plant Physiol. 57, 820-823.
De Polli H Matsui E Döbereiner J and Salati E (1977) Soil Biol. Biochem. 9, 119-123.
Döbereiner J and Day JM (1976) In Newton WE and Nyman CJ, eds, Proc. of the 1st Int. Symp. on Nitrogen Fixation, pp. 518-538, Washington State University Press, Pullman.
Döbereiner J and De Polli H (1980) In Stewart WDP and Gallon GR, eds, Nitrogen Fixation, pp. 301-303, Academic Press, London.
Ela SW Anderson MA and Brill WJ (1982) Plant Physiol. 70, 1564-1567.
Eskew DL Eaglesham ARJ and App AA (1981) Plant Physiol. 68, 48-52.
Gitte R Rai P Patil RR (1978) Plant Soil 50, 553-556.
Goodman RN (1976) In Heitefuss RH and Williams PH, eds, Encyclopedia of Plant Physiology, new series, Vol. 4, Physiological Plant Pathology, pp. 172-196, Springer Verlag, Berlin.
Harper SHT and Lynch JM (1979) J. Gen. Microbiol. 112, 45-51.
Hegazi NA Khawas H and Monib M (1981) In Gibson AH and Newton WE, eds, Current Perspectives in Nitrogen Fixation, pp. 493, Austr. Ac. Sc., Canberra.
Kapulnik Y Okon Y Kigel J Nur I and Henis Y (1981a) Plant Physiol. 68, 340-343.
Kapulnik Y Sarig S Nur I Okon Y Kigel J and Henis Y (1981b) Expl. Agric. 17, 179-187.
Kapulnik Y Sarig S Nur I Okon Y (1983) Can. J. Microbiol. 29, In press.
Klucas RV and Pedersen W (1980) In Newton WE and Orme-Johnson WH, eds, Nitrogen Fixation, Vol. 2, pp. 243-255, University Park Press, Baltimore.
Lethbridge G Davidson MS and Sparling GP (1982) Soil Biol. Biochem. 14, 27-35.
Lin W Okon Y and Hardy RWF (1983) Appl Environ. Microbiol. 45, 1775-1779.
Nur I Okon Y and Henis Y (1982) J. Gen. Microbiol. 128, 1937-1943.
Nur I Steinitz YL Okon Y and Henis Y (1981) J. Gen. Microbiol. 122, 27-32.
Okon Y (1982) Israel J. Bot. 31, 214-220.
Okon Y Heytler PG and Hardy RWF (1983) Appl. Environ. Microbiol. In press.
Patriquin DG (1982) In Subba Rao NS, ed. Advances in Agricultural Microbiology,

pp. 139-190, Oxford and IBH Publishing Co., New Delhi.
Rennie RJ (1980) Can. J. Bot. 58, 21-24.
Rennie RJ deFreitas JR Ruschel AP and Vose PV (1983) Can. J. Bot. 61, In press.
Rennie RJ and Larson RI (1979) Can. J. Bot. 57, 2771-2775.
Reynders L and Vlassak K (1982) Plant Soil 66, 217-223.
Sarig S Kapulnik Y Nur I Okon Y (1984) Expl. Agric. 20, In press.
Schmidt EL (1979) Annu. Rev. Microbiol. 33, 355-376.
Smith RL Bouton JH Schank SC Quesenberry KH Tyler ME Milam JR Gaskins MH and Littell RC (1976) Science 193, 1003-1005.
Subba Rao NS (1981) In Vose PB and Ruschel AP, eds, Associative N_2 Fixation, Vol. 1, pp. 137-144, CRC Press, Boca Raton, Florida.
Suslow TV (1982) In Mount MS and Lacy GH, eds, Phytopathogenic Prokaryotes Vol. 1, pp. 187-223, Academic Press, New York.
Tarrand JJ Krieg NR and Döbereiner J (1978) Can. J. Microbiol. 24, 967-980.
Tien TM Gaskins MH and Hubbell DH (1979) Appl. Environ. Microbiol. 37, 1016-1024.
Umali-Garcia M Hubbell DH Gaskins MH and Dazzo FB (1980) Appl. Environ. Microbiol. 39, 219-226.
Wood LV Klucas RV and Shearman RC (1981) Can. J. Microbiol. 27, 52-56.

ULTRASTRUCTURE AND NITROGENASE ACTIVITY OF FRANKIA GROWN IN PURE CULTURE AND IN ACTINORRHIZAE OF ALNUS, COLLETIA AND DATISCA SPP.

ANTOON D.L. AKKERMANS/FAUZIA HAFEEZ[+]/WIM ROELOFSEN/ASHRAF H. CHAUDHARY[+]/ ROB BAAS
DEPARTMENT OF MICROBIOLOGY, AGRICULTURAL UNIVERSITY, WAGENINGEN, THE NETHERLANDS
+ PRESENT ADDRESS DEPARTMENT OF BIOLOGICAL SCIENCES, QUAID-I-AZAM UNIVERSITY, ISLAMABAD, PAKISTAN

1. INTRODUCTION

Plant species, able to produce N_2-fixing root nodules with actinomycete-like organisms have been discovered in about 200 species distributed over 21 plant genera. In order to distinguish these plants from the nodulated leguminous plants, the term "nodulated non-legumes" was used. Recently, Fessenden has proposed to change this name into "actinorhizal plants, i.e. plants bearing actinorhizas (Fessenden, 1979). The latter term was introduced in order to distinguish the root nodules induced by actinomycetes from those with *Rhizobium* as microsymbiont.
Since the first isolates of the endophytes of actinorhizas became available in 1978 (Callaham *et al.*, 1978; Quispel, Tak, 1978), an increasing number of strains have been obtained from root nodules of *Alnus*, *Casuarina*, *Comptonia*, *Elaeagnus*, *Hippophaë*, *Myrica*, *Purshia*, *Colletia* and *Ceanothus* spp.. Many of these organisms are able to fix N_2 in pure culture and form actinorhizas on their hosts. All these isolates are morphologically very similar actinomycetes which produce typical sporangia and spherical vesicles, which so far known have not yet been observed in other free-living actinomycetes. These microbes are classified in the genus *Frankia*.
Frankia isolates so far obtained have shown to be highly variable with respect of the growth requirements, chemical composition of the cell and the ability to infect various host plants. When grown in media with N_2 as sole N source, many strains are able to induce nitrogenase. Since under these growing conditions the strains produce vesicles it has been suggested that the vesicles are the sites of nitrogen fixation, like heterocysts in cyanobacteria. Additional arguments for this hypothesis have been summarized elsewhere (Akkermans, Roelofsen, 1978; Akkermans *et al.*, 1979, 1982).
Although the formation of spherical vesicles has been observed in root nodules of various plant species as well as in pure cultures, there still are a number of actinorhizas in which the endophyte do not form spherical vesicles under N_2-fixing conditions (e.g. *Casuarina*, *Coriaria*, *Myrica*, *Datisca* and *Comptonia* spp.). This indicates that formation of these structures is not a prerequisited for nitrogen fixation in actinomycetes. In *Casuarina* nodules *Frankia* usually form hyphal clusters although aberrant filamentous "vesicles" may occur. Isolates of these *Frankia* strains however do produce the spherical vesicles when grown in pure culture under N-limiting conditions.
In *Comptonia* nodules the situation is even more complex. In pure culture the isolate formed spherical vesicles when fixing nitrogen. In the nodules, however, the shape of the vesicles is dependent on the host (Lalonde, 1979). When grown in *Comptonia* nodules it formed filamentous vesicles while in *Alnus* spherical vesicles are formed. This indicates that the shape of the vesicles is host-dependent. In *Coriaria* and *Datisca* nodules the structure of the endophyte is even more complex and filamentous "vesicles" are oriented to the centre of the host cell. Although pure cultures of these endophytes have not yet been obtained which fulfil to the postulates of Koch, it is evident that this type of actinomycete-plant symbiosis is too little known in order to draw general

conclusions.
In the present paper a general description will be given of two types of actinomycete-plant symbioses, viz. the actinorhizas of Alnus and Datisca spp. with special attention to the ultrastructure of the endophytes and the optimum conditions for induction of nitrogenase in the in vitro cultures of Frankia strains derived from Alnus nitida and Colletia cruciata.

2. MATERIALS AND METHODS

2.1. Isolation and cultivation of Frankia strain An 1., An 2.10 and Cc 1.17.

Frankia strains used in this study were isolated from actinorhizas of Alnus nitida and Colletia cruciata by the OsO_4-method of Lalonde (Lalonde et al., 1981) and routinely cultivated in a P+N medium, containing propionate as sole organic carbon source and casamino acids as nitrogen source from the field in Pakistan (Hafeez et al., in press). Alnus nodules were collected from the field in Pakistan (Hafeez et al., in press); Colletia nodules were collected from introduced plants cultivated in peat soil in The Netherlands. Vesicles formation and induction of nitrogenase activity were measured after transfer of the cultures from a P+N medium into a P-N medium, containing propionate as organic C-source and N_2 as N source. Strain Cc 1.17 was isolated from nodules of C.cruciata and strains An 1 and An 2.10 originated from A.nitida.

2.2. Scanning and Transmission electron micrographs

Scanning electron micrographs (SEM) were prepared after fixation in glutaraldehyde (2 h, 2 per cent), dehydration and critical point drying as described elsewhere (Hafeez et al., in preparation).
Transmission electron micrographs (TEM) were prepared as described by Hafeez et al. (in preparation).

2.3. Nitrogenase activity

Cells pregrown in P+N medium were harvested by sterile centrifugation and transferred to P-N medium. The cultures were incubated in Hungate tubes (volume 16.5 ml) which were flushed before with argon. Oxygen was added up to different concentrations as mentioned in the Results. Carbon dioxide was added up to a concentration of 0.5%. The tubes filled with different gas mixtures were sterilized by autoclavation before adding the Frankia cultures. The tubes were incubated at 30°C in a water bath and shaken. Acetylene was added to 10% at time: 0.
The nitrogenase (C_2H_2 reduction) activity was followed over a period of ca. two days. After incubation the cells were harvested and used for protein determinations by using Folin reagent. Nitrogenase activity was expressed as nanomol C_2H_4 produced per mg protein per hour.
Duplicate tubes were used for measuring the amount of vesicles.

3. RESULTS AND DISCUSSION

3.1. Ultrastructure of Frankia

3.1.1. Isolates of Frankia

As shown in Fig. 1 Frankia An 1 is a filamentous organisms which produce large numbers of sporangia and spherical vesicles. The shape of the vesicles is rather uniform among different types of Frankia strains and usually are formed at the end of short side branches of the filaments (Fig. 2). Until now it is unknown if other structures than spherical vesicles can be formed by *Frankia in vitro*. Recently *Frankia*-like actinomycetes have been isolated from actinorhizas of *Datisca cannabina* which produced sporangia and spherical vesicle-like structures which resemble those of other *Frankia* strains. These strains however are not infective and it is unknown whether these really are endophytes of *Datisca* or contaminants (Hafeez, *et al*., in preparation).

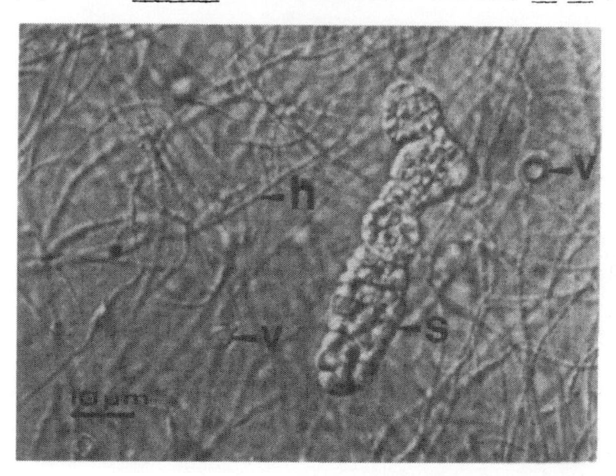

Figure 1. *Frankia* An 1, isolated from root nodules of *A.nitida*, and cultivated in P+N medium. Vesicle (v), sporangium (s) and hyphae (h). Inteference micrograph.

3.1.2. Actinorhizas

Frankia growing in actinorhizas invade the host cell by penetration of the cell wall (Fig. 3) and formation of microcolonies of filamentous cells inside the living host cell. In *Alnus* spp. the hyphen produce terminal spherical vesicles at the periphery of the clusters (Fig. 4). These stuctures are similar to the vesicles produced in the free-living isolates and showed the presence of a complex membrane system around the vesicles (Burggraaf, Meesters, in preparation). These complex double layer systems are probably no artifacts as proposed before (Lalonde *et al*., 1976) and likely to occur in nature as demonstrated in the SEM and TEM pictures of the endophyte in alder nodules (Figs. 5, 6).

In all cases the endophyte is closely connected to the cytoplasm of the plant and vesicles usually are surrounded by several mitochondria (Akkermans and Houwers, 1983, Hafeez *et al*., in press).

In *Datisca* nodules the structure is almost reversed and filamentous vesicles are directed to the centre of the cell (Fig. 7). Of particular interest is the occurrence of congromerates of mitochondria which are embedded in the tissue of the endophyte, indicating the importance of the intime symbioses between plant mitochondria and the nitrogen-fixing endophyte (Akkermans *et al*., 1983). A full description of this nodule type will be published elsewhere (Hafeez, *et al*., in preparation).

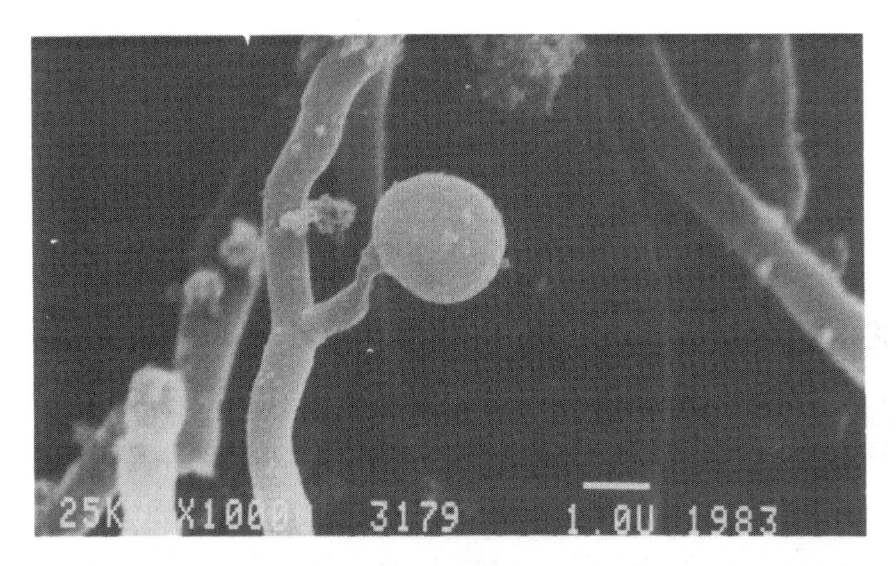

Figure 2. SEM of Frankia An 1 cultivated in P-N medium showing spherica vesicle on side branch of hyphae.

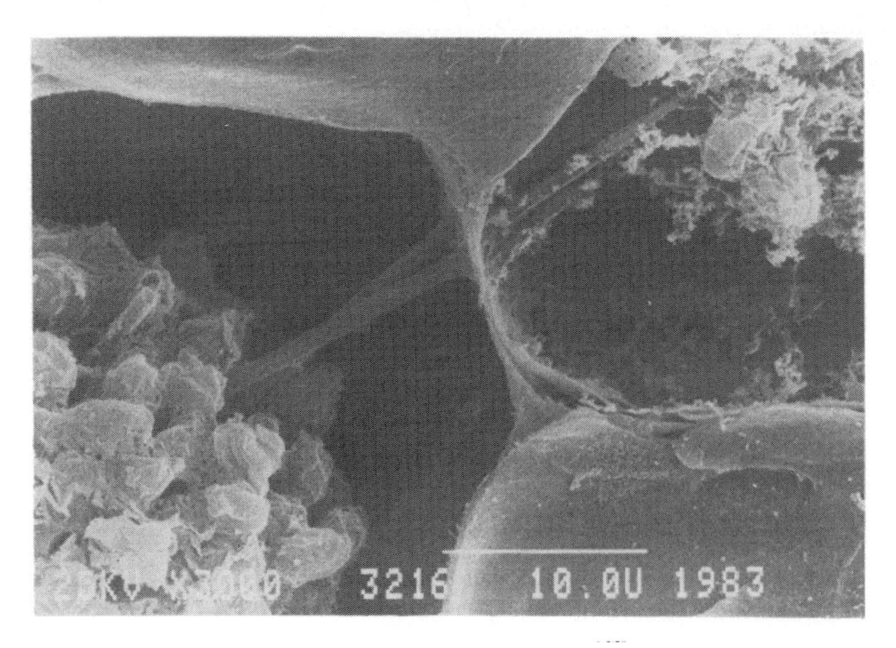

Figure 3. SEM of root nodule of A.glutinosa with spore positive Frankia strain showing penetration of hyphae through the host cell wall and a vesicle cluster inside the cell.

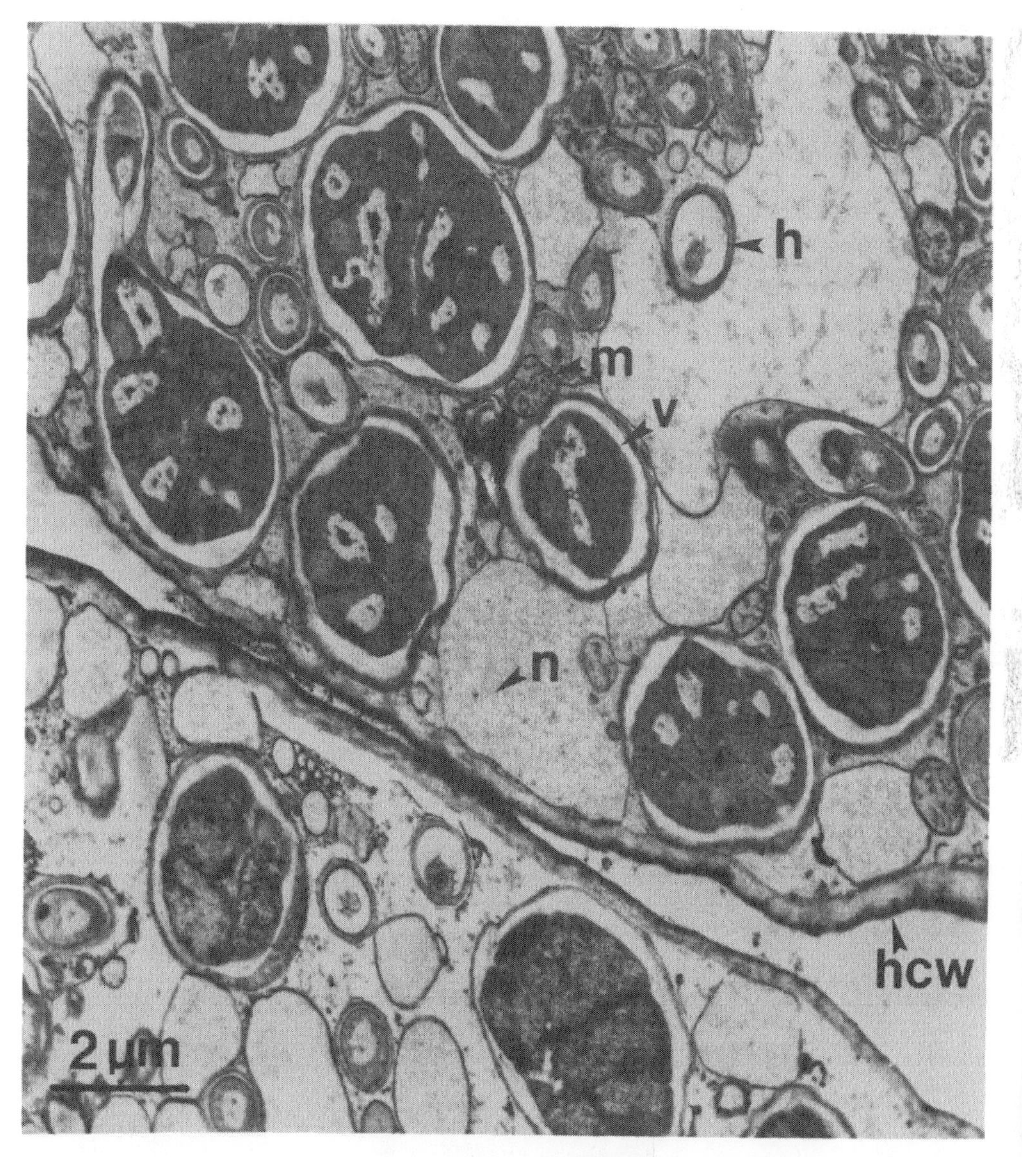

Figure 4. TEM of root nodule of A.nitida showing spherical vesicles (v) near the host cell wall (hcw). Mitochondria (m), hyphae (h), nucleus (n).

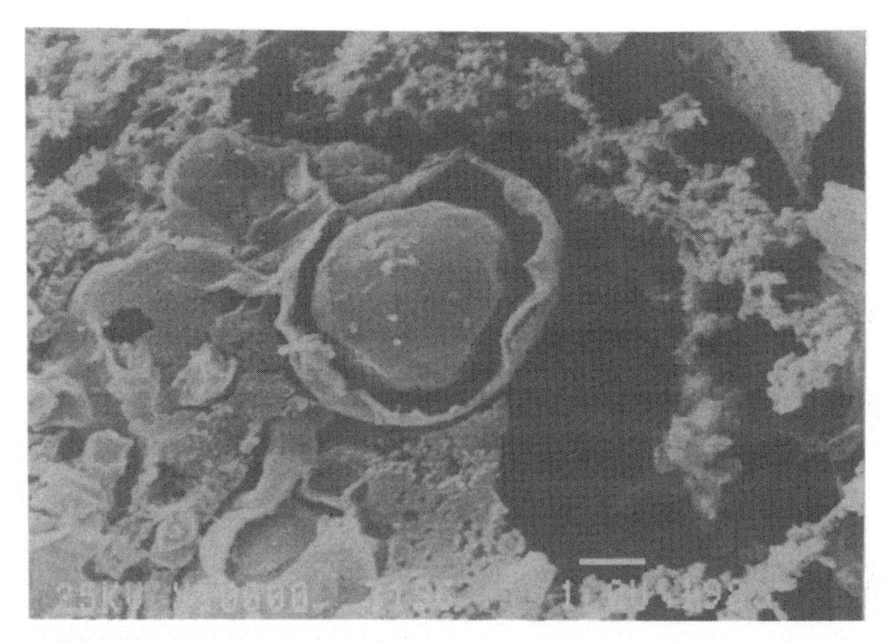

Figure 5. SEM of vesicle of endophyte inside the root nodule of A.glutinosa inoculated with spore positive nodule homogenate (Weerribben). Note membrane system around vesicle.

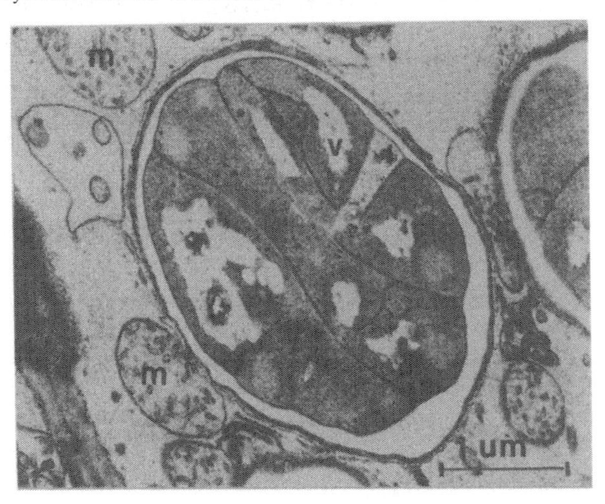

Figure 6. TEM of vesicle of endophyte of A.nitida showing space between two layers around the vesicle. Note the septation of the vesicle and the presence of mitochondria near the vesicle.

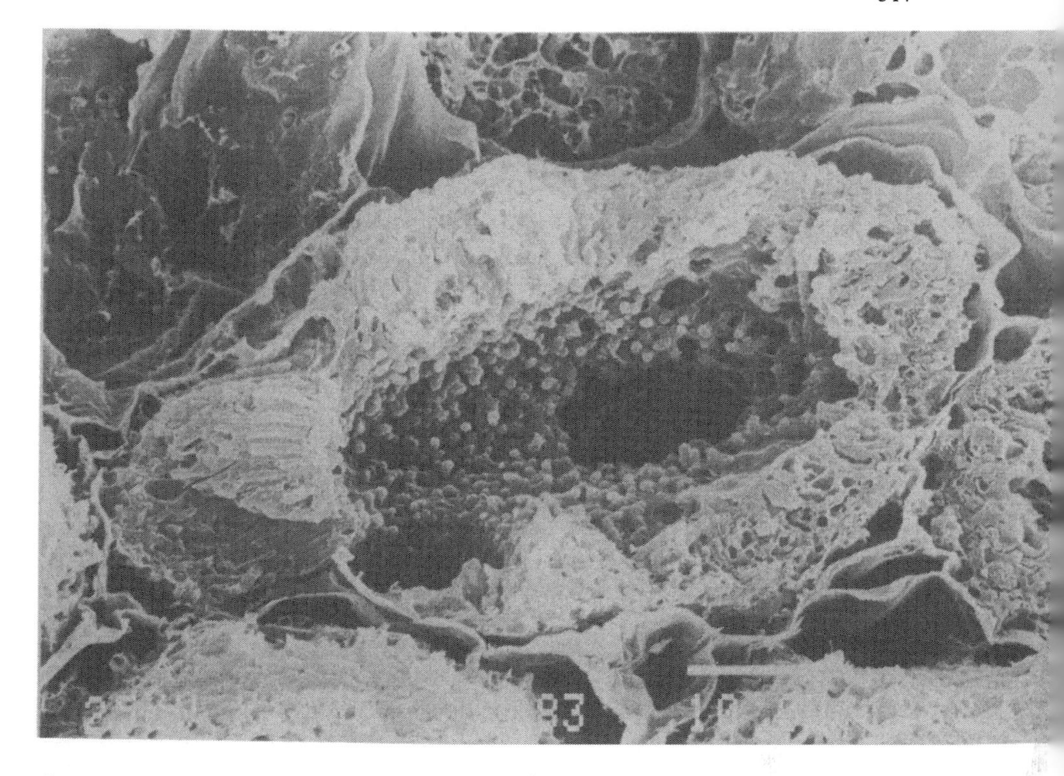

Figure 7. SEM of root nodule of D.cannabina showing vesicle cluster with filamentous hyphae directed towards the centre of the host cell.

4. NITROGENASE ACTIVITY

4.1. Isolates of Frankia

Most Frankia strains are able to grow in media with N_2 as a sole N source. The nitrogenase activity is correlated with the number of spherical vesicles in the culture as has been demonstrated for Frankia strains Cc 1.17 and An 2.10. Cells pregrown in P+N medium and subsequently transferred to P-N medium, showed nitrogenase activity after 20-30 hours of incubation (Fig. 8).
The specific activity was dependent on the O_2 concentration in the gas phase (Fig. 9) and the number of vesicles formed. While strain Cc 1.17 abundantly formed vesicles, strain An 2.10 produced these vesicles only occasionally (data not shown here). The specific activity of strain Cc 1.17 had a maximum of 900 nanomol C_2H_4 produced.mg protein^{-1}.hour^{-1} while the activity of strain An 2.10 was not higher than 20 nanomol.mg protein^{-1}.h^{-1}.

4.2.2. Actinorhizas

The nitrogen-fixing activity of activity usually varied between 1 and 10 µmol C_2H_4 produced. g fresh wt.$^{-1}$.h^{-1}. If one calculates this activity on the amount of Frankia tissue in the nodules it becomes clear that the specific activity of the nitrogenase of the microsymbiont approximates the activity of the pure culture. This can be demonstrated in the following example.
Based on the amount of diaminopimelic acid in the actinorhizas of Alnus glutinosa induced by Frankia Avc 1.1, and similar measurements of the content in

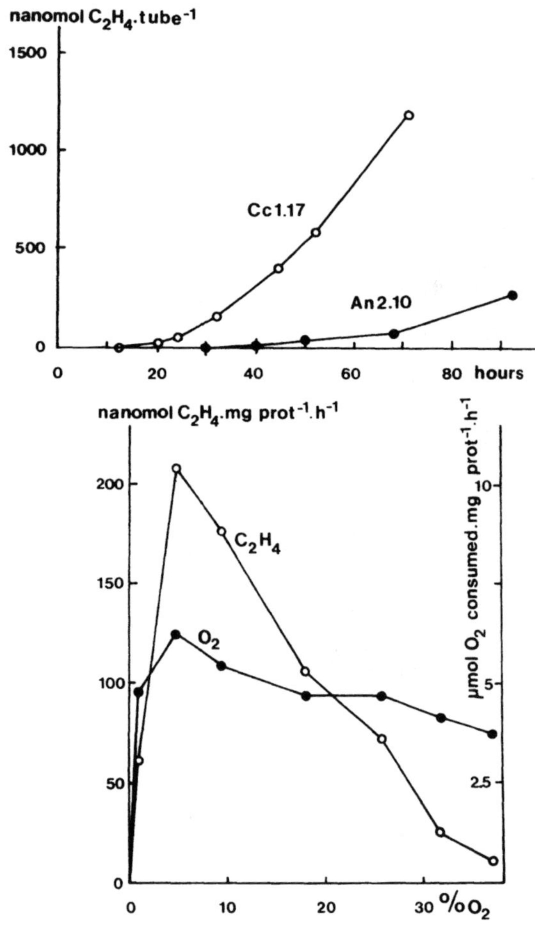

Figure 8. Induction of nitrogenase activity in Frankia strains Cc 1.17 and An 2.10 at 15% O_2 in gas phase.
The amount of cells of strain Cc 1.17 and An 2.10 was 146 µg and 546 µg protein per tube, respectively.

Figure 9. Nitrogenase activity and O_2 uptake by Cc 1.17 at different O_2 concentration.

in the pure culture, it has been computed that ca. 10% of the total nodule mass consists of Frankia tissue (Akkermans et al., in press). If one takes into account the protein content of the cells it indicates that Frankia tissue inside the nodule fixes nitrogen at a rate of 100 - 1000 nanomol. mg Frankia protein^{-1}.h^{-1}. These values are almost similar to the values obtained for the pure cultures. Up to now exact measurements still have to be done with nodules induced by Frankia Cc 1.17. These experiments have not been done becuase strain Cc 1.17 is not (or no more?) infective on Alnus.
Since nitrogenase activity measurements of other strains of Frankia which are infective on Alnus showed similar activities in vitro, it is likely that the statement is correct and that Frankia can fix as much nitrogen in the pure cultures as in the nodule symbioses.

5. REFERENCES

Akkermans ADL, Roelofsen W and Blom J (1979) In Gordon JC, Wheeler CT and Perry DA, eds., Symbiotic Nitrogen Fixation in the Management of Temperate Forests, pp. 160-174, Oregon State Univ., Corvallis.
Akkermans ADL and Roelofsen W (1980) In Stewart WDP and Gallon JR, eds., Nitrogen Fixation, pp. 279-299, Acad. Press, London.
Akkermans ADL, Blom J, Huss-Danell K and Roelofsen W (1982) In Gyllenberg et al., eds., Proceedings Second National Symposium on Biological Nitrogen Fixation, SITRA Report 1, pp. 169-179, Helsinki, ISBN 951-563-069-X.
Akkersmans ADL and Houwers A (1983) in Gordon JC and Wheeler CT, eds., Biological Nitrogen Fixation in Forest Ecosystems, pp. 7-53, Martinus Nijhoff/ Dr. W Junk Publ., The Hague, ISBN 90-249-2849-5.
Akkermans ADL, Roelofsen W, Blom J, Huss-Danell K and Harkink R (1983) Can. J. Bot. 61, in press.
Akkermans ADL, Eijkel L van den, Ham RE and Roelofsen W (in press) Plant and Soil.
Callaham D, Tredici P del and Torrey JG (1978) Science 199, 899-902.
Fessenden RJ (1979) In Gordon JC, Wheeler CT and Perry DA, eds., Symbiotic Nitrogen Fixation in the Management of Temperate Forests, pp. 403-419, Oregon State Univ., Corvallis.
Hafeez F, Akkermans ADL and Chaudhary AH (1984) In Akkermans ADL, Baker D, Huss-Danell K and Tjepkema JD, eds., Proceedings Workshop on Frankia Symbioses, Wageningen, Plant and Soil 77, in press.
Lalonde M, Knowles R and Devoe IW (1976) Arch. Microbiol. 107, 263-267.
Lalonde M (1979) Bot. Gaz. s140, 535-543.
Lalonde M, Calvert HE and Pine S (1981) in Gibson AH and Newton WE, eds., Current Perspectives in Nitrogen Fixation, pp. 296-299, Austr. Acad. Science, Canberra.
Quispel A and Tak T (1978) New Phytol. 81, 587-600.

6. ACKNOWLEDGEMENTS

Electron micrographs were prepared at the Technical and Physical Engineering Research Service (TFDL) Wageningen. Thanks are due to the UNDP for financial support to the first two authors. This investigation was supported by the Netherlands Foundation for Biological Research (BION) with financial support from the Netherlands Organization for the Advancement of Pure Research (ZWO).

POSTER DISCUSSION 5A STEM-NODULATED LEGUMES AND NITROGEN-FIXING TREES

Y. DOMMERGUES, Orstom B.P. 1386 Dakar, Senegal

Only a few among the N_2-fixing systems that exist in nature have been really studied and their use subsequently introduced in agriculture, forestry or agroforestry. Among the non-domesticed systems, two types appear to be most promising: stem-nodulated legumes and N_2-fixing trees (legume or actinorhizal trees).

Studies on stem-nodulated systems (*Sesbania rostrata* and different species of *Aeschynomene*) are rapidly developing in several directions. Some *Rhizobium* strains associated with *Sesbania rostrata* (ORS 571) or *Aeschynomene* sp. are now shown to exhibit the unique characteristic of growing on N_2 as the sole source of nitrogen; this property has obviously triggered investigations on the genetics of ORS571.

The nodulation sites of stem-nodulated legumes are now known to include a dormant root primodium which is infected (and evolves into a stem nodule) only when it can be reached by the competent *Rhizobium* strain. Some stem-nodulated legumes are readily infected (e.g. *Sesbania rostrata*, *Aeschynomene afraspera*), while other hosts nodulate only when the dormancy of the root primodium on the stems is broken. Studies on the nodulation sites of *Sesbania rostrata* have already been completed in at least three laboratories (University of Minnesota, Orstom Dakar and the Australian National University), but, to the best of our knowledge, investigations of the initial stages in the morphogenesis of other stem-nodulated legumes are only just beginning.

From the agronomic point of view, *Sesbania rostrata* is a most promising species in the tropics, since its potential for N_2-fixation is high (200 kg of N_2-fixed per ha in 50 days) provided that soil and climate are favorable. It also exhibits a remarkable tolerance to soil combined nitrogen (which has also been found in *Aeschynomene* sp.). Preliminary field experiments indicate that using *Sesbania rostrata* as a green manure can dramatically increase rice yields in N-deficient paddy soils. *Sesbania rostrata* (and now *Aeschynomene* species) can also be used as a forage.

With the exception of some species (e.g. *Leucaena leucocephala*), our knowledge of nitrogen-fixing trees is still very limited. Up to now, only a few studies have dealt with these N_2-fixing systems that could be harnessed for a variety of purposes in forestry and agroforesty. The relatedness of nitrogen-fixing tree rhizobia to other rhizobia is practically undocumented. We have only very few reliable data on actual N_2-fixation in the field. Our present ignorance can probably be attributed to the difficulties that face the microbial ecologist who studies nitrogen-fixing trees. The most conspicuous obstacles being the high genetic variability of most tree species, the large variations in N_2-fixation that occur throughout the life of trees, the redistribution of N in the plant and the soil profile that occurs under perennial plants. It is thus easy to understand that the assessment of N_2-fixation in the field raises huge technical problems (experimental designs, choice of standard trees, sampling). ^{15}N methods are probably the more reliable, but less expensive techniques should be devised that would facilitate the choice of the most efficient tree species.

More studies on nitrogen-fixing trees are urgently needed especially in developing countries, where the use of fertilizers in forestry would be unrealistic.

THE INFLUENCE OF PHOSPHORUS AND MOLYBDENUM ON NITROGEN FIXATION IN VIGNA UNGUICULATA VAR. IFE BROWN.

ADEWALE ADEBAYO AND JULIUS ADETORO
DEPARTMENT OF SOIL SCIENCE,
UNIVERSITY OF IFE, ILE-IFE, NIGERIA

The application of phosphorus (P) ranging in concentration from 0-60ppm and molybdenum (0-25ppm) to a local cowpea variety Vigna unguiculata Var. 'Ife brown' inoculated with Rhizobium sp. was investigated in a sand culture under greenhouse conditions in a 5x5x3 factorial experiment in Completely Random block design. Some important parameters of N-fixation such as dry matter, yield of roots and shoots, nodule dry weight and the amount of nitrogen fixhd were evaluated. The results of the data were subjected to analysis of variance and Duncan Multiple Range test.

Of all the levels applied, Mo at 10-15ppm and P at 60ppm were found to be highly superior (P=0.01) to all other levels in promoting increased nodule weight, nodule number and dry matter accumulation. The dry weight of nodules increased from 45.06mg/plant at zero levels to a maximum of 282mg/plant at 10ppm Mo and 60ppm P; the dry matter accumulation from 2.3mg to a maximum of 7.0g at 15ppm Mo and 60ppm P. Nitrogen fixation was optimum at 15ppm Mo and 60ppm P. The application of higher rates of Mo (20ppm and above) led to reduced nitrogen fixation and poor crop performance.

From the analysis of the results, it wascconcluded that although tropical legumes have undergone a lot of hybridization, the impact of such hybridization in influencing yield is still very minimal. The increased demand for nutrients as a result of genetic improvement cannot be met from the traditional systems without adequate combination of essential fertilizers such as P and Mo whose contents are very low in unfertilized tropical soils.

EFFECT OF SOIL TYPE AND ADDED NUTRIENTS ON GROWTH AND NITROGENASE ACTIVITY OF SOME TROPICAL FORAGE LEGUMES

BELAL AHMED and JOHN M. KEOGHAN
Caribbean Agricultural Research and Development Institute, University Campus, St. Augustine, Trinidad, W.I.

1. INTRODUCTION

The selection and establishment of productive and well-adapted forage legumes is a first priority towards improvement of pastures in the Caribbean. Towards this objective 13 indigenous and introduced forage legume species were evaluated in pot tests using 12 soils with a wide range of properties from Antigua. W.I.

2. MATERIALS AND METHODS

The legumes were grown with or without added nutrients (full strength N-free Dart, Pate 1959 solution). Uninoculated, scarified legume seeds were planted directly onto undisturbed soil cores in 15 cm plastic pots. Legumes were harvested after nine weeks and dry matter yield and nitrogenase activity determined.

3. RESULTS AND DISCUSSION

Mean values of dry matter yield and nitrogenase activity are presented in Table 1. These parameters differed widely with soil type. Addition of nutrient solution resulted in a significant increase in yield and nitrogenase activity of legumes for all soil types.

TABLE 1. Mean values of dry matter yield and nitrogenase activity of 13 legumes grown in 12 different soils as influenced by added nutrients

Mean	Dry matter yield (g/pot)		Nitrogenase activity μ mole C_2H_4/pot/h	
	No*	N1**	No*	N1**
Soils	1.04-3.68	2.64-5.07	1.46-5.15	4.47-7.43
Legumes	1.30-4.14	1.40-9.28	0.57-11.74	0.99-25.81
Mean	2.43	3.64	2.76	5.71
SE±	0.06		0.20	

*No-No nutrients applied, **N1-extra nutrients applied

Macroptilium lathyroides produced the highest mean yield (6.64 g/pot) while *Centrosema virginianum* exhibited the highest nitrogenase activity (18.77 μ mole C_2H_4/pot/h). The legumes grew best in neutral, free draining soils high in available nutrients (Ottos clay and Yorks clay loam) but not so well in acidic, poor draining, low nutrient soils such as Gunthorpes clay, Shirley loam and Montros clay loam.

4. REFERENCES

Dart PJ and Pate JS (1959) Austr. J. Biol. Sci. 12, 427-444.

NITROGEN-FIXING GROWTH OF *AESCHYNOMENE* STEM-NODULATING RHIZOBIA

D. ALAZARD, and B. DREYFUS, ORSTOM, B.P. 1386, DAKAR, Sénégal

Leguminous plants of the genus *Aeschynomene* form, like *Sesbania rostrata*, nitrogen fixing nodules on their stem. Fifteen strains of *Rhizobium* isolated from stem nodules of seven tropical species of *Aeschynomene (A. afraspera, A. indica, A. sensitiva* spp 1 and 2, *A. schimperi, A. crassicaulis and A. elaphroxylon)* and one strain isolated from root nodules from *A. americana* were first tested for their ability to nodulate roots and stems of all these host plants. According to the cross inoculation test, rhizobia could be divided into three main groups: 1/ rhizobia isolated from stem nodules of *A. afraspera* forming effective stem nodules only on this species.
2/ rhizobia isolated from *A. indica, A. sensitiva* spp 1 and 2, and *A. schimperi* specific to this group. 3/ rhizobia isolated from *A. crassicaulis, A. elaphroxylon* and *A. americana* effectively nodulating the roots of *Macroptilium atropurpureum*, thus related to the cowpea group.
As strain ORS 571 isolated from stem nodules of *Sesbania rostrata* was shown to grow in culture at the expense of N_2 (1), this property was investigated with the rhizobia belonging to the three groups. Experiments were performed in 500ml serum flask containing 20 ml of nitrogen-free medium (MSO) under low O_2 tension (pO_2 0.005 atm). The MSO medium differed from LO medium(1) by replacing sodium lactate by 5 g/l mannitol and 5 g/l disodium succinate. Four vitamins were added: 1 mg/l biotin, 2 mg/l nicotinic acid, 2 mg/l calcium panthotenate, 1 mg/l choline.
Only rhizobia from group 1 and 2 showed significant nitrogenase activity in culture (200 to 300 nmoles C_2H_4/h/mg protein). Furthermore these strains were also able to grow at the expense of N_2. However the O_2 tension required for N_2-fixing growth was lower for these strains than for ORS 571 (pO_2 0.03 atm).
The property to grow on N_2 could thus be a common feature among stem-nodulating rhizobia.

Dreyfus, B., Elmerich, C., and Dommergues, Y. 1983. Appl. Environ. Microbiol. 45, 911-913.

HOST SPECIFICITY, NITROGEN FIXATION AND OXYGEN RELATIONS OF THE *PARASPONIA PARVIFLORA* - *RHIZOBIUM* SYMBIOSIS.

J.H. BECKING
RESEARCH INSTITUTE ITAL, P.O. BOX 48, 6700 AA WAGENINGEN, THE NETHERLANDS.

Cross-inoculation experiments with *Parasponia parviflora* plants and a large number of *Rhizobium* strains isolated from legumes, mainly of temperate origin, showed that some strains of *Rhizobium trifolii* and *R. lupini* can produce root nodulation in *Parasponia*. Root nodulation was also obtained with some tropical *Rhizobium* isolates such as those from *Arachis* and *Albizzia*, but not with *Rhizobium japonicum* strains. Strains of *Rhizobium leguminosarum*, *R. phaseoli*, and *R. meliloti* produced abnormal root nodules or pseudo-root nodules in *Parasponia*. All root nodules induced by foreign *Rhizobium* species in *Parasponia* lacked nitrogenase activity. The reverse combination, testing *Parasponia-Rhizobium* on legumes yielded negative results, except with *Vigna sinensis* and two *Macroptylium* species. Root nodules of the latter legumes showed nitrogenase activity.

$^{15}N_2$ experiments with detached *Parasponia* root nodules showed positive fixation and a mean conversion ratio C_2H_4/N_2 of 6.7. *Vigna sinensis* plants inoculated with *Parasponia-Rhizobium* showed a 3-4 times higher specific nitrogenase activity compared to the same strain on *Parasponia*. Hydrogen evolution and efficiency of nitrogen fixation of the *Parasponia* symbiosis was measured. The relative efficiency $(1 - \frac{H2(air)}{H2\ (Ar)})$ and $(1 - \frac{H2(air)}{C2H4})$ was 0.96 and 0.97 respectively indicating a very efficient recycling of produced hydrogen. The nitrogenase (C_2H_2) activity of detached *Parasponia* root nodules was measured at external O_2 concentrations of 3 - 40%. Nitrogenase activity increased with increasing partial O_2 pressure until 40 or 36% O_2, but then a complete breakdown of the system occurred probably by unsatisfactory protection of the nitrogenase system against O_2 at these O_2 levels.

EFFECTS OF POSTEMERGENCE INOCULATION ON FIELD GROWN SOYBEANS

N. BOONKERD/C. ARUNSRI/W. RUNGRATTANAKASIN AND Y. VASUVAT
DIVISION OF SOIL SCIENCE, DEPARTMENT OF AGRICULTURE, BANGKHEN, BANGKOK 10900, THAILAND

Inoculation of soybean seed with rhizobia has traditionally been performed by applying the inoculum to seed immediately before planting. There are, however, some conditions that the application of rhizobia on seed may fail. The failure of seed inoculation may be remedied by postemergency inoculation. The hypothesis was tested by inoculating field grown soybeans after planting. The experiment was conducted at three locations which were free of native rhizobia. The postemergence inoculation treatments comprised inoculating at 5, 10, 15 and 20 days after planting. Two control treatments of seed inoculation and non-inoculation were included. Methods of postemergence inoculation was performed by suspending peat inoculum at the recommended rate in one liter of water. The water-inoculum suspension was poured over the four planted rows of a treatment of 2.4 x 6 m.
Satisfactory nodulation and plant growth were obtained from inoculum treatments made 5, 10 and 15 days after planting in the two areas where surface soil moisture was excellent. In another area where soil surface was relatively dry at the time of inoculum, application nodulation was greatly reduced. It was also observed that nodules formed by postemergence inoculation treatments were mainly located on lateral roots appearing near surface soil. Inoculation at 20 days after planting was, however, unsuccessful. Under good rainfall distribution, seed yield obtained from seed inoculation was not significantly different from the treatments that inoculated at 5, 10 and 15 days after planting. It can be, therefore, recommended that the failures of seed inoculation can be corrected by postemergence inoculation performed within 2 weeks after planting, provided that surface soil moisture is adequate.

IMPACT OF RHIZOBIA ESTABLISHED IN A HIGH NITRATE SOIL ON A SOYBEAN INOCULANT OF THE SAME STRAIN

J. BROCKWELL[+]/R.J. ROUGHLEY/D.F. HERRIDGE
+ DIVISION OF PLANT INDUSTRY, CSIRO, G.P.O. BOX 1600, CANBERRA, AUSTRALIA.

The interactions between a soybean inoculant and *Rhizobium japonicum* already established in the soil were examined in terms of rhizosphere colonization, nodule formation and nitrogen uptake in a crop of irrigated Bragg soybeans grown in 1982-83 at Breeza, New South Wales, on a vertisol high in soil NO_3^-. Three soil populations of *R. japonicum* CB1809 were established during 1981-82 on the same land which, before that, had been *R. japonicum*-free. Plots with two of these population levels (a. 7.0×10^0, b. 2.7×10^5 per g soil) were used for the 1982-83 crop. For this crop, two rates of inoculation (1. 8.7×10^5, 2. 1.3×10^8 rhizobia per cm row) were supplied by spraying a water suspension of peat inoculant into the seed bed at sowing. Thus, there were four treatments, 1a, 1b, 2a, 2b. The peat inoculant comprised a streptomycin-resistant mutant, CB1809str^r. Therefore, although similar otherwise, the inoculant rhizobia could be distinguished from the soil populations by their resistance to 125 µg per ml of streptomycin. Numbers of rhizobia in the plant rhizospheres were counted. Nodules were examined at intervals and the identity of the strains responsible for their formation was determined. Crop N was calculated using Kjeldahl analysis.

Total populations of *R. japonicum* in the rhizospheres increased from 4.6×10^4 to 4.6×10^5 (means) between 14 and 42 days while inoculant populations decreased from 9.0×10^2 to 2.9×10^2 during the same period. This indicated that populations of rhizobia established in the soil colonized the rhizosphere better than inoculant rhizobia and that this dominance increased with time. The number of nodules formed by inoculant CB1809str^r was proportional to its representation in the rhizosphere but a few nodules were formed by the inoculant even when it was outnumbered >1000 times in the rhizosphere. Small populations of rhizobia in the soil immediately after sowing led to reduced rhizosphere colonization. Rhizosphere populations of 6.6×10^3 (14 days) or 9.5×10^4 (42 days) were not sufficient to maximise nodulation or nitrogen fixation.

TABLE 1. The influence of size and source of *R. japonicum* populations on the development of the symbiosis in field-grown Bragg soybeans.

		1a	2a	1b	2b
Soil rhizobia (CB1809 - $\log_{10}$ per g soil)		0.85	0.85	5.43	5.43
Inoculant rhizobia (CB1809str^r - $\log_{10}$ per cm row)		5.94	8.11	5.94	8.11
Rhizobia in rhizosphere ($\log_{10}$)	14 days - total	2.35	3.82	6.32	6.17
	CB1809str^r	2.57	4.06	2.39	2.80
	42 days - total	2.23	4.98	7.76	7.65
	CB1809str^r	2.40	3.04	2.00	2.42
Nodules due to CB1809str^r (%)	42 days	100	90	0	0
	117 days	92	57	0	4
Plants nodulated (%)	42 days	53	89	100	100
	120 days	77	100	100	100
Crop N (kg per ha)	130 days	156	209	242	248

GROWTH AND NITROGEN FIXATION OF MIXED CULTURES OF *AZOSPIRILLUM BRASILENSE* AND *ARTHROBACTER GIACOMELLOI*

I. CACCIARI/D. LIPPI/S. IPPOLITI/M. DEL GALLO/T.PIETROSANTI/W.PIETROSANTI
IREV-CNR AREA RICERCA ROMA, VIA SALARIA KM 29.3, MONTEROTONDO SCALO, ITALY

Seed inoculation with mixed cultures of diazotrophic free living microorganisms is widely used to improve plant productivity (Tilak K V B R et al. 1982). Little is known however on the factors involved in the interactions among the organisms that can differently respond to nutritional and environmental changes. To study some physiological aspects of mixed cultures a nitrogen fixing association was established with a strain of *A. brasilense* (Mengoni M et al. 1978) and a strain of *A. giacomelloi* (Cacciari I et al. 1971). The organisms were adapted to grow under aerobic conditions in a nitrogen-free mineral medium containing yeast extract (0.08 g/l), biotin (40 µg/l) and sucrose as carbon source. These nutritional conditions supported growth of both microorganisms even though they were not the optimal for growth and nitrogen fixing activity. Acetylene reduction was determined on 5 ml-culture samples incubated standing for 24h at the partial oxygen pressures of 0.2 and 0.02. The microorganisms were well differentiated on agar Phenol Red Dextrose: *A. brasilense* colonies are pinky white while *A. giacomelloi* ones are white-cream. Batch cultures (30 ml) grown on a rotary shaker (100 rpm) showed that nitrogenase activity of the mixed cultures improved by about 35% at pO_2 0.2 and 66% at pO_2 0.02, as compared with the sum of activities of single cultures. The growth of both single and mixed cultures were also followed for 24h in a turbidimetric apparatus. The cultures (20 ml) were incubated with magnetic stirring and aeration at two different flow rates (60 ml/min and 140 ml/min). Single and mixed culture poorly aerated exhibited a higher nitrogen fixing capacity. The improvement in nitrogenase activity of mixed cultures was emphasized under both conditions. To investigate a possible role of bacterial metabolites on the interaction between microorganisms, cultures (150 ml) were grown in the Ecologen, an apparatus that allows flux of metabolites among single cultures but prevents contacts among the cells. The Ecologen was installed in a rotary shaker (100 rpm) and incubated for 24h without or with aeration (500 ml/min). Under both conditions a decrease in nitrogenase activity of *A. brasilense* was observed in the presence of *Arthrobacter* metabolites. This behaviour contrasted with that observed in mixed cultures. Therefore further investigations are necessary to clarify whether this result is due to a real effect of *Arthrobacter* metabolites or it depends on the experimental growth conditions.

REFERENCES

Cacciari I et al. (1971) Ann. Microbiol. 21, 97-105.
Mengoni M et al. (1978) Congr. Naz. Soc. Ital. Microbiol. 2, 1308-1311.
Tilak K V B R et al. (1982) Soil Biol. Biochem. 14, 417-418.

CONTRIBUTION TO THE STUDY OF *RHIZOBIUM* AND *AGROBACTERIUM* GENUS : NUMERICAL TAXONOMY

M. CATTEAU, H. KHANAKA, M.D. LEGRAND and J. GUILLAUME

Laboratoire de Microbiologie SN2 - Université des Sciences et Techniques de Lille I - 59655 Villeneuve d'Ascq - FRANCE

89 strains of the genus *Rhizobium* (comprising 27 strains of *R. meliloti*, 10 strains of *R. leguminosarum*, 9 strains of *R. phaseoli*, 14 strains of *R. trifolii*, 5 strains of *R. lupini*, 12 strains of *R. japonicum* and 12 strains of *R. sp.*), 11 strains of the genus *Agrobacterium* and 3 strains of *Pseudomonas* have been studied. The classification of these strains is realized using numerical analysis of 204 features (30 traditional features, 16 tests of antibiotic sensitivity, 98 tests of carbon substrate utilization and 60 tests of enzymatic reactions). Several tests were quantitative, 322 features have been coded after discarding tests giving identical results for all strains.

The results obtained show a very large heterogeneity. The slow growing *Rhizobium* (*R. japonicum* and *R. lupini* and certain number of *R. sp.*) are heterogeneous and are very different from strains of fast-growing *Rhizobium*. These results substantiate the opinion of JORDAN (1982) who proposed to create for slow growing *Rhizobium* a new genus named "*Bradyrhizobium*".

The fast-growing *R. meliloti* (group M), *Agrobacterium* (group A) and *R. trifolii* (group T) appeared to be more homogeneous than other *Rhizobium* strains. *Rhizobium meliloti* (group M) and *Agrobacterium* (group A) formed closely related species. *R. trifolii* (group T) comprises strains from different cross-inoculation groups. This confirms that the classification which is based on the cross-inoculation concept is unsatisfactory.

All strains of *R. leguminosarum* and *R. phaseoli* studied exhibit large diversity and do not form a homogeneous group. The majority of these strains are placed in the group T and in the group A, others are isolated and remoted from preceding strains.

A comparative study of HU-type proteins isolated from representative strains selected among the homogeneous and heterogeneous groups are presented in the accompagning poster and is in good agreement with our results.

STUDY OF THE PROCESS OF COMPETITION BETWEEN STRAINS OF RHIZOBIUM FOR NODULATION

J.C. CLEYET-MAREL,[1] Y. CROZAT,[2] A. CORMAN[3]

(1) I.N.R.A., Laboratoire de Recherches sur les Symbiotes des Racines - Place Viala - 34060 MONTPELLIER-Cedex (France)

(2) E.S.A., 24 rue Auguste Fontenau - 49004 ANGERS (France)

(3) Laboratoire de Biométrie (LA n° 243) - Université de Lyon I - 43, Bd du 11 Novembre 1918, 69621 VILLEURBANNE (France)

1.- INTRODUCTION.- The factors that affect the competition between strains and the preference for strains by a legume's crop are likely to be considerable practical importance. The mechanisms which determine the competitive ability of a strain are poorly understand. To establish a highly compatible relationship between a legume variety and an effective strain of Rhizobium, as much information as possible should be obtained on the strain's survival and saprophytic competence in soil and rhizosphere.

2.- MATERIAL and METHODS.- We investigated the interactions between four soybean cultivars and six strains of *R. japonicum* (five "slow - growing" - one "fast - growing") used for competition in root nodule formation.

We quantified with fluorescent antibody technique (Schmidt 1974) the population densities of two strains in the rhizosphere (Cleyet-Marel, Crozat 1982) in relation with their nodulation scores. The adsorption kinetic's of these strains on the root's surface were also examinated with the FA technique.

3.- RESULTS .- We observed an effect of the host plant cultivar specific of the mixed strain inoculum used. The "fast growing" strain appears to be more competitive then "slow - growing" strains. Events that occur in the developing legume rhizosphere and its adjacent soil are obviously important to the establishment of the Rhizobium-legume symbiosis. We observed that the more competitive strain for nodulation was more stimulated in the rhizosphere then the other strain. After establishment in the legume rhizosphere, we observed a rapid and important accumulation of the most competitive strain on the host's root surface.

4.- CONCLUSION.- Competition between strains of Rhizobium for nodulation is a complex phenomenon, where several steps are included. It seems necessary to examine how the different compartiments : soil with strain survival and saprophytic competence ; rhizosphere with rhizosphere colonisation and Rhizobial adhesion on roots ; plant tissues with its infection by strains, act in the process of competition between strains.

5.- REFERENCES.-

Cleyet-Marel J.C., and Crozat Y., (1982) Agronomie 2 (3), 243-248.

Schmidt E.L., (1974) Soil Sci., 118, 141-149.

SURVIVAL KINETICS AND ADAPTATION RATES OF *R. japonicum* : DEFINITION OF THE SAPROPHYTIC POTENTIAL OF A STRAIN

Y. CROZAT,[1] A. CORMAN,[2] J.C. CLEYET-MAREL[3]

(1) E.S.A., 24 rue A. Fontenau, 49044 ANGERS (France)

(2) Laboratoire de Biométrie (LA n° 243) Lyon I, Bvd du 11 novembre, 69621 VILLEURBANNE (France)

(3) I.N.R.A., Place Viala 34060 MONTPELLIER-Cedex

1.- INTRODUCTION.- *R. japonicum*, soybean specific bacteria, is naturally absent from French soils. Our aim was to follow survival kinetics of different strains of *R. japonicum* introduced into different soils and to caracterise their saprophytic competence.

2.- MATERIAL AND METHODS.- We study for different inoculation levels the survival kinetics of three strains : G_2 sp (mutant strain from USDA Beltsville 311 B 125) G_{49} (SB 16 IARA New-Dehli) GMB_1 Ka (mutant strain from IRA Madagascar), in three non sterile soils : LAVALETTE (silt calcearous soil, pH 8,2) LYON (heavy soil, pH 6,6), LANDES (sandy soil, pH 6,5). 69 soils samples were incubated in modified Leonard's bottle jar assembly at 22° - 24°C without any plant (CROZAT et al 1982). Rhizobium population change were observed by the fluorescent antibody technique on membrane filters (Schmidt 1974).

3.- RESULTS.- For each soil and strain, whatever the number of Rhizobium introduced, these bacteria reach the same stable level (M). This balance threshold does, not differ very significatively between soils and strains ($10^3 - 10^4$ bacteria.g^{-1} of soil). If we introduced a new strain (y) in a soil with a strain (x) at the balance threshold, strain (y) also reached the same level. From an experimental determination of the ended population levels, the fitting of the survival curves with the GOMPERTZ model (dy/dt = Ky (Log M - Log y) corroborate the existence of a balance level. It also gives a good estimation of mortality rates (or growth) K of each strain. This adaptation rate differs between soils and strains and is independant of the inoculation level.

4.- CONCLUSION.- These data and their good fitting with the GOMPERTZ model would show that their may be at least two regulation types of the growth of Rhizobium in soil : - when the inoculum level is below the balance level, the growth curve is like a population regulated by a growth factor on a relatively large substrat.

- when the inoculum level is superior to the balance level its survival kinetic looks like a "prey" population with a very small growth rate in an environment caracterised by a small predation efficiency. These data also showed that it's possible to measure the saprophytic competence of a strain by its saprophytic potential. The saprophytic potential can be estimated by the balance level and the mortality rate (or growth rate)of a strain obtained from the GOMPERTZ model in a given soil. This latter datum appear statistically relevant to classify the strains in each soil, even if the balance level is not sifnigicatively different between strains.

5.-REFERENCES.-Crozat Y, Cleyet-Marel JC, Giraud JJ, and Obaton M., (1982) Soil Biol Biochem., 14, 401-405.
Schmidt EL, (1974) Soil Sci., 118, 141-149.

PRELIMINARY DATA FROM WORLDWIDE LEGUME INOCULATION TRIALS

R. J. Davis, J. Halliday and F. B. Cady
NifTAL Project, University of Hawaii, P.O.Box"O", Paia, HI, USA
and Cornell University, Ithaca, NY, USA.

The International Network of Legume Inoculation Trials (INLIT) is funded by the United States Agency for International Development (USAID), with the organization, promotion and coordination of the Network assigned to NifTAL. The program was initiated in 1980. The first year, of necessity, was devoted mostly to identifying potential cooperators for the Network, so 1981 was the first year with substantial planting of INLIT experiments. In 1982, results from cooperators started arriving at NifTAL in sufficient quantity for composite analysis.

Cooperators in the Network are individual research scientists or institutions conducting legume inoculation trials. The trials are planned and financed by the cooperator. NifTAL supplies to cooperators a standard experimental design; supplies critical imputs, such as the inoculation package, and seed where necessary; consultation; training, if desireable; and data analysis.

As of mid-1983 experimental results from a total of 110 individual INLIT "A" Experiments from 18 countries had been received at NifTAL, analyzed and the data analysis returned to the cooperator. Data have been received and analyzed for the following crops: *Arachis hypogaea* [6], *Cajanus cajan* [8], *Calopagonium caerotim* [1], *Cicer arietinum* [19], *Glycine max* [22], *Lens culinaris* [15], *Leucaena leucocephala* [1] *Medicago sativa* [1], *Phaseolus vulgaris* [4], *Pueraria phaseoloides* [1], *Vigna mungo* [10], *Vigna radiata* [15], *Vigna unguiculata* [5].

Composite analysis on a wordlwide basis have been performed on 5 crop species: *Cicer arietinum*, *Glycine max*, *Lens culinaris*, *Vigna mungo*, and *Vigna radiata*. The results so far indicate that soybeans (*Glycine max*) give the most consistent evidence of a response to inoculation, being directly associated with prior plantings and inoculation of the crop. However, trends are evident in other species indicating that when more locations have been reported factors will be identified which will indicate whether or not a response can be expected with these species. This information will be extremely useful both to policy makers in developing countries and to donor agencies.

BREEDING SOYBEANS TO EXPLOIT NATIVE RHIZOBIA

A.L. DOTO[1]/M.S. CHOWDHURY[2]
FACULTY OF AGRICULTURE, FORESTRY AND VETERINARY SCIENCE, UNIVERSITY OF DAR ES SALAAM, SUB POST OFFICE-CHUO KIKUU, MOROGORO, TANZANIA[1]
AGRO-FOOD SERVICES LTD, 78 MOTIJHEEL C.A. (THIRD FLOOR), DHAKA-2, BANGLADESH[2]

1. INTRODUCTION

The problems associated with the transfer of inoculation technology to the subsistence farmers in developing countries of the tropics prompted the authors to develop through breeding high yielding soybean cultivars that would nodulate effectively by native rhizobia in Tanzania (1).

2. MATERIALS AND METHODS

A cross between two high yielding cultivars, namely IH/192 (local) and Bossier (exotic) were made at Morogoro, Tanzania in 1979. The crossing was aimed at combining the desirable high yielding potential and effective nodulation by native rhizobia in IH/192 with the earliness, short stature and high yielding potential from Bossier (1). Pedigree selection based primarily on yield, earliness, plant stature along with BNF (nodulation and N_2-ase activity) was done from F_3 to F_7 generation under field conditions.

3. RESULTS AND DISCUSSION

Considerable variation was observed among the segregating lines for all characters studied. A positive but nonsignificant correlation between yield and BNF characteristics was found. Among 50 lines selected from F_5 generation, 3 lines (e.g. $L_{10}R_2P_2$, $L_{13}R_2P_3$ and $L_{16}R_2P_1$) were particularly outstanding in yield and BNF characteristics in F_6 and F_7 generations. All these promising lines had more nodule mass, nitrogenase activity and grain yield per plant than any of their parents. They all matured earlier than their parents and were of intermediate stature.
Two promising segregates (e.g. L_{10} and L_{16}) of the F_3 generation (1) maintained their high yielding characters in F_6 and F_7 generations too. These promising lines from F_7 generation are now in advance stage of development for commercial release in Tanzania, where inoculation technology is still underdeveloped.
Acknowledgements are due to the Research and Publication Committee of the University of Dar es Salaam for financial support.

REFERENCE

1. Chowdhury MS and AL Doto (1982) Biological nitrogen fixation as a criterion for soybean breeding: Preliminary results. In Graham PH and SC Harris (eds) Biological Nitrogen Fixation Technology for Tropical Agriculture. CIAT, Cali, Colombia, pp. 45-48

INOCULATION AND NITROGEN FERTILIZATION OF FIELD GROWN LUPINE

C. DUTHION, N. AMARGER - I.N.R.A. B.V. 1540, 21034 DIJON-CEDEX France

Lupine is being considered as an interesting protein crop which could be developped in different european countries. Its microsymbiont Rhizobium lupini is absent or present in low numbers in soils with pH > 6,5. The present study was made to determine the effect of seed inoculation and nitrogen fertilization on the grain yield and nitrogen content of three species of lupine grown in such soils.

MATERIAL AND METHODS

Experiments were conducted on clay soils, pH between 6.7 and 7.8. The number of R. lupini was between 0.1 and 1 per g. of soil in 1979 and 1980, and < 0.1/ g in 1981 and 1982. Four randomized blocks with individual plots 7.5 x 2 m each, were sown with 6 rows 30 cm apart. The three first years the treatments were : inoculation and no inoculation, 4 levels of nitrogen fertilization (NO_3NH_4) : 0, 40, 80, 160 kg/ ha, 3 species of lupine : Lupinus albus cv Kali, L. luteus cv Sulfa, L. mutabilis cv LM13 in 1979, 1980 and the two first species in 1981. In 1982, 4 cultivars of L. albus Kali, Kievski, Lublanc, Lucky were compared, without and with inoculation, and without and with nitrogen fertilization (160 kg/ha NO_3NH_4).

RESULTS

In 1979 and 1980, although the indigenous population of R. lupini was small,the non inoculated L. albus and L. mutabilis were almost as well in 1979, and as well in 1980, nodulated than the inoculated plants. Consequently there was no significant effect of inoculation on these 2 species in 1980. In 1979, inoculation significantly increased the grain yield of the 3 species. This increase was of 9 %, 18 % and 175 % respectively for L. mutabilis, L. albus and L. luteus. The addition of N-fertilizer increased the yield of the non inoculated yellow and white lupine and of the inoculated white lupine but had no effect on the other treatments. This increase was equivalent for the white lupine and inferior for the yellow to the increase obtained with inoculation.

In 1981, inoculation increased the yield of the 2 species studied L. albus and L. luteus from 2.19 and 1.15 t/ha respectively to 4.17 and 2.19 t/ha. The yields of the non inoculated plants with 160 kg/ha N-fertilizer were 3.54 t/ha and 1.88 t/ha.

With the exception of L. mutabilis in 1980, the grain nitrogen content was always higher in the inoculated plants than in the non inoculated. The increase varied from 2 % in 1980 to 36 % in 1981. The addition of N-fertilizer had no effect on the nitrogen content of either the inoculated or the non inoculated plants.

In 1982, inoculation increased the grain yield of the 4 cultivars, Kali, Kievski, Lublanc, Lucky from 8.3, 10.3, 9.5 and 9.3 t/ha to 24.4, 29.9, 29.5 and 35.8 respectively. The addition of 160 kg/ha fertilizer gave yields of 25.2, 28.9, 27.4 and 29.4 t/ha. The grain nitrogen content increased from an average of 3.8 % for the non inoculated plants to an average of 5.8 % for the inoculated one and to an average of 4.5 % for the N-fertilized non inoculated plants. There was no increase in the nitrogen content of the inoculated plants which received N-fertilizer

CONCLUSION

In neutral or slightly alcaline soils seed inoculation of lupine can give very important yield increases, equivalent or superior to the increases given by the addition of 160 kg/ha of N-fertilizer. Inoculation also increases the grain nitrogen content. This increase is more important than the increase given by N-fertilization.

BEHAVIOUR OF SELECTED FINNISH RED CLOVER *RHIZOBIUM* INOCULANTS UNDER FIELD CONDITIONS

EVA EKLUND / DEPARTMENT OF MICROBIOLOGY, UNIVERSITY OF HELSINKI
SF-00710 HELSINKI 71, FINLAND

1. INTRODUCTION

The effect on red clover dry weight and crude protein production of six Finnish red clover *Rhizobium* inoculants was studied in field experiments carried out at nine geographical sites in Finland. The ability of the inoculants to compete with the native *Rhizobium* populations was studied in three of these field experiments.

2. MATERIALS AND METHODS

Red clover seeds were inoculated with peat based *Rhizobium* preparations simulating commercial preparations used in Finland. The ability of the inoculant strains to compete with the native *Rhizobium* populations and in mixed preparations was studied by two methods:
(a) ELISA (enzyme-linked immunosorbent assay)
(b) IAR (intrinsic antibiotic resistance)

3. RESULTS AND DISCUSSION

All of the inoculants were found well introduced (nodule occupancies 50 - >95 %) in two of the three field experiments studied; the soil types were sand and clay respectively. At the third site, a problematic silt soil, the inoculants were identified in only 10-60% of the root nodules examined. When the field experiment in question was started, the red clover seedlings emerged as late as two months after sowing. The soil surface was nearly impermeable after a rainy period followed by drought (typical for certain areas in Central Finland) ELISA and IAR gave similar results.

The competitiveness between the inoculants in a mixed preparation (60B+7B) was also studied. Ordinarily the introduction of strain 7B was suppressed by 60B. On the other hand, strain 7B seems to be the more persistant one under certain stress conditions (e.g. survival in soil and overwintering in connection with clover roots after waterlogging).

As regards the red clover yields, the relative effectiveness of inoculation and nitrogen fertilization was to some extent dependent on soil structure which in fact involves dependence on various factors like O_2 tension, loss of mineral nitrogen by leaching or denitrification, water stress etc. A marked influence of environmental factors on the competitiveness of inoculants as well as on the effectiveness of the symbiosis was noted. In Finland, due to rather extreme environmental conditions for agriculture, *Rhizobium* strains well adapted to local ecological conditions seem to be an especially important basis for future improvements of inoculants.

DISTINGUISHING RHIZOBIA BY ELECTROPHORESIS OF ISOENZYMES.

K.C. ENGVILD & GUNNAR NIELSEN
AGRICULTURAL RESEARCH DEPARTMENT, RISØ NATIONAL LABORATORY,
DK-4000 ROSKILDE, DENMARK.

Distinguishing rhizobium strains from each other still presents a major problem. There is no universally accepted method. The method of choice depends on preference, experience and equipment in the individual laboratory. We have decided to extend to rhizobium our experience in the identification of plant cultivars by starch gel electrophoresis of isoenzymes.

Rhizobium leguminosarum and other rhizobia, mainly from the Rothamsted culture collection were grown for six days on a slime suppressing substrate with enzyme inducers. The substrate contains tryptone 0.5%, yeast extract 0.3%, $CaCl_2 \cdot 6H_2O$ 0.13%, glucose, galactose, xylose, myoinositol, mannitol, sorbitol, and L-arabinose, each 0.01%, agar 1.5%. The bacterial growth was scraped off the agar (~ 0.1 g) with a spatula into 0.4 ml 0.1 M tris HCl buffer, pH 7.1, and centrifuged to remove slime. The 10-50 mg pellet was sonicated (3 mm tip) in 0.1 ml tris buffer with 0.5% mercaptoethanol for 15 sec in the cold. Centrifuged sonicates were absorbed in 5 x 9 mm filter paper strips and placed in the starch gel.

Buffers: tris-citrate buffer, pH 7.0, for electrophoresis of dehydrogenases over night; discontinuous lithium borate/tris citrate pH 8.3 for electrophoresis of esterases for 4-6 hours. Gels: hydrolyzed starch 13% 18 x 18 cm, 0.8 cm thick. Temperature ~ 4°C. The gels were cut in four slabs with a 3-string "cheese cutter", and each slab stained for individual enzymes. Staining: esterase, fast blue RRsalt and naphtylacetate; dehydrogenases, NAD, MTT, PMS and the substrate in question.

We found useful variation in esterases, 3-hydroxybutyrate dehydrogenase, arabinose dehydrogenase, mannitol dehydrogenase and sorbitol dehydrogenase. Also glutamate-oxaloacetate transaminase and phosphatase showed variation, but were difficult to stain reproducibly, as was sorbitol dehydrogenase sometimes. About 15 other enzyme systems gave too little variation or could not be stained.

Two enzyme systems were rhizobium specific and could be analyzed directly in root nodules: 3-hydroxybutyrate dehydrogenase and L-arabinose dehydrogenase.

REFERENCES

Fottrell PF and O'Hora A (1969) J. Gen. Microbiol. 57, 287-292.
Murphy PM and Masterson CL (1970) J. Gen. Microbiol. 61, 121-129.
Mytton LR McAdam NJ and Portlock P (1978) Soil Biol. Biochem. 10, 79-80.
Shaw CR and Prasad R (1970) Biochem. Genet. 4, 297-320.

MICROCOMPUTER APPLICATIONS

FABRICIUS, B-O.[+], HARPER, R.[+], GYLLENBERG, H.G.[+] and LEPO, J.[++]

+ Department of Microbiology, University of Helsinki, SF-00710 Helsinki 71, Finland
++ Department of Biology, University of Mississippi, University, Mississippi 38677, USA

1. INTRODUCTION

Microcomputers are now readily available and very reasonably priced. Programmes exist for many applications, such as word processing, bookkeeping, graphics, and statistical analysis. Applications also exist for process control in biotechnology, and for identification of microorganisms. Recently, microcomputers have also been used for the storage and retreival of data, whereby microbiologists can use them for the documentation of culture collection information. In addition with the development of worldwide computer networks, fast and efficient exchange of information between scientists who are distant geographically and organizationally, has now become a reality.

2. PROCEDURES

2.1. DATA STORAGE AND RETRIEVAL

A microcomputer and available commercial software (a CP/M-based Nokia MicroMikko M4 computer and the Datastar file handling programme by MicroPro) have been used to enter, store, retrieve, update, and sort Rhizobium strain data from the culture collection of the Department of Microbiology, University of Helsinki. The Wordstar and Mailmerge programmes by MicroPro have been used for making printouts of data. The use of a microcomputer reduces the time spent in maintaining and updating records, and facilitates the entry of new information. It also provides fast and easy access to current details and strain lists.

2.2. COMPUTER CONFERENCING

The microcomputer mentioned above have also been used as a computer terminal for taking part in a computer conference in the COM computer conference system at the computing centre of the University of Stockholm, Sweden.

In a computer conference, the conference participants at different geographical locations are linked to a central computer and are able to communicate with each other at a cost superior to both telephone and face-to-face meetings, and a speed superior to mail. A permanent record of the proceedings is maintained in the memory of the central computer, and can be reviewed at will. The users can belong to open or closed "meetings" and send personal "letters" to each other. The participation is not limited by time or place, and individuals can progress at a pace most convenient to themselves.

In the COM-system there are already more than 400 active meetings. Computer networks, giving accsess to this system cover most of the world. A meeting on the Bioconversion of lignocellulose for the production of fuel, fodder and food has been active from the beginning of 1983, and this meeting is linked to a similar meeting in the north American EIES-computer conference system.

During the 5th International Symposium on Nitrogen Fixation we will collect information to determine if a similar conference can be established for Nitrogen fixation.

AZOSPIRILLUM SPP. ECOLOGY OF ITALIAN SOILS

F. FAVILLI/W. BALLONI/E. CRESTA AND A. MESSINI
INSTITUTO DI MICROBIOLOGIA AGRARIA E TECNICA DELL'UNIVERISTÀ DEGLI STUDI E CENTRO DI STUDIO DEI MICRORGANISMI AUTOTROFI C.N.R. - FIRENZE (ITALY)

Soil samples and roots of cereal crops (wheat, maize, oat and barley), to bacco and prickly pear,collected from 20 localities of northern, central and southern Italy, were examined for the presence of *Azospirillum* spp. The collected data represent the first contribution to the knowledge of the ecology of these diazotrophs in Italy.

23 out of 41 soils and 31 out of 40 root samples showed nitrogen fixing activity and occurrence of *Azospirillum* spp. Appreciable differences in the distribution of *Azospirillum* in the different italian areas examined were not detected. On the whole 62 strains of *Azospirillum* have been isolated in pure culture. Of these 32 came from soils, while the remaining 30 were isolated from roots (12 from wheat, 6 from maize, 1 from oat, 8 from tobacco and 3 from *Opuntia ficus-indica*).

The morpho-physiological properties and the N_2-ase activity of the 30 *Azospirillum* strains isolated from plant roots have been investigated.

In semisolid malate agar under standard conditions the N_2-ase activity ranges widely from strain to strain; the minimum and maximum values were 14 and 366 nmoles C_2H_4 h^{-1} tube $^{-1}$ respectively.

The identification of these strains has been based, as suggested by Tarrand et al. (1978) on cell morphology, glucose utilization, biotin requirement, gas production from nitrate and catalase. Following these criteria, 6 strains were classified as *Azospirillum lipoferum*, 6 as *Azospirillum brasilense* nir - and 6 as *Azospirillum brasilense* nir +. However, 12 strains didn't fit completely the description of either species, since they possess some properties of *Azospirillum lipoferum* while in others they resemble *Azospirillum brasilense*. The results reported here emphasize two points. Firstly, the N_2-fixing bacteria of the genus *Azospirillum* seems to be widely distributed in soils of temperate areas. This conclusion, confirmed by other workers, indicates the need of a detailed investigation on the abundance of these bacteria in non tropical soils. Secondly, it is confirmed that the associative ability of *Azospirillum* is not restricted to the family *Graminaceae*, but includes also other plant taxa not related to them, as *Solanaceae* and *Cactaceae*.This opens new lines of investigations on these ecologically and agronomically important bacteria.

ACKNOWLEDGEMENT: This work was supported by C.N.R, P.F. IPRA.

ROLE OF RHIZOBIUM AND GREEN GRAM GENOTYPES IN DETERMINING COMPETITIVE ABILITY FOR NODULATION

B. S. Ghai and Ashwani K. Gupta
Department of Genetics, Punjab Agricultural University, Ludhiana 141 004, India

In a soil infected with rhizobia the inoculant strain does not give the expected increased yields. Under such condition nodulation is mostly with the native less effective rhizobia. The inoculant strain can give better results only if it can compete better with the native bacteria. Competition for nodulation is reported to be affected by genotype of Rhizobium and soil conditions. In the present studies relative contribution of Rhizobium and green gram genotypes in determining competitive ability for nodulation was studied.

Five effective strains of Rhizobium of the cowpea group which could be identified on the basis of resistance to different antibiotics and seven varieties of green gram constituted the material for this study. The five strains were taken in combination of two strains at a time and also as a mixture of five strains to inoculate the seven varieties of green gram. The experiment was conducted in sterilized soil in pots. The nodules from each treatment were crushed in YEMA media and the rhizobia classified into different strains on the basis of their antibiotic resistance pattern.

The studies showed that there was great variability in the strains of Rhizobium for competitive ability for nodulation. Rhizobium strain 130 showed better competitive ability than other strains. Rhizobium strains 122 and 102 were poor competitors. The varieties of green gram did not show significant differences for competitive ability for nodulation. The Rhizobium-green gram genotype interaction was, however, significant.

On the basis of these studies it is concluded that a Rhizobium strain which has better competitive ability in mixed inoculation and gives high nodule formation with several host strains should be selected.

EFFECT OF TEMPERATURE AND CALCIUM CARBONATE CONTENTS OF SOIL ON THE SURVIVAL AND GROWTH OF RHIZOBIA

E.M. GEWAILY*/M.F.A. KHAN and A.K. KHEDER
Department of Soil Science, College of Agriculture, University of Salah-Din, Arbil, Iraq

The survival and growth of rhizobial suspension mixture (R.leguminosarum 202, R.leguminosarum 1049 and local isolate R.I) prepared with equal ratios were investigated at 4 soil locations. Different temperatures were used such as 20, 25, 30, 35 and 40°C for studying the effect of temperature on survival and growth of rhizobia in sterilized and non-sterilized soils, incubated for 75 days. Also the effect of $CaCo_3$ was studied in other 3 soils containing 12.5, 32.5 and 36.5% $CaCO_3$ pHs from 7.7-7.8 and organic matter varied from 0.58-2.1%. Individual strains and their mixtures were unsed in this experiment and rhizobial counts were carried out at intervals of 15, 30, 45 and 60 days. The study revealed the following:

Soil analysis. The results of these analysis shows that 4 soils were comparatively poor in organic matter content which varied between 0.56-1.06%, while pH-values ranged from 7.4-7.6 and the ECe from 0.65-2.42 mmhos. These soils were non-saline and non-alkaline and its textural grade varied between clay to sandy clay loam.

Temperature effect

1. Incubation at 25, 30 or 35°C caused an increase in rhizobial counts in sterile and nonsterile soils up to the 30th day of incubation, then gradually decreased. However, incubation at 20 or 40°C, the numbers of rhizobial cells increased up to the 45th day.
2. The optimum temperature suitable for rhizobial proliferation in different soils was within 30°C.
3. Multiplication of rhizobia was found to be affected by organic matter, soluble calcium, $CaCO_3$, clay and sand contents.
4. Incubation at 40°C decreased rhizobial number and its ability to infect the host plant roots.
5. Higher number of rhizobia was obtained in nonsterile soils, which showed that such soils were more suitable for growth of rhizobia than sterile ones.

$CaCO_3$ effect

1. Soil containing 12.5% $CaCO_3$ gave the highest number of different strains of rhizobia.
2. It was found that for soils receiving mixed rhizobial suspension having 32.5% $CaCO_3$ was the best.
3. The proliferation of rhizobia strains and its mixture was adversely affected by 63.5% $CaCO_3$ in soil.

The result of our studies on some local and imported rhizobia revealed that the foreign strains proved better than the local one. It is therefore considered that carrying on further studies is necessary to find out a better local strain.

Using rhizobial mixture was quite better for tolerating andverse conditions than the individual strains.

*Present address: Department of Botany, College of Agriculture, University of Zagazig, Zagazig Governorate, Egypt.

SOME ASPECTS OF *Rhizobium*-WOODY LEGUME SYMBIOSIS

M.A. HERRERA/J. OLIVARES
DEPARTAMENTO DE MICROBIOLOGIA, ESTACION EXPERIMENTAL DEL ZAIDIN, GRANADA, SPAIN

1. INTRODUCTION

Woody legumes have a great importance in silviculture, mainly in arid and semi-arid areas. An example of this are represented by the genus *Prosopis* and *Acacia* which have about 900 species distributed in a wide xerophytic zone. In this experience we have studied different *Rhizobium* strains isolated from nodules of woody-legumes growing in Chile.

2. MATERIAL AND METHODS

Source of inocula: *Rhizobium* strains (GRH1; GRH2; GRH3) were obtained from nodules of *Prosopis chilensis, Acadia cyanophylla* and *A. melanoxylon*, respectively. 79 Allen medium has been used for isolation and incubation. For liquid growing it was used a YGT medium.

Plant material: The *Rhizobium* sp. (GRH1) was inoculated to the following legumes: *A. cyanophylla, A. melanoxylon, P. chilensis, Lupinus albus, Medicago sativa, Phaseolus vulgaris* and *Ornitopus sativus*. *Rhizobium* sp. (GRH2 and GRH3) were assayed over *A. cyanophylla. A melanoxylon, P. chilensis* and *M. sativa.* Plants were growing in assay tubes or in small pots with nitrogen-free nutrient solution under controlled temperature and light.

Parameters studied: In bacteria: time of generation; characteristics of culture. In root: nodule production; acetylene reduction activity (ARA); hydrogen evolution.

3. RESULTS

Rhizobium sp. (GRH1) is a slow grower with a generation time of about 10 hours and it is able to infect the same *P. chilensis* and *A. cyanophylla, A. melanoxylon, O. sativus* and *L. albus*. Only in the last species effective nodules were found.
Rhizobium sp. (GRH2 and GRH3) are fast growers with a generation time of about 3 hours. Both *Rhizobium* sp. produce effective nodules in *P. chilensis, A. cyanophylla* and *A. melanoxylon* where a high relative efficiency has been found. In the symbiosis established between *P. chilensis* and *Rhizobium* (GRH2) a strong influence of the time of illumination on the nitrogenase activity and the hydrogen evolution have been observed. According to the values found the relative efficiency varies from 0.03 to near 1.0 along the light period.

EFFECT OF ORGANIC AMENDMENTS ON NITROGEN FIXATION BY ASYMBIOTIC BACTERIA ASSOCIATED WITH WHEAT AND MAIZE PLANTS

Y.Z. ISHAC/M. E. EL-HADDAD/M.A. EL-BOROLLOSY AND M. ISMAIL
UNIT OF BIO-FERTILIZERS, AGRICULTURE EGYPT

Egyptian Soils are rich in asymbiotic N_2-fixers with high potential N_2-fixing activity. Moreover, several investigations reported a high activity of asymbiotic N_2-fixers in the rhizosphere of many economical plants in Egypt.

Two pot experiments were carried out in a net house to study the rhizosphere effect of maize (C_4-plant grown in summer) and wheat (C_3-plant grown in winter) in presence and absence of organic matter wide in C/N ratio (0.2% maize stalks). Changes in the rate of CO_2 evolution, densities of total microbial flora and _Azotobacter_ spp. as well as _Azospirillum_ spp. were recorded under these conditions. Nitrogen balance was also considered as an indication for the potentiality of N_2-fixation process.

The results of this study showed the rhizosphere effect of maize generally gave higher figures as compared with wheat. Such effect was more pronounced on _Azospirillum_ spp. populations.

Addition of maize stalks to the soil was accompanied by an increase in different parameters as compared with the control. Planting of the amended soil with maize or wheat showed a greater influence on different parameters. Such effect was indexed by determining the efficiency of rhizosphere treatment (E.R.T.) i.e., the net effect of rhizosphere treatment divided by the net effect of the same treatment in root-free soil. Data showed that the E.R.T.) of rate of CO_2 evolution, _Azotobacter_ spp. as well as _Azospirillum_ spp. population and nitrogen gain reached 3.32, 5.06, 1.7 and 1.3, respectively under maize plants. On the other hand, the recorded figures under wheat plants were 3.33, 7.57, 6.58, 2.46 and 2.2. It is interesting to note that the higher values of the E.R.T. under maize and wheat plants is correlated with a significant increase in the dry matter and nitrogen contents of both plants being more pronounced in the latter. Such variation was attributed to the difference in the mean temperature during the planting season of maize (28°C and wheat 16°C). The lower temperature in winter resulted in slower rate of maize stalks decomposition and hence higher of N_2-fixation in the amended planted soil in wheat than in maize.

FACTORS AFFECTING WHEAT INOCULATION WITH AZOTOBACTER

Y.Z. Ishac , S.A.Z. Mahmoud , J. Kramer , M. EL-Demerdash and W. Eweda
Unit of Biofertilizers, Ain Shams University, Cairo, Egypt

Greenhouse experiment was carried out to study the effect of wheat seed inoculation with two selected strains of *Azotobacter* or their filtrates (the latter was applied under sterile conditions) on the rhizosphere microflora, seed germination and plant height. Plant dry matter and nitrogen balance were also considered in this investigation. Egyptian silty loam (pH 8.15) and German sandy loam soils (pH 7.9) were used in this work. The C/N ratios of the two soils were 4.9 and 12.6 in respective order. The inoculation has been conducted in absence or presence of compost or wheat straw. The experiment lasted 20 weeks. Day and night temperatures were 28 and 18°C respectively. Day light intensity was ca. 2500 Lux. The selected strains were *A.vinelandii* - A6 (The most active strain in N_2-fixation) and *A.paspali* - A18 (The most active strain in growth promoting substances). Inoculation increased the total bacterial counts in the rhizosphere of wheat. The highest counts were recorded in Egyptian soil amended with straw. Inoculation increased the colonization of *Azotobacter* in wheat rhizosphere and had a variable effect on the colonization of clostridia in rhizosphere. This variation was affected by soil origin, organic matter and Azotobacter strain. *Azotobacter paspali* filtrate significantly affected seed germination and plant length more than those of *A.vinelandii* filtrate. Bacterization significantly affected dry matter of wheat plants. Inoculation with *Azotobacter* in soil amended with compost significantly increased nitrogen gain in Egyptian soil more than German ones, but in case of wheat straw, nitrogen gain was more in German soil. Bacterization with *Azotobacter* increased nitrogen gain than its filtrate, but no marked differences were recorded between the two used strains.

Summing up it might by said that the positive effect of inoculation with *Azotobacter* could be attributed to nitrogen fixation and growth promoting substances.

COMPARATIVE STUDY OF HU TYPE PROTEINS ISOLATED FROM DIFFERENT GROUPS OF *RHIZOBIACEAE*

H. KHANAKA, B. LAINE*, P. SAUTIERE*, M. CATTEAU and J. GUILLAUME

Laboratoire de Microbiologie SN2 - Université des Sciences et Techniques de Lille I - 59655 Villeneuve d'Ascq - FRANCE

*Unité 124 INSERM - Place de Verdun - B.P. 311 - 59020 Lille - FRANCE

The DNA-binding HU type proteins isolated from *Rhizobium meliloti* (strain 2011 Str 3), *Rhizobium leguminosarum* (strains L_{18} and L_{53}), *Rhizobium japonicum* (strain J 5) and *Agrobacterium tumefaciens* (strain B 6) were prepared by affinity chromatography on DNA-cellulose prepared with DNA from *Rhizobium meliloti*. These proteins are respectively called HRm, HRl_{18}, HRl_{53}, HRj and HAt. The HU protein isolated from *Escherichia Coli* is associated with the chromosomal and extrachromosomal DNA and seems to be involved in the condensation of the DNA. HU protein is a low molecular weight (Mr 9500) and sligtly basic protein. Similar proteins have been isolated from other species widely remote in the bacterial kingdom.

The HU-type proteins from *Rhizobiaceae* have been characterized by the following criteria : Electrophoretic mobility on polyacrylamide gel, crossed immunoreactivity, amino acid composition, tryptic peptides mapping performed by reverse phase HPLC and partial sequence determination.

The proteins HRl_{18} and HAt exhibit identical characteristics and show strong structural homologies. It is worth pointing out that the *R. leguminosarum* L_{18} and *Agrobacterium tumefaciens* (B 6) are grouped in the same taxon as evidenced in the accompagning study. By comparison with the protein HRm taken as reference, they show very slight differences in their electrophoretic mobility and amino acid composition and present only conservative substitutions in their amino acid sequence. On the other hand protein HRl_{53} exhibits significant differences particularly in its crossed immuno reactivity against anti protein HRm antibodies, its tryptic peptides mapping and amino acid sequence which differs from that of proteins HRm, HRl_{18} and HAt by non conservative substitutions. Our results emphasize the heterogeneity of the group *R. leguminosarum* ; indeed according to the numerical taxonomy presented in the accompagning poster, the strain L_{53} is related to the group *R. trifolii* whereas the strain L_{18} is connected with the group *Agrobacterium*.

The protein HRj differs strongly from those mentioned above particularly in its electrophoretic mobility on gel containing Triton X-100 and in its amino acid composition. Moreover the protein HRj reacts very weakly against anti protein HRm antibodies. These results substantiate that the strain of slow-growing *R. japonicum* is very remote from the fast-growing strains of *Rhizobium*.

The biochemical characteristics of the HU-type proteins isolated from different groups of *Rhizobiaceae* are in good agreement with the respective position of these bacteria evidenced by numerical taxonomy.

A SEROLOGICAL STUDY OF RHIZOBIUM STRAINS USING DIRECT AND INDIRECT ENZYME-LINKED IMMUNOSORBENT ASSAYS (ELISA).

B. KISHINEVSKY*, A. MAOZ**, DEBORA GURFEL*, CHAJA NEMAS*

*Dept. of Legume Inoculation, Agricultural Research Organization, The Volcani Center, Bet Dagan, 50250, Israel.

**Inst. of Plant Protection, Agricultural Research Organization, The Volcani Center, Bet Dagan, 50250, Israel.

Various forms of ELISA are playing an increasingly important role in studies of serological properties of strains (Ahmad et al., 1981), their competitive ability (Kishinevsky , Bar-Joseph, 1978), the survival of inocula (Jones, Morley, 1981) and as a new inoculant quality control technique (Olsen et al., 1983).

In the present work we describe the relative merits of the direct and indirect ELISA tests used for serological identification of some fast and slow-growing *Rhizobium* strains belonging to different species. Double-antibody sandwich ELISA (Clark, Adams, 1977) enables us to distinguish between some serologically related strains. Thus, there were no or very weak (A_{405}=0.12-0.22) ELISA reactions among four peanut *Rhizobium* strains, whereas the indirect test (Voller et al., 1976) indicated that they are serologically related. Similar results were obtained with two *R. trifolii* strains tested.

It has been shown that antisera adsorbed with antigens of cross-reacting strains failed completely to react in heterologous indirect ELISA reactions without loss of their serological specifity. This may be of practical importance since the indirect ELISA can be used both for defining strain relationships and as a tool for routine rhizobial identification. The indirect ELISA enables the identification of different strains by using a single antiglobulin enzyme conjugate. This eliminates the problems of preparing and storing many different conjugated antisera. Both enzyme-labeled protein A (Engvall, 1978) and goat anti-rabbit immunoglobulin conjugates applied in indirect test enabled good detection of rhizobium antigens in root nodules, but the enzyme-labeled goat anti-rabbit immunoglobulin adsorbed non-specifically to the trapped nodule antigens. In contrast, enzyme-labeled protein A applied in heterologous tests produced no background reactions, whereas its reactivity in homologous antigen-antibody systems was strong. Moreover, enzyme-labeled protein A was always used at considerably higher dilutions than was the anti-immunoglobulin conjugate. For *Rhizobium* strains tested by the indirect assay, 8 µg of nodule tissue added per micro-plate well was sufficient for the accurate detection of the bacterial component of the nodules. This sensitivity was approximately ten times greater than that of the direct ELISA test.

Ahmad MH, Eaglesham ARJ and Hassouna S (1981) Arch. Microbiol. 130, 281-287.
Clark MF and Adams AN (1977) J. gen. Virol. 34, 475-483.
Engvall E (1978) Scand. J. Immunol. 8, 25-31
Jones DG and Morley SJ (1981) Ann. appl. Biol. 97, 183-190.
Kishinevsky B and Bar-Joseph M (1978) Can. J. Microbiol. 24, 1537-1543.
Olsen PE, Rice WA, Stemke GW and Page WJ (1983) Can. J. Microbiol. 29, 225-230.
Voller A, Bidwell D and Bartlett A (1976) In Rose NR and Friedman H, eds. Manual of Clinical Immunology, pp. 506-512, Amer. Soc. Microbiol., Washington, D.C.

FACTORS INFLUENCING GROWTH OF RHIZOBIUM AND ITS BACTERIOPHAGE IN SOIL

K.A. LAWSON and Y.M. BARNET
SCHOOL OF MICROBIOLOGY, UNIVERSITY OF N.S.W., KENSINGTON, N.S.W. 2033, AUSTRALIA

INTRODUCTION. There is little detailed information about the factors which influence the growth, survival and evolution of *Rhizobium* in soil. It has been suggested that rhizobiophages may have a role. However assessment of the reality of phage effects requires knowledge of their numbers, distribution and environmental factors influencing their growth. Only a few studies have sought to provide such information (1,2). This study was undertaken to determine the environmental factors influencing numbers of rhizobiophage and *Rhizobium trifolii* in soils.

MATERIALS AND METHODS. Two adjacent pastures, differing in soil type (a sandy loam containing 10% clay and a silty loam containing 34% clay) were sampled approximately monthly over a period of eighteen months.
Rhizobium trifolii numbers were determined by the most probable number technique (3). Rhizobiophage numbers were determined by plaque counts using a strain of *R.trifolii*, SU91, with a broad phage susceptibility range, and two field isolates, as host indicators. Environmental parameters were determined with appropriate instruments. The resulting data was analysed by a multivariate regression analysis, to determine the association between population levels and environmental variables. The significance (5%) of each regression coefficient was determined by a t-test which compared the coefficient with its standard error.

RESULTS AND DISCUSSION. The rhizobiophage population fluctuated considerably in both soils and reached levels as high as 1.6×10^4 pfu/g soil. These levels are higher than is frequently proposed to exist in soil, but were consistent with recent findings (1,2). The *R.trifolii* population showed less fluctuation than the phage population, but there was no evidence of a positive or negative correlation of phage and rhizobial numbers. Levels of bacteria were positively associated with vegetation height and with amount of solar radiation. Phage numbers showed a negative correlation with soil matric potential and with the silty clay soil, but increased significantly with increasing vegetation height. Phage numbers were thus associated with factors which have the potential to affect their mobility and their host density, while rhizobial numbers were related to factors which would control their nutrient supply.

1. Dhar, B. *et al*. Acta Microbiologica Polonica, 23, 319-324.
2. Dhar, B. *et al*. (1980) Indian Journal Experimental Biology, 18, 1168-1170.
3. Vincent, J.M. (1970) A Manual for the Practical Study of Root Nodule Bacteria. Blackwell Scientific Publications, Oxford, London, Edinburgh, Melbourne.

THE EFFECTS OF NITROGEN FERTILIZER LEVEL AND TRICKLE IRRIGATION REGIME ON THE RATE OF INOCULATION AND NITROGENASE ACTIVITY OF INOCULATED PEANUTS.

RINA LOBEL[1], J. SCHIFFMANN[1], B. BAR YOSEPH[2], J.S. WALLERSTEIN[3], B. SAGIV[2]

Dept. of Legume Inoculation,
Dept. of Soil Chemistry and Plant Nutrition,
Dept. of Industrial Crops, Agricultural Research Organization,
The Volcani Center, Bet Dagan, Israel.

Trickle irrigation of peanuts has not yet been tested extensively in Israel (Wallerstein et al., 1982). Combined nitrogen affects almost all the phases of legume - Rhizobium symbiosis (Reddy, Tanner, 1980; Eaglesham et al., 1983). This research was aimed at studying the combined effects of irrigation and nitrogen rates applied via trickle irrigation on nodulation and N_2 fixation by peanuts.
The experiment was conducted on a sandy soil in the northern Negev, a semi-arid area in Israel. N-fertilizer levels were 0, 60 and 120 ppm N applied through the irrigation system as a compound fertilizer (N:P:K= 5.0:0.9:5.9), at water rates of 0.95 E, 0.65 E, and 0.40 E, every 2 days (E being the evaporation rate from a U.S. Class A pan, in mm). The cumulative E through the growing season was 921 mm.
At 112 days after planting, the addition of 60 to 120 ppm N in irrigation water at 0.65 E and 0.95 E to inoculated peanuts reduced the nodule number to 29 and 20%, respectively, and the nodule dry weight to 22 and 8% of the nodule number and dry weight recorded on inoculated but N-nonfertilized plants.
The N-fertilizer rates of 60 to 120 ppm applied in irrigation water significantly reduced the $N_2(C_2H_2)$ fixation to 6.52 and 1.45µmoles C_2H_4/plant/h respectively, compared with 60µmoles C_2H_4/plant/h on plants only inoculated. Maximum nitrogenase activity in inoculated-nonfertilized plants occurred at the onset of pod-filling stage and continued to be substantial at 112 days (Lobel, Schiffmann, 1982). The inoculated peanuts applied with the 0.95 E, 0.65 E of Class A pan evaporation rates of water showed no substantial differences in nodule number and mass. The lower rate (0.40E) of irrigation water significantly reduced the number and weight of nodules to 32.5 and 44.2% respectively, of the mean values recorded under higher rates of irrigation.
The water stress reduced the $N_2(C_2H_2)$ fixation: at 112 days after planting it reached only 6% of that recorded in higher water rates. The results showed that N-fertilizer application at 0.65 E and 0.95 E did not affect total pod yield (TP), but significantly reduced the export quality pod yields (EP). No differences were recorded between TP and EP of plots applied with the same N-fertilizer level at the two higher rates of irrigation. Water stress (0.40E) decreased the yields (TP and EP) markedly.

Eaglesham ARY, Hassouna S and Seegers R (1983) Agron. J. 75, 61-66.
Lobel R and Schiffmann J (1982) Israel J. Bot. 31, 283-295.
Reddy VM and Tanner JW (1980) Peanut Sci 7, 114-119.
Wallerstein IS, Bar-Yoseph B, Sagiv B, Lobel R, Schiffmann J (1982). Proc. Am. Peanut Res. and Ed. Soc. Inc. (Albuquerque, New Mexico).

RHIZOSPHERE MICROFLORA AND NODULATION OF LICORICE PLANT

S.A.Z. MAHMOUD/M. ABDEL-NASSER/M.F. OUF/A.R. ABDALLAH AND M.A. ATTIA
FACULTY OF AGRICULTURE, SHOBRA, CAIRO, EGYPT

The effect of cutting root-extracts of licorice plant on certain microorganisms *in vitro* was studied. On culture media, the antibacterial effect expressed as "width of inhibition zones" differed according to type of microorganisms and method of extraction. *In vivo* occurrence of certain soil microorganisms in the rhizosphere soil of licorice during different growth phases was also studied. Generally, counts of bacteria (total), sporeformers, actinomycetes, fungi, aerobic nitrogen fixing Azotobacter and of anaerobic nitrogen fixing clostridia, were higher in the rhizosphere soil than in the non-rhizosphere one resulting in positive rhizosphere effects during 120 days after planting. The morphological examination of the isolates from rhizosphere and non-rhizosphere soil showed that there was a marked predominance of Gram-negative types in the rhizosphere soil of licorice; the incidence of Gram-positive types were higher in the control non-rhizosphere one. Number of nodules ranges from 7-10 per plant. Re-inoculation of locorice plant with effective nodule bacteria isolated from locorice plant increased densities of nodules. Cross-inoculation investigation showed that root nodules bacteria isolated from licorice plant formed nodules only on the roots of broad bean (*Vicia faba*) and vice versa.

THE UTILIZATION OF COCONUT WATER AS MEDIUM FOR RHIZOBIA

JUANITA C. MAMARIL/FATIMA T. BEGONIA AND RUBEN B. ASPIRAS
NATIONAL INSTITUTE OF BIOTECHNOLOGY AND APPLIED MICROBIOL.
UNIVERSITY OF THE PHILIPPINES AT LOS BAÑOS, LAGUNA 3720

Coconut water (CW) was obtained from matured nuts and used as culture medium for rhizobial isolates C_{11}, C_4 from *Centrosema pubescens*; L_{15}, L_5 from *Leucaena leucocephala*; M_5, M_4 from *Vigna radiata* (mungbean); P_7, P_3 from *Arachis hypogaea* (peanut) and S_{38}, S_{13} from *Glycine max* (soybean). Growth responses of these isolates in CW media, yeast extract mannitol broth (YEMB) and basal medium (BM) were determined after 4 days incubation at 30°C. The CW media were: aseptically collected CW (aseptic CW), original pH 5.4, aseptic CW, adjusted pH 7; sterilized CW (121°C at 15 psi for 18 min), pH 5.4, sterilized CW, pH 7.0; CW + NH_4Cl (50 mg/100 ml), pH 5.2; and CW + K_2HPO_4 (100 mg/100 ml), pH 6.0. Growth responses were rated according to optical density at 420 nm as follows: $>1.0 = 4$; 0.7 to $0.9 = 3$; 0.4 to $0.6 = 2$; 0.1 to $0.3 = 1$; <0.1 = poor growth. All isolates scored 4 in CW media except C_4, M_5 and P_7 which have slightly lower growth (3) in sterilized CW. Growth responses in YEMB ranged from 2 (C_4, M_5, P_3) to 4 (L_5, L_{15}, P_7) and in BM from 1 (C_4, C_{11}, M_4, M_5, P_3, S_{38}) to 4 (L_5, L_{15}, P_7).

Cell dry weight (mg/10 ml) of *Leucaena* isolates CB81 (Australian strain, L_{15} (native strain) and mungbean isolate M_5 cultured in CW are 141.8, 150.3, and 2.4 while that in YEMB are 141.0, 124.0 and 2.5 respectively. Plate count of viable rhizobium (number of cells/ml) of CB81, L_{15}, and M_5 in CW are 5.5×10^7, 38.5×10^7, and 3.0×10^8 while that in YEMB are 2.0×10^7, 14.0×10^7 and 5.5×10^9 respectively.

CB81 and L_{15}, inoculum from CW, CW + NH_4Cl and YEMB were used to inoculate *Leucaena* and M_5 for mungbean seedlings. Earlier nodulation was observed with CW inoculum. Nitrogenase activity (nM C_2H_2 reduced/mg fresh weight nodule/min) of nodules from 6-week-old seedlings caused by CB81, L_{15} and M_5 inoculum obtained from the following media are: CW - 0.01, 0.04, 0.04; CW + NH_4Cl - 0.01, 0.04, 0.01; YEMB - 0.00, 0.03, and 0.03 respectively.

DISTRIBUTION OF LABELLED NITROGEN FROM N_2-FIXING BACTERIA INTO *Zea mays* AND SOIL.[2]

J. R. MILAM[+]/S. L. ALBRECHT[++]/M. H. GASKINS[++]

\+ DEPARTMENT OF MICROBIOLOGY AND CELL SCIENCE, UNIVERSITY OF FLORIDA, GAINESVILLE, FL 32611, USA.

++ USDA-ARS AND DEPARTMENT OF AGRONOMY, UNIVERSITY OF FLORIDA, GAINESVILLE, FL 32611, USA.

The capacity of nitrogen-fixing root-associated bacteria to provide reduced nitrogen to plants is largely unknown. Unlike the legume-*Rhizobium* symbiosis, there is no direct mechanism for the transfer of fixed nitrogen. The translocation of fixed nitrogen in the rhizosphere, from the bacteria to the plant, is mandatory if associative nitrogen fixation is to make a significant short-term contribution to the nitrogen economy of crop plants. Most nitrogen-fixing bacteria do not excrete appreciable amounts of reduced nitrogen into the environment. This suggests that the nitrogen fixed by these bacteria is released only at cell lysis. This nitrogen is probably incorporated into large molecules that are difficult for plants to assimilate. Many soil organisms have the capacity to metabolize these molecules, giving them a competitive advantage over plants. Soil organic matter, especially compounds that are not readily metabolized, is another sink for reduced nitrogen. Bacterial nitrogen lost from the biomass can be recovered in the humin fraction of the soil. Living cells of *Azospirillum brasilense* (strain JM125A2) and *Klebsiella pneumoniae* (ATCC 15574) were uniformly labelled with ^{15}N (31.5 atom%). The bacteria were inoculated into the rhizosphere of greenhouse-grown *Zea mays* (cv. Funks 509) at two levels of bacterial nitrogen. The plants were harvested at maturity and separated into roots, shoots and grain, dried and weighed. Plant nitrogen and ^{15}N:^{14}N ratios were determined by Kjeldanl analysis and isotope ratio mass spectroscopy, respectively. Bacterial populations in the rhizosphere were estimated by the MPN technique. Dry weight yields, especially grain yields, were improved over the control indicating the bacterial additions were providing some contribution to plant growth. Usually yield increases were directly proportional to the rate of bacterial additions. Large populations of N_2-fixing bacteria were not maintained in the inoculated pots, however, there was an increase of denitrifying bacteria during the growing season. Nitrogen content of the plant tissue generally increased with increased rate of "fertilization." The ^{15}N found in the tissue clearly demonstrates that bacterial nitrogen was incorporated into the plant. The amount of ^{15}N in the plant tissue appears directly proportional to the addition rate. These results suggest that nitrogen from N_2-fixing bacteria can be rapidly incorporated into plant tissue. This study shows that nitrogen in the bacterial biomass, in the soil or rhizosphere, is not completely retained in the soil, and can be made available to plants during the course of a growing season. From our present information, we estimate that 60-70 percent of the bacterial nitrogen introduced into the soil was incorporated into plant tissue during the first growing season.

NEW FORMS OF NITROGEN-FIXING AEROBIC BACTERIA FROM USSR SOILS

E. MISHUSTIN/T. KALININSKAJA/T. REDKINA
INSTITUTE OF MICROBIOLOGY, USSR ACADEMY OF SCIENCES, MOSCOW, USSR

On special media, a set of new cultures of N_2-fixing bacteria were isolated. From soils of the southern zone, particularly in rice fields, *Xantobacter flavus* is numerous. This bacterium can grow on mineral media in the presence of H_2 and CO_2. It does not use carbohydrates, but utilizes organic acids, methanol or ethanol. In the soils of rice fields, N_2-fixing bacteria belonging to *Aquaspirillum* have been found. They utilize carbohydrates and organic acids. Representatives of the genus *Azospirillum* are more common in southern soils; some of them can grow in mineral media using H_2.
Among bacteria utilizing not only organic compounds but also H_2 and CO_2 is *Achromobacter liquefaciens*, which is more common in chernosom soils.
In forest podzolized soils, the acid-tolerant N_2-fixing bacteria belonging to genera *Enterobacter* and *Klebsiella* were often found.

COMPETITION FOR NODULATION BETWEEN STRAINS OF *LEUCAENA* RHIZOBIA IN OXISOL AND MOLLISOL SOILS

H. MOAWAD[+] AND B.B. BOHLOOL
DEPARTMENT OF MICROBIOLOGY, UNIVERSITY OF HAWAII, HONOLULU, HAWAII, U.S.A.
+ PRESENT ADDRESS: NATIONAL RESEARCH CENTRE, DOKKI, CAIRO, EGYPT

The successful nodulation of legume by certain strain of rhizobia is determined by the competition between the desired strain and the mixture of other native and inoculant rhizobia. The competitive ability of six *Leucaena* rhizobia strains in single and multistrain inoculants were studied. For this purpose field inoculation trails were conducted in oxisol and mollisol soils. Strain specific fluorescent antibodies (FA) were used for the identification of the strains in *Leucaena* nodules. Mixture of three *Leucaena* rhizobia strains recommended by NifTAL for INLIT trials (TAL 82, TAL 582, TAL 1145) were used as peat based inoculants either along or with one of the other three strains: B213, B214 and B215. Each of the last three strains was also used as single strain inoculum to study their competition with the native rhizobia in the two soil systems. In oxisol soil, strains B215 and B213, when used as single inocula, outcompeted the native rhizobia and formed 92% and 62% of the nodules respectively. Strain B214 however was the least competitor in oxisol (30% of the nodules) and the best in mollisol soil (70% of the nodules). The most successful competitor for nodulation in mixed strain inoculants was the strain TAL 1145. This strain outcompeted native and inoculant *Leucaena* rhizobia in both soils. None of the strains in single or multistrain inoculants was capable to overwhelm all the resident *Leucaena* rhizobia which gave 4-24% of the total nodules in oxisol and 21-32% in mollisol.

EFFECT OF BARK ASH AND VA-MYCORRHIZAE ON THE GROWTH AND N_2-FIXATION OF TWO LEGUMES

M. NIEMI, M. EKLUND & V. SUNDMAN
DEPARTMENT OF GENERAL MICROBIOLOGY, UNIVERSITY OF HELSINKI
MANNERHEIMINTIE 172, SF-00280 HELSINKI 28, FINLAND

Increasing amounts of ash from bark and wood are annually formed in Finland as an industrial waste, which creates an environmental problem. However, the ash represents a considerable fertilizer potential, since apart from nitrogen, it contains all macro- and micronutrients needed for plant growth and also has a high neutralizing capacity. Previous studies have shown that ash has a good fertilizer effect e.g. on legumes, mainly due to its phosphorus content, the solubility of which is higher than that of apatite (poorly soluble) but lower than that of superphosphate (soluble). The extractability of other ash nutrients in acid acetate buffer varies between 0.5 and 50 %.

Possible improvement of the availability of ash nutrients to plants by inoculation with the VA-mycorrhizal fungus E_3 (*Glomus fasciculatus*) was studied in three pot experiments with two legumes, *Vicia sativa* and *Pisum arvense*. The soil had an original pH of 4.9 and a P-content of 40 ppm Olsen-P. Bark ash as P-fertilizer was compared with apatite and superphosphate; the ash addition was ca. 3000 kg/ha, equivalent to 50 kg P/ha. The liming effect of ash was eliminated by raising the soil-pH to the same level in all treatments.

In steamed soil non-mycorrhizal *Vicia* had poor growth, nodulation and N_2-fixation (ARA) when apatite or no phosphorus was added; after inoculation with E_3 the growth was as good as or better than with superphosphate. This shows the mycorrhizal dependency of legumes under conditions of poor phosphorus availability. The ash phosphorus was soluble enough to sustain non-mycorrhizal plants as well as superphosphate. In unsteamed soil VAM-inoculation of *Vicia* caused only small growth increases, appearently due to the high infectivity and effectivity of the indigenous VAM-flora. As for *Vicia*, bark ash and superphosphate were equally good P-sources for *Pisum* in unsteamed soil. However, ash addition combined with VAM-inoculation gave significantly better growth, nodulation and ARA than did superphosphate, although the P-uptake and VAM-infection levels were equal. This suggests that the positive response to ash fertilization was a micronutrient (e.g. Zn, Cu or Mo) as well as a phosphorus effect, caused by VA-mycorrhizal uptake of ash nutrients important for N_2-fixation.

ON THE ROLE OF NITROGEN-FIXING BLUE-GREEN ALGAE IN THE NITROGEN ECONOMY OF TEMPERATE ZONE SOILS IN THE USSR

E.M. PANKRATOVA
AGRICULTURAL INSTITUTE, KIROV, USSR

Among approximately two thousand species of blue-green algae, about five hundred possess heterocysts and have the ability to fix atmospheric nitrogen. Certain non-heterocystous, filamentous algae also fix nitrogen, but only under microaerobic or anaerobic conditions. Because blue-green algae are photoautotrophic and develop in the upper soil layer, where conditions are usually aerobic, nitrogen fixation by non-heterocystous algae is much slower than by heterocystous blue-green algae. In the soils of the USSR, 133 species of heterocystous blue-green algae, among them about 30 nitrogen-fixing species are widely spread.

In certain soils around Kirov, the biomass (standing crop) of nitrogen-fixing blue-green algae is in range of 2-15 kg/ha of dry weight substance with a yearly productivity of from 66,6 to 577,7 kg/ha of dry weight substance. Annual N_2-fixation (determined by the ^{15}N technique) by algae in a water meadow near the river Vyatka may exceed 26 kg N per hectare, which compares favorable with the considerably lower values (5,1-10,8 kg/ha) reported for arable soils (wheat fields). Input of nitrogen by blue-green algae in highly acidic virgin sodpodzolic soils is minimal or non-existent (3,4-4,0 kg/ha N per annum). Blue-green algae also play an important role in eroded soils on account of their ability to supply organic matter and fixed nitrogen plus a soil-fixing ability. In such situations, the annual production of algae is 266 kg/ha of dry weight with values of the nitrogen contribution usually between 3-13 kg/ha per annum. All these findings lead us to conclude that the total input of nitrogen by blue-green algae to the soils (about 130 millions hectares) of the temperature zone of the USSR is approximately 0,5 million tons per annum, indicating that blue-green algae are an important source of fixed nitrogen in these areas.

Experiments using ^{15}N, ^{13}C and ^{14}C as tracers, indicate that nitrogen-fixing algae provide fixed nitrogen to bacteria, fungi, non-fixing algae and seed plants. Nitrogen-fixing algae are also food for many soil invertebrates. Thus, algal fixation appears to make a substantial contribution to the fertility of soils.

INFLUENCE OF PH ON NITROGEN FIXATION OF CYANOBACTERIA FROM ACID PADDY FIELDS

P.A. REYNAUD
LABORATOIRE DE MICROBIOLOGIE, ORSTOM, B.P. 1386, DAKAR, SENEGAL

When light and moisture are not limiting, pH is the major physico-chemical factor affecting algal growth; occurence of nitrogen fixing blue-green algae (N2BGA) is in close relation with the pH of the soil. In general no growth occurs below pH 5.7 and nitrogen fixation reaches its maximum only in the slightly alkaline range; the fixation decreases markedly below and above the pH range of 7.0-8.0.
Of the different kinds of algae found in paddy fields of Senegal, the blue-green are most abundant; among them N2BGA play an important role for the economy of soils. The present work deals with the response of autochtonous strains to acid conditions. Biomass development and nitrogenase activity as acetylene reducing activity (ARA) are chosen as parameters to estimate the adaptation of N2BGA to different pH regimes.
Biomass estimations are performed by the serial-dilution technic. Unialgal cyanobacteria pellets are suspended in 1ml of medium which is adjusted after autoclaving and cooling at pH values of 4-5-6-7 or 8; from these suspensions ARA is mesured in 10 ml Gravis flasks during one hour.

The mean value for the pH of 79 Senegalese paddy soils is 6.08±0.15, indicating the acidity of a great part of the soils. After four weeks of submersion, pH increases about one unit. When pH is below 6, daily variations of 0.1 unit can have an important effect on nitrogenase activity.
Biomass estimation of N2BGA in relation to soil pH show that: in comparison to the total algal biomass, the portion of N2BGA increases with the pH from 4 to 8; the maximal biomass may increase with the factor 100 for an increase of 4 pH units; for a singular strain (Nostoc punctiforme) the biomass can increase 10,000 times with an increase of 3.5 units.

Six standard curves are obtained from the measure of ARA in the pH range of 4 to 8 on 41 N2BGA isolated from paddy soils mentionned above; they characterized acidophil, acidotolerant, basophil and neutrophil (large or small spectrum of optimum ARA) strains.
Confirmation of these observations is obtained by the isolation in a paddy field of seven N2BGA; their activity is similar to ARA presented in five of the standard curves. The mixture of the seven stains obtained by the addition of equal amounts of protein of each strain, showed a continuous ARA in the pH range 5 to 9.

In situ as *in vitro*, the absolute and relative N2BGA biomass increases with the pH; the inhibition of growth at low pH is more pronounced for a singular strain than for a mixture of strains, showing a buffering effect. This buffering effect is also found for ARA.
The response of ARA to pH variations is not associated to the taxonomic status of the strain. Nostoc, Calothrix and Scytonema strains present generally higher ARA than Anabaena ones.
In a natural biotope, as Senegalese paddy fields, the coexistence of N2BGA with different responses of ARA for pH variations allow theoretically a constant nitrogen fixation between pH 5 to 9.

NODULATION AND NITROGEN FIXATION BY FIVE SPECIES OF LEGUMINOUS TREES GROWN IN SOIL FROM UNDISTURBED AND DISTURBED TROPICAL SITES IN MEXICO

J.P. Roskoski[+]/T. Wood[++]
+ NifTAL Project, P.O. Box 0, Paia, Hawaii 96779, USA
++ Native Plants Inc., 360 Wakara Way, Salt Lake City, Utah 84108, USA

Leguminous trees that grow rapidly, fix nitrogen, and produce fuel-wood and protein-rich foliage and fruits hold potential for use in tropical areas. Many of the valuable characteristics of these plants depend, in part, on the formation of effective nitrogen-fixing symbiosis.

We, therefore, examined the ability of tropical soils from different sites in Mexico to supply rhizobia to Acacia pennatula Schl. and Cham., Albizia lebbek L. Enterolobium cyclocarpum Jacq., Gliricidia sepium Jacq., Leucaena leucocephala Benth. These species were planted in soil from a primary rain-forest (site 1, pH 4.0), two slash-burn fields (sites 2 and 4, pHs 4.1 and 4.8), two pastures (sites 3 and 7, pHs 5.2 and 5.1), a secondary forest (site 5, pH 4.4) and two eroded sites (sites 6 and 8, pHs 4.8 and 4.2). Plant height and biomass, and nodular biomass and specific activity were determined on 2 and 6-month old seedlings. The latter were also used to establish 15N2/C2H2 conversion ratios.

Seedlings of all species grown in soils from sites 3 and 4 exhibited greatest height growth; ranging from a $\bar{X}$ of 22.1 cm for A. lebbek to a $\bar{X}$ of 47.4 cm for A. pennatula, both grown in soil from site 4. Similarly, total above-ground biomass/species was greatest for seedlings grown in soils from sites 3 and 4; ranging from a $\bar{X}$ of 4.36 g D.W./pl for A. lebbek (site 3 soil) to a $\bar{X}$ of 11.86 g D.W./pl for A. pennatula (site 4 soil). All species grew poorly in soils from the primary and secondary forests (sites 1 and 5).

E. cyclocarpum and A. lebbek nodulated in 7 of the 8 soils; the exception being the primary rainforest soil, pH 4.0. In contrast, A. pennatula, G. sepium, and L. leucocephala only nodulated in soils whose pH was 4.8 or greater (sites 3,4,6,7). Species exhibited similar nodulation patterns at 2 and 6 months, with the exceptions of A. lebbek, which nodulated in only 2 soils at 2 months and in 7 soils at 6 months. Nodule biomass for each species in the 8 soils was highly correlated with above-ground biomass and soil pH. Each species yielded greatest nodule biomass when seedlings were grown in soils from either sites 3 or 4. Maximum nodule biomass ranged from .25 g D.W./pl for L. leucocephala to 2.19 g D.W./pl for G. sepium.

Specific nitrogen-fixing activity for 6 month-old seedlings was generally 4 to 10 times greater than that of 2-month-old seedlings and ranged from 1.9 um N/g nod/hr for E. cyclocarpum seedlings grown in site 4 soil to 24.70 um N/g nod/hr for L. lecocephala seedlings grown in site 6 soil. Overall, specific activity was negatively correlated with nodule biomass. For example, seedlings of G. sepium, A. pennatula and L. leucocephala achieved highest nodule biomass but lowest specific activity when grown in site 3 soil.

Rhizobia isolated from nodules of A. pennatula, G. sepium, and L. leucocephala were fast-growing, very sensitive to carbenicillin and cephalothin and moderatley sensitive to chloramphenicol; while isolates from E. cyclocarpum and A. lebbek were slow-growing and sensitive to chloramphenicol and sulfadiazine, respectively. Growth pouch studies revealed that rhizobia were present in all soils and most numerous in soils that had supported greatest nodule biomass production.

Overall, we found that even soils from adjacent areas varied markedly in their ability to support tree legume growth and nodulation, and that conversion of tropical forests to other land uses increased the diversity and abundance of rhizobia specific of leguminous trees.

THE EFFECT OF SEED BACTERIZATION WITH RHIZOBIA AND COINOCULANTS ON RABI PULSES IN INDO-GANGETIC ALLUVIUM OF VARANASI

C.L. SANORIA/A.K. RAWAT/G. RAM/M.K.MALLIK/B.R. MAURYA/B.N. TOSH AND J. PRASED
DEPARTMENT OF SOIL SCIENCE AND AGRICULTURAL CHEMISTRY, BANARAS HINDU UNIVERSITY, VARANASI-221005, INDIA

In statistically designed 14 field experiments, culture suspension of specific Rhizobium and other bacterium (Azotobacter/Beijerinckia/Pseudomonas-PO_4 solubilizer) was used singly and/or combinedly along with gum acacia sticker for plotwise seed inoculation of gram (9;I_0 & I with 3 cultivars; I_0, I_1 & I_2 at 3 P levels; 17; and 13 treatments in trial No. 1&2; 3&4;5;6&7; and 8 respectively), pea (12 treatments in trial No. 9 & 10, 11, and 12 at soil pH 7.5, 9.5 and 8.5 respectively) and lentil (12 treatments in trial No. 13&14). Every plot ($5x2\ m^2$) of each trial was basaly dressed at the rate of 20 kg N(urea) and 50 kg p_2O_5 (super phosphate) per ha. Extra levels of P_2O_5 used in some treatments of the last 4 trials on gram were not beneficial.

Nodulation due to native rhizobia was almost nil in trial No. 6&8 on gram. Inoculant rhizobia for different crops tended to differ among themselves. Rhizobial strain used alone without any coinoculant gave maximum grain yield in only trial No. 8. Promising combined treatments were:
Rh.H_{44}+Az.B_4, Rh.H_{45}+Az.B_5, and Rh.BG_1+Az.B_5+Ps. for gram; Rh.F_{10}+Az.B_5 for pea in all soils; and Rh.F_2+Az.B_4 for lentil. Az.B_5 alone seemed better than Rh. strains for pea in trial No. 12. Yield was significantly reduced by Rh.C_7+Az.B_4 in pea and by Az.B_4 and Rh.C_7+Az.B_5 in lentil. Data on grain protein % were significant in trial No. 5,6,7,10 and 13. Some of the treatments in trial No. 11&12 caused remarkable reduction in soil pH at harvest. As the sites and years (1973-82) of the experiments varied, it is somewhat difficult to explain the mechanism of good/bad effect of coinoculant with particular rhizobial strain. However, there seems possibility of using compatible coinoculant bacterium with specific Rhizobium for greater benefits from pulse crops.

DELONIX REGIA: CONFIRMATION OF ITS NON-NODULATING CHARACTER

Eduardo C. Schroder
Dept. of Agronomy & Soils, Univ. of P.R., Mayaguez, Puerto Rico 00708

1. INTRODUCTION

The subfamily Caesalpinioideae is the most primitive of the leguminosae family. Sixty percent of its species lack root nodules (Allen & Allen, 1981). Delonix regia, a species of this family, is a widely distributed ornamental tree commonly known as flamboyant-tree, or royal poinciana. Results concerning the nodulation status of D. regia have been conflicting. A number of authors have not found any nodules (Allen & Allen, 1936; Bañados & Fernández, 1954; Grobbelaar et al, 1964; and Sen et al, 1980). Lim & Ng (1977); Athar and Mahmood (1980), however, have reported nodulation on this species. The objective of this work was to examine the nodulation status of this species under the tropical conditions of Puerto Rico.

2. MATERIALS AND METHODS

For field surveys, several hundred flamboyant seedlings were excavated and examined for nodulation in different ecological zones as well as diverse soil types. For greenhouse experiments, D. regia seeds were surface sterilized and sown in Leonard jars, following the method described by Vincent (1970).

3. RESULTS AND DISCUSSION

None of the plants examined under field conditions had Rhizobium induced nodules. On the basis of this sampling, and under the natural conditions of Puerto Rico soils, the flamboyant trees do not nodulate.

To determine if the lack of nodulation is due to the lack of the appropriate Rhizobium strain in the soil, seeds of D. regia were inoculated with strains of Rhizobium belonging to different species, including a culture isolated from flamboyant nodules. At harvest, roots were closely examined, but none of the plants had nodules. The "Rhizobium" cultures received from Dr. Lim isolated from D. regia and Brownea ariza, lacked the cultural characteristics of rhizobium (colonies growing on YEM absorb congo red and do not nodulate Macroptilium atropurpureum in tubes). On the basis of these experiments were confirmed the non-nodulating property of D. regia grown in Puerto Rico.

4. REFERENCES

Allen, ON & Allen, EK. 1981. The Leguminosae. Univ. Wis. Press, Madison
Allen, ON & Allen, EK. 1936. Soil Sci. 42;87-91
Athar, M & Mahmood, A. 1980. Trop. Agric. 57;319-324
Bañados, II & Fernández, WL. 1954. Philipp. Agric. 37;529-533
Grobbelaar, N, Beijma, MC van, & Saubert, S. 1964. S. Afr. J. Agri. Sci. 7;265-270
Lim, G & Ng, HL. 1977. Plant and Soil 46;317-327
Sen, R, Sau, A, Naskar, K & Bhattacharjee, A. 1980. Bull. Bot. Sur. India 22;166-172
Vincent, JM. 1970. IBP Handbook No. 15. 164 pp. Oxford, Blackwell Sci. Publ.

5. ACKNOWLEDGMENTS

Work reported was supported by grant from USAID. Strains of Rhizobium were supplied by Dr. G. Lim. The author thanks Mrs. Myrna Gaztambide and Mr. Miguel Rivera for technical assistance.

CHARACTERISTICS OF INDIGENOUS POPULATIONS OF COWPEA RHIZOBIA FROM WEST AFRICA

M.J. SINCLAIR, M.D. STOWERS*, B.J. GOLDMAN, A. AYANABA AND A.R.J. EAGLESHAM
BOYCE THOMPSON INSTITUTE FOR PLANT RESEARCH AT CORNELL, ITHACA, NY, USA

1. INTRODUCTION

The cowpea, *Vigna unguiculata*, is an important source of dietary protein in West Africa, and is grown as a subsistence and a cash crop. Farmers currently depend upon indigenous cowpea rhizobia for nodulation although many cultivars do not obtain optimal amounts of N from symbiotic fixation (Ahmad et al.1981a). Improvement in this situation may be promoted by determining the nature of the native rhizobia. This paper summarizes a program of characterization of three indigenous populations of rhizobia.

2. PROCEDURE

2.1 Materials and Methods

Rhizobia were isolated from cowpeas grown at Onne and Ibadan in Nigeria and at Maradi in Niger (Ahmad et al. 1981a). Symbiotic characteristics, enzyme-linked immunosorbent assay (ELISA) patterns and sodium dodecyl sulphate (SDS)-polyacrylamide gel electrophoresis (PAGE) patterns were examined as described before (Ahmad et al. 1981a; Ahmad et al. 1981b; and Noel, Brill 1980, resp.). Rhizobial stress-tolerances and carbon nutrition were examined on yeast-extract mannitol agar (YEMA). Intrinsic antibiotic resistance (IAR) was examined by a rapid screening method (M.J. Sinclair unpublished).

3. RESULTS

Growth rates in YEM broth and carbon nutritional requirements were typical for slow-growing rhizobia. Colony morphologies on YEMA were of two types. Ibadan strains were 97% the "wet" type (copious slime,often confluent); Maradi strains were 97% the "dry" type (small, discrete); Onne strains were 60% "wet" and 40% "dry". Frequency of highly effective symbioses on cowpea was also influenced by the geographic origin of the strains with Ibadan > Onne > Maradi. Some strains from Maradi and Onne produced unusually dark nodules on cowpea. All of these strains were "dry". Peanut was well-nodulated only by Maradi strains, whereas pigeon pea and mung bean were well-nodulated only by Onne and Ibadan strains. Soybean was generally poorly-nodulated. Chlorosis was induced in soybean seedlings (Eaglesham et al. 1982) with greater frequency among "wet" (74%) than among "dry" (21%) Onne strains. Only the Maradi strains grew well at 40°C and of those that did not grow, only "wet" strains from Onne survived. Tolerance to 0.5% NaCl and to high levels on antibiotics was shown only by "wet" strains. SDS-PAGE and ELISA confirmed fundamental differences between "wet" and "dry" types: greater homology existed between strains of the same colony type from different locations than between strains of different colony type from the same location.

4. REFERENCES

Ahmad MH, Eaglesham ARJ, Hassouna S, Seaman B, Ayanaba A, Mulongoy K, and Pulver EL (1981a) Trop. Agric. (Trinidad) 48, 325-335.
Ahmad MH, Eaglesham ARJ, and Hassouna S (1981b) Arch. Microbiol. 130, 281-287.
Eaglesham ARJ and Hassouna S (1982) Plant and Soil 64, 425-428.
Noel KD and Brill WJ (1980) Appl. Environ. Microbiol. 40, 931-938.

Funded by the United Nations Development Program, grant no. GLO/77/013.

*Present address: Native Plants Inc., Salt Lake City, Utah.

EFFECT OF PLANT-DERIVED AND OTHER CARBON SUBSTRATES ON ASYMBIOTIC N_2-FIXATION

REX L. SMITH*/J. R. MILAM**/S. C. SCHANK*
DEPARTMENT OF AGRONOMY* AND DEPARTMENT OF MICROBIOLOGY AND CELL SCIENCE**, UNIVERSITY OF FLORIDA, GAINESVILLE, FL 32611

Asymbiotic N_2-fixation (also called associative N_2-fixation when plants are involved) is highly variable over different sites and conditions but can occur at magnitudes great enough to be agronomically important. We previously reported finding highly active associative N_2-fixation sites in a statewide search of Florida. Since energy that drives associative N_2-fixation is plant derived and restricting, we questioned whether N_2-fixing bacteria in those highly active soils could utilize a greater proportion of the substrate in fixing nitrogen than those in low activity soils. We further questioned whether those highly active soils could utilize plant structural carbohydrates to fix nitrogen, and also wanted to get estimates of efficiency of substrate use for fixing nitrogen relative to other metabolic uses.

Replicated soil samples from high, medium, and low ARA sites were incubated with glucose, mannitol, sodium succinate, cellulose (non-crystalline) and ground grass-root tissue. Incubation was carried out at 30°C in an atmosphere of 2%O^2-98%N^2, changed daily by flushing. Acetylene reduction analyses were done at one- to four-day intervals by adding 5% acetylene, incubating four hr, then measuring C_2H_2 evolution by gas chromatography. A glucose reference was measured by incubating N_2-fixing bacteria, from a high activity site, in a pre-sterilized system as described above. This was used to make soil efficiency calculations. Counts of N_2-fixing and total bacteria were made using most probable number techniques.

The soils differed greatly in their responses to the substrates, and previous site ARA did not predict substrate utilization for N_2-fixation. Glucose and mannitol stimulated rapid (one day) and maximum ARA responses. Succinate produced rapid but low ARA responses. Cellulose response was intermediate and delayed three to six days. Root tissue ARA responses were very low and also delayed. Estimates of the proportion of the substrate energy going to N_2-fixation were up to 59% for glucose, up to 52% for mannitol and up to 20% for cellulose. Conversions of substrate for root tissue and succinate were very low. Addition of substrate caused an increase in the population of N_2-fixing bacteria of 10^2 to 10^3 while total bacteria either declined or remained constant. Enhancement of N_2-fixation by cellulose was believed to be due to the combined activities of cellulose-digesting and N_2-fixing organisms. Extrapolations from those data (using 4:1::ethylene:nitrogen ratio) show that the glucose reference fixed 21 mg N/g glucose. Calculations show that at maximum soil efficiency 12.6, 11.0 and 4.3 mg of N were fixed per g of glucose, mannitol and cellulose, respectively. At those rates, it would require 470, 900 and 2340 kg of glucose, mannitol and cellulose to fix 10 kg N.

OCCURRENCE OF N_2- FIXING CYANOBACTERIA IN SOMALILAND SOILS

L. TOMASELLI/M.C. MARGHERI/V. BRANCACCIO AND G. FLORENZANO
INSTITUTO DI MICROBIOLOGIA AGRARIA E TECNICA DELL'UNIVERSITÀ DEGLI STUDI E CENTRO DI STUDIO DEI MICRORGANISMI AUTOTROFI C.N.R. - FIRENZE (ITALY)

The occurrence of cyanobacterial communities in three types of Somaliland habitat (alluvional soils and supra-litoral zone),characterized by an high degree of desiccation in a region with a very low rainfall, is described. The samples were found all dominated by cyanobacterial communities. Green algae were practically absent. From the samples examined, 93 filamentous and 2 unicellular cyanobacterial isolates were obtained. The unicellular strains were referred to the genus *Synechocystis* (Section I, according to the taxonomic scheme of Rippka et alii, 1979). The filamentous isolates included 39 non heterocystous strains belonging to the Section III and 54 heterocystous strains included in Section IV. Among the non heterocystous cyanobacteria represented by the genera *Lyngbya*, *Oscillatoria*, *Phormidium*, *Plectonema* and *Schizothrix*, the genus *Oscillatoria* exhibits a remarkable diversity and includes 7 species, according to the "botanical" systematic. Also the genus *Nostoc*, among the heterocystous cyanobacteria represented by *Calothrix* and *Scytonema* too,shows a great species diversity, being recognizable 9 morphological types.

The cyanobacterial pattern is dominated by oscillatoriacean species (*O. acuminata*, *O. proboscidea* and *O. subbrevis*)in the loam soil crusts, by *Scytonema javanicum* in the ferruginous soil crusts and by *Phormidium anomala* in samples from supra-litoral zone. Also in some Nigerian soil *Scytonema* sp. is the dominant organism (Stewart et al.,1978). The structure of the coenosis found in the Somaliland soil crusts differs from that of temperates soils, generally dominated by *Nostoc* species(Tomaselli et al.,1978,1983). In the three types of habitat the amount of cyanobacteria ranged approximately between 1.4×10^4 and 3×10^5 organisms g of dry soil^{-1}. The heterocystous strains showed nitrogen fixation rates ranging between 0.6-3.8 nmoles C_2H_4 mg (d.wt.)$^{-1}$ min^{-1}.

The general deficiency of combined nitrogen and intense illumination favour the selective advantage of cyanobacterial communities in Somaliland soils. The low soil moisture still permitted some cyanobacterial growth but in the rain season the nitrogen fixing activity may be very high:6.8-30.6 Kg N_2 ha^{-1} Yr^{-1}, is fixed by a community where *Scytonema* sp.is dominant(Burris,1976).Nevertheless, also the filamentous strains lacking haterocysts are probably sometimes important contributors to tropical nitrogen fixation(Stah and Krumbein,1981). Potts (1979) found that rates for *Lyngbya aestuarii*,*Schizothrix* sp.and other non-heterocystous species were usually higher than those measured for *Scytonema* and *Calothrix*.

EFFECTIVENESS OF INOCULANT STRAINS AFTER SEVERAL YEARS IN SOIL

H. JANSEN VAN RENSBURG / B.W. STRIJDOM
PLANT PROTECTION RESEARCH INSTITUTE, PRETORIA, SOUTH AFRICA

INTRODUCTION

Although capability of a Rhizobium strain to become part of the soil microbial population is considered desirable for an inoculant strain, there is some concern that its effectiveness in soil might become mediocre with time. The implication that an established inoculant strain may need to be replaced by a more effective strain after a few years prompted this study.

MATERIALS AND METHODS

Rhizobia were isolated from 4 field grown legumes on farms where seed had been inoculated once only, at least 4 years before. The inoculant strains were R. trifolii TA1, Rhizobium XHT1 (lotus), R. meliloti U45 and R. japonicum WB61. It was attempted to isolate from 81 effective nodules (9 from each of 9 plants) of each legume. Effectiveness of isolates on their respective hosts was compared in Leonard jars with that of 20 single colony isolates, as well as whole cultures, of lyophilized inoculant strains. Antisera against the inoculant strains were used in agar gel diffusion tests in an attempt to identify isolates of the inoculant strains serologically.

RESULTS

The percentage isolates with gel diffusion patterns indistinguishable from those of the respective lyophilized inoculant strains were; R. trifolii 64,2, Rhizobium (lotus) 100, R. meliloti 23,4 and R. japonicum 67,5. None of these isolates was less effective than single colony isolates, or whole cultures, of the lyophilized inoculant strains; 2 isolates of R. meliloti (11%) was more effective. With each legume sp. the average effectiveness of the total population (identified and unidentified strains), which agreed closely with the median effectiveness value of the strains, was greater than that of the unidentified population (Table 1).

DISCUSSION

The high percentage strains recovered from some soils where legumes had been inoculated 4 to 8 years before was unexpected. Whereas earlier results indicated the absence from South African soils of strains effective on L. pedunculatus and G. max, the origin of an effective R. japonicum population in addition to the inoculant strain population cannot be explained.

The finding that inoculant strains have survived for several years under field conditions without losing effectiveness, justifies the selection of strains with strong ability to establish in soils. Concern that effectiveness of a strain introduced into soil tend to become mediocre with time seems unfounded, at least in the absence of strong selection pressures. The results also indicate prolonged improvement of the effectiveness levels of the soil rhizobial populations studied as a result of a single seed inoculation.

TABLE 1. Average effectiveness of isolates as percentage of that of 20 single colony isolates of a lyophilized inoculant strain

Legume isolated from	No. of isolates[a]		Years in soil	% effectiveness of isolates[a]		
	I	U		I	U	Total
G. max	56	17	8	94,1	78,1	86,4
M. truncatula	18	62	5	129,7	96,9	113,3
T. repens	52	31	4	95,9	56,1	76,0
L. pedunculatus	88	0	4	113,1	-	113,1

a I = identified, U = unidentified.

NITROGEN FIXATION BY WHITE CLOVER (Trifolium repens L.) IN NOVA SCOTIA PASTURES

J.K. VESSEY and D.G. PATRIQUIN
Biology Dept., Dalhousie University, Halifax, Nova Scotia, Canada B3H-4J1

The objectives of this study were (1) to estimate the amount of nitrogen fixed by white clover in pasture with good clover cover and (2) to determine what are the principle factors influencing the abundance of and N_2 fixation by white clover in permanent pasture.

At three sites, sampled one to two times a month, clover cover increased from less than 10% in April to maxima of 50-70% during anthesis in July. It began to decline in September, reaching 20-30% in December. Acetylene reduction activity (ARA) commenced in April at soil temperature of 5-7°C. Clover-specific ARA (ARA/unit area of clover patch) was high throughout most of May, June and July, and then declined reaching low levels in November. Clover-specific ARA was significantly correlated with rainfall over a period prior to the assay day (highest correlation with 7-28 day period, $r=0.720$, $P<0.01$) and was influenced by phenological development (flowering). Clover cover and the period the temperature was above 5-7°C were the major influences on the total amount of N_2 fixed.

At one site, N_2 fixation was estimated by subtracting the N accumulated in grass patches from the N accumulated in clover patches. Comparison of this to integrated ARA values gave a molar ratio of C_2H_2 reduced to N_2 fixed of 2.78:1. Total N_2 fixation at the three sites was estimated at 66-100 kg N/ha per annum.

Variation in clover cover (2-53%) at eight sites was not significantly correlated with soil pH, cation exchange capacity, Ca, Mg, K, Na, P, C, or N, considered separately. Stepwise multiple regression analysis showed a significant correlation between clover cover and soil Ca and Mg levels considered together ($r=0.914$, $P<0.011$).

Management factors (grazing, N fertilization) appeared to be the principle factors influencing the variation in clover cover at the eight sites.

There appear to be no major climatic or edaphic restrictions to white clover in Nova Scotian pastures. The amount of nitrogen fixed at sites of good clover cover compares favorably to N fertilizer inputs (50-65 kg N/ha) to pasture with poor clover cover.

POPULATION GENETICS OF RHIZOBIUM - A PRELIMINARY STUDY

J.P.W. YOUNG
JOHN INNES INSTITUTE, NORWICH, U.K.

Although we know a good deal about the laboratory genetics of Rhizobium, we know rather little about the structure of natural populations. As a first step, we have looked at natural genetic variation in four species of fast-growing Rhizobium taken from a single site: Rhizobium leguminosarum, R. trifolii, R. phaseoli and R. meliloti.

Isolates of the four species were obtained by sowing the appropriate host plants in a single small area (2m x 2m) within a commercial crop of Phaseolus vulgaris in Norfolk, England. One pure strain was established from each nodule formed. Enzyme electrophoresis was used to detect naturally-occurring genetic variation amongst these isolates. Most of the isolates of all 4 species (about 200 isolates altogether) have been characterized at 3 enzyme loci so far. These are Gpdh (glucose-6-phosphate dehydrogenase), To (tetrazolium oxidase) and Gal (beta-galactosidase). Most isolates had a single allele at each of these loci. All three loci were polymorphic, with allele frequencies that differed amongst species.

R. meliloti isolates were substantially different from those of the other species, with unique alleles at Gpdh and Gal. This is evidence that they are genetically isolated from the other species, as expected from laboratory studies. At each locus, R. leguminosarum, R. trifolii and R. phaseoli share common alleles of the same mobility, and at each locus, R. leguminosarum is more variable than R. trifolii, and R. phaseoli is least variable.

Genotypes do not occur in the frequencies expected for random assortment of alleles at the different loci. For example, the types FF (To-F, Gal-F) and SS are common in both R. leguminosarum and R. trifolii, whereas the corresponding recombinant types FS and SF are rare. It is unlikely that such a strong association between apparently unrelated enzymes would arise independently in each species, the more likely explanation is that R. leguminosarum and R. trifolii share a common pool of genetic variation. In another striking association, the alleles Gpdh-F and To-M always occur together, and furthermore, these alleles are confined to R. leguminosarum. This may be a hint that some parts of the R. leguminosarum gene pool do not exchange freely with R. trifolii.

I should like to thank Palmira Guevara Trejo for preliminary work on R. phaseoli, Annie Rowe for the data on R. meliloti, and Diane Rushbrook and Jackie Roff for technical assistance.

POSTER DISCUSSION 5B FRANKIA SYMBIOSIS

A.D.L. AKKERMANS, Department of Microbiology, Agricultural University, Wageningen, The Netherlands

During the last five years, an increasing number of biologists and foresters became interested in the nitrogen-fixing symbioses between the actinomycete *Frankia* and a number of non-leguminous trees and shurbs. The reason is two-fold. First of all, foresters now realize that input of combined nitrogen by biological nitrogen fixation is often a preriquisite for producing wood at an economically acceptable level in the near future. Fertilization of the forest with nitrogen may be useful at the moment, but doubtless will create economic problems in the future when prices of the fertilizers and transport will increase.
Since growing a forest with an average rotation time of 20-40 years needs a large investment over many years, continuous use of fertilizers may be more risky than using nitrogen-fixing plants as an understory or in interplanting.
A second reason for the interest in *Frankia* is the fact that pure cultures have recently become available. This makes it possible to start detailed studies on the physiology, biochemistry and genetics of these organisms.
Frankia is a filamentous, slow-growing actinomycete which is able to form vesicles and fix nitrogen in pure cultures as well as in the root nodules. Recent observations have shown that these organisms, like cyanobacteria and *Azotobacter*, are free-living nitrogen fixers. The rate of fixation in pure culture approximates that in the nodules. The close correlation between the occurrence of vesicles and formation of nitrogenase suggests that the vesicles probably are involved in nitrogen fixation, like the heterocysts in cyanobacteria. At the moment, special attention is being paid to the morphogenesis and changes in the cell metabolism during the transition of hypha into vesicles. Is nitrogenase really localized inside the vesicles? What is the function of the double membrane systems around the vesicles? Are these involved in protecting nitrogenase against damage by O_2?
In addition to these questions, one has to know whether the vesicles which are formed in the pure culture are physiologically similar to the vesicles which are formed in the nodules. If so, one has to know how the plant is able to change the physiology of the *Frankia* in order to induce exudation of ammonia into the cytoplasm of the plant. In addition to the physiological studies of the nitrogen metabolism, several research groups have initiated studies on the host specificity of *Frankia*. Many strains are now clearly highly promiscuous and form nodules in plant species in different plant families. Other strains, however, are more narrow-minded and have a restricted host specificity. Several strains have now been isolated which are not infective or produce only very few nodules on one host while these regularly produce many nodules on other hosts. During the last few years, a variety of strains have been obtained from the same nodule source. These strains differ in effectivity, host specificity, growth requirements and DNA composition. Moreover, it has been shown that clones and reisolates obtained from one strain may differ considerably. This indicates that *Frankia* is highly variable and that nodules probably contain a mixture of genetically distinct strains.
The recently available information on the biology of *Frankia* is still highly incomplete and does not permit us to draw general conclusions. However, the basic problems of isolation and cultivation of a number of strains now seem to be solved and it is likely in the near future that our knowledge about *Frankia* symbioses will reach the knowledge we have of *Rhizobium* symbioses.
Since *Frankia* is more related to other free-living actinomycetes than to *Rhizobium*, it is likely that actinomycete genetisists and microbiologists can contribute substantially to the understanding of *Frankia* symbioses. For several rea-

sons, actinomycetes have been treated as a separate group of procaryotes. Many of the actinomycetes have great industrial importance, e.g. due to the production of extracellular secundary metabolites, such as antibiotics.
The link between Frankia and the industrially important actinomycetes gives the promise of being able to produce nitrogen-fixing actinomycetes, which can produce antibiotics and grow on simple, highly selective nitrogen-free media. It is tempting to suggest that this potential biotechnological use of Frankia can be explored in the near future.
The great interest in, and potential of, the Frankia symbioses is demonstrated by the posters displayed at this meeting. The abstracts of these posters follow. In addition, further information on research progress in this fiels was presented at the International Workshop on Frankia Symbioses, organized at Wageningen, The Netherlands, 5-6 September 1983., following the 5th International Symposium. The proceedings of this Workshop, containing about 20 original research papers constitutes a separate February 1984 issue of Plant and Soil.

DNA RELATEDNESS OF FRANKIA ISOLATES

AN CHUNG-SUN/W.S. RIGGSBY AND B.C. MULLIN
DEPARTMENTS OF BOTANY AND MICROBIOLOGY, UNIVERSITY OF TENNESSEE - KNOXVILLE, USA

A molecular genetical approach was used to examine the genetic relatedness of 19 *Frankia* isolates by measuring the base composition of their deoxyribonucleic acid (DNA), the size of their genomes, the DNA-DNA homology and thermal stability of the hybrid DNA molecules. The mole % of guanine plus cytosine of *Frankia* DNA ranged from 66.6% to 71.4% with an average of 68.5% as measured by the thermal denaturation method and 69.4% by the buoyant density method. The mean genome size of two *Frankia* strains, Ar14 and Eu11, was 6.2×10^9 and 4.6×10^9 Daltons, respectively (about 1-9 times bigger than that of *E.coli*).

Frankia isolates were divided into two groups based on the results of hybridization tests. Group B, consisting of isolates from *Alnus*, *Myrica* and *Comptonia* host plants, exhibited a high degree (67.4%-94.1%) of homology with Ar14. Group A, consisting of isolates from *Elaeagnus*, *Ceanothus*, *Purshia*, *Casuarina* and Air12 from *Alnus*, did not exhibit significant homology (less than 39%) with Ar14. None of the strains showed a high degree of homology with Eu11 (less than 33%). Among group A strains, subgroupings seemed to exist evidenced by a very high homology (97%) between two isolates from *Casuarina*, DII and G2, but a low homology between other strains in group A and G2 (27%). Thermal stability of the hybrid DNAs which showed a high degree of homology exhibited an average of 3% mismatch while the low homology duplexes exhibited about 5% mismatch. Twelve actinomycetes which share a common cell wall type with *Frankia* were also used to examine the possible genetic relationships of *Frankia* to other actinomycetes by measuring base composition and DNA homology with Ar14 and Eu11 DNA.

STEM-NODULATING LEGUMES

B. DREYFUS, D. ALAZARD, E. DUHOUX, and Y. DOMMERGUES
ORSTOM, B.P. 1386, DAKAR, Sénégal

Only a few legumes form aerial nitrogen-fixing nodules on their stem. Typical stem nodulated legumes belong to three genera: *Sesbania*, *Aeschynomene* and *Neptunia*. All of them are characterized by predetermined stem nodulation sites which always include a dormant root primordium. The rhizobial infection starts at the base of this primordium provided it can be reached by the bacteria. Depending on the host plant, the primordia either pierce the stem cortex, sometimes through a lenticel, or remain embedded in the cortical tissues of the stem. Stem nodulated legumes can be divided, into three groups:

- group 1 with protruding primordia, readily infected by just spraying the stems with a *Rhizobium* culture. E.g.: *Sesbania rostrata*, *Aeschynomene afraspera*.
- group 2 with slightly protruding primordia, less readily infected. E.g.: *A. scabra*, *A. indica* and *A. sensitiva*.
- group 3 with embedded primordia, nodulating only when the dormancy is broken (especially by immersion). There are two subgroups: plants with nodulation sites distributed all along the stems, e.g. *A. elaphroxylon*, like plants from group 1 and 2, and *Neptunia oleracea*, with nodulation sites restricted to the vicinity of stem nodes.

Whereas infection of temperate legumes is through root hairs, infection of stem nodulated legumes does not occur in this way, but by direct intercellular invasion of the base of the root primordium, which is reminiscent of the *Arachis* and *Stylosanthes* infection. Up to now infection threads have only been found in *Sesbania rostrata* (Tsien et al., 1983) and *Neptunia oleracea* (Schaede, 1940). Specific N_2 fixation by stem nodules of *Sesbania rostrata* and *Aeschynomene* spp. is comparable to that of soybean nodules. By virtue of its profuse nodulation(up to 40 g of nodules/plants), *Sesbania rostrata* has a high N_2-fixing potential. Besides this characteristic, most stem-nodulated legumes have the unusual capability of actively fixing N_2 in the presence of high levels of combined nitrogen in the soil. With the exception of *Neptunia oleracea* all stem nodules contain active chloroplasts, which probably contribute to the energy requirements of N_2 fixation. Since root primordia have also been observed on the stem of some other legumes, such as asian cultivars of soybean, *Arachis hypogaea* and *Vicia faba*, one could foresee the possibility of obtaining new legumes bearing on their stems root primordia susceptible to rhizobial infection.

Schaede, R. 1940. Planta (Berlin) 31: 1-21.
Tsien, H.C., B.L. Dreyfus, and E.L. Schmidt, 1983. J. Bacteriol (in press).

MYCORRHIZAL IMPROVEMENT OF THE NON-LEGUMINOUS NITROGEN-FIXING ASSOCIATION IN HIPPOPHAE RHAMNOIDES L.

I.C. GARDNER, D.C. CLELLAND and A. SCOTT
BIOLOGY DIVISION, DEPT. OF BIOSCIENCE & BIOTECHNOLOGY,
UNIVERSITY OF STRATHCLYDE, GLASGOW G1 1XW, GREAT BRITAIN.

Hippophae rhamnoides (sea buckthorn) growing on sand dunes in Scotland possesses actinorhizal nodules and actively fixes atmospheric nitrogen. Examination of root squash preparations and sieved sand samples from such locations reveals that short secondary lateral roots additionally sustain a heavy vesicular arbuscular (VA) mycorrhizal association. Measurement of sporocarps, chlamydospores and vesicles together with consideration of the growth habit of the fungus within the roots led us to identify it as *Glomus fasciculatus* (Thaxter sensu Gerdemann). Gerdemann & Trappe comb. nov.

In physiological terms the most important fungal organ within the host plant is the arbuscule, found intracellularly in the inner cortex of the roots. SEM of such cells, ruptured and cleared of host cytoplasm, shows the extensive dichotomy of the arbuscule. This fine branching increases the host/fungal interface significantly. Uninfected inner cortical root cells are characteristically highly vacuolate. TEM reveals that the cells containing arbuscules have a markedly increased volume of host cytoplasm containing numerous mitochondria, plastids, lipid droplets and E.R. with a membrane of host origin separating the arbuscule from the host cytoplasm. The main trunk hyphae of the arbuscules contain abundant polyphosphate granules and rosettes of glycogen whereas the fine terminal branches are depleted of polyphosphate and have less obvious accumulations of glycogen, consistent with the role of the mycorrhizal arbuscule in metabolite exchange. The polyphosphate granules occur in the extensive vacuolar system of the fungus where alkaline phosphatase has also been located: this last could be involved in the transport of phosphate within the hyphae.

VA mycorrhizal fungi are obligate symbionts but can be conveniently "cultured" in *Zea mays* plants. The mycorrhizal fungus, obtained from field *Hippophae*, "cultured" in this way, was used to inoculate *Hippophae* seedlings grown under sterile conditions in the greenhouse. Experiments, using various inoculation regimes, demonstrated the significant improvement in the mycorrhizal-nodulated plants compared to the nodulated-only and the mycorrhizal-only plants with respect to plant growth, uptake of phosphate and nitrogenase activity, when grown in a medium poor in combined nitrogen and soluble phosphate. The enhancement seen in the experimentally grown *Hippophae* could be expected to be found in the dune habitat where nitrogen and phosphorus levels are low. This, combined with the propensity of the mycorrhizal extra-radical mycelium for sand aggregation and the capacity of the fungus to resist desiccation, makes the triple symbioses a feature of prime importance in the development and stabilisation of dune systems.

A COMPARATIVE ACCOUNT OF ROOT NODULES OF *DATISCA CANNABINA* AND *ALNUS NITIDA* FROM PAKISTAN

F. HAFEEZ/A. H. CHAUDHARY AND A.D.L. AKKERMANS+
DEPARTMENT OF BIOLOGY, QUAID-I-AZAM UNIVERSITY, ISLAMABAD, PAKISTAN
+DEPARTMENT OF MICROBIOLOGY, AGRICULTURAL UNIVERSITY, WAGENINGEN, THE NETHERLANDS

Alnus nitida is a native actionorhizal tree while *Datisca cannabina* a perennial actionorhizal herb has been recently added to the world list of actionorhizas. A comparative study of morphology, anatomy and physiology of root nodules of two species has been investigated. The nodules of both the species form dichotomously branched coralloid clusters typical of the group. Internally the endophyte is uniformly distributed around a central stele in *A. nitida* while it is horse-shoe shaped with acentric stele in *D. cannabina*. The vesicles in *A. nitida* are globose and oriented towards the periphery of host cells while they are cylinderical and directed toward the central vacuole in *D. cannabina*. The young nodules in both the species reduced acetylene at an active rate (*A. nitida*, 36 µmole C_2H_4·g·nodule dry wt-1 h-1; *D. cannabina* 38.0 µmole C_2H_4·g·nodule dry wt-1 h-1). The main transport amino acid in *Datisca* is arginine while in *A. nitida* citrulline is the predominant free amino acid. Root nodules of *D. cannabina* contained relatively large quantities of DAP (5.8 µmole DAP·g^{-1} nodule dry wt). The hydrogenase activity of both summer and winter harvested nodule of *Datisca* is almost constant. Distribution of hydrogenase activity over 20um residue and 20um filterate of nodule homogenate is similar to *A. glutinosa*. The enzymes NADH dehydrogenase and succinate dehydrogenase are present in vesicle clusters of *D. cannabina*. The oxygen uptake is enhanced little bit by succinate in *A. nitida*. The effect of succinate is more and addition of ADP stimulated oxygen uptake in *D. cannabina* nodule homogenate.

AN INVESTIGATION OF THE RESOURCE OF NON-LEGUMINOUS NODULATING TREES IN CHINA

HUANG JIABIN, SHEN SHANMIN, LIU HUICHANG, JIANG JIANDE,
YANG HUIFAN, ZHAO ZHENYING AND YANG BIFANG
INSTITUTE OF FORESTRY AND SOIL SCIENCE, CHINESE ACADEMY OF SCIENCES,
P.O.BOX 417, SHENYANG, CHINA

An investigation of the resource of noduleted non-leguminous trees in different parts of China has been carried out since 1978. Forty-two species belonging to six genera are recorded thus far, they are
Elaeanus angustifolia L., *E. bockii* Diels, *E. commutata* Bernh., *E. conferta* Roxb., *E. crispa* Thunb., *E. glabra* Thunb., *E. henryi* Warb., *E. gonyanthes* Benth., *E. macrophylla* Thunb., *E. mollis* Diels, *E. moorcroftii* Wall., *E. multiflora* Thunb. var. *ovata*, *E. oldhami* Maxim., *E. oxycarpa* Schlecht., *E. pungens* Thunb., *E. stellipila* Rehd. *E. umbellata* Thunb., *E. umbellata* var. *pavifolia* Servet., *E. viridis* var. *delavayi* Lec.,
Hippophae rhamnoides L., *H. rhamnoides* var. *porcera* Rehd.
Alnus cremastogyne Burk., *A. ferdinandi* Makio, *A. formosana* Makio, *A. hirsuta* Turcz., *A. incana*, *A. japonica* Sieb. et Zucc., *A. japonica* var. *microphylla*, *A. mandshurica* Hand-Mzt., *A. nepalensis* D.Don., *A. sibirica* Fisch., *A. tinctoria* Sarg., *A. traboculosa* Hand-Mzt.,
Myrica esculenta Buch.-Ham., *M. nana* Chevat., *M. rubra* Sieb. et Zucc.
Coriaria nepalensis Wall., *C. sinica* Mixam., *C. terminalis* Hemsl.,
Casuarina equisetifolia L., *C. temhissima*, *C. glauca*.
The nodules of all species identified were capable of acetylene reduction activities. The micro-structure of some Chinese native species were examined in both optical and electron microscope and vesicles, which are believed to be responsible for the nitrogen fixation, were observed in all these species.

The Frankia spp. from the nodules of *Alnus cremastogyne*, *A. hirsuta*, *A. japonica*, *A. glutinosa*, *A. nepalensis*, *Elaeagnus oxycarpa*, *E.mollis*, *E. mutiflora*, *E. gonyanthes*, *Hippophae rhamnoides*, *H. rhamnoides* var. *porcera*, *Myrica rubra* and *Casuarina equisetifolia* were isolated and cultured successfully. Infection tests were done to test the nodulation of the strains isolated and results indicated that they were able to nodulate the host plants with the exceptions of the strains isolated from *Myrica* and *Casuarina*. Different methods including sucrose-density fractionation, dilution and other methods were emploied in endophyte isolation from nodules. Among them, the direct isolation from surface-sterilized nodules seemed to be more simple and useful. By which a piece of apical part of a lobe from a fresh nodule, which was sterilized in 0.1 % (v/v) bromine for 5-7 minutes, rubbed on to the Jan Blom's agar medium and the Frankia colonies appeared in two weeks at 28 C. The Frankia spp. grew slowly on the QMOD medium but much faster on the Jan Blom's medium and often with bigger sporangia developed. Vesicles were observed in some of the pure culture and acetylene reduction activities were recorded.

The annual amount of nitrogen fixed by some trees was estimated in field condition by acetylene reduction method. The amount fixed by *Hippophae rhamnoides* (2 years old), *Alnus sibirica* (4 years) and *A. sibirica* (5 years) was 23, 55, and 86 Kg per hecter respectively. Poplar trees (*Populus pseudo-simonii*) grew much better in association with *Hippophae rhamnoides* then in pure stands. The increase in height and width ranged 58-169 % and 106-328 % respectively.

LEVELS OF TREHALOSE AND GLYCOGEN IN FRANKIA sp. HFPArI3

Mary F. Lopez[1], Mark S. Fontaine[2] and John G. Torrey[2]

[1]Department of Botany, University of Massachusetts, Amherst, MA 01002 USA
[2]Cabot Foundation, Harvard University, Petersham, MA 01366 USA

The levels of soluble trehalose were compared in *Frankia* sp. HFPArI3 under various conditions in batch culture.

Trehalose appears to be a specialized carbon reserve which is synthesized upon transfer of cultures to fresh medium; thereafter it is metabolized and broken down. Levels reach a peak of 10-20% cell dry weight depending on cultural conditions, then drop to a minimum of 1-2% cell dry weight. A similar pattern is observed whether cells are grown in a medium containing NH_4 Cl or in a medium lacking nitrogen substrates, however the total levels of trehalose are higher in the latter.

The levels of endogenous trehalose showed an inverse correlation with logarithmic increase in nitrogenase activity. When cells were inoculated into a medium lacking nitrogen to induce fixation of atmospheric nitrogen, trehalose levels reached their peak in 4 days. The nitrogenase activity (measured with the acetylene reduction technique) increased sharply on the fourth to fifth day. Trehalose levels fell sharply from day 4 to day 5; when they fell to half their maximum, nitrogenase activity decreased sharply. Thereafter, the decline in trehalose concentration paralleled that of nitrogenase activity. When cells were washed free of exogenous carbon, the endogenous trehalose levels dropped to half the initial value in one hour. The cells continued to respire for 2 days on endogenous reserves.

In addition to trehalose, glycogen reserves were extracted from *Frankia*. At its maximum, the glycogen totalled approximately 10% of the dry weight of cultures grown on a medium supplemented with NH_4 Cl as exogenous nitrogen source, and decreased to ca. 1.0% dry weight in starved cultures. Glycogen did not accumulate significantly in cultures that were induced to fix nitrogen.

N_2 FIXATION AND N DYNAMICS OF A MYRICA GALE STAND IN MASSACHUSETTS

CHRISTA R. SCHWINTZER
DEPARTMENT OF BOTANY, UNIVERSITY OF MAINE, ORONO, ME, U.S.A. 04469

INTRODUCTION AND METHODS. I examined N_2 fixation, net primary production, litter deposition and leaf litter decomposition in an open peatland dominated by M. gale in central Massachusetts. Nitrogen fixation was measured by acetylene reduction, net primary production by the harvest method, litter deposition in litter traps, and leaf litter decomposition in litter bags.

RESULTS AND DISCUSSION. In 1979 annual net production of M. gale was 550 g $m^{-2}yr^{-1}$ (dry biomass) and contained 8.6 g N m^{-2}. During the same year M. gale fixed about 3.7 g N m^{-2} which supplied approximately 43% of its annual N requirement (Schwintzer 1983a). Biological N_2 fixation by all other agents in the peatland contributed only 0.1 g N $m^{-2}yr^{-1}$ which was divided approximately evenly between the peatland surface (upper 5 cm) and the subsurface peat (Schwintzer 1983b).

M. gale litter excluding large stems added 167 g $m^{-2}yr^{-1}$ (dry biomass). Leaves made up 84% of this litter and small twigs plus fruit and catkins the remaining 16%. The litter contained a total of 3.1 g N m^{-2} with leaves contributing 2.6 g. Newly collected senescent litter leaves contained 1.69% N, 40% lignin and had a C:N ratio of 28.9. During three years of decomposition in the peatland the N concentration in the residue increased to 2.34% and the C:N ratio decreased to 19.7. At the end of three years the leaves were still intact and retained 90% of their original N mass.

The dynamics of Myrica gale leaf litter decomposition follow patterns observed in a wide variety of litter. The rate of decomposition, however, is surprisingly low given the relatively low C:N ratio. The low rate is probably caused by the chemical composition of the litter, in particular its very high initial lignin content. Possible environmental causes of the low decomposition rate can be largely ruled out because equally low rates were observed in an upland forest where the native litter decayed rapidly. At the peatland site slow decomposition of M. gale litter results in the loss of approximately half of its initial N content to peat accumulation. Litter leaves sink below the surface and become incorporated into the peat as fresh litter accumulates above them and older layers compact below them. When litter reaches a depth of 20 cm, the approximate level of the permanently waterlogged zone, it still retains about 32% of its original biomass and 60% of its original N mass. Slow decomposition of litter leads to organic matter accumulation on upland as well as wetland sites. Consequently, presence of M. gale enhances organic matter and N accumulation in all ecosystems in which its litter is retained on the site.

Comparison of decomposition rates and initial N and lignin contents of leaf litter of several woody N_2 fixing species shows that they vary widely. The most rapid decomposition occurs in Alnus glutinosa [>70-90% of initial biomass lost in the first year (Bocock 1964)] and the slowest in M. gale (20-23%). Consequently, no general rule can be formulated for N_2 fixing species.

REFERENCES
Bocock KL (1964) J. Ecol. 52, 273-284.
Schwintzer CR (1983a) Can. J. Bot. 61, in press.
Schwintzer CR (1983b) Amer. J. Bot. 70, 1071-1078.

PRODUCTION OF SPORANGIA BY THE ACTINOMYCETOUS ENDOPHYTE IN ROOT NODULES OF COMPTONIA PEREGRINA: DEVELOPMENT AND CONSEQUENCES FOR NODULE FUNCTION

K. A. VandenBosch[1] and J. G. Torrey[2]

[1]Department of Botany, University of Massachusetts, Amherst, MA 01003

[2]Cabot Foundation, Harvard University, Petersham, MA 01366

Frankia sp., the nitrogen-fixing endophyte in actinorhizal nodules, is a filamentous bacterium which may differentiate two morphologically distinct forms from its hyphae: vesicles and sporangia. In culture and in the nodules of some host genera, vesicles have been identified as the nitrogen-fixing stage of the microorganism. Sporangia, which develop under cultural conditions in most strains, occur only in some nodules. Field collected root nodules of Comptonia peregrina (L.) Coult. (fam.: Myricaceae), infected by the actinomycete Frankia sp., were of two types: those which lacked sporangia entirely, designated spore-minus, and those which showed extensive sporangial development, designated spore-plus. In this study, comparisons were made in the structural ontogeny of the infected host cell and of the endophyte and in the productivity and efficiency of the two symbiotic combinations.

In spore-plus nodules, sporangia began to develop after the differentiation of endophytic vesicles and the onset of nitrogenase activity. Sporangia developed from undifferentiated central hyphae in infected host cells containing mature vesicle clusters. At the onset of sporangial differentiation, host cell cytoplasm and nuclei appeared healthy. However, endophytic vesicles and host cell cytoplasm and nuclei began to senesce rapidly as sporangia developed. Staining of sectioned material with the fluorescent stain Calcofluor White suggested that vesicles, hyphae and young sporangia were enclosed within a host-derived encapsulation, but that mature sporangia were no longer encapsulated. Vesicles were more short-lived in spore-plus than in spore-minus nodules. Spore-minus nodules showed no evidence of sporangial formation and the endophyte was always encapsulated by cell wall material.

Measurements of acetylene reduction (i.e., nitrogenase activity) were coordinated with samplings of nodules for structural studies. Significant differences in acetylene reduction rates were discernable between spore-plus and spore-minus nodules commencing four weeks after nodulation, concommitant with the appearance and maturation of sporangia in the nodule. Spore-plus and spore-minus nodules reached rates of 4.3 and 9.8 μmol C_2H_4 evolved/g fresh weight nodule x h, respectively. Gas chromatographic assays indicated that both spore-plus and spore-minus nodules evolve only small amounts of hydrogen. These data suggest that both types of nodules were equally efficient in recycling electrons lost to the reduction of hydrogen ions by nitrogenase. Respiratory cost of nitrogen fixation, expressed as the quotient of μmol CO_2 to μmol C_2H_4 evolved by excised nodules, was significantly greater in spore-plus than in spore-minus nodules. This value for spore-minus nodules was 6.2 and for spore-plus nodules was 11.7.

Seedlings nodulated with the spore-plus inoculum showed only 60% of the nitrogenase activity and 50% of the net size of their spore-minus counterparts after twelve weeks of culture. The two groups of seedlings had similar numbers and weights of nodules.

The morphogenesis of Frankia in nodules of C. peregrina has thus been seen to influence the symbiotic functioning of the nodule. Nodules containing endophytic sporangia showed less nitrogenase activity than did those without sporangia and did so at a greater respiratory cost to the host. Consequently, spore-plus nodules were capable of supporting less host plant growth.

CHARACTERIZATION OF *FRANKIA* STRAINS IN FINLAND

A. WEBER, A. SMOLANDER, E.-L. NURMIAHO-LASSILA AND V. SUNDMAN
UNIVERSITY OF HELSINKI, DEPARTMENT OF GENERAL MICROBIOLOGY,
SF-00280 HELSINKI 28, FINLAND.

In connection with an energy forest project a number of *Frankia* strains were isolated from *Alnus incana* nodules by the method of OsO_4 sterilization of Lalonde et al. (1981). The nodule material came from a single grey-alder experimental field in southern Finland, where predominantly spore-negative nodules are found. Isolation attempts with nodules collected at other locations were unsuccessful. The isolates were verified for infectivity and effectivity by gnotobiotic inoculation experiments from which they were reisolated. Pure cultures of the isolates were shown to fix nitrogen (reduce acetylene) *in vitro*. All of the isolates grew on acetate and propionate, some of them utilized Tween 80, but none grew on succinate, fumarate, malate or glucose. Optimal growth occurred at about 30 °C; no growth was observed at or below 14 °C or at or above 36 °C. In liquid media the isolates produced compact flocks hard to disintegrate. Neither sonication (Shipton, Burggraaf 1982) nor treatment with 1 N NaOH at 100 °C for 30 min (Akkermans et al. 1983) was sufficient without previous freezing and acid treatments. Electron microscopy revealed sporangia surrounded by a layer 3 to 4 times thicker than the analogous structure of the reference strain *Frankia* Avc I1. The nature of these layers remains unknown.

REFERENCES

Akkermans ADL, Roelofsen W, Blom J, Huss-Danel K and Harkink R (1983) Can. J. Bot. (in press).
Lalonde M, Calvert HE and Pine S (1981) In Gibson AH and Newton WE, eds, Current Perspectives in Nitrogen Fixation, pp. 296-299, Austr. Ac. Sc., Canberra.
Shipton WA and Burggraaf AJP (1982) Plant Soil 69, 149-161.

SPECIFICITY IN SYMBIOTIC DINITROGEN FIXATION

POSTER DISCUSSION 6
RECOGNITION

RECENT IDEAS ON THE PHYSIOLOGY OF INFECTION AND NODULATION

A. Quispel, J.W. Kijne, A.A.N. van Brussel, E. Pees, C.A. Wijffelman, A.J.P. Burggraaf

Department of Plant Molecular Biology, Botanical Laboratory, State University Leiden, The Netherlands

1. INTRODUCTION

The establishment of an endosymbiosis is the result of a complicated series of interactions for which D.C. Smith (1981) postulated a scheme of four stages: a contact, including events leading to contact, b incorporation into host matrix, c integration into regulatory mechanisms and d development of nutrient flows between (endo)symbiont and host. All stages may contain specific interactions which are only possible between homologous partners. However, if such specific interactions, which contain an element of recognition, are part of an essential step, their failure will prohibit all further stages. This failure will be most dramatic if it occurs early in the total sequence of interactions. These interactions form a very subtle, intricate pattern, for a main part operating at a subcellular level. This makes their analysis extremely difficult. Much information may be derived from comparisons with the deviations in the normal development, when less compatible combinations are studied. The progress in the microbial genetics of Rhizobium has markedly improved possibilities for such comparisons and enabled a further analysis of the interactions during infection and nodulation.

2. GENETIC ANALYSIS

In most fast-growing strains essential genes for specific nodulation (nod genes) and genes for N_2-fixation (fix genes) are on large, sometimes highly transmissable sym plasmids. This has enabled the comparison of the host range of Rhizobium and Agrobacterium strains to which sym plasmids were transferred (e.g. Hooykaas et al. 1981, van Brussel et al. 1982, Wijffelman et al. this volume). Recently has been shown that the genes for host-range specificity are on a 10 kb piece of DNA of a cosmid clone which originates from the R.leguminosarum sym plasmid pRL1JI (Downie et al. 1983). The expression of sym plasmids was different when different bacteria were used as hosts. In E.coli and R.meliloti no expression of pRL1JI encoded nod genes could be observed (Schofield et al. pers.comm., Wijffelman et al.in press),in contrast to the kanamycin resistance of the Tn 5 transposon marker. In A.tumefaciens the sym genes of pRL1JI and of the R.trifolii sym plasmid pRtr5a were expressed up to and including bacterial release from the infection thread. Some multiplication of bacteria inside the nodule cells took place, but no bacteroids were formed and no N_2-fixation was measured. The sym genes of pRL1JI were completely expressed in strains of R.phaseoli and R-trifolii. (see also Pühler 1983).

It is not known at this moment whether the different nodulation behaviour of these strains is caused by non-expression of the sym plasmid genes or because other, non-plasmid, genes are needed. Mixed inoculations of Vicia sativa with two non-nodulating strains, the non-mucoid strain RBL 101, which contains a sym plasmid, and the sym plasmid cured strain LPR 5039, resulted in bacteroid-containing ineffective nodules. Both strains could be re-isolated from the nodules and were non-infective, while no changes in plasmid patterns were observed. This indicates a role of the non-plasmid encoded mucoid in bacteroid formation (van Brussel pers.comm.). A similar conclusion was based on a study of non-mucoid R.trifolii mutants by Djordjevic et al. (pers.comm.).

Such results combined with the difference in infectivity and effectivity towards Vicia sativa and V.hirsuta by some Rhizobium transconjugants (van Brussel et al. 1982) illustrates the subtle interactions between sym plasmids, other parts of rhizobial genome and specific plant functions.

Though the production of a nodule up till and including the stage of bacterial release, needs a series of different phenotypic steps (Vincent 1980), the small 10 kb nod region of pRL1JI appears to contain only few essential genes (Downie et al. 1983). After transposon mutagenesis van Brussel et al. (this volume) could distinguish three classes of nod$^-$ mutants according to their induction of root hair curling (hac) and the induction of thick and short roots on Vicia sativa (tsr). The genes for hac and tsr were mapped in the nod region of pRL1JI by transduction experiments (Pees et al. in prep). Hac$^+$,tsr$^-$,nod$^-$ mutants were found, but no hac$^-$,tsr$^+$,nod$^-$ mutants. Assuming that hac and tsr are on one operon this leads to the conclusion of a hac $\rightarrow$ tsr gene sequence. It is suggested that the tsr gene, which induces the teratological formation of thick and short roots in V.sativa, will play an as yet unknown role somewhere in normal nodulation. A transduction map of the pRL1JI plasmid (Pees et al. in prep) shows the nod region within two fix regions, and at another site a second cluster which contains genes for plasmid transfer (tra) and genes coding for the repression of small bacteriocin excretion (rps), small bacteriocin sensitivity (sbs), medium bacteriocin production (mep) and immunity (mei). Deletion mutants missing the genes rps, sbs, tra and mep, still form effective nodules so that these genes code for functions which are not essential for nodulation (Priem et al. this volume). However, this does not exclude that they have some secondary function e.g. during competition between strains. The gene cluster has a striking resemblance to a part of the A.tumefaciens plasmid pTiC58 studied by Ellis et al. (1982) (Pees et al. this volume). The genes tra, sbs and rps of pRL1JI can be compared with the gene tra, the gene for sensitivity to the bacteriocin agrocine 84 and the gene that prevents the excretion of agrocine 84 on pTiC58. The gene for sensitivity to agrocine 84 codes for a transport system for agrocine 84 and for the opines agrocinopine A and B, which are produced by crown-galls. This production is induced by these agrocinopines. It is tempting to extend this resemblance and to suggest that gene sbs of pRL1JI also codes for a transport system which is induced by an opine produced by root nodules. Such opines could function as selective substrates for Rhizobium strains, as has been observed for Agrobacterium. Indeed Tempé et al. (1982) already succeeded in extracting opine-like substances from alfalfa root nodules.

3. PHYSIOLOGICAL ANALYSIS

It is the challenge for the plant and microbial physiologist to translate the gene functions postulated by the genetical research into biochemical and physiological terms. For that purpose we will follow the four stages in the development of endosymbiosis by Smith (1981) and the sequence of Vincent (1980) for the root nodule symbiosis, extended by Rolfe et al. (1981).

3.1. Contact, including events leading to contact: This stage consists of root colonization (Roc), root adhesion (Roa), root hair branching (Hab), root hair (marked) curling (Hac). We propose to precede this sequence by chemotaxis (Cht) of motile strains (Bowra and Dilworth 1981). Though chemotaxis is not specific (Gaworzewska and Carlile 1982) and not essential under axenic conditions there is a competitive advantage for motile above non-motile strains in co-infection experiments (Ames and Bergman 1981). Most attention has been paid to root adhesion as here the lectin-recognition hypothesis offers a plausible explanation for specific recognition between plant roots and homologous bacteria. We refer to the critical remarks of Robertson et al. (1981) and restrict ourselves to some recent additional information. The most convincing evidence has been obtained with the combination of Trifolium repens cv. Ladino and Louisiana with R.trifolii strain 0403 used by Dazzo and his group (recent review by Dazzo 1981 and Dazzo and Truchet 1983). A new evidence in support of the theory has been added by the observation that a strain of Agrobacterium tumefaciens after incorporation of a sym-plasmid from R.trifolii binds to root hairs of clover, induces root-hair curling and infection thread formation and binds Trifoliin A (Dazzo et al. 1983). With other plant-Rhizobium combinations besides much supporting evidence conflicting observations have been reported. Here we must distinguish between observations where non-specific root hair binding (e.g. Solheim 1983) or lectin-binding of heterologous Rhizobium strains (Pueppke et al. 1980, Wong, 1983) raise doubt to the role of lectins as specific recognition system and those observations where the role of lectins in root adhesion as such is disputed. Observations about Glycine varieties which lack the 120.000 D soybean lectin (SBL) (Pull et al. 1978) but still can be nodulated are no longer conflicting since more sensitive methods still show an albeit low content of SBL (Tsien et al. 1983). The demonstration of another lectin with a different hapten-specificity (4-0-methyl-D-glucuronic acid) (Dombrinck-Kurzman 1983) explains the infectivity of strains which lack galactose (a hapten for the 120.000 D SBL) but do have 4-0-methyl-D-glucuronic acid in their capsular polysaccharides (Dudman 1978).

It is generally accepted that lectin receptors may be found in cell wall LipoPolySaccharide (LPS), Capsular PolySaccharide (CPS), or Extracellular PolySaccharide (EPS). Many contradictory results may be explained by the different extraction procedures or the different methods to determine the presence of lectin receptors. Especially the CPS fraction may contain different components while we can not exclude parts of the LPS e.g. 0-antigens which are not found with the usual tests for heptoses of KDO as LPS indicators. Though much attention has been paid to correlations between infectivity and composition of surface polysaccharides like CPS and EPS it is by no means certain that these polysaccharides are still present or functional in the rhizosphere, the less so because in root exudates CPS-degrading enzymes have been found (Dazzo et al. 1982). As to the lectin binding-tests it appeared that binding of FITC-labelled pea-lectin was not specific for R.leguminosarum RNL1, but could be observed with most heterologous Rhizobium strains and even with

A.tumefaciens (Van der Schaal 1983,this volume). However, pea-lectin mediated agglutination was only observed with R.leguminosarum and a pea nodulating R.phaseoli :: pRL1JI transconjugant. This latter reaction was only observed when, due to oxygen limitation in the batch cultures, the logarithmic phase stopped and, after a short growth-interphase, was superseded by a linear growth phase. An enzyme-linked lectin-binding assay (ELBA) showed that exactly at this end of the logarithmic phase a sharp increase was observed of lectin-receptors in a CPS containing extract, though the total amount of CPS per cell decreased. This is an indication that a special fraction in the CPS-containing extract is involved in binding of pea-lectin. At a somewhat later moment of the growth curve a temporary decrease of lectin receptors in the CPS extract coincided with an increase in the EPS of the nutrient solution. It is well possible that this EPS is merely a soluble form of the lectin-binding polysaccharide in the CPS.

In R.trifolii the increase of binding of the Trifolium-lectin Trifoliin A at early stationary phase coincides with an increase in Trifoliin A binding LPS and a change in the content of the hapten-sugars quinovosamine and N acetyl-quinovosamine, (Hrabak et al. 1981). Binding of lectins to LPS has been observed in other Rhizobia as well, though there are no simple correlations between specificity and LPS composition (Zevenhuizen et al. 1980). Kamberger (1979) suggested that extracellular receptors, like those in CPS, are responsible for adsorption of Rhizobia, after which a secundary recognition to LPS triggers further reactions in the infection process. This idea of biphasic absorption has been supported by E.M.studies of the binding of Rhizobium to the root hairs (Dazzo 1981). During a phase I adherence docking of the bacteria is initiated by contact of the fibrillar capsule of R.trifolii with electron-dense globular aggregates on the outer periphery of the clover root-hair wall. This is followed by a phase II attachment in which the bacterial cells are anchored to the root hair by a fibrillar material (cellulose? pili?) (Dazzo and Truchet 1983). As compared to the bacteria, far less attention has been paid to the plants. Plant lectins, originally isolated from seeds, are really present at the root surface. In Trifolium repens immunological methods detected Trifoliin A at the roots, while nodulation-inhibiting concentrations of nitrate reduced the detectable level of Trifoliin A on the clover root (Truchet and Dazzo 1982, Dazzo and Hrabak 1982). An immunolatex technique detected soybean lectin at the root surface (Stacey et al. 1980). By means of a sensitive ELISA technique pea lectins were demonstrated in cell-walls and slime of pea roots (Diaz et al. this volume).

While the lectin-recognition hypothesis is based on a possible role of plant lectins there is all reason to consider a function of bacterial lectins as well. In the study of intestinal infections by E.coli considerable attention is paid to pili with lectin activity as adhering agents to host cells (Gaastra and de Graaff 1982). In batch cultures of R.leguminosarum strain RBL1 the phase of lectin-mediated agglutination was succeeded by a autagglutination, which was prevented by addition of mannose but not by glucose. In the autagglutinates many pili were observed which frequently cross-bridged the distance between the Rhizobia (Kijne et al. 1983 and this volume). A correlation between binding of early log phase cells of R.japonicum to soybean root hairs with the development of pili was found by Vesper and Bauer (1983). It is too early to tell whether bacterial lectins in pili play a role during infection and, if they do, whether they are involved in phase I or phase II.

The next stages root hair curling and root hair branching will be extensively discussed by Bhuvaneshwari (next paper). Root hair branching apprears to be induced by bacterial oligosacharides, perhaps products of host-symbiont interaction. No new evidence is available about the factors inducing root hair (marked) curling, which is found to occur about 24 hours after adherence of bacteria to the root hairs (Turgeon and Bauer 1982). The enclosure of a sufficient amount of bacteria to induce further infection is mostly obtained by root hair curling, but is as well possible between adjacent root hairs (Dazzo et al. 1982).

3.2. Incorporation into host matrix: This stage compromises the sequence: infection (Inf), infection thread growth (Ing) and infection thread branching (Inb), nodule initiation (Noi), nodule development (Nod) and bacterial release(Bar). We propose to add a phase 'infection thread growth' as distinct from the primary infection. Nod is the last phase which in any case has to be developed before a non-effective nodule is formed, e.g. by nod^{+} fix^{-} Rhizobia.

There is no doubt that infection is one of the most decisive and specific steps in the process. Even though the bacterial phase I adhesion, as discussed before, may have specific aspects, the most specific limitation is phase II attachment which triggers infection. (Kamberger 1979). The factors involved in triggering infection are unknown though there are some candidates. We refer to the binding to LPS described by Dazzo et al. this volume), the effects of the oligosaccharides from R. japonicum CPS (Bhuvaneshwari this volume), the periplasmatic cyclic 1-2 glucan from R. trifolii (Abe et al. 1982) or, from the plant side, the factors in a cowpea root exudate studied by Bhagwat and Thomas (1982).

From the work of Callaham and Torrey (1981) we must conclude that infection is not an invagination but a loosening of the structure of the root-hair cell wall. This makes the old controversy about the induction of cell wall degrading enzymes again highly actual. Hubbell (1981) suggested that an interaction of biochemically related and functionally analogous components of the surfaces of root hairs and Rhizobia leads to the control of infection. Though it is not clear why his hypothesis proposes that plant lectins must be enzymes similar to those of Rhizobium, a study of the enzymatic activities and their inducing substrates is essential for our further understanding of both the plant as the bacterial roles. Galactosidase activity has been shown for mung bean lectin (Hankins and Shannon 1978), in Rhizobium the induction has been observed of pectinases (Hubbell et al. 1978) and cellulases (Hubbell pers. comm.). The remarkable structure of the cell wall at the root-hair tip described by Goosen-de Roo et al. (this volume) for pea roots where, contrary to root hairs of non-leguminous species, the fibrillar arrangement appears to start at the outside of the cell wall, might give a clue to the suitability for infection of leguminous plants.

Infections appear to be limited to very young root hair tips, either of those root hairs, which started to develop at the moment of inoculation or in fully developed root hairs after induction of branching (Bhuvaneshwari next paper). It is at this point that the first comparisons with Frankia infections are possible. Normally the pattern of nodulation along the root axis confirms the infection of very young root hairs at the time of inoculation. The infection of root hairs, which must have developed far later may be explained by ripening and germination of spores. If mildly sonicated Frankia is used for inoculation mature root hairs are infected. This might be explained if sonication released substances which stimulate root hair branching (Burggraaf et al. this volume).

3.3. Integration into regulatory mechanisms: This is one of the most essential stages for our understanding of an endo-symbiosis. In phytopathogenic infections the reactions after penetration of the parasite are characteristic for a more-or-less severe disregulation. Elicitor-induced synthesis of phytoalexins and hyper sensitive reactions are common defense mechanisms.

Though R. leguminosarum is sensitive to the pea phytoalexin pisatin, the concentration of pisatin in the young nodules is so low that it could not have any effect. In old nodules a role during senescence might be possible (Van Iren et al. this volume). The most typical aspect of the nodulation process is not a dis-regulation but a re-regulation so that a functional organ is formed in which the bacteroids functions as N_2-fixing organells. Bacteroid development (Bad) might be the result of the chelating effect of succinate (Urban and Dazzo 1982). Nitrogen fixation (Nif) and complementary functions (Cof) are the result of different regulations. At the bacterial side the derepression of sym plasmid genes (Krol et al. 1982) among which the derepression of the nif genes and the repression (and inactivation) of both types of glutamin synthetase, besides the stimulation of heme synthesis. On the plant side we find the de-repression of the synthesis of leghemoglobin and other nodulines, the stimulation of the synthesis and/or activity of plant glutamin synthetase and other N-assimilatory enzymes (recent reviews by Verma and Long 1983).

When we compare the root-nodule symbiosis with other endo-symbioses, like symbioses with photosynthesizing algae, we are struck by a general tendency (Smith 1981, Richmond and Smith 1981). In all cases the first assimilatory steps of CO_2-assimilation and N_2-fixation are stimulated, while the subsequent steps of assimilation are repressed and/or inhibited so that the first assimilatory product (sugars, NH_3) are excreted to the host. It is tempting to look after a common mechanism for this general aspect of endosymbioses.

3.4. Developments of nutrient flow between plant and bacteroids: This stage can hardly be separated from the preceding stage. Here as well we see the stimulation and reduction of transport processes, through the complicated membrane system between bacteroids and host cell. As the peribacteroid membrane will be discussed by Robertson, we will restrict ourselves to the bacterial transport. As compared to cultivated bacteria bacteroids no longer have an active transport of sugars (de Vries et al. 1982) though sugars may be used at low oxygen pressures (Trinchant et al. 1981, 1983) or taken up by passive transport (Hooymans et al. this volume), there is a good transport of C_4-Carboxylic acids (de Vries et al. 1982), whose uptake and metabolism are essential for effective N_2-fixation (e.g. Finan et al. 1983), a reduced though still active transport of amino acids with changed competition pattern (H. Staal et al. in prep.), no uptake of (methyl) ammonium but an equilibration by passive or facilitated diffusion leading to efflux of synthesized NH_3 (Dillworth and Glenn 1982). Such differences may be explained by the loss of periplasmatic binding proteins, the (de)repression by plant metabolites or by a direct effect on the cytoplasmic membrane. It is not too speculative to remind here of the intensively studied effects of lectins on animal cell membranes and the presence of lectins inside roots (Gade et al. 1983).

Suggestions about the effect of host factors on the viability of bacteroids in vivo have been related to an old remark of D.C. Jordan in 1962 about selective interference in normal synthesis of cell wall or cytoplasmic membranes or both (Sutton and Paterson 1983). In this context it is suggestive to make a comparison with the endophytic phase of Frankia. Here it appears that the successful isolation of Frankia in many nodules within a short period of only a few weeks is only possible after addition of lipids from an Alnus root extract. While we originally believed that these lipids belonged to the growth requirements of certain strains it now became evident that these lipids enable Frankia to adapt from the endophytic to the free-living way of life (Quispel et al., this volume). An obvious hypothesis is that these lipids will play a role in effecting membrane structures. We suggest that direct research on membrane systems in endosymbionts will be most promising in elucidating essential aspects of endosymbiotic interactions.

The final adaptation of metabolic and transport systems of both host plant and endosymbionts leads to a regulated system in which aminoacids are excreted in the xylem, while the use of carbohydrates functions as a sink for the phloem. In this way the root nodules are integrated in the total transport system of the plant till the onset of senescence.

REFERENCES

Abe M, Anemura A, Higashi S (1982) Plant Soil 64, 315-329

Ames P, Bergman K (1981) J. Bacteriol. 148, 728-729

Bhagwat AA, Thomas J (1982) Appl. Environm. Microbiol. 43, 800-805

Bøg-Hansen TC, Spengler GA (eds) (1983) Lectins Vol. III, Walter de Gruyter, Berlin, New York

Bowra BJ, Dillworth MJ (1981) J. Gen. Microbiol. 126, 231-233

Brussel AAN, Tak T, Wetselaar A, Pees E, Wijffelman CA (1982) Plant Science Letters 27, 317-325

Callaham DA, Torrey JG (1981) Can. J. Bot. 59, 1647-1664

Dazzo FB (1981) J. Supramol. Struct. & Cell Biochem. 16,29-41

Dazzo FB, Hrabak EM (1982) Plant Soil 69, 259-264

Dazzo FB, Truchet GL, Hooykaas P (1983) Abstract P52 in 9th North Amer.Conf. Rhizobium Boyce Thompson Inst. Ithaca New York, p 26

Dazzo FB, Truchet GL, Sherwood JE, Hrabak EM, Gardiol AE (1982) Appl. Environm. Microbiol. 44, 478-490

Dazzo FB, Truchet GL (1983) J. Membran Biol. 73, 1-16

Dazzo FB, Truchet GL, Kijne JW (1982) Physiol. Plant 56, 143-147

Dillworth MJ, Glenn AR (1982) J. Gen. Microbiol. 128-29-37

Dombrick-Kurtzmann MA, Dick WE, Slodki MW (1983) Biochem. Biophys. Res. Comm. 111, 798-803

Downie JA, Hombrecker G, Ma Q-Sh, Knight CD, Wells B (1983) Mol. Gen. Genet. in press

Dudman WF (1978) Carbohydrate Research 66, 9-23

Ellis JG, Murphy PJ, Kerr A (1982) Mol. Gen. Genet. 186, 275-281

Finan TM, Wood JM, Jordán DC (1983) J. Bacteriol. 154, 1403-1413
Gaastra W, Graaf FK de (1982) Microbiol. Rev. 46, 129-161
Gade W, Schmidt EL, Wold F (1983 Planta 158, 108-110
Gaworzewska ET, Carlile MJ (1982) J. Gen. Microbiol. 128, 1179-1188
Hankins CN, Shannon LM (1978) J. Biol. Chem. 253, 7791-7797
Hooykaas PJJ, Brussel AAN van, Dulk-Ras H den, Slogteren GMG van, Schilperoort RA (1981) Nature 291, 351-353
Hrabak EM, Urbano MR, Dazzo FB (1981) J. Bacteriol. 148, 697-671
Hubbell DH (1981) Bio Science 31, 832-837
Hubbell DH, Morales VM, Umali-Garcia M (1978) Appl. Environm. Microbiol. 35, 210-213
Kamberger W (1979) FEMS Microbiol. Lett. 6, 361-365
Kato G, Maruyama Y, Nakamura M (1980) Agric. Biol. Chem. 44, 2843-2855
Kijne JW, Schaal CAM van der, Diaz CL, Iren F van (1983) In Bøg-Hansen TC, Spengler GA (eds.) Lectins Vol III 521-529 Walter de Gruyter. Berlin, New York
Krol AJM, Hontelez JGJ, Kammen A van (1982) J. Gen. Microbiol. 128, 1839-1847
Pueppke SG, Freund TG, Schulz BC, Friedman HP (1980) Can. J. Microbiol. 26, 1489-1497
Pühler A(ed) Molecular Genetics of the plant-bacteria interaction. Springer, Berlin-Heidelberg in press
Pull SP, Pueppke SG, Hymowitz Th, Orf JH (1978) Science 200, 1277-1279
Richmond MH, Smith DC (eds) (1981) Proc. Roy Soc. London Sci B 204, 115-287
Robertson JG, Lyttleton P, Pankhurst CE (1981) In Gibson AH, Newton WE (eds) Current perspectives in nitrogen fixation Austral. Acad. Sci. Canberra 280-291
Rolfe BG, Djordjevic M, Scott KF, Hughes JE, Badenoch Jones J, Gresshoff PM, Cen Y (1981) In AH Gibson, WE Newton (eds) Current Perspectives in Nitrogen fixation pp 142-145 Austral. Acad. Science, Canberra
Schaal CAM van der,Thesis Univ. Leiden 1983
Smith DC (1981) Ber. D. Bot. Ges. 94, 517-528
Solheim B (1983) In Bøg-Hansen TC, Spengler GA. Lectins Vol III p 539-548. Walter de Gruyter, Berlin, New York
Stacey G, Paau A, Brill WJ (1980) Plant Physiol. 66, 609-614
Sutton WD, Paterson AD (1983) Plant Science Letters 30, 33-41
Tempé J, Petit A, Bannerot H (1982) Compt. Rend. Acad. Sci. Paris t. 295 ser. III 413-416
Trinchant JC, Birot AM, Denis M, Rigaud J (1983) Arch. Microbiol. 134, 182-186
Truchet GL, Dazzo FB (1982) Planta 154, 352-360
Tsien HC, Jack MA, Schmidt EL, Wold F (1983) Planta 158, 128-133
Turgeon BG, Bauer WD (1982) Can. J. Bot. 60, 152-161
Urban JE, Dazzo FB (1982) Appl. Environm. Microbiol. 44, 219-226
Verma DP, Long S (1983) Int. Rev. Cytol. Suppl. 14, 211-245
Vesper SJ, Bauer WD Abstract P5 in 9th N. Amer. Rhiz. Conf. Ithaca p.3
Vincent JM (1980) In WE Newton, WH Orme-Johnson Nitrogen fixation Vol II Univ. Park Press Baltimore pp 103-129
Vries GE de, Brussel AAN van, Quispel A (1982) J. Bacteriol.149,872-879
Wong PP (1983) Plant Physiology in press
Zevenhuizen LPTM, Scholten-Koerzelman I, Posthumus MA (1980) Arch. Microbiol. 125, 1-8

PREINVASION EVENTS IN LEGUME/RHIZOBIUM SYMBIOSIS

T.V. BHUVANESWARI
CHARLES F. KETTERING RESEARCH LABORATORY, YELLOW SPRINGS, OHIO 45387.

1. INTRODUCTION

The symbiotic association between bacteria of the genus *Rhizobium* and leguminous hosts is intimate, intricate and well regulated. The process begins on the root surface and follows a specific sequence of interdependent steps which culminate in the formation of nitrogen fixing nodules (Dart 1977; Robertson, Farnden 1980; Vincent 1980; Bauer 1981). A wide variety of extracellular products synthesized by rhizobia in culture have been tested for their role in this course of events; for their ability to induce root hair deformation and enhance infection and nodulation in particular (Sahlman, Fahraeus 1963; Fahraeus, Ljunggren 1967; Ljunggren 1969; Higashi, Abe 1980a; Bauer 1982). In this paper I intend to evaluate the available information on the substances that cause root hair deformation and describe our recent experiments designed to understand the significance of root hair branching in the nodulation of white clover. The importance of *R. japonicum* capsular polysaccharide in the nodulation of soybean will be discussed briefly.

2. ROOT HAIR DEFORMATION - ITS SIGNIFICANCE IN THE SYMBIOTIC INFECTION

Root hair deformation is a consequence of host interaction with the symbiont. Deformed hairs may be branched, lobed, intertwined or curled. Depending on the degree of deformation, Yao and Vincent (1969) characterized the deformations into branching (lateral branches and short out-growths), moderate curling (root hairs curled less than 360°) and marked curling (hair curled 360° or more). Marked curling occurs only when the root hairs interact with live cells of homologous rhizobia. Infections occur frequently (but by no means exclusively) in such markedly curled root hairs. Rhizobia enter these hairs at the point of most acute curling. Recent cytological evidence, however, suggests that infections can originate wherever the bacteria get trapped between closely appressed walls and that such an entrapment is achieved most frequently by the marked curling of root hairs (Napoli, Hubbell 1975; Higashi, Abe 1980b; Callaham, Torrey 1981; Robertson 1981; Turgeon, Bauer 1982).

Root hair branching and moderate curling do not require host interaction with live cells of nodulating bacteria. Such deformations can be induced in root hairs of white clover and alfalfa by exposing the roots to culture fluids of nodulating and in some cases even non-nodulating rhizobia (Sahlman, Fahraeus 1963; Yao, Vincent 1969). Infections do occur in such moderately curled hairs (Nutman 1959; Callaham 1979; Higashi, Abe, 1980b; Dazzo, Truchet 1983). At present the significance of root hair branching and moderate curling in the establishment of the symbiosis is in general poorly understood (Nutman 1959; Sahlman, Fahraeus 1963; Hubbell 1970; Solheim, Raa 1973; Yao, Vincent 1969, 1976; Shimakanova 1978). In the following sections I describe a few of our recent experiments with white clover that indicate that root hair branching is essential for infections to occur in hairs that are developmentally mature at the time of inoculation.

2.2 CHARACTERIZATION OF SUBSTANCES THAT CAUSE ROOT HAIR DEFORMATION

The nature of the substances that cause deformation of root hairs in legume hosts is not established. Hubbell (1970) tested a cell free preparation of the extracellular polysaccharide (EPS) from *R. trifolii* cultures for its infection related biological activity and reported that it induced curling and deformation of *T. fragiferum* (clover) root hairs. The extent of deformation was proportional to the concentration of the crude polysaccharide in the growth medium. EPS prepared from a non-infective strain of *R. trifolii* only induced slight deformation of root hairs. Addition of the EPS from an invasive strain to seedlings inoculated with a non-invasive strain increased the number of deformed hairs in these plants. Erwin and Hubbell (1983) tested the cell free culture fluids, EPS, capsular polysaccharide (CPS), lipopolysaccharide (LPS) and the cell envelopes of *R. trifolii* for root hair deforming activity and found that all the fractions except LPS deformed the root hairs. The EPS preparations of Hubbell (1970) and that of Erwin and Hubbell (1983) contained some protein. An earlier attempt by Fahraeus and Ljunggren (1967) to induce root hair deformation with an EPS preparation, which had been extensively treated to remove protein, was unsuccessful. The role of *R. trifolii* EPS in the deformation of clover root hairs thus remains a controversial issue (Solheim, Raa 1973; Yao, Vincent 1976). Solheim and Raa (1973) purified substances that caused deformation of white clover root hairs from cell free filtrates of *R. trifolii* cultured in the presence of host roots and reported that the deforming activity was associated with two fractions; a heat labile protein fraction and a heat stable polysaccharide or nucleic acid fraction. The presence of both thermolabile and thermostable deforming factors in the culture fluids was later confirmed by Yao and Vincent (1976). These authors also confirmed the earlier observations of Ljunggren (1969) that in addition to high molecular weight substances there are heat stable and dialyzable substances in the culture fluids of *R. trifolii* that induce deformations in clover root hairs. Yao and Vincent (1976), also for the first time, succeeded in separating the factors that induced root hair branching from those that caused moderate curling. These studies suggest that the deformation of root hairs is caused by more than one substance present in culture filtrates.

Results of recent attempts (T.V. Bhuvaneswari, B. Solheim, unpublished) to characterize the low molecular weight root hair branching factors (BF) present in the culture fluids confirms this possibility. The BF was isolated from filtrates of *R. trifolii* cultured for 48h in the presence of white clover seedlings using root hair branching as the bioassay. A brief account of the purification and characterization is given here. A detailed description of the bioassay and the purification procedures will be published elsewhere. Freeze-dried, cell-free filtrates were extracted sequentially with 95% ethanol, 60% ethanol and water. The BF was insoluble in 95% ethanol but highly soluble in 60% ethanol. Some residual activity, however, was retained in the water soluble fraction even after repeated extraction with 60% ethanol. The active substance in the 60% ethanol soluble fraction consistantly voided a DEAE column (pH 7.5, 50 mM Imidazole-HCl). Further separation of the pooled void fractions by gel chromatography (Biogel P4) and HPLC yielded several peaks of biological activity. All the biologically active peaks contained neutral sugars. The branching activity was dialyzable and stable to autoclaving at neutral pH. The active fractions lacked detectable quantities of uronic acids and UV absorbance. These general properties suggest that the root hair branching factor(s) may be oligosaccharides of various sizes (T.V. Bhuvaneswari, B. Solheim, unpublished).

The significance of the heat labile protein fraction in root hair deformation still remains unclear. No attempt has been made to purify the protein from the culture fluids to test for its deformation activity. The possibility that this factor may be an enzyme that degrades the CPS and EPS of R. trifolii to low molecular weight oligosaccharides, needs to be explored.

3.3 THE SIGNIFICANCE OF ROOT HAIR BRANCHING IN WHITE CLOVER INFECTIONS.

White clover root hairs branch in response to inoculation. Infections occur quite frequently in these branched hairs. The question of whether the infections occur in branched hairs before or after they develop the branch was first raised by Fahraeus (1957). Nutman (1959) categorized those infections that originate in the branches or outgrowths of a root hair as lateral and these that originate at the curled apex as apical. He further stated "Although infection at the curled apex of root hair has been described more often in literature than infections of a lateral branch, the latter is commoner..". Our results indicate that infections occur through lateral branches more frequently only in root hairs that are developmentally mature at the time of inoculation (mature root hair, MRH zone) and that apical infections are more frequent in root hairs that differentiate and/or mature after inoculation (immature root hair, IRH zone) (T.V. Bhuvaneswari, B. Solheim, unpublished; Table 1.). The higher frequency of lateral infections in the MRH zone indicates that

Table 1. Infection Pattern (Trifolium repens) inoculated with R. trifolii TA 1.

Infection Thread Origin	Number of Infections	
	MRH Zone	IRH Zone
Branched Hairs (Lateral Infections)	94 (69%)	4 (7%)
Root Hair Apex (Apical Infections)	9 (7%)	21 (38%)
Not Traceable	33 (24%)	31 (55%)
Total Number of Infections Observed in 12 Root Samples	136	56

fully developed root hairs branch before infection. More significantly it also implies that root hairs which are developmentally mature at the time of inoculation do not become infectible until and unless they interact with the rhizobia in the inoculum and develop new growing points. Induction of branching by R. trifolii thus appears to be an early and essential event for the initiation of symbiosis in these hairs.

3.4 EFFECT OF ROOT HAIR BRANCHING FACTOR (BF) ON NODULATION

Exposing clover roots to substances synthesized by R. trifolii in culture has been shown to enhance infection and nodulation (Ljunggren 1969; Higashi and Abe 1980a). Exposing the roots to isolated BF influenced the time course of nodule development only in the MRH zone. Purified BF did not enhance nodulation per se (Fig. 1) (T.V. Bhuvaneswari, B. Solheim, unpublished).

The fractions that voided the DEAE column (Section 3.2) were pooled, freeze dried, dissolved in water, and autoclaved. Plants were treated with 50 μl of a 25 μg/ml (glucose equivalent) solution of this fraction 6 h before inoculation with R. trifolii TA1. Control plants were exposed to sterile water. Nodulation in the MRH and IRH zones of these plants was followed over a period of one week. The results presented in Fig. 1 indicate that nodule development in the MRH zone reached its peak in the BF pretreated plants 2 days before the water treated controls. The time course of nodule development in the IRH zone remained unaltered by this pretreatment.

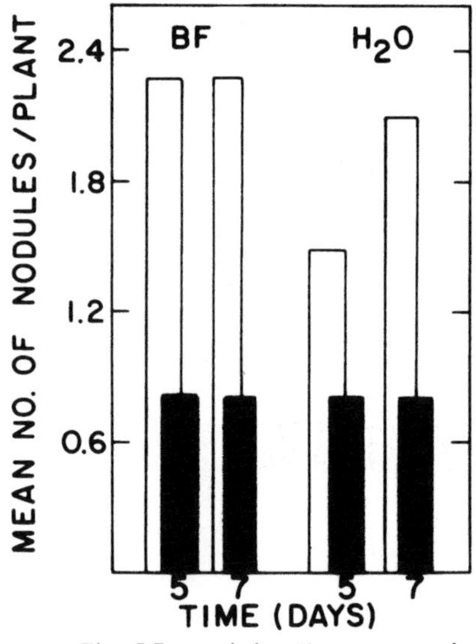

Fig. 1. Effect of BF pretreatment on the nodulation of white clover cv. Regal Ladino. Each bar represents the mean from at least 90 plants. Plants were grown in plastic growth pouches (Bhuvaneswari et al. 1981). Six hours before inoculation each plant was treated with 50 μl of BF or sterile water. Inoculum was prepared from midlog phase cultures of R. trifolii TA 1 (Bhuvaneswari et al. 1981). Plants were inoculated with 100 μl of bacterial suspension containing 1 X 10^7 cells/ml.

☐ - Nodulation in the MRH and IRH zones.

■ - Nodulation in the IRH zone only.

The BF used in these experiments at a concentration of 25 μg·glucose equivalent/ml caused approximately 30% of the mature hairs to branch. Less than 5% of the mature hairs were branched in the control plants. At 6.3 μg/ml, root hair branching and the effect on the nodulation were reduced. Neither of these activities were detectable when plants were pretreated with 1.6 μg/ml of the BF (data not presented). Based on these results it is possible to conclude that nodules developed sooner in the MRH zone of the BF pretreated plants because the mature root hairs of these plants were already branched (and hence were infectible) at the time of inoculation; and that in control plants, nodule development was delayed in MRH zone because the root hairs in this zone became branched and infectible only after they interacted with the BF generated by the rhizobia present in the inoculum (Bhuvaneswari et al. 1981; T.V. Bhuvaneswari, B. Solheim, unpublished).

4.1 ROLE OF CAPSULAR POLYSACCHARIDE IN SOYBEAN/R. JAPONICUM SYMBIOSIS

Capsular polysaccharide (CPS) synthesized by many strains of R. japonicum binds to a lectin present in the seeds and roots of soybean (Bal et al. 1978; Calvert et al. 1978; Tsien, Schmidt 1981; Gade et al. 1981; 1983). Bohlool and Schmidt (1974) were the first to postulate that the binding of soybean lectin to *R. japonicum* cell surface molecules could determine the host specificity in this association. Since then the binding of *R. japonicum* CPS to soybean lectin has become well documented (Mort, Bauer 1980, 1982; Tsien, Schmidt 1981), but the significance of this interaction in determining host specificity still remains controversial (Pull et al. 1978; Schmidt 1979; Bauer 1981; Bhuvaneswari 1981; Gade et al. 1983; Graham 1981; Robertson 1981; Pueppke 1983; Tsien et al. 1983). Based on the lectin mediated model of attachment proposed by Dazzo and Hubbell (1975), Stacey et al. (1980) suggested that the attachment of *R. japonicum* to soybean root hairs may involve the binding of CPS to the lectin molecules present on the surface of the root. Law et al. (1982) have shown that mutants of *R. japonicum* that lack the lectin binding capsules can adhere to soybean roots as well as strains which do form capsules. These mutants, however, were reported to be less efficient than the capsulated parent in nodulating soybean cultivar Williams. Subsequent studies in our laboratory have revealed a strong correlation between the ability of various *R. japonicum* strains to synthesize the lectin binding CPS in vitro and their efficiency in nodulating soybean (Bhuvaneswari et al. 1983; T.V. Bhuvaneswari, W.D. Bauer, unpublished). In addition, we have recently found that it is possible to increase the efficiency of *R. japonicum* cells by supplementing the inoculum with either a crude preparation of the CPS or with highly purified preparations of its oligosaccharide components.

4.2 EFFECT OF *R. JAPONICUM* CPS AND ITS COMPONENTS ON SOYBEAN NODULATION.

Susceptibility to *Rhizobium* infection is developmentally regulated in soybean root cells (Bhuvaneswari et al. 1980; Pueppke 1983). Consequently at any given time, *Rhizobium* infections can occur only in a small segment of the root. The cells in this area remain susceptible to infection for only a brief period of 4-6 h. The time period during which these root cells stay infectible is inadequate for bacterial cells present in the inoculum to proliferate significantly because most *R. japonicum* strains possess a doubling time of 8-10 h in culture. The number of nodules that develop in the zone infectible at the time of inculation thus accurately reflects the efficiency of the cells in the initial inoculum in initiating these nodules (Bhuvaneswari et al. 1983). We measured the efficiency of *R. japonicum* strain 3I1b 138 to nodulate soybean cultivar Williams in the presence and absence of a crude preparation of *R. japonicum* CPS by measuring nodulation in the initially infectible zone.

To obtain a crude preparation of CPS, a cell free bacterial exudate (BE) of *R. japonicum* was prepared as described in Bhuvaneswari et al. (1981). The procedure solubilizes the CPS of *R. japonicum* cells. CPS is thus likely to be a major component of the BE used in these experiments. Soybean plants were grown in plastic growth pouches and the zone infectible at the time of inoculation marked as described in Bhuvaneswari et al. (1980). Plants were inoculated with 250 µl of inoculum (containing 10^3 cells, ml^{-1}) prepared either in water or in BE (exudate from 4 X 10^8 cells, ml^{-1}). The nodules that developed in the marked infectible zone of the roots were counted one week after inoculation. The results are presented in Table II.

Plants that were exposed to inoculum prepared in BE developed more nodules in the infectible zone than the plants than were inoculated with cells prepared in water. We find that it is critical to use a suboptimal dose of inoculum to see this additive effect of the BE on nodulation. The active fraction in the BE is most probably the solubilized CPS or its components because addition of 5 μg per plant of the purified oligosaccharide fragments of the CPS (Mort, Bauer 1982) with the inoculum also increased nodulation in the infectible zone of the root (Table II).

Table II.

Experiment #	Treatment[a]	Average No. of Nodules/Plant in the Infectible Zone[b]
1	Inoculum in water	1.96
	Inoculum in BE	3.18[c]
2	Inoculum in water	1.70
	Inoculum in oligosaccharide solution (20 μg/ml)	2.57[c]

[a]Inoculum, R. japonicum strain 3I1b138. Viable cell count in the inoculum. 1.5 X 10^3/ml in Expt 1 and 2.5 X 10^3/ml in Expt 2.
[b]Average of at least 55 plants/treatment.
[c]Statistically significant from respective controls at 99% confidence level.

5. SUMMARY AND CONCLUDING REMARKS

From our studies on white clover we conclude that 1) the root hair branching is a necessary initial event in the infection of white clover root hairs that are fully developed at the time of inoculation and that 2) the substances that induce root hair branching and infectibility in mature root hairs may be oligosaccharides of various sizes.

The origin of the BF is not known at present. The BF is recovered at high concentrations only from filtrates of R. trifolii cultured in association with its host (Solheim, Raa 1973). It is thus possible that the substances which cause branching are products of host-symbiont interaction. The recent report of Dazzo et al. (1982) that white clover root exudates contain enzymes capable of degrading the CPS of R. trifolii may be significant in this regard.

Our results on soybean suggest that the CPS of R. japonicum plays some role in the initial interactions leading to nodule formation in soybean. How the CPS and its components influence the soybean root cells and whether their biological activity is related to their lectin binding property is not known at present. Future experiments will be directed towards answering these questions.

6. ACKNOWLEDGEMENTS

I am grateful to Drs. D.D. Baker, W.D. Bauer, B. Solheim and S. J. Vesper for critically reading this manuscript and to Ms S. Kannagi for expert technical assistance. The culture of R.japonicum 138 and R. trifolii were obtained from Dr. D. Weber of USDA, Beltsville, Maryland and Dr. A.H. Gibson of CSIRO, Australia, respectively. Purified preparations of R. japonicum oligosaccharide were a generous gift from Dr. A.J. Mort, Department of Biochemistry, OSU, Stillwater, Oklahoma. Work presented here was supported in part by a grant from the National Science Foundation (PFR 7727269) and in part by a grant from the Norwegian Agricultural Research Council.

7. REFERENCES

Bal AK, Shantaram S and Ratnam S (1978) J. Bacteriol. 133, 1393-1400.
Bauer WD (1982) in Asada Y et al., eds, The Physiological and Biochemical Basis of Plant Infection, Japan Sci. Soc. Press, Tokyo, Springer Verlag, Berlin.
Bauer WD (1981) Ann. Rev. Pl. Physiol. pp. 407-448.
Bhuvaneswari TV (1981) Econ. Bot. 35, 204-223.
Bhuvaneswari TV, Bhagwat AA and Bauer WD (1981) Plant Physiol. 66, 1044-1049.Bhuvaneswari TV, Mills K, Crist DK, Evans WR and Bauer WD (1983) J. Bacteriol.
Bhuvaneswari TV, Turgeon BG and Bauer WD (1980) Plant PHysiol. 66, 1027-1031.
Bohlool BB and Schmidt EL (1974) Science 185, 269-271.
Callaham DA (1979) A Structural Basis for Infection of Root Hairs of _Trifolium repens_ by _Rhizobium trifolii_. MS Thesis, Univ. of Mans, Amherst, 41 pp.
Callaham DA and Torrey G (1981) Can. J. Bot. 59, 1647-1664.
Calvert HE, LaLonde M, Bhuvaneswari TV and Bauer WD (1978) Can. J. Microbiol. 24, 785-793.
Dart PJ (1977) in Hardy RWF and Silver WS, eds., A Treatise on Dinitrogen Fixation, Wiley, New York.
Dazzo FB and Hubbell DH (1975) Appl. Microbiol. 30, 172-177.
Dazzo FB and Truchet GL (1983) J. Membrane Biol. 73, 1-16.
Dazzo FB, Truchet GL, Sherwood JE, Hrabak EM and Gardiol AE (1982) Appl. Environ. Microbiol. 44, 478-490.
Erwin S and Hubbell DH (1983) 9th North American _Rhizobium_ Conference. Abst. No. L26.
Fahraeus G (1957) J. Gen. Microbiol. 16, 374-381.
Fahraeus G and Ljunggren H (1967) in Gray TRG and Parkinson D, eds., Ecology of Soil Bacteria, pp. 396-421.
Gade, W, Jack MA, Dahl JB, Schmidt EL and Wold F (1981) J. Biol. Chem. 256, 12905-12910.
Gade W, Schmidt EL and Wold F (1983) Planta _158_, 108-110.
Graham TL (1981) in Giles KL and Atherley AG, eds., Biology of the Rhizobiaceae, pp. 127-148, Academic Press, New York.
Higashi S and Abe M (1980a) Appl. Environ. Microbiol. 39, 297-301.
Higashi S and Abe M (1980b) Appl. Environ. Microbiol. 40, 1094-1099.
Hubbell DH (1970) Bot. Gaz. 131, 337-342.
Law IJ, Yamamoto Y, Mort AJ and Bauer WD (1982) Planta 154, 150-156.
Ljunggren H (1969) Physiol Plant. Suppl. V, pp. 1-84.
Mort AJ and Bauer WD (1982) J. Biol. Chem. 257, 1870-1875.
Mort AJ and Bauer WD (1980) Plant Physiol. 66, 158-163.
Napoli CA and Hubbell DH (1975) Appl. Microbiol. 30, 1003-1009.
Nutman PS (1959) J. Exp. Bot. 10, 250-262.
Pueppke SG (1983) Can. J. Microbiol. 29, 70-75.
Pull SP, Pueppke SG, Hymovitz T and Orf JH (1978) Science 200, 1277-1279.
Robertson JG (1981) in Gibson AH and Newton EW, eds, Current Perspectives in Nitrogen Fixation, pp. 280-291.
Robertson JB and Farnden KFJ (1980) in Stumpf PK and Conn EE, eds., The Biochemistry of Plants, A Comprehensive Treatise, Vol. 5, pp. 65-113, Academic Press, New York.
Sahlman K and Fahraeus G (1963) J. Gen. Microbiol. 33, 425-427.
Schmidt EL (1979) Ann. Rev. Microbiol. 33, 355-376.
Shimakanova NM (1978) Microbiologia 48, 341-345.

Solheim B and Raa J (1973) J. Gen. Microbiol. 77, 241-247.
Stacey G, Paau AS and Brill WJ (1980) Plant Physiol. 66, 609-614.
Tsien HC, Jack NA, Schmidt EL and Wold F (1983) Planta 158, 128-153.
Tsien HC and Schmidt EL (1981) J. Bacteriol. 39, 1100-1104.
Turgeon BG and Bauer WD (1982) Can. J. Bot. 60, 152-161.
Vincent JM (1980) in Newton WE and Orme-Johnson WH, eds, Nitrogen Fixation Vol. II, pp. 103-129, University Park Press, Baltimore.
Yao PY and Vicent JM (1969) Austral. J. Biol. Sci. 22, 413-423.
Yao PY and Vincent JM (1976) Plant Soil 45, 1-16.

THE POLYSACCHARIDES AND OLIGOSACCHARIDES OF RHIZOBIUM AND THEIR ROLE IN THE INFECTION PROCESS

WILLIAM F. DUDMAN
DIVISION OF PLANT INDUSTRY, CSIRO, G.P.O. BOX 1600, CANBERRA, A.C.T., AUSTRALIA

INTRODUCTION

Recent advances in knowledge about Rhizobium polysaccharides make it essential to review the subject before considering their possible biological role. We now realize that rhizobia are capable of producing a wide range of different saccharides and it is necessary to be aware of these. A representative scheme for fractionating a culture to obtain the various saccharides is shown in Fig. 1.

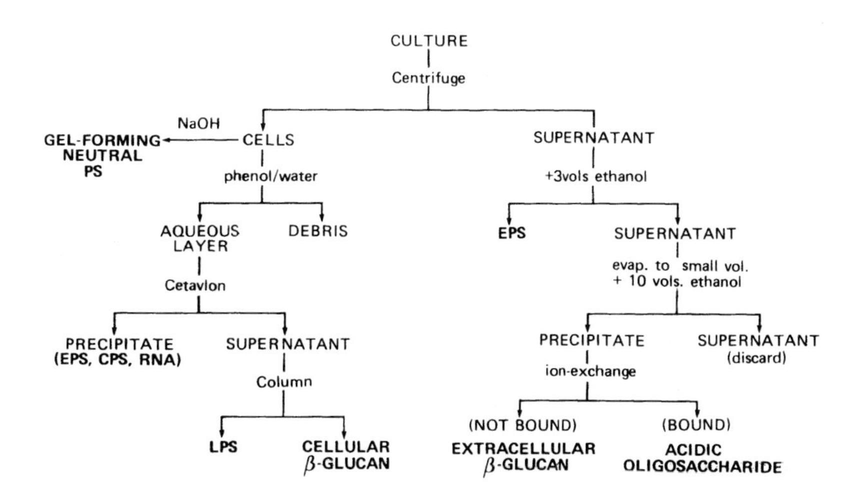

FIGURE 1. Representative fractionation

2. EXTRACELLULAR POLYSACCHARIDES (EPS)

The structures of the EPSs of some two dozen strains of Rhizobium of all species have been determined and interesting relationships can be discerned between them. Among strains of the fast-growing species, there are two groups of EPS based on the type of backbone chain. One group, mainly of R. meliloti and

FIGURE 2.

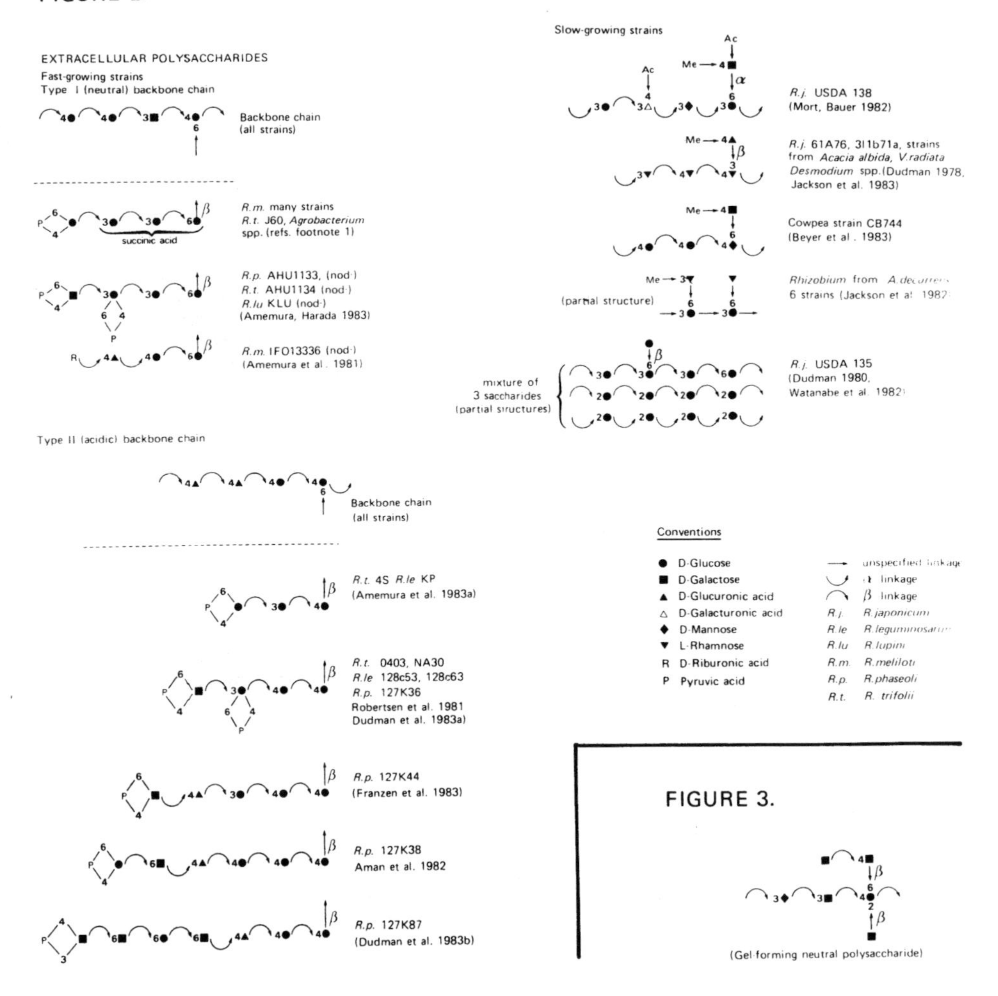

FIGURE 2. Structure of Rhizobium EPSs. The side chains of the repeating units for the fast-growing strains are shown separated from their respective backbone chains.

FIGURE 3. Structure of the gel-forming neutral polysaccharide found in some fast-growing strains

Footnote 1. Jansson et al. 1977; Hisamatsu et al. 1980 Åman et al. 1981; Ghai et al. 1981.

some strains of other species (which are all non-nodulating except for R. trifolii J60, which is Nod$^+$ (A. Amemura, personal communication)) has a common neutral backbone composed of glucose (Glc) and galactose (Gal), with different side-chains attached (Fig. 2). The three side chains are all the same length and are similar to each other in having two acidic groups of one type or another; the first side chain has a succinic acid residue attached by an ester linkage but its location is not known. In addition, two strains of R. meliloti (201, 206) have been found which make more than one EPS. Strain 201 has two polysaccharides, containing respectively Glc, mannose (Man), glucuronic acid (GlcA) (3:3:2) and Glc, Gal, Man and pyruvic acid (PyrA) (4:3:2:1) (Yu et al. 1983).

All strains in the second group (R. leguminosarum, R. phaseoli and R. trifolii) share an identical backbone which has the unusual feature of two adjacent GlcA residues per repeating unit. In this group the side chains are of varying lengths (Fig. 2), terminated with PyrA groups and all (except the shortest side chain) have a second acidic group. One of these structures is common to the EPSs of strains of R. leguminosarum, R. phaseoli and R. trifolii that are all infective on their hosts and specific in their nodulation. Species-linked differences in non-sugar substitutents have been found recently in these otherwise identical polysaccharides (M. McNeil et al., personal communication); this observation shows that differences may be present at a very subtle level.

The EPSs of R. japonicum and the cowpea miscellany are distinctly different from those of the fast-growers. Their repeating units are smaller and the side chains consist of single sugar residues, which are methylated (Fig. 2). R. japonicum USDA135 is unique among rhizobia in seeming to produce only glucans and yet it is effective on a wide range of soybean cultivars. Other strains of R. japonicum and the cowpea miscellany are known which produce EPSs containing sugars different from those above but nothing is known of their structures (Kennedy 1980; Footrakul et al. 1981).

3. ACIDIC OLIGOSACCHARIDES

Oligosaccharides of the same size and structure as the single repeating units of the respective EPS have been found to be produced by all strains of R. meliloti and by the strains of R. trifolii and R. lupini with Type I EPSs (Fig. 2) (Amemura, et al. 1983b; Zevenhuizen, van Neerven 1983). These acidic oligosaccharides are secreted into the medium in substantial amounts (0.3 - 1.6 g/L) and are present in the culture supernatants (Fig. 1). Although the sugar moieties are identical with the repeating units of the EPSs, not all the oligosaccharide molecules are fully substituted with pyruvyl or succinyl groups.

4. EXTRACELLULAR CYCLIC (1→2)-β-D-GLUCANS

These glucans, first discovered in culture supernatants after precipitation of the EPSs (York et al. 1980) appear to be produced by all strains of the fast-growing species, in amounts ranging from 0.1 - 0.4 g/L (Zevenhuizen 1981; Amemura et al. 1983b). Little is known of their occurrence among the slow-growing rhizobia. Cultures contain mixtures of these glucans composed of from 17 to 23 glucose residues (McNeil et al. 1982); the distribution of glucans of various sizes in cultures of different strains is strain-dependent and appears to be correlated with the structure of the EPS of the strain (Hisamatsu et al. 1983).

5. INTRACELLULAR CYCLIC (1→2)-β-D-GLUCANS

Similar glucans are present also in the cell walls of fast-growing rhizobia (Planqué, Kijne 1977; Zevenhuizen, Scholten-Koerselman 1979). Some strains which produce intracellular glucans also secrete them into solution (Zevenhuizen, van Neerven 1983); these give identical size distribution chromatograms upon HPLC analysis (Hisamatsu et al. 1983).

6. LIPOPOLYSACCHARIDES (LPS)

Rhizobium LPSs are unusual in containing uronic acids, fucose, rhamnose, methylated 6-deoxyhexoses, methylated hexosamines and methylated heptoses (Carlson et al. 1978, Planqué et al. 1979; Zevenhuizen et al. 1980). The uronic acid in *R. leguminosarum* LPS has been identified as GalA (Carlson, Lee 1983). The LPSs are the most heterogeneous of all the various saccharides produced by rhizobia. In addition, the LPS produced by single strains of fast-growing strains appear to be mixtures of incomplete LPS (lacking the O-antigen component) and a heterogeneous range of complete LPS molecules varying either in the number of O-antigen repeating units attached or in the degree of aggregation (Carlson 1983).

7. GEL-FORMING POLYSACCHARIDE

A completely new type of polysaccharide has been found in *R. leguminosarum*, *R. trifolii* and *R. phaseoli* (total of 8 strains) (Zevenhuizen, van Neerven 1984) and may be ubiquitous among fast-growing rhizobia. This neutral polysaccharide is not soluble and is extracted in *M* NaOH at room temperature or in water at 100°C. It is present in large amounts in some strains: 1 g of washed cells yielded 0.2 - 0.3 g of this polysaccharide. Its structure is shown in Fig. 3. The polysaccharide forms firm gels in water at concentrations as low as 0.2 per cent.

8. POLYSACCHARIDES AND SYMBIOTIC INFECTION

There is contradictory evidence in the literature for and against the involvement of *Rhizobium* polysaccharides in infection (e.g. Sanders et al. 1978; Napoli, Albersheim 1980; Rolfe et al. 1981; Sanders et al. 1981; Chakravorty et al. 1982; Law et al. 1982; Raleigh, Signer 1982) and it will be difficult to be certain that the evidence is more than circumstantial until the molecular basis for the involvement is understood. In assessing results obtained using mutants with impaired EPS production, it is important to establish the nature of the saccharide being detected: e.g. the small amount of anthrone-positive material produced by the non-nodulating mutant Exo-1 of *R. leguminosarum*, previously presumed to be EPS, has been found to be extracellular LPS and low-MW glucan (Carlson, Lee 1983). However, non-nodulating mucoid strains capable of copious EPS production, and conversely, non-mucoid strains producing no detectable EPS and which are yet capable of nodulation, are known.

Interactions with lectins are an obvious starting point in discussing the role of saccharides. This topic is well known and has been dealt with in recent reviews (e.g. Bauer 1981; Dazzo 1981; Graham 1981; Schmidt 1979) and can be discussed only briefly here. The decreased ability of cells of R. japonicum 110 and 138 to bind soybean lectin when cultures become older has been correlated with increased methylation of the Gal side chains of their EPSs (Mort, Bauer 1980). A similar increase in methyl content of EPS with culture age has been found also with rhizobia from Acacia decurrens (Jackson et al. 1980) but its effect on lectin binding was not reported. Changes in the lectin-binding properties of R. trifolii 0403 EPS at different stages of culture age are related to changes in pyruvyl and acetyl substitution (Sherwood et al. 1983; see also Cadmus et al. 1982); different abilities to react with clover lectin in the LPS isolated from cultures of this strain at mid-log and early stationary phases are also correlated with differences in composition (Hrabak et al. 1981). Such changes help to account for the transient nature of the lectin-binding properties of rhizobia.

The discovery of a new lectin specific for 4-O-methyl-D-glucuronic acid (4-Me-GlcA) in every variety of soybean examined (Dombrink-Kurtzman et al. 1983) is good support for the lectin hypothesis and for the involvement of the EPSs which contain this component. Other Rhizobium strains producing EPSs containing 4-Me-GlcA have been isolated from Acacia albida, Vigna radiata and Desmodium spp., and have been found to nodulate soybeans effectively (Jackson et al. 1983). When considering mechanisms involving lectins, it is important to remember that lectin-mediated binding is a threshold phenomenon dependent on the density of both the lectin and sugar/receptor molecules on the surfaces of the cells in question (Rando et al. 1979; Weigel et al. 1979). When the density of either reactant is less than the required threshold level, binding does not occur despite the presence of many binding sites.

It is difficult to see a role in specificity for the β-glucans but there is evidence that they influence nodulation. The unusually enhanced nodulating ability of two mutants of R. japonicum with diminished EPS production was correlated with the presence of an additional saccharide containing mostly or only glucose (Law et al. 1982). The most direct demonstration of the activity of glucans was by Abe et al. (1982) who used a purified preparation of β-glucan from R. trifolii 4S and obtained increased infection thread formation in clover seedlings inoculated with R. trifolii cells. EPS and cyclodextrin, separately, also stimulated infection threads but to a lesser degree.

The mechanism by which these glucans might exert their biological effects could be related to their uniform linkage and cyclic structure. Glucans of α-1,4-linkage (cyclodextrins) have a wide range of chemical and biological activities derived from their property of forming inclusion complexes with many types of molecules (Saenger, 1980). Although containing twice as many Glc residues as cyclodextrins, β-glucans may be coiled in such a way as to have cavities similar in size and function as those of cyclodextrins. The nature of the guest molecule for which such a β-glucan cavity is intended is beyond even speculation at present but these glucans have an affinity for the linear acidic oligosaccharides of rhizobia (Amemura et al. 1983b). They may stabilize EPSs in certain configurations.

It is easier to see a role for the EPSs in specificity, through effects in addition to reactions with lectins. The differences in number and distribution of acidic groups among the various classes of EPS (Fig. 2) suggest the possibility in differences in chelating or cation-binding activity and there is some experimental evidence for this (Sømme 1974). The requirement for the acidic EPS to be electrochemically neutral may be relevant to infection when one considers that a feature of infection threads is that they often appear to start where rhizobia are trapped in pockets (Higashi, Abe 1980; Callaham, Torrey 1981; Dazzo et al. 1982). If rhizobia continue to synthesize EPSs when enclosed in such environments, it is conceivable that they affect the distribution of cations on the surfaces of the plant cells through the synthesis of new acidic groups requiring counter-ions, the closest source of which are the counter-ions of the GalA residues of the cell wall pectin. The withdrawal of cations from membranes which require cations for dimensional stability may result in altered permeability (Young et al. 1982) and ultimately in penetration. Regions of the root surface where the pectic portions of the walls are methylated would be less susceptible to this effect (Rao et al. 1982) and this could be one factor in the transient and localized nature of the susceptibility of roots to infection. For this hypothetical mechanism to have any relevance to specificity, there would have to be differences in the responses of root surfaces to the cation-attracting activities of the various types of EPS -about which nothing is known.

The ion-binding properties of the EPSs of R. japonicum and cowpea strains would be much less than those of the others; some EPSs (from CB744 and USDA135) are neutral, and no such mechanism can be invoked for these. The most striking features of this group of EPSs (except for the all-glucan 135) is the presence of the methyl group on the side chains, which would contribute some hydrophobic character to the EPS. The significance of this is unclear but biological activity in another group of bacterial polysaccharides with methyl groups is related to their ability to assume conformations with hydrophobic cavities and thereby form inclusion complexes with hydrophobic molecules (Bergeron et al. 1975). It may be significant that the slow-growing rhizobia produce EPSs with this character and concomitantly do not appear to produce or to require β-glucans, which are postulated to also have the ability to form inclusion complexes.

But perhaps polysaccharides play a less important part than we tend to give them. By transposon mutagenesis, Hooykaas et al. (1981) transferred the Sym plasmid of R. trifolii into Agrobacterium tumefaciens, giving cells of the transconjugated strain the ability to form ineffective nodules on clover roots. No differences could be detected between the EPS of the parent or the transconjugated A. tumefaciens strains (Dudman, unpublished). Therefore, if any changes occurred to the saccharides as a result of insertion of the Sym plasmid, it must have been to one of the fractions other than the EPS. The role of any Rhizobium poly- or oligosaccharide must be regarded as being still open to speculation.

9. REFERENCES

Abe M, Amemura A and Higashi S (1982) Plant Soil 64, 315-324.

Aman P, Franzén L-E, Darvill JE, McNeil M, Darvill AG and Albersheim P (1982) Carbohydr. Res. 103, 77-100.

Aman P, McNeil M, Franzén L-E, Darvill AG and Albersheim P (1981) Carbohydr. Res. 95, 263-282.

Amemura A and Harada T (1983) Carbohydr. Res. 112, 85-93.

Amemura A, Harada T, Abe M and Higashi S (1983a) Carbohydr. Res. 115, 165-174.

Amemura A, Hisamatsu M, Ghai SK and Harada T (1981) Carbohydr. Res. 91, 59-65.

Amemura A, Hisamatsu M, Mitani H and Harada T (1983b) Carbohydr. Res. 114, 277-285.
Bauer WD (1981) Annu. Rev. Plant Physiol. 32, 407-449.
Bergeron R, Machida Y and Bloch K (1975) J. Biol. Chem. 250, 1223-1230.
Beyer R, Melton LD and Kennedy LD (1983) Carbohydr. Res. 111, 195-203.
Cadmus MC, Burton KA and Slodki ME (1982) Appl. Environ. Microbiol. 44, 242-245.
Callaham DA and Torrey JG (1981) Can. J. Bot. 59, 1647-1664.
Carlson RW (1983) J. Bacteriol. in press.
Carlson RW and Lee R-P (1983) Plant Physiol. 71, 223-228.
Carlson RW, Sanders RE, Napoli C and Albersheim P (1978) Plant Physiol. 62, 912-917.
Chakravorty AK, Zurkowski W, Shine J and Rolfe BG (1982) J. Mol. Appl. Genet. 1, 585-596.
Dazzo FB (1981) J. Supramol. Struct. 16, 29-41.
Dazzo FB, Truchet GL and Kijne JW (1982) Physiol. Plant. 56, 143-147.
Dombrink-Kurtzman MA, Dick WE, Burton KA, Cadmus MC and Slodki ME (1983) Biochem. Biophys. Res. Commun. 111, 798-803.
Dudman WF (1978) Carbohydr. Res. 66, 9-23.
Dudman WF (1981) In Gibson AH and Newton WE, eds, Current Perspectives in Nitrogen Fixation, p. 427, Austr. Ac. Sc., Canberra.
Dudman WF, Franzén L-E, Darvill JE, McNeil M, Darvill AG and Albersheim P (1983a) Carbohydr. Res. 117, 141-156.
Dudman WF, Franzén L-E, McNeil M, Darvill AG and Albersheim P (1983b) Carbohydr. Res. 117, 169-183.
Footrakul P, Suyanandana P, Amemura A and Harada T (1981) J. Ferment. Technol. 59, 9-14.
Franzén L-E, Dudman WF, McNeil M, Darvill AG and Albersheim P (1983) Carbohydr. Res. 117, 157-167.
Ghai SK, Hisamatsu M, Amemura A and Harada T (1981) J. Gen. Microbiol. 122, 33-40.
Graham TL (1981) Int. Rev. Cytol., Suppl. 13, 127-148.
Higashi S and Abe M (1980) Appl. Environ. Microbiol. 40, 1094-1099.
Hisamatsu M, Abe J, Amemura A and Harada T (1980) Agric. Biol. Chem. 44, 1049-1055.
Hisamatsu M, Amemura A, Koizumi K, Utamara T and Okada Y (1983) Carbohydr. Res. in press.
Hooykaas PJJ, van Brussel AAN, den Dulk-Ras H, van Slogteren GMS and Schilperoort RA (1981) Nature (London) 291, 351-353.
Hrabak EM, Urbano MR and Dazzo FB (1981) J. Bacteriol. 148, 697-711.
Jackson LK, Keyser HH, Burton KA and England RE (1983) Am. Chem. Soc. 17th Great Lakes Regional Meet. (St Paul) in press.
Jackson LK, Slodki ME, Cadmus MC, Burton KA and Plattner RD (1980) Carbohydr. Res. 82, 154-157.
Jackson LK, Slodki ME, Plattner RD, Burton KA and Cadmus MC (1982) Carbohydr. Res. 110, 267-276.
Jansson PE, Kenne L, Lindberg B, Ljunggren H, Lönngren J, Rudén U and Svensson S (1977) J. Am. Chem. Soc. 99, 3812-3815.
Kennedy LD (1980) Carbohydr. Res. 87, 156-160.
Law IJ, Yamamoto Y, Mort AJ and Bauer WD (1982) Planta 154, 100-109.
McNeil M, Darvill AG, Spellman MW, Sharp JK, Waeghe TJ, Dudman WF, Franzén L-E, Aman P, York WS, Nothnagel E, Lau JM, Valent BS, Albersheim P and Dell A (1982) Int. Carbohydr. Symp. 11th, Vancouver, Abst. II-1.
Mort AJ and Bauer WD (1980) Plant Physiol. 66, 158-163.

Mort AJ and Bauer WD (1982) J. Biol. Chem 257, 1870-1875.
Napoli C and Albersheim P (1980) J. Bacteriol. 141, 1454-1456.
Planqué K and Kijne JW (1977) FEBS Lett. 73, 64-66.
Planqué K, van Nierap JJ, Burger A and Wilkinson SG (1979) J. Gen. Microbiol. 110, 151-159.
Raleigh EA and Signer ER (1982) J. Bacteriol. 151, 83-88.
Rando RR, Orr GA and Bangerter FW (1979) J. Biol. Chem. 254, 8318-8323.
Rao SS, Lippincott BB and Lippincott JA (1982) Physiol. Plant. 56, 374-380.
Robertsen BK, Aman P, Darvill AG, McNeil M and Albersheim P (1981) Plant Physiol. 67, 389-400.
Rolfe BG, Shine J, Greshoff PM, Scott KF, Djordjevic M, Cen Y, Hughes JE, Bender GL, Chakravorty A, Zurkowski W, Watson JM, Badenoch-Jones J, Morrison NA and Trinick MJ (1981) Proc. North Am. Rhizobium Conf., 8th.
Saenger W (1980) Angew. Chem. Int. Ed. Eng. 19, 344-362.
Sanders RE, Carlson RW and Albersheim P (1978) Nature (London) 271, 240-242.
Sanders RE, Raleigh E and Signer E (1981) Nature (London) 292, 148-149.
Schmidt EL (1979) Annu. Rev. Microbiol. 33, 355-376.
Sherwood JE, Truchet GL, Dazzo FB and Vasse J (1983) Proc. North Am. Rhizobium Conf., 9th, Abstract L24.
Sømme R (1974) Carbohydr. Res. 33, 89-96.
Watanabe T, Kamo Y, Matsuda K and Dudman WF (1982) Carbohydr. Res. 110, 170-175.
Weigel PH, Schnaar RL, Kuhlenschmidt MS, Schmell E, Lee RT, Lee YC and Roseman S (1979) J. Biol. Chem. 254, 10830-10838.
York WS, McNeil M, Darvill AG and Albersheim P (1980) J. Bacteriol. 142, 243-248
Young DH, Köhle H and Kauss H (1982) Plant Physiol. 70, 1449-1454.
Yu N-X, Hisamatsu M, Amemura A and Harada T (1983) Agric. Biol. Chem. 47, 491-498.
Zevenhuizen LPTM (1981) Antonie van Leeuwenhoek 47, 481-497.
Zevenhuizen LPTM and Scholten-Koerselman HJ (1979) Antonie van Leeuwenhoek 45, 165-175.
Zevenhuizen LPTM, Scholten-Koerselman I and Posthumus MA (1980) Arch. Microbiol. 125, 1-8.
Zevenhuizen LPTM and van Neerven ARW (1983) Carbohydr. Res. 118, in press.
Zevenhuizen LPTM and van Neerven ARW (1984) Carbohydr. Res. in press.

10. ACKNOWLEDGEMENTS

The author wishes to thank Drs. P Albersheim, A Amemura, RW Carlson, FB Dazzo, M McNeil, ME Slodki and LPTM Zevenhuizen for their kindness in entrusting him with information prior to publication.

POSTER DISCUSSION 6 RECOGNITION

F.B. DAZZO, Department of Microbiology, Michigan State University, East Lansing, Michigan 48824, U.S.A.

The poster session on Recognition had many contributions of great importance to plant-microorganism symbiosis. The evening discussion session was equally exciting, so much so that it continued for two additional hours beyond the formal conclusion.

Posters were selected for discussion under three main categories: analysis of symbiotically defective rhizobia; dynamics of lectins and lectin receptors; and the activation of the root-hair infection process by *Rhizobium* polysaccharides.

Carlson summarized the differences in carbohydrate compositions of LPS, CPS and EPS of symbiotically defective mutants of *R.trifolii* obtained by Rolfe and colleagues. These studies will give important insight into how these surface polysaccharides play a role in the infection process and also the molecular basis for lectin-polysaccharide interactions with the *R.trifolii*-white clover symbiosis.

Stacey presented exciting results of the phenotypic reversal of a late-nodulation mutant strain of *R.japonicum* HSIII by its binding to soybean lectin present *in situ* in seedling exudate. This effect of soybean lectin was specific and did not occur with cowpea root exudate. This study is significant since it illustrates that the interaction of excreted lectin with the bacteria in the simulated rhizosphere plays a role in the nodulation process.

Ugalde compared two colony variants of *R.meliloti* 102F51 which differed in competitiveness on alfalfa. These variants differed in degree of interaction with alfalfa agglutinin, phage sensitivity, and the exclusive presence in the more competitive strain of a soluble factor required for the *in vitro* incorporation of galactose from NDP-galactose into a surface antigen. This study is significant since mutant analysis of these strains implicated the role of this surface antigen in competition.

There were 3 posters which reported the *de novo* synthesis of *Rhizobium*-binding lectins in roots of white clover (J. Sherwood), pea (C. Diaz), and soybeans (C. Sengupta-Gopalan). In the related clover and pea systems, the lectin is synthesized by roots grown under N-free conditions and with 15 mM KNO_3, but the level of lectin on the root cell walls in the latter case is significantly lower for both legumes. These results are consistent with the previous studies by the Leiden and East Lansing Groups, that the root wall chemistry changes with nitrate supply and binds less lectin under conditions where nitrate severely inhibits root infection. The poster on soybean lectin synthesis is significant since it shows that nodulating soybean lines, which lack this lectin in seed, nevertheless make this lectin in roots.

Kijne showed electron microscopic evidence for three types of piliation in autoglutination of *R.leguminosarum* and presented evidence from growth studies that they are produced under oxygen-limiting conditions. In view of their lectin-like activities, the role of pili in autoagglutination and attachment to plant cells should be thoroughly examined.

Van der Schaal developed a new enzyme-linked lectin binding assay (a modified ELISA) which allowed a high resolution of the transient appearance of pea lectin receptors on *R.leguminosarum*. This poster is very significant since it indicates that the cells in broth culture momentarily cease to grow when they can bind pea lectin during a brief period preceeding early stationary phase.

Several posters independently reported that a variety of polysaccharides and oligosaccharides isolated from *Rhizobium* enhance infection of root hairs of the legume host.

Dazzo presented results which indicated that highly purified trifolii-binding LPS from 2 R.trifolii strains significantly increased root hair infection of white clover when the roots were pretreated with a narrow range of concentration (6-60 nM). This effect was dose-dependent, time-dependent, inhibitory above 60 nM concentration, Rhizobium specific, less apparent with non-lectin binding LPS from R.trifolii, but incapable of compromising the inhibitory effect of 15 mM KNO_3 on root hair infection. This result is significant, since it is the first report of a carefully purified, lectin-binding polysaccharide from Rhizobium shown to play a role in root hair infection.
Solheim demonstrated root hair branching activity with several partially purified oligossacharides isolated from the root exudates of white clover inoculated with R.trifolii. These substances contained various neutral hexoses and also could enhance subsequent root hair infection of clover. Olivares presented evidence that crude extracellular polysaccharides from R.meliloti specifically enhanced infection of alfalfa. At the high concentration of EPS added, root hair adsorption by R.meliloti was inhibited. This work will be easier to interpret once the active material is fractionated to homogeneity, but nevertheless suggests that similar effects may occur in the R.meliloti-alfalfa symbiosis.
The session convenor acknowledged the various speakers and the audience for the lively discussions and reminded them that there were several other significant posters in the 'Recognition' section that should be studied in detail during the rest of the meeting. All-in-all, the area of recognition and the infection process over the last two years has had a very exciting level of development.

STUDY OF THE LECTIN-BINDING CAPSULAR POLYSACCHARIDE OF *RHIZOBIUM TRIFOLII* 0403 USING A BACTERIOPHAGE-DEPOLYMERASE.

M. ABE, J.E. SHERWOOD, AND F.B. DAZZO
DEPARTMENT OF MICROBIOLOGY, MICHIGAN STATE UNIVERSITY, EAST LANSING, MICHIGAN, 48824, USA.

The ability of the capsular polysaccharide (CPS) of *Rhizobium trifolii* 0403 to bind with the clover lectin, trifoliin A, appears transiently with cell culture age (Dazzo et al. 1979). A similar phenomenon has been observed with the lipopolysaccharide of *R. trifolii* 0403 (Hrabak et al. 1981). Previously, a polysaccharide depolymerase had been isolated from the phage lysate of *R. trifolii* 4S (Higashi, Abe 1978) was used to hydrolyze the extracellular polysaccharide to its repeating unit for structural analysis (Amemura et al. 1983). In this study, a polysaccharide depolymerase has been isolated from a phage lysate of *R. trifolii* 0403 to identify the lectin-binding site of the CPS of *R. trifolii* 0403.

MATERIALS AND METHODS *Polysaccharide Depolymerase Isolation.* The phage lysate was obtained by increasing the amount of $CaCl_2$ in B3 broth medium from 30 µM to 700 µM. The enzyme was partially purified by precipitation with ammonium sulfate and DEAE-cellulose column chromatography. Enzyme activity was measured by viscosity decrease in 30 min using CPS from 5-day culture as substrate. *CPS Isolation.* CPS was extracted from cells cultured on B3 agar medium at 30 C with 0.5 M NaCl in phosphate-buffered saline. *Preparation of Oligosaccharide.* The reaction mixture of CPS and enzyme was chromatographed on Bio-Gel P10 with 20 mM Tris-HCl buffer, pH 7.2. *Deacylation of Polysaccharide.* CPS was mixed with 10 mM NaOH for 3 h at room temperature under nitrogen and dialyzed against distilled water. *Lectin-Binding Assay.* Lectin binding activity was assayed by bacterial agglutination inhibition (Hrabak et al. 1981).

RESULTS AND DISCUSSION The molecular weight of the enzyme is about 540,000. The optimum temperature is 40 C and the optimum pH is 5.4. Enzyme activity, using CPS from 5- or 7-day cultures as substrate, is greatly stimulated by the addition of 1 mM Ca^{2+}, but inhibited by 1 mM Mg^{2+}, 1 mM Hg^{2+}, and 0.1 mM EDTA. The CPS from 5- and 7-day cultures showed higher viscosity than the CPS from 3- or 14-day cultures, and had greater decreases in viscosity when treated with the enzyme. The CPS was composed of glucose, galactose, and glucuronic acid, with pyruvyl and O-acetyl substitutions. Chemical analyses indicated that the pyruvate and O-acetyl levels increased in the CPS from 5- and 7-day cultures. The CPS hydrolyzed by the enzyme chromatographed in 3 peaks. However, deacylated CPS was hydrolyzed completely and eluted at the repeating unit fraction. CPS from the 5-day culture and its oligosaccharide showed the highest affinity for trifoliin A. However, the oligosaccharide from deacylated CPS showed lower binding activity than its polysaccharide. From these results, it is assumed that some component and/or structural changes have occurred with the CPS from 5- to 7-day old cells which have high lectin-binding activity, and that some alkaline-labile substitution may affect both the enzyme and lectin-binding site of the CPS.

REFERENCES Amemura A et al. (1983) Carbohydr. Res. 115, 165-174. Dazzo FB et al. (1979) Current Microbiol. 2,15-20. Higashi S and Abe M (1978) J. Gen. Appl. Microbiol. 24,143-153. Hrabak EM et al. (1981) J. Bacteriol. 148,697-711.

INTERACTION BETWEEN *Rhizobium meliloti* AND ALFALFA AGGLUTININ

M. T. ADLER, A. T. DE MICHELI, D. A. SORGENTINI and G. FAVELUKES
Química Biológica I, Facultad de Ciencias Exactas, Universidad Nacional de La Plata, 1900-La Plata, Argentina.

Rhizobium meliloti (Rme) is specifically agglutinated at pH 4.0 by alfalfa agglutinin, and this protein has been proposed as the root lectin hypothetically involved in the mutual recognition of both symbionts (Paau et al., 1981). However, early events in this association which precede infection and nodulation are completely inhibited at pH 4 (Munns, 1968). In this study we have explored whether agglutinin can interact with Rme in conditions which are permissive for preinfection.

Partially purified alfalfa seed agglutinin (AGL) produced a direct agglutination of Rme U-45, U-210 and L5-30 (in divalent cations-saline solution, DCSS) at pH 4.0, but not at pH 5.0 or higher. In another type of approach, bacteria were incubated with AGL at different pH's in the range 4.0-7.0, then washed, resuspended, and finally reincubated at pH 4.0 without AGL. In these conditions agglutination occurred during the second incubation when the pH of the first had been 5.5 or lower. On the other hand, using bacteria pretreated with AGL at pH 5.0 (Where no direct agglutination took place) and washed as before, they became agglutinated if the pH during reincubation was 4.8 or lower. These results indicate that i) stable, soluble AGL-Rme complexes can be formed without agglutination at pH's as high as 5.5, and ii) these complexes are able to agglutinate but only at pH's below 5.0. Thus, simple complexing of AGL to Rme is distinct from agglutination, as shown by its requirement of less acid pH.

Although the interaction of AGL with Rme could not be detected at pH 7.0, a slow but intense agglutination was obtained when the incubation was done for 20 hours in the presence of glucose (5-25 mM) in a pH 7.2, poorly buffered DCSS solution. This effect, which required AGL, was accompanied by a gradual decrease in pH, reaching values below 5.0. Controls without glucose, with fructose instead of glucose, with heat-killed bacteria, with azide, or with increased buffering capacity (in all of which cases acidification was inhibited) did not agglutinate. These experiments indicated that agglutination induced by glucose is dependent on acidification of the medium produced by aerobic metabolism of glucose (Courtois et al., 1979).

Our results show that the interaction of AGL with Rme takes place at pH's as high as 5.5, which is compatible with the minimal pH required for successful development of preinfection in the alfalfa system (Munns, 1968). They also suggest that, in the nearly neutral medium optimal for nodulation of alfalfa, the low pH in the bacterial microenvironment required for the interaction of Rme with AGL, may be generated at some stage during preinfection, by its own metabolism supported by root exudates. In this way, rhizobial activity might turn to be an important factor in driving the system through this putative recognition step towards symbiotic association.

Courtois, B., Hornez, J. P., Dérieux, J. C., (1979) Can. J. Microbiol. 25:1191-1196.
Munns, D. N., (1968) Plant Soil 28:129-147.
Paau, A. S., Leps, W. T., Brill, W. J., (1981) Science 213:1513-1515.

EFFECT OF IRRADIATION: NITROGEN FIXATION AND MUTAGENESIS IN RHIZOBIUM MELILOTI

J. AL-MAADHIDI
AGRICULTURE AND BIOLOGY DEPARTMENT, NUCLEAR RESEARCH CENTRE, BAGHDAD, IRAQ

Nitrogen, phosphorus and potassium are the essential macroelements for plant growth. Nitrogen is the most important one, nevertheless its quantity which is available for plants is not enough to cover crops requirements. It is well known that dinitrogen fixation is carried out by several groups of micro-organisms in symbiotic or free forms (Englund, B. 1977, Steward 1977).

The objective of this study is to induce mutations in *Rhizobium meliloti* by gamma rays characterized by high nodulations and nitrogen fixation.

The methods were carried out as follows:

1. The bacteria was irradiated in all experiments at late exponential phase using gamma cell 220 with a ^{60}Co source.
2. The experiments were performed in a green house under controlled conditions (temperature, humidity, illumination).
3. The irradiated bacteria was screened for mutants using replica plating method.
4. The efficiency of semi - mutant bacteria and irradiated one were examined for nitrogen fixation using tube method (Gibson 1963).
5. Total nitrogen and protein were estimated using Büchi, 320 N_2 distillation unit.

Experiment results showed that:

1. The bacteria which was isolated from root nodules of Alfalfa plants identified morphologically and physiologically as *R. meliloti*.
2. Reinfectivity test showed a 100% infection of seedling roots.
3. The bacteria was sensitive to gamma rays, LD_{50} was ranging between 5 to 10 krad.
4. No auxotrophic mutant was obtained, while incomplete auxotrophic mutants were isolated and identified by weak growth on a minimal medium, comparing with intensive growth on a complete medium (YMA).
5. The treated seedling of Alfalfa plants with the same number of bacterial cells irradiated with 10 krads, incomplete auxotrophic mutants as well as wild type as a control, separately showed no significant differences between treated and untreated plants on fresh weight, number of leaves, number of nodules, dry weight and total nitrogen.
6. Alfalfa seedling treated with bacteria irradiated with 5, 10, 15 krads or with strain 4 of semi - auxotrophic mutant or with wild type showed that 5 krads and strain 4 are the most effective ones in comparison with the wild type.

References

1. Englund, B. 1977 Studies on nitrogen fixation by free-living and Symbiotic blue-green algae using the acetylene reduction technique. Ph.D. thesis, Uppsala.
2. Steward, W.D.P. 1977. Present - day nitrogen - fixing plants. Ambio 6 (2-3): 336-348.
3. Gibson, A.H. 1963. Physical environment and Symbiotic nitrogen fixation, I. The effect of root temperature on recently nodulated *Trifolium subterraneum* L.Plants. Astral. J. Biol. Sci. 16, 28-42.

ROLE OF ABSCISIC ACID IN ROOT NODULATION AND NITROGEN FIXATION IN FABA VULGARIS AND ALNUS GLUTINOSA

A. BANO, J.R. HILLMAN AND C.T. WHEELER
BOTANY DEPARTMENT, GLASGOW UNIVERSITY, GLASGOW G12 8QQ, U.K.

The role of abscisic acid (ABA) in the correlative control of root-nodule development and functioning has been evaluated in *Faba vulgaris* and *Alnus glutinosa*. Endogenous free and conjugated ABA were detected in *Alnus* fruit and *Faba* cotyledons at the time of cotyledon excision as well as in seedlings of both the species at the time of inoculation: the analyses were carried out by GC-MS.

In *Faba*, cotyledonary excision prior to germination modified the radial location of nodules on the primary root such that there was an increase in the number of nodules opposite the protophloem radii.

Exogenous ABA at 10^{-8} mol m^{-3} applied either at the time of inoculation of *Faba* seedlings with *Rhizobium leguminosarum* (WPBS 29d), or at the stage of nodule-primordia development, retarded the growth and development of the host plant, particularly modifying root growth. These effects were apparently dependent on the growth rate of the host plant, the endogenous ABA levels and the % metabolism of the applied ABA. At the nodule-primordia development stage when the endogenous ABA levels were low in the root and cotyledons, and only 6% of the applied ABA was recovered, the growth of the primary root was inhibited by exogenous ABA forming a compact root system with abundant root hairs. At the functional-nodule stage, which was associated with an enhanced ABA content of the root and reduced metabolism of applied ABA, superficial browning of the laterals was observed in response to ABA treatment. ABA treatment significantly reduced N_2-ase activity in *Faba* by delaying initiation of the nodule and bacteroid tissue and by decreasing bacteroid-tissue volume. *Alnus* seedlings, which possessed twice the level of endogenous free ABA required 10-fold more exogenous ABA than *Faba* to exhibit similar effects; nevertheless, ABA treatment led to a stimulation of primary-root elongation and a reduction of lateral-root formation. At later stages there was an increase in the volume of bacteroid tissue per nodule apparently compensating for the initial reduction in N_2-ase specific activity in ABA-treated plants.

Treatment with 10^{-8} mol m^{-3} ABA in *Faba* for 24 h during the active growth phase of nodules stimulated N_2-ase specific activity. No significant differences were observed in N_2-ase either in *Faba* or in *Alnus* fed with ABA for 14d. However, early senescense of lower leaves occurred in response to ABA in both species and an apparent effect on root and nodule permeability was observed in *Faba*. ABA treatment decreased H_2 evolution in air and apparently decreased the total electron flux through N_2-ase, as estimated by H_2 evolution in an Ar/O_2 atmosphere.

An initial stimulation of H_2/D_2 exchange followed by a subsequent decrease was observed in *Faba* nodules treated with ABA.

INFECTIVITY OF PURE FRANKIA CULTURES FROM ALNUS GLUTINOSA

A.J.P. Burggraaf, A. van Vianen, J. van der Linden and T. Tak.
Dept. Mol. Plant Biol., Univ. of Leiden, The Netherlands

The infectivity of Frankia was studied with respect to a) infectivity of hyphal cells, spores and crude homogenates, b) the relation between growth and infectivity and c) the nodulation pattern of A.glutinosa seedlings.

Material and Methods. Frankia strains: isolated from spore positive (LDAgp1, LDAgp1r) or spore negative (LDAgn2r,AGNIg) alder root nodules (Burggraaf and Shipton, 1983). Studied parameters: 1) Most Probable Number (MPN) of infective units (Iu) 2) number of viable units (Vu) on agar plates, 3) Infectivity (If) is reciprocal value of the number of Vu necessary to generate one nodule, 4) nodule zonation pattern along alder roots (according to Bhuvaneswari et al. 1980, 5) Frankia growth - total protein. Inocula: spores collected from 1 - 2 months old agar plates, hyphae from stationary liquid cultures grown at 35°C (inhibition of sporulation). Media: plants - (modified) Jensen agar (Burggraaf et al. 1983) in tubes and petri dishes, Frankia - mineral medium with propionic acid (0.5 gr/l) and NH_4Cl (0.1 gr/l) (Burggraaf and Shipton, 1983).

Results and discussion. a) The infectivity (If) of spore and hyphal suspensions and crude homogenates ranges between 1/100 - 1/2000. Low values ($1/10^5$) are found with LDAgp1 (spores and hyphae) and LDAgp1r (spores) and high values for LDAgn2r (1/11, spores). b) Iu and Vu are well correlated over a 30 day growth batch culture period with LDAgn2r. The maximum of Iu is reached after 2-3 weeks (10^4Iu,10^6Vu). Low values of Iu (10^2) are found with LDAgp1r and AGNIg, Vu reaches 10^5-10^6. LDAgp1 is hardly infective. c) Pure spore and hyphal suspensions show a clear zonation in nodulation. Differences in nodulation pattern arise when crude homogenates are used as inoculum, sometimes encompassing the entire nodulation class interval (160 mm). Mild inoculum sonication (10-20 sec, 90 W) results in a more narrow and proximal situation of the nodulation pattern. This results with LDAgn2r in 35% nodulation in the maturising root hair region (at the time of inoculation).

The broad nodulation zone in crude homogenates is probably due to the irregular release of spores from maturizing sporangia. Thus no antagonistic effect of the plant is to be expected after initial nodulation. Mild sonication releases most of the spores simultaneously. The more proximal pattern of nodulation after mild inoculum sonication could mean that certain particles (immature spores?) bear a different infective potential. This concept can be coupled with the release of certain substances upon sonication (polysaccharides), which stimulate root hair branching and recognition (Bhuvaneswari, 1983).

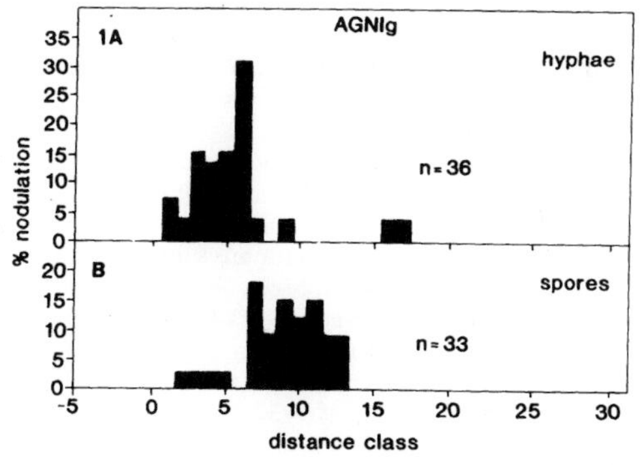

Distance class = 5mm; n = no. root nodules class; 0 = root tip mark (at inoculation); -2mm = shortest root hair mark.

References

Burggraaf AJP and Shipton WA (1983) Can. J. Bot. (in press).
Bhuvaneswari TV et al. (1980) Plant Physiol. 66, 1027-1031.
Burggraaf AJP et al. (1983) Plant and Soil (in press).
Bhuvaneswari TV (1983) this volume.

ANALYSIS OF THE SURFACE POLYSACCHARIDES ISOLATED FROM TRANSPOSON GENERATED SYMBIOTIC MUTANTS OF RHIZOBIUM TRIFOLII.

R.W. Carlson/E.A. Turnbull/J. Duh
Chemistry Dept., Eastern Ill. Univ., Charleston, Ill. 61920 USA

1. INTRODUCTION

Rhizobia contain extracellular (EPS), capsular (CPS), lipopolysaccharides (LPS), neutral and small molecular weight (SmPS) glucans. We have been examining the polysaccharides from transposon generated mutants of *R. trifolii* which are defective in certain symbiotic steps. These mutants were supplied by Dr. Barry Rolfe of The Australian National University.

2. PROCEDURE

All strains were grown to early stationary phase in YEM media. The polysaccharide isolation and analysis were as described (Carlson and Lee, 1983). Combined GC/MS was done at Washington University, St. Louis, Mo.

3. RESULTS AND DISCUSSION

3.1 Mutant 851. This mutant has a Tn5 insert in the sym plasmid and fails to cause root hair curling (hac$^-$). Its EPS has the same composition as 843 EPS and CPS, being gal/glc/glcA=1/4/2. Its CPS is reduced in amount and has a different composition than 843 CPS, being gal/glc/galA+glcA/hept=1/4/5/1. EM studies show that 851 lacks a capsule (B. Rolfe, pers. comm.). NMR of cells show that 851 is lacking in acetyl groups (A. Jones, The Australian National University). Our interpretation is that 851 is defective in the adherence of EPS to the outer membrane.

3.2 The 845 mutant. Strain 845 lacks the sym plasmid and is hac$^-$. The EPS and CPS are normal in amount and composition. However the viscosity of 845 EPS is greatly reduced. The cause of this reduced viscosity is under investigation. EM studies show that 845 lacks a capsule (B. Rolfe, pers. comm.). This conflicts with the fact that we obtain CPS from 845. The reason for the descrepancy in these results is unknown.

3.3 The 845(pBR1AN) strain. This is strain 845 which contains another R. trifolii sym plasmid, is nod$^+$fix$^+$ and produces normal EPS and CPS in amount, composition and viscosity. It should be mentioned at this point that the LPS from all the mutants were altered in that they all contained glcA and galA instead of just galA. Our interpretation is that the LPS change is not important for symbiosis. Genes controlling EPS viscosity are on the added sym plasmid.

3.4 The 437 mutant. This mutant contains a Tn5 insert, not in the sym plasmid, and is nod$^+$fix$^-$. All polysaccharides are altered when compared to 794. The EPS, CPS and LPS of 794 have the same composition as 843. Strain 437 produces 10% of 794 EPS and the major sugar is mannose. This EPS may be a yeast mannan and not a bacterial polysaccharide. The 437 CPS contains LPS sugars, heptose and KDO. There is more KDO in the CPS than the LPS. The LPS from 437 vary from one batch to the next, but several preparations are identical to 794 LPS. Interpretation of these data is difficult until more details are available. It is apparent that the cell surface of 437 has been altered. Analysis of one batch of 437RP4, 437 with a 4kb DNA segment from the wild-type strain, has shown that all the polysaccharides return to normal. We are examining these strains further. The SmPS fractions of all strains are under investigation.

4. REFERENCES

R.W. Carlson and R. Lee. 1983. Plant Physiol. 71: 223. (This work is supported in part by grants from the NSF [PCM-8104481] and EIU.

SPECIFIC ENHANCEMENT OF CLOVER ROOT HAIR INFECTIONS BY TRIFOLIIN A-BINDING LIPOPOLYSACCHARIDE FROM RHIZOBIUM TRIFOLII

F. B. DAZZO, G. L. TRUCHET, and E. M. HRABAK
DEPARTMENT OF MICROBIOLOGY, MICHIGAN STATE UNIVERSITY, EAST LANSING, MICHIGAN 48824 U.S.A.

Lipopolysaccharide (LPS) from Rhizobium trifolii 0403 in early stationary phase binds specifically to the clover lectin, trifoliin A (Hrabak et al. 1981). Kamberger (1979) proposed that the specific interaction of host lectin with LPS of fast-growing rhizobia triggers successful infection of root hairs after attachment via cross-bridging of lectin and capsular polysaccharides. In this study, we examined the effect of trifoliin A-binding LPS on the infection of white clover root hairs by R. trifolii 0403.

PROCEDURE Louisiana Nolin seedlings were germinated for 2 days from surface-sterilized seeds, then incubated with filter-sterilized, purified LPS in Fahraeus medium, rinsed, and inoculated with 2.5×10^7 cells of R. trifolii 0403 in slide cultures. The location of the root tip at the time of inoculation was marked on the microscope slide with a felt-tip pen. Cultures were grown in a growth chamber for 4 days and then scored by phase contrast microscopy for root hair infections (infection threads).

RESULTS AND DISCUSSION Immunofluorescence microscopy showed that LPS from R. trifolii 0403 in early stationary phase (K90) bound to root hair tips. LPS pretreatment significantly increased root hair infections (Table 1). This enhancement was dose- and time-dependent, less apparent with non-lectin binding LPS of exponentially growing cells (K50), restricted to the root region present at the time of exposure to LPS (above the root tip mark), and incapable of reversing the inhibition of infection by NO_3^- and NH_4^+. Root hair infections by R. trifolii 0403 were similarly enhanced by pretreatment of seedlings with LPS (1-5 µg/ml) from the serologically unrelated strain R. trifolii 2S-2, but not by LPS from R. meliloti 102F28 or Escherichia coli 0127:B8. These results support the hypothesis that trifoliin A-binding LPS plays a specific role in white clover root hair infection by R. trifolii.

REFERENCES Hrabak EM, Urbano MR, Dazzo FB (1981) J. Bacteriol. 148, 697-711.
Kamberger W (1979) FEMS Microbiol. Lett. 6, 361-365.

TABLE 1 Effect of LPS on root hair infection by R. trifolii 0403

Pretreatment of Seedlings (LPS source, µg/ml, min exposure)	Root Hair Infections per Plant	
	Above Root Tip Mark	Below Root Tip Mark
None	11.5 + 2.9	0.5 + 0.7
K50, 0.1, 60	12.8 + 3.1	1.4 + 1.3
K50, 1.0, 60	15.8 + 10.3	1.0 + 1.4
K50, 5.0, 60	14.5 + 5.2	0.3 + 0.5
K90, 0.1, 60	35.8 + 11.7	0.5 + 1.0
K90, 1.0, 60	50.5 + 5.6	2.0 + 2.1
K90, 5.0, 1	31.3 + 11.1	1.5 + 1.9
K90, 5.0, 30	59.4 + 5.0	3.5 + 2.1
K90, 5.0, 60	61.3 + 10.4	2.3 + 0.5
K90, 5.0, 30, then 15 mM KNO_3 continuous	0.0 + 0.0	0.0 + 0.0
K90, 5.0, 30, then 1 mM NH_4Cl continuous	0.5 + 0.7	0.0 + 0.0
K90, 5.0, 30, then 15 mM KCl continuous	48.5 + 9.2	2.0 + 1.4

DETERMINATION OF LECTINS IN ROOT SLIME AND ROOT CELL WALL PREPARATIONS OF *Pisum sativum L.*

Clara L. Díaz, Ton J. van Driel, Ineke A.M. van der Schaal and Jan W. Kijne.
Dept. of Plant Molecular Biology, Research Group of Nitrogen Fixation.
Botanical Laboratory, Nonnensteeg 3, Leiden, The Netherlands.

Introduction.

Root lectins are believed to participate in the recognition between *Rhizobium* and its host plant. Among other factors, testing of this hypothesis is difficult due to the very low amounts in which root lectins are produced. A variant of the enzyme-linked immunoassay, ELISA, was used to determine nanograms of pea lectin in the root slime and root cell wall preparations of 4 and 7 days old peas growing in presence or absence of a NO_3^- concentration inhibiting nodulation.

Methods.

Peas seeds (cv Finale) were inoculated with *R. leguminosarum RBL1*. After 21 days growth nodules were not observed on plants growing in gravel soaked with rooting medium suppied with 20 mM NO_3^-. Root fractions were prepared from 4 and 7 do peas growing aseptically with or without 20 mM NO_3^-. Water washed slime, (RS), was concentrated and lyophilized. Cell walls (Buchala, Franz, 1974) were extracted with buffered 1.5 M NaCl, pH 7. The extract was precipitated at 80% $(NH_4)_2SO_4$, dialyzed and lyophilized, (CWE).
Pea seed and root lectins were previously found to give an immunological reaction of identity (Hosselet et al, 1983). Specific rabbit antibodies against chromato focused pea seed isolectin 2 were isolated. A conjugate was produced by attaching alkaline phosphatase to a part of the purified antibodies. The remainder was used to coat microtiter plates (overnight, 4°C). The coating was fixed with 1% glutaraldehyde (10 min, 4°C). The plates were subsequently incubated with known lectin quantities and samples, followed by the conjugate (3h, 37°C, each). The amount of bound lectin was proportional to the reaction of the enzyme with p-nitrophenylphosphate (1h, 37°C), measured at 405 nm, and was linear between 20-100 ng lectin.

Results and Discussion.

	4do	4doN	7do	7doN
mg RS, 100 roots	15	12	46	79
µg lectin/ mg RS	0.47	0.75	0.06	0.07
µg lectin, RS 100 roots	7	9	3	6
mg CWE, 100 roots	3	2.3	17	8
µg lectin/ mg CWE	0.61	0.28	0.50	1.01
µg lectin,CWE 100 roots	1.8	0.65	8.6	8.1

The amount of pea root lectin present in the slime and associated with cell walls of young roots changes in time. In the 4do roots, lectin seems to be secreted into the slime rather than being retained in the cell walls. This trend is reversed as the root ages.
NO_3^-, in a concentration inhibiting nodulation, appears to increase slime lectin production and decreases (2-3 fold) salt extractable cell wall lectin in the 4 do roots. An increase in loosely bound slime lectin might be correlated with a decrease in bacterial attachment to the root cell wall.

References.

Buchala A and Franz G (1974) Phytochemistry 13,1887-1889.
Hosselet M, van Driessche E, van Paucke M and Kanarek L (1983) In Bøg-Hansen TC and Spengler GA, eds, Lectins, Vol 3, pp 549-558, W de Gruyter, Berlin.

ANALYSIS OF SURFACE INFECTIVITY DETERMINANTS IN *RHIZOBIUM*

M.R. ESPUNY/R.A. BELLOGÍN/A. TORRES AND J.E. RUIZ
DEPARTAMENTO DE MICROBOLOGÍA, FACULTAD DE BIOLOGÍA, UNIVERSIDAD DE SEVILLA, SEVILLA, SPAIN

Recent work has focused attention to the study of interactions between host lectins and rhizobial surface carbohydrates. We have examined the possible role of cell wall lipopolysaccharides (LPS) on the infective capability of *Rhizobium*.

R. trifolii RS 169 is an effective strain on *Trifolium alexandrinum* and other clovers. This strain has two megaplasmids (M.W. higher than 200 megadaltons). A non-infective mutant (nod^-) from the RS 169 parental strain has been obtained by treatment with acridine orange. Agarose electrophoresis showed that this mutant (RS 169-NA3) has lost the smallest plasmid of the parental strain. SDS-polyacrylamide gel electrophoresis has been used to study LPS of these strains. The obtained data demonstrate that LPS from the strain RS 169 presents only one band, while that of the mutant (RS 169-NA3) has two additional bands with lower mobility.

Host-range plasmid pJB5JI has been transferred from *R. leguminosarum* T83K3 to the mutant RS 169-NA3. Transconjugants harbouring pJB5JI are effective on peas. LPS from these transcojugants and that from *R. leguminosarum* T83K3 showed a similar polyacrylamide electrophoretic pattern, consisting in two bands.

STRUCTURE OF THE EXOPOLYSACCHARIDES OF THREE STRAINS OF RHIZOBIUM

M. FERNÁNDEZ PASCUAL AND E. CABEZAS DE HERRERA
INST "JAIME FERRÁN", C.S.I.C. JOAQUIN COSTA, 32, MADRID-6 SPAIN

The molecular basis of specificity Rizobium-legume has been objetc of multiple controversy (Wolper and Albersheim 1979).

The carbohydrate of bacterial cell wall can play an importan paper in the process of recognition (Kato et al. 1980). On the other hand, several reports showed that host Rhizobium recognition might be based on the interaction between leguminous lectins and rhizobial cells in the infection process (Bohlool and Schmidt 1974, Mort and Bauer 1980). The lectin hypothesis in Rhizobium-clover symbiosis has been supported by some experimental results presented by Dazzo et al., 1979. However there was not enongh evidence, in other host symbiont systems.

This work shows the results obtained with regard to the chemical composition and structure of the exopolysaccharides of three strains of Rhizobium belonging three different taxonomic groups.

Analysis of the EPS were performed with descending paper and gas liquid chromatography, glucose, galactose and manose were the major components detected and glucuronic and galacturonic acids in a lesser quantity. The precipitation of the EPS with cetyltrimethylammonium bromide shows the acidic nature of these polysaccharides.

After desacilation of the EPS with KOH the presence of volatile and non-volatile acids were investigated by paper chromatography. Structural analysis were obtained by the KBr technique on a Perkin Elmer 457 infrared spectrophotometer. The infrared spectrum is characteristic of a polysaccharide having at 850 cm^{-1} but, it appears a band at 820 cm^{-1} characteristic of -configuration. Periodate oxidation was performed according to Aspinall and Ferrier (1957). The formic acid was titrated with 0.99 mM NaOH. All this demostrated the presence of 1-- 6, 1-- 4 and 1-- 3 bonds in different porcentages in every strain of Rhizobium proportions and the presence of esther and B-glicosyl linkages.

REFERENCES:

ASPINALL, G.O., FERRIER, R.J. (1957). Chem. and Industry 1957, 1216.
BOHLOOL, B.B., SCHMIDT, E.L. (1974). Science, 185, 269-271.
DAZZO, F.G., HUBBELL, D.H. (1975). Appl. Microbiol. 30, 172-177.
KATO, G.Y.M., NAKAMURA, M. (1980). Agric. Biol. Chem. 44, 2843-2855.
MORT, A.J., BAUER, W.D. (1980). Plant. Physiol. 66, 158-163.
WOLPER, J.S., ALBERSHEIM, P. (1976). Biochem. Biophys. Res. Commun. 70, 729-737.

THE ULTRASTRUCTURE OF PEA ROOT HAIR CELL WALLS

L. GOOSEN-DE ROO, A.M. MOMMAAS-KIENHUIS, J.W. KIJNE
DEPARTMENT OF PLANT MOLECULAR BIOLOGY, UNIVERSITY OF LEIDEN,
NONNENSTEEG 3, 2311 VJ LEIDEN, THE NETHERLANDS

Introduction

Most known species and isolates of *Rhizobium* are able to infect developing root hairs of their leguminous host plants. The mechanism of root hair curling and infection thread formation is unknown; however, a rhizobial interference with root hair cell wall metabolism seems obvious. Little is known of the ultrastructure and the composition of leguminous root hairs. Using transmission electron microscopy, we studied the ultrastructure of uninfected developing pea root hairs, with special attention to the cell wall.

Material and methods

Surface sterilized seeds of *Pisum sativum* L. cv. Finale were germinated and the seedlings grown on sterile agar (3%) in Petri dishes, or in sterile coarse gravel in tubes, with use of the N-free medium of Raggio and Raggio (1956), in case supplemented with 20 mM KNO_3.
The zone with the developing root hairs was cut from seven-days old seedling roots, and immediately fixed. The fixation- and embedding-procedure was essentially the same as described by Kijne (1975, fixation method 3) with a reduced glutaraldehyde-fixation time, viz. 2 hrs.

Results and discussion

The general cytological characteristics of growing pea root hairs are similar to those described for root hairs of other plant species (e.g. Belford, Preston, 1961). The ultrastructure of the developing cell wall, however, differs. The fibrillar arrangement appears to start at the *outside* of the cell wall, which is opposite to the situation in other root hairs, in which the so-called fibrillar β-layer develops along the plasmamembrane. The fibrillar structure in pea root hair cell walls presumably is a network in a right-angular arrangement.
Presence of 20 mM nitrate in the culture medium (inhibiting nodulation almost completely) has no influence on the root hair ultrastructure in comparison with root hairs grown in a N-free medium.
A particular cell wall structure in growing leguminous root hairs might directly be correlated with the susceptibility of these hairs to infection by *Rhizobium*.

References

Belford, DS and Preston, RD (1961) J. Exp. Bot. 12, 157-168.
Kijne, JW (1975) Physiol. Plant Pathol. 5, 75-79.
Raggio, N and Raggio, M (1956) Phyton 7, 103-119.

PHENOTYPIC REVERSAL OF A MUTATION IN RHIZOBIUM JAPONICUM CAUSING DEFECTIVE NODULE INITIATION

LARRY J. HALVERSON AND GARY STACEY,
DEPARTMENT OF MICROBIOLOGY, UNIVERSITY OF TENNESSEE, KNOXVILLE, TN, U.S.A.

I. Introduction. Lectins, proteins that bind carbohydrates, have been implicated as important in determining the specificity of the legume-*Rhizobium* interaction. The case for lectin involvement in nodulation would be greatly strengthened if a direct effect of lectin presence on nodulation by *Rhizobium* could be demonstrated. The following data demonstrate just such an effect.

II. Results. The mutants of *Rhizobium japonicum* used in this study are those reported earlier by Stacey et. al., 1982. One mutant, strain HS111, is characteristic of a class of mutants that is slow-to-nodulate producing visisble nodules on inoculated plants 7-8 days later than those produced by the wild-type. Nodulation by mutants of this class was examined using the methods developed by Bhuvaneswari et. al. (1980, 1981) in which nodulation is scored with respect to the area between the root tip (RT) and the smallest emergent root hair (SERH) visible with a 10x dissecting microscope. This is the area in which most of the nodules are normally found.

The wildtype *R. japonicum* strain 3I1B110 when inoculated onto soybean formed nodules with an average distance above the RT mark made at the time of inoculation of 3.5±0.3 mm. In contrast, nodules were formed by mutant strain HS111 at an average distance of 18.4±4.0 mm below the RT mark. These data suggest that strain HS111 is delayed in its ability to initiate nodulation in the RT-SERH zone and must have a period of "conditioning" within the root environment before nodulation can be initiated. To test this, cells of strain HS111 were preincubated in soybean root exudate for various times prior to inoculation onto soybean roots (Fig. 1). By this treatment, the average distance of the nodules formed by strain HS111 was shifted from -18 mm below the RT mark to 0.2±1.2 mm above the RT mark. This difference is statistically significant to a confidence level of $p=0.01$.

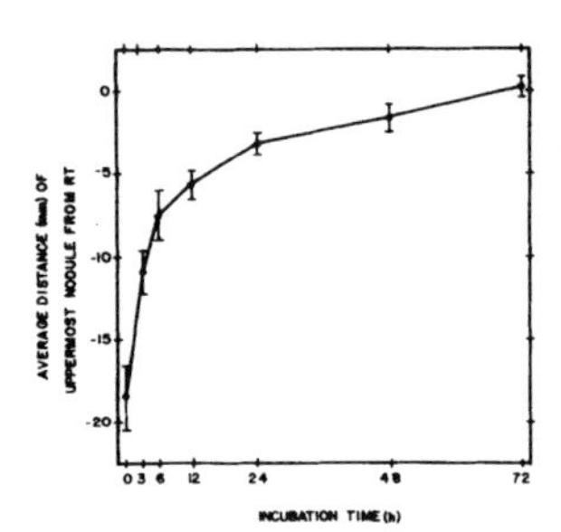

Experiments indicate that the change in the nodulation profile of strain HS111 brought about by incubation in soybean root exudate is due to a phenotypic change rather than a genotypic change.

Treatment of soybean root exudate with trypsin or by boiling for 30 min prior to the addition of cells of strain HS111 prevents the enhancement of nodulation. Table 1 shows the effect of pretreating cells of strain HS111 with soybean lectin prior to inoculation onto plants.

Table 1. Effect of *Rhizobium japonicum* pretreatment with SBL.

Treatment	Position of Uppermost Nod From RT	% Nod Only Below RT	Ave. # of Nod Per Plant
HS111 RE	-1.6 ± 1.9 (SEM)	39	4.7 ± 2.4 (SD)
HS111 PNS/BSA	-14.6 ± 2.4	73	4.1 ± 1.3
HS111 PNS	-16.2 ± 3.2	73	3.3 ± 1.8
HS111 PNS/SBL	-2.9 ± 1.6	39	4.4 ± 2.5

PNS = plant nutrient solution; RE = soybean root exudate; BSA = bovine serum albumin; SBL = soybean lectin

The data indicate that SBL can enhance the nodulation of strain HS111 and is likely to be the active factor in soybean root exudate.

Bhuvaneswari, T.V., Bhagwat, A.A., and Bauer, W.D. 1981. Plant Physiol. 68: 1144-1149.

Bhuvaneswari, T.V., Turgeon, G., and Bauer, W.D. 1980. Plant Physiol. 66:1027-1031.

Stacey, G., Paau, A.S., Noel, K.D., Maier, R.J., Silver, L.E., and Brill, W.J 1982. Arch. Microbiol. 132:219-224.

ANOMALOUS NODULATION OF SUBTERRANEAN CLOVER BY *RHIZOBIUM LEGUMINOSARUM* 1020.

E.M. HRABAK, G.L. TRUCHET AND F.B. DAZZO.
DEPT. OF MICROBIOLOGY AND PUBLIC HEALTH, MICHIGAN STATE UNIVERSITY, E. LANSING, MICH 48824 USA

Some naturally occurring strains of pea-nodulating rhizobia, e.g. *Rhizobium leguminosarum* 1020, also nodulate subterranean clover (Hepper 1978; Hepper,Lee 1979). The purpose of this work was to determine if *R. leguminosarum* 1020 is a pure culture capable of nodulating both peas and subterranean clover and how this anomalous nodulation process differs from normal clover nodulation by *R. trifolii* 0403.

Bacteria used were *R. leguminosarum* 1020 and *R. trifolii* 0403 from Rothamsted Experimental Station, U.K. Pea seeds (line 9888F) were from Canners Seed Corp., IN. Subterranean clover seeds (vars. Clare and Woogenellup) were from M. George, Univ. of CA-Davis. White clover seeds (vars. Ladino and Louisiana Nolin) were obtained commercially. Rhizobia were grown on BIII defined medium.

Culture Purity. *R. leguminosarum* 1020 had a uniform colony morphology on BIII agar before and after passage through peas or clover, and reisolates retained the ability to nodulate both peas and clover. Spontaneous rifampicin resistant (rif^r) mutants of *R. leguminosarum* 1020, each theoretically arising from a single cell, were able to nodulate both peas and subterranean clover. Nodule isolates were still rif^r. Thus, the culture was assumed to be pure.

Nodulation Characteristics. We confirm the reports of Hepper 1978 and Hepper and Lee 1979 that anomalous nodulation of subterranean clover by *R. leguminosarum* 1020 was delayed and occurred primarily at lateral root emergence. Clover inoculated with *R. leguminosarum* 1020 did not reduce acetylene and produced as many nodules as the effective strain, *R. trifolii* 0403. Nodulation by both strains was inhibited by nitrate.

Infection Process. In contrast with *R. trifolii* 0403, *R. leguminosarum* 1020 did not bind the white clover lectin, trifoliin A, either after 6 days incubation in slide cultures or when grown on BIII plates for 2-12 days. *R. leguminosarum* 1020 attached to white clover root hairs only at the background level (2.6 cells/200 µm root hair length after 24 hr). On subterranean clover, root hair deformation with either *R. leguminosarum* 1020 or *R. trifolii* 0403 was seen by day 5. Microscopic studies indicated that *R. trifolii* 0403 infected subterranean clover by formation of infection threads in root hairs. However, this is not an obvious route of infection for *R. leguminosarum* 1020. Subterranean clover nodules incited by *R. leguminosarum* 1020 have typical meristematic-type zonation with bacteria released from infection threads in the 'infection zone' and transformed into enlarged, 'Y'-shaped bacteroids in the 'bacteroid zone'. Nodule ultrastructure appeared to be normal. Leghemoglobin was detected by SDS-PAGE in soluble extracts of pea nodules formed by *R. leguminosarum* 1020 and in subterranean clover nodules formed by *R. trifolii* 0403 but not in soluble extracts of clover nodules incited by *R. leguminosarum* 1020, even after concentration by ammonium sulfate fractionation. Therefore, although the morphology of subterranean clover nodules incited by *R. leguminosarum* 1020 appears normal, perhaps they remain ineffective because no leghemoglobin is present. Further study of this anomalous nodulation system will aid in understanding how *Rhizobium* can overcome the host barriers to infection.

Hepper C M (1978) Ann. Bot. 42,109-115.
Hepper C M and L Lee (1979) Plant Soil. 51,441-445.

PILIATION AND AUTAGGLUTINATION OF RHIZOBIUM LEGUMINOSARUM RBL 1

J.W. KIJNE /C.A.M. VAN DER SCHAAL /G. VAN DER PLUIJM /J.B. DE KORTE /
G.J. MEDEMA /A.J.P. DE HAAS /L. VAN DE OEVER
DEPARTMENT OF PLANT MOLECULAR BIOLOGY, UNIVERSITY OF LEIDEN,
NONNENSTEEG 3, 2311 VJ LEIDEN, THE NETHERLANDS

INTRODUCTION

Rhizobium leguminosarum RBL 1 shows autagglutination in batch culture on several media. Cell clustering appears characteristically at two growth stages: end lag-phase and, particularly, end log-phase. In a yeast extract/mannitol medium the end log-phase autagglutination could be prevented by addition of 0.2 M mannose; glucose, galactose and mannitol were inactive (Kijne et al. 1983). This interesting observation led us to a more detailed study of autagglutination in a defined medium, in view of a possible role of this clustering behaviour in sugar-specific attachment of Rhizobium to legume root hairs.

MATERIAL AND METHODS

strain: R.leguminosarum RBL 1 (= A171)
medium: B^{-}glu: glucose 10 g; glutamate 750 mg; $MgSO_4.7H_2O$ 540 mg; KCl 40 mg; $CaCl_2.2H_2O$ 64 mg; biotin 100 µg; thiamin 100 µg; NaFe EDTA 33 mg; KH_2PO_4 993 mg; K_2HPO_4 320 mg; microelements; demineralized water 1000 ml; pH 6.2.
culture conditions: 50 ml medium in 100 ml erlenmeyer flasks (180 rpm, $28^{o}C$), inoculated with 1% (v/v) liquid-precultured rhizobia.
Oxygen was measured with the YSI 53SA oxygen monitor system, glutamate with the ninhydrin method. Negative staining visualized capsulation (light microscope, Indian ink) and piliation (EM, 1% PTA).

RESULTS

(Electron)microscopical study revealed the dominant presence of both piliated and capsulated cells in the rhizobial agglutinates. Pili frequently crossbridge the distance between rhizobia, which suggests (not more than that) a role for pili in cell-to-cell binding. Three types of pili or pili-like extensions have been observed, the most abundant being long, very thin filaments, which only can be clearly visualized on rhizobia growing under low carbon conditions.
R.leguminosarum RBL 1 grows logarithmically on B^{-}glu-medium up to an OD_{620} of 0.48, at which cell density the oxygen in the culture medium becomes limiting. The percentage of piliated cells increases significantly under oxygen-limited conditions (up to 50-75%). The percentage of capsulated cells decreases during growth, but capsulation is induced again under late stationary growth conditions. The increase in cell number stops at an OD_{620} of 0.62, when the glutamate is exhausted; shortly afterwards the rhizobia autagglutinate. Other growth limitations (e.g. Ca, Fe, Mg or P limitation) also induce autagglutination.

DISCUSSION

The observations indicate that at least one factor involved in rhizobial autagglutination (the pilus-receptor?) is produced under growth-limiting conditions. As the leguminous rhizosphere provides batch culture conditions, the importance of growth limitation in the attachment-behaviour of Rhizobium is stressed (see also Van der Schaal et al., this issue).

REFERENCES

Kijne JW, Van der Schaal CAM, Díaz CL, Van Iren F (1983) in Bøg-Hansen TC and Spengler GA, eds, Lectins Vol 3, pp. 521-529, W. de Gruyter, Berlin.

ANTIGENIC ANALYSIS OF *ANABAENA AZOLLAE* AND PRESENCE OF LECTIN IN *AZOLLA-ANABAENA* ASSOCIATION

J. K. LADHA/I. WATANABE
THE INTERNATIONAL RICE RESEARCH INSTITUTE
P.O. BOX 933, MANILA, PHILIPPINES

Azolla has worldwide distribution and is represented by six recognizable species. The algal symbiont belongs to the Nostocaceae and generally referred to as *A. azollae*. It is not clear whether the symbiont is the same in the various *Azolla* species and specimens or if there are several strains of the symbiont. The present study was undertaken to study the antigenic relationship among symbiotic *A. azollae* (separated from different species and specimens of *Azolla*), with its host *Azolla* and few selected free-living *Anabaena* species.

Fluorescent antibodies (FAs) prepared against five strains of symbiotic *A. azollae* strongly cross-reacted with symbiotic *Anabaena* of the 32 specimens of *Azolla* (belonging to six species) tested, but not with any of the free-living blue-green algae. FAs against two strains of free-living *A. azollae* (Newton, Bai) did not cross-react with any of the symbiotic *A. azollae*. These results suggest that: (i) symbiotic *Anabaena* from different specimens of *Azolla* share identical and highly specific antigens and (ii) free-living *A. azollae* are either not true isolates, or their antigenic properties were altered during isolation and culturing. The possibility that the symbiotic *A. azollae* cells have on their surfaces antigens that are cross-reactive with *Azolla* cells was confirmed from the following results: (i) on absorbing the FA against *A. azollae* with leaf extract suspension of *Anabaena*-free *Azolla*, the absorbed FA reduced the reactivity with all symbiotic *A. azollae* tested and, (ii) FA against *Anabaena* free *Azolla* showed 2+ fluorescence with several symbiotic *A. azollae* tested while no fluorescence was observed with free-living *Anabaena* species tested.

We then examined the lectin in cell extracts of *Azolla-Anabaena* symbiosis and a few free-living *Anabaena* species. The whole *Anabaena-Azolla* symbiosis and of *Anabaena* free *Azolla* plants both caused agglutination of human and rat erythrocytes whereas extracts of symbiotic *A. azollae* and free-living *A. azollae* (Newton) did not cause haemagglutination. α-D-galactose was found to be most effective carbohydrates tested in preventing haemagglutination.

A. azollae (Newton) did not cross-react when stained periodically with FA's against symbiotic *A. azollae* and *Anabaena* free *Azolla* thus ruling out possibility that *A. azollae* (Newton) could develop *Anabaena-Azolla* cross-reactive antigens at some stage of its growth phase. These results strongly suggest that *A. azollae* isolated by Newton is not a true isolate of *Azolla*.

In *Azolla-Anabaena* symbiosis there is a continuity of association between the algal symbiont and *Azolla* host during the sexual cycle and therefore recognition and infection of host by free-living *Anabaena* probably are not required. Thus the possible significance of cross-reactive antigens and lectin in *Azolla Anabaena* symbiosis is not fully understood.

COMPETITION STUDIES WITH FAST-GROWING R. JAPONICUM 'PRC' STRAINS

THOMAS J. McLOUGHLIN/ANN DAMEWOOD/SCOTT ALT
AGRIGENETICS ADVANCED RESEARCH LABORATORY, 5649 EAST BUCKEYE ROAD, MADISON, WISCONSIN 53716 USA.

INTRODUCTION

We have recently shown that a fast growing R. japonicum strain USDA 191 can form a partially effective symbiosis on commercial cultivars of Glycine max (1). Some fast growing strains were reported to be more competitive in forming nodules when challenged at the time of inoculation with equal numbers of slow growing strains on cv. Peking but less competitive on cv. Lee (2).

MATERIALS AND METHODS

Fast growing R. japonicum strains 191, 192, 193, 194, 201, 205, 206, 208, 217, 257, and slow growing 110 were obtained from USDA, Beltsville, Maryland; Ag-39 (a member of 122 serogroup) was obtained from D. Bauer, Kettering.

Genetically marked strains and serology (3) were used for identification of those strains. Competition studies were carried out in growth pouches by mixing fast:slow growers in the desired proportion (1:1, 1:10 and 10:1). Eight replications were used per treatment and 40 nodules per treatment were identified after 3 weeks' growth.

Ten 'PRC' strains were inoculated at 10^6 and 10^9 cells/ml on cv. Peking and cv. J130 in soil pots containing 3.5×10^5 indigenous Rhizobium/gm of soil. Ten replications were used per treatment; 100 nodules per treatment were identified 5 weeks later.

RESULTS AND DISCUSSION

When equal proportions of fast:slow growers were mixed on cv. Peking in growth pouches, greater than 80% of the nodules were formed by the slow grower. When the fast growers were added at a 10:1 ratio in their favor, again the majority of the nodules were formed by the slow growers. These results are in disagreement with the results of (2), who reported that the fast growing strains (201, 205 and 208) were more competitive when added in equal proporation with USDA 110 and 122 on cv. Peking in a Rhizobium free soil. On cv. J130 the slow growers occupied all the nodules.

In soil pot experimments, three fast growing strains USDA 257, 193 and 206 formed > 65% of the nodules on cultivar Peking. The other strains competed poorly against the indigenous population. On cv. J130 none of the nodules were formed by the introduced strains.

In terms of strain competitiveness, the ranking would be USDA 257 > 193 > 206 > other fast growers. After screening competitive strains in the laboratory, it is important to test those strains under field conditions. We are currently testing these fast growing strains in different mid-Western soils.

REFERENCES

1. McLoughlin TJ et al. (1983) 9th N. American Rhizobium Conf., Cornell.
2. Sanagho S and Keyser H (1981) Agronomy abstracts.
3. Johansen E (1983) Ph.D. Thesis, M.I.T.

ACKNOWLEDGEMENTS

We would like to thank Drs. E. Appelbaum and E. Johansen for providing mutant strains of USDA 191 and Ag-89, also, J. Adang, J. Pregler and C. Hopka for assistance in the preparation of this poster.

EXTRACELLULAR POLYSACCHARIDES AND *RHIZOBIUM MELILOTI* INFECTIVITY

J. OLIVARES/E. MARTINEZ-MOLINA
DEPARTMENTO DE MICROBIOLGIA, ESTACION EXPERIMENTAL DEL ZAIDIN, GRANADA, SPAIN

1. INTRODUCTION

Extracellular polysaccharides (EPS) that are produced by *Rhizobium* seem to be involved in the earlier steps of the root hair infection. Their actual role is not yet well understood although their significance is out of doubt. The aim of this work has been to study the effect of different EPS preparations on the infectivity of several *R. meliloti* strains and to know the possible implication of these substances in the infection process.

2. MATERIAL AND METHODS

The microorganisms and the plant: *R. meliloti* (strains Rm4, Rm4c, GRC 60, AK631) showing a different degree of infectivity, *R. leguminosarum* (strain GR024) and *Medicago sativa* have been used.

EPS preparations: Crude EPS preparations were obtained following the techniques of Amarger et al. (1967) and Robertsen et al. (1981). Crude preparations were either dialyzed, extracted with ethyl acetate to separate hormone-like substances or hydrolyzed with 1N Hcl. Aqueous fraction of each treatment was precipitated with acetone and dissolved in distilled water to be assayed.

Test using plants: To test the effect of the different EPS preparations on nodulation rate and on root hair adsorption of bacteria the techniques of Olivares et al. (1980) and Dazzo et al. (1976) were followed, respectively.

3. RESULTS AND DISCUSSION

Results have shown that when EPS obtained from a high infective *R. meliloti* strain is added to a poor infective one its infectivity increases as EPS concentration does. This positive effect can not be attributed to hormone-like substances present in the crude preparations. When EPS are submitted to an acid hydrolysis a decrease or even loss of their activity occurs. Studies at microscopical level have shown an increase in the adsorption of bacteria to root hair surface. Hydrolyzed EPS causes a total inhibition of the adsorption.

4. REFERENCES

Amarger N, Obaton M and Blanchere H (1967) Can. J. Microbiol. 13, 99-105.
Dazzo FB, Napoli CA and Hubbell DH (1976) Appl. Environ. Microbiol. 31, 166-171.
Olivares J, Casadesús J and Bedmar EJ (1980) Appl. Environ. Microbiol. 39, 967-970.
Robertsen BK, Aman P, Darvill AG, McNeil M and Albersheim P (1981) Plant Physiol. 67, 389-400.

EFFECT OF *AZOSPIRILLUM* STRAINS ON RHIZOBIUM-LEGUME SYMBIOSIS

JACEK PLAZINSKI/ELENA GÄRTNER/JAN MCIVER/ROLAND JAHNKE* AND BARRY G. ROLFE
GENETICS DEPARTMENT AND *NEUROBIOLOGY DEPARTMENT, RESEARCH SCHOOL OF BIOLOGICAL SCIENCES, AUSTRALIAN NATIONAL UNIVERSITY, CANBERRA, AUSTRALIA 2601

The interaction of five *Azospirillum* strains of the legume nodulation capacity of 14 *Rhizobium* strains was studied by using the rapid plant screening assay (Rolfe 1980). All *Azospirillum* strains showed an ability to inhibit or enhance nodulation of *Rhizobium trifolii*, *R. meliloti*, *R.leguminosarum* and *Rhizobium* "cowpea" strains on their respective plant hosts. An inhibition of nodulation was observed when *Azospirillum* and *Rhizobium* strains were mixed at a precise cell ratio and inoculated onto plants. Stimulation of nodule formation occurred when plants were inoculated first with an appropriate *Rhizobium* strains, and an *Azospirillum* strain added at least 24h later. In addition, the same phenomenon was observed when plants were inoculated first with an *Azospirillum* strain and *Rhizobium* added at least 24h later. Another *Azospirillum-Rhizobium* phenomenon was observed when a particular combination of an *R.trifolii* strain and an *Azospirillum* strain formed no nodules but gave a stimulation of clover plant growth on nitrogen free media. When *Azospirillum* addition caused an increase in nodule number there was a concomitant decrease in the effectiveness of the *R.trifolii* strains. All *Azospirillum* strains (SP7,SP59,SP107,SP242 and SP245) showed variation in their ability to inhibit or stimulate *Rhizobium* nodulation. Our root-segment-squash and nodule isolation methods showed that *Azospirillum* cells were present in both root segments and nodules of all investigated test legume plants. A detailed study of this interaction has shown that when nodulation was blocked (for those cases of mixed cultures containing the appropriate ratios of *Azospirillum* and *R.trifolii*) a stimulation of root hair growth was observed and an overproduction of mucigel by clover plants occurred. This plant response decreased root hair susceptibility to *Rhizobium* infection, and was similar to that observed when plants were inoculated with *Rhizobium* in the presence of different concentrations of phytohormones.

References

Rolfe B.G., Gresshoff P.M., Shine J. (1980). Plant Sci.Lett.19,277-284

CHARACTERIZATION AND LOCALIZATION OF AUXOTROPHIC SYMBIOTIC MUTANTS IN RHIZOBIUM TRIFOLII. CORRELATION WITH THEIR EXTRACELLULAR POLYSACCHARIDES.

A.J. PALOMARES, M.A. CAVIEDES, M. MEGIAS and RUIZ-BERRAQUERO, F.

DEPARMENT OF MICROBIOLOGY. FACULTY OF PHARMACY. SEVILLE. SPAIN.

The symbiotic relationship between a species of nitrogen-fixing Rhizobium and a specific legume which results in the development of a functional nitrogen-fixing root nodule must involve the interaction of a number of genes of bacterial and plant origin. Mutants of Rhizobium that affect formation of normal functional nodules are likely to be useful in the analysis of such interactions.

To obtain auxotrophic symbiotic mutants we have used Tn5 transpositions. We used plasmid pJB4JI which confers gentamicin and kanamicin resistance. Escherichia coli 1830 (containing pJB4JI) was conjugated with -- R. trifolii (Str^r or Rif^r) and Km^rStr^r or Km^rRif^r exconjugants were obtained at frequencies about $1x10^{-4}$. On the average 98 to 99% of the Km^r exconjugants were sensitive to gentamicin and these were candidates for isolates containing Tn5 transpositions to R.trifolii replicons. Growth requirements of auxotrophic mutants were determined using the test of Holliday and were assigned into 10 phenotipic groups (14 Ade^-, 11 Trp^-, 9 Met^-, 6 His^-/Gln^-, 5 His^-, 3 Met/Cys^-, 3 Pyr^-, 1 Leu^-, 1 Pan^-, 1 Thr^-). Ten were unidentified Frequency of appearance of auxotrophs after Tn5 mutagenesis was 0.4%. To identify Tn5 induced symbiotic mutants, auxotrophic mutants were tested -- individually for infectivity on clover after 4-5 weeks and their symbiotic effectivity was stimated by their acetylene reduction ability in the nodule. A total of 3 Nod^- and 12 Fix^- symbiotic mutants have been isolated.

The linkage map of R.trifolii chromosome (Megias et al.,1982) -- contains auxotrophic and resistance markers. These markers provide a frame for the localization of any other chromosomal mutations. To map a symbiotic mutation, plasmid pJB3JI was introduced into R.trifolii RS637 and RS 661 (both Nod^+ Fix^+) and then mated with 3 Nod^- and 7 Fix^- infividually -- recipients. Exconjugants of 3 Nod^- were all Nod^-. Exconjugants of 3 Fix^- - were Fix^+ while the others were Fix^-. From the results of this experiment we preliminary located three fix markers in the chromosome of R.trifolii.

Rhizobia characteristically synthesize copious amount of extracellular polysaccharides (EPS). R.trifolii produces an EPS rich in fructose, glucose and galactose (Caviedes et al.,1982). It has been suggested that the nodulation by several mutants was linearly proportional to the amount of acidic exopolysaccharide that they released into the culture medium -- during the growth phase, indicating that such polysaccharide synthesis is important and perhaps required for nodulation.

We have studied the production and composition of EPS from some symbiotic mutants (3 Nod^- and 3 Fix^-). Glucose and fructose are present in all the mutants. The three Fix^- lack glucuronic acid which is present in - the Nod^-. Galactose is present in the Fix^- and in one Nod^-.

ADAPTATION FROM THE ENDOPHYTIC FORM TO THE FREE-LIVING FORM OF FRANKIA

A. Quispel, A.J.P. Burggraaf, T. Tak. Dept. of Plant Mol. Biol., Univ. of Leiden, The Netherlands.

During isolation of Frankia from root nodules of Alnus glutinosa the endophyte has to adapt to the free-living form. The growth-requirements during isolation and subsequent cultivation are studied. Special attention is given to the role of root lipids, propionate and amino acids.

Material and methods. Alder root nodules (watercultures) were desinfected (alcohol and OsO_4 1.5%), fragmented and fragments inoculated on bottom agar and covered by a top agar (Burggraaf et al. 1981). The percentage of nodule fragments showing Frankia growth was counted after 10 to 20 and 41 days. Total lipid (TL) extraction from Alnus roots was according to Kates (1972). TL, silicagel chromatographic (Hirsch and Ahrens 1958) or thin layer chromatographic (Skipski et al. 1965) fractions were applied to the top agar-layer.

Results

1) TL is essential for outgrowth of hyphae within 20 days. Ocassional hyphal outgrowth is found without TL presumably because lipids are present inside the nodule fragments.

2) The percentage of nodule fragments showing hyphal outgrowth and sporangia is initially inhibited by propionic acid, though after several subcultures propionate is an excellent carbon source.

3) Clear differences exist in the isolation success with different amino acids.

4) The active factors in the TL are on thin layer chromatograms cochromatographing with the mono- and diglycerides and hydroxy-fatty acids.

5) For isolations these lipids can not be replaced by Tween 80, Triton X-100 or lecithin (5mg/l) which after isolation stimulate growth.

6) Reisolates from nodules formed after infection with the cultivated strains retain the characteristics of these cultivated strains.

Hypothesis: Lipids from TL extract influence the functioning of the endophytic cytoplasma membranes. Thanks to this influence the cells can adapt to the requirements of a free-living form.

References

Burggraaf AJP et al. (1981) Plant and Soil 61, 157-168.
Hirsch J and Ahrens EH (1958) J. Biol. Chem. 233(2), 311-320.
Kates M (1972). Techniques in Lipodology. North Holland/American Elsevier, Amsterdam, London, New York.
Skipski et al. (1965) Biochem. Biophys. Acta 106, 386-396.

1=P-N (Propion.med.-NH_4Cl)
2=P-N/TL
3=P/TL (Propion.med.)
4=M/TL (Mineral med.)
5=M-N/Cas/TL
6=P-N/Cas/TL
7-11= M-N/0.005,0.01,0.05, 0.08 or 0.1 gr NH_4Cl
12-16=M-N/0.1,0.5,1.0,3.0 gr/l Casamino acids (Cas)
Striated area:Frankia with few or no sporangia, Black area: Many sporangia and extensive hyphal growth.

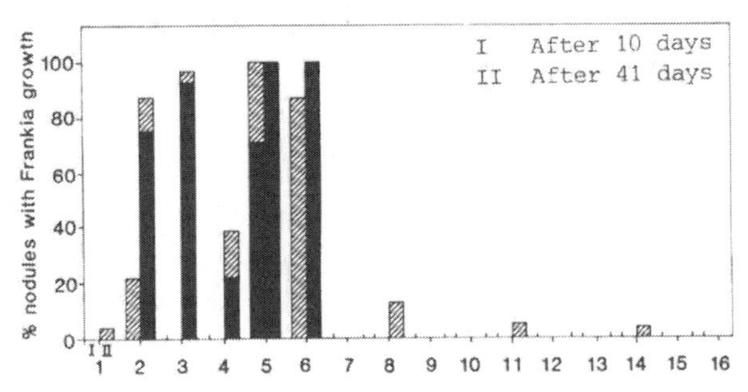

RE-EXAMINATION OF THE ROLE OF LECTIN IN RHIZOBIUM-GLYCINE SYMBIOSIS

CHAMPA SENGUPTA-GOPALAN/JAN W. PITAS/TIMOTHY C. HALL
Agrigenetics Advanced Research Laboratory, 5649 East Buckeye Road, Madison, Wisconsin 53716 USA

Introduction

It has been suggested that seed lectin may be involved in the specific recognition of symbiotic rhizobia by legumes. However, the two major problems with this hypothesis has been the finding that cultivars of soybean lacking the seed lectin (SBL) nodulate normally and the fact that all the circumstantial evidence in favor of the hypothesis is based on the interaction between seed lectin rather than root lectin and *R. japonicum*. Towards a better understanding of the recognition mechanism, we have used the sensitive techniques described below to critically test for the presence of SBL in the so-called 'lectinless' lines. We have also used these techniques to determine if root lectins are similar or different from SBL.

Materials & Methods

Detection of lectin in the tissues was done by immunoprecipitation and Western blot hybridization using antisera to SBL. mRNA was detected by immunoprecipitation of proteins translated *in vitro* with anti-SBL antibody and by Northern blot hybridization with a ^{32}P-labeled cDNA clone of SBL mRNA. In-gel proteolysis of the lectin protein was done using *Staphylococcus aureus* V8 protease.

Results & Discussion

By Western blot hybridization, a protein co-migrating with the 35kD SBL could be detected in the seeds and cotyledons of 'lectinless' lines (lec^-) at a 10 to 20 fold lower level than in the lectin-containing lines (lec^+). The cotyledons of lec^- lines also showed the presence of an immunoreactive ~43kD protein. Poly(A) RNA from cotyledons of lec^- and lec^+ lines directed the translation *in vitro* of a 32kD polypeptide that was immunoprecipitable with antibody to SBL. The level of this protein was about 10 fold lower in the lec^- lines than in the lec^+ lines. However, the lectin protein from lec^- and lec^+ lines showed a different peptide map suggesting different amino acid sequences. Blots of cotyledon RNA when hybridized to a SBL cDNA clone, showed strong hybridization to a 1.1kb RNA molecule in the lec^+ lines and showed no hybridization in the lec^- lines. A 10,000 fold decrease in transcription of the functional lectin gene in the lec^- lines due to the presence of an insertion sequence in the gene has been reported (1).

Immunoprecipitation of labeled *in vivo* root exudates showed two major immunoreactive polypeptides (~45kD and a 35kD), one of which co-migrated with SBL. Immunoprecipitation of *in vitro* translation products to poly(A) RNA from roots with anti-SBL antibody showed a major 32kD polypeptide. However, the cDNA clone to SBL mRNA did not show any hybridization to root mRNA, suggesting that the seed and root lectin mRNA have little homology.

Our data suggest that there are at least two classes of lectin genes in soybean, one under strict developmental control in the cotyledons and the other being constitutively expressed in roots and cotyledons. Expression of the latter set accounts for the presence of lectin in the seeds of lec^- lines and in roots of both lec^- and lec^+ lines. However, the two gene products are immunologically, and probably functionally, similar.

References

1. Goldberg RB *et al.* (1983) Cell 33, 465-475.

DE NOVO SYNTHESIS AND MOBILIZATION OF TRIFOLIIN A BY CLOVER SEEDLING ROOTS

J.E. SHERWOOD, G.L. TRUCHET, AND F.B. DAZZO
DEPARTMENT OF MICROBIOLOGY, MICHIGAN STATE UNIVERSITY, EAST LANSING, MICHIGAN, 48824, USA

The interaction between legume lectins and the Rhizobium symbiont has been implicated in the specific recognition of these bacteria by host root hairs (Bohlool, Schmidt 1974, Dazzo, Hubbell 1975). The study of the synthesis and transport of lectins has been made easier and more sensitve when labeled amino acids could be incorporated (Crispeels, Bollini 1982, Roberts, Lord 1981). We are studying the synthesis and mobilization of trifoliin A, a white clover lectin which binds R. trifolii, in order to better elucidate the regulation of this system.

MATERIALS AND METHODS. Trifolium repens var. Louisiana Nolin seeds were surface sterilized, germinated, and grown in sterile Fahraeus medium (Fahraeus 1957). ^{3}H- and ^{14}C-amino acids in Fahraeus medium were added, and, after incubation, the medium ("root exudate") was removed. The seedlings were washed with Fahraeus medium and then rinsed with the lectin hapten, 2-deoxyglucose, to remove lectin bound to the "root surface". After additional washing, the roots were excised and homogenized ("root homogenate"). All three fractions were assayed for total protein (precipitation in trichloroacetic acid or acetone) and lectin (immunoprecipitation with antibody prepared against the seed lectin).

RESULTS AND DISCUSSION. Labeled amino acids were incorporated into protein synthesized by intact clover seedlings and mobilized to the root surface and root exudate. Incorporation into lectin was shown by immunoprecipitation and confirmed by fluorography following SDS polyacrylamide gel electrophoresis (SDS PAGE). Incorporation of label (cpm/mg root fresh weight) into protein and lectin in the root exudate and root surface samples was greatest with 2 day-old seedlings. The amount of root surface lectin dropped sharply with older seedlings, suggesting that the root receptor sites for trifoliin A had become saturated. The amount of label incorporated into lectin in the root homogenate fractions remained relatively constant with seedling age. As compared with seedlings grown in N-free Fahraeus medium, the addition of 1% sucrose or 1 mM KNO_3 caused an increase in labeled root surface lectin, while a 15 mM KNO_3 amendment caused no increase. Although the addition of the high level of nitrate resulted in greater cpm/mg seedling fresh weight in the root exudate, there were several proteins from the root exudate samples which coprecipitated with the lectin, as shown by SDS PAGE/fluorography, making quantitation unreliable in this case. More labeled amino acids were incorporated into protein when seedlings were grown in the dark than with a 14 h photoperiod. Labeled protein, but not labeled lectin, was detected in plant tops (cotyledon and hypocotyl) following incubation with the amino acids.

REFERENCES. Bohlool BB and EL Schmidt (1974) Science 185,269-271.
Crispeels MJ and R Bollini (1982) Plant Physiol. 70,1425-1428.
Dazzo FB and DH Hubbell (1975) Appl. Microbiol. 30,1010-1033.
Fahraeus G (1957) J. Gen. Microbiol. 16,374-381.
Roberts LM and JM Lord (1981) Eur. J. Biochem. 119,31-41.

CHARACTERIZATION OF ROOT HAIR BRANCHING FACTOR IN WHITE CLOVER/*RHIZOBIUM TRIFOLII* SYMBIOSIS.

B. SOLHEIM[+]/T.V. BHUVANESWARI[++]
+ INSTITUTE OF BIOLOGY AND GEOLOGY, UNIVERSITY OF TROMSØ, TROMSØ, NORWAY AND
++ C.F. KETTERING RESEARCH LABORATORY, YELLOW SPRINGS, OHIO 45387.

In white clover *Rhizobium* infections that lead to nodule formation occur frequently in root hairs that are developmentally mature at the time of inoculation. These mature root hairs become susceptible to infections only after they develop lateral branches (Bhuvaneswari, T.V. this volume). Cell free filtrates of *R. trifolii* cultured in the presence of white clover plants induced branching in mature root hairs. A bioassay for branching was developed to test the fractions purified from the filtrate. Each fraction was tested on a minimum of 14 plants growing in 1.5 ml chambers (7 plants/chamber) made up of a microscope slide and a 22 X 60 mm cover slip. Roots were scored after 14 h on a scale of 0-4 (0 is no root hairs branched and 4 is almost all root hairs branched). Control roots exposed to water averaged a score of 0.5. Freeze dried cell free filtrates were sequentially extracted with 95%, 60% ethanol and water. The 60% ethanol soluble fraction which was biologically most active was concentrated by evaporation at reduced pressure, freeze dried and applied to a DEAE Sephadex column (imidazole/HCl 0.05M pH 7.5). Active unbound components were pooled, dialysed (1200 dalton cut off), freeze dried and applied to a Biogel P4 column. Three active fractions were collected. Further separation of these fractions by HPLC with a silica column yielded eight active fractions ranging from 1200-10,000 daltons in molecular weight. All these fractions contained neutral sugars and their biological activity was stable to autoclaving above pH 6.0. The sugar composition of the individual peaks was determined by GC-MS of the alditol acetate derivatives (Table 1).

Table 1. Relative sugar composition of HPLC purified fraction.

	Fraction Number							
Sugar	1	2	3	4	5	6	7	8
Glucose	0.24	0.11	1.24	6.06	4.52	2.53	1.54	1.26
Galactose	1.00	1.00	1.00	1.00	1.00	1.00	1.00	1.00
Mannose	1.04	1.40	1.51	0.65	1.68	1.71	1.37	3.37
Xylose	0.33	tr	0.20	tr	+	tr	0.81	0.91
Arabinose	0.41	0.21	0.68	0	+	tr	tr	0.23
Fucose	0	0	0	0	0	0	1.16	tr
Rhamnose	0	0	0	0	0	0	0.96	tr
Biological activity	1.0	1.8	1.4	1.4	1.1	0.6	0.8	1.0

tr = trace amounts; + = overlapping peaks, not calculated.

The results suggest that small polysaccharides produced when rhizobia interact with the host roots, induce branching in mature root hairs. Further characterization of the active fractions is needed to identify any possible common determinant responsible for the observed biological activity.

This work was supported in part by a National Science Foundation Grant (PFR 7727269) and in part by a grant from Norwegian Agricultural Research Council.

GALACTOSE INCORPORATION AND COMPETITIVENESS IN RHIZOBIUM MELILOTI

R. A. UGALDE, J. HANDELSMAN, W. J. BRILL
DEPT. OF BACTERIOLOGY, UNIVERSITY OF WISCONSIN, MADISON, WI 53706, U.S.A.

In cultures of R. meliloti strain 102F51, two spontaneous cell-surface variants were distinguished. One is agglutinated at high diluting titer of an alfalfa agglutinin (HA), sensitive to phage F20 ($F20^s$) and resistant to phage 16B ($16B^r$), the other is agglutinated at lower agglutinin titer (LA), is resistant to phage F20($F20^r$) and sensitive to phage 16B($16B^s$).

Alfalfa plants inoculated with a mixture of equal number of these two phenotypes are predominantly nodulated by the LA, $F20^r$, $16B^s$ strains.

In vitro studies showed that the more competitive strains LA, $F20^r$, $16B^s$, have a galactosyl-transferase activity that is absent in the HA, $F20^s$, $16B^r$ less competitive strains. Spontaneous phage 16B resistant mutants were selected from LA, $F20^r$, $16B^s$ cells. In all the cases HA, $F20^s$, $16B^r$ strains were isolated. The galactosyl-transferase activity and competitiveness were lost in these phage selected strains, suggesting a direct implication of this enzymatic activity in the cell-surface modifications that determine the competitive ability.

The galactosyl-transferase that distinguishes one phenotype from the other is inner-membrane bound and has an absolute requirement for a phosphorylated small molecular-weight factor recovered from the cells during its permeabilization with EDTA. This factor is alkaline-phosphatase sensitive and is present in both phenotypes of R. meliloti and other species of Rhizobium; however, it was not recovered from Escherichia coli, Azotobacter vinelandii, Klebsiella pneumoniae or Agrobacterium rhizogenes cells. This reaction requires Mg^{2+}, has an optimum pH at 8.2, an optimum reaction temperature of 25°C and an apparent K_m for UDP galactose of 1.6 μM. It transfers galactose from UDP galactose to a high molecular weight water-insoluble product that can be solubilized by alkaline treatment (0.1 N NaOH at 65°C for 10 minutes) and partially purified by gel chromatography in Sephadex G200. Antibodies prepared against whole cells of LA, $F20^r$, $16B^s$ strains precipitate this partially purified product while antibodies raised against HA, $F20^s$, $16B^r$ cells do not. These results indicate that the product prepared in vitro is present on the surface of the more competitive LA, $F20^r$, $16B^s$ strains and absent on the HA, $F20^s$, $16B^r$ less competitive strains.

All the evidences support that the reaction described here is involved in the synthesis of a R. meliloti cell surface antigen required for the attachment of phage 16B and its presence on the surface effects the interaction of the cells with phage F20 and with the alfalfa agglutinin. In all the cases studied, these changes result in strains with more competitive ability to nodulate alfalfa.

THE INFLUENCE OF NITRATE AND BACTERIAL POLYSACCHARIDES ON WALL-BOUND GLYOSIDASES OF WHITE CLOVER ROOTS.

JANKEES VAN DER HAVE AND FRANK DAZZO
DEPARTMENT OF MICROBIOLOGY AND PUBLIC HEALTH, MICHIGAN STATE UNIVERSITY, EAST LANSING, MICHIGAN 48824, U.S.A.

A hypothesis for the involvement of enzymatic breakdown of plant material in the infection process has been presented by Fahraeus and Ljunggren (1). Since the infection occurs at the cell wall (2) and since bacterial components, like lipopolysaccharides, which may play a specific role in the infection change significantly during bacterial growth (3), we investigated the influence of several bacterial LPS on the activity of cell wall associated exoglycosidases in white clover (*Trifolium repens*) roots. The level of activity of these enzymes is also investigated in plants grown in 15 mM nitrate, when hardly any infection occurs.

Material and Methods

White clover seedlings were grown axenically in Fahraeus medium containing 1 mM nitrate. The medium was enriched with 15 mM nitrate, or with LPS from *Rhizobium trifolii*, from *R.meliloti* or from *Escherichia coli*, in a final concentration of 1 μg/ml. After 5 days the roots were cut off, ground, sonicated and washed 5 times with acetate buffer pH 5.5. From the remaining cell walls enzymes were extracted with 1 M NaCl at 30°C. Enzyme activities were measured by using p-nitrophenyl substituted glycosides as substrates (4).

Results and Discussion

Enzyme assayed	*	activity following growth with supplements				
	1 mM NO_3	15 mM NO_3	*R.trifolii* K 90 LPS	*R.trifolii* K 50 LPS	*R.meliloti* LPS	*E.coli* LPS
α-L-arabinosidase	61.1	-59.2%	0	-47.2%	-40.7%	0
β-D-fucosidase	92.9	+50.6%	+60.7%	+24.1%	+59.1%	0
α-D-galactosidase	23.4	0	+162.0%	0	0	0
β-D-galactosidase	89.0	+124.8%	+126.3%	+38.9%	+84.5	+111.1%
β-D-glucosidase	261.1	+44.7%	+46.9%	0	+63.5%	+26.4%

table 1. Results of the protected t-test of the glycosidase activity measurements in the cell wall preparations. *=activity in units per mg protein (one unit hydrolyses 1 μMol of substrate per min. at 30°C.), 0=no significant change K 90=Trifoliin A-binding LPS, K 50=non-Trifoliin A-binding LPS, test performed on 8 replicates of each treatment.

It has been shown that very low concentrations of LPS can have significant influence on the activity or extractability of the enzymes described above. This effect in white clover is not specific for *R.trifolii* LPS, except for α-D -galactosidase. High levels of enzyme activity are also found when plants are grown in 15 mM nitrate.

The trifoliin A-binding LPS (K 90) seemed to cause the greatest increase in the activity of β-D-fucosidase and α- and β-D-galactosidase. It should be determined how these enzymes function in cell wall metabolism and root hair formation.

1. Fahraeus, G. and Ljunggren, H. (1959) Physiol Plant 12, 145-154
2. Callaham, D.A. and Torrey, J.G. (1981) Can. J. Bot. 59, 1647-1664
3. Hrabak,E.M. et al (1981) J. Bact. 148, 687-711
4. Dazzo, F.B. et al (1982) Appl. Envir. Microbiol. 44, 478-490

GROWTH-PHASE-DEPENDENT PEA LECTIN RECEPTORS ON *Rhizobium leguminosarum* AS DETERMINED BY AN ENZYME-LINKED LECTIN BINDING ASSAY (ELBA)

Ineke A.M. van der Schaal, Trudy J.J. Logman, Clara L. Díaz, Jan W. Kijne
Dept. of Plant Molecular Biology, Research Group of Biological Nitrogen Fixation, Botanical Laboratory, Nonnensteeg 3, Leiden, The Netherlands

Introduction. Agglutination of Rhizobium with pea lectin (PL) was restricted to pea nodulating strains and only occurred during a short period after exponential growth had stopped. Binding of PL to Rhizobium was observed with several species and was not restricted to a particular growth phase. Hot saline extraction of the cells removed their light microscope visible capsule and their PL binding capacities (Van der Schaal et al. 1983). We have developed an enzyme-linked lectin binding assay (ELBA) to be able to quantitate PL receptors in these extracts and in extracellular polysaccharides (EPS) of Rhizobium.

Methods. EPS were prepared from cell free culture supernatant by addition of 3 volumes of ethanol 96% (-20°C). Precipitated material was collected by centrifugation. Capsules were extracted from washed bacteria by shaking them for 1h at 80°C in 0.9% NaCl. Bacteria were removed by centrifugation and capsular polysaccharides (CPS) were prepared from the extracts as written for EPS. ELBA: Ovalbumine (PL-binding protein) was adsorbed onto microtiter plates. Afterwards PL, labeled with horseradish peroxidase (HRPO), was added, together with EPS or CPS. Bound enzyme activity was colorometrically determined after addition of enzyme substrate. Lectin affinity (K_L) was expressed as the concentration of EPS or CPS necessary to inhibit 50% binding of HRPO-PL conjugate to ovalbumine, divided by the concentration of D-mannose (PL hapten) necessary to inhibit 50% binding.

Results. (see figure). CPS of R. leguminosarum (▲—▲ A, production; B, PL-affinity) always showed PL binding activity of which the strenght was dependent on the age of the culture. Highly active CPS were obtained between A_{620} = 0.6-0.7, which coincides with the period of PL-agglutinability of R. leguminosarum. In this period no cell multiplication occurred (●—●), probably due to the then limiting oxygen concentration (○—○). Soon after growth had resumed, cells autoagglutinated (vertical line) and PL binding activity appeared in EPS (△—△ A, production; B, PL-affinity).

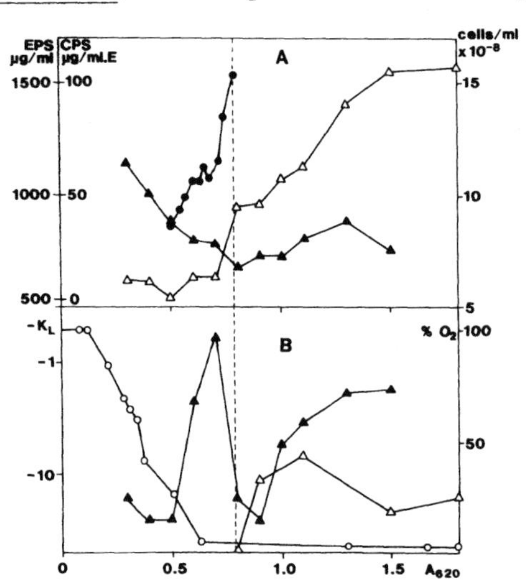

Conclusion. We propose that R. leguminosarum produces two PL receptors: one receptor which is always present, not correlated with host specificity and also present on other rhizobial species, and a second receptor with higher PL-affinity, only present under certain conditions and correlated with the pea nodulating capacities of the species.

Reference. Van der Schaal IAM, Díaz CL, Kijne JW, Van Iren F (1983) In Bøg-Hansen TC and Spengler GA, eds, Lectins, vol. 3, pp 531-538, W. de Gruyter, Berlin.

PHYTOALEXINS AND NODULATION IN PISUM SATIVUM

F. VAN IREN /M. VAN DER KNAAP /J. VAN DEN HEUVEL* /J.W. KIJNE
DEPT. OF PLANT MOLECULAR BIOLOGY, UNIVERSITY OF LEIDEN, NONNENSTEEG 3
2311 VJ LEIDEN, THE NETHERLANDS.
*WILLIE COMMELIN SCHOLTEN PHYTOPATHOLOGICAL LABORATORY, JAVALAAN 20,
3742 CP BAARN, THE NETHERLANDS.

1. INTRODUCTION

When thinking of the evolutionary history of root nodule symbioses we wondered whether the present state of things might represent a perfectly balanced pathogenesis. As in many plant infections phytoalexins appear to play a role in the control of the intruder, we studied whether the major phytoalexin of pea, viz. pisatin, is involved in controlling the growth of Rhizobium leguminosarum during pea root nodule formation.

2. EXPERIMENTAL AND RESULTS

2.1. Materials and Methods: We used R.leguminosarum A 171 and Pisum sativum cv. Rondo. Pisatin was prepared from Rhizoctonia solani infected pea seedlings. Pisatin was estimated in plant tissues and culture media by UV spectroscopy and GLC, after extraction and TLC (Pueppke, VanEtten, 1974 and 1976). Antimicrobial compounds were detected on TLC plates by growing Cladosporium herbarum on the plates (Bailey, 1973).

2.2. Influence of pisatin on bacterial growth in culture: In a glucose/nitrate medium, pH 6, generation time of R.leguminosarum increased linearly from 6 h (control) to 20 h at 100 mg l^{-1} pisatin. This indicates that growth regulation by pisatin is well possible. Cruickshank, 1962 and Pankhurst and Biggs, 1980 found other values. After 4 d of culture, bacterial degradation of pisatin appeared negligible.

2.3. Pisatin synthesis during nodulation: First traces of pisatin (tissue concentration about 0.1 mg l^{-1}) in nodules were found 30 d after inoculation of the seeds. By that time, N-fixation activity has passed already its maximum and scenescence of the bacteroids has already begun. Concentration raises to a level of about 10 mg l^{-1} after 35 d. This concentration will not account for a significant inhibition of bacterial growth. So, unless the endosymbiont is much more sensitive to pisatin than free living bacteria, pisatin is not involved in the arrest of multiplication, in bacteroid formation, and in scenescence of Rhizobium during symbiosis.

2.4. Other antimicrobial substances: The Cladosporium-test (see 2.1.) revealed such a substance in the 28 d nodule extract with a Rf value of 0.95. It failed from the 22 d extract. Its role remains to be studied.

3. CONCLUSIONS

Probably, pisatin plays no role in normal nodulation of pea. We found another substance that might be involved. It is not excluded that pisatin is involved in abnormalities, resulting in ineffective nodules (Pankhurst, Jones, 1979).

4. REFERENCES

Bailey JA (1973) J. Gen. Microbiol. 75, 119-123
Cruickshank AIM (1962) Aust. J. Biol. Sci. 15, 147-159
Pankhurst CE and Biggs DR (1980) Can. J. Microbiol. 26, 542-545
Pankhurst CE and Jones WT (1979) J. Exp. Bot. 30, 1095-1107
Pueppke SG and VanEtten HD (1974) Phytopathol. 64, 1433-1440
Pueppke SG and VanEtten HD (1976) Physiol. Pl. Pathol. 8, 51-61

GEL-FORMING CAPSULAR POLYSACCHARIDE OF *RHIZOBIUM LEGUMINOSARUM* AND *RHIZOBIUM TRIFOLII*

L.P.T.M. ZEVENHUIZEN AND A.R.W. VAN NEERVEN
LABORATORY OF MICROBIOLOGY, AGRICULTURAL UNIVERSITY, HESSELINK VAN SUCHTELENWEG 4, 6703 CT WAGENINGEN, THE NETHERLANDS

Extraction of cell pellets of several *R. leguminosarum* and *R. trifolii* strains with M NaOH at room temperature released an anthrone-positive material which, on neutralisation of the extract, precipitated as a gel. The polysaccharide could also be extracted with hot water. On cooling to room temperature, the clear filtrate solidified to a stable gel, already at a poly-saccharide concentration as low as 0.2%. The sol-gel transition temperature lies between 50-55°C.
Hydrolysis of the polysaccharide released D-galactose, D-glucose and D-mannose in the molar ratios 4:1:1 suggesting a hexasaccharide repeating unit. Methyl-ation analysis of the polysaccharide revealed two D-galactosyl end-groups, one (1→3)-linked D-mannosyl, one (1→3)-linked D-galactosyl, and one (1→4) linked D-galactosyl residue, and a doubly branched D-glucosyl residue linked through O-1,2,4,6.

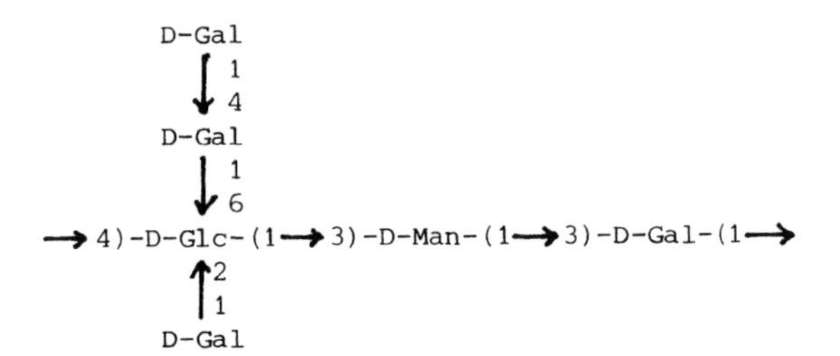

Periodate oxidation, followed by Smith degradation removed three D-galactose residues, present as side chains, and afforded a degraded polymer being the unbranched main chain, consisting of equal amounts of D-glucose, D-galactose, and D-mannose.
Sequence of the hexose residues in the main chain was studied by a second periodate oxidation, as the now (1→4)-linked D-glucose residues were suscep-tible to periodate attack.

PHYSIOLOGY OF INFECTED PLANTS

POSTER DISCUSSION 7A
PHOTOSYNTHESIS AND HYDROGEN METABOLISM IN RELATION TO NITROGENASE ACTIVITY OF LEGUMES

POSTER DISCUSSION 7B
ENVIRONMENTAL EFFECTS ON NODULATED PLANTS

CARBON METABOLISM IN THE LEGUME NODULE

T.A.LARUE, J.B.PETERSON AND S.TAJIMA
BOYCE THOMPSON INSTITUTE FOR PLANT RESEARCH
ITHACA, N.Y. 14853

The nodule uses photosynthate for its growth and maintenance, to provide energy and reductant for nitrogenase, and for the incorporation and transport of newly fixed nitrogen to the shoot. There is ample evidence that symbiotic fixation is demanding of energy, and imposes a respiratory burden on the legume plant. An increase in nitrogen fixation by crops might be achieved by increasing the supply of photosynthate to nodules, or by a more efficient use of carbon compounds within the nodule. This will require more knowledge than we now have on how carbon compounds are used by the symbionts.

The pathways by which the carbohydrate is metabolized in nodules are known only in part. There is little information on metabolism in the uninfected plant cells. We have only begun to determine the relative contributions of the bacteroid, peribacteroid space, plant cytoplasm, membranes and plant organelles in the infected cells.

In the past, metabolic pathways in the bacteroid could only be inferred from studies of free living bacteria. In recent years two experimental techniques have added greatly to our knowledge of bacteroid metabolism. One is the use of rhizobial mutants, and examination of their ability to infect, differentiate and to fix nitrogen. The other is improved rapid methods of isolating bacteroids. These preparations provide greater confidence that their contents and enzymatic activities are representative of the situation in the nodule.

In this paper we will summarize recent reports on the metabolism of carbohydrates and organic acids in nodules. We will also review work showing that the nodule contains enzymes and substrates typical of anaerobic metabolism.

1. SUCROSE

Studies of the slow growing rhizobia indicate that their growth cannot be supported by disaccharides as sole carbon source, and that they lack invertase and sucrose phosphorylase. Sucrose can support growth of the fast growing rhizobia (e.g. Martinez-de-Drets *et al*., 1974). The fast growing rhizobia have an active disaccharide uptake system and an inducible invertase. However bacteroids of *R. leguminosarum* from pea and of *A. lupini* from yellow lupin are unable to accumulate disaccharides (Glenn, Dilworth, 1981). While sucrose in phloem is the major carbon source for the nodules (Pate, 1975) it has generally been assumed that sucrose is not important in bacteroid metabolism.

Recent work will force a reexamination of this assumption. Soybean shoots were exposed to $^{14}CO_2$ and the distribution of label in the nodule was followed for several hours (Reibach, Streeter, 1983). Two hours after the pulse, about 60% of the radioactivity in bacteroids was in the neutral fraction, and of this about half was in the form of sucrose. The decrease in radioactivity in sucrose was accompanied by an increase in radioactivity of trehalose. The radioactivity in sucrose in bacteroids declined more rapidly than did the radioactivity in sucrose in the cytosol fraction. These results indicate not only that bacteroids of *R. japonicum* can degrade sucrose, but that they either accumulate it or reconstitute it rapidly from labelled precursors.

2. HEXOSES

A bacteroid preparation from soybean nodules had an energy dependent glucose uptake and phosphorylating mechanism. Uptake is only partly inhibited by 2-deoxyglucose, indicating two mechanisms are operative (Jacobsen, San Francisco, 1983).

Bacteroids of *R. leguminosarum* isolated from pea nodules are incapable of transporting glucose (Hudman, Glenn, 1980; de Vries *et al.*, 1982) or oxidizing it (Glenn, Dilworth, 1981). Nevertheless they contain glucokinase, and the Entner-Doudoroff pathway enzymes 6-phosphogluconate dehydratase and 2-keto-3-deoxy-6-phosphogluconate aldolase (Glenn *et al.*, in press). The specific activity of those enzymes in bacteroids was about half that found in free living bacteria grown on glucose. The bacteroids contained 6-phosphogluconate dehydrogenase, and hence may have a pentose phosphate pathway. The bacteroids, like glucose grown cells, lacked phosphofructokinase and thus do not have the Embden-Meyerhof pathway.

The presence of these enzymes is hard to reconcile with the reported inability of bacteroids to use glucose to support respiration or fixation. It is proposed (Trinchant *et al.*, 1983) that the reported inertness of sugars for bacteroids was due to inadequate oxygen tensions in the test systems. Bacteroids of *R. phaseoli* were isolated from nodules of french beans. Glucose stimulated respiration and C_2H_2 reduction to about twice the endogenous activities when the dissolved oxygen was about 3 µM. By contrast, when succinate was substrate, the optimum oxygen concentration for nitrogenase activity was 9 µM, and respiration was increased five fold above endogenous. Hooijmans and Logman (these proceedings) find that bacteroids of *R. leguminosarum* have an increased respiration in the presence of high concentrations (0.3M) of glucose and fructose. It remains to be determined if the peribacteroid membrane can accumulate such concentrations in the peribacteroid space.

The use of hexoses by bacteroids is moot, but there is strong evidence that it is not essential. Two transposon induced mutants of *R. leguminosarum* were obtained which lacked fructokinase, and failed to grow on fructose or mannitol (Glenn *et al.*, in press). Another mutant lacked glucokinase and did not grow on glucose. The three mutants nevertheless formed red nodules capable of C_2H_2 reduction. Isolated bacteroids were found to lack the enzymes, so reversion had not occurred in the nodule. This strongly suggests that the use of glucose or fructose is not essential either to establish the symbiosis or to support fixation.

Mutants of *R. trifolii* were prepared deficient in fructose uptake, in glucokinase, and in both activities (Ronson and Primrose, 1979). All mutants formed an effective symbiosis on red clover indicating that neither sucrose, glucose or fructose are essential for the symbiosis.

A mutant of *R. meliloti* low in phosphoglucose isomerase grew on glucose but not fructose. The strain accumulated toxic levels of fructose-6-phosphate when grown in a mixture of glucose and fructose. The mutant formed nodules, but the C_2H_2 reduction activity, on a whole plant basis, was less than that of the parental strain.

3. CYCLITOLS

In the labelling experiment described above (Riebach and Streeter, 1983) little ^{14}C appeared in cyclitols in soybean nodules. It is unlikely that the cyclitols have a major role in the energy metabolism of the nodule.

4. ORGANIC ACIDS

Formic acid can serve as sole carbon source for growth of *R. japonicum* (Manian *et al.*, 1982) and is oxidized by bacteroids isolated from soybean (Riuz-Argueso *et al.*, 1979). In the presence of other carbon compounds (glutamate, ribose or aspartate), but not alone, formate supports nitrogenase activity in free living *R. japonicum*.

A mutant strain unable to grow on formate was found to lack formate dehydrogenase (Manian *et al.*, 1982). The mutant did not reduce acetylene *ex planta* when formate was a carbon source. The strain formed effective nodules, indicating that formate does not play an essential role in the bacteroids metabolism.

Malonic acid is a major constituent of soybean nodules. Alone, it can be oxidized by *R. japonicum* bacteroids if it is at a high concentration (10 mM) (Werner *et al.*, 1982). Its oxidation was unaffected by the presence of arabinose or xylose (1 mM) but was depressed 85 to 100% by the presence of 0.1 to 1 mM succinate. Conversely the oxidation of 0.1 mM succinate by bacteroids was reduced by only 10 to 30% by the presence of 1 to 10 mM malonate. These results indicate that, at the concentrations found in the soybean nodule, malonate probably is not oxidized by bacteroids, and does not prevent the oxidation of succinate.

That malonate is not an important metabolite is also indicated by the labelling experiments of Reibach, and Streeter (1983), where only traces of label were found in malonate after $^{14}CO_2$, feeding to the soybean shoot.

The presence of the tricarboxylic acid cycle in bacteroids has long been established and, at least in *R. japonicum* bacteroids, the anapleurotic partial glyoxylate cycle (Stovall, Cole, 1978). That dicarboxylic acids are necessary for the complete functioning of bacteroids is indicated by experiments using mutants of three rhizobial species.

A mutant of *R. meliloti* unable to grown on L-arabinose, acetate or pyruvate was found to lack α-ketoglutarate dehydrogenase (Duncan, Fraenkel, 1979). The strain forms nodules on alfalfa which do not have nitrogenase activity.

Mutants of *R. trifolii* defective in C_4 dicarboxylate transport were unable to grow on or transport succinate, fumarate or malate (Ronson *et al.*, 1981). The mutants formed white ineffective nodules on white and red clover. The bacteroids of mutant strains were smaller than the normal parental strain. The mutants formed pleomorphic rods and spherical cells similar to those observed in the normal nodule.

Transposon mediated C_4-dicarboxylic acid transport mutants were obtained of *R. leguminosarum*. Two strains formed small white ineffective nodules on pea, while a third formed nodules with reduced nitrogenase activity. A succinate dehydrogenase deficient mutant also formed ineffective nodules.

A succinate dehydrogenase mutant of *R. meliloti* showed delayed nodulation and formed white ineffective nodules on alfalfa (Gardiol *et al.*, 1982).

These results strongly suggest that a complete TCA cycle is required for effective nitrogen fixation. The results do not however resolve the problem of what carbon compounds the rhizobia obtain during the infection process. The fact that the mutants are capable of infection, division and pleomorphic change indicates that they are capable of obtaining adequate carbon compounds for growth.

5. ANAEROBIC METABOLISM IN THE NODULE

The interior of the nodule is essentially microaerobic. Measurements by leghemoglobin spectroscopy (Bergeson, Turner 1975) and oxygen micro electrodes (Tjepkema, Yocum, 1974) show that the oxygen partial pressure within soybean nodules is only about 10 NM.

Early workers recognized that the nodule contained "fermentative" enzymes e.g. (Burris, Wilson, 1939). However, the implications of the low oxygen tension to carbohydrate metabolism remained unconsidered until recently. De Vries *et al.* (1980) found that the presence of 3.4 mM malate in pea nodules could be explained by the action of PEP carboxylase and malate dehydrogenase in the nodule cytosol, and they observed that the nodule contained alcohol dehydrogenase. The activities of these enzymes were higher in nodules than in the root. If plants were cultured with their roots in suboptimal oxygen concentrations, the enzyme activities in roots then resembled those in nodules. De Vries *et al.*, proposed that because of its essentially anaerobic nature, the nodule cytosol might have an anaerobic metabolism like that in roots of flooded plants.

We have found that the soybean nodule is "flood tolerant (Huang, LaRue, unpublished). Soybeans growing on N-free nutrient in fine sand in the greenhouse were flooded when 5 weeks old. Control plants were not flooded. Plants were uprooted at two day intervals and nitrogenase activity (C_2H_2) determined on excised roots in air. At two day intervals, the water was drained from some of the flooded plants, and nitrogenase activity of those plants determined one or two days after draining.

Nitrogenase activity is essentially absent after two days flooding. If the pots are drained, however, activity returns within two days to the same level as unflooded controls. The nodules can survive eight days flooding; the rapid return of activity with two days suggests that there is little if any loss of nodule structure.

During the flooding, there is an increase in alcohol dehydrogenase activity in the nodule. That the nodule can maintain its intregity and synthesize enzyme shows that it can obtain at least some energy under anaerobic conditions. It may be, therefore, that the fermentation enzymes observed in nodules are a protective mechanism against flooding or root anaerobiosis. Their presence does not prove they function under normal conditions.

In some flood tolerant plants, the "branch point" for aerobic or anaerobic metabolism is pyruvic acid. In oxidative metabolism, pyruvate is converted to acetyl CoA which enters the tricarboxylic acid cycle. In anaerobic metabolism, pyruvate may be converted to oxalacetate, malate and succinate (via PEP carboxylase), to lactate (via lactic dehydrogenase), to alanine (via transaminase) and to acetaldehyde and ethanol (via pyruvate decarboxylase). There is evidence that all these pathways are operative in the legume nodules.

5.1 Lactate dehydrogenase

Lactate dehydrogenase activity was detected in the cytosol from nodules of faba bean, lima bean, cowpea, pea, lupin, red clover and soybean (Tajima, LaRue, 1982). A possible fate of lactate was demonstrated by Jackson, Evans (1966), who found by labelling experiments that extracts of *R. japonicum* bacteroids converted lactate to propionate, which in turn was a precursor of heme.

5.2 Pyruvate decarboxylase

Soybean nodules contain low levels (40-60 nM) of acetaldehyde (Tajima LaRue, 1982). That compound seldom is found in high concentration in biological systems, for it acts as a feedback inhibitor of the enzymes producing it. The nodule cytosol fraction of seven legume species contain pyruvate decarboxylase. The enzyme was not found in the *R. japonicum* bacteroid or in the soybean roots. As the soybean plant ages the activity of pyruvate decarboxylase in nodules rises and falls in a similar fashion to nitrogenase activity, peaking about flowering.

5.3 Alcohol dehydrogenase

The soybean nodule contains ethanol. Labelled PEP and pyruvate are precursors to ethanol; and presumably the alcohol arises *via* acetaldehyde and alcohol dehydrogenase. By contrast with pyruvate decarboxylase, alcohol dehydrogenase activity is found in the bacteroids and in the root as well as in the cytosol of soybean nodules. We have failed to find alcohol dehydrogenase in bacteroids isolated from nodules of pea or sweet clover.

5.4 Aldehyde dehydrogenase

We demonstrated that acetaldehyde is converted to acetate in extract of *R. japonicum* bacteroids. The acetaldehyde may be metabolized by one of at least three enzyme systems in the bacteroid. There is a soluble NAD preferring aldehyde dehydrogenase, a soluble NADP preferring aldehyde dehydrogenase and an uncharacterized aldehyde stimulated respiration inhibited by azide but not by 6-cyano purine (Peterson, LaRue, 1982).

The soluble NAD preferring aldehyde dehydrogenase was studied in some detail. It occurs both in bacteroids and free living *R. japonicum* cultured on mannitol. The specificity is broad; acetaldehyde, propionaldehyde, butyraldehyde, and succinic semialdehyde are substrates. The aldehyde dehydrogenase was inhibited by 6-cyano purine, but not metronidazole. Both compounds inhibit C_2H_2 reduction by isolated bacteroids with several carbon substrates.

6. USE OF FERMENTATIVE PRODUCTS BY THE BACTEROID

Alcohols and aldehydes can increase respiration and C_2H_2 reduction above endogenous levels by isolated R. japonicum bacteroids. They are not, however, as active as malate or succinate in these regards. Like dicarboxylic acids, ethanol and acetaldehyde shift upward the concentration of O_2 at which nitrogenase activity is maximum. This effect may be due to a respiratory protection: compounds supporting oxidative phosphorylation reduce the intracellular concentration of O_2 which would otherwise inhibit nitrogenase.

The role of ethanol and acetaldehyde in the nodule is not known. They are attractive as putative substrates for the bacteroid; being neutral molecules they should pass through the peribacteroid and bacteroid membranes without the energy cost associated with transport of organic acids. The midpoint potentials of the ethanol-acetaldehyde and acetaldehyde-acetate couples are -197 and -586 mV respectively. These oxidations might be the source of low potential reductant needed for nitrogenase.

If ethanol is a major carbon source for bacteroids, and if most of the bacteroids metabolism is to support nitrogenase activity, then blocking nitrogenase activity might lead to an accumulation of substrate.

We have found that a brief treatment with 100% O_2 is a convenient way of reducing nitrogenase activity without apparent other damage to the soybean plant (Patterson, Peterson, LaRue, 1983a, Patterson, LaRue, 1983b). Decapitated roots or excised nodules were analyzed for C_2H_2 reduction and CO_2 evolution before and after flushing with 100% O_2. The $\Delta CO_2/\Delta C_2H_2$ is an estimate of the amount of carbon used to support nitrogenase activity. In the same or parallel experiments, using root segments as nonfixing controls, we measured by gas chromatography the accumulation of ethanol on the gas phase.

TABLE I

Activity μmole h^{-1} gm nodule^{-1}	Before O_2 Treatment	After O_2 Treatment	Change in Activity
Nitrogenase	3.8	0	3.8
CO_2 evolution	35	28	7
Ethanol evolution	0	.042	.042

Oxygen eliminated nitrogenase activity and decreased nodule respiration (Table I). However ethanol did not accumulate at a concentration similar to the decrease in CO_2. Our observations, of which this is typical, is that ethanol is not likely a major C source for the soybean bacteroid.

References

Arias A, Cervenansky C, Gardiol A. and Martinez-Drets G. (1979) J. Bacteriol 137, 409-414
Bergersen FJ and Turner GL (1975) J. Gen. Microbiol 89, 31-47
Burris RH and Wilson PW (1939) Cold Spring Harbor Symp Quant Biol 7, 349-361
De Vries GE, In't Veld P and Kijne JW (1980) Plant Sci Lett 20, 115-123
De Vries GE, van Brussel AAN and Quispel A (1982) J. Bacteriol 149, 872-879
Duncan MJ and Fraenkel DG (1979) J. Bacteriol 37, 415-419
Finan TM, Wood JM and Jordan DC (1983) J. Bacteriol 154, 1403-1413
Gardiol A, Arias A, Cervenansky C and Martinez-Drets G. (1982) J. Bacteriol 151, 1621-1623
Glenn AR and Dilworth MJ (1981b) Arch Microbiol 129, 233-239
Glenn AR and Dilworth MJ (1981b) J. Gen Microbiol 126, 243-247
Glenn AR, McKay IA, Arwas R and Dilworth MJ. J. Gen Micrbiol, in press
Hooijmans JJM and Logman GJJ (1983) These proceedings
Hudman JF and Glenn AR (1980) Arch. Microbiol 128, 72-77
Jackson EK and Evans HJ (1966) Plant Physiol 41, 1673-1680
Jacobson GR and San Francisco M (1983) 9th North American Rhizobium Conference, Abstr. p56
Manian SS, Gumbleton R and O'Gara F (1982) Arch Microbiol 133, 312-
Martinez de Drets G, Arias A, Rovira de Cutinella M (1974) Can. J. Microbiol 20, 605-609
Patterson TG, Peterson JB and LaRue TA (1983) Plant Physiol 72, 695-700
Patterson TG, and LaRue TA (1983) Plant Physiol 72, 701-705
Peterson JB and TA LaRue (1982) J. Bacteriol 151, 1473-1484
Reibach PH and Streeter JG (1983) Plant Physiol
Ronson CW and Primrose SB (1979) J. Gen Microbiol 112, 77-88
Ronson CW, Lyttleton P and Robertson JG (1981) Proc Natl Acad Sci 78, 4284-4288
Ruiz-Argueso T, Emerich DW and Evans HJ (1979) Arch Microbiol 121, 199-206
Smith AM and ApRees T (1979) Phytochemistry 18, 1453-1458
Stovall I and Cole M (1978) Plant Physiol 61, 687-690
Stowers MD and Elkan GH (1983) Can J. Microbiol 29, 398-406
Tajima S and LaRue TA (1982) Plant Physiol 70, 388-392
Tjepkema JD and Yocum CS (1974) Planta 119, 351-360
Trinchant JC, Birot AM and Rigaud J (1981) J. Gen Microbiol 125, 159-165
Werner D, Dittrich W and Thierfelder H (1982) Z. Naturforsch 37c, 921-926

THE CELLULAR AND INTRACELLULAR ORGANIZATION OF THE REACTIONS OF UREIDE BIOGENESIS IN NODULES OF TROPICAL LEGUMES

KAREL R. SCHUBERT AND MICHAEL J. BOLAND
MONSANTO AGRICULTURAL PRODUCTS, ST. LOUIS, MISSOURI 63167, U.S.A.

Ureides - Major Products of Recently-Fixed Nitrogen in Tropical Legumes.

Although the ureides, allantoin and allantoic acid, were first detected in leguminous plants in the 1930's (Fosse et al., 1930; Umbreit, Burris, 1938), the importance of the ureides to the nitrogen economy of many legumes has only recently been defined. These compounds represent the major products of recently-fixed nitrogen exported from the nodules of many tropical legumes including soybeans, cowpeas, mung bean, winged bean, and garden bean (Matsumoto et al., 1977a; Herridge et al., 1978; McClure, Israel 1979; Streeter 1979; Pate et al., 1980) Rainbird (1982) has conducted an extensive survey of ureide-exporting and amide-exporting legumes. The ureides account for 60 to 99% of the total nitrogen in the xylem exudate of soybeans (Schubert 1981).

Pathways of Ureide Biogenesis in Nodules

Two pathways were proposed for ureide biogenesis (Bollard 1959; Reinbothe, Mothes 1962). The first involved the condensation of urea and a two-carbon compound such as glyoxylate while the second involved the oxidative catabolism of purines. To date, there is little evidence to support the occurrence of the first pathway. The presence of significant levels of the enzymes of purine oxidation, anthine dehydrogenase, urate oxidase, and allantoinase, during nodule development support the existence of the latter pathway (Atkins et al., 1980; Schubert 1981). The effects of allopurinol, a specific inhibitor of xanthine dehydrogenase, on the accumulation or ureides *in vivo* and on the conversion of purine nucleotides and nucleosides *in vitro* supported this proposal (Fujihara, Yamaguchi 1978; Atkins et al., 1980; Triplett et al., 1980; Atkins, 1981).

Involvement of de novo Purine Synthesis

Ureides can be formed via the turnover of preexisting nucleic acids or from purines synthesized *de novo*. The rapid labeling of ureides in nodules with $^{15}N_2$ (Ohyama, Kumazawa 1978; Tajima, Yamamoto 1975) or $[^{13}N]N_2$ (Schubert, Coker 1981) argue for the latter possibility. This conclusion is supported by several lines of evidence. To begin, activities of several key enzymes in purine biosynthesis and the synthesis of purine precursors were elevated in soybean nodules in relation to the activities in lupin nodules, an amide-exporter (Reynolds et al., 1982a). Second, the activities of these enzymes including PRPP (Phosphoribosylpyrophosphate) synthetase and PRPP amidotransferase were specifically induced during nodule development (Schubert 1981; Reynolds et al., 1982b). The *in vitro* activity of PRPP synthetase, which catalyzes the synthesis of PRPP used in purine synthesis, was correlated with the estimated rates of N_2 fixation (based on C_2H_2 reduction) and the apparent rates of ureide export from soybean nodules (Schubert 1981). Based on these estimates, the rates of PRPP synthesis were more than sufficient to account for the rates of purine synthesis necessary for the flux of recently-fixed N_2 being exported as allantoin and allantoic acid. The third line of evidence for the involvement of *de novo* purine synthesis is based on labeling studies using purine precursors. Atkins et al., (1980) first demonstrated that labeled glycine, a purine precursor, was incorporated into allantoic acid in cowpea nodule slices. The majority (70%) of the label was present in carbons which would be derived from the incorporation of glycine into the purine ring. These findings were consistent with the action of glycinamide ribonucleotide synthetase, a purine biosynthetic enzyme, although they do not exclude the possible condensation of glycine or a two carbon compound derived from glycine with urea to form allantoic acid. Boland and Schubert (1981) confirmed that purines are synthesized de novo in intact nodules by labeling the purine ring at position 6 with ^{14}C from $^{14}CO_2$. The insertion of $^{14}CO_2$ into the purine ring is catalyzed by aminoimidazole ribonucleotide carboxylase. Because C_6 is oxidized in the urate oxidase-dependent oxidation of the purine ring to produce allantoin, these experiments were conducted in the presence of allopurinol. Under these conditions, xanthine labeled specifically in C_6 accumulated. All of these results support the involvement of de novo purine synthesis in the biogenesis of ureides in nodules.

Intracellular Organization of the Reactions of Ureide Biogenesis.

Urate oxidase and allantoinase are reportedly present in subcellular organelles in both plants and animals (Tolbert 1980). These reports prompted a number of investigators to investigate the subcellular location of these enzymes in nodules. Atkins et al. (1980) reported that the enzymes of purine oxidation and catabolism were in the soluble fraction of extracts of cowpea nodules fractionated by sucrose gradient centrifugation, although they did not exclude the possibility of subcellular compartmentation. Similar findings were reported by Triplett et al. (1980) for soybean nodules. Hanks et al. (1981) developed extraction procedures and gradient separations to maximize the recovery of intact peroxisomes from nodules. Using these techniques, Hanks et al. (1981) demonstrated that urate oxidase and catalase were, in fact, localized in peroxisomes in soybean nodules. The localization of xanthine dehydrogenase and allantoinase, two other enzymes involved in purine catabolism, was examined. Xanthine dehydrogenase was cytoplasmic. In contrast to other reports which indicated that allantoinase was associated with microbodies (see Tolbert 1980), Hanks et al. (1981) found that allantoinase was associated with the microsomal fraction originating from the endoplasmic reticulum. These observations were just the beginning of the complex pattern of compartmentalization of the reactions of ureide biogenesis which has emerged.

Boland et al. (1982) conducted similar studies to examine the subcellular localization of enzymes involved in ammonium assimilation, purine biosynthesis, and the synthesis of the carbon-containing purine precursors. Glutamine synthetase, the first enzyme of ammonium assimilation was located almost entirely in the soluble fraction. A similar location has been reported for nodules of Phaseolus vulgaris (Awonaike et al., 1981) and in roots of maize, rice, bean, pea, and barley (Suzuki et al., 1981). Glutamate synthase and aspartate aminotransferase are both associated, at least in part, with the plastid fraction. Glutamate synthase apparently is in the plastids of bean nodules (Awonaike et al., 1981) and soybean roots (Suzuki et al., 1981). Aspartate amino transferase has also been associated with plastids in bean nodules (Miflin 1974). This enzyme exists as two isozymes in soybean nodules (Ryan et al., 1972) as well as in the nodules of lupin (Reynolds, Farnden 1979). The two isozymes can be separated and identified by gel electrophoresis; the isozyme with higher mobility is induced in the nodule along with other assimilatory enzymes and presumably is the major enzyme responsible for aspartate synthesis (Reynolds, Farnden 1979). Boland et al. (1982) found that the faster-moving isozyme was localized within the plastid. Asparagine synthetase activity, present in young nodules at low levels, was also located in the plastid fraction. These observations suggest that many of the reactions of ammonia assimilation and amino acid synthesis are compartmentalized within the plastid.

The synthesis of purines requires glycine and one-carbon units. Results of enzymological studies presented by Reynolds et al. (1982b) suggested that these compounds may be formed from serine via the phosphoglycerate-phosphoserine pathway. The presence of other amino acid synthesizing enzymes in plastids suggested that a similar location would be logical for enzymes involved in the synthesis of serine and ultimately glycine and C1-units. This notion was substantiated by Boland et al. (1982). Phosphoglycerate dehydrogenase, serine hydroxymethylase and N^{15}-N^{10} methylene tetrahydrofolate dehydrogenase were all associated with the plastid. A specific phosphatase for phosphoserine was detected in nodule extracts and was enriched in the plastid fraction (Schubert, Boland 1982). A specific aminotransferase has not been detectable using glutamate or alanine as an amino donor (Boland, Schubert, unpublished).

The distribution of two key enzymes of purine synthesis in different subcellular fractions were determined in these studies (Boland et al., 1982). PRPP synthetase activity was primarily soluble but was highly labile and a large proportion of the activity was lost during the centrifugation. Small amounts of activity were associated with the organelles. PRPP amidotransferase on the other hand was localized within the plastid. Based on these results Boland et al. (1982) proposed a model for the intracellular organization of the reactions of ureide biogenesis in which the synthesis of purines and purine precursors starting from glutamine and the glycoytic intermediate, triose phosphate, occurred in the plastids of soybean nodules.

To substantiate this proposal, Boland and Schubert (1983) isolated a plastid fraction on sucrose step gradients and incubated this fraction with labeled precursors. This fraction was capable of incorporating ^{14}C from [^{14}C]glycine into purines in the presence of added PRPP, glutamine, aspartate, ATP, bicarbonate, methenyl tetrahydrofolate, $MgCl_2$ and KCL. Ribose-5-phosphate and ATP could be substituted for PRPP in the reaction mixture and still maintain purine synthesis. This observation suggests that PRPP synthetase activity may be located within or associated with the plastid. Label was incorporated from [U-^{14}C]serine and [3-^{14}C]serine into purines in the presence of NADP+ and tetrahydrofolate, supporting the presence of a functional pathway for glycine and C1 synthesis from serine. The major product was IMP. Atkins and Shelp (1983) have recently reported preliminary findings which indicate that a similar compartmentalization of the purine biosynthetic pathway exists in cowpea nodules.

The oxidative conversion of IMP to uric acid can occur via several routes. The conventional pathway involves the conversion of IMP to hypoxanthine, with subsequent oxidation to xanthine and then uric acid in reactions catalyzed by xanthine dehydrogenase. Alternatively IMP can be oxidized to XMP through the action of IMP dehydrogenase and then XMP can be used to form xanthine. The former was favored because of the inability to detect IMP dehydrogenase activity in crude extracts of cowpea or soybean nodules (Atkins, Blevins, 1981). Results of studies in which allopurinol was used to inhibit xanthine dehydrogenase contradicted this conclusion. In the presence of allopurinol, xanthine, and not hypoxanthine as expected, accumulates in nodules. The detection of IMP dehydrogenase in the proplastid fraction of soybean nodules provided the first evidence that the pathway through XMP was operational. Boland and Schubert (1983) have suggested that this is the normal pathway for ureide biogenesis. Shelp and Atkins (1983) have recently reported the presence of IMP dehydrogenase in cowpea nodule extracts. This enzyme functions at the branch between funneling purines into ureide synthesis or into synthesis of purine nucleotides for cellular needs and, therefore, may be expected to be a very important regulatory protein. The enzyme, however, is highly labile and has been very difficult to purify.

Cellular Organization of Reactions of Ureide Biogenesis.

Using electron microscopy of soybean nodules Newcomb and Tandon (1981) observed that peroxisomes were present primarily in the uninfected cells of soybean nodules. These cells represent approximately half of the cells within the central nodule tissue. Smooth endoplasmic reticulum was also more prevalent in the uninfected cells. On this basis, Newcomb and Tandon (1981) suggested that part of the ureide biogenic pathway may be located in uninfected cells. Hanks et al. (1983) have isolated protoplasts of uninfected and infected cells of soybean nodules and have shown that urate oxidase and allantoinase are enriched in the uninfected cells. Plastids are present in both cell types in Newcomb and Tandon's electromicrograph. Although structural differences in the plastids are evident, enzymes involved in ammonium assimilation were distributed more evenly within cell types. The plastid isozyme of aspartate amino transferase, however, was slightly enriched in the infected cells. The difficulty in detecting the purine biosynthetic enzymes has hindered the establishment of the cellular location of purine synthesis in the two cell types.

Atkins (1983) has suggested that a similar pattern of cellular compartmentalization exists in cowpea nodules. Based on the results discussed above, the following model for the cellular and subcellular organization of the reactions of ureide biogenesis in nodules of tropical legumes is proposed (Fig. 1).

The proposed localization of plastids involved in purine synthesis within the infected cells may offer several advantages. First, the enzymes involved in the synthesis of purines and purine precursors as well as reaction intermediates are oxygen labile. The low partial pressure of oxygen in infected cells would help circumvent this sensitivity. Second, the purine pathway requires a large flux of ATP to sustain activity and is normally very sensitive to inhibition by ADP. The higher metabolic activity associated with infected cells would favor purine synthesis in these cells.

Urate oxidase has been purified from cowpea (Rainbird, Atkins, 1981) and soybean nodules (Lucas et al., 1983). The enzyme from both sources has a very high Km for oxygen and on this basis, Rainbird and Atkins (1981) proposed that the urate oxidase reaction could be a rate-limiting step in ureide synthesis and N_2 fixation. Assuming that urate oxidase is present in the leghemoglobin-containing cells, urate oxidase would be virtually non-functional in the presence of the low oxygen-tensions present within infected cells. In contrast, localization of uricase in uninfected cells could circumvent this apparent limitation.

Figure 1. Proposed Model for the Cellular and Subcellular Organization of the Reactions Involved in Ureide Biogenisis

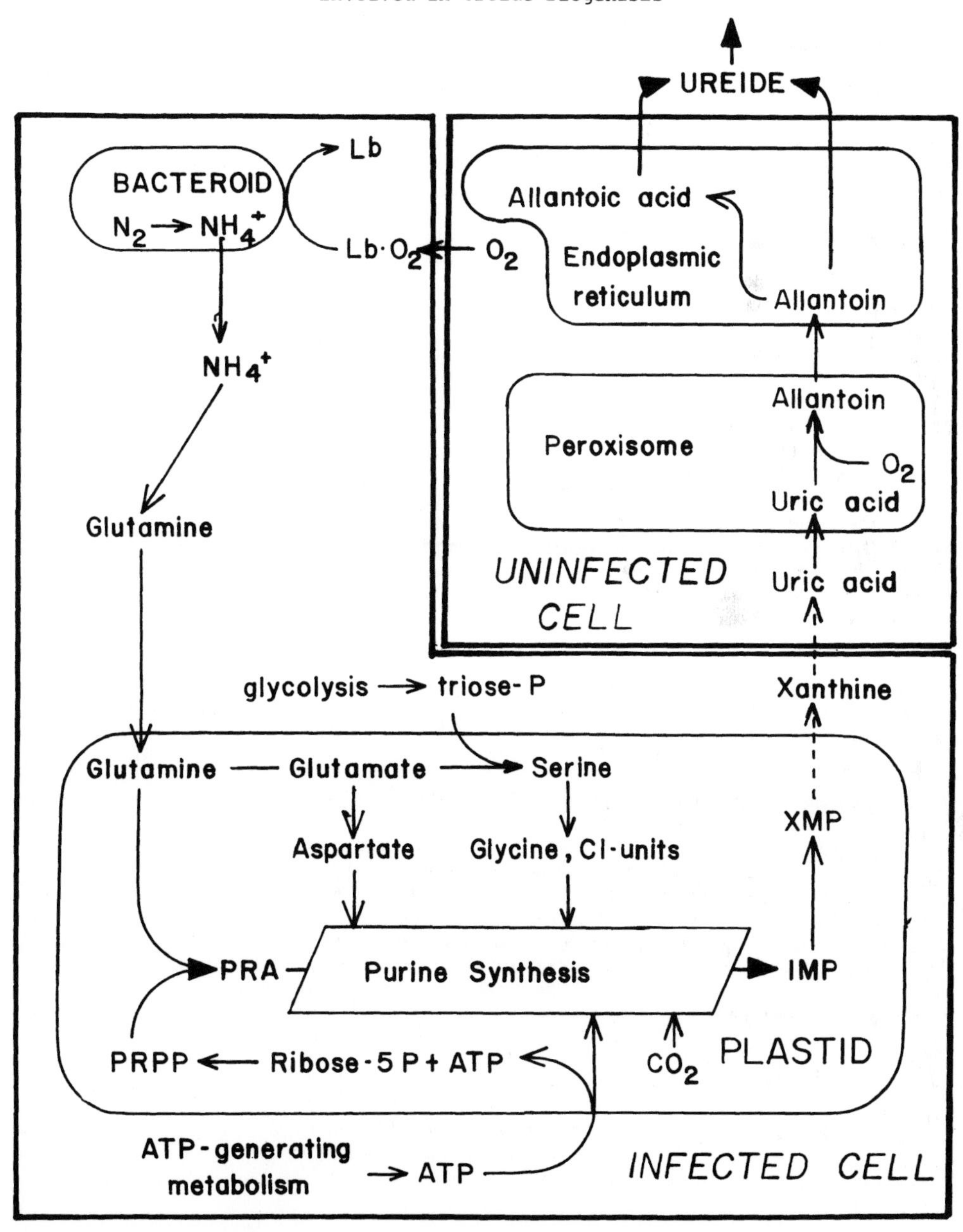

Uninfected cells may also play a role in the loading of ureides into the xylem stream. Presumably, a steep concentration gradient of nitrogenous intermediates from the infected cells through uninfected cells to the xylem would be necessary to drive the transport of these intermediates into the various subcellular and cellular compartments and to maintain the flow of precursors through the pathway. The energetic cost of maintaining this gradient is difficult to estimate. Based on the value of one ATP equivalent per molecule transported across a membrane, however, the cost could be substantial.

The model proposed is speculative and further research using enzymological, immunological, molecular, and tracer techniques is required to substantiate or refine the model, Key areas requiring further experimentation include the determination of the exact location of plastids involved in purine synthesis, the identification of the nitrogenous intermediate transported between the infected and uninfected cells, and the localization of PRPP synthetase and xanthine dehydrogenase.

Areas of Future Research

An understanding of the molecular and biochemical regulation of ureide synthesis is one important area for future research. The mechanism regulating the differentiation of cortical cells into uninfected and infected cells is unknown. Likewise, very little is known about the control of the enzymes involved in purine synthesis and catabolism. Because of the essential nature of purines for cellular functions, the balance of purine synthesis for ureide production and for nucleotide synthesis must be highly regulated. Key points at which regulation may occur include the reactions catalyzed by PRPP synthetase, PRPP amidotransferase, and IMP dehydrogenase. The possible existence of different isoenzymes and/or compartments for purine biosynthesis for cellular purine pools needs addressed.

Other areas requiring additional investigation include the biochemical mechanism of ureide utilization within leaves and pods and the effects of nitrate on the ureide biosynthetic pathway.

Acknowledgement

The authors would like to thank all those who have contributed scientifically to the rapid increase in our knowledge of ureide biogenesis, especially J. F. Hanks, C. A. Atkins, N. E. Tolbert, P. H. S. Reynolds, D. G. Blevins. Special thanks are due to G. M. DeShone for her technical assistance and support and to B. J. Smith and M. H. Zimmer for their careful preparation of this manuscript. Much of the work reported was supported by grants from the National Science Foundation and the United States Department of Agriculture Competitive Research Grants Program.

LITERATURE CITED

Atkins CA (1981) FEBS Lett. 125, 89-93.
Atkins CA and Blevins DG (1981) In Gibson AH and Newton WE, eds, Current Perspectives in Nitrogen Fixation, pp. 271-272, Austr. Ac. Sc., Canberra.
Atkins CA, Rainbird RM and Pate JS (1980) Z. Pflanzenphysiol. 97, 249-260.
Atkins CA and Shelp BJ (1983) Plant Physiol. 72 S, 113.
Awonaike KO, Lea PJ and Miflin BJ (1981) Plant Sci. Lett. 23, 189-195.
Boland MJ, Hanks JF, Reynolds PHS, Blevins DG, Tolbert NE and Schubert KR (1982) Planta 155:45-51.
Boland MJ and Schubert KR (1981) Arch. Biochem. Biophys. 213, 486-491.
Boland MJ and Schubert KR (1983) Arch. Biochem. Biophys. 220, 179-187.
Bollard EG (1959) Symp. Soc. Exp. Biol. 13:304-329.
Fosse R, Brunel A, de Draeve P, Thomas PE and Sarazin J (1930) C.R. Acad. Sci. 191, 1153-1155.
Fujihara S and Yamaguchi M (1978) Plant Physiol. 62, 134-138.
Hanks JF, Schubert KR and Tolbert NE (1983) Plant Physiol. 71, 869-873.
Hanks JF, Tolbert NE and Schubert KR (1981) Plant Physiol. 68, 65-69.
Herridge DF, Atkins CA, Pate JS and Rainbird RM (1978) Plant Physiol. 62, 495-498.
Lucus K, Boland MJ and Schubert KR (1983) Arch. Biochem. Biophys. 225, in press.
Matsumoto T, Yatazawa M and Yamamoto Y (1977a) Plant and Cell Physiol. 18, 353-359.
Matsumoto T, Yatazawa M and Yamamoto Y (1977b) Plant and Cell Physiol. 18, 459-462.
McClure PR and Israel DW (1979) Plant Physiol. 64, 411-416.
Miflin BJ (1974) Plant Physiol. 54, 550-555.
Newcomb EH and Tandon SR (1981) Science 212, 1394-1396.
Ohyama T and Kumazawa K (1978) Soil Sci. Plant Nutr. 24, 525-533.
Rainbird RM (1982) Ph.D. Thesis, Univ. Western Australia, Nedlands.
Rainbird RM and Atkins CA (1981) Biochem. Biophys. Acta. 659, 132-140.
Reinbothe H and Mothes K (1962) Ann. Rev. Plant Physiol. 13:129-150.
Reynolds PHS, Blevins DG, Boland MJ, Schubert KR and Randall DD (1982a) Physiol. Plantarum 55, 255-260.
Reynolds PHS, Boland MJ, Blevins DG, Schubert KR and Randall DD (1982b) Plant Physiol. 69, 1334-1338.
Reynolds PHS and Farnden KJF (1979) Phytochemistry 18, 1625-1630.
Ryan E, Bodley F and Fotrell PF (1972) Phytochemistry 11, 957-963.
Schubert KR (1981) Plant Physiol. 68, 1115-1122.
Schubert KR and Boland MJ (1982) Plant Physiol. 69 S, 112.
Schubert KR and Coker GT (1981) In Root JW and Krohn KA, eds, Short-lived Radionucleides in Chemistry and Biology, Adv. Chem. Ser. 197, 317-339.
Shelp BJ and Atkins CA (1983) Plant Physiol. 72 S, 113.
Streeter JG (1979) Plant Physiol. 63, 478-480.
Suzuki A, Gadal P and Oaks A (1981) Planta 151, 457-461.
Tajima S and Yamamoto Y (1975) Plant and Cell Physiol. 16:271-282.
Triplett EW, Blevins DG and Randall DD (1980) Plant Physiol. 65, 134-138.
Tolbert NE (1980) In Stumpf PK and Conn EE, eds, The Biochemistry of Plants - A Comprehensive Treatise, Vol 2 - Metabolism and Respiration, pp. 488-523, Academic Press, New York.
Umbreit WW and Burris RH (1938) Soil Sci. 45, 111-126.
Woo KC, Atkins CA and Pate JS (1981) Plant Physiol. 66, 735-839.

ENERGY METABOLISM IN NODULATED ROOTS

HANS LAMBERS AND RIES DE VISSER
DEPARTMENT OF PLANT PHYSIOLOGY, UNIVERSITY OF GRONINGEN, P.O. BOX 14, 9750 AA HAREN, THE NETHERLANDS

1. INTRODUCTION

This paper discusses some aspects of the energy metabolism in nodulated roots. It is the aim to address the problem if the supply of photosynthates to the nodulated roots or the production of metabolic energy in respiratory pathways limits the rate of N_2-fixation in nodulated roots of vegetative *Pisum sativum* plants. Our approach involves the determination of the activity and/or capacity of the electron transport pathways in pea roots. Hence some information is included on the non-phosphorylating CN-resistant electron transport pathway -the alternative path- and the regulation of electron transport via the phosphorylating cytochrome path. For further details, see Laties 1982, Lambers 1982 and Day, Lambers 1983.

The alternative path, which branches from the cytochrome path at the level of ubiquinone, is not coupled with ATP formation. It is only engaged upon saturation of the cytochrome path and should be considered as an "overflow" rather than as a "leak" for electrons. Saturation of the cytochrome path occurs when the activity of glycolysis plus TCA-cycle is so high that more NADH is generated than can be accommodated by the cytochrome path plus biosynthetic reactions requiring NADH, e.g. NO_3-reduction. Engagement of the alternative path is generally associated with a high input of carbohydrates into a sink organ. Saturation of the phosphorylating path is determined either by the capacity of (one of) the carriers in this pathway, or by the availability of ADP.

As will be shown below, the alternative path is only marginally engaged in nodulated pea roots, whereas the cytochrome path operates without any constraint by ADP. The implications of these observations for the regulation/limitation of the rate of N_2-fixation will be discussed.

2. MATERIAL AND METHODS

2.1. Germination and growth: Seeds of *Pisum sativum* L. cv. Rondo were germinated and inoculated as described by De Visser, Lambers (1983). Six days old seedlings were transferred into aerated nutrient solutions without combined N; all other conditions were as described before, unless stated otherwise. NO_3-grown plants were cultivated in a solution as described by De Visser, Lambers (1983). The age of the pea plants used in the present experiments varied between 20 and 30 days. Germination and growth of *Lupinus albus* L. cv. Ultra was as described by Lambers et al. 1980.

2.2. Root respiration: O_2-uptake was measured polarographically (Lambers et al. 1983); inhibitors were added as described elsewhere (Day, Lambers 1983). CO_2-production was measured by infrared gas analysis.

2.3. N_2-fixation: C_2H_2-reduction and H_2-evolution were measured by gas chromatography (De Visser, Poorter 1983).

2.4. Carbohydrate assays: Soluble and insoluble (starch) carbohydrates were extracted and analyzed with the anthrone reagent.

3. RESULTS AND DISCUSSION

3.1. Respiratory characteristics of NO_3-fed and N_2-fixing roots

3.1.1. Respiratory rates: It is often assumed -on the basis of a high C-requirement of N_2-fixation- that the rate of root respiration is higher in N_2-fixing plants than in NO_3-fed plants. When the rate of N_2-fixation equalled that of NO_3-assimilation, O_2-uptake in roots of *Pisum sativum* was slightly, and CO_2-evolution was significantly higher in N_2-grown plants (Table 1). However, the rate of root respiration -both O_2-uptake and CO_2-production- in *Lupinus albus* was highest in NO_3-grown plants (Table 1).

Table 1. Respiratory characeristics of the roots of N_2-fixing and NO_3-grown Pisum sativum. For comparison data on 28 days old Lupinus albus (Lambers et al. 1980) are included. Rates in µmoles.h^{-1}.g^{-1} (dry wt); relative contribution of the alternative path (%) in brackets. All rates were measured during the light period. For L. albus the rates were somewhat lower during the dark period.

Species	N-source	Total O_2-uptake	RQ	O_2-uptake via alternative path
Pisum sativum	N_2	110	1.3	0 (0)
Pisum sativum	NO_3	98	0.9	20 (20)
Lupinus albus	N_2	119	1.6	10 (8)
Lupinus albus	NO_3	184	1.4	50 (27)

A significant portion of the respiration of nodulated roots resides in the nodules. On a dry weight basis the O_2-uptake in the nodules of *Pisum sativum* was ca. 80% higher than that of the rest of the roots.

3.1.2. The respiratory quotient (RQ): The RQ of root respiration in N_2-fixing plants is generally higher than that in NO_3-grown plants, also when the rate of N-assimilation is the same (Table 1). This is not to be expected from the basic equations describing the reduction of NO_3 and N_2:

$$1\ NO_3 + 4\ \text{"NADH"} \rightarrow 1\ NH_3 \qquad (1)$$

$$\tfrac{1}{2}\ N_2 + 1\tfrac{1}{2}\ \text{"NADH"} \rightarrow 1\ NH_3 \qquad (2)$$

The production of each mole of "NADH" (i.e. reducing power such as NAD(P)H, ferrodoxin, etc.) is linked to the evolution of ½ mole of CO_2. The rate of CO_2-production associated with other respiratory processes being the same in NO_3-grown and N_2-fixing roots, eqns (1) and (2) lead to a higher RQ in NO_3-grown plants. However, a significant part of NO_3^--reduction may occur in the leaves (Wallace, Pate 1965), whilst a part of the reducing equivalents in nodules is donated to H^+ (Schubert, Evans 1976):

$$2\,H^+ \;+\; 1\ \text{"NADH"} \;\rightarrow\; 1\,H_2 \qquad (3)$$

Both phenomena offer an explanation for the lower RQ in NO_3-grown plants, in comparison with N_2-fixing plants. There is one more process in CO_2-exchange which needs to be considered. The high activity of PEP-carboxylase in some nodules, which is responsible for a high rate of dark CO_2-fixation, serves as an anaplerotic route. If glutamate or glutamine are the final products of N_2-fixation, an equivalent amount of CO_2 is produced by isocitrate dehydrogenase as fixed by PEP-carboxylase, so that there is no net effect on RQ. If asparagine or aspartate are the final products, RQ will tend to be reduced. However, since a similar anaplerotic route must also occur in NO_3-fed roots, PEP-carboxylase activity will not seriously affect the difference in RQ between N_2-fixing and NO_3-fed plants.

3.1.3. The CN-resistant, alternative path: A significant difference in respiration between N_2-fixing and NO_3-assimilation roots is the engagement of the alternative path (Table 1). The contribution of the alternative path in respiration of NO_3-grown roots is ca. 30%, whilst that in N_2-fixing roots is not significantly different from 0%. In roots of N_2-fixing plants of *Vigna angularis*, *V. unguiculata* and *Phaseolus aureus* similarly low activities of the alternative path were found (Table 2).

Table 2. The rate of root respiration in µmoles $.h^{-1}.g^{-1}$ (dry wt) and the relative contribution of the alternative pathway (%) of some N_2-fixing legume-Rhizobium combinations. One of the Rhizobium strains was isolated from Desmodium (Rh.ex Dem.). Culture conditions were as described by Lambers et al. (1980) for N_2-fixing plants of Lupinus albus. The plants were in the vegetative growth phase during the experiments, which were conducted by one of us (H. Lambers) and Dr. R.M. Rainbird at the Department of Botany, University of Western Australia.

Legume-Rhizobium combination	Total O_2-uptake	O_2-uptake via alternative path
Vigna angularis x CB756	130	1
V. angularis x Rhizobium sp. ex. Desmodium	120	0
Vigna unguiculata x CB756	120	4
V. unguiculata x Rhizobium sp. ex. Desmodium	150	4
Phaseolus aureus x CB756	160	7

3.2. The limitation of N_2-fixation by photosynthates

3.2.1. Short term effects of an increased supply of carbohydrates: The alternative path in roots of many higher plants is found to be engaged only when the input of carbohydrates into the roots exceeds their demand for growth and energy production (Lambers 1982). Therefore, the very small contribution of the alternative path in root respiration of N_2-fixing plants (Tabs 1 and 2) indicates that these roots do not have an excess of carbohydrates. Consequently, these results might point to the limitation of N_2-

fixation by the supply of carbohydrates, as suggested by many authors (e.g. Bethlenfalvay et al. 1979; Hardy, Havelka 1975; Herridge, Pate 1977).

To test the hypothesis that N_2-fixation is limited by carbohydrates, the level of sugars in the tissue was manipulated by addition of 25 mM sucrose to the medium and by increasing the light intensity from 225 to 375 $\mu E.m^{-2}.s^{-1}$. C_2H_2-reduction and H_2-evolution were slightly inhibited upon addition of sucrose. Upon an increase of the light intensity, C_2H_2-reduction and H_2-production were initially unaffected; only after prolonged exposure to 375 $\mu E.m^{-2}.s^{-1}$ these activities commenced to increase. The concentration of carbohydrates in the roots increased much earlier: 5 h after increasing the light intensity.

The lack of response of N_2-fixation in the first 5 h, despite the increased concentration of carbohydrates in the roots, indicates that carbohydrates per se were not limiting N_2-fixation. Two possibilities remain. (1) The capacity of the respiratory pathways, in particular that of the cytochrome path, was not sufficient to oxidize the extra input of sugars, so that the energy could not become available for N_2-fixation. And, (2) the capacity of N_2-fixation per se was not sufficient to allow increased activity of C_2H_2-reduction and/or H_2-evolution. A combination of (1) and (2) can also be envisaged (Fig. 1).

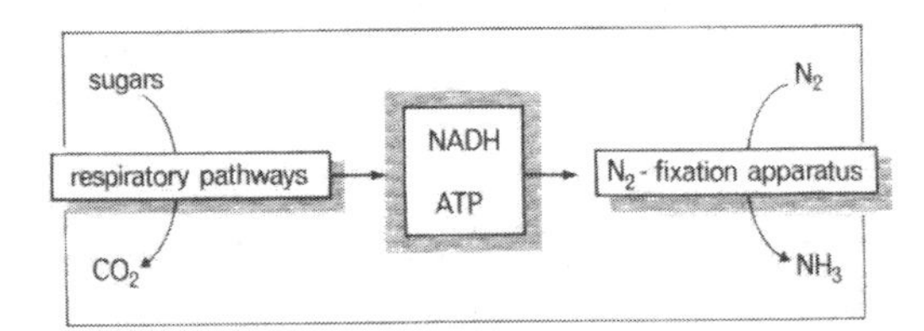

Fig. 1. Respiratory metabolism and N_2-fixation and their link through NADH and ATP.

There is good evidence that in nodulated pea roots the capacity of (one of) the carriers of the cytochrome path restricts the flux of electrons through this pathway. Upon addition of uncoupler to nodulated roots, respiration increases. However, it is only the activity of the alternative path which increases, whilst the flux of electrons through the cytochrome path is not affected (Table 3). This demonstrates that the flow of electrons through the cytochrome path is not restricted by the availability of ADP, such as in roots of wheat and other non-legume crop plants (Day, Lambers 1983), but rather by the capacity of (one of) the cytochromes. A limitation of the activity of the cytochrome path by the capacity of (one of) the carriers of this pathway was also found for roots of non-nodulated NO_3-grown legumes, including *Phaseouls vulgaris* (Day, Lambers 1983) and *Pisum sativum* (De Visser, Blacquière 1983).

Table 3. Root respiration in µmol $O_2.h^{-1}.(g\ dr\ wt)^{-1}$ of 20 d old nodulated roots of Pisum sativum, grown in nutrient solution without combined N. Root respiration was first measured in the absence of inhibitors (control) and then in the presence of 25 mM SHAM, to determine the activity of the cytochrome path (v_{cyt}, i.e. respiration in the presence of SHAM) and that of the alternative path (v_{alt}, i.e. control - v_{cyt}). The capacity of the cytochrome path (V_{cyt}) as determined by the carriers of the cytochrome chain was determined by measuring respiration in the presence of 25 mM SHAM and 1 µM CCCP. For comparison data on root respiration in a NO_3-fed non-nodulated legume (bean) and a non-legume crop plant (wheat) are included (Day, Lambers 1983).

Species	Control	v_{cyt}	V_{cyt}	v_{alt}	v_{cyt}/V_{cyt}
Pea	170	140	140	30	1.0
Bean	320	300	300	20	1.0
Wheat	200	140	170	60	0.8

Thus, an increased supply of sugars to the roots could not immediately lead to enhanced ATP-production, coupled to electron transport through the cytochrome path. And, consequently, N_2-fixation was unaffected by the increased supply of sugars. After prolonged exposure to high light intensity, the capacity of the cytochrome path increased, thus allowing an increase in the production of ATP (see section 3.2.2.).

Van Mil (1981) compared the actual rate of N_2-fixation, i.e. the in vivo activity of C_2H_2-reduction, with the potential rate, i.e. the C_2H_2-reduction rate in the presence of ATP and reducing power. He concluded that in the root nodules of 28 days old *Pisum sativum* plants the actual rate was ca. 80% of the potential rate of N_2-fixation. This suggest that possibility (2) can be eliminated.

Combination of the information on the effects of light intensity on N_2-fixation with that on the capacity of the phosphorylating cytochrome path in *Pisum sativum* roots (see above) leads to the suggestion that the rate of N_2-fixation in nodulated pea roots is not limited by the supply of photosynthates. Rather, it is the capacity of the nodulated roots -and more specifically presumably that of the bacteroids- to generate ATP in the respiratory processes which constrains N_2-fixation.

3.2.2. Long term effects of an increased supply of carbohydrates: After exposure of pea plants to 375 $\mu E.m^{-2}.s^{-1}$ for 24 hours, C_2H_2-reduction (expressed per g dry wt of nodules) increased to 400% of control plants, kept at 225 $\mu E.m^{-2}.s^{-1}$. Root respiration (in µmoles of CO_2 per g dry wt) increased to 180% of the control during the same period.

Growth of *Pisum sativum* under high light conditions (i.e. 375 vs. 225 $\mu E.m^{-2}.s^{-1}$) or at elevated concentration of CO_2 in the atmosphere (700 vs. 350 μl CO_2 (l of air)$^{-1}$) enhanced the rate of C_2H_2-reduction per plant. However, this was solely due to increased production of nodule weight and not to an increased specific activity of the nodules. Growth at high light intensity enhanced the rate of root respiration by 40%. However, an elevated concentration of CO_2 had no such effect (Table 4).

Table 4. The rates of acetylene reduction (μmoles.h^{-1}.(g dry wt of nodules)$^{-1}$ and root respiration (μmoles $CO_2.h^{-1}$. g dry wt^{-1}) and the nodule dry weight (mg. (g dry wt of plant)$^{-1}$) in control plants (225 μE.$m^{-2}.s^{-1}$, 350 μlCO_2.(l air)$^{-1}$) and in plants grown at high light intensity (375 μE.$m^{-2}.s^{-1}$) or an elevated atmospheric CO_2 concentration (700 μl CO_2 (l air)$^{-1}$) for 7 days.

	Control	High light	Control	High CO_2
C_2H_2-reduction	120 ± 15	118 ± 11	148 ± 13	126 ± 14
Root respiration	108 ± 6	153 ± 1	80 ± 6	85 ± 3
Nodule weight	40 ± 2	49 ± 1	33 ± 3	54 ± 3

It is concluded that an increased supply of carbohydrates to nodulated roots only increases the rate of C_2H_2-reduction, when the capacity of the cytochrome path has increased. Growth under conditions where the supply of carbohydrates to the nodulated roots is high leads to an increased production of nodule weight. This leads to enhanced N_2-fixation.

4. GENERAL DISCUSSION

The activity of the cytochrome path (v_{cyt}) in nodulated pea roots was demonstrated to closely match the carrier capacity of this path in nodulated roots of *P. sativum*. Consequently, upon on increased input of carbohydrates into the nodulated roots, extra ATP could not rapidly be generated to allow enhanced N_2-fixation. Upon prolonged exposure to conditions of increased carbohydrate input into nodulate roots, the capacity of the cytochrome path (V_{cyt}) increases, allowing increased respiration via the cytochrome path, increased ATP-generation and thus an enhancement of N_2-fixation.

It is concluded that N_2-fixation in nodulated pea roots grown under the present conditions is constrained by the capacity of (one of) the carriers of the cytochrome path. Thus, enhancement of N_2-fixation in nodulated pea roots can only occur if more of the fixed N is invested in the respiratory pathways in roots. Naturally, this will primarily be at the expense of N-investment into other proteins, e.g. those involved in photosynthesis. However, the following calculation suggests that there is scope for optimization of the rate of N_2-fixation in pea.

Consider a "realistic" pea plant (De Visser, Lambers 1983) with 4.5 and 3.5% N in its shoot and roots, respectively, a shoot to root ratio of 2 and a relative growth rate of 75 mg.g^{-1}.d^{-1} (both roots and shoot). One gram of these roots will accumulate

2.63 mg N per day; ca. 25% of this 0.68 mg) will be invested in mitochondria (R. de Visser, T. Blacquière, pers. comm.). The shoot in this plant will accumulate 6.75 mg N, approxiamtely half of which (3.4 mg) will be invested in the photosynthetic apparatus.

Consider now a "hypothetical" plant. This plant still fixes 2.63 + 6.75 = 9.38 mg N, but invests 10% more in its mitochondria in the nodulated roots (0.75 vs. 0.68 mg N), at the expense of an investment in the photosynthetic apparatus (now 6.68 vs. 6.75 mg N). This plant is likely to fix proportionally (ca. 2% less CO_2 and fix 10% more N (0.77 vs. 0.70 mg N), provided some stored reserves of carbohydrates can be used to generate sufficient metabolic energy and C-skeletons. (The latter proviso is likely to be met: De Visser, Lambers 1983). This extra N can now be invested into photosynthesis to replenish the carbohydrate pool used in this day and in addition be used for further growth. Fig. 2 shows the difference in photosynthesis and N-accumulation between a "realistic" and a "hypothetical" pea plant.

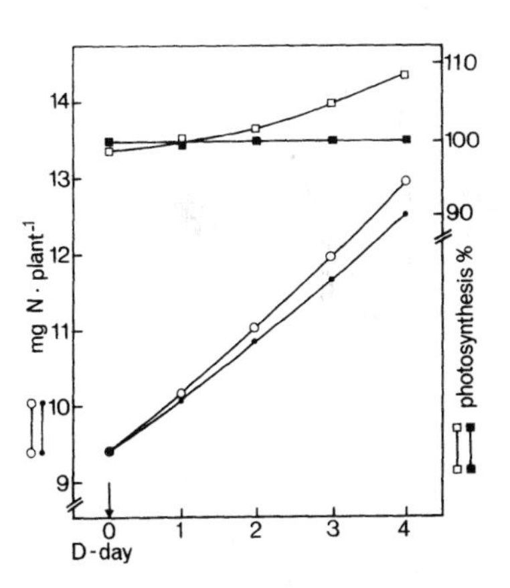

Fig. 2. N_2-fixation and photosynthesis in a "realistic" and a "hypothetical" pea plant. See text.

Our calculation with a hypothetical plants is a rather crude one. However, it does suggest that a change in investment of N, more N utilized for mitochondria, might lead to increased yields in legumes.

ACKNOWLEDGEMENTS

This research was supported in part by the Foundation for Fundamental Biological Research (BION) which is subsidized by the Netherlands Organization for the Advancement of Pure Research (Z.W.O.).

REFERENCES

Bethlenfalvay GJ, Norris RF and Philips DA (1979) Plant Physiol. 63, 213-215.
Day DA and Lambers H (1983) Physiol.Plant. 58, 155-160.
De Visser R and Blacquière T (1983) Plant Physiol., in press.
De Visser R and Lambers H (1983) Physiol.Plant. 58, 533-543.
De Visser R and Poorter H (1983) Physiol.Plant., submitted.
Hardy RWF and Havelka UD (1975) Science 188, 633-643.
Herridge DF and Pate JS (1977) Plant Physiol. 60, 759-764.
Lambers H (1982) Physiol.Plant. 55, 478-485.
Lambers H Layzell DB and Pate JS (1980) Physiol.Plant. 50, 319-325.
Laties GG (1982) Annu. Rev.Plant Physiol. 33, 519-555.
Schubert KR and Evans HJ (1976) Proc.Nat.Acad.Sci. 73, 1207-1211
Van Mil M (1981) PhD Thesis, Wageningen.
Wallace W and Pate JS (1965) Ann.Bot. 29, 655-671.

THE UNIQUE PROPERTIES OF THE SYMBIOTIC *ANABAENA AZOLLAE* IN THE WATER FERN *AZOLLA*: METABOLISM AND INTERCELLULAR RECOGNITION

ELISHA TEL-OR[1], TAL SANDOVSKY[1], HANNA ARAD[1], AVI KEYSARY[2] and DAVID KOBILER[2]

[1]Dept. of Agricultural Botany, The Hebrew University of Jerusalem, Rehovot 76100 and [2]Israel Institute for Biological Research, Ness Ziona 70450, Israel.

ABSTRACT

Cultured isolates of *Anabaena azollae* provide a suitable tool for studying the metabolism, transport, cell composition and antigenic properties of the cells. The isolates of *Anabaena azollae* from *Azolla caroliniana* and *Azolla filiculoides* exhibit antigenic homology in the quantitative immunoassays employed, suggesting that they may stem from the same parental *Anabaena*. These isolates were also similar in respect to the fructose supported N_2-fixation activity: the fructose is taken up actively by the cells of *Anabaena azollae* and although it does not support the growth of *Anabaena*, it carries a series of changes involving cell composition, differentiation and metabolism. Respiration was facilitated, the amount of storage products increased, and the frequency of heterocysts was increased. Sucrose and glucose stimulated respiration, but showed limited effect on N_2-fixation. The uptake of fructose, dependent on ATP synthesis, is recovered faster in starved cells than N_2-fixation activity. Fructose carrier seems to be constitutive in the cultured isolate of *Anabaena azollae* grown autotrophically. The lectins identified in *Anabaena azollae* Newton cells are constitutive as well, and their actual role in the symbiosis with *Azolla* has not been verified yet. They may be involved in the regulation and control of the development of *Anabaena* in the leaf cavity and, possibly, trigger the exchange of metabolites between the host *Azolla* and the N_2-fixing *Anabaena*.

INTRODUCTION

The heterosporous aquatic fern *Azolla* is associated with the N_2-fixing phycobiont *Anabaena azollae*. The phycobiont is accommodated in the leaf cavity of the host fern and the exchange of metabolites between the two partners is assumed to be an active process of photosynthate translocation from the host fern and translocation of fixed nitrogen compounds from the *Anabaena* (Peters et al., 1979). The development of *Anabaena azollae* in the leaf cavity is synchronized with the development of the accommodating leaf (Peters et al., 1982; Hill, 1977).

Few cultured isolates of *Anabaena azollae* were obtained for comparative studies with the fresh isolates (Newton, Herman, 1979; Gates et al., 1980; Ladha, Watanabe, 1982; Tel-Or et al., 1983). The cultured isolate of the phycobiont may retain the original metabolic properties, expressed *in situ*, thus being suitable for the investigation of its heterotrophic nature. On the other hand, antigenic differences were observed between fresh *Anabaena azollae* and cultured cells of the phycobiont (Gates et al, 1980; Ladha, Watanabe, 1982), suggesting additional modifications of the properties of the phycobiont in culture.

A comparison of cell antigens of the cultured isolates of *Anabaena azollae* may provide a tool for the determination of the degree of homology of the phycobiont cells in the different *Azolla* species.

A phytophaemagglutinin was observed in cells of cultured *Anabaena azolla*

(Kobiler et al., 1981) which may be involved in the recognition in situ between the phycobiont and the host cells forming the leaf cavity (Tel-Or et al., 1983).

This report provides a survey of the recent studies carried out with the cultured isolates of Anabaena azollae and the relevance of these studies to the properties of the phycobiont in situ.

1. Effect of Fructose on Growth, Cell Structure and Composition of Anabaena azollae

The activity of N_2-fixation in the dark with cells of Anabaena azollae var. caroliniana was supported by fructose, while sucrose, glucose, citrate, malate, pyruvate, succinate and acetate did not stimulate the activity of N_2-fixation (Tel-Or, Sandovsky, 1982). The effect of fructose on the growth of Anabaena azollae var. caroliniana was tested in the dark and in the light. Fructose facilitated the growth of cells only in the light in the presence of DCMU, where photosystem 2 activity is blocked. Fructose did not facilitate the growth of cells in the light or in the dark, but frequency of heterocysts was higher in the presence of fructose in the dark, light and light+ DCMU. The incorporated fructose caused an increase in the C/N ratio in the filament which was suggested to enhance differentiation of heterocysts (Stewart, 1976). Cells of Anabaena azollae grown in the dark or in the light in the presence of fructose undergo major modifications in ultrastructure: cell size of the fructose-grown cells was much larger and the cell surface was rough, as compared with the smooth cell surface of control cells. The intracellular structure, observed through the transmission electron microscope, is characterized by few typical modifications in fructose-grown cells. The number of cyanophycin granules and of glycogen granules is much higher in the vegetative cells grown in the presence of fructose, as compared with the control. The fructose-grown cells contain more polyhedral bodies and the heterocysts of the fructose-grown cells contain enlarged polar nodules and cyanophycin granules. The presence of cyanophycin granules in heterocysts is uncommon in free-living cyanobacteria. The cells of Anabaena azollae were found to contain concentric lamellar structures resembling the thylakoid structure (Lang, 1965). These bodies may carry out photosynthesis at localized sites in the cell which produce nonhomogeneous gradients of oxygen in the cell.

The structural modifications in the fructose-grown cells are related to the supported activity of nitrogenase, producing excess fixed nitrogen compounds which are accumulated in the form of storage protein products. The cultured Anabaena azollae do not release free amino acids or ammonia to the growth medium, and fructose does not control the release of fixed nitrogen compounds. The situation in situ must be different and the host may operate the appropriate signal to induce the relase of fixed N_2 products (Stewart, 1978).

2. The Effect of Sugars on Photosynthesis and Respiration

Fructose, glucose and sucrose are produced in high quantities by the host (Peters, Kaplan, 1981) and may affect additional physiological activities of Anabaena azollae such as photosynthesis and respiration. As shown in Table 1, fructose enhanced respiration by a two-fold stimulation and slightly inhibited photosynthetic O_2 evolution. Glucose and sucrose enhanced respiratory activity to the same extent as fructose.

These results provide indirect evidence that the three sugars are taken up by the cells of Anabaena azollae to facilitate respiration.

3. Fructose-Supported N_2-Fixation and Its Uptake by Anabaena azollae

The effect of fructose on N_2-fixation of Anabaena azollae was studied with three isolates from A. caroliniana, A. pinnata and A. filiculoides. The studies were extended in order to compare the effect of fructose to that of sucrose and of glucose at higher concentrations, and to follow the effect of the sugars in the light and in the dark in the presence of DCMU. Fructose supported acetylene reduction activity of the three isolates of Anabaena azollae in the dark, and in the light in the presence of DCMU, to the level of N_2-fixation in the control light experiment. Sucrose and glucose were less effective and a limited support of N_2-fixation in the dark, and in the light in the presence of DCMU, was observed.

These results provide a similar pattern of the effect of fructose on N_2-fixation in the three cultured isolates of Anabaena azollae, which suggests that fructose is preferential to other photosynthetates as a general phenomenon in Anabaena azollae. Fructose uptake was followed, in parallel to its effect on nitrogen fixation. Fructose is taken up in the light and in the dark, and in the light+DCMU (Table 2). The uncoupler CCCP inhibited fructose uptake in the dark and in the light, indicating that fructose uptake is an active ATP-dependent process. The uptake of fructose is also inhibited by fluoride, which is known to inhibit aldolase (Table 2). Aldolase participates in glycolysis and in the pentose phosphate pathway, therefore the energy required for fructose uptake may derive from these pathways. Cells of Anabaena azollae which were incubated for 16 hours in the dark, are deprived of storage material and cannot generate sufficient energy for N_2-fixation. The rate of fructose uptake in these cells is similar to the rate of uptake of the control cells, left in the light, suggesting that energy required for fructose uptake is more easily available in the cells than the energy required for N_2-fixation. The uptake of fructose was not affected by glucose and sucrose, suggesting that the cells of Anabaena azollae take up the fructose by a separate carrier.

4. Cell Antigens of Anabaena azollae

The enzyme-linked immunosorbent assay and the Fahrr radioimmunoassay were used to test the degree of homology between cell antigens of the three cultured isolates of Anabaena azollae: Newton cells from A. caroliniana, Arad cells from A. filiculoides and Pinnata cells from A. pinnata. The Newton and Arad cells cross-reacted identically with the anti-Newton serum and silimarly with the anti-Arad serum, suggesting that their cell antigens are very closely related. The Pinnata cells interacted only partially with the anti-Newton and anti-Arad sera, and are antigenically different. The fresh isolate of Anabaena azollae from A. filiculoides was found different from the three cultured isolates of Anabaena azollae. The Arad cells, which are cultured isolates of fresh cells, differ to a great extent from the parental fresh isolate, indicating that changes in cell antigens occurred as a result of cultivation separated from the symbiotic state. A similar resolution of cross-reaction between the Anabaena azollae isolates was obtained with rabbit antisera. The close similarity between the Arad and Newton cells is evident, and the difference between the Pinnata cells and the other two types was clearly demonstrated.

The relations between the surface antigens of the three isolates of Anabaena azollae were tested employing the Fahrr technique. The antibodies produced against Newton cells interacted similarly with intact cells of the Newton and Arad types, and the same tendency was observed with cell wall preparations of both cell types, as shown in Table 3. Cell surface antigens of Pinnata cells were found significantly different from the Newton cells. The studies on cell antigens of Anabaena azollae comprise an appropriate tool for the comparison of cultured isolates, but are not suitable for a comparison between the fresh isolate of Anabaena azollae and the cultured cells.

5. The Recognition Between Anabaena and Azolla

A lectin isolated and characterized from the host Azolla was found to specifically bind D-galactose, and was suggested to be involved in the recognition between Anabaena and Azolla (Mellor et al., 1981). A specific haemagglutinin, binding D-fucose and L-rhamnose, was characterized in Newton cells of Anabaena azollae. The haemagglutination was inhibited by an extract of isolated leaf cavities of Azolla (Tel-Or et al., 1983). The presence of fucose at the cell walls of the leaf cavity envelops of Azolla was demonstrated with a fucose-binding protein, suggesting that the leaf cavity envelop of the host Azolla is enriched in fucose and may recognize the lectin of Anabaena.

TABLE 1: The effect of sugars on photosynthesis and respiration of Anabaena azollae var. caroliniana.

Conditions	O_2 Evolution	O_2 Consumption
Fructose (1mM)	90	189
Glucose (4mM)	108	178
Sucrose (4mM)	117	193

Values are expressed as percent of control activities: 100% activity were 201 μmoles O_2 evolved and 27 μmoles O_2 consumed x mg chlorophyll^{-1} x hr^{-1}.

TABLE 2: Fructose incorporation and fructose-supported N_2-fixation in Anabaena azollae var. caroliniana.

Conditions	N_2-Fixation		Fructose Incorporation	
	Light	Dark	Light	Dark
Control	100	100	-	-
Fructose	100	400	100	100
Fructose+DCMU	95	-	83	-
Fructose+CCCP	0	0	11	1
Fructose+fluoride	53	55	29	0

Values are expressed as percent of control activity in the light or in the dark: 1mM fructose, 10μM DCMU, 40μM CCCP, 10mM fluoride.

TABLE 3: The interaction of rabbit anti-Newton sera with Newton and Arad cells.

	Intact Cells*	Cell Walls**
Newton cells	12693	7691
Arad cells	13092	6295

* 9 x 10^6 cells of Newton and Arad types.

**3 x 10^6 cells of Newton and Arad types.

Numbers represent the values of triplicate determinations.

REFERENCES

Gates JE Fischer RW, Goggin TW and Azrolan NI (1980) Arch. Microbiol. 128, 126-129.

Hill DJ (1977) New Phytol. 78, 611-616.

Kobiler D, Cohen-Sharon A and Tel-Or E (1981) FEBS Lett. 133, 157-160.

Ladha JK and Watanabe I (1982) Biochem. Biophys. Res. Commun. 109, 675-682.

Lang NJ (1965) J. Phycol. 1, 127-134.

Mellor RB, Gadd GM, Rowell P and Stewart WDP (1981) Biochem. Biophys. Res. Commun. 99, 1348-1353.

Newton JW and Herman AI (1979) Arch. Microbiol. 120, 161-165.

Peters GA, Calvert HE, Kaplan D, Ito O and Toia RE (1982) Isr. J. Bot. 31, 305-323.

Peters GA and Kaplan D (1981) Plant Physiol. 67, S-37.

Peters GA, Mayne BC, Ray TB and Toia RE (1979) in: Nitrogen and Rice, IRRI, Los Banos, Laguna, Phillines, pp. 324-344.

Stewart WDP (1976) in: Hardy RW and Silver WS, eds, A Treatise on Dinitrogen Fixation, Section III - Biology, John Wiley and Sons, New York, pp. 63-193.

Stewart WDP (1978) in: Encyclopedia of Plant Physiology, Vol. 6, Photosynthesis II, pp. 457-471.

Tel-Or E and Sandovsky T (1982) Isr. J. Bot. 31, 329-336.

Tel-Or E, Sandovsky T, Kobiler, D, Arad H and Weinberg R (1983) in: Papageorgiou GC and Packer L, eds, Photosynthetic Prokaryotes: Cell Differentiation and Function, Elsevier Science Publishing Co., New York, pp. 303-314.

OXYGEN, HEMOGLOBINS, AND ENERGY USAGE IN ACTINORHIZAL NODULES

JOHN D. TJEPKEMA
DEPARTMENT OF BOTANY AND PLANT PATHOLOGY, UNIVERSITY
OF MAINE, ORONO, ME 04469, U.S.A.

INTRODUCTION

The nitrogen-fixing symbioses between higher plants and Rhizobium, Frankia, and cyanobacteria differ in a number of important ways. Identification of the nature and range of these differences may help in understanding how these systems function and the possible means by which new symbioses may be created by genetic engineering. My own recent studies have emphasized comparisons between actinorhizal plants where Frankia is the nitrogen-fixing symbiont, and legumes and Parasponia where Rhizobium is the symbiont

The agronomic importance of legumes is well known, and actinorhizal plants, although less well known, are also of economic importance, with increasing use in forestry, land reclamation, and horticulture. In this paper I will compare energy usage, protection of nitrogenase from oxygen, and the occurrence of hemoglobin in Rhizobium and Frankia induced nodules.

EXPERIMENTAL

Seedlings of Myrica gale were germinated in sand, inoculated with a crushed nodule suspension immediately after germination, and were watered with a one fourth strength -N Hoagland's solution supplemented with 1 mM urea for the next 20 days. They were then transplanted to vermiculite and watered with -N solution without urea. They were grown at a constant 23°C root and air temperature, with a 17 hour day, and 7 hour night. Total plant CO_2 evolution and/or C_2H_2 reduction were measured by placing the intact plants in a closed container in the dark and measuring the rate of CO_2 and C_2H_4 accumulation by gas chromatography (Winship, Tjepkema, 1982). These rates were constant throughout a 24 hour cycle. Carbon dioxide evolution by plant parts was measured in the same way, immediately after shaking the vermiculite from the roots and separating the plants into shoot, root, and nodule fractions. Most nodules were clustered together in the upper part of the root system. The nodules were left attached to roots extending from the nodules towards the stem, but all other roots were removed except for the nodule roots. Total CO_2 evolution rate of the separated root, nodule, and shoot fractions averaged 20% more than CO_2 evolution rate of the intact plant, while C_2H_2 reduction was 5% greater. Nodule respiration in Table 2 was calculated by multiplying intact plant respiration by the per cent nodule respiration (Table 1). Further dissection of the nodule fraction showed that the nodule roots and other attached roots contributed about 10% of the total CO_2 evolution by the nodule fraction. The values in Tables 1 and 2 are for the total nodule fraction and thus include nodule roots and some other roots. The rate of carbon accumulation as plant dry matter was calculated from the growth rate, which was logarithmic, and the carbon content of field plants of Myrica gale (Schwintzer, 1983).

ENERGY USAGE

The annual rate of nitrogen fixation by legume and actinorhizal plants is similar, as are rates of nitrogen fixation per gram of nodules and nodule weight as a fraction of total plant weight (Bond, 1958; Torrey, 1978). This suggests that energy usage for nitrogen fixation is similar in legumes and actinorhizal plants, but does not quantify the amount of respiratory energy used by the nodules.

Isolated nodules. We have approached the question of energy usage for nitrogen fixation by measuring the ratio between respiration and nitrogenase activity in root nodules isolated from the plant by excision or by enclosing them in a gas-tight chamber. Nitrogenase activity is confined to the nodules, and the only respiratory activity known to occur in the nodules is for support of nitrogen fixation and directly related metabolism, such as growth, maintenance respiration, and amino acid synthesis and export. We have measured respiration primarily from the rate of CO_2 evolution, but get very similar results when electron flow from respiratory substrates is measured as the sum of oxygen uptake, acetylene reduction, and hydrogen evolution (Schwintzer, Tjepkema, 1983). We usually measure the ratio between CO_2 evolution and C_2H_2 reduction because this can be rapidly and accurately measured with a single gas sample by gas chromatography (Winship, Tjepkema, 1982). Using this method we find that legume and actinorhizal nodules give a similar range of values for the ratio between CO_2 evolution and acetylene reduction (Tjepkema, Winship, 1980). A variety of factors might increase the ratio between respiration and nitrogen fixation. Examples are rapid growth, drought and flooding stresses, and ineffective endophyte. Thus the minimum ratio between respiration and nitrogenase activity is of greatest interest. For legume nodules we found a minimum value of 2.5 moles CO_2 per mole of C_2H_2 reduced in soybean nodules, while in actinorhizal nodules this minimum ratio was 2.8 in nodules of *Elaeagnus umbellata*. Using different methods other workers have estimated similar ratios between respiration and nitrogenase activity in legume nodules (Schubert, 1982). In limited measurements of nodules of *Parasponia*, a non-legume that is nodulated by a *Rhizobium*, we found a minimum ratio of 3.4 moles CO_2 per mole of C_2H_2 reduced (Tjepkema, Cartica, 1982). As has been previously noted (Bond, 1958; Torrey, 1978) we found that nitrogenase activity per gram of nodule was also similar in legume and actinorhizal nodules. Thus the photosynthate required for both nodule growth and respiration is similar in both kinds of nodules and one cannot be recommended over the other for economic use in forestry or other applications, on these grounds. Also there is no indication of a general occurrence of a high degree of respiratory protection of nitrogenase from oxygen in either kind of symbiosis, since at maximum efficiency nitrogenase requires about 1.2 moles CO_2 respired per mole of C_2H_2 reduced, and this does not include nodule growth, amino acid synthesis, or maintenance respiration (Tjepkema, Winship 1980).

Whole plants. Energy usage by root nodules should be considered in the context of the total carbon and nitrogen budget of the host plant. Although extensive studies of this nature have been made of legumes, there is little data for actinorhizal plants. Some of our initial data for *Myrica gale* are presented in Tables 1 and 2. If judged by nodule size and dry weight, *Myrica gale* seedlings appear to use only a moderate amount of photosynthate for nitrogen fixation. But nodule respiration is a very large fraction of total respiration, being twice the size of root respiration. Considering carbon use in both nodule growth and respiration we estimate that nitrogen fixation consumes 17.6% of gross photosynthesis (or about 20% of net photosynthesis). Although other data are not available for nodule respiration as a fraction of total respiration in actinorhizal plants, our data agrees with that of others for plant growth rate and nodule fraction of total plant weight. Bond (1951) found that nodules averaged 7.4% of dry weight for seedlings of *Myrica gale* grown in -N water culture, while Ingestad (1980) found that nodules were 5.0% of dry weight for young seedlings of *Alnus incana*. Ingestad's plants were growing logarithmically, as were ours, and he found a relative growth rate of 10.4% per day which is close to our value of 9.8% per day for *Myrica gale*.

TABLE 1. Partitioning of dry weight and CO_2 evolution in 52 day-old seedlings of *Myrica gale*. Average dry weight of 6 plants = 0.173g each.

	Shoot	Roots	Nodules
Dry weight	76.1%	17.5%	6.4%
CO_2 evolution	50%	15%	36%

TABLE 2. The relationship between carbon usage by nodules and gross photosynthesis at 52 days after germination.

	μ moles C day^{-1} per plant	Percentage of gross photosynthesis
Whole plant growth	701	62
Whole plant respiration	435	38
Gross photosynthesis	1136	100
Nodule growth	45	4.0
Nodule respiration	155	13.6
Nodule growth and respiration	200	17.6

Our values for photosynthate consumption by nitrogen fixation in *Myrica gale* are somewhat higher than comparable data for legumes. For the vegetative phase of pea growth, Minchin and Pate (1973) found that 4.4% of net photosynthate was used for nodule growth and 10.6% for nodule respiration. For vegetative growth in cowpea, Herridge and Pate (1977) found that 3% of net photosynthate was used for nodule growth and 5% for nodule respiration. One reason for greater energy usage by *Myrica gale* nodules may have been their logarithmic growth which required high investment in nodule growth a number of days in advance of nodule ability to fix nitrogen. Another factor is the presence of nodule roots in *Myrica gale*, which add moderately to both nodule growth and respiration. Finally, the plants were inoculated with crushed nodules that contained spore (+) endophyte which may cause less efficient nitrogen fixation (Hall et al., 1979; Normand, Lalonde, 1982; K. A. VandenBosch, personal communication). In fact the ratio between nodule CO_2 evolution and C_2H_2 reduction for the plants in Tables 1 and 2 was 5.3 moles CO_2 per mole of C_2H_2 reduced, which is high compared to many actinorhizal nodules.

Field nodules. Energy usage for nitrogen fixation in field plants may be rather different than in young seedlings in the laboratory. One difference is that substantial fixed nitrogen may be absorbed from the soil. Schwintzer (1983) estimated that only 43% of the nitrogen used in the net annual production of a stand of Myrica gale in Massachusetts came from nitrogen fixation. The carbon used for nitrogen fixation in this same stand was measured during a growing season (Schwintzer, Tjepkema, 1983). Nodule growth contained 3.0g carbon m^{-2} out of a net production of 281g C m^{-2} while CO_2 evolution by the nodules during a year contained 18.0g C m^{-2}. Thus nodule growth plus respiration was 7.5% of the carbon in the annual production. The corresponding value calculated from Table 2 is 28.5%. Thus nitrogen fixation by the field plants was considerably more energy efficient, even when one considers that they were fixing only 43% of their nitrogen. The small nodule mass relative to net production accounts for part of the greater efficiency and may be partly due to a relatively linear rather than logarithmic growth rate by the plants. Another factor may be that these nodules had spore (-) rather than spore (+) endophyte (Schwintzer et al., 1982). But the field nodules were only moderately more efficient in terms of moles of CO_2 evolved per mole of C_2H_2 reduced. This ratio was 4.9 for the field plants compared to 5.3 for the nodules in Tables 1 and 2. Finally, the nitrogen content of the more woody biomass produced by the field plants was lower, thus reducing the energy used in nitrogen fixation per unit of production by the field plants.

PROTECTION OF NITROGENASE FROM OXYGEN

Free-living cultures. From the proceeding there is no evidence of marked differences in the rate and energy cost of nitrogen fixation in legume and actinorhizal nodules, but obviously more work is needed. But marked differences do exist in the means by which nitrogenase is protected from oxygen. This is demonstrated most clearly by the difference in the optimal pO_2 for nitrogen fixation by cultures of free-living Rhizobium and Frankia. For Rhizobium the optimal pO_2 is 0.1 kPa (0.1%) or less (Keister, Evans, 1976) while for Frankia the optimal pO_2 is in the range of 5 to 20 kPa (Tjepkema et al., 1981; Ganthier et al., 1981). It seems most likely that the ability of Frankia to fix nitrogen at atmospheric pO_2 is due to the presence of structures called "vesicles" which are terminal swellings at the ends of branch hyphae. These vesicles have laminated outer layers similar in appearance to those of the heterocysts of cyanobacteria (Torrey, Callaham, 1982). These laminated layers may restrict oxygen diffusion into the vesicle and combined with high metabolic rates this could result in a low pO_2 at the site of nitrogen fixation in the vesicle. This possibility is supported by the recent findings that oxygen uptake by nitrogen-fixing cultures of cyanobacteria and Frankia is not oxygen saturated at atmospheric pO_2, in a fashion consistent with diffusion-limited uptake (Rhodes, 1981; M. A. Murry, personal communication). These observations agree with studies of acetylene reduction by nodules of Alnus rubra as a function of acetylene concentration (Winship, Tjepkema, 1983). The results could not be explained by enzyme kinetics alone, but did fit an equation combining enzyme kinetics and diffusion. The calculated diffusion coefficients were quite temperature dependent, which is consistent with diffusion limited flux through a lipid such as may surround the vesicles of *Frankia*.

A related difference between *Frankia* and *Rhizobium* may be in the rate of oxygen uptake per unit volume of nitrogen-fixing cells. The rates of nitrogen fixation and oxygen uptake by actinorhizal and legume nodules appear to be quite similar (Tjepkema, Winship, 1980). But inspection of nodule cross sections suggests that infected host cells are a smaller fraction of nodule volume in actinorhizal nodules than in legume nodules. Furthermore, the vesicles, the presumed site of nitrogen fixation (Tjepkema et al., 1980), are a smaller fraction of infected cell volume than are rhizobia. This suggests that the rate of oxygen uptake per unit volume of *Frankia* vesicles is substantially greater than oxygen uptake of an equal volume of *Rhizobium* bacteroids. If so, both high intensity of oxygen uptake and low permeability of the vesicle wall to oxygen may be factors that allow *Frankia* to fix nitrogen at atmospheric pO_2. The importance of the rate of oxygen uptake is emphasized by the observation that the optimum pO_2 for nitrogen fixation by *Frankia* is reduced if carbon supply is limited (M. A. Murry, personal communication).

Oxygen diffusion in nodules. In view of the ability of *Frankia* to fix nitrogen at atmospheric pO_2, it is not surprising to find that there is a negligible pO_2 gradient between the soil atmosphere and the infected cells in alder nodules. Oxygen enters the nodule through lenticels and diffuses through a continuous network of intercellular air spaces to the surfaces of the infected cells (Tjepkema, 1979; Wheeler et al., 1979). Calculations show that the pO_2 gradient in air spaces such as these is negligible (Tjepkema, Cartica, 1982), so that the surfaces of the infected cells are at nearly the same pO_2 as the soil atmosphere. In fact, it may be important that actinorhizal nodules be well aerated, since we observed a marked reduction in nitrogenase activity by cultures of *Frankia* at 0.5 kPa O_2 (Tjepkema et al., 1981).

Such free access of atmospheric O_2 to the infected cells does not occur in legumes and *Parasponia* nodules (Tjepkema, 1979; Tjepkema, Cartica, 1982). In these nodules, the infected cells are completely surrounded by the inner nodule cortex, and this layer of cells lacks air spaces. Calculations show that the rate of oxygen uptake by the infected cells combined with the high resistance to oxygen diffusion through the layer of cortical cells that lacks intercellular air spaces can account for the low oxygen concentrations observed in the infected cells of legume and *Parasponia* nodules. But intercellular air spaces do occur between the cells in the region of infected tissue. These spaces are essential, since *Rhizobium* generates the ATP needed for nitrogen fixation by oxidative phosphorylation. Without the air spaces the center of the nodule would become anaerobic, and the pO_2 in the outer infected cells would rise and approach atmospheric pO_2 due to reduced oxygen consumption by the infected cells as a whole. As it is, calculations show that there is a negligible pO_2 gradient in the air spaces of the infected tissue between the outer edge and the center of the tissue (Tjepkema, Cartica, 1982).

Leghemoglobin may serve to minimize the pO_2 gradient from the intercellular air spaces to the centers of the infected cells. At the low pO_2 found within the infected cells the concentration of oxyleghemoglobin is very much greater than the concentration of free oxygen. Under these conditions most of the flux of oxygen along a concentration gradient would occur via leghemoglobin (Appleby, 1981). The absence of leghemoglobin might cause a situation analogous to the lack of air spaces. Thus the bacteroids in the centers of the cells might be anaerobic, while those adjacent to the air spaces might be at a pO_2 that is inhibitory to nitrogen fixation. But the expected pO_2 gradient in the absence of hemoglobin would be relatively small in absolute terms, and its effect on nitrogen fixation is not entirely clear (Tjepkema, Cartica, 1982). This is because of imprecision in the calculation and incomplete knowledge about the sensitivity to oxygen of nitrogen fixation by Rhizobium.

The recent demonstration of hemoglobin in Parasponia (Appleby et al., 1983) suggests that hemoglobins may be essential for all symbioses between Rhizobium and higher plants. All evidence suggests that oxygen transport is similar in legumes and Parasponia (Tjepkema, Cartica, 1982), so the simplest hypothesis is that hemoglobin functions in the same way in Parasponia and legume nodules. The properties of the hemoglobin in Parasponia are consistent with a function in oxygen transport (Appleby et al., 1983). But it should also be noted that Parasponia hemoglobin is somewhat different from legume hemoglobin; for example there is evidence that it is a readily dissociable dimer.

HEMOGLOBINS IN ACTINORHIZAL NODULES

Since hemoglobins cannot facilitate oxygen flux at atmospheric pO_2, and Frankia is able to fix nitrogen at atmospheric pO_2, one might expect that hemoglobins would have no function in actinorhizal nodules and thus would not occur in these nodules. But Davenport (1960) presented convincing evidence for hemoglobins in nodules of Casuarina cunninghamiana, which led me to further investigate this question. Absorbance spectra of nodule slices from C. cunninghamiana and Myrica gale showed distinct absorption bands corresponding to oxyhemoglobin when measured in an atmosphere of pure O_2, which were reversibly replaced by the absorption band of deoxyhemoglobin when assayed in N_2 (Tjepkema, 1983b). Partial oxygenation was observed at atmospheric pO_2. But such spectra indicated only a low concentration of hemoglobin in nodules of Alnus rubra and little or none in Ceanothus americanus and Datisca glomerata. The nature of these actinorhizal hemoglobins is not yet known, but contrary to the results of Davenport (1960), a hemoglobin has been extracted in soluble form from Casuarina cunninghamiana (Tjepkema, 1983; Appleby et al., 1983). Based on total heme content and also the absorbance of the carboxyhemoglobin form, Davenport (1960) found that the hemoglobin content of C. cunninghamiana nodules was equal to that of pea nodules.

The function of hemoglobin in actinorhizal nodules is unknown, but there are a number of possibilities. One is that hemoglobin is an adaptation that permits greater nitrogen fixation in soils of low pO_2. Myrica gale, Casuarina cunninghamiana, and many genera of Alnus frequently grow in wet soils where the pO_2 at the nodule surface may be very low. Under these conditions hemoglobin might allow a much greater rate of oxidative phosphorylation than possible in its absence, by facilitating oxygen flux from the intercellular air spaces through the host cytoplasm to the endophyte.

A second possibility is that nitrogen fixation by Frankia is more energy efficient at subatmospheric pO_2 and that some actinorhizal nodules provide an environment of low pO_2 for nitrogen fixation as do the nodules of legumes and Parasponia. In fact, oxygen microelectrode measurements show that there are regions of very low pO_2 in the nodules of Myrica gale (Tjepkema, 1983a). Hemoglobin might minimize pO_2 gradients in these regions by facilitated transport of oxygen. However, there is no evidence that nitrogen fixation is more energy efficient in actinorhizal nodules with high hemoglobin concentrations (Myrica gale) than in nodules with low or zero concentration (Ceanothus americanus) judging by nodule CO_2/C_2H_4 ratios (Tjepkema, Winship, 1980).

The high concentration of hemoglobin in the nodules of Casuarina cunninghamiana is especially interesting in view of the absence of Frankia vesicles or at least typical vesicles in these nodules (Tyson, Silver, 1979; Berg, 1983). Berg has presented evidence that the walls of the infected cells are suberized and/or lignified, and it is possible that such deposits might restrict oxygen diffusion to the endophyte, thus creating a low pO_2 environment where hemoglobin could facilitate oxygen flux. Vesicles might thus be absent because they are not needed to restrict oxygen diffusion. But Schaede (1938) observed lignification in the walls of infected cells in Myrica gale, and in this case vesicles are present. Also Davenport (1960) reported high concentrations of total heme in nodules of Alnus glutinosa and Hippophaë rhamnoides, and there are vesicles and no reports of a special host cell wall in these nodules.

REFERENCES

Appleby CA (1981) In Gibson AH and Newton WE, eds, Current Perspectives in Nitrogen Fixation, pp. 265-270, Austr. Ac. Sc., Canberra.
Appleby CA, Tjepkema JD and Trinick MJ (1983) Science 220, 951-953.
Berg RH (1983) Can. J. Bot. (in press).
Bond G (1951) Ann. Bot. N.S., 15, 447-458.
Bond G (1958) In Hallsworth EG, ed, Nutrition of the Legumes, pp. 216-231, Butterworths, London.
Davenport HE (1960) Nature (Lond.) 186, 653-654.
Gauthier D, Diem HG and Dommergues Y (1981) Appl. Environ. Microbiol. 41, 306-308.
Hall RB, McNabb HS, Maynard CA and Green TL (1979) Bot. Gaz. 140, S120-S126.
Herridge DF and Pate JS (1977) Plant Physiol. 60, 759-764.
Ingestad T (1980) Physiol. Plant. 50, 353-364.
Keister DL and Evans WR (1976) J. Bacteriol. 127, 149-153.
Minchin FR and Pate JS (1973) J. Exp. Bot. 24, 259-271.
Normand P and Lalonde M (1982) Can. J. Microbiol. 28, 1133-1142.
Rhodes KS (1981) Oxygen sensitivity of Nitrogen Fixation in the cyanobacterium Anabaenopsis arnoldii. Ph.D. thesis, Univ. of Michigan.
Schaede R (1938) Planta 29, 32-46.
Schubert KR (1982) The Energetics of Biological Nitrogen Fixation, 30 p, Workshop Summaries-I, Amer. Soc. Plant Physiol., Rockville, MD.
Schwintzer CR (1983) Can. J. Bot. (in press).
Schwintzer CR, Berry AM and Disney LD (1982) Can. J. Bot. 60, 746-757.
Schwintzer CR and Tjepkema JD (1983) Can. J. Bot. (in press).
Tjepkema J (1979) In Gordon JC, Wheeler CT and Perry DA, eds, Symbiotic Nitrogen Fixation in the Management of Temperate Forests, pp. 175-186, Forest Res. Lab., Oregon State Univ., Corvallis.
Tjepkema JD (1983a) Amer. J. Bot. 70, 59-63.
Tjepkema JD (1983b) Can. J. Bot. (in press).
Tjepkema JD and Cartica RJ (1982) Plant Physiol. 69, 728-733.
Tjepkema JD, Ormerod W and Torrey JG (1980) Nature (Lond.) 287, 633-635.
Tjepkema JD, Ormerod W and Torrey JG (1981) Can. J. Microbiol. 27, 815-823.
Tjepkema JD and Winship LJ (1980) Science 209, 279-281.
Torrey JG (1978) BioSci. 28, 586-592.
Torrey JG and Callaham D (1982) Can. J. Microbiol. 28, 749-757.
Tyson JH and Silver WS (1979) Bot. Gaz. 140 (Suppl.), S44-S48.
Wheeler CT, Gordon JC and Ching TM (1979) New Phytol. 82, 449-547.
Winship LJ and Tjepkema JD (1982) Plant Physiol. 70, 361-365.
Winship LJ and Tjepkema JD (1983) Can. J. Bot. (in press).

THE ROLE OF PERIBACTEROID MEMBRANE IN LEGUME ROOT NODULES

JOHN G. ROBERTSON[+], PAMELA LYTTLETON[++] AND BRIAN A. TAPPER[+]
+ APPLIED BIOCHEMISTRY DIVISION, DEPARTMENT OF SCIENTIFIC AND INDUSTRIAL RESEARCH, PALMERSTON NORTH
++DEPARTMENT OF CHEMISTRY, BIOCHEMISTRY AND BIOPHYSICS, MASSEY UNIVERSITY, PALMERSTON NORTH, NEW ZEALAND.

INTRODUCTION

The function of legume root nodules is to provide an environment in which the bacteroid form of rhizobia can fix N_2 and the ammonia so produced can be efficiently assimilated for plant growth. The peribacteroid membrane plays a key role in the establishment of this environment by allowing the rhizobia to function, albeit temporarily, in combination with this plant membrane as an intracellular organelle.

The presence of a plant membrane surrounding the bacteroids in legume nodules (Figs 1,2,3) was first reported in studies of soybean by Bergersen and Briggs (1958) who called it the membrane envelope. Robertson et al. (1974, 1978a) coined the term peribacteroid membrane (pbm) as a more specific name which has now gained wide acceptance (Newcomb, 1981; Verma, Long, 1983). The pbm was isolated from lupin and soybean nodules by first preparing membrane-enclosed bacteroids (Fig.4) and then removing the pbm by osmotic shock (Robertson et al., 1978b) or mechanical action (Verma et al., 1978). In thin sections of nodule tissue or isolated membranes, stained with uranium and lead, the pbm appears as a trilaminar membrane about 5.3 nm from midpoint to midpoint of the densely staining layers (Robertson et al., 1978a, b). The pbm stains with phosphotungstic acid (Robertson et al., 1974, 1978a, b; Verma et al., 1978) and with the Thiéry (1967) stain for polysaccharides (Fig.13) and can be conveniently identified using the freeze-fracture technique (MacKenzie et al., 1973; Robertson et al., 1978a, b; Tu, 1979). Detailed studies on the composition of the pbm have yet to be published. Robertson et al. (1978b) reported a lipid-to-protein ratio of 6.1:1 which is high compared with most biological membranes (Korn, 1969). Analysis of the protein composition of the pbm using SDS gel electrophoresis revealed a unique pattern compared with other nodule fractions (Robertson et al., 1978b; Verma et al., 1978). Cytochemical studies of nodule tissue have led to the suggestions that acid phosphatase (Truchet, Coulomb, 1973), adenyl cyclase (Tu, 1974) and ATPase (Verma et al., 1978) activities are associated with this membrane. ATPase activity was also detected in isolated pbm preparations (Verma et al., 1978; Robertson et al., 1978b) and subsequent tests have shown that activity is stimulated by Mg^{++} at pH 5.25 and inhibited by ADP (Robertson, unpublished results).

THE PERIBACTEROID MEMBRANE AS A BACTEROID CARRIER IN PLANT CYTOPLASM

It is well established that conditions of low pO_2 are essential for N_2-fixation to occur in bacteroids (Bergersen, 1974, 1977; Laane et al., 1978; Sutton et al., 1981). Within nodules such conditions are apparently created by a combination of the restriction of O_2 diffusion through the cortical layers (Tjepkema, Yocum, 1973, 1974) and consumption of O_2 by both bacteroids (Bergersen, 1962, 1978a) and mitochondria in the central tissues. The development of an infected nodule cell should therefore be considered in relation to declining pO_2 from the time the first rhizobia are taken

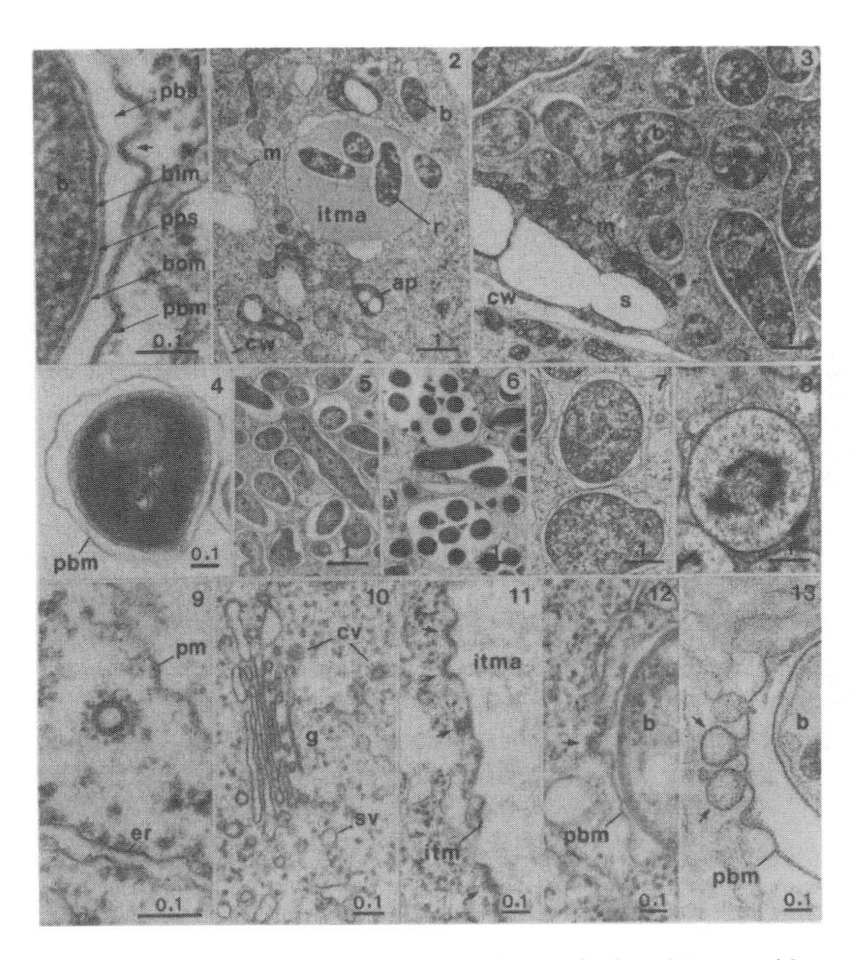

Figs.1-13. Thin sections of legume nodules or isolated bacteroids stained with uranium and lead or (Fig.13) with silver. Abbreviations: ap, amyloplast; b, bacteroid; bim, bacteroid envelope inner membrane; bom, bacteroid envelope outer membrane; cv, coated vesicle; cw, cell wall; er, endoplasmic reticulum; g, golgi body; itm, infection thread membrane; itma, infection thread matrix; m, mitochondria; n, nucleus; pbm, peribacteroid membrane; pbs, peribacteroid space; pm, plasma membrane; pps, periplasmic space; ps, polysome; r, rhizobium; s, starch; sv, smooth vesicle. (Scale in µm).
Fig.1. Portion of a mature bacteroid showing the various membranes. Note the adherence of the pbm to the bacteroid envelope outer membrane (arrowhead). Fig.2. Early infection zone showing rhizobia in infection thread; a bacteroid, mitochondria and amyloplasts occur throughout the cytoplasm. Fig.3. Late infection zone with many bacteroids in cytoplasm; mitochondria and amyloplasts occur at the cell periphery. Fig.4. Isolated bacteroid with pbm. Figs 5-8. Mature bacteroids in legume nodules. Fig. 5. Lupin.

Fig.6. Adzuki bean. Fig.7. White clover. Fig.8. Peanut (courtesy Alan Craig). Figs 9-13. Early infection zone of white clover nodules. Fig.9. Coated vesicle close to plasma membrane and endoplasmic reticulum; note relative membrane widths. Fig.10. Coated vesicles close to or associated with a Golgi body. Fig.11. Coated vesicles (arrowheads) associated with infection thread membrane. Fig.12. Coated vesicle (arrowhead) associated with pbm. Fig.13. Smooth vesicles (arrowheads) associated with pbm.

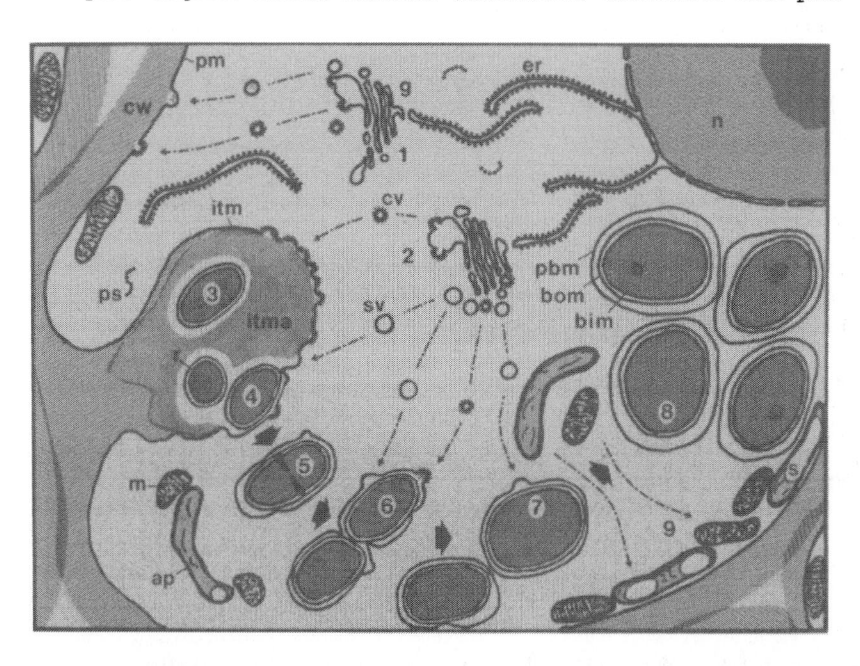

Fig.14. Diagram of the role of the pbm as a carrier for bacteroids in a nodule cell in white clover. Stages 1-2: plant endomembrane system involved in plasma membrane, cell wall, infection thread and pbm biogenesis. Stages 3-8: endocytosis of rhizobia and multiplication, distribution and maturation of bacteroids enclosed by pbm in the plant cytoplasm. Stage 9: migration of mitochondria and amyloplasts.

---

into the cell from the tip of the infection thread (Fig.2) to the time when the cell is packed with bacteroids (Fig.3). That such a decline in pO_2 has considerable effect on the organisation of an infected cell is suggested by the fact that mitochondria, which are known to move in living cells in relation to O_2 gradients (Thyberg et al., 1982), collect at the periphery adjacent to air spaces as the bacteroids multiply (Fig.3,14). It has not been established whether the distribution or size of bacteroids in the plant cytoplasm is a critical element influencing nodule efficiency (Sen, Weaver, 1980; Bal et al., 1982). However, it might be predicted that dispersion of the bacteroids would offer the greatest opportunity for 1) their synchronous development 2) synchronous induction of plant proteins (Robertson, Farnden, 1980) and 3) a final arrangement in which the greatest number of bacteroids are sited adjacent to plant cytoplasm.

For the pbm to function as a carrier of the rapidly dividing bacteroids in the plant cytoplasm, the plant cell must have considerable capacity to synthesise membranes since the pbm occurs in amounts (per cell) up to 20-40 fold that of the plasma membrane (Verma, 1982). Although it is accepted that the pbm is initially derived from the plasma membrane by endocytosis (Goodchild, Bergersen, 1966; Bergersen, 1974; Goodchild, 1977; Robertson et al., 1978a; Werner, Mörschel, 1978; Newcomb, 1981; Verma, 1983) the system of membrane biogenesis, which enables endocytosis and the subsequent growth and division of the pbm to occur, has not received wide attention. Verma et al. (1978) have shown, using incorporation of radioactive amino acids, that synthesis of pbm proteins is under the control of the plant. This is consistent with studies of thin sections of both lupin (Robertson et al., 1978a) and white clover nodules (Robertson, Lyttleton, 1982) from which it was proposed that the synthesis of the infection thread and the pbm involves the functional modification of the plant endomembrane system (Robertson, Farnden, 1980). The endomembrane concept as proposed by Morré and coworkers (Robinson, Kristen, 1982) integrates the endoplasmic reticulum (Fig.9) and the Golgi apparatus (Fig.10) into a functional unit in which the Golgi are the main station of membrane transformation from an "endoplasmic reticulum-like" to a "plasma membrane-like" membrane (Fig.9). Although the endomembrane concept is unproven there is considerable indirect evidence to support it at least in broad outline (Robinson, Kristen, 1982). The formation of the infection thread and the pbm is seen therefore as a modulation of a plastic system for plant membrane and cell wall biogenesis. The observation (Robertson, Lyttleton, 1982) that coated vesicles (Fig.9,10) and smooth vesicles (Fig.10) are involved in the biogenesis of infection thread membranes and possibly infection thread matrix material (Fig.11) and also pbm (Figs 12,13) supports this concept. It appears therefore that the pbm is a modified or incomplete plasma membrane without an overlying plant cell wall.

The endocytotic system of uptake of rhizobia operates for both effective and ineffective strains of rhizobia although cell senescence occurs at an earlier stage with ineffective strains since plant cell development is not sustained (Ronson et al., 1981). Distribution of the bacteroids in the plant cytoplasm requires that the pbm divides in step with division of the bacteroids. The mechanism by which division of pbm occurs (Fig.14) appears to involve adherence (Fig.1) of the pbm to the outer surface of the bacteroid envelope outer membrane (Robertson, Lyttleton, in preparation).

THE PERIBACTEROID MEMBRANE AS A PARTITION

By completely enclosing the bacteroids the pbm forms a partition between plant and bacterial systems which may be essential if the plant is to protect itself against hydrolases of bacterial origin. We have observed in thin sections of nodule tissue containing otherwise normal bacteroids and plant cytoplasm, an occasional bacteroid which has undergone lysis within a pbm sac. Such lytic events would almost certainly involve protease activity, known to occur in nodule tissue (Malik et al., 1981), from which the plant cytoplasm would need protection.

The nature of the material in the pbs, which varies in volume (Figs 5-7) and is virtually nil in peanuts (Fig.8), is controversial. The space often contains electron dense material (Fig.6,8,12,13) (Dart, 1977) and debris from the bacteroid envelope outer membrane (Bal, Wong, 1982). It is not known whether the pbs contains proteins which are confined solely to this space although unique patterns on SDS gels have been obtained for a fraction which should contain such proteins (Robertson et al., 1978b; Melik-Sarkisyan et al., 1981). Melik-Sarkisyan et al. (1982) suggested a direct relationship between the pbs and N_2-fixing activity in bacteroids. However N_2-fixation by bacteroids is known to be dependent upon an intact cytoplasmic membrane (Laane et al., 1978) which would almost certainly be damaged by osmotic shock (Melik-Sarkisyan et al., 1982) causing loss of nitrogenase activity. Controversy exists as to whether leghemoglobin (Lb) is present in the pbs (Goodchild, 1977; Bergersen, 1978b; Robertson, Farnden, 1980; Verma, Long, 1983). In recent studies Bergersen and Appleby (1981) and Livanova et al. (1979) presented evidence which they claim supports such a location. By contrast Melik-Sarkisyan et al. (1981) reported no Lb associated with membrane-enclosed bacteroids and Verma et al. (1979) suggested from studies of *in vitro* protein synthesis that Lb is synthesised in the cytoplasm and does not cross the pbm. Ongoing experiments in our laboratory fail to find Lb in the pbs in lupin nodules.

The presence of enzymes on the plant side of the pbm associated with carbon and nitrogen metabolism which are directly associated with the N_2-fixing system in legume nodules is well established (see Robertson, Farnden, 1980). Recent reports have assigned enzymes of these pathways to a variety of compartments within the plant cell (Boland et al., 1982; Reynolds et al., 1982).

THE PERIBACTEROID MEMBRANE AS A TRANSPORT BARRIER

Localisation of various enzymes of importance to bacteroid function in legume nodules coupled with studies of specific mutants of rhizobia (Ronson et al., 1981) has given an insight into what metabolites might be transported across the pbm. Materials supplied to the bacteroids by the plant will include amino acids, sugars, organic acids and a range of trace elements and vitamins (Sutton et al., 1981; Verma, Long, 1983). Ronson et al. (1981) concluded that organic acids in particular must be supplied by the plant to support N_2-fixation in bacteroids in white clover. Both O_2 and N_2 must also cross the pbm and it has been suggested that the high lipid content of the pbm might be important in terms of the solubility of these gases (Robertson et al., 1978b). Sutton and Patterson (1983) have suggested that factors from the plant cytoplasm affect the development of bacteroids by interfering with bacteroid cell wall development.

Materials which originate in the bacteroids and which are passed to the plant cytoplasm will include ammonia and carbon dioxide (Laing et al., 1979; Robertson, Farnden, 1980) as well as other materials of possible importance to the symbiosis (Truchet et al., 1980). It is not known whether macromolecules are transported in either direction across the pbm. Both smooth and coated vesicles have been reported in association with the membrane (Kijne, Planque, 1979; Robertson, Lyttleton, 1982) and these vesicles (Figs 12,13) may be involved in transport of macromolecules as well as membrane biogenesis. The suggestion that heme is synthesised by the bacteroids for Lb formation by the plant (see Robertson, Farnden, 1980) has yet to be understood in terms of transport across the bacteroid envelope membranes or the pbm.

THE PERIBACTEROID MEMBRANE AND LEGHEMOGLOBIN - AN OVERVIEW

The pbm is in a position to play a regulatory role in nodules by controlling transport of metabolites to the bacteroids. Recent studies of plant plasma membranes have revealed that they possess transport properties which require that they be energised (Hodges, 1976; Leonard, Hodges, 1980; Stout, Cleland,

1982). These functions focus attention on the availability of ATP in the plant cytoplasm and on the importance of mitochondria (Figs 2,14). In infected cells of legume nodules not only will ATP be required for maintenance of activities of the pbm and plasma membrane it will also be required for ammonia assimilation in the plant cytoplasm. Thus in an infected cell fixing N_2 the demand for ATP in the cytoplasm may well be higher than in other plant cells. If this is so a parallel might be drawn with muscle cells in which ATP is required for muscle action. In these cells myoglobin occurs in the cell cytoplasm and Cole (1982) has proposed that this protein plays a role as an oxygen store, as an agent involved in enhancing O_2 flux through the tissue, or possibly as an oxygen buffer maintaining cell pO_2 constant where there are changes in O_2 supply and demand. In the infected cells of legume nodules O_2 is required for oxidative phosphorylation by bacteroids as well as mitochondria and therefore the function of both these organelles must be considered in relation to the O_2 status of the plant cell. Thus it appears there is greater value for Lb to be sited in the plant cytoplasm which is continuous throughout the infected cell, rather than simply as an O_2 transporter in the isolated compartments formed by the peribacteroid membrane.

ACKNOWLEDGEMENTS

We thank many colleagues in Applied Biochemistry, Grasslands and Plant Physiology Divisions DSIR for helpful discussions and growth room facilities, Professor R.D. Batt for generous support, Douglas Hopcroft and Raymond Bennett for photography and Cathy Isles for typing the manuscript.

REFERENCES

Bal AK, Shantharam S and Wong PP (1982) App. Env. Microbiol. 44, 965-971.
Bal AK and Wong PP (1982) Can. J. Microbiol. 28, 890-896.
Bergersen FJ (1962) J. Microbiol. 29, 113-125.
Bergersen FJ (1974) In Quispel A, ed, Biology of Nitrogen Fixation, pp. 473-498, North Holland Publ. Co., Amsterdam.
Bergersen FJ (1977) In Hardy RWF and Silver WS, eds, A Treatise on Dinitrogen Fixation, Section III. Biology, pp. 519-555, John Wiley and Sons, New York.
Bergersen FJ (1978a) In Loutit MW and Miles JAR, eds, Microbial Ecology, Proceedings in Life Sciences, pp. 367-383, Springer-Verlag, Berlin.
Bergersen FJ (1978b) In Döbereiner J, Burris RH and Hollaender A, eds, Limitations and Potentials for Biolocial Nitrogen Fixation in the Tropics, pp. 247-261, Plenum Press, New York.
Bergersen FJ and Appleby CA (1981) Planta 152, 534-543.
Bergersen FJ and Briggs MJ (1958) J. gen. Microbiol. 19, 482-490.
Boland MJ, Hanks JF, Reynolds PHS, Blevins DG, Tolbert NE and Schubert KR (1982) Planta 155, 45-51.
Cole RP (1982) Science 216, 523-525.
Dart P (1977) In Hardy RWF and Silver WS, eds, A Treatise on Dinitrogen Fixation, Section III. Biology, pp. 367-472, John Wiley and Sons, New York.
Goodchild DJ (1977) Int. Rev. Cytol. Suppl. 6, 235-288.
Goodchild DJ and Bergersen FJ (1966) J. Bact. 92, 204-213.
Hodges TK (1976) In Lüttge U and Pitman MC, eds, Encyclopedia of Plant Physiology, Transport in Plants II, Vol IIA, pp. 260-283, Springer-Verlag, Berlin.
Kijne JW and Planqué K (1979) Physiol. Plant Path. 14, 339-345.
Korn ED (1969) Fed. Proc. 28, 6-11.
Laane C, Haaker H and Veeger C (1978) Eur. J. Biochem. 87, 147-153.

Laing WA, Christeller JT and Sutton WD (1979) Plant Physiol. 63, 450-454.
Leonard RT and Hodges TK (1980) In Stumpf PK and Conn EE, eds, The Biochemistry of Plants, A Comprehensive Treatise, Vol. 1, pp. 163-182.
Livanova GI, Zhiznevskaya GYa and Andreeva IN (1979) Doklady Biochem. 245, 739-742.
Malik NSA, Pfeiffer NE, Williams DR and Wagner FW (1981) Plant Physiol. 68, 386-392.
MacKenzie CR, Vail WJ and Jordan DC (1973) J. Bact. 113, 387-393.
Melik-Sarkisyan SS, Tikhomirova AI and Kretovich VL (1981) Doklady Biochem. 259, 1498-1501.
Melik-Sarkisyan SS, Tikhomirova AI and Kretovich VL (1982) Doklady Biochem. 264, 157-160.
Newcomb W (1981) Int. Rev. Cyt. Suppl. 13, 247-298.
Reynolds PHS, Boland MJ, Blevins DG, Randall DD and Schubert KR (1982) TIBS 7, 366-368.
Robertson JG and Farnden KJF (1980) In Stumpf PK and Conn EE, eds, The Biochemistry of Plants, A Comprehensive Treatise, Vol. 5, pp. 65-113, Academic Press, New York.
Robertson JG and Lyttleton P (1982) J. Cell Sci. 58, 63-78.
Robertson JG, Lyttleton P, Bullivant S and Grayston GF (1978a) J. Cell Sci. 30, 129-149.
Robertson JG, Taylor MP, Craig AS and Hopcroft DH (1974) In Bieleski RL, Ferguson AR and Cresswell MM, eds, Mechanisms of Regulation of Plant Growth, Bulletin 12, pp. 31-36, Roy. Soc. NZ, Wellington.
Robertson JG, Warburton MP, Lyttleton P, Fordyce AM and Bullivant S (1978b) J. Cell Sci. 30, 151-174.
Robinson DG and Kristen U (1982) Int. Rev. Cytol. 77, 89-127.
Ronson CW, Lyttleton P and Robertson JG (1981) Proc. Natl. Acad. Sci. 78, 4284-4288.
Sen D and Weaver RW (1980) Plant Sci. Lett. 18, 315-318.
Stout RG and Cleland RE (1982) In Marmé D, Marré E and Hertel R, eds, Plasmalemma and Tonoplast: Their functions in the plant cell, Developments in Plant Biology, Vol. 7, pp. 401-407, Elsevier Biomedical Press, Amsterdam.
Sutton WD, Pankhurst CE and Craig AS (1981) Int. Rev. Cytol. Suppl. 13, 149-177.
Sutton WD and Patterson AD (1983) Plant Sci. Lett. 30, 33-41.
Thiéry JP (1967) J. Microscopie 6, 987-1018.
Thyberg J, Sierakowska H, Edstrom J-E, Burvall K and Pigon A (1982) Develop. Biol. 90, 31-42.
Tjepkema JD and Yocum CS (1973) Planta 115, 59-72.
Tjepkema JD and Yocum CS (1974) Planta 119, 351-360.
Truchet G and Coulomb Ph (1973) J. Ultrastruct. Res. 43, 36-57.
Truchet G, Michel M and Dénarié J (1980) Differentiation 16, 163-172.
Tu JC (1974) J. Bact. 119, 986-991.
Tu JC (1979) Physiol. Plant Path. 15, 35-41.
Verma DPS (1982) In Smith H and Grierson D, eds, The Molecular Biology of Plant Development, Botanical Monographs, Vol. 18, pp. 437-466, Blackwell Sc. Pub., Oxford.
Verma DPS, Bal S, Guerin C and Wanamaker L (1979) Biochemistry 18, 476-483.
Verma DPS, Kazazian V, Zogbi V and Bal AK (1978) J. Cell Biol. 78, 919-936.
Verma DPS and Long S (1983) Int. Rev. Cytol. Suppl. 14, 211-245.
Werner D and Mörschel E (1978) Planta 141, 169-177.

SYMBIOTIC NITROGEN FIXATION INVOLVING RHIZOBIUM AND THE NON-LEGUME PARASPONIA

PETER M. GRESSHOFF, SUSAN NEWTON, SHYAM S. MOHAPATRA, KIERAN F. SCOTT SUSAN HOWITT, G. DEAN PRICE, GREGORY L. BENDER, JOHN SHINE and BARRY G. ROLFE.

BOTANY AND GENETICS DEPARTMENTS
THE AUSTRALIAN NATIONAL UNIVERSITY CANBERRA, ACT, AUSTRALIA

1. INTRODUCTION

In classical terms Rhizobium is defined as that bacterial soil organism which can elicit nodule formation and if optimal, symbiotic nitrogen fixation on the roots (or at stems as in the case of Sesbania) of legumes. This definition may require alteration in view of the discovery 10 years ago that the tropical trees (or shrubs) belonging to the Parasponia genus are capable of nodulation and efficient nitrogen fixation in symbiosis with Rhizobium strains.

Parasponia, a tropical pioneer plant genus found in the Malay Archipelago, Papua and New Guinea, Fiji and Tahiti, was first recognised to be nodulated as early as 1909 (Ham, 1909). However, as Parasponia belongs to the Ulmaceae, which in Europe and North America have several species which nodulate with the actinomycete Frankia, this observation was not pursued further. It was Michael Trinick who in 1973 recognised that the microsymbiont was Rhizobium. The Rhizobium characteristic was ascertained as single colony isolates could nodulate and fix nitrogen in both the legume and non-legume host. More importantly Koch's postulates of reisolation and reinfection were tested positively. Trinick (1973) identified that the host was Trema aspera, which was later amended to Trema cannabina (Trinick and Galbraith 1976) and later to Parasponia rugosa (Akkermans et al., 1978). Trema is a close relative of Parasponia, but repeated attempts either in the field or in the laboratory have failed to demonstrate nodulation by Rhizobium (Becking, 1979). This in itself is of interest. Why does the Parasponia genus being made up of P. andersonii, P. parviflora, P. rugosa and P. rigida possess this general susceptibility to nodulation by Rhizobium? Are other Ulmaceae in the same habitat such as Celtis, Aphananthe, and Gironniera which as yet have not been tested, capable of non-legume Rhizobium induced symbiotic nitrogen fixation? Are there even other plant species which interact with Rhizobium, that do not belong to the classical legume host group? In the last meeting of this conference Athar and Manood (1981) reported that 5 related species of Zygophyllum, namely Tribulus terrestris, T. longipetalous, Zygophyllum simplex, Z. pripinquium and Fagonia indica found in Pakistan were nodulated by a Rhizobium-like organism. The nodules were similar in morphology to those found in Parasponia or actinomycete induced nodules on non-legumes (i.e. they were coralloid in nature with a central vascular system and reminiscent of a modified lateral root). We have found nodules on Tribulus (catseye, three prong jack) in central and northern New South Wales in Australia; however, we have no evidence for nitrogen fixation or the type of bacterial (if any) occupant.

Little is known about the mode of infection and nodule initiation in Parasponia. It appears that invasion is via the root hair as in most legumes. The information available so far is, however, that on agar plates Parasponia andersonii seedlings develop nodules within 14 days. By 3-4 weeks such seedlings express measurable nitrogenase. A similar rate of nodule formation and nitrogenase development was seen on inoculated cuttings of *Parasponia rigida* (Fig. 1). It is thus possible to classify a *Parasponia*-symbiosis within the same period of time that is required in parallel legume experiments. Part of our overall strategy, therefore, is to utilise the "wide" host range of certain *Parasponia* isolates to determine what functions (genes) are necessary for symbiotic nitrogen fixation in general and which are legume or non-legume specific.

To understand and classify the symbiotic steps (for example as outlined by Vincent 1980 and Rolfe et al. 1982) it was important to describe the symbiotic phenotype of a legume and *Parasponia* nodule produced by the same bacterial inoculant.

Previous studies especially by Trinick and Galbraith (1976), Becking (1979) and Trinick (1979) were confirmed in our laboratory using *P. rigida* and *Rhizobium* strain ANU289 (Price et al., 1984). Before demonstrating these features as we recognise them in *P. rigida* and siratro (*Macroptilium atropurpureum*), it is essential to clarify the reasons for selection of strain ANU289. The papers by Shine et al., Weinman et al., Newton et al., (all this volume) and Scott et al. (1983) outline the molecular genetic studies carried out with this strain. Studies by Morrison et al. and Badenoch-Jones et al. (this volume) focus on the bacterial genetics of a fast-growing *Rhizobium* strain (ANU240), which also nodulates *Parasponia*.

Trinick and Galbraith (1980) tested 15 different bacterial isolates from *P. andersonii* nodules for their host range and effectiveness on a large number of legumes. Strain CP283 stood out inasfar as it was highly effective on *Parasponia* and showed a wide and effective host range on legumes such as *Lablab purpureus*, *Glycine wightii*, *Stylosanthes humilis*, *Vigna luteola*, *V. marina*, *V. unguiculata*, *Flemingia congesta*, *Macroptilium lathyroides* and as mentioned siratro (*Macroptilium atropurpureum*). We received strain CP283 from M. Trinick and quickly discovered that the isolate contained two near isogenic forms. These are characterised by a differential amount of mucoid exopolysacharide excretion on mannitol based growth medium. Strain ANU289 is the streptomycin resistant derivative of the non-mucoid form of CP283. It is further characterised by a spontaneous rifampicin resistance (up to 500 g/ml). The mucoid and non-mucoid forms (ANU288 and ANU289 respectively) both nodulate and fix nitrogen in *Parasponia* and siratro. In all tests (including EcoR1 restriction patterns) both strains are identical. We have used this difference to explore the role of exopolysacharide production on *in vitro* nitrogenase derepression.

2. THE *PARASPONIA* NODULE

Nodules on the roots of different *Parasponia* species appear to be similar (Becking, 1979; Trinick, 1979; Price et al., 1984). They are coralloid in structure (Trinick and Galbraith, 1976), arising from single cylindrical nodules with apical meristems, which tend to bifurcate giving rise to nodule clusters. On 8 month old plants it is not rare to find nodule clusters of about 15-20 mm diameter. Nodulation is prolific and nodules seem to be restricted to those parts of the root system, which are well aerated.

Contrary to Trinick (1976) we found the vegetative propagation of Parasponia very easy. Success rates of 80% can be obtained by the simple dusting of hardened-off cuttings (about 10 cm in length) with indolebutyric acid (as in commercial rooting powder) and incubation of the basal region in moistened mineral wool under mist spray (see Fig. 1).

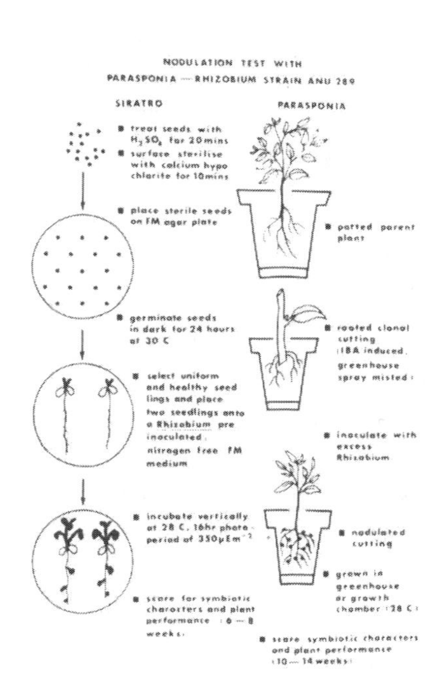

Fig. 1 Symbiotic tests of Rhizobium strain ANU289 on siratro & Parasponia

Inoculation of such cuttings with a specific strain at day 0 results in nodulation and nitrogen fixation as early as 6 weeks. Isolation of 10 colonies each from 10 different nodules showed that 100% of the original inoculant can be recovered. Parasponia plants yield large nodule masses, e.g. a plant of 35-40 cm height (being about 6-8 months after propagation) can support as much as 25 gm of nodules.

The Parasponia nodule resembles a modified lateral root. One finds a central vascular cylinder, a region of meristematic growth on the apex and a zone of invaded cortical tissue. The invaded cells show the ontogeny of Rhizobium invasion, with the more mature (and more densely filled) cells being furthest from the apical primary invasion zone. These cells are hypertrophic, as one finds in invaded legume cells. A comparison of the nodule morphology of Parasponia rigida and siratro (both elicited by ANU289) points to more significant differences than the macroscopic differentiation between the coralloid and determinate nodule type.

Rhizobia in Parasponia rigida nodules remain within the infection thread. Electronmicrographs, however, show that the infection threads vary. There are those in which the bacteria are very tightly packed, obviously in division and with a more electrondense infection thread wall. This wall loosens as the ontogeny progresses, bacteria are less organised as the space within the thread increases. Bacteria now show increased presence of poly-β-hydroxybutyrate (PHB). We would propose that this latter stage is the site of nitrogen fixation and no longer invasion and infection. It thus seems more appropriate to label those structures "fixation threads" (Fig. 2, Price et al., 1984).

Parasponia nodules cells are less densely invaded than corresponding siratro cells, as we found 295 and 654 bacteroids per 1000 cubic microns respectively in cells of the most advanced stage of infection within a nodule. Yet acetylene reduction rates of siratro can be similar to

P. rigida. Direct comparison, of course, is difficult as specific activities depend strongly on the age of the nodule being tested. Parasponia and siratro nodule bacteroids can be fractionated using step-wise sucrose gradients (Ching et al., 1977; Mohapatra et al., this volume). In both cases three nearly equally massed fractions are obtained, varying mainly in the amount of PHB, (but not in size). Pleomorphy as described by Becking (1979) and Appleby et al., (1981) is minimal and is mainly attributable to PHB deposits.

3. OXYGEN EFFECTS ON NITROGENASE IN PARASPONIA SYMBIOSIS

The recognition of the role of leghemoglobin in legume nodules was a significant advance in nodule physiology, but it presented a (yet unsolved) evolutionary dilemma. How can a hemoglobin exist within this isolated group of plants (the legumes) and numerous animal species, when none of the intermediate genera show this trait? Did one see divergent, convergent or horizontal transfer type of evolution? With the elucidation of the gene structure for soybean hemoglobin (see Verma and Long (1983) for a review) showing that the intron-exon arrangement is essentially conserved between mammalian and soybean globin genes, this question became more pressing.

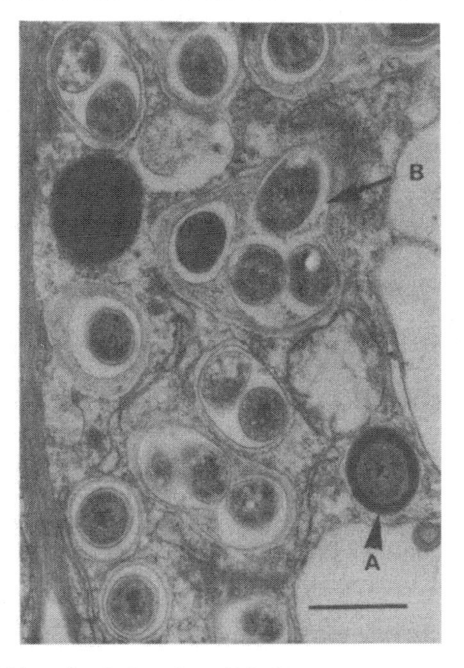

Fig. 2 Infection (A) & fixation threads (B) of Rhizobium strain ANU289 in Parasponia rigida nodule. Bar represents 1 micrometer.

With the discovery of Rhizobium nitrogen fixation in Parasponia came the search for a mechanism of oxygen protection and supply of oxygen to the nitrogenase in that genus.

Parasponia nodules, when sliced open across the invaded zone, fail to show the blood-red or even brownish colours of the siratro nodule. The invaded zone is clearly recognisable by "pinker" colouration compared to the yellow nodule mantle, but this zone progressively decolours to brownish upon exposure to air. Clearly polyphenyloxidases do their damage. Coventry et al. (1976) looked at Parasponia nodules and found no soluble hemoglobin. Tjepkema and Cartica (1982) used direct micro-measurements to illustrate that the oxygen level in the invaded zone is as low as that found in legume nodules. Visual observations after India ink infusion suggested a role of a physical boundary layer of cells which restricts oxygen access.

In our laboratory it was possible to derepress nitrogenase of strain ANU289 using defined *in vitro* conditions (Mohapatra et al., 1983). In agitated liquid cultures with media containing 3 mM glutamate, 50 mM succinate and oxygen at 0.25% the high oxygen sensitivity of the derepressed nitrogenase system was verified (Mohapatra and Gresshoff, 1984). A comparison of the oxygen sensitivity between ANU289 bacteroids isolated from siratro and *P. rigida* again illustrated the high oxygen sensitivity of the *Parasponia* derived cells (R. Sandeman, pers. comm.; Gresshoff et al., this volume). Furthermore siratro bacteroids in the presence of their nodule supernatant (mainly hemoglobin) are more oxygen tolerant for longer periods of time than corresponding *Parasponia* bacteroids. This suggests that the *Parasponia* nodule supernatant is not as efficient in handling the external oxygen supply as the siratro nodule.

Elegant work by Appleby et al. (1983) illustrated the presence of a hemoglobin in *P. andersonii* nodules. Using related techniques we have isolated hemoglobin from nodules of *P. rigida* infected with ANU289. The *P. rigida* hemoglobin has a molecular weight of about 19000 which contrasts that of *P. andersonii* (MW 21000, Appleby et al., 1983). Using spectral

TABLE 1

Comparison of *Parasponia* cDNA clone derived protein sequence with internal leghemoglobin residues of broadbean and soybean

		Sequence
Parasponia	:	-ala-ser——tyr-phe-ile-val-arg-gly-glu-ile-leu-
Broadbean	:	-tyr-ser-val-leu-phe-tyr-thr-ile——ile-leu-
Soybean	:	-tyr-ser-val-val-phe-tyr-asn-ser——ile-leu-
		30 exon 1 ← → exon 2
Parasponia	:	-gly-phe-met-lys-asp-glu-gln-thr-ala-lys-ala-
Broadbean	:	-glu——lys-ala-pro——thr-ala-lys-ala-
Soybean	:	-glu——lys-ala-pro——ala-lys-asp-
		40
Parasponia	:	——phe-ala——lys-asp-val-phe-ile-asn-
Broadbean	:	met-phe-ser-phe-leu-lys-asp-ser-ala-gly-val-
Soybean	:	leu-phe-ser-phe-leu-ala-asn-gly-val-asp-pro-
		50
Parasponia	:	-lys-glu-arg——lys-leu-gly-ala-arg-arg-arg-
Broadbean	:	-val-asp-ser-phe-lys-leu-gly-ala-his-ala-glu-
Soybean	:	-thr-asn-pro——lys-leu-ile-gly-his-ala-glu-
		60

The screening of cDNA clones derived from poly A-RNA of Parasponia nodules using a soybean leghemoglobin probe obtained from Dr. D.P. Verma gave weak hybridisation of one clone (see Newton et al., this volume). The nucleotide sequence of the insert suggests more partial homology with the leghemoglobin of broadbean than that of soybean (Table I). The region shown represents the junction of exon 1 and exon 2 in the soybean leghemoglobin gene. Homology of amino acid residues is 16 out of 45 (broadbean and Parasponia) and 10 out of 45 (soybean and Parasponia). In contrast, illustrating the variability of leghemoglobins, the broadbean and soybean sequences share only 22 out of 45 residues. Ten amino acids are shared by all three sequences. The Parasponia sequence shows one methionine residue, which supports the finding by Appleby (pers. comm.) of methionine incorporation into in vitro translation products of Parasponia nodule RNA.

Appleby et al. (1983) outlined the properties of the P. andersonii hemoglobin (assumed to be quite similar to that of P. rigida). The observations of reversible oxygenation and those of nodules slices (as described in Appleby et al., 1983) support the view of an oxygen carrier. At pH 6.8 at 20°C the O_2 OFF rate constant of 0.3 sec^{-1} is about one-thirteenth that of oxy-leghemoglobin. Appleby et al. (1983) suggested that the hemoglobin might be bound to the bacteroid surface, thereby facilitating a high, localised concentration. In view of the observation that the bacteria in Parasponia are within the fixation thread, and since the hemoglobin stems from the plant cytoplasm, this presupposes a transport across the plant membrane and the thread wall. This may be possible, especially in view of the general lack of knowledge regarding the distribution of leghemoglobin in the legume nodule. Alternatively, one might assume that the hemoglobin is tightly associated with the plant side of the plasma membrane, or perhaps (if secreted) is present in the region between the thread membrane and the thread wall. Such high degree of localisation would optimise the effectiveness of the oxygen carrier, especially if coupled with an anatomical protective mechanism as outlined by Tjepkema and Cartica (1982).

Thus the question of the phylogeny of the plant hemoglobins becomes more puzzling. With the Ulmaceae and Leguminoseae possessing hemoglobins and similar observations having been reported in Casuarina (Davenport, 1960; Appleby et al., 1983), one wonders whether all angiosperms possess such a protein. One has to ask then, what is its function in a plant that does not nodulate and hence use the hemoprotein in symbiotic nitrogen fixation? The possible answer, perhaps, lies in the thought that hemoglobin like all other previously thought of nodule specific proteins, represents only a nodule-amplified protein (like uricase, glutamine synthetase and allantoinase) which already was part of the angiosperms repertoire evolved for its own, nodulation unrelated, physiological function. Perhaps our study of the non-legume, Rhizobium induced nodule will help us to understand how Rhizobium triggers this amplification of gene expression and perhaps what the functions of seemingly nodule specific genes are.

REFERENCES:

Akkermans ADL et al. (1978) Plant and Soil 49, 711-716.
Appleby CA et al., (1981) In: Gibson AH and Newton W, eds Current prespectives in nitrogen fixation research pp 369 Aust.J.Acad.Sci. Canberra.

Appleby CA et al. (1983) Science 220, 951-954.
Becking JH (1979) Plant and Soil 51, 289-296.
Ching TM et al. (1977) Plant Physiol. 60, 771-775.
Coventry DR et al. (1976) Biochim Biophy Acta 420, 105-111.
Davenport HE (1960) Nature 186, 653-654.
Ham SP (1909) Handel 10. Congress Nederl.Indisch Landbouw. Syndicaat II, p. 26.
Mohapatra SS et al., (1983) Arch.Microbiol. 134, 12-16.
Mohapatra SS and Gresshoff PM (1984) Plant Sci.Lett. (in press).
Price GD et al. (1984) Bot.Gazette (submitted).
Rolfe BG et al. (1982) Aust.Microbiol. 3, 33-37.
Scott KF et al. (1983) DNA 2, 141-148.
Tjepkema JD and Cartica RJ (1982) Plant Physiol. 69, 728-733.
Trinick MJ (1973) Nature 244, 459-460.
Trinick MJ (1976) In: Newton W and Nyman CJ, eds Proc.I.Inst.Symp.Nit.Fix. Vol. 2 pp 507-517 Wash.State Univ.
Trinick MJ (1979) Can.J.Microbiol. 25, 565-578.
Trinick MJ and Galbraith J (1976) Arch.Microbiol. 108, 159-166.
Trinick MJ and Galbraith J (1980) New Phytol. 86, 17-26.
Verma DPS and Long S (1983) Int.Rev.Cytol.supp. 14, 211-245.
Vincent JM (1980) In: Newton W and Orme-Johnson W, eds Nitrogen Fixation Vol. II pp 103-129 Univ. Park Press, Baltimore.

POSTER DISCUSSION 7A PHOTOSYNTHESIS AND HYDROGEN METABOLISM IN RELATION TO NITROGENASE ACTIVITY OF LEGUMES

F.R. MINCHIN, The Grasslands Research Institute, Maidenhead, Berkshire, S16 5LR, United Kingdom

As there were nearly 30 posters relevant to this topic, it was decided that the discussion would not deal with the details of the posters but range generally over the area covered by them. This area was subdivided into 4; carbon supply to nodules, carbon use for nitrogenase, hydrogen metabolism and ammonia assimilation.

Carbon supply to nodules.

There is still no general consensus on whether the reductions in N_2 fixation associated with applied N and water stress reflect either direct effects on the nodules or indirect effects on carbon supply. These are areas where more extensive physiological measurements need to be made. The question as to whether carbohydrate levels in nodules are good indications of carbon supply was considered but not resolved.

Responses to short-term deprivation of carbon supply depended on the buffering effect of available carbohydrate pools within the nodules and supporting roots. These pools are affected by the prevailing environmental conditions prior to the deprivation treatment. Consequently, results from these types of experiments are difficult to interpret.

Carbon use for nitrogenase

There have recently been 5 additions to the list of available techniques for measuring the quantity of carbon used for nitrogenase and/or the efficiency of its use. However, these aroused little controversy and there did not appear to be any desire to standardize on one or two techniques, although the use of flow-through systems is becoming an increasingly common feature.

Efficiency values within the range of 3-5 gC(gN fixed)$^{-1}$ are becoming accepted as the "normal" for legume nodules. There was no disagreement with the suggestion that efficiency values lower than the theoretical estimate of 2.57 gC.(gN fixed)$^{-1}$ are likely to be erroneous.

Suggested reasons for variations in efficiency values within the 3-5 gC.(gN fixed)$^{-1}$ range included the costs of associated transport and assimilation processes (although these are separated with some measurement techniques), variations in bacteroid P/O ratios, and variation in the ratio of the two components of the nitrogenase enzyme. An understanding of this variation in efficiency may well allow for genetic manipulation resulting in increase growth and economic yield of legumes.

There was general agreement that O_2-supply limited nitrogenase activity in the short-term, due to the need for nitrogenase protection by an O_2 diffusion resistance. However, the long-term consequences of this protection for carbon use were not clear, probably reflecting the paucity of critical experimentation in this area. Carbon metabolism studies provide indirect evidence that carbon supply is not limiting under normal growth conditions (see discussion session 4B). Nevertheless carbon supply will almost certainly be limiting under extreme conditions, such as severe defoliation.

Hydrogen metabolism

The consensus of the discussion was that the recycling of H_2 through hydrogenase offered no great advantage in terms of carbon saving. Indeed , the amount saved is relative to the efficiency of carbon use for nitrogenase and in many cases could be too small to produce a significant yield response. It

is even possible that H_2-recycling could be detrimental in O_2-limited nodules. However, the use of Hup in genetic manipulation studies was defended on the basis that it provides a useful model. This is particularly true as the expression of Hup can be regulated by the host plant.
The reasons for Hup-expression in rhizobia were briefly discussed and it was suggested that it could be an advantage in conferring autotrophic capacity to bacteria within the host rhizosphere. However, such an advantage requires Hup expression to be repressed in bacteroids to allow for release of H_2.
Most delegates seemed to consider that measurements of relative efficiency of H_2-production were reasonably accurate, despite errors in the measurements of total electron flow using either acetylene or H_2-production under argon. Therefore, variations in RE induced by changes in carbon supply were considered to be genuine reflections of variations in electron allocation between H^+ and N_2. If this is true, then RE cannot be used for symbioses expressing Hup, because two variables (electron allocation and rate of hydrogenase activity) are covered by a single measurement.

Ammonium assimilation

A short discussion on the possible role of PEP carboxylase in ureide producing legumes suggested that it provides organic acids for export in the xylem sap, although the possible contribution of anaerobic respiration by the cytosol of infected cells was not discounted. The presence of high quantities of organic acids in the xylem of ureide producers makes the C:N ratio of their sap comparable to that of amide producers. The reason for ureide production in tropical legumes is, therefore, still unresolved.

PARASPONIA HEMOGLOBIN: IMPLICATIONS FOR THE ORIGIN OF PLANT HEMOGLOBINS

C.A. APPLEBY[1], Y.P. CHEN[1 2], W.F. DUDMAN[1], T.J. HIGGINS[1], A.S. INGLIS[3], A.A. KORTT[3], W.D. SUTTON[1 4], M.J. TRINICK[1].

[1]Divn. of Plant Industry, CSIRO, Canberra, Australia; [2]Shanghai Institute of Plant Physiology, Academia Sinica, China; [3]Divn. of Protein Chemistry, CSIRO, Melbourne, Australia; [4]Plant Physiology Divn., DSIR, Palmerston North, N.Z.

The purification of hemoglobin (Hb) from nitrogen fixing root nodules formed by Rhizobium on a non-leguminous plant, Parasponia (Appleby et al. 1983), casts doubt on hypotheses that legume nodule leghemoglobin (Lb) might have arisen by a unique act of horizontal gene transmission, and that Lb is a sophistication rather than essential for the functioning of symbiotic nitrogen fixation in plants. Parasponia Hb combines reversibly with oxygen (Appleby et al. 1983) which means that it could act as an oxygen carrier in vivo, as does Lb in legume symbioses. Hence, the possible requirement of an Hb presence in new, genetically-engineered nitrogen-fixing symbioses prompts the asking of several questions:-

* Is Parasponia Hb a plant or a bacterial product; what factors control its expression?
* Do it and legume nodule Lb have the same genetic origin?
* Do other higher plant families also possess (ancient?) Hb gene(s) related to those responsible for Lb and Parasponia Hb expression.

When the same Rhizobium strain CP283 is used to nodulate Parasponia andersonii (Planch) and cowpeas, isoelectric focusing analysis of nodule extracts shows the production of very different Hb patterns (Appleby et al. 1983). Hence the plant host must have substantial control over Parasponia Hb synthesis. The total or poly(A)-enriched RNA from Parasponia nodules is efficiently translated by a wheat germ system containing ^{35}S methionine, yielding Parasponia Hb as the major labelled product (with identity established by molecular weight and immuno-selection analyses). These results indicate that Parasponia Hb mRNA is a plant gene product rather than a bacterial product. Differences between Parasponia Hb and soybean Lba are emphasized by the absence of immunological cross reactivity when they are examined using rabbit antibodies specific for each of the two proteins, and lack of cross-hybridization between Parasponia nodule RNA and a full length cDNA for soybean Lba (kindly provided by K.A. Marcker). The similarity between Parasponia Hb and soybean Lba is emphasized by an extraordinarily-close amino acid sequence homology between their respective N-terminal and C-terminal regions. Sequence homology between soybean Lb a and animal myoglobins (proteins known to have the same genetic origin) is much less. We tentatively conclude that Parasponia Hb has the same evolutionary origin as soybean Lb and lies close to each it in evolutionary history. This conclusion, and the demonstration (J.D. Tjepkema, these proceedings; A.I. Fleming, C.A. Appleby, unpublished observations) of Hb presence in Casuarina and other actinorhizal nodules, imply the widespread distribution of ancient Hb genes within the plant kingdom. We have, however, been unable to detect Hb gene expression in un-nodulated plants.

Appleby, C.A., Tjepkema, J.D., Trinick, M.J. (1983) Science 220, 951-953.

HOST PLANT CONTROL OF RHIZOBIUM HYDROGENASE ACTIVITY IN PISUM SATIVUM L.

EULOGIO J. BEDMAR AND DONALD A. PHILLIPS
DEPT. OF AGRONOMY & RANGE SCIENCE, UNIV. OF CALIFORNIA, DAVIS, CA 95616 U.S.A.

Total H_2 evolved from legume root nodules represents the sum of H_2 produced by nitrogenase and H_2 oxidized by any uptake hydrogenase in *Rhizobium*. The host plant can alter relative H_2 production by nitrogenase in rhizobia lacking uptake hydrogenase (Hup^- phenotype) (Edie, Phillips, 1983), and even larger differences in H_2 released from the nodule can be found between Hup^+ and Hup^- rhizobial strains (Ruiz-Argüeso et al., 1978). Potential advantages of using Hup^+ rhizobia have been discussed (Evans et al., 1981), and symbiotic performance of rhizobia has been improved by incorporating the plasmid pIJ1008, which carries *hup* and other symbiotic traits (DeJong et al, 1982). Without postulating that the Hup^+ phenotype is related to better symbiotic performance, it is reasonable to study factors affecting its expression. It is known that different species of host plants produce Hup^+ or Hup^- phenotypes in single strains of *R. leguminosarum* (Dixon, 1972) and *R. japonicum* (Keyser et al., 1982). Such findings are informative, but effects on plant growth of such host control of *Rhizobium* cannot be compared between two species.

The difficulty of comparing physiological parameters between species was overcome when we found that cultivars within *Pisum sativum* L. produced quite different Hup phenotypes in *R. leguminosarum* (Bedmar et al., 1983). In symbiosis with strain 128C53, root nodules of the pea line JI1205 had 12 times as much uptake hydrogenase activity on a nodule fresh weight basis in the $^3H_2$-incorporation assay as nodules formed by the same bacteria on the cultivar Feltham First. On a whole-plant basis, those differences resulted in nearly 24 times more H_2 being evolved from the Feltham First/128C53 association than from the JI1205/128C53 symbiosis at similar rates of C_2H_2 reduction (25 $\mu mol \cdot pl^{-1} \cdot h^{-1}$). Another pea cultivar, Alaska, produced intermediate levels of total H_2 evolution and H_2 uptake with strain 128C53, relative to JI1205 and Feltham First.

Subsequent studies have further characterized host pea control of Hup in *Rhizobium*. A survey of 14 different pea cultivars showed that the phenotypes observed in JI1205, Alaska, and Feltham First are not unique. Of eight cultivars tested with the putative Hup^- strain 300, none produced convincing uptake hydrogenase activity, but two lines that produced the greatest uptake hydrogenase activity in strain 128C53 had significantly lower relative efficiencies with strain 300 than other cultivars tested. Grafting studies showed that a transmissible factor supplied by Alaska and JI1205 shoots produced 5 to 7 fold increases in uptake hydrogenase activity of Feltham First root nodules, but Alaska and JI1205 roots maintained uptake hydrogenase activity at least 4 weeks after being grafted to a Feltham First shoot. The H_2 evolution phenotypes of F_2 progeny from the six possible crosses among Feltham First, Alaska, and JI1205 show complex inheritance patterns. Hopefully it will be possible to identify nearly isogenic lines still segregating for H_2 evolution phenotypes in the F_6 generation.

REFERENCES

Bedmar EJ, Edie SA and Phillips DA (1983) Plant Physiol. 72, In press.
DeJong TM, Brewin NJ, Johnston AWB, and Phillips DA (1982) J. Gen. Microbiol. 128, 1829-1838.
Dixon ROD (1972) Arch. Mikrobiol. 85, 193-201.
Edie SA and Phillips DA (1983) Plant Physiol. 72, 156-160.
Evans HJ, Purohit K, Cantrell MA, Eisbrenner G, Russell SA, Hanus FJ and Lepo JE (1981) In: Gibson AH and Newton WE, eds, Current Perspectives in Nitrogen Fixation, pp. 84-96, Austr. Ac. Sc., Canberra.
Keyser HH, Van Berkum P and Weber DF (1982) Plant Physiol. 70, 1626-1630.
Ruiz-Argüeso T, Hanus J and Evans HJ (1978) Arch. Microbiol. 116, 113-118.

EFFECT OF SPLIT APPLICATIONS OF UREA AND THE NITRIFICATION INHIBITOR DWELL ON NITROGEN FIXATION AND YIELD OF PEANUTS (ARACHIS HYPOGEA VAR. TAINUNG)

WALTER CANESSA AND CARLOS RAMIREZ
CENTRO DE INVESTIGACIONES AGRONOMICAS, UNIVERSIDAD DE COSTA RICA, SAN PEDRO, COSTA RICA

Excessive amounts of starter nitrogen inhibit nodulation. Nitrate is apparently mediating this inhibition. Split nitrogen fertilization can avoid this problem or, another approach, is the use of nitrification inhibitors which prevents the formation of NO_3^-. These two possibilities were investigated in two separate adjacent field trials, using peanuts as a test plant.
Seeds of peanuts (*Arachis hypogea* var. Tainung) were planted at "Finca Experimental Fabio Baudrit, La Garita, Cost Rica", in a soil (Typic Dystrandept) in furrows 60 cm apart and a 20 cm between plants. Inoculation was carried out with a water suspension of peat inoculum. Urea was applied as a water solution whenever Dwell (5-Ethoxy-3-trichloromethyl-1,2,4 thiadiazole), the nitrification inhibitor, was used it was thoroughly mixed with the urea solution before its application to the soil.

Results were summarized in tables 1,2. The application of 30 $kg \cdot ha^{-1}$ of urea at preflowering or at after flowering had a better effect on yield than a similar amount applied at planting. Increasing the rate of application of urea beyond 30 $kg \cdot ha^{-1}$ did not increase yield. There was no difference between treatments in regards to % nitrogen. The data on nodulation (table 2) suggest that the nitrification inhibitor prevented urea (as a potential source of NH_4^+ and NO_3^-) at a rate of 50 $kg \cdot ha^{-1}$ from inhibiting nodulation. Ammonium may well be a less drastic nodulation inhibitor than NO_3^-. There was an increase of the nitrogen content of the plants upon urea fertilization regardless of the use of Dwell. Neither dry matter yields nor grain yields were significantly different between treatments.

TABLE 1. YIELD OF PEANUTS (ARACHIS HYPOGEA VAR. TAINUNG) UNDER DIFFERENT RATES AND TIME OF APPLICATION OF UREA

TREATMENT Nº	UREA RATES $kg \cdot Ha^{-1}$			YIELD $TONS \cdot Ha^{-1}$
	PLANTING	PREFLOWERING	POSTFLOWERING	
(Uninoculated check)	0	0	0	2,2 a
2	0	0	0	2,6 b
3	30	0	0	2,4 ab
4	30	30	0	3,1 c
5	30	30	30	3,0 c
6	30	0	30	2,9 cb
7	0	30	30	3,2 c
8	0	0	30	3,3 c
9	0	30	0	3,3 c

Figures with same letter do not differ at 0.05% level.

TABLE 2. EFFECT OF UREA ± DWELL ON CONTENT AND NODULE DRY WEIGHT

Nº	TREATMENTS		% NITROGEN	NODULE DRY WEIGHT/ 5 PLANTS
	Urea $kg.Ha^{-1}$	Dwell 2 $l.Ha^{-1}$		g
1	0	-	3,5 a	0,46 b
2	0	+	3,6 ab	0,60 a
3	25	-	3,8 b	0,73 a
4	25	+	3,6 ab	0,67 a
6	50	-	4,0 b	0,34 c
7	50	+	3,8 b	0,62 a

HYDROGEN EVOLUTION AND ENERGY METABOLISM IN NODULATED PEA ROOTS

R. DE VISSER, B. FRISO, AND R. SPLINT
DEPT. OF PLANT PHYSIOLOGY, UNIVERSITY OF GRONINGEN, P.O. BOX 14, 9750 AA HAREN, THE NETHERLANDS

1. INTRODUCTION

The control of the relative efficiency (RE) of N_2 fixation in legume root nodules has been the subject of many investigations and hypotheses (Davis et al.1975, Bethlenfalvay, Phillips 1977, Edie 1983, Hageman, Burris 1980). We report on the RE of pea root nodule nitrogenase activity as affected by energy supply and mean age of the bacteroids.

2. MATERIAL AND METHODS

H_2 production, C_2H_2 reduction, nodule growth, and respiration rate were measured on intact nodulated roots (De Visser, Lambers 1983) of Pisum sativum L.cv Rondo x Rhizobium leguminosarum strain PF2 (Hup^-;Dr.TA Lie) during early ontogeny and as influenced by increasing the light intensity with 50% to 330 $\mu E\ m^{-2} s^{-1}$. The activity of the alternative path was considered to be an indicator of the availability of photosynthates in the root (Lambers 1982).

3. RESULTS AND DISCUSSION

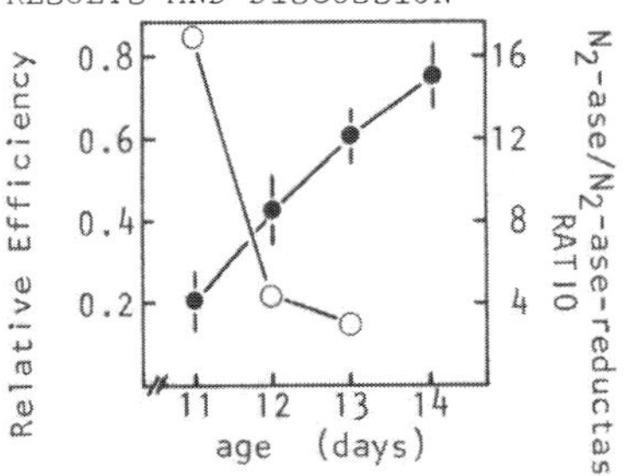

FIGURE 1. RE (●) and component ratio of N_2-ase (O; Bisseling et al.1980) in pea.

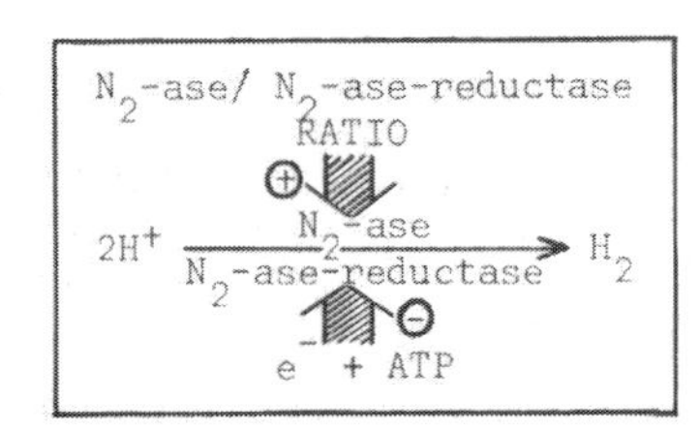

FIGURE 2. Model for the control of N_2-ase proton reduction (in vitro data of Davis et al. 1975 and Hageman, Burris 1980) in intact nodulated roots of pea.

The RE increased from 0.2 to 0.8 during early ontogeny (Fig.1) in accordance with data of Bethlenfalvay, Phillips (1977). The RE and the alternative path activity were not correlated. A positive correlation was expected on the above-mentioned hypothesis. The RE increased from 0.74 to 0.86 ($P<0.05$) one day after increasing the light intensity to 50%. However, a decrease followed to 0.72, ultimately (control: 0.86; $P<0.01$). This decrease of the RE was accompanied by a higher relative growth rate of the nodules, leading to a higher fraction of young nodule tissue. This tissue shows a higher N_2-ase/ N_2-ase-reductase ratio(Fig.1), which favours H^+ reduction (Davis et al.1975). Conclusion: The N_2-ase component ratio controls the RE in situ (Fig.2).

4. REFERENCES: Bethlenfalvay GJ, Phillips DA (1977) Plant Physiol. 60, 419-421. Bisseling T et al.(1980) J.Gen.Microbiol. 118, 377-381. Davis LC et al. (1975) BBA 403, 67-78. De Visser R, Lambers H (1983) Physiol. Plant.58, 533-543. Edie SA (1983) Can.J.Bot. 61, 780-785. Hageman RV, Burris RH (1980) BBA 591, 63-75. Lambers H (1982) Physiol. Plant.55, 478-485.

FACTORS AFFECTING THE ELEMENTAL CHEMICAL COMPOSITION OF NODULATED PLANTS. MOLYBDENUM CONCENTRATION AS A NOVEL PARAMETER OF NITROGEN FIXATION

C.E. DOUKA[+]/A.C. XENOULIS[++]/T. PARADELLIS[++]
NUCLEAR RESEARCH CENTER DEMOKRITOS, AGHIA PARASKEVI, ATHENS, GREECE
[+] Department of Biology
[++] Accelerator Laboratory

The concentration of inorganic elements in plants grown under different conditions with and without inoculation with effective Rhizobia was measured, by the X-ray fluorescence technique, and compared in order to determine whether and to which extent the concentration is affected by biological nitrogen fixation. Relevant results, which have been already reported (Douka et al. 1982), show that either the elemental uptake by the plant or its elemental composition is affected by inoculation.

Plants of Medicago sativa were grown either in the laboratory on nitrogen-free agar and soil-agar medium (Douka 1979) or in field experiments with or without inoculation with an effective Rhizobium meliloti strain. In all cases of plant growth, it has been observed that the concentration of several elements, such as Ca, Mn, Fe, Co, Zn, Rb, Sr and Mo was significantly affected by inoculation. Relevant field experiments have been also carried out with soybean. Of particular interest has been seen to be the behavior of molybdenum, the concentration of which in a nodulated plant is reduced by inoculation almost by an order of magnitude.

A large number of relevant experimental data have suggested that a correlation may exist between the degree of induced biological nitrogen fixation and the Mo concentration of a nodulated plant. An explicit form of such a correlation has been actually identified. Specifically, in an association of the logarithm of the molybdenum concentration of a plant with either the corresponding number of nodules, or the dry weight plant yield or the nitrogen yield, a systematic trend emerges, illustrating a nearly linear dependence of all the cases investigated. This correlation reveals quantitatively that as the plant yield, the nitrogen yield or the degree of nodulation of a plant increases the Mo concentration of the plant decreases.

The results demonstrate that, in addition to the well established procedures of measuring plant and nitrogen yields, measurements of molybdenum concentration in a plant could be used as a criterion of the effectiveness of biological nitrogen fixation.

Douka CE (1979) An unsterilized soil-agar technique for studying Rhizobium plant relationships. J. Appl. Bacteriol. 216, 615.
Douka CE, Xenoulis AC and Paradellis T (1982) Effect of inoculation on elemental uptake by plants grown on saline soils. Folia Microbiol. 27, 278.

INFLUENCE OF THE MACROSYMBIONT ON RELATIVE EFFICIENCY OF NITROGEN FIXATION BY LEGUME ROOT-NODULES

J.J. DREVON, P. TILLARD, L. SALSAC

Laboratoire de Recherches sur les Symbiotes des Racines - INSTITUT NATIONAL de la RECHERCHE AGRONOMIQUE, 9, Place Viala, 34060 Montpellier-Cedex France.

The relative efficiency is defined as $R.E. = 1 - \frac{\text{rate of } H_2 \text{ evolution}}{\text{rate of } C_2H_2 \text{ reduction}}$

1. Influence of the host-plant on the R.E. of nodules formed by a Hup^- (H_2 uptake negative) strain of Rhizobium (glycine max L. MERRIL associated with Rhizobium japonicum).

The R.E. varies during growth cycle. It is optimal at flowering stage and decreases at the end of the cycle.

The R.E. of nodules formed by a same strain, varies from one cultivar to another. In field, among 11 cultivars assayed, the lowest R.E. was 0,44 and the highest was 0,65 ;

Therefore, the host plant may influence the allocation of electrons between H^+ and N_2. This could be due to the quantity and nature of carbohydrates supplied to bacteroids ; to the conditions of diffusion of H_2 and O_2 in nodules.

2. Influence of the host-plant on the R.E. of nodules formed by Hup^+ (H_2 uptake positive strain) (Vigna unguiculata and radiata associated with Rhizobium cowpea CB 756).

The nodules formed by strain CB 756 on Vigna unguiculata have a R.E. close to 1.0, though the nodules formed by the same strain on Vigna radiata have a R.E. of 0,75. This difference is widely explained by a high hydrogenase activity of bacteroids from V. unguiculata, though bacteroids of V. radiata have no detetable hydrogenase activity.

The host plant would therefore play a role in derepression of synthesis or in activation of the strain CB 756 hydrogenase. Similar results are available in GIBSON et al., 1981.

Conclusion : There would be two means of increasing the Relative Efficiency of symbiotic nitrogen fixation :

. decrease the electron allocation by nitrogen to protons (to increase the relative efficiency of Hup^- nodules from 0,50 to 0,75 should theoretically save 16 ATP per N_2 reduced).

. recycle the hydrogen by mean of an hydrogenase (to increase the relative efficiency from 0,50 to 1,0 by mean of hydrogenase would supply 6 ATP).

References :

GIBSON A.H., DREYFUS B.L., LAWN R.J., SPRENT J.I., TURNER G.L. (1981) in Current perspectives in nitrogen fixation. A.H. GIBSON - W.E. NEWTON Ed. p. 373 Griffin press Ltd Netley.

SCHUBERT K.S., EVANS H.J., 1976 - Proced Natl. Ac. Sc. (U.S.A.) 73, 1207-1211

IS HYDROGENASE ACTIVITY BENEFICIAL FOR NITROGEN FIXATION BY SOYBEAN ROOT NODULES

DREVON Jean-Jacques[+] - SALSAC Louis[+]

+ Laboratoire de RECHERCHES sur les SYMBIOTES DES RACINES, I.N.R.A. 9, Place Viala, 34060 MONTPELLIER-Cedex FRANCE.

1 - ATP GENERATION FROM HYDROGEN OXYDATION VERSUS CARBOHYDRATE OXYDATION

In an experiment described previously, it appeared that in Hup^+ soybean nodules, the consumption of carbohydrates was 9,5% less than in Hup^- isogenic nodules, presumably because of H_2 recycling (DREVON et al 1982). This economy saves 5,4 micromoles of glucose saved per hour and per gram fresh nodules of which the oxydation would have supplied 19,5 micromoles of ATP (36 ATP/Glucose oxydized). The oxydation of the amount of H_2 that is synthetised in Hup^- nodules, which is 5.15 micromoles H_2 per hour per gram fresh nodules would have supplied 10,30 micromoles of ATP (2 ATP/H_2 oxydized). So the oxydation of H_2 in Hup^+ nodules would have generated less ATP than the oxydation of carbohydrates in Hup^- nodules with the same amount of O_2.

2 - OXYGEN VERSUS CARBOHYDRATES AS A LIMITING FACTOR OF SPECIFIC NITROGENASE ACTIVITY IN SHORT TERM EXPERIMENTS

In the short term experiment refered above the amount of carbon available in soybean nodules would not be a limiting factor of nitrogenase activity. Indeed raising the partial presure of O_2 from 20 to 40% increased C_2H_2 reduction by more than 60%. So nodules had enough carbon to support a higher level of nitrogenase activity; subsequently in air in which pO_2 is 21%, the nitrogenase activity of soybean nodules is limited by the amount of O_2 supplied to the bacteroids.

Conclusion

With free living bacteria or bacteroid suspension, hydrogenase activity has been shown to protect nitrogenase against O_2. Inside nodules efficient mechanisms maintain the partial pressure at such a low level, when nodules are incubated in air, that the oxyhydrogenation reaction would be a not necessary extra mechanism of decreasing O_2 pressure in bacteroids. On the opposite, by consumming O_2 on a less efficient ATP generating process than carbohydrates oxydation this reaction would decrease the availability of O_2 for respiration. Consequently, in nodules conditions where the O_2 concentration is yet limiting nitrogenase activity, the hydrogenase reaction would not be beneficial for nitrogen fixation by Rhizobium-legume symbiosis. This would explain why Hup^+ nodules did not fix more N_2 than Hup^- nodules in this experiment refered above.

This might be the selection pressure causing that Hup^- strains are the majority among wild strains of various groups of Rhizobium (EVANS et al 1981), which seems opposite to what is observed among free living fixing bacteria. This would explain why the hydrogenase nitrogenase correlation proposed by WILSON (WILSON, 1969) which applies well to various fixing procaryots, does not fit with Rhizobium species.

References

DREVON J.J., FRAZIER L., RUSSELL J., EVANS H.J., (1982) Plant Physiol., 70, 1341-1346

EVANS H.J., PRUROHIT K., CANTRELL M.A., EISBRENNER G., RUSSELL S.A., MANUS F.J., LEPO J.E., (1981) in Current perspectives in nitrogen fixation - A.H. GIBSON, W.E.NEWTON. Ed. pp. 84-96 Giffin press Ltd. Netley

WILSON P.W. (1969) Proc R. Soc London Ser B - 172 ; 319-325

UREIDE N AS INDEX OF BIOLOGICALLY FIXED N IN ISOLINES OF SOYBEANS HAVING *Rj* AND *rj* GENES

J.N. DUBE AND B.B. SINGH
DEPARTMENTS OF SOIL SCIENCE, AND OF AGRICULTURAL BOTANY, JAWAHARLAL NEHRU AGRICULTURAL UNIVERSITY, JABALPUR, 482004, INDIA

Two isolines of soybean cultivar Clark 63 were grown in N-free sand in Leonard jars with applied NO_3-N at 0, 30 and 60 ppm, and the isoline carrying *Rj* was inoculated with *R. japonicum* strain CB 1809. Quarter-strength Jensen's nutrient solution was used for irrigation till emergence, when the NO_3-N was applied. The chlorate anion (1 ppm) was used prophylactically to prevent any provident denitrification. At 2-week growth one set of plants was decapitated, a transfusion tube was fixed in a grip over the stump, and under a mild negative pressure cell sap was drawn, and frozen till analyzed for NO_3-N and ureide N (allantoin + allantoic acid). When the cotylendonary reserve of N exhausted, another sampling was done. Finally the third sampling was done when the flower buds appeared. The NO_3-N was measured colorimetrically through p-DMBA method, and the ureide-N by Young and Conway's method. The ratio of NO_3-N to Ureide N varied as high as 1, and as low as one-tenth. Symbiotically grown plant showed the lowest ratio, and exhibited the highest ureide-N. Physiological stages of growth may very will govern this ratio.

LEGUME STEM NODULES

ALLAN R. J. EAGLESHAM, ADRIENNE KOERMENDY AND FRED SACK
BOYCE THOMPSON INSTITUTE FOR PLANT RESEARCH AT CORNELL, ITHACA, NY, USA

1. INTRODUCTION

Legume stem nodules were described in 1928 by the Danish scientist Hagerup, on *Aeschynomene aspera* growing in shallow waters of the upper River Niger. The nodules were predominantly on submerged parts of the stems, but occurred also well above the water line. However, not until 1979 was it proven that *Aeschynomene* stem nodules do fix nitrogen (Yatazawa and Yoshida). Shortly after, nitrogen-fixing upper-stem nodules were reported in a second legume genus, on *Sesbania rostrata* (Dreyfus and Dommergues, 1981). Eaglesham and Szalay (1983) induced nodules on dry stems of seven *Aeschynomene* species on four of which stem nodules had not previously been reported.

2. PROCEDURE

2.1. Materials and Methods

Aeschynomene and *Sesbania* plants were raised from seed or cuttings in the greenhouse, and maintained in waterlogged sand culture (Eaglesham and Szalay, 1983). *Aeschynomene* plants were inoculated with rhizobial strain BTAi 1 and *Sesbania* with BTSr 1. BTAi 1 was cultured in yeast extract mannitol broth and BTSr 1 in yeast extract succinate broth.

3. RESULTS

Nodules were induced on the upper stems of several species of *Aeschynomene* and on *Sesbania rostrata*. Time to visible detection of stem nodules varied with conditions; they were observed as early as three days post-inoculation on *A. indica* and at four days on *S. rostrata*. Stem nodule morphology differed among species from shallow circular swellings on *A. sensitivia* to well pronounced swellings on *A. indica* and spherical easily removed nodules on *S. rostrata*. Transmission electron micrographs show a single bacteroid per envelope in stem nodules of *A. scabra*. Cortical cells with starch-filled chloroplasts surround the centrally-located bacteroid zone. The close juxtposition of energy source and sink may contribute to the high specific ARAs of *A. scabra* stem nodules (up to 448 μmoles/g dry nodule/h). We are examining the possibility that stem nodules may be energy sufficient; four days after defoliation ARAs were not significantly different from untreated controls (P. Harrison and A.R.J. Eaglesham, unpublished). As with *S. rostrata* (Duhoux and Dryefus, 1982) adventitious root initials appear to be the sites of rhizobial infection of *Aeschynomene* stems. On. *A. scabra* these root initials only just penetrate the stem epidermis and are approximately 0.25 mm in diameter, much smaller than those of *Sesbania*. Direct inoculation with *R. japonicum* did not result in stem nodules on a soybean cultivar which forms adventitious root initials on the lower stem. However after flooding to induce root growth followed by inoculation, axillary nodules formed. Draining and excision of the roots did not disturb the integrity or function of these "stem" nodules.

Dreyfus BL and Dommergues Y (1981) FEMS Microbiol. Lett. 10, 313-317.
Duhoux E and Dreyfus B (1982) Co-ptes Rendues 294, 407-411.
Eaglesham ARJ and Szalay AA (1983) Plant Sci. Lett. 29, 265-272.
Hagerup O (1928) Dansk Botanisk Arkiv 5, 1-9.
Yatazawa Y and Yoshida S (1979) Physiol. Plant. 45, 293-295.

PHOTOSYNTHESIS INTENSITY AND NITROGENASE ACTIVITY IN ROOT ZONE OF CEREALS

V.T. EMTSEV
MICROBIOLOGY DEPARTMENT OF THE TIMIRYAZEV AGRICULTURAL ACADEMY, TIMIRYASEV ST., 50 BUILDING 9, 127550, MOSCOW, U.S.S.R.

The correlation between the photosynthesis intensity of oats (C_3) and corn (C_4) and the nitrogenase activity in their root zones has been investigated. The observations conducted during 1980 to 1982 show that the nitrogenase activity in the root zone of corn (C_4) was higher than that in oats (C_3) and correlated with the photosynthetic activity of the plants. In addition, the nitrogenase activity of the rhizosphere of the plants as well as their photosynthesis were the highest during the first period of vegetation.
A closer parallelism between the photosynthesis of the plants and the nitrogenase activity was observed by us in a 24 hr. study of these processes. The results obtained have shown that there are two levels of the nitrogenase activity, one being higher (at about 12 AM) and characteristic of the day-time activity and the other being lower (at about 12 PM) and characteristic of the night-time period. These two levels are divided by periods of lower nitrogenase activity.
The investigation on the effect of different lighting of plants on the plant photosynthesis and the nitrogenase activity in the root zone of oats and corn has shown that the nitrogenase activity in the rhizosphere of the plants resembles in all cases the change of the photosynthetic activity of oats and corn, that is, the nitrogenase activity is in direct dependence on the intensity of the photosynthesis and hence on the level of lighting, all other conditions of the environment being equal.
Differences have also been determined in the levels of the nitrogenase activity in the rhizosphere and rhizoplane of the plants.

EFFECT OF PROLONGATION OF DARK PERIOD ON RATE OF NITROGEN FIXATION AND SOME ASSOCIATED METABOLIC FACTORS IN COWPEA

O.P. GARG[+] AND K. SWARAJ[++]
+ DEPARTMENT OF BOTANY, PANJAB UNIVERSITY, CHANDIGARH, INDIA
++DEPARTMENT OF BOTANY, HARYANA AGRICULTURAL UNIVERSITY, HISSAR, INDIA

Plants of cowpea (Vigna unguiculata Walp.)cv HFC 42-1 were raised from nocu lated seeds in sand culture supplied periodically with N-free nutrient olution. Nodule senescence was induced by placing well grown plants in continuous dark. A darkening period of 18 hr reduced the rate of N-fixation (moles of ethylene produced) to 43% of the light controls, of 24 hr to 15% and of 48 hr to almost negligible level. Plants exposed to 42 hr of darkness and returned to daily photoperiods did not show even partial recovery of N-fixation. Total recovery took place weeks later only on production of new leaves and nodules (Fig. 1).

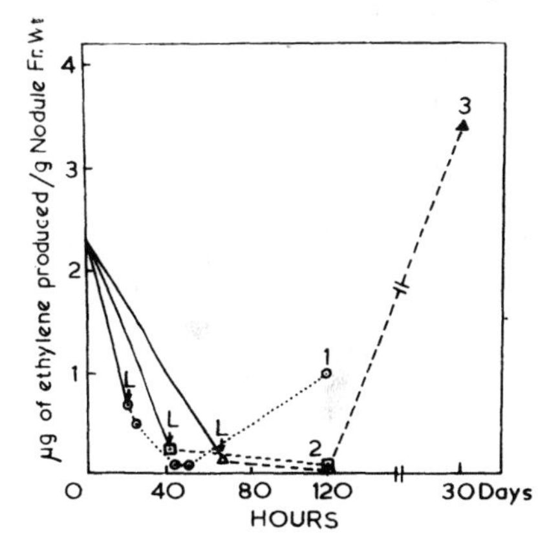

Fig. 1 Revival of nitrogen fixation in nodules of cowpea plants exposed to predarkening periods of different durations and reshifted to normal alternation of light and dark. L depicts exposure to light- solid lines depict predarkening period;
Curve 1- Predarkening 20 hr;
Curve 2-Predarkening 42 hr;
Curve 3-Predarkening 66 hr.

The fall in ascorbic acid content of nodules on dark-retained plants to 80% of light control in 24 hr is not commensurate with the precipitate decline in N-fixation. Earlier studies by us (1970, 1977) had highlighted the importance of this vitamin in maintaining N-fixation under steady state conditions. There was a very small reduction in soluble carbohydrate and respiratory rate, but free amino acids accumulated.

Total heme proteins in the water extract of nodules (unfractionated) did not show any decrease even up to 42 hr in darkness but those precipitating between 50-80% of $(NH_4)_2SO_4$ saturation (mainly leghemoglobin) showed only a small decline.

None of the biochemical parameters studied is quantitatively commensurate with the early sharp decline in the rate of N-fixation, suggesting triggering by some more critical and faster acting factor(s).

REFERENCES

Swaraj K and Garg OP (1970) Physiol. Plant 23, 889-97.
Swaraj K and Garg OP (1977) Proc. Nat. Symp. N-Assimilation and Crop Productivity, 164-170.

CALORIMETRY OF NITROGENASE EFFICIENCY IN SOYBEAN NODULES

P.G. HEYTLER AND R.W.F. HARDY
E.I. DUPONT DE NEMOURS & COMPANY
CENTRAL RESEARCH & DEVELOPMENT DEPARTMENT
WILMINGTON, DE 19898 USA

Assessment of the metabolic cost of nitrogenase in intact, detached nodules, was made by a calorimetric technique which we designed for the purpose. This approach avoided some uncertainties of respirometric methods and allowed mathematical treatment to separate the requirement of reduction *per se* vs. that of enzyme activation.

Nodules were harvested from ∿40 day, hydroponically grown soybean plants inoculated at germination with a Hup^- strain of *R. japonicum*. Our cuvette held 1 g of tissue + .3 ml buffered malate medium, and was slowly flushed with a pre-equilibrated gas mixture. Typically, this was 20% O_2 in argon mixed with 12% of substrate (N_2, C_2H_2 or N_2O). In some runs, air was the main component. Heat production was measured continuously and the exiting gas was sampled every 4-6 minutes for CO_2, H_2 and C_2H_4 analysis by GC.

Nitrogenase was selectively inhibited, stepwise, with brief pulses of pure O_2. Both heat evolution and nitrogenase activity decreased. The metabolic cost of the particular nitrogenase-mediated reduction was reflected by the slope obtained in plotting cal/h against μmol product/h. This cost averaged -171 Kcal/mol H_2 (under Ar-O_2); -250 Kcal/mol C_2H_2; -784 Kcal/mol N_2. The inherent heats of reduction are derivable from thermodynamic data. The remaining ΔH, associated with activation of nitrogenase e^- transfer, is 180-190 Kcal/$2e^-$ in all cases.

The calorimetric data were used to calculate the net photosynthate-carbon cost of N_2 reduction in nodules at 3.6 g C/g N. Simultaneous CO_2 evolution data were in reasonable agreement (4.3 C/N).

References

Brown HD 1969 Biochemical Microcalorimetry, Academic Press, NY.

Pate JS, CA Atkins, RM Rainbird 1980 Theoretical and experimental costing of nitrogen fixation and related processes in nodules of legumes. In Current Perspectives in Nitrogen Fixation, (Gibson AH, WE Newton, eds) Elsevier, Amsterdam. pp 105-116.

Rainbird RR, WD Hitz, RWF Hardy 1983 Experimental determination of respiration associated with soybean/ Rhizobium nitrogenase function, nodule maintenance and total nodule fixation (submitted for publication).

Schubert KR 1982 The energetics of biological nitrogen fixation. Workshop Summaries - I. Amer. Soc. Plant Physiol.

SEASONAL VARIATIONS ON NODULE METABOLISM OF *Phaseolus vulgaris*

MARIANGELA HUNGRIA[+]/M.C.P. NEVES[++]
+ EMBRAPA/PNPBS, Km 47 SEROPÉDICA, 23460, RIO DE JANEIRO, BRASIL
++ DEPTº SOLOS, UNIV. FEDERAL RURAL DO RIO DE JANEIRO, 23460, SEROPÉDICA RIO DE JANEIRO, BRASIL

INTRODUCTION

A three-fold increase in the Brazilian average seed yield of *Phaseolus vulgaris* can be achieved without fertilizer nitrogen (Neyra, Pollack, 1982; Franco *et al.*, unpublished) if plants are well nodulated with efficient *Rhizobium* strains. As inoculation becomes a common practice, the physiological factors affecting nodule functioning and limiting seed production in the bean/*Rhizobium* system are of great concern. Here we report some effects of *Rhizobium* strains and bean cultivars on nodule metabolism of glasshouse grown plants.

MATERIALS AND METHODS

Bean cultivars Negro Argel and Venezuela 350 were inoculated with either strain C05 Hup^+ (CENA, Brasil) or 127 K17 Hup^- (Nitrargin, USA) and grown in sterilized sand and vermiculite mixture receiving complete nutrient solution devoid of N. Mean temperatures throughout plant growth were 28.3/23.2^{o}C (day/night, respectively). Weekly measurements were made from germination to seed maturation, and involved C_2H_2 reduction activity, H_2 evolution in air and xylem sap analysis (Thomas *et al.*, 1979; Vogel, Drift, 1970).

RESULTS AND DISCUSSION

During plant growth there was marked variation on C_2H_2 reduction, H_2 evolution, relative efficiency (RE) and xylary transport of fixed N_2, which were also affected by plant cultivar and *Rhizobium* strain. The resulting seed yield was 38% increased by the best cultivar/strain combination.

Maximum rates of xylary transport of N compounds were recorded at early flowering (35 days after germination), however , acetylene reduction activity and H_2 evolution peaked 7 days later. Cultivar and strain affected the N concentration of xylem exudates but differences were greater when calculated as the rate of N transported.

Both strain and cultivar affected the percentage of N transported as ureide in the xylem sap and was positively correlated with seed yield ($r = 0.800^{**}$) and harvest index ($r = 0.851^{**}$), reflecting a better utilization in seed yield of the N fixed by the best cultivar/strain combination.

The R.E. measured throughout plant growth was positively correlated with the total xylary N transport and percentage of N as ureides ($r = 0.624^{**}$). Linear correlations were also obtained between R.E. and both seed yield ($r = 0.880^{**}$) and harvest index ($r = 0.710^{**}$).

The results have showed that R.E. and % N transported as ureides are physiological factors that should be considered when selecting bean cultivars and *Rhizobium* strains combinations for higher seed production.

Neyra CA and Pollack BL (1982) New Jersey Agric. Exp. Station Publication
Thomas RJ, Feller U and Erismann KH (1979) New Phytol 82, 657-669.
Vogels GD and Van der Drift C (1970) An. Biochem. 33, 143-157.

NITROGENASE ACTIVITY IN ROOT NODULE HOMOGENATES OF *Alnus incana*

KERSTIN HUSS-DANELL AND ANN-SOFI AHLQVIST
DEPARTMENT OF PLANT PHYSIOLOGY, UNIVERSITY OF UMEÅ
S-901 87 UMEÅ, SWEDEN

Nitrogenase activity in symbiotic *Frankia* is mostly measured on excised root nodules, nodulated roots or intact nodulated plants while the use of nodule homogenates is less common. Root nodule homogenates of actinorhizal plants represent *Frankia* in a symbiotic stage but released from the environmental influence of the host plant.

We have studied the nitrogenase activity of *Alnus incana* (L.) Moench at various intactness of the symbiosis by comparing the activity of root nodule homogenates with that of excised nodules, nodulated roots and intact potted plants. Two techniques for preparing nodule homogenates were compared. Anaerobic homogenization with a blender in buffer supplied with sucrose, polyvinylpyrrolidone and reducing substances gave higher yields than crushing the nodules in liquid nitrogen. The blender technique was therefore used throughout the study.

The activity in the homogenates was very reproducible and was, on average, nearly twice as high as the activity in excised nodules and c. 10 % of the activity in intact plants. The difference in activity between nodule homogenates (and excised nodules) and intact plants was, roughly by halves, due to removal of the root system from the pot and to the excision process in itself. The activity in the homogenates was slightly higher when nodule excision was done in Ar or under water as well as after treatment of the homogenates with osmotic shock, toluene or Triton X-100, but lysozyme gave decreased activity. The gainings in activity obtained by the treatments were considered too small to outweigh the increased complications of preparing homogenates for routine studies. Due to the reproducible recovery of nitrogenase activity in the homogenates the technique seems useful for physiological studies on nitrogen fixation in *Alnus incana*.

LEGHEMOGLOBIN REDUCTION IN SOYBEAN NODULES

R. KLUCAS/ L. SAARI AND K. LEE
DEPARTMENT OF AGRICULTURAL BIOCHEMISTRY,
UNIVERSITY OF NEBRASKA, LINCOLN, NEBRASKA 68583-0718, USA

We present spectrophotometric evidence that ferric leghemoglobin (Lb) is reduced in nodules and suggest that ferric Lb reductase may be the enzyme responsible for the reduction. Thin nodule slices (1-2 mm) from fresh nodules were washed and maintained in isotonic solutions containing various oxidants and ligands. Nodule slices isolated under anaerobic conditions exhibited typical ferrous Lb absorption band which was not changed by the addition of ferricyanide. Slices of nodules incubated with hydroxylamine exhibited a 627 nm absorption peak which shifted to 608 nm or 622 nm if the slices were exposed to fluoride or acetate respectively and disappeared if exposed to nicotinate which are typical responses of ferric Lb. With longer incubation in nitrogen or air, the 627 nm peak disappeared with the appearance of typical ferrous Lb band or if nicotinate was present, a sharp peak at 554 nm appeared which is diagnostic for ferrous Lb nicotinate complex. These data substantiate that ferric Lb can be reduced *in situ* to ferrous Lb.

An NADH: (acceptor) oxidoreductase (EC 1.6.99.3) from the cytosol of soybean root nodules was purified by ammonium sulfate fractionation, hydroxylapatite adsorption and Sephacryl S-200 Superfine chromatography. The native molecular weight of the reductase was found to be 100,000 by analytical gel filtration and 83,000 by equilibrium ultra centrifugation. The subunit molecular weight was 54,000 as determined by SDS-polyacrylamide slab gel electrophoresis. The pI of the enzyme was 5.5.

Nearly identical initial velocities were obtained using either CO or O_2 to ligate the enzymatically produced ferrous leghemoglobin (Lb). With CO as the ligand in the reaction, the product of the enzyme-catalyzed, NADH-dependent reduction of ferric Lb was spectrally identified as LbCO. Initial velocity was a linear function of increasing protein concentration. NADPH was only 31% as effective an electron donor as NADH. The Michaelis constants (K_M) for ferric Lb*a* and NADH were 9.5 μM and 18.8 μM, respectively. Myoglobin, Lb*a*, Lbc_1, Lbc_2, Lbc_3 and Lb*d* were reduced at similar rates by the reductase. At pH 5.2, acetate-bound ferric Lb and nicotinate-bound ferric Lb were reduced by the enzyme at 83% and 5%, respectively, of rates observed in the absence of ligand. The rate of enzymatic reduction of ferric Lb was constant between pH 6.5 and 7.6 but increased approximately 3-fold at pH 5.2. The results indicate that the NADH: (acceptor) oxidoreductase could be identified as a ferric Lb reductase.

TEMPERATURE & RHIZOBIA STRAIN EFFECTS ON RE & COST OF N_2 FIXATION IN SOYBEANS

DAVID B. LAYZELL, PAULA ROCHMAN AND DAVID T. CANVIN
DEPT. OF BIOLOGY, QUEEN'S UNIV., KINGSTON, ONTARIO, CANADA. K7L 3N6

INTRODUCTION

Little is known of the effects of low root temperatures on the interrelationships between photosynthate supply, photosynthate utilization and N_2 fixation in nodulated legumes. In this paper, we report the effects of low temperature on two measures of carbon use efficiency in nodulated soybean roots: a) the apparent respiratory cost and b) the relative efficiency of N_2 fixation.

METHODS

Soybean plants (Harosoy 63, 28-37 d old) infected with either USDA 16, 35 or 110 or Type S inoculant were connected to an open gas exchange system and rates of CO_2 and H_2 evolution and C_2H_2 reduction continuously monitored while pot temperatures were dropped from the growing temperature (25°C) to 10°C at 5°/hour. Uptake hydrogenase activity was measured in excised nodules by both the incorporation of $^3H_2$ into nodules and the depletion of H_2 from the gas phase (10% C_2H_2 in air) of sealed vials.

RESULTS & DISCUSSION

At 25°C, rates of CO_2 evolution, and C_2H_2 reduction were from 185-270 µmole. CO_2/gDW nod rt.hr and 21-42 (same units) respectively. While no H_2 was evolved from the USDA 110 symbioses, the other 3 symbioses displayed rates ranging from 10 to 18 µmoles/gDW nod rt.hr. In the H_2 evolving symbioses, rates of C_2H_2 reduction were least affected by temperature (Q_{10} = 1.34 to 2.35) while in USDA 110, the Q_{10} was a high 15.2. Measurements of CO_2 evolution revealed Q_{10} values between 2.03 and 2.87 while H_2 evolution was markedly affected by temperature change (Q_{10} = 3.16 to 3.70). When roots of one symbioses (Type S) were held at 10°C for 96 hours, the rates of CO_2 and H_2 evolution remained low, while the C_2H_2 reduction rates increased within 24 hours to 80% of the initial rates at 25°C (cf Duke et al., 1979, Plant Physiol 63: 956-62).

Using these measurements, values for the relative efficiency (RE= 1-H_2 evolution in air/C_2H_2 reduction) and the apparent respiratory cost (CO_2 evolved/C_2H_2 reduced) of N_2 fixation were computed at the various temperatures. In the three H_2-evolving symbioses, RE increased from 0.57 to 0.62 at 25°C to 0.85 to 0.92 at 10°C. Longer term exposure (24 to 96 hrs) to low root temperatures resulted in even higher RE values (to 0.99).

Attempts to account for the observed changes in RE by increases in uptake hydrogenase activities were unsuccessful. Net H_2 uptake was not detectable at 15 to 25°C and measurements of $^3H_2$ uptake were, in all cases, similar to, or greater than, the observed changes, with temperature, in C_2H_2 reduction activity. These results, although not conclusive, suggest that low temperatures directly affect nitrogenase electron allocation.

The results also suggested that longer term (24 to 96 hr) exposure to low root temperatures decreases the apparent respiratory cost of N_2 fixation from ca 8.0 (at 25°C) to 3.5 (at 10°C) moles CO_2 evolved/mole, C_2H_2 reduced. Consequently, the findings of this study indicated that low root temperatures may have important effects on both the pattern of assimilate partitioning and the efficiency of photosynthate use in nodulated legumes.

Acknowledgements: The technical assistance of Glenn Weagle and Kerry Walsh is appreciated. This work was funded by the NSERC (Canada) (DBL) and Agriculture Canada (DTC).

CYTOLOGY AND MORPHOGENESIS OF SEXUAL ORGANS OF *AZOLLA*

LE DUY THANH
DEPARTMENT OF GENETICS, UNIVERSITY OF HANOI, HANOI, VIETNAM

In natural conditions in North Vietnam, the sexual organs of *Azolla pinnata* usually appear during the period from November to March. Sporocarps are borne in pairs. They may be male and female or of the same sex. In general, most of varieties, especially in greenhouse, produce only microsporocarps. All the phases of morphogenesis of micro- and macrosporocarps for several varieties of *A. pinnata* and *A. mexicana*, from initial stages until the production of a functional macrospore and sixty-four microspores, have been described. Special attention was paid to the organization and microstructure of the mature micro- and macrosporangium and their spores as well as to their further development. Differences in structure of the complex swimming apparatus of the macrosporangium between species *A. pinnata* and *A. mexicana* were noted. While in the macrosporangium of *A. pinnata*, 9 floats are symmetrically arranged in groups of 3, *A. mexicana* had only 3 macrospore floats. In addition, the sexual reproduction of *Azolla* in relation to the transfer of the algal symbiont, *Anabaena azollae* during its whole life cycle has been probed. In order to obtain a hybrid between different species of *Azolla*, a series of experiments inducing gametophytes were performed.

AN ULTRASTRUCTURAL STUDY OF THE *AZOLLA* - *ANABAENA AZOLLAE* RELATIONSHIP

LE DUY THANH
DEPARTMENT OF GENETICS, UNIVERSITY OF HANOI, HANOI, VIETNAM

Light- and electron-micrographs reveal the following aspects of the organization and ultrastructural characteristics in the relation between *Azolla pinnata* and *Anabaena azollae*. Particular attention was paid to leaf cavity formation and the association of algae with hair cells during leaf development. The findings reveal that the stem tip is the site of initiating the infection of the plant by algae. A large increase of the leaf cavity occurs in the young leaf. Then, the algal filaments fill the cavity completely. The subsequent development of the leave shows enlargement of the leaf cavity with the algae lining the leaf cavity. Some pictures show that the algae form a slime layer covering the wall of the leaf cavity. The plant cells produce hairs protruding through the slime layer into the cavity. The hairs can contain one, two or more cells. But usually they consist of two or more cells. At most stages of development, the *Anabaena* cells are associated with hair cells. The hairs produce labyrinthine wall ingrowths and have cytoplasm with numerous organelles. Many of them appear to be high vacuolated and filled with a dense substance. The new hairs initially appear as a small bud on an epidermal cell which then undergoes a transverse division. This morphology and ultrastructure of the hair cells demonstrate and support the notion that they may either have a function in absorption of algal metabolites or in the possible metabolic interchange between *Azolla* and *Anabaena azollae* (Duckett, Toth and Soni, 1975).

DIVERSION BETWEEN C_2H_2-REDUCTION AND HETEROCYST FREQUENCY IN A CYCAD ROOT

PETER LINDBLAD
INSTITUTE OF PHYSIOLOGICAL BOTANY, BOX 540, S-751 21 UPPSALA, SWEDEN

All cycads so far investigated contain heterocystous cyanobacteria (*Nostoc*/*Anabaena*) in coralloid roots. An increased heterocyst frequency towards the basal parts of these roots was earlier reported (Grilli Caiola 1975, 1980). Similarly cyanobacteria of some other symbiotic systems showed increased heterocyst frequency in older parts (Becking, Donze 1981; Englund 1977; Hill 1977; Peters et al. 1982; Silvester 1976). However, three different patterns in nitrogenase activity have been described. First, a parallel increase in heterocyst frequency and nitrogenase activity (C_2H_2-reduction) in the water fern *Azolla* (Peters et al. 1982). Second, two "peaks" in the nitrogenase activity; one at approximately 20% and the second at approximately 40% heterocysts, in *Azolla pinnata* (Becking, Donze 1981). Third, in *Gunnera* and the lichen *Peltigera aphthosa* nitrogenase activity (C_2H_2-reduction) initially increased and then decreased in older parts (Silvester 1976; Englund 1977).
In this study a diversion between heterocyst frequency and nitrogenase activity (C_2H_2-reduction) in coralloid roots from the cycad *Zamia floridana* A.DC. is described.

Materials and Methods

Coralloid roots from the cycad *Zamia floridana* A.DC. were collected in greenhouse, cut into sections and then prepared for acetylene reduction assay, heterocyst count and filament length estimation. Acetylene reduction was determined according to Stewart et al. (1967) using 7 ml serum bottles and 6 hours incubation time. Heterocyst frequency was determined according to Hitch, Millbank (1974) and the same technique was used to determine filament length.

Results

The results are summarized in figure 1, means of three coralloid roots.

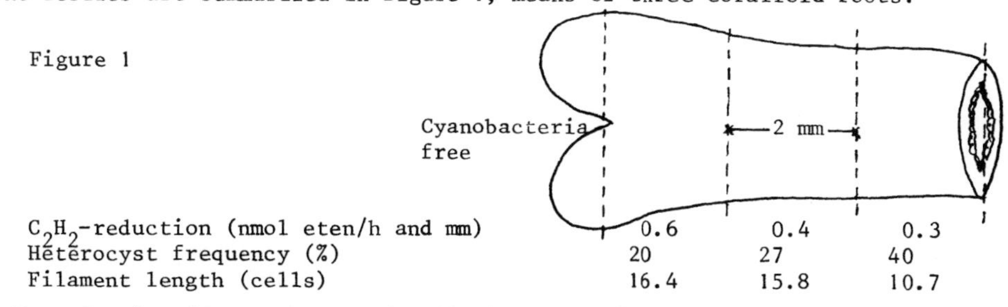

Figure 1

C_2H_2-reduction (nmol eten/h and mm)	0.6	0.4	0.3
Heterocyst frequency (%)	20	27	40
Filament length (cells)	16.4	15.8	10.7

C_2H_2-reduction decreases towards older parts, while heterocyst frequency increases towards older parts of the coralloid root.

References

Becking, J.H., Donze, M. 1981. Plant & Soil 61:203-226
Englund, B. 1977. Physiol.Plant. 41:298-307.
Grilli Caiola, M. 1975. Phycologia 14:25-33.
Grilli Caiola, M. 1980. New Phytol. 85:537-544.
Hill, D.J. 1977. New Phytol. 78:611-616.
Hitch, C.J.B., Millbank, J.W. 1974. New Phytol. 74:473-376.
Peters, G.A. et al. 1982. Isr. J. Bot. 31:305-323.
Stewart, W.D.P. et al. 1967. Proc.Natl.Acad. 58:2071-2078.

CHANGES IN C_2H_2 REDUCTION AND H_2 EVOLUTION FOLLOWING DISRUPTION OF THE PHOTOSYNTHATE SUPPLY TO PEA NODULES

J.D. MAHON
NATIONAL RESEARCH COUNCIL OF CANADA
PLANT BIOTECHNOLOGY INSTITUTE, SASKATOON, SASKATCHEWAN, S7N OW9, CANADA

When the supply of photosynthate to actively fixing nodules was disrupted by darkening or excising the shoots or by chilling the stem, the rates of both C_2H_2 reduction and H_2 evolution in the intact roots and nodules decreased. The changes in C_2H_2 reduction after these disruption treatments were similar in plants inoculated with a low H_2 evolving strain (R.F.=0.75) and a mixed strain commercial inoculant (R.E.=0.59), but the changes in H_2 evolution differed. With both inoculants, H_2 evolution was a continuous function of the rate of acetylene reduction. However, while H_2 evolution was almost proportionally related to C_2H_2 reduction with the low efficiency mixed strain, the relationship was distinctly non-linear and the rates were reduced with the low H_2 evolving strain.

The decreases in C_2H_2 reduction after disruption were kinetically different depending on the mechanism of disruption, the age of plant, time of day and root temperature. The rate decreased rapidly after both darkening and excising the shoots if the treatment was initiated early in the photoperiod. Later in the day, only excision reduced the rate of C_2H_2 reduction unless the irradiance had been decreased during the photoperiod before darkening. Thus photosynthesis was important in supporting C_2H_2 reduction not only during the photoperiod, but also through its effects on storage pools, during the night. Roots and nodules were also able to supply substrate for nitrogenase activity. Although the substrate available from this source was considerably less than that supplied by either photosynthesis or shoot storage pools, it was able to support an increasing amount of C_2H_2 reduction per gram of root and nodule tissue as the plant aged.

The results indicated that nitrogenase activity could be supported by substrate from at least three different sources. They also showed that the rate of H_2 evolution was related to the total nitrogenase activity, regardless of the source of substrate. However, because of the non-linearity of this relationship, estimates of relative efficiency tended to increase as the rate of C_2H_2 reduction decreased.

EXPOSURE OF COTYLEDONS TO LIGHT AFFECTS RHIZOBIUM INFECTION AND NODULE DEVELOPMENT IN SOYBEANS

N.S.A. MALIK/H.E. CALVERT/M.K. PENCE/W.D. BAUER
CHARLES F. KETTERING RESEARCH LABORATORY, YELLOW SPRINGS, OHIO 45387

In soybean seedlings the exposure of cotyledons/hypocotyls to light strongly affects the initiation and development of root nodules. Two day-old soybean seedlings in plastic growth pouches were inoculated with *R. japonicum* in very dim light. At the time of inoculation, the position of the root tip (RT) of each seedling was marked on the face of the pouch. Upper portion of seedlings (cotyledons/hypotocotyl) were exposed to light at various times before and after inoculation. The number and position of nodules on the primary roots relative to RT mark were scored 7-8 days after inoculation or first exposure to light whichever came last. A set of 100 seedlings were used for each treatment. For anatomical studies representative root samples fixed in paraffin were serially sectioned, stained and then examined under microscope to score the location and stage of infection development.

If the cotyledons/hypocotyls of young seedlings were exposed to light 2 days prior to inoculation, nodulation above the RT mark was reduced by 50% compared to plants which were first exposed to light 1 day after inoculation (Fig. 1). However, the exposure of cotyledons to light after inoculation was necessary for normal nodulation. Seedlings kept in the dark for 7 days after inoculation formed only 25% as many nodules above the RT mark as seedlings kept in the dark for 1 day after inoculation (Fig. 1). Anatomical studies revealed that the exposure of cotyledons to light prior to inoculation reduced the number of infection centers with visible infection threads. If seedlings were kept in the dark for 7 days after inoculation, normal number of infection threads formed above the RT mark, but few of these developed further into mature nodules. It appears that light may stimulate soybean cotyledons to produce substances which can both inhibit the formation of infection threads and enhance the development of nodules from established infection threads. The effects of light appear to be independent of the rapid self-regulatory response elicited in the host plant by exposure of the roots to *R. japonicum*.

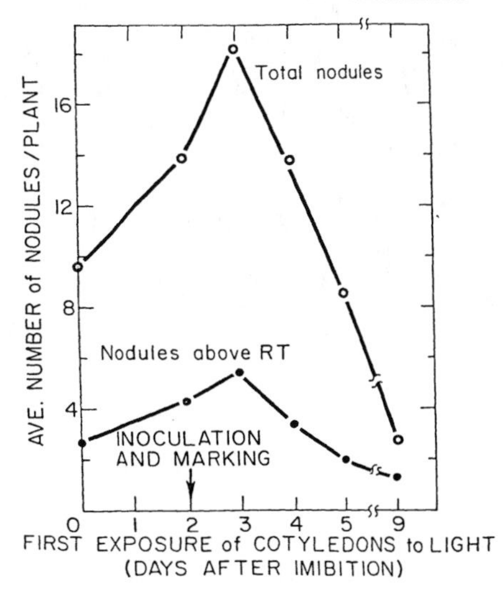

Figure 1. Effect of exposing soybean cotyledons at different time intervals before and after inoculation on nodulation. Open circles (oo) represent total number of nodules on the primary root and closed circles (●●) represent the number of nodules formed above the RT mark. Seedlings were marked (RT) and inoculated with 2.5 X 10^4 rhizobia/plant 48 h after imbibition. The cotyledons were exposed to light (1200 Ft-c) in the growth chamber (14 h photoperiod) either from the time of imbibition or at various time intervals post imbibition.

IN VITRO NITROGENASE ACTIVITY OF RHIZOBIUM ASSOCIATED WITH CAROB EXCISED ROOTS

M. A. Martins-Loução* / C. Rodríguez-Barrueco**
* Dep. Botany. Fac. Sciences. 1294 Lisboa codex. Portugal.
** Unidad de Fijacion de Nitrogeno. CSIC: Salamanca. Spain.

1. INTRODUCTION

Stimulation of nitrogenase activity in rhizobia by carob (*Ceratonia siliqua* L.) callus cultures has been obtained (3). This paper shows that both seedlings and roots "in vitro" were also able to induce nitrogenase activity in two different rhizobia.

2. MATERIAL AND METHODS

After 6-8 days germination (1), seedlings and excised roots were associated with either *R. meliloti* or *R. lupini*. Seven days of association were passed before the samples were assayed for nitrogenase activity by gaschromatographical detection.

3. RESULTS

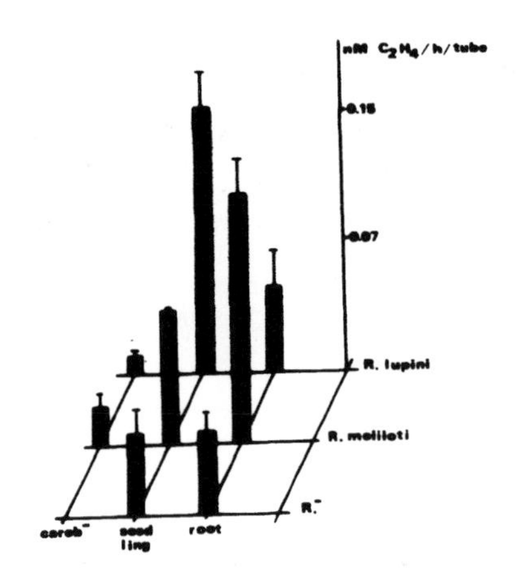

Nitrogenase activity in *R. meliloti* and *R. lupini* can be induced either by carob seedlings or excised roots (fig.). This activity is higher in carob seedlings with *R. lupini* and higher in excised roots with *R. meliloti*. It seems that the presence of some difusible factors can be stimulatory for *R. lupini* and, on contrary, inhibitory for *R. meliloti*. Although nitrogenase activity showed considerable variation from assay to assay, like in others (3), there is no doubt that the seedlings or excised roots from carob were able, like witn callus (2), to induce nitrogenase activity in two different strains of rhizobia. These results can be very interesting for the understanding of the interactions of plants and nitrogen-fixing bacteria.

4. REFERENCES

1. Martins-Loução & Rodríguez-Barrueco (1981) Z.Pfl. 103, 297-303
2. - - - - (1982) Proc. V Int. C. Plant Cell Cult., 671-672
3. Schetter, C., Hesse, D. 1977. Plant. Sci. Lett. 9, 1-5

THE INTERACTION BETWEEN NITROGEN AND CARBON METABOLISM IN NITROGEN FIXING SOYBEAN BACTEROIDS.

D.L. MCNEIL/B.J. CARROLL/P.M. GRESSHOFF
BOTANY DEPT, AUSTRALIAN NATIONAL UNIVERSITY, PO BOX 4, CANBERRA, AUSTRALIA

Bacteroids were prepared from 6-8 week old soybean nodules (after Eady, 1980). Large particles were removed by a low speed centrifugation. Two washes using high speed centrifugation were then carried out and the bacteroids resuspended at 1 mg bacteroid protein per ml. The washing buffer was 0.15M NaCl, 50mM Phosphate, 2mM $MgCl_2$, 0.5mM $CaCl_2$, pH7.3. C_2H_2 reduction rates were measured after 100 & 300 minutes in vials containing 25ml of 5% C_2H_2 in N_2, 1ml buffer, 0.5mg bacteroid protein and O_2 levels and C-sources as required. Generally, results given are for the C_2H_2 reduction rate over the first 100 minutes from duplicate or triplicate samples at the optimum of 4 O_2 levels.

At low levels of O_2 (0.3-0.6%) bacteroids had high autonomous rates of acetylene reduction (up to $5x10^{-9}$ mol C_2H_2 reduced (mg bacteroid protein)$^{-1}$ min^{-1}). With added succinate (1.5-50mM) rates were slightly higher (up to 7nmol mg^{-1} min^{-1}) but had a higher O_2 optima (1.0-1.5%) and lower efficiency of O_2 consumption (7.3 compared to 3.6 mol O_2 consumed / mol C_2H_2 reduced). Addition of glucose or sucrose (50mM) did not increase bacteroid C_2H_2 reduction rates in contrast to the findings of Trinchant & Rigaud (1979) with _R. phaseoli_.

Autonomous acetylene reduction by bacteroids was lower in plants that had been pregrown in the dark for 3^1/2 days before use (75% reduction). The intact nodules had C_2H_2 reduction reduced by 95%. However full bacteroid activity was restored with the additon of succinate (1.5 or 50mM). This suggested that bacteroids were C limited 'in vivo' without loss of function when light was limiting. Glucose and pyruvate (50 & 10mM) gave low initial bacteroid C_2H_2 reduction rates (1.25 & 3.4nmol mg^{-1} min^{-1} but the activity lasted longer (40 & 60% of the original activity remained between 100 & 420 minutes compared to 10 & 8% for bacteroids supplied succinate or no C-source). The high O_2 consumption and short fixation duration of succinate supplied bacteroids suggests that succinate is not the sole source of C for bacteroids 'in vivo'.

Acetylene reduction was substantially reduced in nodules treated with 20mM KNO_3 for 2 days (70% reduction), or 20mM NH_4Cl for 1 day (40% reduction). However bacteroids from these nodules were unaffected in acetylene reduction rates when supplied with succinate (1.5mM) (cf Houwaard, 1980). Autonomous acetylene reduction also remained high (150 & 60% of the controls) suggesting that C deprivation alone did not cause the reduced activity. After longer periods of treatment (4 & 6 days respectively giving 87% and 40% reductions in intact nodule acetylene reduction rates) the bacteroids failed to recover fully upon the addition of succinate (67% and 61% decline in C_2H_2 reduction rates). This was in spite of the maintenance of bacteroid numbers and protein levels in the nodules. Neither did the loss of bacteroid function correlate well with nodule levels of leghaemoglobin, NO_2^-, NO_3^- or intact nodule acetylene reduction rates. It appears that several factors lead to the loss of nodule fixing ability in combined N inhibited nodules.

Eady, R.R., (1980) In F.J. Bergersen ed., Methods for evaluating Biological Nitrogen Fixation pp213-264. J Wiley and sons, New York.
Trinchant J., J Rigaud (1979) Physiol Veg., 17, 547-556.
Houwaard F., (1980) Plant & Soil, 54, 271-282.

SERINE HYDROXYMETHYLASE FROM SOYBEAN NODULES

MICHELLE K. MITCHELL, PAUL H.S. REYNOLDS AND DALE G. BLEVINS
DEPARTMENT OF AGRONOMY, UNIVERSITY OF MISSOURI, COLUMBIA, MISSOURI USA.

Serine hydroxymethylase (SHM, EC 2.1.2.1) catalyzes the reaction serine + tetrahydrofolate (FH_4) glycine + methylene·FH_4. It has been proposed that SHM may be important in the C flow from glycolysis to purine biosynthesis in ureide transporting soybean nodules. Through its products glycine and methylene·FH_4, this enzyme has the potential to provide four C to the purine ring. This paper reports a partial purification and some of the kinetic properties of SHM from soybean root nodules.

Methods: Soybean plants (*Glycine max* L. Merr. cv Williams) were grown and a crude plant fraction prepared from the nodules (Reynolds et al., 1982). SHM activity was measured by following the transfer of radiolabel from [3-^{14}C]serine to FH_4 (Taylor & Weissbach, 1965). SHM was purified using ammonium sulphate (AS) fractionation, Sepharose CL-6B column chromatography and Affi-gel Blue (AGB) affinity chromatography. In the AGB step SHM activity was stabilized by the addition of 10% (v/v) glycerol and 4 mM serine. The enzyme bound to the column in the presence of 10 mM NAD+ and was eluted by 500 mM KCl.

Results and Discussion: SHM was purified 750-fold (Table 1). The high recovery of activity was due to stabilization of the enzyme by glycerol and serine.

Table 1. Purification of serine hydroxymethylase from soybean root nodules.

Fraction	Total Protein mg	Total Activity nmol min^{-1}	Specific Activity nmol min^{-1} mg^{-1}	Fold Pure	Recovery %
Crude Plant	144	1160	8.1	1	100
AS pellet	27.5	843	31	4	73
CL-6B enzyme	4.3	1368	318	39	118
AGB enzyme	0.128	804	6385	788	69

The activity was stable at -20° C for several weeks. An apparent M_r of 280 000 was obtained using gel filtration. The enzyme exhibited optimal activity at pH 8.5. K_m values of 1.5 mM and 0.25 mM were obtained for serine and dl-L-FH_4, respectively. There was a second serine K_m at 40 mM. Glycine was a competitive inhibitor with respect to serine and methylene·FH_4 inhibited noncompetitively with FH_4. The K_i values obtained were 2.4 and 0.5 mM, respectively. These data are consistent with an ordered sequential mechanism in which serine is the first substrate to bind and glycine is the last product released. Methyl·FH_4, a major storage form of FH_4 in plants (Cossins, 1980), was a noncompetitive inhibitor with FH_4. SHM activity was inhibited 16% by 2.5 mM NADPH and stimulated 20% in the presence of 2.5 mM NADP+. Two end products of purine biosynthesis, IMP and GMP, had no effect on SHM activity. The enzyme was also unaffected by 2.5 mM phosphoglycerate, HCO_3^- or 10 mM NH_4Cl. In contrast to the enzyme from mung bean seedlings (Rao and Rao, 1982) soybean nodule SHM exhibited Michaelis-Menton kinetics. Care was taken to avoid the use of purification methods known to interfere with the regulatory properties of allosterically modulated enzymes. SHM was not affected by products or substrates of purine biosynthesis, but may be subject to feed-back control by intermediates and cofactors of one-carbon metabolism.

Cossins E (1980) In Davies DD, ed, The Biochemistry of Plants, Vol 2, pp 366-412, Acad. Press, New York.
Rao DN and Rao NA (1982) Plant Physiol. 69, 11-18.
Reynolds PHS Boland MJ Blevins DG Randall DD and Schubert KR (1982) Plant Physiol. 68, 894-898.
Taylor RT and Weissbach J (1965) Anal. Biochem. 13, 80-84.

ISOLATION AND CULTURE OF ANABAENA AZOLLAE IN VITRO

NGUYEN THANH HIEN AND NGUYEN NAM HOA
DEPARTMENT OF GENETICS, UNIVERSITY OF HANOI, HANOI, VIETNAM

Anabaena azollae is a symbiont of the small aquatic fern, *Azolla*. *Azolla* containing *Anabaena* is capable of assimilating atmospheric nitrogen. In order to investigate the mechanism of nitrogen fixation, *Anabaena azollae* has been isolated from *Azolla* and cultured separately in synthetic media.
Anabaena was isolated by "the gentle roller method" (Peters GA, Mayne BC, 1974), followed by filtering through cheesecloth. Cyanobacteria were grown in Erlenmeyer flasks without agitation. The optimal inoculum varies between 10^{-5} - 10^{-6} cells/ml. We tried to grow *Anabaena* on BG-11, a free-living blue-green algal and bacterial media. Frond extract was also tried. The viability was quickly reduced on BG-11 and free-living blue-green algal media, like 'C' medium (Kratz and Myers, 1955), De medium (De, 1939) and Uspenskaja medium (Uspenskaja, 1966). Frond extract gave a positive effect, especially when it was added to BG-11 with glutamate as a nitrogen source. After 5 days, the viability reduced by a factor of two.
Anabaena azollae could be maintained much better on the bacterial minimal medium for *E. coli*. We have drawn a growth curve for *Anabaena*. The concentration of cells was counted after each 5 days. During the first five days, the viability reduced 1.3-times and afterwards, slower. Sometimes it increased. The concentration of cells of a 3-month population was still quite high, it was about 5 x 10^5 cells/ml.
Anabaena grows not only on glucose as carbon source in the light, but also on fructose. On fructose, it grows in the light as well as in the dark and forms photosynthetic pigments.

LEGHEMOGLOBIN PROTECTION IN SOYBEAN NODULES.

A. PUPPO/L. DIMITRIJEVIC/J. RIGAUD
LABORATOIRE DE BIOLOGIE VEGETALE, FACULTE DES SCIENCES, 06034 NICE CEDEX, FRANCE.

The main function of leghemoglobin (Lb) in vivo appears to be ensuring an adequate flux of oxygen to the bacteroids (Bergersen 1980). The ferrous form only is able to bind oxygen. As other oxygen-carrying hemoproteins, Lb is still sensitive to autoxidation which results in the formation of both ferriLb and superoxide anion (O_2^-). This reaction is favoured by :

- acidic pH values : at the nodule pH (6.4) $t_{1/2}$ is 1.5 day (Puppo et al. 1981)
- presence of nitrite : KNO_2 (62.5 µM) produces a 50 % oxidation of oxyLb (24 µM) after 6 min (Rigaud, Puppo 1977)
- presence of O_2^- ; a superoxide dismutase, located in the host cells, is able to scavenge O_2^- ; avoiding Lb oxidation (Puppo et al. 1982).

However ferriLb can be reduced in vivo by two different pathways :

- by a NADH dependent reductase found in the nodules. The enzyme has an optimal activity at acidic pH values and nicotinic acid inhibits the reduction (Puppo et al. 1980a)
- during the simultaneous oxidation of indole-3-acetic acid. This type of reaction constitutes a model for similar systems involving other substrates (Puppo, Rigaud 1979).

Furthermore, ferriLb is oxidized to Lb(IV) by hydrogen peroxide (H_2O_2) and this irreversible reaction leads to a complete inactivation of the hemoprotein. Nodule host cells contain two enzymic systems which avoid this reaction :

- a catalase which is able to degrade 6.8 mmol of H_2O_2 per min and per mg (Puppo et al. 1982)
- a peroxidase whose activity is comparable to that of low-efficiency plant peroxidases. The rate constant of the reaction with H_2O_2 is $3.10^5\ M^{-1}s^{-1}$ (Puppo et al. 1980b).

The conjugate actions of these mechanisms can contribute to explain the presence of high levels of ferrous Lb in vivo (Nash, Schulman 1976). They can allow an optimal working of the nitrogen fixation process.

Bergersen FJ (1980) In Stewart WDP and Gallon JR. eds. Nitrogen Fixation pp: 139-160, Academic Press, London.
Nash DT and Schulman HM (1976) Biochem. Biophys. Res. Comm. 68, 781-785.
Puppo A and Rigaud J (1979) FEBS Lett. 108, 124-126.
Puppo A, Rigaud J and Job D (1980a) Plant Sci. Lett. 20, 1-6.
Puppo A, Rigaud J, Job D, Ricard J and Zeba B (1980b) Biochim. Biophys. Acta 614, 303-312.
Puppo A, Rigaud J and Job D (1981) Plant Sci. Lett. 22, 353-360.
Puppo A, Dimitrijevic L and Rigaud J (1982) Planta 156, 374-379.
Rigaud J and Puppo A (1977) Biochim. Biophys. Acta 497, 702-706.

5-PHOSPHORIBOSYLPYROPHOSPHATE AMIDOTRANSFERASE FROM SOYBEAN NODULES.

PAUL H.S. REYNOLDS, DALE G. BLEVINS AND DOUGLAS D. RANDALL*
AGRONOMY & BIOCHEMISTRY* DEPARTMENTS, UNIVERSITY OF MISSOURI, COLUMBIA, MO. USA.

The proposed pathway of ureide biogenesis directs the flow of recently fixed N from initial assimilation into amino acids to allantoic acid, via purine synth esis and subsequent partial purine oxidation (Reynolds et al. 1982). 5-Phospho ribosylpyrophosphate amidotransferase (PRAT) catalyzes the first committed step of *de novo* purine biosynthesis and could be important in the metabolic control of nitrogen assimilation in ureide plants. This paper reports on the kinetic and regulatory properties of PRAT from soybean nodules.

Methods: Plants (*Glycine max* L. Merr. var. Williams) were grown, their nodules harvested and a crude plant fraction prepared (Reynolds et al. 1982b). PRAT activity was assayed by measuring PRPP-dependent glutamate formation from [U-^{14}C] glutamine (Holmes et al. 1973). Experiments in which NH_3 replaced glutamine used the [^{35}S]cysteine assay of King and Holmes (1976). PRAT activity was purified from the plant fraction of the nodule by high speed centrifugation followed by chromatography on a Sepharose CL-2B column. The activity could be further purified and an estimate made of M_r by centrifugation down a 33 ml linear 10-40% (v/v) glycerol gradient.

Results and Discussion: PRAT activity has been demonstrated in soybeans and shown to increase in specific activity concomitant with the onset of N_2 fixation (Reynolds et al. 1982b). Further, the enzyme is localized in the proplastid fraction of the nodule (Boland et al. 1982) and this nodule fraction can synthesize purines *de novo* (Boland, Schubert 1983). PRAT was purified 1500-fold and is the first purificaiton of this activity from a plant source. The apparent M_r was 8.0×10^6, and maybe a purine biosynthetic complex. There was an optimum for V_m at pH 8, though there was only a 30% variation in the V_m values over the pH 6 to 10 range tested. K_m values for glutamine and PRPP were 18 mM and 0.4 mM, respectively. The reciprocal plots for these substrates both intersected at points on, or very close to, the X-axis which suggested nonconsecutive binding of glutamine and PRPP at independent binding sites. Glutamate was a competitive inhibitor with glutamine, and PP_i inhibited uncompetitively with PRPP. These data were consistent with ordered product release-PRA, PP_i, glutamate. NH_4Cl was a substrate for PRAT and inhibited glutamine utilization by the enzyme with a K_i of 16 mM. The K_m for NH_4Cl at pH 8 was 16 mMl, and the V_m was twofold higher than when glutamine was used as a substrate. NH_4Cl inhibition of glutamine utilization was greater at pH 9 than pH 7 for a given NH_4Cl concentration, which suggested that NH_3, rather than NH_4^+, was the reactive molecular species. The enzyme was unaffected by 30 mM glycine or HCO_3^-. Inhibition of PRAT activity was observed in the presence of end-products of purine biosynthesis. XMP was a linear competitive inhibitor with PRPP and had a K_i value of 1.2 mM. IMP, on the other hand, was a non-linear inhibitor of PRAT activity and promoted cooperativity in PRPP binding, with a Hill coefficient of 1.7. There was evidence for only one binding site for IMP per molecule of enzyme. These data are consistent with PRAT occupying a key regulatory point in ureide biogenesis.

Boland MJ Hanks JM Reynolds PHS Blevins DG Tolbert NE & Schubert KR (1982) Planta 155, 45-51.
Boland MJ and Schubert KR (1983) Arch. Biochem. Biophys. 20, 179-187.
Holmes EW McDonald JA McCord JM Wyngaarden JB & Kelley WN (1973) J. Biol. Chem. 248, 144-150.
King GL and Holmes EW (1976) Anal. Biochem. 75, 30-39.
Reynolds PHS Boland MJ Blevins DG Randall DD & Schubert KR (1982) Trends in Biochem. Sci 7, 366-368.
Reynolds, PHS Boland MJ Blevins DG Randall DD & Schubert KR (1982b) Plant Physiol. 68, 894-898.

ACETYLENE REDUCTION ACTIVITY AND NON-STRUCTURAL CARBOHYDRATE CONTENT OF HEMARTHRIA ALTISSIMA CV. BIGALTA, AFTER DEFOLIATION.

S.C. SCHANK[+] /R.L. SMITH[+] /K.H. QUESENBERRY[+] /R.C. LITTELL[++] /
UNIV. OF FLORIDA
[+] Dept. Agron., 2183 McCarty Hall, Gainesville, FL. USA 32611
[++] Dept. Stat., 524 Nuclear Sci., Gainesville, FL. USA 32611

1. ABSTRACT

Cut vs. uncut plants of Hemarthria altissima cv. Bigalta were studied to ascertain if the defoliated plants had significantly lower acetylene reduction activity (ARA) than unclipped plants. (Fig. a). Non-structural carbohydrate determinations were made of dried root samples. Roots sectioned with a freezing microtome were observed to contain starch in the cortical cells in the center of the root. ARA per gram of root, was significantly lower in the cut plants (p=.02), and percent carbohydrate content was also significantly lower in the cut plants 17 days after clipping (p=.001). A summary of percent carbohydrates found in the stems, crowns and roots of cut vs. uncut Hemarthria altissima is also shown. (Fig. b). Observations of root and shoot development indicated that cut plants had increased shoot growth, but initiated fewer new roots than the uncut plants. The ARA results are similar to, but at lower rates than reported with legume seedlings (Cralle, Heichel 1981) (Vance et al. 1979) and in field pastures (Halliday, Pate 1976).

2. REFERENCES

Cralle HT and Heichel GH (1981) Plant Physiol. 67, 898-905
Halliday J and Pate JS (1976) J. Brit. Grass. Soc. 31, 29-35
Vance CP, Heichel GH, Barnes DK, Bryan JW and Johnson LE (1979) Plant Physiol. 64, 1-8

3. ACKNOWLEDGEMENT

This research was supported by contract AID/ta-C-1376 and by the Florida Agric. Expt. Sta.

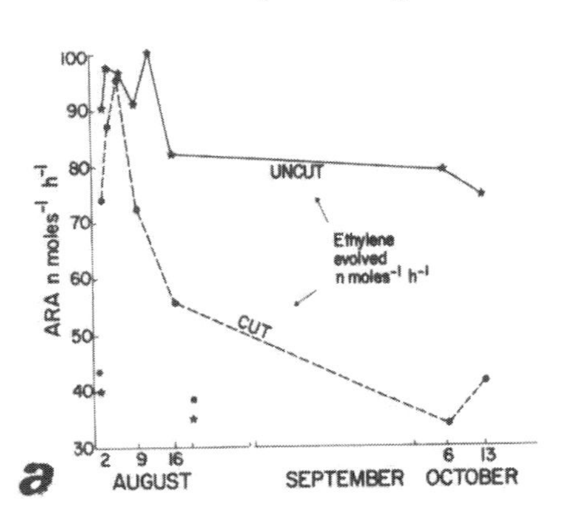

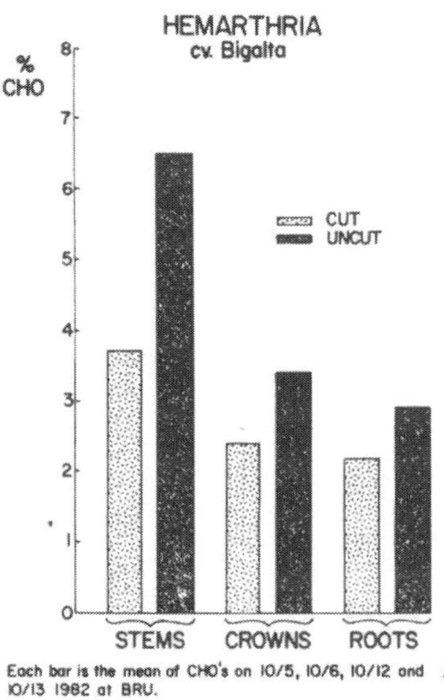

Each bar is the mean of CHO's on 10/5, 10/6, 10/12 and 10/13 1982 at BRU.

COMPARISON OF GROWTH, NITROGEN FIXATION AND RELATIVE EFFICIENCY OF NITROGENASE IN *ALNUS INCANA* GROWN IN DIFFERENT CULTIVATION SYSTEMS

ANITA SELLSTEDT AND KERSTIN HUSS-DANELL
DEPARTMENT OF PLANT PHYSIOLOGY, UNIVERSITY OF UMEÅ, S-901 87 UMEÅ, SWEDEN

For studies of physiological aspects of nitrogen fixation as, e.g. energy demand of the nitrogen fixation process, it is important to have a cultivation system where neither water nor nutrients are limiting or fluctuating. Three cultivation systems for growing cloned plants of nitrogen-fixing *Alnus incana* (L.) Moench were tested. In one system the alders were grown hydroponically in an aerated weak nutrient solution. In the two other systems the plants were in pots with gravel and either received water twice a day and concentrated nutrient solution once a day or were continuously supplied with an aerated diluted circulating nutrient solution.

Alders grown in the continuously circulating system showed significantly better growth and produced more biomass than alders in the other cultivation systems. Nitrogen fixation and relative efficiency of nitrogenase were repeatedly measured during the growth periods. Nitrogen fixation was highest in the continuously circulating system. The relative efficiency of nitrogenase held a constant value of about 0.80 in alders continuously supplied with water and nutrients, while it decreased during the growth period of alders given water and nutrients at intervals. No strict relationship was found between RE and nitrogen content or between RE and plant productivity.

METABOLISM OF CARBOHYDRATES IN SOYBEAN (GLYCINE MAX (L) MERR) NODULES

J. G. STREETER AND P. H. REIBACH
Dept. of Agronomy, Ohio State University (OARDC), Wooster, OH 44691, USA

Carbohydrates previously identified in soybean nodules include sucrose, α,α-trehalose, maltose, glucose, fructose and 3 cylitols: D-pinitol, D-chiro-inositol, and myo-inositol. The concentrations of all of these compounds in nodules increase rapidly during the onset of N_2 fixation (Streeter, 1980). We assumed that sucrose is the major carbohydrate entering nodules because sucrose is the major compound exported from soybean leaves (Vernon, Aronoff, 1952). The objective was to begin elucidation of the principal routes for utilization of carbohydrates by studying the distribution of ^{14}C-photosynthate and the distribution of enzymes of carbohydrate metabolism.

Plants were supplied with $^{14}CO_2$ (ca 70 μCi/plant) for 30 min and nodules were harvested at 0.5- or 1-hr intervals from 1 to 5 hr after the $^{14}CO_2$ labeling period. Bacteroids and cytosol were separated with Percoll density gradients and ^{14}C in carbohydrates, phosphate esters, organic acids, and amino acids were determined. There were three $^{14}CO_2$-labeling experiments each involving 4 to 6 sampling times and 2 or 3 replicates. Enzymes were extracted from nodules formed by R. japonicum USDA 110 or USDA 138. After separation of bacteroids and cytosol, bacteroids were ruptured and the following enzymes were assayed in crude, Sephadex G-25-filtered preparations: P-glucomutase, glucose-dependent hexokinase, fructose-dependent hexokinase, P-glucoseisomerase, P fructokinase, fructose-1,6-bis P aldolase, glucose-6-P-dehydrogenase, 6-P-gluconate dehydrogenase (NAD or NADP-dependent). Attempts to detect the following enzymes in bacteroids or bacteroid fragments (after sonication) were unsuccessful: fructokinase, glucose dehydrogenase, PPi-dependent P-fructokinase, and pyruvate formation from 6-P gluconate (Entner-Doudoroff pathway).

Results/Conclusions: (1) Cyclitols (myo-inositol, D-pinitol, D-chiro-inositol) and malonate were not significantly labeled at any sampling time up to 5 hrs. These compounds, which are major carbon compounds in soybean nodules, are not rapidly formed from incoming photosynthate and do not appear to serve as carbon substrates for bacteroids. (2) The major fate of incoming ^{14}C-sucrose was conversion to organic acids (mainly malate). The appearance of ^{14}C in organic acids was much more rapid in cytosol than in bacteroids. Key enzymes of glyolysis and the oxidative pentose-phosphate pathway were present in cytosol. (3) ^{14}C in bacteroids increased with time up to 30% of total ^{14}C in nodules after 5 hrs. Sixty to 65% of the ^{14}C in bacteroids was in sugars at all sampling times. Most of the neutral fraction ^{14}C was in sucrose and significant ^{14}C was present in α,α-trehalose at later sampling times. Bacteroids appeared to absorb some carbon as carbohydrate (probably sucrose) directly from the cytosol. (4) However, bacteroids lack invertase, P-fructokinase and NAPD-dependent 6-P-gluconate dehydrogenase. Thus, mechanisms for catabolism of sucrose and glucose in bacteroids are presently unclear. (5) Bacteroids did possess enzymes for the formation of hexose phosphates from glucose or fructose, and glucose-P would be required for trehalose systhesis. Since ^{14}C did accumulate in trehalose in bacteroids, trehalose systhesis may be a major fate of sugars entering bacteroids.

References:

Streeter JG (1980) Plant Physiol. 66,471-476.
Vernon LP and Arnonoff S (1952) Arch. Biochem. Biophys. 36,385-398.

THE INFLUENCE OF PROLONGED DARK PERIOD ON FUNCTIONING OF SYMBIOTIC NITROGEN FIXING SYSTEM IN SOYBEAN

K.Swaraj*, I.N.Andreeva+, G.I.Kozlova+

*Botany Department, Haryana Agricaltural University, Hissar 125004, India.
+K.A.Timiriazev Institute of Plant Physiology, USSR Academy of Sciences, Moscow 127276, USSR.

Present investigations were undertaken to probe into physiological events accompanying fall in nitrogen fixation in dark. Decrease in leghemoglobin content and accumulation of amino acids in nodules (Roponen,1970), decrease in nodule energy charge (Ching et al.1975), decrease in nodule ascorbic acid content (Swaraj and Garg,1976) and accumulation of poly-β-hydroxybutyrate (PHB) in bacteroids (Romanov et al.1978) in dark have been reported.

Material and Methods. Soybean (*Glycine max* L.Merr. cv.Amurskaya-310) seeds were inoculated with *Rhizobium japonicum* strain 648. Plants were raised in sand culture in glass house with day and night temperature 25°C and 16 hr day length. Four weeks old plants were shifted to a dark room with temperature 25°C and relative humidity 70% at 5 PM after 9 hr exposure to normal light conditions. Nodule ultrastructure was studied by usual methods of electron microscopy. C_2H_2, CO_2 and H_2 evolution were estimated gas chromatographically.

Results. Prolongation of dark period resulted in a sharp decline in nitrogenase activity which was accompanied by a parallel reduction in respiration upto 42 hr. After this period, fall in nitrogenase activity was sharper than that in respiration. Thus, after 90 hr in dark, energy cost of acetylene reduction (expressed as ratio of CO_2 evolved to C_2H_2 evolved) showed a considerable increase over control. H_2 evolution showed an even steeper decline in dark. As a result nitrogen fixing efficiency showed values higher than the control upto 42 hr in dark and later on declined. Decline in nitrogenase activity in dark upto 4 days is completely reversible on shifting the plants back to normal conditions.Electron microscopic studies showed numerous changes in nodule ultrastructure in dark. Within 18 hr bacteroids showed a large accumulation of PHB, which however declined after 90 hr. A long dark period also resulted in considerable increase in size of the peribacteroid space and number of bacteroids enclosed in it. Enlargement of peribacteroid space in dark was accompanied by appearence of network of unknown nature,which occupied almost whole peribacteroid space exept immediately around bacteroids. The peribacteroid space in control was electron transparent. On shifting the plants after 2 days in dark back to normal light conditions PHB granules in bacteroids declined after 48 hr recovery, but network in peribacteroid space remained unaffected.

References

Chihg Te May et al. (1975) Plant Physiol. 55, 796-798.
Romanov VI et al. (1978) Plant Soil 56, 379-390.
Roponen I (1970) Physiol. Plant. 23, 452-460.
Swaraj K, Garg OP (1976) In Sen et al. eds, Proc. Nat. Symp."Nitrogen Assimilation and Crop Productivity", pp. 164-170, Assoc. Publ.Co., New Delhi.

ONONITOL AND O-METHYL-*SCYLLO*-INOSITOL IN PEA ROOT NODULES

LEIF SKØT AND HELGE EGSGAARD
AGRICULTURAL RESEARCH DEPARTMENT AND CHEMISTRY DEPARTMENT
RISØ NATIONAL LABORATORY, DK-4000 ROSKILDE, DENMARK.

1. INTRODUCTION

Cyclitols (hexahydroxycyclohexanes) and their methyl ethers occur in legumes (1). In soybean root nodules a correlation between D-pinitol (3-O-methyl-*chiro*-inositol) and nitrogen fixation has been observed suggesting a role for pinitol in N_2 fixation (2). Here we report on ononitol (4-O-methyl-*myo*-inositol) and O-methyl-*scyllo*-inositol as major components of the soluble carbohydrates in pea root nodules and that the cyclitol pattern depends on the Rhizobium strain inhabiting the nodules.

2. MATERIALS AND METHODS

Pisum sativum cv. Safir nodulated by *Rhizobium leguminosarum* strain la or 1045 was grown in a leonard jar pot system in a growth room with a 16 h photoperiod. Quantitation of ethanol soluble carbohydrates from the plant material was done by GLC of trimethylsilyl (TMS) derivates using a packed column. Identification of cyclitols was established by GC-MS utilizing TMS- and acetyl-derivatives.

3. RESULTS AND DISCUSSION

The most abundant component of the ethanol soluble carbohydrates in pea root nodules formed by 1045 was not any of the 'common' sugars (sucrose, glucose etc.). GC-MS of acetylderivatives revealed fragmentation pattern with two major ions at m/z 182 and 140 which is 28 mass units below the characteristic inositol ions at m/z 210 and 168, respectively ($R-O-COCH_3 \rightarrow R-O-Me$: 28 amu), these ions posses aromatic ion structures and are very characteristic of the cyclitols. When demethylated and reacetylated the product formed co-chromatographed with *myo*-inositol-hexaacetate and the fragmentation pattern was identical with that of authentic *myo*-inositol-hexaacetate (3).

The acetylated component from the nodules formed by 1045 co-chromatographed with ononitolpentaacetate, and could be separated from sequoyitol and bornesitol. Therefore we conclude that the compound is ononitol. Using a similar procedure O-methyl-*scyllo*-inositol was shown to occur in nodules formed by la. The cyclitols were present in the denodulated roots in smaller amount, but was not detected in shoot material.

The role of cyclitols in pea root nodules is unknown. The information available at present suggests that O-methyl-inositols are not carbon sources in plants, but perhaps play a role in osmoregulation.

4. REFERENCES

Smith AE and Phillips DV (1980) Crop Sci. 20, 75-77.
Streeter JG and Bosler ME (1976) Plant Sci. Lett. 7, 321-329.
Sherman WR, Eilers NC and Goodwin SL (1970) Org. Mass. Spectrom. 3, 829-840.

NITROGENASE AND HYDROGENASE IN *R.LEGUMINOSARUM* AND N-CONTENT OF PEA PLANTS.

T.A.Truelsen and R.Wyndaele
Institute of Molecular Biology and Plant Physiology, University of Aarhus, 8000 Aarhus C, Denmark.

The hydrogen production from nitrogenase represents an energy loss in the *Rhizobium*-legume symbiosis.Some *Rhizobium* strains contain an oxidizing hydrogenase(Hup$^+$ strains)and are supposed to be more yielding symbionts than Hup$^-$ ones. 23 strains of *R.leguminosarum* were examined for the presence of the Hup character and its influence on hydrogen recycling and nitrogen fixation in symbiosis with pea plants cv.Bodil grown in a N-free nutrient solution.Nitrogenase and hydrogenase activities of the nodules were measured once weekly from 4 to 7 weeks after inoculation.Nitrogenase was measured as acetylene reduction(Ac) and H_2production in 80%Ar,20%O_2(H_2ArO_2).The latter gas analysis as well as H_2evolution in air(H_2air) and H_2 uptake(Hup) -measured on halved nodules- were performed in a standardized Clark-type electrode. Four strains were found to be Hup$^-$.Three strains were Hup$^+$ with a clear recycling efficiency (Hup$^+$Ref$^+$)and able to recycle all of the H_2 produced by the nitrogenase.The remaining 16 strains were Hup$^+$, but without any discernible recycling capacity *in situ*.They were thus Hup$^+$Ref$^-$.Figure 1.shows the regression line between H_2air and Ac for the 23 *R.leguminosarum* strains.

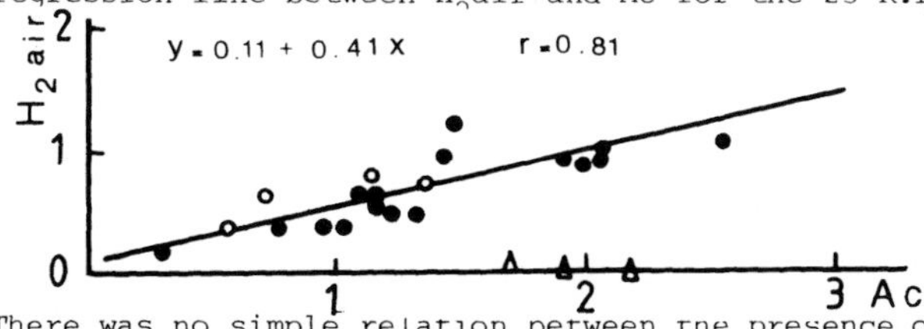

Fig.1.Regression line between H_2air and nitrogenase activity measured as acetylene reduction.
● Hup$^+$Ref$^-$
Δ Hup$^+$Ref$^+$
o Hup$^-$

There was no simple relation between the presence of Hup character and the yield of the host plants measured as N-content and dry matter after 7 weeks. The correlation between N-content of aerial plant parts and the average nitrogenase activity during the four weeks(4) or two weeks(2) with significant N-fixation was fairly low for all methods used in measuring enzyme activity as shown in Tabel 1.

Regression-(b) and correlationcoefficients(r) for N-content(mg N/plant) in aerial plant parts after 49 days and different parameters of nitrogenase activities in nodules.The latter are either averages of 4 or 2 (weekly) measurements in weeks 4-7 or 5-6 after inoculation respectively.

	Ac		H_2ArO_2		Ac-H_2air		H_2ArO_2-H_2air	
	(4)	(2)	(4)	(2)	(4)	(2)	(4)	(2)
b	0.033	0.032	0.051	0.053	0.017	0.014	0.035	0.035
r	0.65	0.53	0.56	0.53	0.37	0.23	0.56	0.50

Those bad correlations were partly due to bacterial induced differences as :
- distribution of dry matter and N-content between aerial plant parts and infected roots
- assimilation and transport products
- metabolic efficiencies.

This work was financially supported by a grant of the Danish SVLF.

^{15}N-NATURAL ABUNDANCE OF TRIFOLIUM SPP. INOCULATED WITH DIFFERENT RHIZOBIUM STRAINS.

CHRISTOPHER VAN KESSEL
UNIV. OF WISCONSIN-MADISON, DEPT OF BIOCHEMISTRY, MADISON, WI 53706

Recently, attention has been given to the ^{15}N-natural abundance of N_2-fixing plants as an indicator for calculating the % of total N derived from N_2-fixation (1,3). N_2-fixing plants should show a lower ^{15}N-natural abundance than non-N_2-fixing control plants (2). Differences in atom % ^{15}N should give an indication about the % of total N derived from N_2-fixation. This paper summarizes the effect of *Rhizobium* on the ^{15}N-natural abundance of *Trifolium* spp.

Materials and Methods

Trifolium pratense, *Trifolium hybridum*, and *Trifolium repens*, nodulated with one of the following *Rhizobium* strains; *R. trifolii* 162 BB1, *R. trifolii* 162X6, *R, trifolii* 162P17, were grown in a sand-vermiculite medium, placed in a green house. A N-free Hoagland solution was added twice a week. At time of harvest, nodules, roots, and stem+leaves were analyzed for total N by steam distillation. Atom % ^{15}N of plants parts was determined with a MAT 250 isotope ratio mass spectrometer. As standard was used atmospheric N_2.

Results and Discussion

Leaves+stems of all 9 different symbionts did show a negative value for atom % ^{15}N-excess. Roots did show positive as well as negative values whereas nodules only showed positive values.

Plant	R. strain	nMol $^{15}NH_4^+$ excess				atom % ^{15}N-excess plant[1]
		leaves	roots	nod.	total	
T. pratense	162BB1	-64.6	-0.8	+2.6	-62.9±21.8	-0.0008±0.0004 cd
T. pratense	162P17	-103.9	-11.3	+4.1	-111.2±28.2	-0.0017±0.0003 ab
T. pratense	162X6	-70.2	+1.1	+2.3	-65.4±17.1	-0.0013±0.0002abc
T. hybridum	162BB1	-64.8	-9.9	+0.6	-44.2±21.0	-0.0010±0.0002bcd
T. hybridum	162P17	-30.0	+12.9	+0.8	-21.7± 2.7	-0.0004±0.0001 d
T. hybridum	162X6	-63.5	-6.3	+0.6	-69.3±20.0	-0.0010±0.0003bcd
T. repens	162BB1	-58.9	-1.7	+0.1	-60.6±17.8	-0.0011±0.0002bcd
T. repens	162P17	-84.7	-2.3	+0.6	-85.5±23.2	-0.0012±0.0002abc
T. repens	162X6	-95.0	-1.0	+1.6	-94.4±43.1	-0.0018±0.0004 a

[1]Values followed by the same letter are not significant different at the 95 % confidence level.

Significant differences in atom % ^{15}N-excess were observed among clover plants inoculated with different *Rhizobium* strains as well as within clover species,all dependent on atmospheric N_2 as their sole N-source. When the same results are obtained with other leguminous species, the ^{15}N-naturel abundance method for calculating the % of total N derived from N_2-fixation will not achieve a high accurancy, especially when is known that the differences in atom % ^{15}N between test plant and control plant are rather small.

(1) Amarger et al., 1977 C.R.Acad.Sc.Paris, t. 284 Serie D;2179-2182. (2) Delwiche et al., 1979 Bot.Gaz. 140 (Suppl);S65-S69. (3) Kohl et al., 1980 Plant Physiol. 66;61-65.

REVERSIBLE-DARK INDUCED SENESCENCE OF SOYBEAN ROOT NODULES

FRED W. WAGNER/C.S. SODHI AND N.E. PFEIFFER
DEPARTMENT OF AGRICULTURAL BIOCHEMISTRY
UNIVERSITY OF NEBRASKA-LINCOLN
LINCOLN, NE 68583 USA

1. Introduction

In previous communications (1,2) it was demonstrated that dark stress of soybean plants caused a deterioration of both nodule and shoot tissue which was markedly similar to that of naturally senescing plants. Specifically, plants kept in total darkness for 4 days completely lost their capacity to reduce acetylene. A significant reduction in cytosolic protein was evident in 4 days and in 8 days total darkness leghemoglobin was absent and total protein was only 35% the level of controls. Concomitant with the loss of cytosol protein was a 4 to 5 fold increase in cytosolic endopeptidase activity. When plants were dark-stressed for 8 days, then returned to a 14 hr photoperiod, they recovered and all the metabolic parameters measured returned to normal levels. Separate experiments demonstrated that detached nodules from recovered plants reduced acetylene at normal rates, but that detached nodules from 8 day dark stressed plants did not reduce acetylene (1). While these experiments demonstrated dramatic metabolic changes in cytosolic constituents during stress, changes in bacteroid proteins and enzyme activities, other than whole plant acetylene reduction, were not measured. This report presents results which suggest that bacteroids from nodules of dark stressed plants do not senesce concomitant with tissue of plant origin.

2. Procedure. Thirty day old soybean plants (*G. max* L. Merrill cv. Woodworth) were placed in total darkness for up to 8 days then returned to a 14 hr photoperiod (1). Isolation of bacteroids and assay of bacteroid enzymes were the procedures of Pfeiffer et al. (2). Acetylene reduction and respiration rates of isolated bacteroids were performed by the procedure of Klucas (3). The experiments reported were designed to determine if pronounced changes in bacteroid constituents accompany changes in cytosol due to dark stress. We measured the content of protein per fresh weight of nodule, examined bacteroid proteins by electrophoresis and measured a number of bacteroid enzymes during dark stress and recovery. There were no appreciable changes in any of these parameters. Moreover, while acetylene reduction by bacteroids was lost during dark stress, bacteroid respiration remained high. Preliminary evidence suggests bacteroids recover their ability to reduce acetylene. These results are consistent with the thesis that nodule senescence involves degeneration of the plant tissue and re-differentiation of the bacteroid to return to the free living state.

1. Pfeiffer, N.E., N.S.A. Malik and F.W. Wagner (1983) *Plant Physiol.* *71*, 393-399.

2. Pfeiffer, N.E., C.M. Torres and F.W. Wagner (1983) *Plant Physiol.* *71*, 797-802.

3. Klucas, R.V., S.R. Koch and H.J. Evans (1968) *Plant Physiol.* *43*, 1906-1912.

CARBON ECONOMY IN RELATION TO DINITROGEN FIXATION IN GLYCINE MAX (L) Merr. cv. Hodgson USING DOUBLE LABELLING TECHNIQUES $^{14}CO_2$ AND $^{15}N_2$

F.R. WAREMBOURG[1], M. FERNANDEZ[2], D. MONTANGE[2], R. BARDIN[2]
[1] C.E.P.E., C.N.R.S., B.P. 5051, 34033 Montpellier Cedex, FRANCE
[2] Université Claude Bernard, Lyon, FRANCE

This is a collection of data gathered in situ during one season with soybean plants.

The study was done with plants grown in natural conditions under the mediterranean climat of south of France. At four different phenological stages, series of 6 containers holding 3 plants each were exposed to an atmosphere containing $^{14}CO_2$ during one day. Among them, and during the same time, 4 sets had their root systems under an atmosphere containing $^{15}N_2$. The 2 remaining sets (nodulated and non-nodulated plants) were analysed for soil respiration including measurements of $^{14}CO_2$ evolution. These measurements lasted 7 days after labelling. Sampling of containers and analysis of ^{14}C, C, ^{15}N and N contents of plant parts were done after completion of respiration measurements. Part of the containers were sampled later during the season.

The data collected concerned : assimilated carbon distribution and fixed N distribution within the plants during the various phenological stages, fate of C and fixed N in plant parts and estimations of root respiration according to plant age and dinitrogen fixing activity. They can be summarized as follow :

- The distribution of assimilated carbon 7 days after incorporation into the plants showed marked differences according to the state of organs development. During the vegetative period, the aerial parts were the major sinks with as much as 75 % located in leaves and stems. With the formation of reproductive organs, more and more assimilates were directed toward them with as much as 70 % near maturity, this mainly to the detriment of the leaves.
- The distribution of fixed nitrogen was very similar except in the beginning of nodulation when the nodules kept almost 30 % of the amount incorporated.
- As obvious, part of the carbon and nitrogen incorporated into structural components were remobilized and directed toward other plant parts. Near maturity, 90 % of the total amount of nitrogen fixed during the season was recoveded in the seeds. It should be noted that around day 90, stems and petioles acted as storage organs for nitrogen.
- Detailled analysis of $^{14}CO_2$ evolution by root systems of nodulated and non-nodulated plants showed marked differences mainly when the nodulated plants were actively fixing N_2. Calculations of the net cost of the nitrogen fixation process indicated values ranging between 2.5 (until day 80) and more than 7 mg (after day 80) of carbon lost per mg of N_2 fixed.

On the overall, the double labelling techniques applied on soybean has proven very successfull in estimating coefficients for distribution, transfers and losses of C and N that can be very usefull in a model simulating the C and N economy of legumes grown under near natural conditions.

THE USE OF X-RAY FLUORESCENCE TECHNIQUE IN THE STUDY OF BIOLOGICAL NITROGEN FIXATION

A.C. XENOULIS[+]/T. PARADELLIS[+]/C.E. DOUKA[++]/T. PRASSAS[++]
NUCLEAR RESEARCH CENTER DEMOKRITOS, AGHIA PARASKEVI, ATHENS, GREECE
[+]Accelerator Laboratory
[++]Department of Biology

The utilization of X-ray fluorescence technique in the measurement of elemental concentration of plant samples has been seen to be especially useful in the study of biological nitrogen fixation. This technique is well suited for the analysis of biological samples since it demands small amount of sample (down to 5 mg) thus permitting analysis of plants available in small quantities, such as those obtained in controlled laboratory experiments. Furthermore, the preparation of the sample for analysis is very simple, the technique is non destructive and permits the simultaneous measurement of all elements with atomic number between 19 and 42 percent in a sample.

Employment of the technique has already helped to identify the existence of interaction between microorganisms, chemical composition and environment in salt affected soils (Douka et al. 1983). Furthemore, it has been obtained that inoculation with effective Rhizobia affects the elemental uptake and the elemental composition of nodulated plants (Douka et al. 1982). Of particular importance is considered to be a significant effect of biological nitrogen fixation on the concentration of molybdenum in a plant, which has been observed to decrease almost by an order of magnitude by inoculation.

The results obtained until now suggest that the elemental composition of a nodulated plant constitutes a very promising novel parameter in the study of biological nitrogen fixation.

Douka CE, Xenoulis AC and Paradellis T (1982) Effect of inoculation on elemental uptake by plants grown on saline soils. Folia Microbiol. 27, 278.
Douka CE, Xenoulis AC and Paradellis T (1983) Interaction between microorganisms, chemical composition and environment in salt-affected soils. Folia Microbiol. 28, 57.

POSTER SESSION 7B ENVIRONMENTAL EFFECTS ON NODULATED PLANTS

J.I. SPRENT, Dept of Biological Sciences, University of Dundee, Dundee, UK

The posters covered many aspects of the responseS of legumes to environmental variations. The majority included some data on combined nitrogen effects and the discussion centred on this topic.

It was clear from the discussion that the effects of combined nitrogen on nodulated plants cannot be considered alone: responses at all stages of growth are strongly conditioned by environment. For example, some workers criticised pot experiments on the requirements for "starter" N on the grounds that, in field conditions, soils may contain sufficient combined N to render a starter dose unnecessary. Others, in particular those working with low latitude, low N soils, where growth following germination is rapid, reported N-deficiency symptoms in inoculated plants before nodules become active.
Whether or not a given dose of combined N depresses nodulation and nodule activity is related to species, plant age (Koemendy and Eaglesham), temperature and irradiance (Silsbury and Catchpole). Generally, the better the growing conditions, the more combined N can be applied without depressing N-fixation (see also Herridge *et al*.)
It was agreed that the poster showing that water stress affects N-fixation more than nitrate reduction (Obaton) was generally typical of stress responses. There is still no universally accepted explanation for the effects of nitrate on nodule activity. However, several posters (Nelson; Harper & Gibson; Deignon *et al*) and discussants indicated that there is scope for selecting both rhizobia and legume cultivars (but particularly the latter) which can tolerate a moderate level of combined N in soil without showing reduced nitrogen fixation. The need for more field experimentation using a greater range of species and geographical locations was very evident.

THE GROWTH OF *VIGNA RADIATA* L. AFTER INFECTION WITH DIFFERENT ANTIBIOTIC RESISTANT MUTANTS OF *RHIZOBIUM* COWPEA GROUP

F.Y. AL-ANI/A.A. AL-SAADI and A.T. ABU-TABIKH
BIOLOGY DEPARTMENT, SCIENCE COLLEGE, BAGHDAD UNIVERSITY, IRAQ

In the present work we aimed to find out the effects of different antibiotic resistant markers in *Rhizobium spp*. (cowpea group) on, total nitrogen, dry weight, and nodules number of the host plant (green gram) at two periods of growth, after 50 days (before flowering) and after 90 days (maturation time).

For this purpose twelve mutants resistant to a single antibiotic and three resistant to two antibiotics were isolated. The methods used in this work were described before (Al-Saadi, 1978). The results were subjected to statistical analysis, method of analysis of variance.

The main results can be summarized in the following points:

1. The least value observed in the three measurements are in the group of negative control (non-inoculated plants).
2. The highest increase in N% was detected in plants infected with mutant resistant to chloramphenicol (350 μg/ml). The increase % in the first and second periods of growth was 17% and 39% respectively as compared with wild type. The dry weight of these plants in both periods were higher than that of the positive control. Previous studies on this aspect (Schwinghamer, 1967; Pankhurst, 1977) contradict our results.
3. No correlation was observed between nitrogen fixation, dry weight and number of nodules.

REFERENCES

1. Al-Saadi AA (1978) M.Sc. Thesis, Baghdad University, Iraq.
2. Pankhurst CE (1977) Can.J.Microbiol. 23, 1026-1033.
3. Schwinghamer (1967) Ant. van Leeuw. 33, 121-136.

LEGHAEMOGLOBIN FROM ROOT AND STEM NODULES OF *SESBANIA ROSTRATA*

D. BOGUSZ
Microbiologie des Sols, ORSTOM, BP 1386, Dakar, Senegal

The tropical legume *Sesbania rostrata* forms both root and stem nodules in symbiotic association with a fast-growing *Rhizobium* (1). In contrast with root nodules, stem nodules harbour chloroplasts in their cortex (2) and exhibit a higher nitrogenase activity than the root nodules (3). We hypothesise that these differences between root and stem nodules properties might reflect differences in the leghaemoglobins.
Using the chromatographic procedure described by Appleby (4) we fractionated root nodule leghaemoglobin into three components and stem nodule leghaemoglobin into four components. One of the components was found to be specific to stem nodules, the other ones presented the same elution volume and electrophoretic mobility in both types of nodules. A nicotinate binding assay was made using the root or stem purified total leghaemoglobins. Absorption spectra of oxidized and reduced root and stem leghaemoglobins were modified in the presence of nicotinate which indicated that leghaemoglobins were able to bind nicotinate in either their ferric or ferrous forms.
Antiserum prepared against purified total stem nodule leghaemoglobin did not cross-react with nodule extract from soybean and alfalfa. Investigations are under way to determine the cross-reactivity of *S.rostrata* leghaemoglobins with other taxonomically related legumes namely *Aeschynomene* spp.

1. Dreyfus *et al*. (1983) Appl.Environ.Microbiol. 45, 711-713.
2. Dreyfus and Dommergues (1981) FEMS Microbiol. Lett. 10, 313-317.
3. Dreyfus (1982) PhD thesis, University of Paris VII.
4. Appleby *et al* (1975) Biochemistry 14, 4444-4450.

PARTITIONING OF NITROGEN FROM UPTAKE AND FIXATION IN THE TRIFOLIUM REPENS/ RHIZOBIUM TRIFOLII SYMBIOSIS.

M.T. DEIGNAN, J.E. COOPER AND A.J. HOLDING
DEPARTMENT OF AGRICULTURAL AND FOOD BACTERIOLOGY, THE QUEEN'S UNIVERSITY OF BELFAST, NEWFORGE LANE, BELFAST BT9 5PX, N. IRELAND.

INTRODUCTION

In agriculture *Trifolium repens* is often supplied with mineral N in various forms, concentrations and combinations, but the effect of mineral N on N fixation and the relative contributions of these two sources to the total N content of plants is unclear. This study used indirect (acetylene reduction, nitrate reductase, remaining N concentration) and direct (^{15}N) methods to partition N from uptake and fixation and to identify *Rhizobium trifolii* strains which continue to fix N in the presence of mineral N.

METHODS

T. repens was grown under microbiologically controlled conditions with a mineral solution containing either no N, or various levels and combinations of NH_4-N or NO_3-N. Plants remained uninoculated or were inoculated with *R. trifolii* strains. Assays were performed for %^{15}N content, remaining N concentration, acetylene reduction or nitrate reductase activity.

RESULTS

(a) Plant yield

Combining data from inoculated and uninoculated plants, all N treatments increased plant yield by 10 weeks. Addition of 50 ppm NO_3-N or NH_4-N gave similar increases in yield. Overall, inoculated plants produced significantly higher shoot yields than uninoculated plants. *R. trifolii* strain R5 gave the largest yields across the treatments.

(b) C_2H_4 evolution

Mineral N caused a reduction in C_2H_4 evolution. 50 ppm NO_3-N or NH_4-N produced equal reductions. A 5.5 ppm combined N treatment had a smaller effect although it significantly reduced production compared to an N-free treatment. Strain R5 was least affected by N addition, suggesting that increased yields of plants inoculated with R5 may be due to N fixation.

(c) Nitrate reductase activity and remaining N concentration

Plants inoculated with strains of varying effectiveness (P3, R5, F73) displayed no significant difference in total plant nitrate reductase activity but a significantly higher activity was recorded for plants inoculated with the ineffective strain R4. Highest remaining NO_3-N concentration was found in the rooting solution of plants supplied with 50 ppm NO_3-N and inoculated with the effective strain R5.

(d) ^{15}N analyses

The N content of plants supplied with 50 ppm NO_3-N alone or in combination with an ineffective strain (R4) or an effective strain (R5) was the same. However, 60% of the N was derived from fixation in plants nodulated by strain R5, indicating that R5 fixed 83% of the N which it was capable of fixing under N-free conditions.

CONCLUSIONS

While some of the indirect methods for assessing N fixation in the presence of mineral N can indicate host/strain combinations which warrant further study, only the ^{15}N technique can distinguish precisely between different nitrogen sources. With this technique it was established that strain R5 was capable of maintaining a high level of symbiotic effectiveness in the presence of 50 ppm NO_3-N.

Role of Ni in nitrogen metabolism

D.L. Eskew and R.M. Welch
U.S. Plant, Soil and Nutrition Lab. Ithaca NY 14853 U.S.A.

Urease is a Ni-metalloenzyme and Ni has been shown to be involved in several other biological processes (Welch 1981). Recently we reported that Ni deficient soybeans [Glycine max (L) Merr] accumulated 2.5% urea in necrotic lesions at their leaflet tips (Eskew et al. 1983). Nodulation was delayed by 2-3days, growth was reduced and leaflet tip necrosis was more severe with N_2 fixing plants. Both leaf and seed urease activity were increased by addition of $1ugNi\ell^{-1}$ and leaflet tip necrosis was prevented. Here we present further evidence that Ni has an essential role in plant N metabolism.

Although our previous experiments showed that 1 or $10ugNi\ell^{-1}$ was sufficient to prevent tip necrosis, it is still desirable to show that this was due to Ni in the plant and not to the presence of Ni in the solution. It is also necessary to show that other elements cannot substitute for Ni. Therefore the effects of carryover of Ni in the seed and the effects of adding $10ug\ell^{-1}$ of Al, Cd, Sn and V (A-V solution) were investigated. Seeds from plants which had been given 0,1 or $10ugNi\ell^{-1}$ supplementation were used to grow a second generation. The A-V solution, or A-V plus $10ugNi\ell^{-1}$, were provided to plants from $1ugNi\ell^{-1}$ seed.

Table 1: Occurence of tip necrosis

Pretreatment $ugNi\ell^{-1}$	Current Supplement	% Tip necrosis
0	0	25.6 a[+]
1	0	20.0 b
10	0	1.0 c
1	A-V	11.0 d
1	A-V + Ni	0 c

+ values followed by the identical letter are not significantly different

Carryover of Ni in the seed was sufficient to reduce the incidence of tip necrosis (Table 1). The A-V solution reduced the incidence of tip necrosis but did not prevent it, whereas Ni addition prevented it completely.

Experiments were also conducted with cowpeas [Vigna unguiculata (L) Walp]. The results with nodulated, nitrogen fixing cowpeas were similar to those previously reported for soybeans. Nodulation was delayed by 2-3days without added Ni and plant development was visually observed to be retarded. Plant dry weight was reduced from 0.75 to 0.55 at 30days, although the differences were not statistically significant. Contrary to the results with soybeans, however, both + and- Ni plants had a high frequency of tip necrosis. This suggests that $1ugNi\ell^{-1}$ is not sufficient for N_2 fixing cowpeas.

An experiment with cowpeas supplied NO_3 and NH_4 was performed with 7uM NaEDTA above that added as FeEDTA. This treatment lowers the ion concentration of several metal ions, including Ni. After the plants began to flower the characteristic necrosis developed on 63.8% of the leaflet tips. The sympton was completely absent on cowpeas supplied 1uM NiEDTA.

The data presented here establish that Ni has a similar effect in two plant species, that presence of Ni in the plant alone is sufficient and that none of the 4 metals tested was able to substitute for Ni.

References:

Eskew DL, Welch RM, and Cary EC (1983) Science, In press
Welch RM (1981) J. Plant Nutr. 3:345-356

NITRATE INHIBITION OF NODULATION AMONG SOYBEAN CULTIVAR X *RHIZOBIUM* STRAIN COMBINATIONS.

J.E. Harper and A.H. Gibson, USDA & Dep of Agron, U of Illinois, URBANA, IL, U.S.A., and CSIRO, Div of Plant Ind, CANBERRA, A.C.T., Australia.

Studies were conducted in the CERES phytotron to determine the effect of external NO_3^- concentration and rate of NO_3^- uptake on nodulation of various soybean cultivar x *Rhizobium* strain combinations. Pregerminated seedlings were transplanted on day 7 to inoculated nutrient solutions containing NO_3^- levels of 0, 0.5, and 1.0 mM (resupplied daily to initial levels) and 1.0, 2.0, and 4.0 mM (allowed to deplete with plant growth). Nodulation was observed, pH was adjusted, and nitrate uptake was monitored daily over the subsequent 14-day growth period. Plants were then harvested for measurement of nitrogenase (acetylene reduction) activity of nodulated roots.

Rates of NO_3^- uptake were similar among all treatments in spite of an initial eight-fold range in NO_3^- level. Higher external NO_3^- levels were more inhibitory to nodule appearance even though NO_3^- uptake rates were similar. The external concentration of NO_3^- rather than the rate of NO_3^- uptake appeared to have a major effect on the initial stages of nodulation. Maintaining the concentration of NO_3^- in the solution following appearance of nodules greatly retarded, or prevented, the development of nitrogenase activity.

Eight strains (selected from preliminary evaluation of 46 strains) of *R. japonicum* were compared for tolerance to NO_3^- when inoculated on a single soybean cultivar (Williams). All strains, in the absence of added NO_3^-, nodulated soybean between 6.0 and 6.3 days after inoculation. With the 1.0 mM NO_3^- treatment (rundown), a rather narrow range in time to the first nodulation was observed (8.3 to 10.2 days). Time to detection of 50 (at least) nodules per pot of four plants ranged from 7.0 to 8.5 days for controls, and the NO_3^- treated plants were delayed 3.5 to 5.1 days relative to respective controls; again a rather narrow range. The 1.0 mM NO_3^- treatment was quite inhibitory to acetylene reduction activity and the inhibition ranged from 84 to 98%, depending on *Rhizobium* strain. This indicated that only limited variability existed among strains for tolerance to NO_3^-.

Twelve soybean cultivars (selected from preliminary evaluation of 24 cultivars) were compared for tolerance to NO_3^- when inoculated with a single strain (USDA 110) of *R. japonicum*. Nodulation of 0 NO_3^- controls ranged from 5.7 to 6.4 days after inoculation while the delay (relative to controls) due to NO_3^- treatment ranged from 1.3 to 5.3 days. The cultivar Dickie which had the greatest delay in nodulation (5.3 days) also had the greatest inhibition (99%) of C_2H_2 reduction activity. In contrast, Elf and Avoyelles which showed 1.5 and 1.3 days delay in nodulation, in response to NO_3^- treatment, were inhibited in C_2H_2 reduction activity by only 30 and 48%, relative to 0 NO_3^- controls. No relationship could be established between rates of NO_3^- uptake by the various cultivars and delay in nodule initiation or inhibition of acetylene reduction activity. However, more variability in nodulation tolerance to NO_3^- appeared to exist among soybean cultivars than among *Rhizobium* strains. From this it was deduced that the host plant may play a greater role in nodulation tolerance to NO_3^- than does the *Rhizobium* strain.

HIGH SOIL NITRATE REGULATES SOYBEAN SYMBIOSIS

D.F. HERRIDGE[+]/R.J. ROUGHLEY/J. BROCKWELL
+N.S.W. DEP. AGRICULTURE, AGRICULTURAL RESEARCH CENTRE, TAMWORTH, AUSTRALIA.

A field experiment was done to determine whether the relative dependence of a soybean crop on soil N and symbiotic N could be manipulated by varying the levels of inoculation. Irrigated Bragg soybeans were grown on a vertisol, high in soil NO_3^- (27 µg per g, 0-60 cm) and previously free of *R. japonicum*. Three levels of inoculation, nil (O), 1.0 x 10^6 (n), 1.0 x 10^8 (100n) rhizobia per cm of row, were applied by spraying a water suspension of peat inoculant into the seed bed at sowing. Counts were made of the number of rhizobia in the inoculant and the populations that subsequently developed in the plant rhizospheres. At intervals during the growth of the soybeans, measurements were made of soil NO_3^- through the profile, the nitrogenous solutes in vacuum-extracted xylem sap and crop N. The level of ureides relative to total nitrogenous solutes was used as an index of symbiotic N-fixation.
High soil NO_3^- impaired early nodule formation and the onset of N-fixation (Table 1). Colonization of the rhizospheres by *R. japonicum* and the amount of nodulation were related to level of inoculation (Table 1). Acceleration of nodulation between 42 and 62 days coincided with depletion of NO_3^- from the soil (Table 2). Xylem sap ureides increased also during this time and were related directly to the levels of inoculation and nodulation. By day 126 ureide values were 19% (O), 57% (n) and 71% (100n). Crop N was improved by inoculation despite interference to early nodulation by high soil NO_3^-. It was concluded that the uptake of soil NO_3^- and N-fixation were complementary in meeting the N requirements of the crop.

TABLE 1. Influence of level of inoculation on the development of the symbiosis of Bragg soybeans grown in a high NO_3^- soil.

		0	n	100n
Inoculant rhizobia (log_{10} per cm of row)	0 day	nil	6.02	8.02
Rhizobia per rhizosphere (log_{10})	14 days	0	3.00	4.16
	42 days	0	5.33	5.85
Nodule mass (mg per plant)	42 days	0	13	32
	62 days	0	122	269
Xylem sap ureides (%)	42 days	10	10	10
	63 days	13	24	29
	76 days	9	11	19
Crop N (kg per ha)	126 days	170	280	280

TABLE 2. Levels of soil NO_3^- (µg per g, 0-30 cm) measured periodically under Bragg soybeans compared with NO_3^- levels in adjacent fallow.

Treatment	day 15	day 35	day 63	day 78	day 112
Soybeans, 0 inoculation	31	29	12	5	1
Soybeans, 100n inoculation	30	29	10	8	1
No plants (fallow)	30	32	27	39	16

THE EFFECT OF STARTER N ON SOYBEAN NODULATION

ADRIENNE KOERMENDY AND ALLAN R. J. EAGLESHAM
BOYCE THOMPSON INSTITUTE FOR PLANT RESEARCH AT CORNELL, ITHACA, NY 14853, USA

1. INTRODUCTION

The effect of nitrate on soybean (Glycine max cv. Wilkin) growth, nodulation and nitrogen fixation were studied, addressing the following questions. (i) How do starter N levels affect nodulation? (ii) Does starter N influence the relative effectiveness of strains? (iii) Are some strains less sensitive to high levels of starter N?

2. MATERIALS AND METHODS

Sterile nutrient solution containing potassium nitrate (0 to 450 mg N/pot) was added to sand-filled styrofoam pots just prior to planting. When the first trifoliates appeared, seedlings were thinned to one/pot. Plants were grown in either a greenhouse or high intensity lightroom (16 h photoperiod) with 30/20 or 30/25 C day/night temperatures respectively and were watered by adding sterile -N solution to the saucers.

3. RESULTS AND DISCUSSION

(i) The effects of N depended on the amount applied, Rhizobium strain and plant age at harvest. At 14 days after planting (DAP), 90 mg caused a reduction of nodule number with 4 out of 7 strains, whereas at 24 DAP, 90 mg increased nodule number by more than 100% with 6 of the strains. At 24 DAP, nodule number with 180 mg was equal to that of the -N controls (5 strains), and with 270, 360 and 450 mg, nodule number was less than in the -N controls (all strains). At 45 DAP, highest nodule number and dry weight occurred with 180 and 270 mg. With several strains, nodule number with 360 and 450 was twice the -N control. Clearly, the older the plants were when harvested, the higher was the level of starter N required to maximize nodule number, nodule dry weight and ARA. This can be interpreted as follows, a plant provided with high N becomes nitrogen-deficient as a large plant with a large N demand; this demand is met by a burst of nodulation.

(ii) Strains which were judged as poorly effective based on plant growth in the absence of N, were as effective as strains considered as very effective when grown in the presence of high levels of starter N. For example, USDA I110 formed 22 nodules at 24 DAP and IRj 2179AI resulted in only 5 nodules at the -N level. A significant difference persisted between the 2 strains at the -N and 90 mg levels with final total plant dry wt. significantly higher for plants inoculated with I110. However, with higher N levels (180, 270, 360 and 450 mg), nodule number, nodule dry wt. and ARA levels were closely similar with these strains at 45 DAP and final harvest. Final yields were not significantly different.

(iii) Forty strains of Rhizobium were screened for the ability to nodulate early in the presence of 360 mg N/pot. They fell into 2 categories in terms of the nodulation at 21 DAP: (a) those which failed to nodulate or produced significantly fewer nodules than the -N controls (8), (b) those producing nodules equal in number, but smaller, to the -N controls (32). Strains in category (a) were less effective than those in category (b) as judged at the -N level.

CHLOROSIS-INDUCING TOXINS PRODUCED BY RHIZOBIUM AND THEIR RELATIONSHIP TO NODULATION OF NON-NODULATING SOYBEAN

JEFFREY S. LA FAVRE AND ALLAN R. J. EAGLESHAM
BOYCE THOMPSON INSTITUTE FOR PLANT RESEARCH AT CORNELL, ITHACA, NY, USA.

1. INTRODUCTION

Some strains of Rhizobium japonicum are known to produce a chlorosis-inducing toxin called rhizobitoxine, an amino acid which reacts with ninhydrin to form a yellow product (Owens et al., 1972). A chlorosis-inducing toxin has also been found in the supernatant of broth cultures of some West African "cowpea" rhizobia (Eaglesham, Hassouna, 1982). Rhizobitoxine-producing strains of R. japonicum have been shown to possess a limited ability to nodulate non-nodulating (rj_1) soybean (Devine, Weber, 1977). This report summarizes results of a search for strains of rhizobia capable of producing chlorosis-inducing toxins and a comparison of nodulating ability on rj_1 soybean.

2. MATERIALS AND METHODS

A soybean (Glycine max) seedling bioassay was developed to detect chlorosis-inducing toxin(s) present in broth cultures of Rhizobium using an Asian soybean (cv. TGm 344) grown in 180 ml polypropylene bottles containing vermiculite (10 ml late log broth culture/bottle added at planting, one plant/bottle).

3. RESULTS

Chlorosis was first observed in the unifoliate or first trifoliate leaves. High levels of toxin resulted in completely yellow or even necrotic leaves while moderate levels resulted in a distinctive net pattern of green on a yellow background. Chlorosis was the result of toxin uptake by roots rather than production of toxin in nodules. Of 39 strains of R. japonicum and 43 strains of "cowpea" Rhizobium assayed for toxin production in broth, 51% of both groups produced detectable levels. The geographic origins of toxin producers were diverse, including West Africa, Panama, and the United States. Toxin produced by a West African "cowpea" Rhizobium (IRc 291) was partially purified. Paper chromatography was used to detect amino acids present in fractions collected from a cation exchange column. Fractions capable of inducing chlorosis in soybean contained an amino acid which reacted with ninhydrin to yield a yellow product. The chlorosis pattern induced by the toxin was identical to that induced by R. japonicum strain USDA 76 which is known to produce rhizobitoxine. Three toxin-producing strains of R. japonicum and one toxin-producing "cowpea" Rhizobium were compared with four strains that do not produce toxin. The nodulation capability on rj_1 soybean (cv. Harosoy) was not correlated to the ability of the rhizobia to produce detectable levels of toxin in culture. The rj_1 soybeans required high inoculum levels (about 1×10^{11} cells/plant) for good nodulation (48 nodules/plant using strain USDA 76).

4. REFERENCES

Devine TE and Weber DF (1977) Euphytica 26, 527-535.
Eaglesham ARJ and Hassouna S (1982) Plant Soil 65, 425-428.
Owens LD, Thompson JF, Pitcher RG and Williams T (1972) J.C.S. Chem. Comm. 1972, 714.

THE ROLE OF NITROGEN FIXATION IN SEED DEVELOPMENT OF SOYBEAN

D. R. NELSON, R. J. BELLVILLE, C. A. PORTER
Monsanto Company, St. Louis, Missouri, USA

1. INTRODUCTION

Seasonal nitrogen fixation profiles of soybean (*Glycine max* L. Merr) were published by Harper (1974), Thibodeau and Jaworski (1975) and others using acetylene reduction (AR) with washed, detached roots. They observed a decline a AR activity coincidential with increasing seed size. In a heavily cited article, Thibodeau and Jaworski hypothesized that the proximity of the pods to the leaves deprived the roots of carbohydrate and caused the decline in AR activity. Coupled with the observation that nitrate reductase activity ceases after flowering (Harper 1974), Thibodeau and Jaworski proposed a utilization model which concluded that nitrogen redistribution, rather than assimilation, satisfied the nitrogen requirement late in the season.

2. MATERIALS AND METHODS

A non-destructive AR assay was developed for soybean field plots. The system consisted of 120x150x20 cm chambers containing 65 plants in 5 rows spaced 25 cm apart. There were 3 replications (chambers) per treatment. Plants were grown in a medium grade sand; nutrient, moisture and root temperature were controlled. At flowering, combined nitrogen was removed from the nutrient. To begin the assay, acetylene at 10 cc/l was circulated through manifolds in the sand; AR activity equilibrated in 5 min and was linear with time. Plant growth and yield were comparable to soil grown companion plots.

3. RESULTS AND DISCUSSION

Using the above AR assay and omitting combined nitrogen after flowering, AR peaked 3 weeks after the beginning of seed development, not coincident with it. A correlation between seasonal AR and yield representing 4 yr of data was R=.999. This suggested a cooperative relationship between nitrogen fixation and yield. Consistent with these results, total nitrogen data of Pal and Saxena (1976), Hashimoto (1971) and others demonstrated that both nodulated and non-nodulated soybeans accumulated more than 60% of the seasonal total after a stage at which nitrate reductase activity had ceased and acetylene reduction activity was minimal. This suggests that developing seeds depend primarily on currently fixed nitrogen.

Evidence for nitrogen assimilation during seed growth does not preclude competition between seeds and roots for carbohydrate. If carbohydrate is partitioned preferentially to the seeds, then a partial defoliation should create a carbon shortage and decrease AR activity more than yield. Continuous removal of all lateral leaflets from each trifoliolate (60% defoliation), from flowering to maturity, resulted in a 23% reduction in yield and no effect on AR activity. The defoliation demonstrated that neither a carbon shortage during podfilling nor the proximity of the pods to the source was invariably detrimental to nitrogen fixation.

4. REFERENCES

Harper JE (1974) Crop Sci. 14,255-260.
Hashimoto K (1971) Hokkaido Nat Agr Exp Sta Res Bull 99,17-29.
Pal UR and Saxena MC (1976) Agron.J. 68,927-932.
Thibodeau PS and Jaworski EG (1975) Planta (Berl) 127,133-147.

VARIATION IN THE RESPONSE OF N_2-FIXING RHIZOBIUM LEGUMINOSARUM ISOLATES TO THE APPLICATION OF NH_4NO_3.

L.M. NELSON
NATIONAL RESEARCH COUNCIL OF CANADA, PLANT BIOTECHNOLOGY INSTITUTE, SASKATOON, SASKATCHEWAN, S7N 0W9, CANADA.

Selection of effective *Rhizobium* strains is generally carried out in N-free conditions, yet low levels of combined nitrogen are usually present in the field. The contribution of N_2 fixation to plant nitrogen could be increased if the symbiotic system were more tolerant of combined nitrogen. There is some evidence that the rhizobial strain is a source of variation with respect to effectiveness in the presence of combined nitrogen. The present study was undertaken to determine whether *R. leguminosarum* isolates varied in their ability to fix N_2 in the presence of NH_4NO_3. A second objective was to determine whether this variability in response to NH_4NO_3 was affected by time of application of NH_4NO_3, NH_4NO_3 concentration, and plant age.

In an initial experiment 38 isolates of *R. leguminosarum* which were effective in N-free conditions were inoculated onto *Pisum sativum* L. cv. Homesteader plants and grown for 4 weeks in N-free conditions or with weekly additions of 2mM NH_4NO_3. Acetylene reduction was inhibited relative to N-free controls in all isolates but the extent of inhibition ranged from 60 to 100%. Ten isolates whose C_2H_2 reduction rates ranged from 0 to 9% of the N-free controls were selected for further study.

The effect of NH_4NO_3 concentration (0,1,2, and 5mM) and plant age (3,4, and 5 weeks) on the N_2 fixation response to NH_4NO_3 by the 10 isolates was determined when NH_4NO_3 was applied at seeding. Plant age, NH_4NO_3 treatment, and the isolate were significant sources of variation for C_2H_2 reduction but interactions between these factors were also significant. There were significant differences between isolates in the rate of decrease in C_2H_2 reduction with increasing NH_4NO_3 (responsiveness to NH_4NO_3) as well as in overall C_2H_2 reduction rates. C_2H_2 reduction responsiveness to NH_4NO_3 was correlated with C_2H_2 reduction rates at 2 and 5mM NH_4NO_3. This implys that ranking of the isolates with respect to C_2H_2 reduction will vary dependent on NH_4NO_3 level. C_2H_2 reduction rates were correlated with shoot dry weights at 0,1, and 2mM NH_4NO_3.

When plants inoculated with each isolate were grown in N-free conditions for 3 weeks prior to addition of NH_4NO_3 there were significant differences between isolates in C_2H_2 reduction responsiveness to NH_4NO_3 4 and 7 days after application of NH_4NO_3. The ranking of the isolates with respect to C_2H_2 reduction responsiveness to NH_4NO_3 at 7 days after NH_4NO_3 application was significantly correlated with the rankings obtained when NH_4NO_3 was applied at seeding.

These results suggest that *R. leguminosarum* isolates do vary in their response to NH_4NO_3 and this response appears to be a stable character of the isolate regardless of time of application of NH_4NO_3, NH_4NO_3 concentration and plant age. It may be possible to enhance N_2 fixation in peas through selection of rhizobia with maximum effectiveness at low levels of combined nitrogen.

INFLUENCE OF A WATER DEFICIT ON NITRATE REDUCTASE AND NITROGENASE ACTIVITIES ON FIELD GROWN SOYBEAN (*Glycine max*)

M. OBATON, Geneviève CONEJERO, Anne-Marie DOMENACH

I.N.R.A., Place Viala, 34060 MONTPELLIER-Cedex

Université de Lyon, Bat.741, 43, Boulevard du 11 Novembre 1918
69622 VILLEURBANNE

(FRANCE)

1.- INTRODUCTION.- The effect of a water deficit on the activities of nitrate reductase and nitrogenase is well documented, but the relative importance of the two enzymatic activities at the field level was not studied.

2.- MATERIAL and METHODS.- Soybean (*Glycine max*), cultivar Hodgson, is grown in a field and submitted to four types of irrigation (watering at field capacity or interruption of this irrigation during 28, 33 or 53 days - see graphs). The nitrate reductase activity, acetylene reduction and 15N natural abondance are mesured as previously indicated (Obaton 1982).

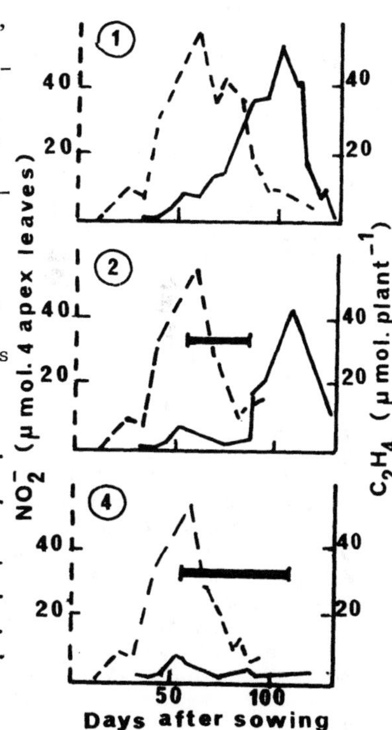

3.- RESULTS.- In these conditions, the cultivar Hodgson grown without water constraint fixes 43% of the nitrogen contained in the aerial parts. But in presence of a water deficit this rate decline to 25, 16 or less than 5 p. 100.

TABLE : Rate of nitrogen fixation under different watering conditions

Treatment	1	2	3	4
Total amount of water received (mm)	676	457	531	242
Days of drought	0	33	28	53
% N fixed	43	25	16	<5

The 15N mesure supports the enzymatic study.

4.- CONCLUSION.- When Soybean (cv. Hodgson) suffers of a water deficit, the reduction of nitrate in the leaves is less affected than the nitrogen fixation. Under a drastic stress this cultivar grows almost without nitrogen fixation.

Work is presently in progress to determine the causes of this differential effect of a water deficit on these two enzymatic activities and if there is a variability between cultivars to increase by breeding the capacity to fixe nitrogen under water deficit.

5.- REFERENCE.- Obaton M., Miquel M., Robin P., Conejero G., Domenach A.M., and Bardin R., (1982), C.R. Acad. Sci., 294, Série D, 1007-1012.

EFFECTS OF $[NO_3^-]$ ON N_2-FIXATION BY TRIFOLIUM SUBTERRANEUM

J.H. SILSBURY & D. CATCHPOOLE
DEPARTMENT OF AGRONOMY, UNIVERSITY OF ADELAIDE.

INTRODUCTION

The adverse effects of external mineral nitrogen on the N_2-fixing legume - *Rhizobium* symbiosis is well documented. Theories proposed to explain the suppression of nitrogenase activity by exogenous NO_3^- involve: (i) a direct effect; (ii) photosynthate deprivation; (iii) a feed-back mechanism. We have studied the dynamics of nitrogenase activity of swards of *Trifolium subterraneum* (subterranean clover) as measured by acetylene reduction assay after the application of nutrient solutions containing NO_3^- up to 16 mM.

METHODS

Small swards of clover (cv. Mt. Barker) were grown in 2 l pots containing fritted clay. Surface sterilized seed was inoculated with *Rhizobium trifolii* WU/95 after sowing. Growth temperature was 20 ± 0.4°C and the photoperiod, 12 h. The sward was contained within the area of the pot, so that after the canopy had closed, net daily CO_2-uptake was near constant. Swards were irradiated at PPFD values ranging from 300 to 1000 μmol quanta/m²/s. Plants were watered daily with nutrient solutions. KNO_3, K_2SO_4, $CaNO_3.4H_2O$ and $CaSO_4.2H_2O$ were used to provide variable $[NO_3^-]$ from 0 to 16 mM keeping $[K^+]$ and $[Ca^{2+}]$ constant. NO_3^- treatments were imposed by flushing the pots with the appropriate solution. Plants were assayed for H_2 evolution and C_2H_2 reduction at the end of the dark period.

RESULTS AND DISCUSSION

The initial response in AR assay after application of NO_3^- was a rapid stimulation of activity within 12 hours, followed by inhibition in an asymptotic trend with time. H_2 evolution followed a similar trend. The degree and the rate of inhibition were correlated with $[NO_3^-]$, 7.5 mM and 15 mM giving almost total inhibition within 4 days. However, the degree to which a lower $[NO_3^-]$ inhibited nitrogenase activity depended on the irradiance. At a PPFD of 300 μmol quanta/m²/s, 0.5 mM and 1.0 mM $[NO_3^-]$ inhibited fully within 7 days but at 1000 μmol quanta/m²/s only a partial inhibition occurred. This was sustained for 35 days. If NO_2^- is produced by NO_3^--reductase in the nodules, the amount and the degree of binding to the Mo/Fe component must, at least initially, be proportional to $[NO_3^-]$ since in the whole plant system AR activity was not immediately arrested by $[NO_3^-]$. Inhibition of the synthesis of nitrogenase over several days is consistent with the data, as is interference by NO_3^- with the capacity of leghaemoglobin to regulate O_2 to the bacteroids. Some support for the assimilate deprivation hypothesis is provided by evidence of an interaction between $[NO_3^-]$ and the PPFD.
It is probable that N_2-fixation in the actively nodulated plant is strongly integrated with all growth processes such that the rate of biosynthesis of new material (of given N- content) is regulated primarily by the rate of photosynthesis. When such a system is disrupted by the induction of NO_3^--reductase, the rate of N_2-fixation is initially accelerated in response to a new demand for N (synthesis of NO_3^--reductase) and is then depressed by a 'flooding' of the transport system with reduced N produced at sites closer to the sinks than when the only source is from the nodules.

THE EFFECT OF SALINITY ON SURVIVAL OF RHIZOBIUM, NODULE FUNCTION AND NODULE FORMATION IN THE SOYBEAN-RHIZOBIUM JAPONICUM SYMBIOSIS

P.W. SINGLETON[+]/ B.B. BOHLOOL[+]/ UNIVERSITY OF HAWAII NifTAL PROJECT
[+] P.O. BOX O, PAIA, HAWAII 96779, USA

Symbiotic nitrogen fixation may be adversely affected by saline environments. This paper describes three experiments that assess the salt sensitivity of: 1) *Rhizobium* as free living organisms; 2) soybean nodule function; and 3) soybean nodule formation. In addition, a split-root plant growth system is described which can be used to separate the effects of salinity stress on host yield potential from the effects of salinity on nodule processes.

The growth rate of *Rhizobium* in culture media was slowed by the addition of NaCl. Some strains were incapable of growth at the highest level of salt used (120 mM NaCl). However, two sensitive and two tolerant strains withstood substantial osmotic shock and survived for extended periods in saline solutions equivalent to the concentration of sea water. Survival of *Rhizobium* in salinized soil at different moisture tensions was a function of the additive effects of osmotic and matric potentials.

By independently subjecting nodules and shoots to salinity stress in a split-root growth system it was possible to demonstrate that the soybean-*Rhizobium japonicum* nodule system was not greatly affected by exposure to 120 mM NaCl. The main reduction in nitrogen fixation from salinity stress was the indirect effect of salinity on leaf expansion, shoot yield potential and the sink for nitrogen. Nitrogenase activity of salt stressed nodules was relatively unaffected when shoots remained unstressed.

The early processes of nodule formation however, were extremely sensitive to NaCl in the rooting medium. When only 26.3 mM NaCl was added to the nutrient solution two hours prior to inoculation of one half of a split-root system, nodule number and mass were reduced by 50% and 79.9 mM NaCl reduced nodule number, mass and nitrogen fixation to less than 10% of the controls. *Rhizobium japonicum*, reisolated from nodules from the high salt treatment did not form more nodules under saline conditions than isolates from controls.

EFFECTS OF SALINITY ON GROWTH AND NODULATION OF *Arachis hypogaea*

J.I.SPRENT and S.G.McINROY
DEPARTMENT OF BIOLOGICAL SCIENCES, UNIVERSITY OF DUNDEE, DD1 4HN SCOTLAND

INTRODUCTION

The numerous studies which have been made of the effects of salinity on nodulation, have all used species which are infected *via* root hairs (see Sprent, 1983 for discussion). *Arachis hypogaea* is infected at the junctions of branch roots (Chandler, 1978 and others). We have begun to investigate whether this mode of infection is more or less resistant to stress than root hair infection.

MATERIALS AND METHODS

Arachis hypogaea cv J11 was germinated for 48h on moist paper before being transplanted into pots of sand. At the time of transplanting seedlings were inoculated with a dense suspension of *Rhizobium* strain 3824 from the Rothamsted collection. Dry sand was moistened with either water, 50 mol m^{-3} or 100 mol m^{-3} NaCl to maintain water holding capacity. Pots were weighed daily and given water or N-free nutrient to maintain their water content. After 21 days, 9 plants from each treatment were harvested. They were photographed, weighed, nodules were counted and 3 from each group were fixed in glutaraldehyde so that roots and nodules could be examined by scanning electron microscopy.

RESULTS

Growth, but not nodulation was reduced by 50 mol m^{-3} NaCl. Double this concentration severely depressed both growth and nodulation.

There were no obvious differences in the appearance of nodules on 0 and 50 mol m^{-3} NaCl treated plants. Both showed a clear ring of hairs at the root junction (see plate 1A in Chandler, 1978). These hairs were also visible on the 100 mol m^{-3} plants, even though the degree of root branching was much reduced.

Macroscopically, more sand was seen to adhere to roots of the 100 mol m^{-3} plants than in either of the other two treatments. Microscopically, this was seen to be due to the presence of hairs along the roots surface. These hairs are superficially like root hairs, though of more scattered distribution. Hairs of this type were not seen on roots of plants from the other treatments (nor were they observed by Chandler, 1978).

DISCUSSION

Arachis hypogaea appears to be amongst the more salt tolerant of legumes with respect to nodulation (see Sprent, 1983, Table 4). Its production of root hairs, apparently as a specific response to high salinity is a surprising and interesting result, which we shall be investigating further.

REFERENCES

Chandler MR (1978) J. exp. Bot. 29, 749-755.
Sprent JI (1983) Proceedings of this conference.

PHYSIOLOGICAL STUDIES ON THE LUPIN-RHIZOBIUM SYMBIOSIS: pH AND MINERAL NITROGEN EFFECTS

P. SUBRAMANIAM/ R.M. PARADINAS/ C. RODRIGUEZ BARRUECO
UNIT OF NITROGEN FIXATION. CENTRO DE EDAFOLOGIA Y BIOLOGIA APLICADA. C.S.I.C. SALAMANCA. SPAIN.

1. INTRODUCTION

Nodulation and nodule function of L. angustifolius have been studied in order to evaluate its tolerance to soil pH and mineral nitrogen, two important nodule stress-inducing factors.

2. MATERIAL AND METHODS

L. angustifolius cv. Unicrop was grown under water culture using N-free Crone salts under artificial illumination. Nodulation response was studied over 7 pH levels, while nodule function was studied over 5 pH levels in the presence of KNO_3 using nodulated plants. Nodule formation, function and growth of plants supplied with differential quantities of KNO_3 and $(NH_4)_2SO_4$ were also evaluated (See graphs).

3. RESULTS AND DISCUSSION

3.1. pH effects:

pH levels markedly influenced nodulation, while the symbiotic performance was less affected over pH 6-9, indicating that the nodulation is more pH-sensitive than nodule function. This could be attributed to the effect of pH on the survival and virility of rhizobia and the deficiency of certain nutrients essential for infection (fig. 1).

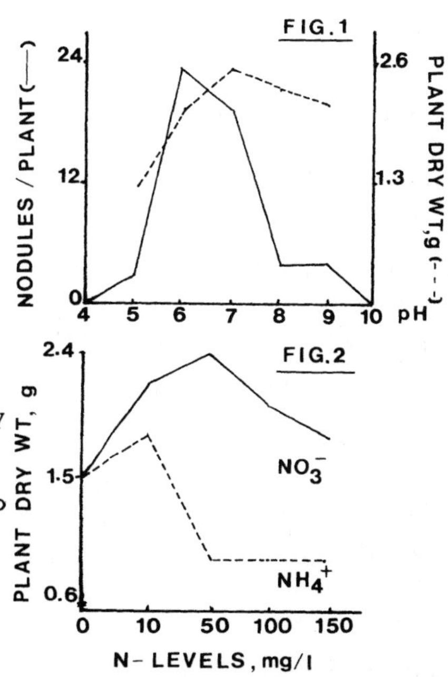

3.2. Mineral nitrogen effects:

NH_4 nutrition affected more severely nodulation and nodule function of lupins than NO_3 nutrition. Low levels of NO_3-N resulted beneficial to the whole plant growth and improved the N status owing to stimulatory effect on early growth by compensating for a suboptimal rate of symbiotic fixation during that period (fig. 2).

pH EFFECTS ON *Casuarina cunninghamiana* SYMBIOSIS

P. SUBRAMANIAM / C. RODRIGUEZ-BARRUECO
UNIT OF NITROGEN FIXATION. CENTRO DE EDAFOLOGIA Y BIOLOGIA APLICADA. C.S.I.C. SALAMANCA. SPAIN.

1. INTRODUCTION

pH effects on the nodule function of *Casuarina cunninghamiana* has been studied with the object of understanding the physiology of nitrogen fixation in this economically important tree species.

2. MATERIAL AND METHODS

Seedlings of *C. cunninghamiana* were raised from surface sterilized seeds and were inoculated with a crushed nodule inoculum after 45 days growth. Nodulated seedlings were maintained in half-strength nitrogen-free Crone nutrient solution for two months prior to the imposition of different pH treatments. Six pH levels were tested, three in the acid range (4, 5 and 6), and three in the alkaline range (8, 9 and 10). The pH levels of the solutions were monitored twice daily. The plants were grown under artificial illumination for six months, at the end of which observations were made on growth and acetylene reduction.

3. RESULTS AND DISCUSSION

pH dependance of nodule function in *C. cunninghamiana* was evaluated in terms of growth and acetylene reduction. pH levels 9 and 10 affected adversely plant height, dry wt. accumulation and nitrogenase activity, while pH 4 to 8 caused little variation (Fig. 1), indicating a higher acid pH dependance of nodule function.

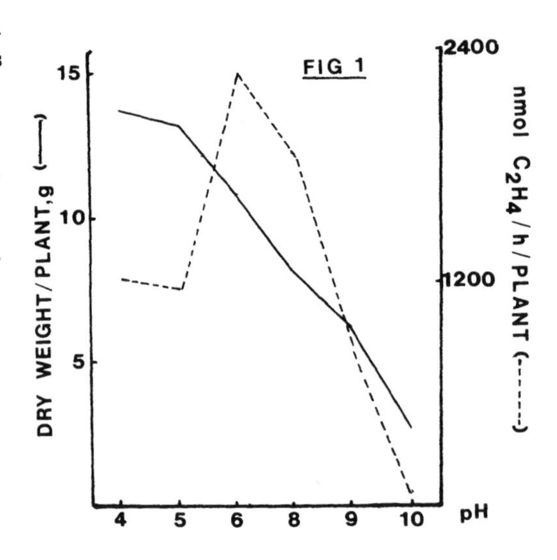

4. CONCLUSION

Moderate acid-dependance for nodule function in *Casuarina cunninghamiana* is indicated.

CALCIUM OXALATE CRYSTALS IN DETERMINATE ROOT NODULES OF THE LEGUMINOUS TRIBE PHASEOLEAE

J.M.SUTHERLAND AND J.I.SPRENT
DEPARTMENT OF BIOLOGICAL SCIENCES, UNIVERSITY OF DUNDEE, DD1 4HN SCOTLAND

Crystalline structures have been found in the outer cortex of nodules of the leguminous tribe Phaseoleae. These structures have been examined by various microscopic methods. During fixation and embedding for light microscopy the crystals were extracted, but their shape and distribution were determined by examination of the crystal-shaped holes which remained. Direct examination of the crystals was carried out by scanning electron microscopy of freeze-dried sections of *Phaseolus* nodules. The crystals were found to be similar in structure to the calcium oxalate crystals found in the leaves of leguminous species by Horner and Zindler-Frank (1982).

Calcium oxalate crystals were said by Spratt (1919) to occur in the cortex of *Phaseolus* nodules, but no evidence was given for their identification as such and recent structural studies (see Newcomb, 1982) have not followed up this earlier report. The lack of positive evidence for their identification as calcium oxalate is probably due to the presence (around the crystals) of cell walls which are resistant to both acid and alkali. These walls were retained when nodules were cleared in 5% sodium hydroxide followed by hydrogen peroxide. Though the presence of these resistant walls prevented extraction and purification of the crystals for examination by X-ray diffraction, it proved a useful means of examining their distribution. Staining with toluidine blue and sudan black suggested the presence of lignin and of either lipid or suberin. Electron microscopy revealed that the crystals were surrounded by 2 and sometimes 3 layers of wall material: an irregularly thick fibrillar outer layer; a thin electron translucent layer and sometimes an intermediate homogeneous layer. The crystals stained positively for calcium oxalate by a method considered diagnostic by Horner and Zindler-Frank (1982). X-ray microprobe analysis, though not quantitative, showed that the crystals contained very high relative levels of calcium. Overall the evidence is considered sufficient for the identification of these crystals as calcium oxalate.

Similar crystals were found in nodules of *Phaseolus*, *Vigna*, *Lespedeza*, and *Desmodium*, but they were absent from nodules of *Vicia*, *Pisum*, *Ononis* and *Dorycnium*. Sprent (1980) has described structural and physiological differences between the nodules of these two groups of genera; The most important physiological difference is in their nitrogenous export products. Determinate nodules - those which also contain calcium oxalate crystals- export chiefly ureides, while the remainder, which have indeterminate growth, export amides. The possible relationships between these differences and calcium oxalate production are currently under investigation.

Horner HT and Zindler-Frank E (1982) Can. J. Bot. 60, 1021-1027.
Newcomb W (1981) Int. Rev. Cyt. Suppl. 13, 247-297.
Spratt ER (1919) Ann. Bot. 33, 189-199.
Sprent JI (1980) Pl. Cell Environ. 3, 33-43.

THE EFFECT OF TEMPERATURE STRESS ON N_2-FIXING *PHASEOLUS VULGARIS* L.

R.J. THOMAS, J.I. SPRENT, M. HUNGRIA*, J. DOBEREINER*
DEPARTMENT OF BIOLOGICAL SCIENCES, UNIV. OF DUNDEE, DUNDEE DD1 4HN, SCOTLAND
*EMBRAPA/PNPBS, KM 47, 23460 SEROPEDICA, RIO DE JANEIRO, BRASIL.

When grown in temperate climates *Phaseolus* beans are subjected to suboptimal growth temperatures throughout their life cycle. On the other hand, when grown in their more native sub- or tropical environments they can be subjected to supraoptimal temperatures. Common beans grown in a mixed cropping system with maize in Brasil have been observed to grow better in the mixed crop than in a monocrop during vegetative growth. Shading of bean plants by maize and hence lower soil temperatures has been suggested as a possible cause of this effect. Here we examine some physiological responses of beans to 1) low temperature stress as experienced by the U.K. crop and 2), high temperature stress when grown under Brasilian conditions.

1) Low temperature stress

A cold tolerant (194) and a sensitive (Seafarer, SF) line of *P. vulgaris* were grown in cabinets with a 14/10 h photo- and thermo-period at 15/10, 20/15 and 25/15°C. All plants were inoculated with *Rhizobium phaseoli* strain 3622 and no combined N was fed to the plants.

Both lines produced functional nodules at a mean temperature of 12.9°C (15/10) but 194 produced more and larger nodules which fixed more N (acetylene reduction and total plant N) and also had greater rates of xylary N transport than SF. At 15/10°C during vegetative growth 194 accumulated greater amounts of dry wt. than SF but had similar RGR's and NAR's. RLGR and RGR_N, expressed in terms of total plant N, on the other hand were markedly greater in 194 than in SF. The results suggest that the ability to nodulate well, increase nodule size in response to low temperatures and achieve maximum leaf expansion via increased rates of N transport from the roots, are important factors for the superior growth of 194 over SF.

2) High temperature stress

Nodulated beans (cv. Negro Argel) were grown outdoors in Brasil with full, 70 and 50% sunlight in uncovered or covered pots. Covering pots with aluminium foil resulted, on average, in a 2°C lower soil temperature (2-3 cm below surface). Soil temperature ranged from 27-39°C and was generally highest in full sunlight. Shading plants by as much as 50% with nylon netting did not affect net photosynthesis (dry matter accumulation) or nitrogenase activity (acetylene reduction). Dry matter accumulation and acetylene reduction were up to 15 and 50% lower, respectively in uncovered pots compared with covered pots in full sunlight. Only acetylene reduction was lower in uncovered pots compared with covered pots in the 70 and 50% sunlight treatments. Unlike low temperature stress, high temperatures did not affect nodule size or number. It is concluded that high soil temperatures can inhibit bean growth under Brasilian conditions and one beneficial effect of shading the crop would be a reduction in soil temperature.

ALUMINIUM TOXICITY AND NODULATION OF WHITE CLOVER

M. WOOD, J. E. COOPER AND A. J. HOLDING
DEPARTMENT OF AGRICULTURAL AND FOOD BACTERIOLOGY, THE QUEEN'S UNIVERSITY OF BELFAST, NEWFORGE LANE, BELFAST BT9 5PX, NORTHERN IRELAND.

White clover (*Trifolium repens*) responds poorly to inoculation with elite strains of *Rhizobium trifolii* in limed acid mineral soils in the United Kingdom. Initially these soils contain a high level of exchangeable aluminium, therefore the possible role of aluminium in limiting nodulation of white clover has been investigated using an axenic solution-culture technique. At pH 4.3 - 5.0 root elongation and root hair formation are limited by aluminium rather than by acidity, whereas Rhizobium multiplication and nodulation are limited by acidity alone. With 50 µM aluminium the pH must be raised to 5.0 for optimum root growth and development, but pH 6.0 is required with 10 µM phosphate for optimum Rhizobium multiplication and nodulation. The effect of 50 µM aluminium at pH 5.5 in inhibiting Rhizobium multiplication and in reducing nodule formation contrasts with the lack of any effect on root growth and development and occurs despite the precipitation of aluminium from solution under these conditions.

The same response by Rhizobium to pH calcium and aluminium was found in a defined arabinose-galactose-glutamic acid medium supplemented with vitamins. Screening of 80 strains of *R. trifolii* isolated from soils in Northern Ireland showed that aluminium is generally toxic at pH 5.5, and the selection of acid-tolerant strains indicated that aluminium imposes an additional stress at pH 4.5.

The precipitate formed by aluminium at pH 5.5 is likely to be an amorphous hydroxy-aluminium-phosphate polymer; the aluminium toxicity was overcome by increasing the phosphate concentration from 10 to 100 µM, but evidence indicates that the toxicity is not indirect in reducing the concentration of phosphate. It is suggested that the aluminium toxicity is due to the negatively-charged Rhizobium cells being adsorbed to the positively-charged hydroxy-aluminium polymers, thereby preventing multiplication and nodulation. This hypothesis could explain the poor response by white clover to inoculation in acid mineral soils which have been limed as recommended to pH 5.2 - 5.8.

^{15}N STUDY ON THE PARTITIONING AND METABOLISM OF THE NITROGEN DERIVED FROM ATMOSPHERIC N_2 AND COMBINED N IN SOYBEANS

Tadakatsu Yoneyama and Junji Ishizuka
Division of Plant Nutrition, National Institute of Agricultural Sciences, Kannondai, Tsukuba, Ibaraki 305, Japan

1. Partitioning of N in intact plants

Soybean plants can utilize atmospheric N, and combined N (nitrate and ammonium) as N source for growth. Differences in the utilization of these N sources in water-cultured plants were investigated using ^{15}N.

Partitioning of the ^{15}N from ^{15}N-labeled three N sources showed distinct characteristics. The N from ammonium preferentially distributed to the developing organs in comparison with N from nitrate which distributed relatively more to mature organs. N from dinitrogen showed a distribution pattern in shoots similar to that of N from ammonium. These differences in partitioning may be related with the N forms in the transport from roots to shoots

2. Partitioning of N in detached vegetative shoots

The main forms of N transport from roots to shoots in the plants depending on nitrate are nitrate and amides (asparagine and glutamine), and those in the plants depending on fixed N are ureides (allantoin and allantoic acid) and amides. ^{15}N-labeled nitrate, glutamine (amide label), asparagine (amide label) and allantoin were fed to the cut shoots via transpiration streams.

2.1. Partitioning

Nitrate-N distributed to both growing and mature leaves intensively with little retention in the stem + petioles, while glutamine-N and asparagine-N distributed relatively more actively to the growing leaves than to the aged leaves with large retention in the stem + petioles. Allantoin-N partitioned by an intermediate pattern of those of nitrate-N and amides-N.

2.2. Metabolism

Nitrate-N was assimilated into amino acids in both growing and mature leaves more actively than amide-N of asparagine and allantoin-N. Glutamine synthesis may be operating in the different compartments, chloroplasts and cytoplasm, using nitrate-N and ammonia from other sources, respectively. Asparagine synthesis was more active in aged leaves than growing leaves, while utilization of asparagine was more active in the growing leaves than in mature leaves. The amide N of asparagine was transferred to the amide of glutamine followed by glutamic acid and alanine. Allantoin was synthesized using added nitrate and asparagine (amide) as the N source in both growing and mature leaves, and the utilization of allantoin-N to form amino acids was very active in the growing leaves.

A SIGNIFICANT ROLE OF BACTERIAL *Rhizobium japonicum* ASPARTATE AMINOTRANSFERASE IN AMMONIA ASSIMILATION IN *Glycine max* ROOT NODULES.

Zlotnikov K., Marunov S., Khmelnitsky M.
Institute of Biochemistry and Physiology of Microorganisms, USSR Academy of Sciences, I42292,Pushchino, USSR

An assimilation of ammonia from N_2 fixation in legume nodules is carried out in reactions catalyzed by GS, GOGAT, AAT, and AS enzymes which, according to literature (Boland, et al.,I980), have plant origins. In our work we checked the possibility of a bacterial origin of one of these enzymes, aspartate aminotransferase (AAT, EC2.6.I.I.). For this purpose *asp4* mutant of *R.japonicum* IIO strain was isolated and tested in symbiosis with soybean plants.

Materials and Methods.

The wild type strain *R.japonicum* IIO, the procedure for isolation of *asp4* mutant, method for testing the symbiosis with plants, and growth conditions were all as described (Khmelnitsky, et al.,I98I). The AAT activity was assayed using the malate dehydrogenase-linked reaction system (Decker,Rau,I963).

Results and Discussion

The *asp4* mutant cells required aspartate or asparagine for growth on minimal media, induced pink normal morphology nodules on soyabean plants, and could be recovered from the nodules as the original stable auxotrophic mutants. The activity of AAT enzyme in wild type bacteria was around 35 μmol/min/mg of protein, in the cells of *asp4* mutant it was undetectable. Nodules formed by the mutant had the increased level of nitrogenase activity (acetylene reduction). At the same time plants inoculated with *asp4* mutant were yellow and had one half the quantity of nitrogen (Kjeldahl analyses) in their leaves. Therefore our results provide evidence supporting the principal role of bacterial AAT for ammonia assimilation in soyabean root nodules.

References.

Boland MJ, Farnden KJF, Robertson JG (I980) in Newton WE and Orme-Johnson H, eds, Nitrogen Fixation, v.II.pp.33-52, U.P.P., Balt.
Decker LE and Rau EM (I963) Proc.Soc.Exp.Biol.Med., I22,I44-I49.
Khmelnitsky MI,Zlotnikov KM,Bayev AA(I98I)Dokl.Acad.Nauk SSSR. 256, I9I-I95.

PLANT GENES INVOLVED IN DINITROGEN FIXATION

POSTER DISCUSSION 8
PLANT GENES INVOLVED IN NODULATION AND SYMBIOSIS

LEGHEMOGLOBIN AND NODULIN GENES: TWO MAJOR GROUPS OF HOST GENES INVOLVED IN SYMBIOTIC NITROGEN FIXATION

D.P.S. Verma, J. Lee, F. Fuller and H. Bergmann

Plant Molecular Biology Laboratory
Department of Biology
McGill University
Montreal, CANADA H3A 1B1

INTRODUCTION

The legume root nodule, an organ Sui generis, is a product of the interactions of two genomes (plant and Rhizobium) culminating in the fixation of atmospheric nitrogen symbiotically. It is apparent that these interactions occur at several nested levels (see Verma and Long 1983) and that any perturbation of these interactions aborts the symbiotic route and pathogenesis ensues. Our knowledge of these interactions at both genetic and molecular levels is very limited, which is primarily due to the fact that no suitable mutations in the host are available and the genes that affect nodule development are not yet identified and characterized. Classical genetic experiments have revealed a number of host genes (see for reviews, Holl and LaRue 1976; Nutman 1981; Verma and Nadler 1983) which play a role in symbiotic nitrogen fixation. However, the mode of interactions of these gene products and their function in symbiosis is not known. Moreover, most of these genes are recessive and conditional (i.e., Rhizobium strain specific). Recent studies using molecular approaches have identified a group of host gene products which are produced specifically in nodules and may play a role in the process of morphogenesis or symbiotic functions of the nodule.

MOLECULAR ANALYSIS OF HOST GENES INVOLVED IN SYMBIOSIS

Towards the identification and isolation of specific host genes which may play a role in symbiotic nitrogen fixation, immunological and RNA/cDNA hybridization techniques have been employed (Legocki and Verma 1979, 1980; Auger and Verma 1981; Bisseling et al. 1983). These studies led to the identification of a major group of host gene products, termed nodulins, which are expressed in moderate abundance and are specific to nodule tissues. Since these sequences are induced before the commencement of nitrogen fixation activity in nodules and their accumulation is influenced by various mutations in Rhizobium which render nodules ineffective, these host gene products are believed to be involved in symbiotic nitrogen fixation.

Phytohormones have been implicated in nodule development (see Verma 1982; Verma and Long 1983) and are known to be produced by Rhizobium. Accordingly, a subfraction of moderately abundant mRNAs common to nodule and root tissues has been found to be more abundant in the nodule and its level in the root can be increased by treatment with the plant hormone, IAA (Auger and Verma

1981; Verma et al. 1983a). In addition, a small group of proteins was observed which appeared three days after infection with Rhizobium and similarly appeared after IAA treatment (Verma et al. 1983). Phytohormones may also repress some host genes normally expressed in the root since the application of auxins is known to reduce the concentration of several mRNA sequences in soybean hypocotyl (Baulcombe and Key 1981).

Taken together, these results suggest that nodule development requires a highly interactive developmental program leading to the establishment of a symbiotic state. Figure 1 summarizes various groups of host genes that have been postulated to be involved in this process. At present, there are no molecular probes for the host genes believed to be involved in the process of early infection and repression of the host defence mechanism. Such genes cannot be identified with the techniques employed for the study of nodulins and leghemoglobins.

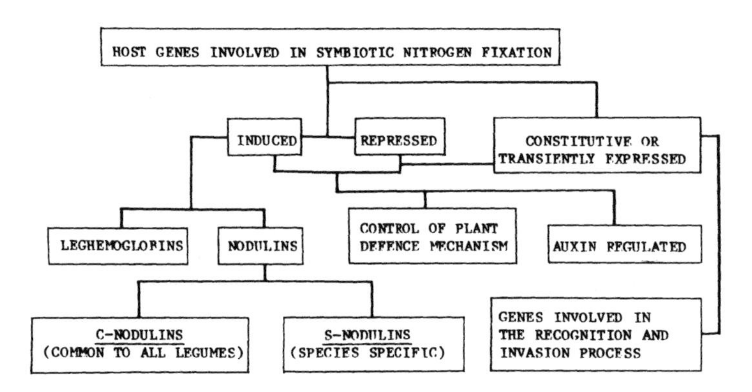

Figure 1. Summary of various host genes postulated and/or identified to be involved in symbiotic nitrogen fixation.

LEGHEMOGLOBIN GENES

A direct correlation between the presence of leghemoglobin (Lb) and the effectiveness of legume nodules in nitrogen fixation, as well as the presence of a similar protein in non-legume nodules (Appleby et al. 1983) suggests that this host protein is essential for symbiosis. Its function in oxygen transport in nodules at low pO_2 is now well recognized. The leghemoglobins are monomeric while the globin identified in non-legume nodules appears to be dimeric (Appleby et al. 1983).

Leghemoglobins have been shown to be induced prior to and independent of nitrogen fixation (Verma et al. 1979, 1981a). Therefore, it appears that this molecule is produced early in infection as a host response to invading Rhizobium. In soybean, there are four major leghemoglobins, Lba, $-c_1$, $-c_2$ and c_3, each of which is post-translationally modified into minor components referred to as Lbb, d_1d_2 and d_3 (Whittaker et al. 1981). The major Lb components are encoded by separate genes of a polymorphic gene family

(Brisson and Verma 1982; Wiborg et al. 1982). In addition to normal genes, there are pseudo and truncated genes in this family (Brisson and Verma 1982). These genes are arranged on the chromosome in more than four regions (Lee et al. 1983). The main locus (Lba locus) consists of a Lba gene and two other normal Lb genes (Lbc_1 and c_3) as well as a pseudogene, ψLb_1 (Fig. 2). All these genes are arranged in the same transcriptional orientation. This locus is flanked on each end by genes that appear to be expressed preferentially in leaf and root tissues. The second locus consists of the Lbc_2 and another pseudogene, ψLb_2. The two truncated genes are found on two different genomic fragments not linked to each other. One feature of all Lb containing regions is the presence of a common sequence at the 3' end of the region. This sequence may be involved in duplication of these loci.

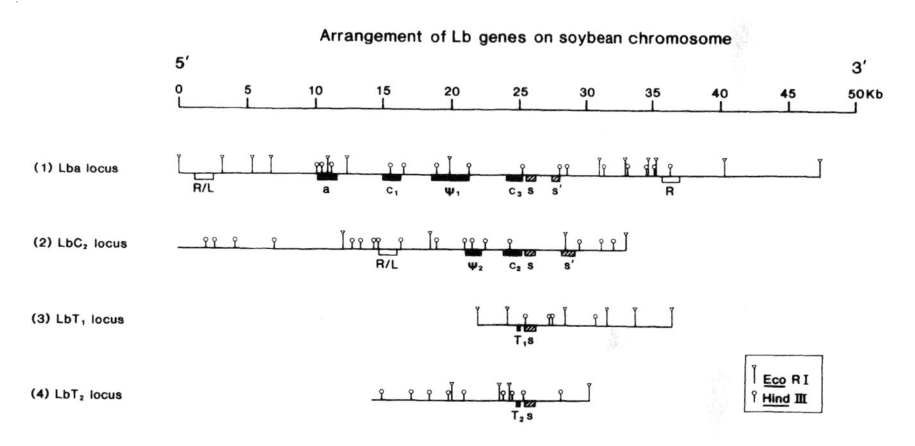

Figure 2. Organization of four Lb containing regions in soybean chromosome R, sequence expressed in root, R/L sequence expressed in root and leaf, s and s' are two repeat elements of which s is common to all four regions (see Lee et al. 1983).

Some evidences suggest that the expression of the Lb genes in the main locus may be temporally regulated; Lbc_3 being transcribed before Lba. The Lba protein eventually predominates in mature nodules (Verma et al. 1974; 1979). The specific sequences on these genes that act as cis-acting regulatory elements have not yet been identified. However, a comparison with mammalian globin genes shows (Brown et al. 1983) that Lb genes have all the potential regulatory sequences found in animal globin genes including a sequence known to be involved in enhanced transcription in vivo and located -100 bp upstream of the transcription initiation site (Dierks et al. 1983).

NODULIN GENES

Nodulins have been identified as products of plant mRNAs in soybean (Legocki and Verma 1980) and as nodule accumulation products in pea (Bisseling et al. 1983), in soybean (Legocki and Verma 1979; Verma et al. 1983b) and in alfalfa (Ausubel personal communication). The mode of induction of these proteins and their function (except that of nodulin-35) is largely unknown. In order to isolate nodulin genes, we prepared a cDNA library to nodule mRNA and screened the library for moderately abundant, nodule-specific sequences. This led to the identification of five nodulin clones in addition to leghemoglobin coding clones (Fuller et al. 1983; Fuller and Verma, 1983). These nodulin sequences encode proteins ranging from 23.5 to 100 kilodaltons (Table 1) which are immunoreactive with antibodies against nodule proteins.

TABLE I. SOYBEAN NODULE-SPECIFIC SEQUENCES

Sequence	Mole Fraction mRNA*	Size of Genomic EcoRI Fragments (kb)	Translation Product (MW x 10^{-3})
Leghemoglobin	12.0	(0.6 to 11.5)	16
Nodulin-44 (pNodA15)	6.0	7.0	44
Nodulin-23.5-24.5 (pNodA25)	-	7.0**	23.5/24.5
Nodulin-27 (pNodB45)	0.8	5.8	27
Nodulin-24 (pNodC60)	0.5	11.6	24
Nodulin-100 (pNodD24)	0.5	2.9 & 3.4	100-120

* These values are based upon relative hybridization slopes (Fuller *et al.* 1983).

** This is the same size fragment as for Nodulin-44 but these two nodulins are encoded on different sized *Hin*dIII fragments (Nguyen and Verma unpublished).

Hybridization of cDNA clones representing Lb and five nodulin species to RNA isolated from effective and ineffective soybean nodules (Figure 3) showed that all these nodulin sequences are present very early in both effective and ineffective nodules; however, their accumulation levels are much different in the ineffective nodules. Two genetically different, effective strains of *R. japonicum* (61A76 vs USDA110) appear to have a different affect on the early accumulation of nodulin as well as leghemoglobin mRNAs. The consequence of this, if any, on the ability of nodules to begin nitrogen fixation needs to be explored. These results suggest that the induction of nodulins following invasion of the host by specific *Rhizobium* is a part of the nodule developmental program which is moderated by *Rhizobium*.

Cross-hybridization experiments with RNA from *Phaseolus* nodules indicated that nodulin -100 may be present in *Phaseolus* as well while nodulins-44,-27, 24 and-24.5-23.5 may be specific to *Glycine max* (Fuller and Verma 1983). The genes encoding these nodulin sequences have been isolated from a soybean genomic library (Katinakis, Nguyen, Purohit and Verma unpublished data). In contrast to the Lb genes which are found to be linked (Lee et al. 1983), these nodulin genes do not appear to be linked to each other or to Lb loci at this stage of analysis.

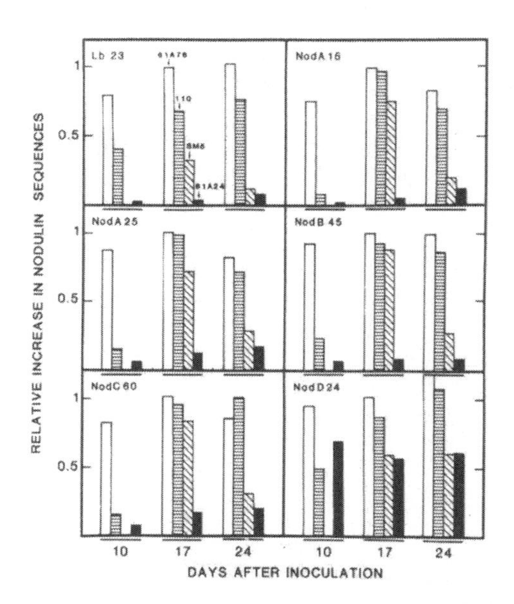

Figure 3. Accumulation of various nodulin mRNAs in effective and ineffective nodules at different stages of development.

Nodulin-35

An analysis of the nodule-specific host proteins revealed a protein with an apparent MW of 35,000 in soybean nodules (Legocki and Verma 1979). Since a major part of the enzymatic activity in the nodule is carbon and nitrogen metabolism, it was anticipated that some nodulins may encode specific enzymes involved in these assimilation pathways (Verma et al. 1983b; Fuller et al. 1983). Consequently, assays of major enzymes which are known to be induced in nodule tissues led to the identification of uricase activity associated with nodulin-35 (Jochimson and Rasmussen 1983; Verma et al. 1983b). This enzyme (uricase II) is nodule-specific, and we have localized it in the uninfected cells of the soybean nodules (Fig. 4). This location is consistent with the role of this enzyme in ureide biosynthesis which has been postulated to take place in peroxisomes of uninfected cells (Schubert 1981; Hanks et al. 1983; Newcomb and Tandon 1981). Nodulin-35 is induced very early after infection; however, significant uricase II activity was not detected until the commencement of nitrogen fixation (Bergmann et al. 1983). Thus, nodulin-35 may undergo some modifications before it is active as uricase II. This enzyme consists of four subunits, has an optimum pH of 9.5, contains Cu and exhibits some unusual physical properties and amino acid composition. Recent studies on Phaseolus nodules have also revealed a nodule-specific glutamine synthetase (Cullimore et al. 1983), an enzyme

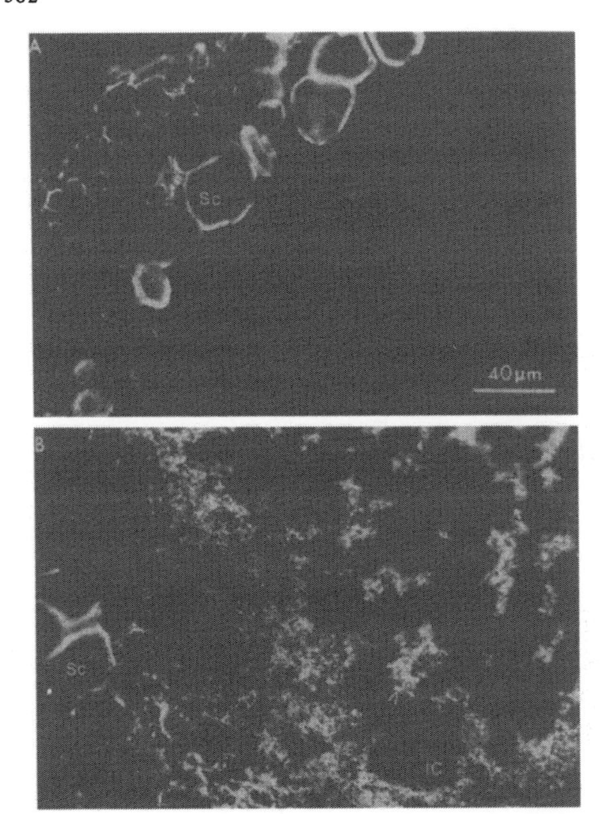

Figure 4. Localization of nodulin-35 (uricase II) in soybean root nodules. Note that the fluorescence is specific to the uninfected cells (arrowhead). Sclerenchyma (Sc) and some cells of the outer cortex autofluoresce (see Bergmann et al. for details).

which appears to be located in the infected cells of the nodules (Shelp et al. 1983). It would thus appear that even though many enzymatic activities required for ammonia assimilation already exist in other tissues, legumes have evolved separate set(s) of genes encoding these activities for their specific use in symbiotic nitrogen fixation. Whether this reflects a requirement for coordinate regulation of these and other nodulin and leghemoglobin genes or subtle differences in required enzymatic activities (as is suggested by the difference in pH optimum for the two uricase forms), remains to be determined.

Attempts to assign functions to other nodulins are being made via a variety of approaches which include the expression of the individual nodulin clones in E. coli followed by the preparation of antibodies which can be used for enzyme inhibition studies of various enzyme activities known to be induced in nodules and for subcellular localization of nodulins.

PERSPECTIVES

Nodule development can be divided into three overlapping phases, invasion, nodule organogenesis and achievement of effective symbiotic nitrogen fixation. While success at the first phase, invariably leads to the second, the second phase does not ensure effectiveness (see Verma and Nadler 1983). Therefore, a coordination of both host and Rhizobium gene expression is essential during the second phase in order to develop an effective symbiotic state. It is during the initiation of the second phase that induction of leghemoglobin and nodulin genes appears to take place. Since some nodulins are localized in one or the other kind of cells, the function of the nodule will be affected if nodulins are not present or induced at appropriate levels in the proper cell types. Nodule organogenesis is directly affected by phytohormones and the type of Rhizobium invading the host. The cells of this organ may remain uninfected (pseudonodule) partly or fully infected and yet remain ineffective. In a fully effective nodule, no more than 50% of the cells are infected. This suggests that uninfected cells have a specific role to play in this organ. Nodulin-35 (uricase II) has been specifically localized in these cells. It is also now apparent that irrespective of the nodule effectiveness, both nodulin and leghemoglobin genes, two major groups of nodule-specific host genes, are induced as a part of the nodule developmental program (see Fig. 3 and Verma et al. 1981b). The level of induction may vary with the type of Rhizobium and the extent of infection as well as the effectiveness of the nodule (Legocki and Verma 1980; Auger and Verma 1981; Fuller and Verma 1983). A lesion resulting in an ineffective nodule need not be apparent at a morphological level. For example, the process of transformation of Rhizobium into bacteroids may be directly affected and this could explain why some fully competent strains are unable to form effective nodules on other varieties or species of plants. Thus, specific markers for nodule development are essential for our understanding of effective nodule symbiosis. Detailed analysis of leghemoglobin and nodulin genes may provide us some clues about the way in which these genes are coordinately regulated during nodule development. Attempts to induce these genes by substances other than infection will also be valuable in understanding the mechanism of regulation of these genes in root nodule symbiosis. Finally, the understanding of the function of nodulins is of paramount importance if the symbiosis is to be analyzed at a molecular level.

ACKNOWLEDGEMENTS

This study was supported by grants from the NSERC of Canada and FCAC Quebec. We wish to thank Dr. M.G. Zelechowska for her assistance in immunological studies and Miss Y. Mark for the preparation of this manuscript.

REFERENCES

Appleby CA, Tjepkema JD and Trinick MJ (1983) Science 220, 951-953.
Auger S and Verma DPS (1981) Biochemistry 20, 1300-1306.
Baulcombe D and Key J (1980) J. Biol. Chem. 255, 8907-8913.
Bergmann H, Preddie E and Verma DPS (1983) The EMBO Jour. (submitted).

Bisseling T, Been C, Klugkist J, van Kammen A and Nadler K (1983) The EMBO Jour. 2 961-966.
Brisson N and Verma DPS (1982) Proc. Natl. Acad. Sci. USA 79, 4055-4059.
Brown G, Lee JS, Brisson N and Verma DPS (1983) J. Mol. Evol. (submitted).
Cullimore JV, Lara M, Lea PJ and Miflin BJ (1983) Planta 157, 245-253.
Dierks P, van Doyen A, Cochran MD, Dobkin C, Reiser J and Weissman C (1983) Cell 32, 695-706.
Fuller F, Künstner PW, Nguyen T and Verma DPS (1983) Proc. Natl. Acad. Sci. USA 80, 2594-2598.
Fuller F and Verma DPS (1983) Plant Mol. Biol. (submitted).
Holl FB and LaRue TA (1976) In Proc. First Int. Symp. on Nitrogen-Fixation Vol. II (Newton WE and Nyman CJ, eds) Washington State University Press, Pullman, Washington. pp. 391-399.
Hanks JF, Schubert KR and Tolbert NE (1983) Plant Physiol. 71, 869-873.
Jochimsen B and Rasmussen D (1983) In Molecular Genetics of the Bacteria-Plant Interactions (Pühler A. ed). Springer-Verlag, Berlin Heidelberg (in press).
Lee JS, Brown G and Verma DPS (1983) Nucl. Acids Res. (in press).
Legocki RP and Verma DPS (1979) Science 205, 190-193.
Legocki RP and Verma DPS (1980) Cell 20, 153-163.
Newcomb EH and Tandon SR (1981) Science 212, 1394-1396.
Nutman PS (1981) In Current Perspectives in Nitrogen Fixation. Gibson AH and Newton WE, eds). Aust. Acad. Sci., Canberra. p. 194.
Schubert KR (1981) Plant Physiol. 68, 1115-1122.
Shelp BJ, Atkins CA, Storer PJ and Canvin DT (1983) Arch. Biochem. Biophys. (in press).
Sullivan D, Brisson, N. Goodchild B, Verma DPS and Thomas DY (1981) Nature 289, 516-518.
Verma DPS and Long S (1983) In International Review of Cytology (Jeon K, ed), Suppl. 14, pp. 211-245, Academic Press, New York.
Verma DPS, Haugland R, Brisson N, Legocki R and Lacroix L (1981a) Biochim. Biophys. Acta 653, 98-107.
Verma DPS, Legocki RP and Auger S (1981b) In Current Perspectives in Nitrogen Fixation (Gibson AH and Newton WE, eds). The Aust. Acad. Sci., Canberra. pp. 205-208.
Verma DPS and Nadler K (1983) In Genes Involved in Plant-Microbe Interactions (Verma DPS and Hohn T, eds) Springer-Verlag (in press).
Verma DPS, Bewley JD, Auger S, Fuller F, Purohit S and Künstner P (1983) In Genetic Engineering: Application to Agriculture (ed) Owens LD, Academic Press pp. 236-245.
Verma DPS, Ball S, Guérin C and Wanamaker L (1979) Biochemistry 18, 476-483.
Verma DPS, Nash DT and Schulman HM (1974) Nature (London) 251, 75-77.
Verma DPS, Bergmann H, Fuller F and Preddie E (1983b) In Molecular Genetics of the Bacteria-Plant Interactions (Pühler A, ed). Springer-Verlag, Berlin, Heidelberg (in press).
Whittaker RG, Lennox S and Appleby CA (1981) Biochem. Int. 3 117-124.
Wiborg O, Hyldig-Nielsen JJ, Jensen EO, Paludan K and Marcker KA (1982) Nucl. Acids Res. 10, 3487-3494.

Nitrogen and Carbon Assimilation in *Medicago sativa* Mediated by Host Plant Genes for Nodule Effectiveness.

C.P. Vance, G.H. Heichel, D.K. Barnes, USDA-ARS and the Department of Agronomy and Plant Genetics, University of Minnesota, St. Paul, MN 55108.

Root nodule formation is mediated by both the host plant and *Rhizobium* genomes (Nutman, 1981; Holl and LaRue, 1976; Vance, 1983). Numerous studies with *Rhizobium* have led to substantial progress in identification of specific areas on either plasmids or chromosomes which regulate infection, nodule formation, nitrogenase expression and hydrogenase development (Ausubel, 1982; Beringer et al., 1979; Brewin et al., 1980, Nuti et al., 1979). In contrast to the proliferation of genetic information regarding bacterial control of root nodule symbiosis, relatively little progress has been made in understanding plant genes that regulate symbiosis. Several specific genes are known to control root nodule formation and effectiveness in pea, clover, soybeans and alfalfa (Holl and LaRue, 1976; Nutman, 1956; Vest and Caldwell, 1972; Peterson and Barnes, 1981). Yet, the physiological and biochemical manifestations of these genes are poorly understood. Studies with soybean (Verma et al., 1981; Legocki and Verma, 1979) and pea (Bisseling, this volume) have shown that plant genes code for nodule specific proteins (nodulins). However the physiological and biochemical function of most of these nodulins is not understood.

Our approach to studying genetic control of nodulation in alfalfa has been to: 1) find naturally occurring non-nodulating or ineffectively nodulating lines; 2) examine the inheritance of those traits; 3) describe the anatomical, physiological and biochemical manifestions of those traits; and 4) utilize them in field and greenhouse studies to ascertain how carbon and nitrogen assimilation are interrelated in the nodule and in the intact plant.

At least five genes in alfalfa designated in_1-in_5 condition ineffective nodule formation [Table 1] (Peterson and Barnes, 1981). All genes controlling ineffectiveness are recessive with tetrasomic inheritance. These plants are ineffective with all strains of *R. meliloti*

Table 1. Designation, proposed genotype and nodule phenotype of ineffective and non-nodulating alfalfa genotypes.

Designation	Genotype	Phenotype
Mn Saranac (In)[1]	$in_1in_1in_1in_1$	numerous, large pale
Mn Agate (In)		numerous, small white
MnNC-3226 (In)	$in_2in_2in_2in_2$	tumor-like
MnNC-3811 (In)	$in_3in_3in_3in_3$	tumor-like
MnPL-480 (In)	$in_4in_4in_4in_4in_5in_5in_5in_5$	tumor-like
MnNC-1008 (NN)[1]	$nn_1nn_1nn_1nn_1nn_2nn_2nn_2nn_2$	non-nodulating

[1]In = ineffective nodules, NN = no nodules

tested to date. A non-nodulation trait (NN-1008) is conditioned by two tetrasomically inherited recessive genes (nn_1 and nn_2). The in_1 genotype has numerous small nodules relatively normal in external morphology with early senescence of bacteroids (Vance and Johnson, 1983). In contrast, the in_2 - in_5 genotypes have normal invasion and release of bacteria with premature senescence of bacteria. Nodules then continue to proliferate forming a tumor-like structure. Although we have utilized all of the genotypes, our most extensive studies have been with Mn Saranac (In) [in_1] because it appears most normal in nodule morphology and displays herbage and root characteristics similar to those of effective plants.

The ineffective genotypes and NN-1008 have been utilized in field studies to assess which genotype(s) might be most useful as non-fixing controls for field measurements of N_2 fixation. The effective control had significantly higher herbage dry weight, N concentration and N yield than the ineffective genotypes and NN-1008 (Table 2). Significant differences were noted between some ineffective genotypes and NN-1008 for herbage dry weight and N yield. However, the herbage N concentrations

Table 2. Yields of herbage and N, and the N concentration of herbage, for one normal and six non-N_2-fixing genotypes of alfalfa averaged over four harvests of hill plots in 1980. Values are means $\pm$ one s.e. of five replicates.

Designation	Herbage Dry Weight	Herbage N Concentration	Herbage N Yield
	(g/plant)	(%)	(mg/plant)
Saranac (Eff)	1.61 $\pm$ 0.33	3.51 $\pm$ 0.10	56.1 $\pm$ 12.0
MnPL-480 (In)	0.08 $\pm$ 0.02	2.86 $\pm$ 0.11	2.2 $\pm$ 0.6
MnNC-3811 (In)	0.13 $\pm$ 0.05	2.50 $\pm$ 0.20	3.1 $\pm$ 1.4
MnNC-3226 (In)	0.32 $\pm$ 0.09	2.81 $\pm$ 0.18	9.0 $\pm$ 2.6
Mn-Saranac (In)	0.21 $\pm$ 0.08	2.57 $\pm$ 0.18	5.5 $\pm$ 2.3
Mn-Agate (In)	0.17 $\pm$ 0.06	2.70 $\pm$ 0.21	4.6 $\pm$ 1.9
MnNC-1008 (NN)	0.28 $\pm$ 0.10	2.97 $\pm$ 0.15	8.2 $\pm$ 2.9

of these genotypes were similar. These data suggest that the non-N_2-fixing genotypes differ in the uptake of soil N and in the efficiency of herbage production under conditions of limiting N. Additional data indicate that ineffectiveness and non-nodulation traits were accompanied by differences in correlated traits that involve plant growth and development other than through nitrogenase expression (Barnes et al., 1983). The genotype MnPL-480(In) exemplifies this in that it has shorter internodes, a prostrate stature, and is prone to flower abcission compared to effectively nodulating alfalfa. Interestingly, MnPL-480(In) ineffectiveness is controlled by two genes, and it is the poorest performer of all the non-fixing genotypes (Table 2).

We also evaluated the recovery of ^{15}N-labeled fertilizer from soil by the non-fixing genotypes and reed canarygrass. There were significant differences between genotypes for uptake of ^{15}N-labeled fertilizer. All alfalfa genotypes were more efficient at ^{15}N uptake than reed canarygrass. The uptake of ^{15}N by a non-fixing control species is used to calculate the amount of N derived from symbiosis by the isotope dilution technique. Our data suggest that overestimates or underestimates of N derived from symbiosis are likely depending upon the non-fixing control utilized in the experiment.

The observation that some non-fixing genotypes had alterations in metabolism was substantiated by studies of nodule enzymes associated with C and N metabolism (Table 3). In contrast to actively-fixing effective nodules, nodule GS, GOGAT, and PEPC specific activities in ineffective

Table 3. Comparisons of nitrogenase and ammonia assimilating enzymes between effective and ineffective alfalfa genotypes.

Designation	Nitrogenase[1]	GS	GOGAT	GDH	PEPC
	(% effective control)	--------(nmoles·min^{-1} mg protein^{-1})----			
Saranac (Eff)	100	85 ± 9	90 ± 11	65 ± 5	636 ± 50
Mn Saranac (In)	0	25 ± 4	18 ± 2	84 ± 12	68 ± 20
Mn Agate (In)	0	35 ± 6	7 ± 1	34 ± 8	119 ± 30
MnNC-3226 (In)	0	34 ± 5	8 ± 1	74 ± 9	34 ± 7
MnNC-3811 (In)	0	12 ± 2	4 ± 1	95 ± 8	83 ± 10

[1]Nitrogenase is based on acetylene reduction activity after 120 min incubation period.

[2]Each value is mean ± SE of three determinations.

nodules were very low. These results support previous studies of lupine (Rawsthorne et al., 1980), soybeans (Werner et al. 1980) and alfalfa (Groat and Vance, 1981) which indicated that GS and GOGAT are the major enzymes of ammonia assimilation in nodules. GDH is ubiquitious in legume nodules and a definite role for this enzyme in N metabolism has not been established. High specific activities of GDH in ineffective alfalfa nodules indicate, however, that GDH is not associated with assimilation of symbiotically fixed N. Nodule PEPC was greatly reduced when compared with effective nodules, suggesting that PEPC may be integrally involved in active N_2-fixation. These results demonstrate the close link between bacteroid N_2-fixation and nodule cytosol ammonia and carbon assimilation. These enzymes may be components of the nodulins described by Verma et al., 1981. Our data on total protein in ineffective nodules indicate that effectiveness requires a substantial increase in either the activity or the presence of many proteins. These alfalfa genotypes might provide ideal systems for studying and identifying nodulins and determining their importance to both nodule morphology and effectiveness. Our ineffective genotypes might also provide a useful source for isolating plant genomic components regulating symbiosis.

Resolution of the importance of nodule CO_2 fixation to symbiotic N_2-fixation can be approached by comparing nodule CO_2 fixation and transport of nodule-fixed CO_2 in both effectively and ineffectively nodulated plants (Table 4).

Table 4. Nodule CO_2 fixation, respiration, acetylene reduction and xylem sap characteristics of ineffectively (MnSa[In]) and effectively (Saranac) nodulated alfalfa.[1/]

Variable		MnSa(In)	Saranac
Nodule CO_2 fixation	($\mu g \cdot kg^{-1} \cdot s^{-1}$)	96[1/]	394[2/]
Nodule respiration	($mg \cdot kg^{-1} \cdot s^{-1}$)	0.5	4
Acetylene reduction	($nmol \cdot mg^{-1} \cdot h^{-1}$)	0	172
Asparagine + Aspartate in xylem sap	($nmol \cdot ml$ xylem sap^{-1})	150	6500
Nodule fixed $^{14}CO_2$	($dpm \cdot \mu l$ xylem sap^{-1})	16	121
Percent of xylem sap radioactivity in amino acids	(% of total)	3[3/]	82

[1/]Adapted from Maxwell et al. 1983.

[2/]Each value is the mean of 3 replicates.

[3/]Remainder is in organic acid fraction.

Nodule CO_2 fixation and respiration are much higher in effective than in ineffective nodules (Table 4). Active nitrogenase as measured by acetylene reduction was not detectable in the ineffective nodules. These data indicate that nodule CO_2 fixation is closely associated with functionally effective nodules. Also, the high rates of effective nodule respiration indicate that internal nodule CO_2 concentrations are high enough to maintain the CO_2 fixing enzyme phosphoenolpyruvate carboxylase at saturating levels. If we assume a 3:1 ratio for acetylene reduction to N reduction, then nodule CO_2 fixation may provide up to 25% of the carbon required for the assimilation of fixed N (Vance et al., 1983).

Evaluation of xylem sap of effective and ineffective plants after exposing intact nodulated roots to $^{14}CO_2$ demonstrated that: 1) the major transport products of effective alfalfa are asparagine and aspartate; 2) xylem sap radioactivity is much higher in effective nodules than in ineffective nodules; and 3) the major portion of radioactivity in effective xylem sap was in the amino acid fraction while in xylem sap of ineffective plants the major radioactive fraction was organic acids (Table 4). The data show that CO_2 fixed by effective nodules is incorporated into amino acids for transport in the xylem sap. The data further support a direct connection between nodule CO_2-fixation, N_2-fixation and N assimilation. The observed CO_2 fixation in ineffective nodules supports a role for CO_2 fixation being involved in aspects of nodule metabolism other than N-assimilation. These may include pH stat (Davies, 1979) nodule maintenance (Rawsthorne et al., 1980) and ion balance (Israel and Jackson, 1982).

The availability of ineffectively nodulating alfalfa has also provided a unique opportunity to investigate the role of nodule effectiveness in the interorgan partitioning of current photosynthate (Boller and Heichel, 1983), and in canopy structure and photosynthesis under field conditions (Boller and Heichel, 1984). Pulse-chase experiments with $^{14}CO_2$ were used to measure the allocation of photosynthate to nodules and to four other plant parts by utilizing the concept of relative specific activity, RSA (Mor and Halevy, 1980).

Significant differences among organs in photosynthate allocation were apparent during vegetative growth and at 5% flower (Fig. 1). Across the

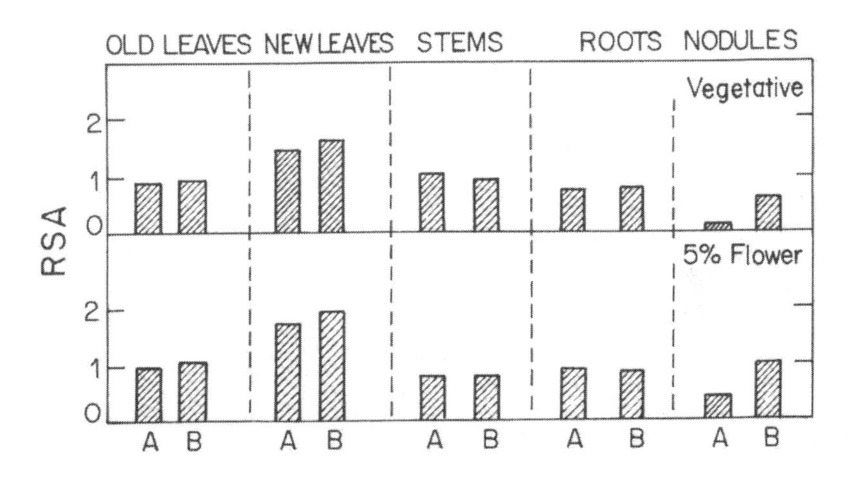

Figure 1. Relative specific activity (RSA) of five organs of ineffectively nodulating alfalfa, Mn Saranac [In] (A), and effectively nodulating MnNC alfalfa (B) 24 h after pulse-labeling shoots of intact field-grown plants with $^{14}CO_2$. Measurements are for the vegetative (upper panel) and 5% flower (lower panel) growth stages of the second harvest-regrowth cycle. Adapted from Boller and Heichel (1983).

ineffectively and effectively nodulating populations, the RSA of new leaves was always significantly greater than that of any other organ. Old leaves, stems, and roots had a similar priority for photosynthates at both growth stages. The RSA of nodules was significantly less than that of all other organs for vegetative plants and is similar to that of old leaves, stems, and roots at 5% flower.

Two features of photosynthate partitioning in response to differences in nodule effectiveness were noteworthy. First, allocation of photosynthate to nodules of ineffectively nodulating alfalfa was greatly reduced at both growth stages in comparison to nodules of effectively nodulating alfalfa. In the absence of N_2 fixation, the RSA of nodules of ineffectively nodulating alfalfa probably reflected the partitioning of recent photosynthate for use in maintenance of the nodule in the absence of the energy needs of functional bacteroids.

Second, the presence of functional bacteroids in nodules resulted in substantially increased photosynthate partitioning to nodules, but the presence of effective nodules did not change the priority rankings for recent photosynthate among the five organs of effectively nodulating alfalfa (Fig. 1, A vs B). This shows that even effective and ineffective alfalfa nodules are relatively weak sinks for recent photosynthate. The results clearly show that recent photosynthates are not as important a source to sustain the comparatively high carbohydrate requirements of nodules as they are for other sinks.

Canopy structure and leaf area distribution were affected more by ineffective nodulation than were rates of apparent photosynthesis per unit leaf area (Fig. 2). Ineffectively nodulating alfalfa grown in soil low in available N had mean leaf N concentrations 40% less than those of effectively nodulating alfalfa. Nevertheless, rates of apparent photosynthesis were similar in upper sections of canopies (nos. 3, 4, & 5, Fig. 2) lacking mutual shading among leaves. Addition of 100 kg/ha supplemental N decreased leaf photosynthesis in both geneotypes. Thus, ineffectively nodulated alfalfa maintained rates of apparent photosynthesis similar to those of normal, effectively nodulated alfalfa at the expense of leaf area growth.

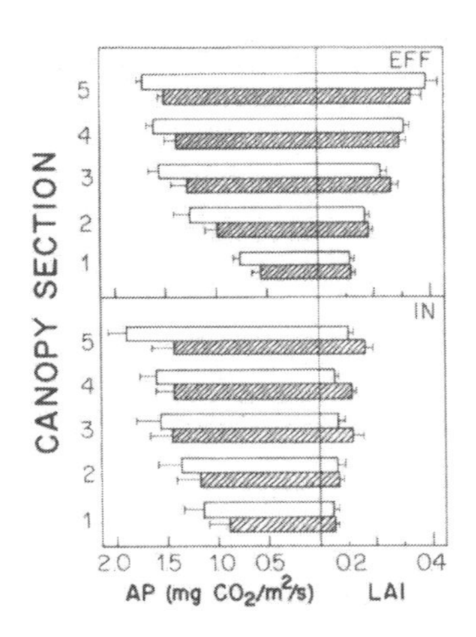

Figure 2. Rate of apparent photosynthesis per leaf area (AP) and leaf area index (LAI) in five canopy sections (1=base, 5=top) of effectively (EFF) and ineffectively (IN) nodulating alfalfa grown at zero (open bar) and 100 kg/ha (hatched bar) supplemental N. Values are averaged across the three growth-harvest cycles of the season for vegetative canopies 2 weeks before harvest. Adapted from Boller and Heichel (1984).

Our studies have demonstrated that plants selected for genes controlling effectiveness are useful as tools to evaluate plant physiological and biochemical constraints to effective symbiosis and growth from the cellular to the crop community level. These ineffective plants displayed biochemical and morphological alterations in a number of characteristics. This implies that genes for effectiveness are either closely linked to genes for a number of correlated traits, or that ineffectiveness may result in a cascade effect by which other plant parameters are altered. The alfalfa genotypes described in these studies might also be useful to molecular biologists in assessing plant genes and proteins that regulate nodule formation and function.

REFERENCES

Ausubel FM (1982) Cell 29, 1-2
Barnes DK et al. (1983) Agron. Abstracts. p. 59
Berringer JE et al. (1979) Proc. R. Soc. (London) Ser. B. 204, 219-233
Boller BC and Heichel GH (1983) Crop Sci. 23, in press
Boller BC and Heichel GH (1984) Crop Sci. 24, in press
Brewin NJ et al. (1980) J. Gen. Microbiol. 1203, 413-420
Davies DD (1979) Ann. Rev. Plant. Physiol. 30, 131-158
Groat RG and Vance CP (1981) Plant Physiol. 67, 1198-1203
Holl FB and LaRue TA (1976) In Newton WE and Nyman CJ eds, 1st Int. Symp. Nitrogen Fixation, pp 391-399, Washington State Univ. Press
Israel DW and Jackson WA (1982) Plant Physiol. 69, 171-178
Legocki RP and Verma DPS (1979) Science 205, 190-193
Maxwell CA et al. (1983) Crop Sci. 24, in press
Mor Y and Halevy AH (1980) Plant Physiol. 66, 990-995
Nuti MP et al. (1979) Nature 282, 533-535
Nutman PS (1956) Biol. Rev. Cambridge Phil. Soc. 31, 109-51
Nutman PS (1981) In Gibson AH and Newton WE eds., Current Perspectives in Nitrogen Fixation, pp 194-205, Austr. Ac. Sc., Canberra
Rawsthorne, S et al. (1980) Phytochemistry 19, 341-355
Vance CP (1983) Ann. Rev. Microbiol. 37, in press
Peterson MA and Barnes DK (1981) Crop Sci. 21, 611-616
Vance CP et al. (1983) Plant Physiol. 72, 469-473
Vance CP Johnson LEB (1983) Can. J. Bot. 61, 93-106
Verma, DPS et al. In Gibson AH and Newton WE eds. Current Perspectives in Nitrogen Fixation pp. 205-209, Austr. Ac. Sc., Canberra
Vest G and Caldwell BE (1972) Crop Sci. 12, 692-694
Werner D et al. (1980) Planta 147, 320-329

THE SOYBEAN LEGHEMOGLOBIN GENES

K.A. MARCKER/K. BOJSEN/E.Ø. JENSEN/K. PALUDAN
DEPARTMENT OF MOLECULAR BIOLOGY AND PLANT PHYSIOLOGY, UNIVERSITY OF AARHUS, DK-8000 ÅRHUS C, DENMARK

INTRODUCTION

Leghemoglobins (Lbs) are myoglobin-like proteins synthesized exclusively in the root nodules which develop through the symbiotic association of Rhizobium with legumes. The function of the Lbs is to facilitate the diffusion of oxygen at a low oxygen tension, thus making the oxygen available for metabolic processes without inactivating the nitrogenase which is very oxygen-sensitive (Bergersen et al., 1973).

Soybean nodules contain four major components of Lbs denoted Lba, Lbc_1, Lbc_2 and Lbc_3, respectively, and some less characterized minor components (Fuchsman, Appleby, 1979). The minor components are most likely post-translational modification products of some of the major components. The differences in the amino acid sequences among the various Lb components are small and correspond to about 10 amino acids (Sievers et al., 1978).

The Structure of the Lb Genes

The Lbs are encoded in the soybean genome by a small family of genes (Sullivan et al., 1981; Marcker et al., 1981). An Lb cDNA clone hybridizes to at least six EcoR1 genomic restriction fragments which all have been isolated from various soybean DNA libraries. Subsequent DNA sequence analysis of the isolated clones revealed that four genes coded for Lba, Lbc_1, Lbc_2 and Lbc_3, respectively (Hyldig-Nielsen et al., 1982; Wiborg et al., 1982; Brisson, Verma, 1982). A fifth Lb gene has an unusual number of replacement site substitutions in the coding sequence and corresponds to no known soybean Lb. In addition the 5' flanking sequence of this particular Lb gene is mutated in two regions which are of importance for transcription, and may thus be an Lb pseudogene (ψ_1Lb) (Wiborg et al., 1983). A sixth isolated Lb gene has a stop codon in the second exon and is therefore a pseudogene (ψ_2Lb). In addition one or two truncated Lb genes have been isolated (Brisson, Verma, 1982; Bojsen et al., 1983).

All Lb genes are interrupted by three intervening sequences (IVS) at identical positions corresponding to codon 32 (IVS-1), codons 68-69 (IVS-2) and codons 103-104 (IVS-3) (Fig. 1). The positions of IVS-1 and IVS-3 correspond precisely to the psitions of the two intervening sequences found in all α-, β- and myo-globin genes sequenced so far (Jeffreys, 1981; Blanchetot et al., 1983). This supports the hypothesis that all globin genes including the Lb genes are derived from a common ancestral gene. The position of IVS-2 in Lb genes is in agreement with a hypothesis proposed by Gō (1981) which suggests that the central exon in animal globin genes might be the result of a fusion between two ancestral exons with a division somewhere between amino acids 66-71. All

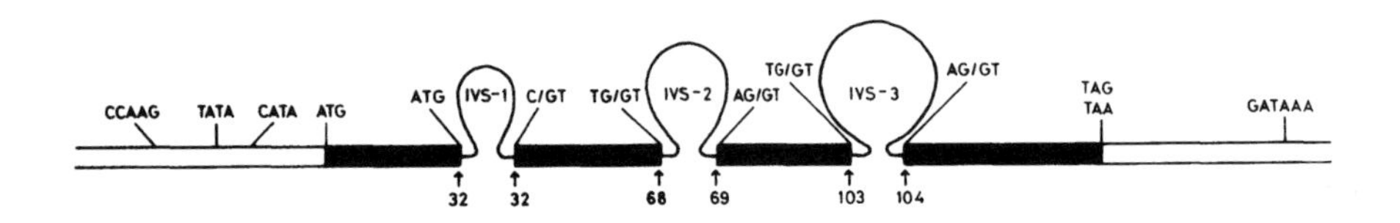

FIGURE 1. General DNA sequence organization of the LB genes. Solid boxes indicate coding sequences. Open boxes indicate flanking regions, while single lines indicate intervening sequences. Numbers refer to the number of the corresponding amino acid in the protein sequence. The partial DNA sequences above the gene are putative regulatory sequences. "CAT" box (CCAAG), "TATA" box (TATA), initiator codon (ATG), intron/exon junctions (GT/AG), terminator codons (TAA, TAG) and a putative Poly(A) addition signal (GATAAA).

complete Lb genes contain putative regulatory signals identical to or very similar to the corresponding signals in other eukaryotic genes. This implies that the mechanisms for gene transcription in plants are very similar to those found in other eukaryotes.

The Chromosomal Arrangement of the Soybean Leghemoglobin Genes

The chromosomal organization of the Lb genes has been determined by means of genomic blotting analysis and isolation of overlapping clones from two independent soybean genomic libraries (Bojsen et al., 1983). Six Lb genes (four functional and two pseudogenes) are arranged in two independent clusters.

Four genes are very closely linked in the order: 5' Lba - Lbc_1 - ψ_1Lb - Lbc_3 3' (cluster I), while two other genes are linked in the order: 5' ψ_2Lb - Lbc_2 3'(cluster II). The distances between the genes in both clusters are about 2-3 kb (Fig. 2).

No clone linking the two clusters have so far been obtained. DNA sequences extending 11-15 kb at each side of the two clusters do not contain Lb sequences and the distance between the two clusters, if placed on the same chromosome, must thus be at least 25-30 kb.

The ψ_2Lb and Lbc_2 genes (cluster II) are located within a 10 kb EcoR1 restriction fragment. During phage propagation the ψ_2Lb gene is very often deleted leaving a phage with a 6 kb fragment containing only the

Lbc_2 gene. The Lbc_2 gene was therefore isolated on a 6 kb EcoR1 fragment (Wiborg et al., 1982). An Lb cDNA clone does hybridize to a 6 kb EcoR1 fragment in genomic blots of soybean DNA. This genomic 6 kb EcoR1 fragment now seems to be the result of incomplete EcoR1 digestion of soybean DNA. The EcoR1 restriction site between the 4.2 kb EcoR1

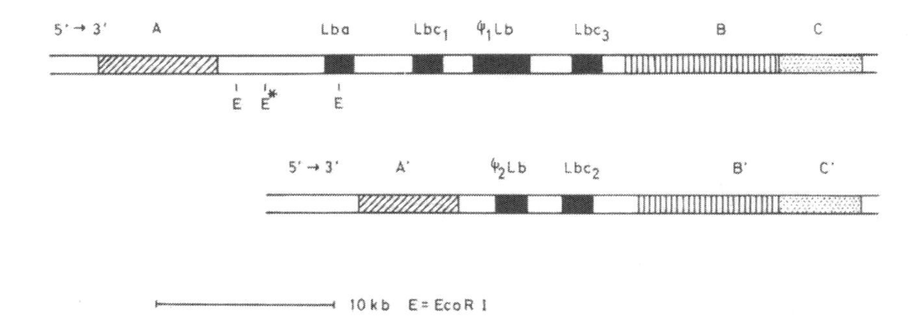

FIGURE 2. Chromosomal arrangement of six soybean Lb genes. Solid boxes indicate the position of the Lb genes (the three introns present in all Lb genes are not shown). Identical shadings (A and A', B and B', C and C') represent cross hybridizing non-Lb regions. The transcriptional polarity of the Lb genes is indicated by the two arrows. The EcoR1 restriction sites proximal to the Lba gene are indicated. E^* denotes the EcoR1 site which is difficult to cleave in soybean genomic DNA.

fragment containing the 5' end of the Lba gene and a 1.6 kb EcoR1 fragment located 5' to the 4.2 kb fragment is partly resistant to EcoR1 digestion (Fig. 2). EcoR1 digestion of genomic soybean DNA thus gives a 6 kb and a 4.2 kb fragment both containing the 5' end of the Lba gene.

Non-Lb Genes are Present in Each Lb Gene Cluster in Corresponding Positions

Molecular hybridization experiments revealed extensive cross hybridization between several DNA regions outside the Lb genes in the two clusters (Bojsen et al., 1983) (Fig. 2). A DNA region (A) in cluster I cross hybridizes strongly to a corresponding region (A') in cluster II. Furthermore sequences in regions B and C in cluster I cross hybridize strongly to sequences in regions B' and C'. Southern blotting experiments showed that DNA sequences from regions B and B' are repeated 10-20 times in the soybean genome while the DNA sequences in regions A and C are only present in the A' and C' regions of the soybean genome. Thus regions B and B' contain a repetitive DNA sequence, while the DNA sequences present within regions A (A') and C (C') probably encode structural sequences. Northern blotting analysis using subcloned DNA fragments from regions A and C demonstrated hybridization to specific mRNAs in poly A^+ RNA extracted from soybean roots, leaves and nodules, while no transcripts could be detected in any tissue when using a subcloned DNA fragment from region B'. No amplification of mRNA encoded in region A and C could be detected in nodules, indicating that the infection of roots with Rhizobia has no effect on the expression of these genes. Consequently DNA sequences within regions A, A', C and C' most likely encode structural soybean genes regulated in a manner different from that of the Lb genes.

The presence of two very similar Lb gene clusters in the soybean genome indicates that a duplication of a large area of the chromosome has occurred. Soybean is a tetraploid plant (Datta, Saha, 1973) and the two Lb gene clusters may therefore have arisen by genome duplication in an ancestral soybean species. However, it cannot be ruled out that only a duplication of a part of the chromosome containing the Lb genes has occurred. Another example of the occurrence of two independent globin gene clustes has been found in the tetraploid organism Xenopus laevis (Jeffreys et al., 1980). A diploid relative, Xenopus tropicalis, has only one cluster, thus indicating that the occurrence of two globin gene clusters in the tetraploid organism is due to tetraploidization. We have recently suggested (Jensen et al., 1983) that the Lb genes arose by single duplication events followed by a genome duplication. Subsequently a recombination event in one of the clusters resulted in the formation of a fused Lb gene (Lbc_2) in addition to an unequal distribution of Lb genes in the two clusters.

Regulation and Expression of the Lb Genes

Animal globin genes are differentially regulated during development (Jeffreys, 1981). The general rule is that these genes are arranged in the order of their expression during development. An exception to this rule is the chicken β-globin gene cluster, where embryonic genes are placed both 5' and 3' to the adult genes (Villeponteau, Martinson, 1981). Sequences within or very close to the animal globin structural sequences are important for the regulation of the differential expression of the genes within the clusters (Weatherall, Clegg, 1982; Chao et al., 1983).

Changes in relative Lb contents in soybean nodules during development indicate that Lb genes are also induced at different times during nodule development (Fuchsman, APpleby, 1979; Verma et al., 1979, 1981). The synthesis of Lbcs precedes the synthesis of Lba, and it is likely that the activation of the Lbc_3 gene precedes the activation of the Lbc_1 gene (Fuchsman, Appleby, 1979). The synthesis of Lbc_2 probably starts at the same time as the synthesis of Lbc_3, but the amount of Lbc_2 in nodules is only half the amount of Lbc_3 (Fuchsmann, Appleby, 1979) indicating that the Lbc_2 gene is less active than the other Lb genes. It is not known at present whether activation of the Lbc_1 and Lba gene is followed by deactivation of the Lbc_3 and Lbc_1 genes, respectively, or whether all Lb genes are active during the later stages of nodule development. However, it is clear that the chromosomal arrangement of the Lb genes is in the opposite order of their expression.

In conclusion the soybean Lb genes are closely linked to other soybean genes that are regulated in a way different from that of the Lb genes. In addition the latter genes most likely are differentially regulated during nodule development. This implies that the soybean DNA element(s) which respond to the Rhizobial signal activating the Lb genes most likely are located close to the structural Lb sequences.

Lb Gene Activation and Methylation Variance

Most active eukaryotic genes are hypomethylated when compared to the corresponding inactive genes (Felsenfeld, McGhee, 1982). Unfortunately the soybean Lb gene region contains very few restriction sites that are sensitive to methylation, thus making it difficult to detect a possible hypomethylation during activation of these genes. There are for example only 4 MspI and 4 AvaI sites close to the Lb genes (Fig. 3). Southern blotting analysis of nodule and seedling DNA digested with the following restriction enzymes XhoI, HaeIII, HhaI, BamH1 and PvuII using an Lb cDNA clone as hybridization probe gave essentially similar results. However, digestions with three enzymes: Sau3A, MspI and AvaI of nodule and seedling DNA indicated methylation variance in the Lb gene region. Thus

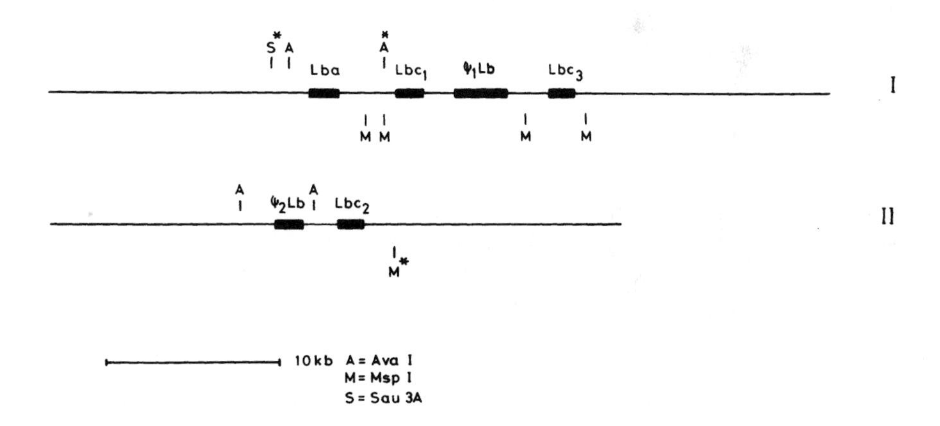

FIGURE 3. Physical map of the two Lb gene clusters showing locations of the MspI and AvaI sites. Lb genes are indicated by solid boxed. Asterisks indicate the restriction sites which are hypomethylated in nodule DNA but not in seedling DNA. S* is the Sau3A site which is hypomethylated in nodule DNA.

a Sau3A site located 5' to Lba, a MspI site 3' to Lbc_2 and probably an AvaI site 5' to Lbc_1 are hypomethylated in nodule DNA when compared to seedling DNA (Fig. 3). These results therefore indicate that in the soybean Lb gene region an inverse correlation between the level of methylation and the transcriptional activity of a gene exists in analogy with other eukaryotic genes.

Acknowledgements

We thank Drs. R. Goldberg and R. Fischer, UCLA, for providing the limited EcoR1 and Alu/HaeIII soybean libraries. This research was supported by grants from De Danske Sukkerfabrikker A/S and the Danish Natural Science Research Council.

References

Bergersen FJ, Turner GL and Appleby CA (1973) Biochim. Biophys. Acta 292, 271-282.
Blancetot A, Wilson V, Wood D and Jeffreys AJ (1983) Nature 301-732-734.
Bojsen K, Abildsten D, Jensen EØ, Paludan K and Marcker KA (1983) EMBO J. 2, 1165-1168.
Brisson N and Verma DPS (1982) Proc. Natl. Acad. Sci. USA 79, 4055-4059.
Chao MV, Mellon P, Charnay P, Maniatis T and Axel R (1983) Cell 32, 483-493.
Datta PC and Saha N (1973) Genét. Ibér. 25, 37-62.
Felsenfeld G and McGhee J (1982) Nature 296, 602-603.
Fuchsman WH and Appleby CA (1979) Biochim. Biophys. Acta 579, 314-324.
Gō M (1981) Nature 291, 90-92.
Hyldig-Nielsen JJ, Jensen EØ, Paludan K, Wiborg O, Garrett R, Jørgensen P and Marcker KA (1982) Nucl. Acids Res. 10, 689-701.
Jeffreys AJ (1981) In Williamson R, ed, Genetic Engineering, vol 2, pp. 1-48, Academic Press, London.
Jeffreys AJ, Wilson V, Wood D, Simons JP, Kay RM and Williams JG (1980) Cell 21, 555-564.
Jensen EØ, Hein J, Paludan K and Marcker KA (1983) UCLA Symposia in Molecular and Cellular Biology.
Marcker KA, Gausing K, Jochimsen B, Jørgense P, Paludan K and Truelsen E (1981). In Panopoulos NJ, ed, Genetic Engineering in the Plant Sciences, pp. 63-71. Praeger Publishers, New York.
Sievers SG, Huhtala, M-L and Ellfolk N (1978) Acta Chem. Scand. B32, 380-386.
Sullivan D, Brission N, Goodchild B, Verma DPS and Thomas DY (1981) Nature 289, 516-518.
Verma DPS, Ball S, Guérin C and Wanamaker L (1979) Biochemistry 18, 476-483.
Verma DPS, Haugland R, Brisson RN, Legocki RP and Lacroix L (1981) Biochim Biophys Acta 653, 98-107.
Villeponteau B and Martinson H (1981) Nucl. Acids Res. 9, 3731-3746.
Weatherall DJ and Clegg JB (1982) Cell 29, 7-9.
Wiborg O, Hyldig-Nielsen JJ, Jensen EØ, Paludan K and Marcker KA (1982) Nucl. Acids Res. 10, 3487-3494.
Wiborg O, Hyldig-Nielsen JJ, Jensen EØ, Paludan K and Marcker KA (1983) EMBO J. 2, 449-452.

EXPRESSION OF NODULIN GENES DURING NODULE DEVELOPMENT FROM EFFECTIVE AND INEFFECTIVE ROOT NODULES

TON BISSELING/FRANCINE GOVERS/RITA WYNDAELE/JAN-PETER NAP/JAN-WILLEM TAANMAN and ALBERT VAN KAMMEN
DEPARTMENT OF MOLECULAR BIOLOGY, AGRICULTURAL UNIVERSITY, DE DREIJEN 11, 6703 BC WAGENINGEN, THE NETHERLANDS

1. INTRODUCTION

Root nodule formation in the Rhizobium-legume symbiosis is the result of a series of plant-Rhizobium interactions, in which each induces gene expression in the other. One of the steps in this process is the release of rhizobia into the cytoplasm of host cells, whereafter the infected cells differentiate into non-dividing ones, which form the nodule tissue. Furthermore, the rhizobia differentiate into the characteristic bacteroid forms. During the formation of a root nodule several host genes are specifically induced. Immunological techniques have been used to detect the occurrence of nodule specific polypeptides, the so-called nodulins in soybean and pea root nodules (Legocki, Verma 1980; Bisseling et al., 1983). Leghemoglobins (Lbs) are of course an example of nodule specific proteins, but many more (about 25) nodulins have now been identified. These findings were confirmed by hybridization analysis of nodule mRNA, which indicated the presence of several nodule specific mRNA sequences (Auger et al., 1981).
In this paper we describe the use of antisera raised against purified nodulins and the use of nodule specific cDNA clones respectively, to study nodulin gene expression at the protein and mRNA level during nodule development of effective and ineffective root nodules.

2. RESULTS

Pea nodulins were identified with a nodule-specific antiserum. This preparation was made from an antiserum raised against total proteins from the cytoplasmic fraction of pea root nodules by titration with proteins from uninfected roots. After titration it showed only a faint reaction with root proteins, whereas in the protein pattern from root nodules several polypeptides still showed a distinct reaction (Bisseling et al., 1983).
Nodulin patterns were analysed with this antiserum preparation at different stages of development. Cytoplasmic plant proteins from root nodules and uninfected roots were isolated, separated by SDS-gel electrophoresis and blotted onto nitrocellulose. The blots were incubated with the nodule specific antiserum and then, with ^{125}I protein A to visualise the specific immune complexes of nodulins with the respective antibodies (for details see Bisseling et al., 1983). An example of such a blot is shown in Fig. 1. By this method we have been able to identify about 30 nodulins, with molecular weights ranging from 120 K to 15 K. The nodulins were numbered according to size, N-1 being the largest nodulin detected. The smallest nodulin, N-30, is Lb. We concluded that some nodulins e.g. N-3 and N-8 are detectable before Lb in the course of nodule development while others appeared at the same time as Lb and accumulated similarly. Furthermore we could demonstrate that some nodulins are located in the plant cytoplasm (e.g. Lb) whereas others are specifically found within the peribacterial membrane (Bisseling et al., 1983).
The nodule specific antiserum used, however, is too complex to allow study of the expression of individual nodulin genes. Therefore we have prepared antisera against purified nodulins and we isolated several nodule specific cDNA clones.

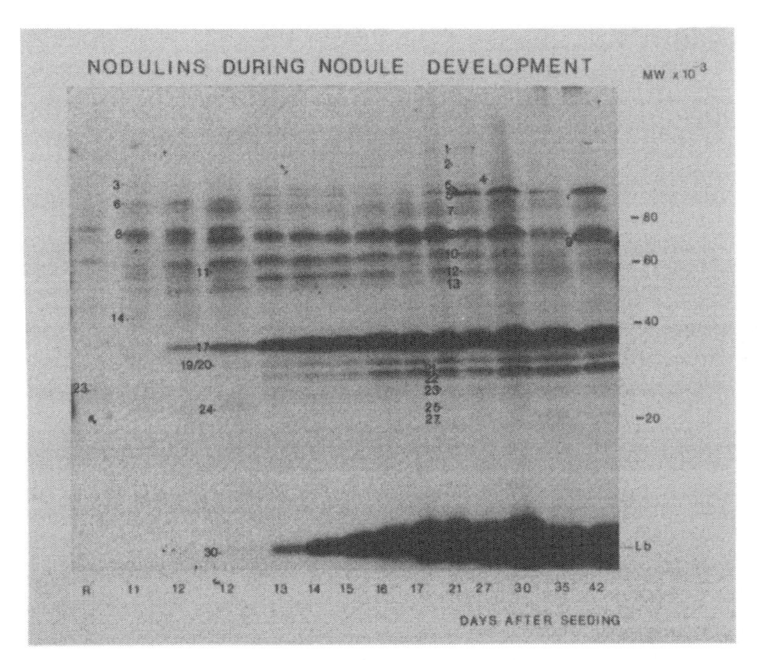

FIGURE 1. Autoradiograph of a protein blot of soluble pea proteins from uninfected roots and root nodules of pea and R.leguminosarum (PRE) harvested at different times after seeding as indicated. The blot was incubated with a nodule specific antiserum preparation and [^{125}I]protein A.

2.1. Anti-nodulin sera

Nodulins were purified by $(NH_4)_2SO_4$ precipitation, DEAE chromatography and preparative gel electrophoresis. The antisera were raised in rabbits. (Details will be published elsewhere, F. Govers et al., in preparation).
In this paper we will present the results obtained with three different antisera. One of the antisera we used was anti-Lb serum, in order to be able to compare the synthesis of other nodulins with that of Lb. Another antiserum was raised against N-19, a nodulin located in the peribacteroid space (Bisseling et al., 1983). The third antiserum was raised against a fraction containing three nodulins N-17a, N-27 and N-29 (Fig. 2). As it became clear during purification that more than one nodulin migrated at the position of N-17 we have numbered this one N-17a. N-17a and N-29 are cytoplasmic nodulins, whereas N-27 is found in the peribacteroid space (Fig. 2). None of these nodulins could be detected in uninfected roots.

2.2. Nodule development

In the development of an effectively nitrogen fixing nodule several steps can be distinguished, e.g. nodule initiation - infection-thread branching - bacterial release - induction of Lb genes - induction of nif genes. In our system the first visible root nodules appear 10 days after seeding and Lb was first detected at day 12, followed one day later by nitrogenase (Bisseling et al., 1980).
Root nodules were harvested at 13, 16, 20 and 24 days after seeding, whereas at days 6 and 9, pieces of main root, where root nodules were expected to be formed, were collected. Equal amounts of total protein from the host cytoplasm were separated on polyacrylamide gels and protein blots were incubated with anti-Lb

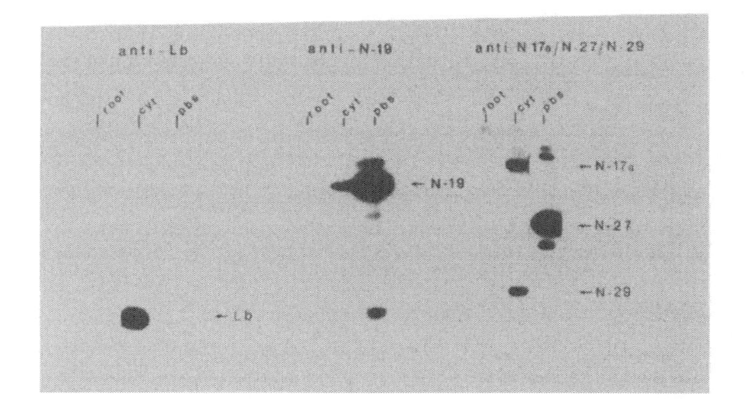

FIGURE 2 Autoradiograph of protein blots containing proteins from root nodules of pea - R.leguminosarum (PRE) and uninfected roots. Root nodule proteins were fractionated in proteins located within the peribacteroid membrane and the host cytoplasm. The protein blots were incubated with anti-Lb, anti-N-19 and anti-N-17a/N-27/N-29 as indicated, which was followed by $[^{125}I]$protein A.

and the anti-nodulin sera.

In Fig. 3 the appearance of N-17a, N-19, N-27, N-29 and Lb during the course of nodule development is shown. Lb appears at day 13 and accumulates during further nodule development. The two peribacteroid space proteins N-19 and N-27 both behave in a similar way as Lb, whereas the course of appearance of N-17a and N-29 markedly differs from that of Lb. N-29 can already be identified in infected roots at 6 days after seeding; it accumulates till day 17 and then its concentration starts to decrease. N-17a appears at day 17, so the gene is expressed after the Lb genes during nodule development.

We selected these four nodulins because of their different locations in the cell and their different time course of appearance during nodule development. Two of them are examples of respectively an "early" and a "late" nodulin occurring in the cytoplasm, whereas the two other nodulins are located within the peribacteroid membrane and appear at the same time as Lb in the course of nodule development.

The results obtained with the immunological detection of nodulins on protein blots strongly suggest that the different nodulins appear in the order indicated. However, exact quantification of the respective nodulin concentration may be essential to prove this order of appearance.

2.3. Ineffective root nodules

Nodulin gene expression was also studied in ineffective nodules, which are formed upon inoculation with nod^+fix^- R.leguminosarum strains. Three different strains were used:

- R.leguminosarum (PRE)2(Tn5::nif D) is a mutant of the parental strain R.leguminosarum (PRE), with Tn5 inserted in nif D (Van den Bos et al., 1983). This mutant differentiates into normal y-shaped bacteroids. With antisera against the nitrogenase components a small amount of CII can be detected whereas CI was not detectable (Fig. 4)
- R.leguminosarum (1062)116(pop^-) is a mutant of R.leguminosarum 1062, obtained from K. Nadler. It is pop^- which has the effect that the bacteroids of 116 excrete only a low amount of heme (Nadler, 1981). This mutant differentiates into long rods and y-shaped bacteroids, containing a reduced amount of CI, whereas CII is below the detection level. The wild type strains PRE and 1062

induce similar nodulin patterns (Bisseling _et al._, 1983).
P8 is a wild nod^+fix^- isolate which is not capable to differentiate into bacteroids with a characteristic shape. With CI and CII antisera no nitrogenase

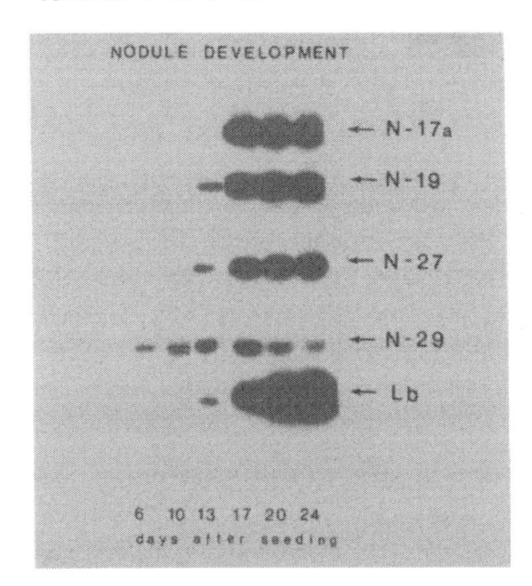

FIGURE 3 Autoradiograph of a protein blot containing soluble pea proteins from root nodules, formed after inoculation with _R.leguminosarum_ (PRE), harvested at different times after seeding as indicated. The autoradiograph shown is composed of two blots, one incubated with anti-N-19 and the other successively with anti-Lb and anti-N-17a/N-27/N-29 followed by $[^{125}I]$protein A.

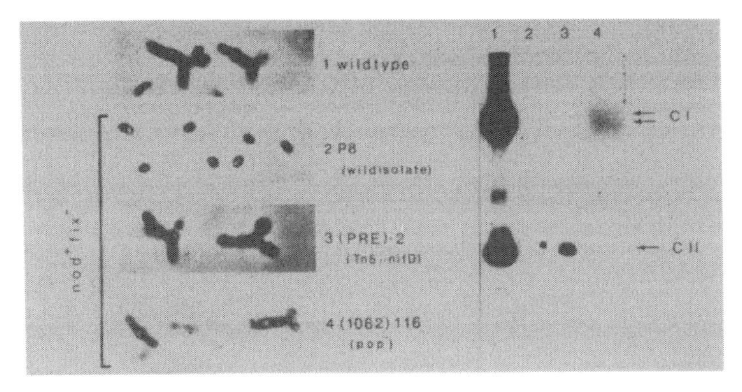

FIGURE 4 Autoradiograph of a protein blot containing proteins from bacteroids of three nod^+fix^- strains and wild type PRE. Bacteroids were isolated from 25 day old pea plants. The blots were incubated successively with anti-CI, anti-CII and $[^{125}I]$protein A.

components could be detected (Fig. 4).
The synthesis of nodulins N-17a, N-19, N-27, N-29 and Lb in the course of the development of the ineffective nodules induced by these three different nod^+fix^- strains was compared with that in effective nodules produced by _R.leguminosarum_ (PRE).

These nod$^+$fix$^-$ strains were selected as they seem to disturb nodule development at different stages. (PRE)2 is blocked in CI synthesis which is one of the last steps before nitrogen fixation can start. (1062)116 has a defect in heme synthesis which will affect the synthesis of Lb. Since Lb synthesis preceeds the synthesis of the nitrogenase components, this mutant will probably disturb nodule development at an earlier stage than (PRE)2. In contrast to these two strains P8 does not differentiate into y-shaped structures, indicating that the interaction P8 - Pisum sativum is disturbed even earlier during nodule development.
Nodulin patterns of the effective and the three types of ineffective nodules were studied at 17, 20 and 24 days after seeding and inoculation. The appearance and accumulation of Lb and the other four nodulins is shown in Fig. 5. In all three types of ineffective nodules some Lb can be detected. In (PRE)2-induced nodules Lb appears and accumulates as in the wild-type till day 20, whereafter its concentration decreases markedly. The (1062)116-induced nodules behave in a similar way but the Lb concentration decreases already after 17 days, whereas in P8-induced nodules only a reduced Lb concentration can be detected at day 17.
Markedly reduced levels of the late nodulin N-17a can be detected in (PRE)2 and (1062)116 nodules, whereas in P8 nodule N-17a is absent.
N-19 is expressed in (PRE)2 nodules and it follows the Lb pattern since its concentration decreases after 20 days. N-19 cannot be detected in (1062)116 and P8 nodules.
The N-27 gene is expressed in (PRE)2 and (1062)116 nodules and follows the respective Lb patterns, whereas in P8 nodules N-27 is not found.
N-29 is absent in all three ineffective nodules.

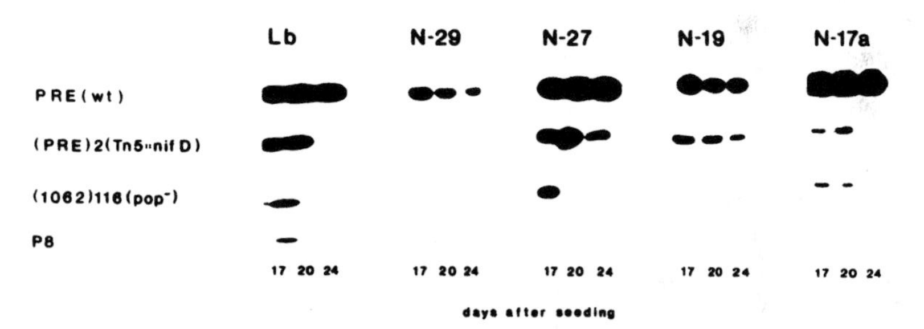

FIGURE 5 Autoradiographs of protein blots containing soluble pea protein from root nodules formed after infection with three nod$^+$fix$^-$ R.leguminosarum strains and the wild type PRE respectively, harvested at different times after seeding as indicated. These blots were incubated with anti-Lb, anti-N-19 and anti-N-17a/N-27/N-29, respectively, followed by [^{125}I]labeled protein A.

2.4. Nodule specific cDNA clones

A cDNA library was constructed from poly(A)+ RNA from pea root nodules (de Vries et al., 1982) in pBR322 by the dG-dC homopolymer tailing method as described by Land et al. (1981). Details about the construction will be published elsewhere (F. Govers et al., in preparation, see also poster 8-8, these Proceedings). By differential hybridization we selected 30 nodule specific clones which did not hybridize to a previously obtained Lb cDNA clone. This Lb-cDNA clone (pPsLb101) was identified with anti-Lb serum after hybrid selected translation and by comparing the nucleotide sequence with the aminoacid sequence of

pea Lbs (Lehtovaara et al., 1980). From the 30 clones, we have so far studied two clones: pPsNod201 and pPsNod301. By hybridization it was shown that none of the other selected clones contained similar nucleotide sequences.
Northern blots of poly(A)+ RNA from root nodules and from uninfected pea roots hybridized with pPsNod201 and pPsNod301 as a probe confirmed that the two clones contained DNA inserts complementary to nodule specific mRNA, just as is the case for Lb mRNA (Fig. 6). pPsNod 201 hybridized to mRNA with a molecular weight of 2 x 10^5 D, similar in size to Lb mRNA, whereas pPsNod301 hybridized to mRNA with a molecular weight of 6 x 10^5 D.

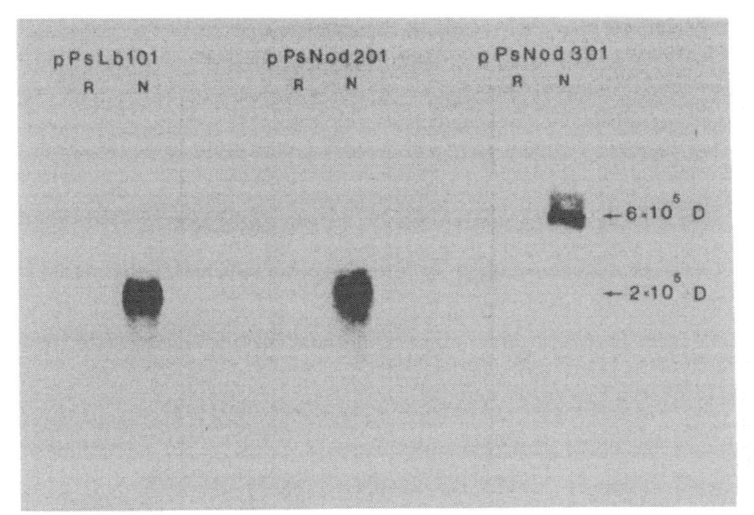

FIGURE 6 Autoradiograph of Northern blots of poly(A)+ RNA from uninfected roots and root nodules, harvested at 15 days after seeding. Blots were hybridized with pPsLb101, pPsNod201 and pPsNod301.

Southern blots of EcoRI digests of genomic pea DNA and total DNA of free living rhizobia hybridized with pPsNod201 and pPsNod301 further demonstrated that both clones contain sequences which are pea genome borne and contain DNA copies of mRNA transcribed from pea DNA. Whereas at least 8 EcoRI restriction fragments of pea DNA hybridized with pPsLb101, illustrating that, comparable to soybean (Sullivan et al., 1980; Jensen et al., 1981), the pea genome carries a small family of Lb genes, only one EcoRI fragment hybridized with pPsNod201 indicating that the gene corresponding to pPsNod201 is present in a single copy (Fig. 7).

3. DISCUSSION

With anti-nodulin sera and cDNA clones it appears possible to study the expression of nodulin genes. We have not yet determined for which nodulins the nodule specific cDNA clones are coding, but we intend to characterize these clones using the anti-nodulin sera. The fact that different nodulins become detectable at different times after infection, indicates differential expression of nodulin genes, which confirms that nodule formation can be divided into different stages. One of the nodulins we have studied, N-29, appears already 6 days after seeding and inoculating the peas which implies that the gene for this nodulin is induced shortly after the infection has occurred. The fact that its

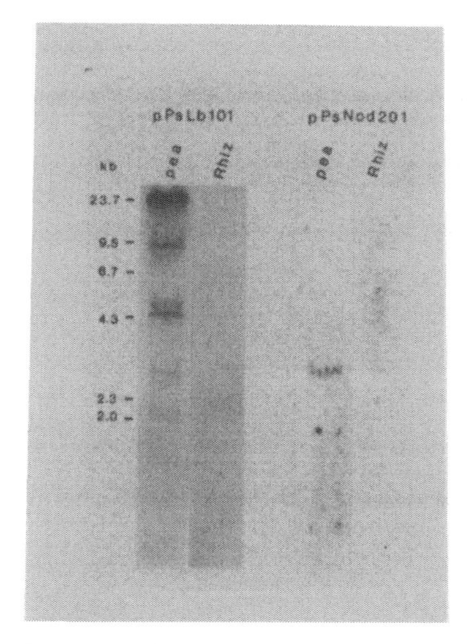

FIGURE 7 Autoradiograph of Southern blots of total *R.leguminosarum* DNA and genomic pea DNA digested with EcoRI. The blots were hybridized with pPsLb101 and pPsNod201.

concentration decreases after 16 days indicates that it will be mainly involved in the early steps that lead to effective nodule formation.
We have studied ineffective nodules which arise after inoculation with three different nod^+fix^- *R.leguminosarum* strains. These strains were chosen since they seem to be blocked at different stages of development. P8 nodules appear to be disturbed in an earlier stage of the development of an effective nitrogen fixing root nodule than PRE(2) nodules which might have a defect in one of the final steps of effective nodule development.
Our analysis of the ineffective nodules of each of these strains confirms to a certain extent this order of disturbance in nodule development. The four nodulins we studied are absent in P8 nodules, whereas two of these are present in nodules produced by (1062)116 and even three after inoculation with (PRE)2. Furthermore, the decrease in concentration of N-27 and Lb in (1062)116 nodules preceeds that in (PRE)2 nodules. These results suggest that nodule formation is much stronger disturbed upon infection with strain P8 than with strain (PRE)2. In all ineffective nodules the early nodulin N-29 is absent, which is very curious. It means that mutations in genes essential for heme or nitrogenase synthesis repress expression of this early nodulin gene. However, we cannot yet exclude, that the gene for N-29 is expressed before day 13 in ineffective nodules.
The function of most nodulins is still unclear. Some nodulins appear to be involved in NH_3 metabolism: nodulin-35 (Legocki, Verma, 1979) in soybean nodules probably is the enzyme uricase (Jochemsen, Rasmussen, 1982) and in *Phaseolus* a nodule specific form of glutamine synthetase has been identified (Cullimore *et al.*, 1982). We think that the analysis of the course of nodule development from effective and ineffective nodules will be a step toward the elucidation of the functions of nodulins.

4. ACKNOWLEDGEMENTS

We thank I. ten Velde and Th. Schetgens for their contribution to some of the experiments, P. Madern for making illustrations and M. van Neerven and J.C. Toppenberg-Fang for typework. This work is financially supported by a grant from the Netherlands Organisation for Biological Research (B.I.O.N.) to F.G. and a fellowship from the OECD to R.W.

5. REFERENCES

Auger S and Verma DPS (1981) Biochemistry 20, 1300.
Bisseling T, Moen AA, Van den Bos RC and Van Kammen A (1980) J.Gen.Microbiol. 118, 377.
Bisseling T, Been C, Klugkist J, Van Kammen A and Nadler K (1983) EMBO Journal 2, 961.
Cullimore JV, Lara M, Miflin BJ (1982) In 2st Int.Symp.Mol.Genet.Bacteria-Plant interaction, Bielefeld, ed. Pühler A, p. 14.
Jensen EO Paludan K, Hyldig-Nielsen JJ, Jorgensen P and Marcker KA (1981) Nature 291, 677.
Jochimsen B and Rasmussen O (1982) In 1st Int.Symp.Mol.Genet. of the Bacteria-Plant Interaction, Bielefeld, ed. Pühler A, p. 26.
Land H, Grez M, Hauser H, Lindenmaier W, Schatz G (1981) Nucleic Acids Res. 9, 2251.
Legocki RP and Verma DPS (1979) Science 205, 190.
Legocki RP and Verma DPS (1980) Cell 20, 153.
Lehtovaara P, Lappalainen A, Ellfolk N (1980) Biochem.Biophys.Acta 623, 98.
Nadler KD (1981) In Gibson AH and Newton WE (eds.), Proceedings of the 4th International Symposium on Nitrogen Fixation, Australian Academy of Sciences, Canberra, p. 143.
Sullivan D, Brisson N, Goodchild B, Verma DPS and Thomas DY (1981) Nature (London) 289, 516.
Van den Bos RC, Schetgens ThMP, Bisseling T, Hontelez JGJ and Van Kammen A (1983) In Proc. 1st Int.Symp.Mol.Genet. Bacteria-Plant Interaction, Bielefeld, ed. Püler A, in press.

POSTER DISCUSSION 8 PLANT GENES INVOLVED IN NODULATION AND SYMBIOSIS.

A. VAN KAMMEN, Department of Molecular Biology, Agricultural University,
De Dreijen 11, 6703 BC Wageningen, The Netherlands.

Considering the great interest in the results presented in both the papers and the posters on plant genes involved in nodulation and symbiosis, several groups will likely initiate research programs on plant proteins involved in the development of root nodules. Agreement on the terminology and notations to be used in describing the results of such research is necessary to minimize confusion and misunderstanding and to facilitate the comparison of results obtained with different legume nodules as much as possible.

At the poster discussion session, there was a call for care in using similar terms for different nodule-specific proteins. On the last day of the Symposium this matter was discussed by D.P.S. Verma (Montreal, Canada), N. Lang-Unnasch (Boston, MA, USA) and T. Bisseling, F. Govers and A. van Kammen (Wageningen, The Netherlands). They agreed upon the following definitions and decided to press other researchers to use the following terminology and notations: The development of legume root nodules is dependent on genetic functions of both the _Rhizobium_ bacterium and the host plant. In root nodules, a number of _nodule-specific proteins_ are found which can be of plant as well as bacterial origin. Nodule-specific proteins can be divided into two classes:

1. _bacteroidins_ , which occur in the endosymbiotic forms of _Rhizobium_, but not in the free living bacteria; and 2. _nodulins_ encoded by host plant genes which are expressed during the development of the symbiosis. Nodulins are plant proteins found only in root nodules and not in uninfected roots or other parts of the host plant.

If, in an extract of root nodules, nodule-specific proteins are detected, it must be determined whether they are nodulins or bacteroidins. Since nodulins are plant genome borne, mRNA for nodulins will most probably be polyA(+)RNA. If, therefore, translation of polyA(+)RNA from root nodules in an _in vitro_ protein-synthesizing system results in the production of polypeptides identical to one or more of the nodule-specific proteins, this can be considered evidence for the nodule-specific proteins being nodulins. Further evidence can be obtained by cDNA cloning of mRNA from root nodules and probing the host plant genome with a specific cDNA clone for the occurrence of nodulin specific sequences.

Bacteroidins are encoded by _Rhizobium_ genes which are expressed during differentiation of free-living bacteria into bacteroids. Nodule-specific proteins can be identified as bacteroidins, if identical proteins can be shown to occur in intact bacteroids. Examples of bacteroidins are the components I and II proteins of nitrogenase, which are found in bacteroids but not in free-living _Rhizobium_ bacteria under normal free-living conditions. Bacteroidins like the nitrogenase proteins may be produced in cultures of free-living rhizobia under defined conditions imitating the endosymbiotic state.

One should be aware that the plant cytoplasmic fraction of a root nodule extract may contain bacterial proteins due to leakage or lysis of bacteroids. Such proteins contaminating the plant cytoplasmic fraction may be a mixture of bacteroidins and other bacterial proteins also occurring in free-living _Rhizobium_ bacteria. In the future, proteins encoded by _Rhizobium_ may be found in cells of root nodules outside the bacteroid, but with a specific function in the _Rhizobium_-plant symbiosis.If such proteins are demonstrated, these may be called _symbactins_. So far, however, there are no examples of nodule-specific proteins which deserve this name. In the peribacteroid space, bacterial proteins are found,

some of which are also excreted by free-living *Rhizobium* bacteria. Some of these proteins in the peribacteroid space prove to be symbactins,. But, such conclusion cannot be drawn before an essential function for them in the symbiosis has been established.
There is as yet no evidence for the occurrence of nodulins in the peribacteroid space. Leghemoglobin, the best-known example of a nodulin, has long been thought to occur within the peribacteroid membrane in close contact with the bacteroids. But there is now good evidence that leghemoglobin occurs in the plant cytoplasm and has its function there. Convincing data that leghemoglobin also occurs in the peribacteroid space are lacking.
Nodule specific proteins should be indicated by the letter N and their molecular weight as determined by SDS polyacrylamide gel electrophoresis and should not be indicated by numbers or letters. For example, N-45 stands for a nodule specific protein with a molecular weight of 45 kilodaltons; N-16 for a nodule specific protein with a molecular weight of 16 kilodaltons, etc. These notations should be used until a defined function has been assigned to the protein. If the protein appears to be a nodule-specific form of an enzyme also occurring elsewhere in the host plant, the use of the letter n as a prescript to the name of the protein is suggested. In soybean root nodules, for example, one of the nodulins previously indicated as N-35, has now shown to be n-uricase. Similarly, in root nodules of *Phaseolus vulgaris*, one of the nodulins has been proved to be n-glutamine synthetase (nGS). Leghemoglobins, which are found solely in root nodules do not require the prescript n-. This rule will also apply to other proteins, when identified, which are found only in root nodules.
Unfortunately, it has not been possible to use the proposed terminology and notation in the papers and poster abstracts published in these Proceedings. We hope, however, that the recommendations, which have come from the deliberations of D.P.S. Verma, N. Lang-Unnsal, T. Bisseling, F. Govers and myself, will be found acceptable and that authors will use the proposed terminology and notation in forthcoming papers and reviews.

SELECTION FOR, AND CHARACTERIZATION OF, DIFFERENCES BETWEEN LEGUME GENOTYPES IN THE EXTENT OF INHIBITION OF NITROGEN FIXATION BY EXTERNAL NITRATE

B.J. CARROLL/K.S. SCHULLER/D.L. MCNEIL/P.M. GRESSHOFF
BOTANY DEPT, AUSTRALIAN NATIONAL UNIVERSITY, PO BOX 4, CANBERRA, AUSTRALIA

Genetic approaches to overcome NO_3 inhibition of nodulation and nitrogen fixation have been generally confined to the micro-symbiont. These studies have implied an epistatic contribution by the host genome. This paper assesses the feasibility of selecting host genotypes for increased nodulation in the presence of NO_3.

Three clover species (Trifolium repens, T. dubium & T. subterraneum) (cultured as in Carroll & Gresshoff (1983)) showed significant variation in their ability to nodulate and fix nitrogen in the presence of NO_3 during early plant development (0-3 wks). In Hawaii, visual screening of 270 soybean cultivars for increased nodulation in NO_3-fertilized soil yielded a spectrum of phenotypes. Subsequent testing of good and poor nodulating cultivars in sand pots, indicated heritable differences in symbiotic sensitivity to NO_3. Seeds of Davis (a promising cv.) were mutagenized with rays and the M_2 screened for increased nodulation on NO_3. Two lines had increased acetylene reduction activity compared to parent cv. Davis when cultured on 8mM KNO_3. This demonstration of heritable variation provided the impetus for a larger scale attempt to isolate (Bragg) soybean mutants.

mutagen	dose	chl[-a]%
rays	25 krad	0.3
NaN_3	1.0mM, 2 hrs	0.2
EMS	.44%, 4 hrs	0.9
EMS	.5%, 6 hrs	2.8

[a] chlorophyll deficient mutants in M_2

The M_2 populations from EMS mutagenesis were used in subsequent screening. Analysis of families segregating for chlorophyll deficiency indicated that a plausible value for the genetically effective cell number in soybean is 3 or 4.

Eleven thousand M_2 seedlings were inoculated with USDA110 and scored for extent of nodulation on 4-5 mM NO_3:

Score	description	number of selections
0	nod⁻	18
1	average nodulation	-
2	marginally increased nodulation	43
3	substantially increased nodulation	3

Siblings and progeny of the three class 3 variants are being retested.

Eleven thousand seedlings were screened for lack of constitutive NO_3 reductase activity (NRA) in the primary leaves and several selections were made. One M_2 family, T-356, had depressed activity at the primary-leaf growth stage under glasshouse conditions (mean temp, 17°C). However, when plants were grown to the same stage in a growth cabinet (28°C) constitutive NRA wasn't significantly different from Bragg. Siblings of two other selected lines showed reduced activity ($^2/47$ for C-345 and $^2/8$ for C-258)

20,000 M_2 plants were screened for lack of nodulation in the field. 54 plants were selected, reinoculated and grown to seed. Two of these were nod⁻ at harvest and both of these produced nod⁻ progeny in Leonard jars.

Carroll, B., P. Gresshoff (1983) Z. Pflanzenphysiol. Bd., 110. S., 77-88.

MOLECULAR CLONING OF PLANT GLUTAMINE SYNTHETASE FROM PHASEOLUS ROOT NODULES

JULIE V. CULLIMORE and B. J. MIFLIN
Biochemistry Department, Rothamsted Experimental Station, Harpenden, Hertfordshire, U.K.

cDNA clones have been prepared from mRNA from the plant fraction of *Phaseolus* root nodules and a clone coding for glutamine synthetase (GS) has been preliminarily identified. Poly A^+ RNA was prepared from root nodules of *Phaseolus vulgaris* and size fractionated on dimethyl sulphoxide and formamide containing sucrose density gradients. The fractions containing mRNA coding for GS were identified by *in vitro* translation and immunoprecipitation of the products with specific anti-GS-antiserum [1]. cDNA was prepared from these fractions, made double stranded and inserted into the plasmids pUC8 and pBR322. *Escherichia coli* was transformed and over 1000 recombinant clones were obtained. These clones were screened by colony hybridization using as a probe labelled mRNA highly specific for GS which was obtained by immuno-purification of polysomes. A clone which hybridized strongly to this probe produced a single polypeptide band identical in mobility to authentic GS on hybrid release translation. This clone hybridized to nodule, root and leaf mRNA of *Phaseolus*. In Southern blot analysis a copy number of at least two genes per genome were estimated for *Phaseolus* and hybridization was also observed to DNA from soybeans, alfalfa and peas.

[1] Cullimore JV and Miflin BJ (1983) FEBS Letters 158, 107-112.

CHARACTERISTICS OF NODULE AND ROOT GLUTAMINE SYNTHETASES OF *PHASEOLUS VULGARIS*

JULIE V. CULLIMORE, M. LARA, P. J. LEA and B. J. MIFLIN
Biochemistry Department, Rothamsted Experimental Station, Harpenden, Hertfordshire, U.K.

Phaseolus vulgaris nodules contain two separable glutamine synthetases (GS_{n1} and GS_{n2}) in the plant cytoplasm; one of these (GS_{n1}) is specific to nodule tissue whereas GS_{n2} resembles the form of GS present in uninoculated roots (GS_r). Both forms have been purified to apparent homogeneity and their properties examined. They have many characteristics in common, with similar K_m or $S_{0.5}$ values for the synthetase-reaction substrates and identical molecular weights for the holoenzyme (380,000) and subunits (41,000) but they differ in their behaviour on ion-exchange chromatography and native polyacrylamide gel electrophoresis and in the ratio of transferase to synthetase activity [1]. The two forms can also be differentiated immunologically [2]

Free-living *Rhizobium phaseoli* also contain two forms of GS both of which have different molecular weights from the plant enzymes. Bacteroids contain only the higher molecular weight form of rhizobial GS. During nodulation, GS activity increased 16 fold and this is due almost entirely to the production of GS_{n1}; GS_{n2} activity remains almost constant [3]. This increase in GS_{n1} occurs concurrently with the synthesis of nitrogenase and leghaemoglobin. In non-fixing nodules produced by infection with mutants of *R. phaseoli* only trace amounts of GS_{n1} and leghaemoglobin were detected.

[1] Cullimore, J. V., Lara, M., Lea, P. J., and Miflin, B. J. (1983) Planta 157, 245-253.
[2] Cullimore, J. V. and Miflin, B. J. (1983) J. Exp. Bot.(in press)
[3] Lara, M., Cullimore, J. V., Lea, P. J., Miflin, B. J., Johnston, A. W. B., and Lamb, J. W. (1983) Planta 157, 254-258.

BREEDING FOR A HIGHER NITROGEN FIXATION ABILITY IN FABA BEAN (*VICIA FABA* L.): METHODOLOGY STUDIES

G. DUC/H. PATRIAT/J. PICARD[1]/A. MARIOTTI[2]
(1) STATION D'AMELIORATION DES PLANTES, BV 1540, 21034 DIJON CEDEX FRANCE
(2) LABORATOIRE DE GEOLOGIE DYNAMIQUE, UNIVERSITE PIERRE ET MARIE CURIE
4, PLACE JUSSIEU 75230 PARIS CEDEX 05 FRANCE

Several techniques were tested in order to be introduced in a breeding program aiming at improving N_2 fixation of leguminous plants.

1. ^{15}N dilution technique with a low enrichment of the soil (A. MARIOTTI, N. AMARGER, 1979) was tested at the level of the field.

2. Acetylene reducing activity (ARA) was measured on plants extracted from the field several times during growing season.

3. Glasshouse experiments were performed during winter. Growth of innoculated cultivars was compared under 0 meq or 12 mq of mineral nitrogen in the nutrient solution.

4. We have been looking for morphological characters of plants linked to fixating ability.

^{15}N Technique appeared of interest as it was in good correlation with ARA and it allowed to detect genetic variability with a good precision. However, this technique is associated to some problems : its cost, the need for a regular enrichment of soil in ^{15}N in space and time, the need of good control plant and the need of higher precision to discriminate among high fixating cultivars.

Results of glasshouse experiments were not so much consistant with field results. Some limiting factor as light could explain these discrepancies.

A good positive correlation between yield and % of fixed N_2 was measured. However this correlation has to be controled on a shorter scale of yields among high yielding cultivars.

PROTEIN SYNTHESIS IN THE PLANT FRACTION OF LUPIN AND PEA NODULES

K.J.F. FARNDEN/N.S. FERNANDO/M.R. GRANT/A.L. JOHNSON/V.S. NYONI AND C.W. RONSON[a]
DEPARTMENT OF BIOCHEMISTRY, UNIVERSITY OF OTAGO, DUNEDIN, N.Z. AND
[a]GRASSLANDS DIVISION, DSIR, PALMERSTON NORTH, N.Z.

Our studies of enzymes and proteins in the plant fraction of developing legume nodules demonstrated changes in the levels of leghemoglobin and the enzymes involved in ammonia assimilation which correlated with the appearance of the bacteroid nitrogenase activity (Robertson and Farnden, The Biochem. of Plants 5, 65-113, 1980). We have now initiated a study of the synthesis of these proteins in the plant fraction of legume nodules. A total RNA fraction has been prepared from *Lupinus angustifolius* and *Lupinus luteus* developing nodules and roots using a guanidine hydrochloride extraction procedure. These RNA preparations have been translated *in vitro* using a wheat germ cell-free system and the protein products analysed on SDS-polyacrylamide gels and detected by fluorography. *In vivo* labelled proteins have been analysed similarly.
A C_4-dicarboxylate transport mutant of *Rhizobium leguminosarum* has been isolated using the procedures of Ronson, Lyttleton and Robertson, PNAS 78, 4284-88, 1981. Assays of nodule extracts from the mutant-inoculated plants indicated that leghemoglobin and glutamine synthetase were present at very low levels or absent (compared to wild-type); while the specific activities of malate dehydrogenase and phosphofructokinase were slightly lower than in the wild-type inoculated nodules. We are currently isolating and translating mRNA from both the effective and ineffective nodules.

EXPRESSION OF HOST SPECIFIC SEQUENCES DURING DEVELOPMENT OF ROOT NODULES IN PEA.

Francine Govers, Jan-Peter Nap, Ton Bisseling, Rita Wyndaele and Albert van Kammen.
Department of Molecular Biology, Agricultural University, De Dreijen 11, 6703 BC WAGENINGEN, The Netherlands.

In root nodules obtained after infection of leguminous plants with Rhizobium a number of host-encoded nodule-specific proteins, nodulins, are found. Previous studies of the symbiosis between pea (Pisum sativum) and Rhizobium leguminosarum have shown that during nodule development about 30 nodulins can be detected by means of a nodule specific antiserum (Bisseling et al., 1983). Some of these nodulins appear before nitrogen fixation starts, whereas the majority is synthesized during stages of active nitrogen fixation.

In order to study the expression of the corresponding genes involved in nodule development of effective and ineffective root nodules we constructed cDNA clones of nodule specific sequences (I) and raised antisera against some purified nodulins (II).

I. PolyA(+)RNA was isolated from active pea root nodules, harvested 14 days after seeding and inoculation. The RNA was reverse transcribed and the double-stranded cDNA was size-selected, dC-tailed and inserted into the dG-tailed Pst1 site of pBR322. We obtained a cDNA library consisting of approximately 8000 clones. A small part was screened by differential colony filterhybridization and 60 apparently nodule-specific cDNA clones were selected. They hybridized to nodule polyA(+)RNA but failed to hybridize to root polyA(+)RNA and to pPsLb101, a previously obtained leghaemoglobin (Lb) cDNA clone. This clone was identified with anti-Lb serum after hybrid selected translation and by comparing the nucleotide sequence with the amino acid sequence of pea Lbs. Dot hybridization of isolated plasmid DNA's showed that 30 clones were indeed nodule specific. Just two clones, pPsNod201 and pPsNod301, were further analyzed. They did not show homology to any of the other selected clones. Hybridization of pPsNod201 and pPsNod301 to DNA from pea and from R.leguminosarum revealed that these sequences are encoded by the plant genome. Transcripts corresponding to both clones having M.W.s of 2.10^5D for pPsNod201 and 6.10^5D for pPsNod301 were only represented in root nodule RNA and not in root RNA. The latter results confirmed the nodule specific character of these clones.

II. Expression of host genes involved in nitrogen fixation was also studied by protein analysis of root nodules with specific antisera against Lb and four other nodulins (N29, N27, N19 and N17a; nomenclature according to Bisseling et al., 1983). The results show that the corresponding genes are expressed at different stages during nodule development. In infected roots N29 appears 6 days after seeding and infection while N19 and N27 can just be identified in 13 day old nodules, similar to the Lb pattern. N17a is a "late" protein appearing at day 17. By using Rhizobium mutants (nod^+fix^-) it was shown that repression of nodulin genes in ineffective nodules depends on the stage in which nodule development is disturbed.

Detailed methods and results are described by Bisseling et al. (these proceedings) and F. Govers et al. (in preparation).

This work was financially supported by a grant from the Netherlands Foundation for Biological Research.

Reference: Bisseling T, Been C, Klugkist J, Van Kammen A, Nadler K (1983) The EMBO Journal 2, 961-966.

ENVIRONMENTAL AND GENOTYPIC EFFECTS ON DINITROGEN FIXATION OF CONTRASTING ALFALFA CLONES

G.H. HEICHEL[+]/D.K. BARNES[+]/C.P. VANCE[+]/G. HARDARSON[++]
+ USDA-ARS, 1509 GORTNER AVENUE, ST. PAUL, MN 55108, USA
++ IAEA SEIBERSDORF LABORATORY, A-2444 SEIBERSDORF, AUSTRIA

INTRODUCTION. Improvement of N_2 fixation by legumes depends upon development of methods which allow accurate identification of genetic variability for this trait and accurate measurement of its expression in the field. Our objectives were to investigate N_2 fixation of clones, initially selected from adapted alfalfa (*Medicago sativa* L.) germplasm, over 4 years of growth in the field, and to evaluate their rhizobial preference in specific years.

METHODS. Eleven effectively nodulating clones that had previously been selected in a breeding program for contrasting nitrogenase activity (Viands et al., 1981) and one ineffectively nodulating clone (Viands et al., 1979) were vegetatively propagated in the glasshouse. The clones were inoculated with a mixture of rhizobial strains (except in 1979, when two antibiotic resistant mutant strains were used), and transplanted to field plots in each of four successive years. Dinitrogen fixation was measured by the isotope dilution method (Heichel et al., 1981). Serology was used in 1978 and antibiotic resistant mutants (Hardarson et al., 1982) in 1979 to test rhizobial preference.

RESULTS. The average N_2-fixation ranged several-fold across the clones in each year. Except for clone 10 in 1979, clones 8, 10, and 11 ranked consistently low across years. In comparison, clones 2, 3, and 5 ranked consistently high. Nitrogen fixation of clones 1, 4, 6, 7, and 9 showed the greatest interaction with year of growth. Dinitrogen fixation was highly correlated ($r = 0.70$ to 0.99) with herbage yield within each year and across years. Clone 12, which was incapable of N_2 fixation, exhibited the lowest herbage yield. The clones also differed significantly among years in the proportion of herbage N derived from fixation. Clones 8, 10, and 11 always ranked consistently lower than the remainder except for the ineffectively nodulating control. In years with normal weather, most clones averaged about 80% N from symbiosis over a two-fold range of herbage yields.

There were significant differences in nodulation among the clones in 1978 and in 1979. Number of nodules/plant was significantly correlated with mg N_2 fixed/plant in both years ($r = 0.63, 0.65$), with proportion of N from fixation ($r = 0.57, 0.65$), and with herbage yield ($r = 0.66, 0.83$). Use of antibiotic-resistant mutants showed significant differences among clones in preference for indigenous strains and for the mutant 102F51 $str^r a$. Most of the nodules were formed by indigenous strain(s) other than those inoculated.

These results suggest that the N_2 fixation capability of adapted alfalfa germplasm can be reliably determined in the field by the isotope dilution method without specific knowledge of the rhizobial preference of the host. Identification of plant material potentially useful in a breeding program should be accomplished more easily and reliably than has previously been possible.

REFERENCES

Hardarson G, Heichel GH, Barnes DK and Vance CP (1982) Crop Sci. 22, 55-58.
Heichel GH, Barnes DK and Vance CP (1981) Crop Sci. 21, 330-335.
Viands DR, Barnes DK and Heichel GH (1981) USDA Tech. Bull. 1643
Viands DR, Vance CP, Heichel GH and Barnes DK (1979) Crop Sci. 19, 905-908.

PLANT GENOTYPE AND THE CONTROL OF NITROGEN FIXATION

John Imsande
Dept. of Genetics, Iowa State University, Ames, Iowa 50011 USA

Recently a nondestructive assay for nitrogenase activity was described that permits the monitoring of nitrogen fixation periodically throughout the lifetime of the nodulated soybean plant (Imsande, Ralston, 1981). This procedure, which relies upon hydroponic growth and the repeated measurement of acetylene reduction by the nitrogenase complex, is as follows: 1) seeds are sprouted aseptically in rolls of moistened germination paper; 2) approximately 3 days after emergence, the seedlings are transferred, aseptically, to hydroponic growth in a plant growth chamber; 3) after 21 days of aseptic hydroponic growth, the entire root system of each plant is immersed for 10 minutes in a pure culture of Rhizobium japonicum, rinsed, and returned to hydroponic growth; 4) twice each week for approximately 8 weeks individual plants are weighed, assayed for acetylene reduction activity, and returned to hydroponic growth. Growth medium supplemented with 1.0 to 3.0 mM nitrate is changed twice each week (Ralston, Imsande, 1983). Using this procedure I attempted to identify individual soybean plants that fix nitrogen at a relatively high rate and to determine whether or not the capacity for enhanced nitrogen fixation is controlled by the plant genotype.

For routine screening experiments, plants of the same genotype are grown in sets of 12, 24, 36, or 96. In one such screening, 36 wild-type 'Harosoy' plants were screened for nitrogenase activity (acetylene reduction) for 10 weeks. One of these 36 plants, a plant designed "0-25", was selected because of its high rate of acetylene reduction and because of its extended duration of acetylene reduction. Seeds from this plant were subsequently grown in the field and each plant (F_1) was thrashed individually yielding F_2 progeny. Twelve seeds from each of 3 F_2 plants and 12 wild-type 'Harosoy' seeds were then grown individually under laboratory conditions. Some of the F_2 progeny of plant "0-25", when grown in the presence of 2.0 mM nitrate (i.e., high nitrate), reduced acetylene at a much higher rate than the control group. One F_2 progeny, designated F_2-21, reduced acetylene at a rate approximately 4 times greater than the control (4 std deviations from the mean) and, furthermore, the total, measured acetylene reduced (671 μmoles) was more than 4 times greater than the control plants (6.8 std deviations from the mean of the control group). The probability that such a plant would occur by chance alone is less than one out of 10,000 plants. The probability of finding one such plant in a group of 36 experimental plants is less than 0.004. This strongly suggests that the ability of plant "0-25" to reduce acetylene at a high rate has been passed on to some of its progeny. The fact that plants "0-25" and F_2-21 were each identified from a relatively small sample suggests that at least some aspects of nitrogen fixation can be enhanced by the action of a relatively small number of alleles, possibly as few as 3 or 4 pairs. Thus these preliminary data indicate that, using this screening system, plants of the desired phenotype can be identified, and furthermore, these data also indicate that partial resistance of nitrogen fixation to repression by nitrate is a heritable trait.

References

Imsande J and Ralston EJ (1981) Plant Physiol. 68, 1380-1384.
Ralston EJ and Imsande J (1983) J. Exp. Bot. 34 (In press).

A NEW PEA MUTANT EFFICIENTLY NODULATING IN THE PRESENCE OF NITRATE.

E. Jacobsen, Department of Genetics, Biological Centre, University of Groningen, Haren, The Netherlands.

Nitrate administered at the moment of sowing inhibits nodulation; when added after nodulation acetylene reduction is impaired. In this respect no genetical variability from the part of the plant has been described in literature. Here, the selection of a, on nitrate containing medium, persistently nodulating pea (Pisum sativum) mutant is reported.
Seeds of parent variety Rondo, and of M_2-families harvested from EMS-treated M_1-seedlings, were nodulated with Rhizobium leguminosarum strain PF_2 on an aerated standard mineral solution (SMS) (Feenstra and Jacobsen, 1980) complemented with 15 mM KNO_3. Nodulation of cv Rondo on SMS + 15 mM KNO_3 is strongly inhibited (>70%). Under these culture conditions one nodulating mutant, designated as nod_3, among 222 M_2-families was found which showed to be monogenic and recessive.

Table 1: Nodulation and acetylene reduction of cv Rondo and mutant nod_3 cultured on SMS or on SMS + 15 mM KNO_3

	number of nodules	nodule fresh weight (mg)	acetylene reduction* per plant	acetylene reduction* per g of nodule fresh weight	culture conditions
cv Rondo	59.3	125	2.0	16.1	SMS
	16.6	26	0.2	8.5	SMS + KNO_3
nod_3	> 300	589	6.8	11.6	SMS
	> 250	682	4.1	5.6	SMS + KNO_3

* = umoles of C_2H_4 produced per h.

The most important data are shown in Table 1. Mutant nod_3: 1. nodulates persistently on nitrate containing medium, 2. is highly nodulating, 3. reduces about 3 times more acetylene after nodulation on SMS, and on nitrate containing medium acetylene reduction is still more than that of cv Rondo nodulated on SMS, and 4. is as cv Rondo, inhibited by nitrate, in nodule activity per g of fresh weight.
In further investigations with nod_3 more questions have to be solved about strain specificity, the effect of other forms of bound nitrogen, the physiological and/or morphological basis of the nodulating ability. The mutant character of nod_3 could be of importance for agricultural application.

References

Feenstra, W.J. and E. Jacobsen, 1980. Theor. Appl. Genet. 58, 39 - 42.

INTERACTIONS OF GLYCINE SOJA GENOTYPES WITH FAST AND SLOW GROWING SOYBEAN RHIZOBIA

HAROLD KEYSER AND PERRY CERGAN
USDA-ARS, BARC-W, Bldg. 011, HH-19, Beltsville, MD 20705, USA

Fast growing soybean rhizobia were reported to be ineffective in N_2-fixation with several N. American soybean cultivars, though they were effective with cv. Peking and a *Glycine soja* (wild soybean) line from China (Keyser et al. 1982). In further investigation of their effectiveness with *Glycine* species, they were compared with effective slow growing soybean rhizobia on *G. soja* lines from China, Korea and Japan.

Material and Methods

Seven *G. soja* genotypes of diverse origin were initially screened in a greenhouse by inoculation with one slow (USDA 122) and two fast (USDA 191 and 192) growers in modified Leonard jars. Plants were harvested after 35 days and top dry weights determined. Two lines (PI 342.434 and PI 468.397) with the most marked and opposite rhizobia preferences were further tested in a growth chamber with six strains of each rhizobia group. At 42 days, plants were harvested and top dry weights, nodule dry weights, and total N accumulation in tops were determined. Treatments were replicated four times.

Results

Results from the initial trial showed that the fast growers were significantly more effective than the slow grower with two *G. soja* lines, while the opposite was found with four other lines, and no interaction for another line. Results from the expanded second trial verified that marked interactions for N_2-fixation occurs between the two rhizobia groups with different *G. soja* genotypes. In symbiosis with PI 342.434, the slow growers fixed more than three times as much N_2 as the fast growers, whereas, with PI 468.397 the fast growers fixed more than twice as much as the slow growers.

Discussion

Glycine soja is considered to be the wild ancestor of the domesticated soybean, *G. max* (Hymowitz 1970). Our results demonstrate that rhizobia strain and group (fast or slow) effectiveness interactions exist in the *G. soja* germplasm. Genotypes as PI 342.434 have symbiotic behavior similar to several *G. max* genotypes in their effectiveness with slow growers. However, it is apparent that there are *G. soja* genotypes which are much more effective with fast growers than with slow growers. Line 468.397 is the genotype from which the fast grower USDA 193 was isolated and demonstrates the usefulness of paired collections.

References

Hymowitz T (1970) Econ. Bot. 24, 408-421.
Keyser, HH (1982) Science 215, 1631-1632.

EMS DERIVED MUTANT OF *PISUM SATIVUM* RESISTANT TO NODULATION

Barbara E. Kneen/Thomas A. LaRue.
Boyce Thompson Institute, Tower Road, Ithaca, NY 14853 USA

A non-nodulated pea plant was obtained by screening M_2 progeny from plants arising from EMS-treated seeds. The nodulation resistant character is stable, and has been maintained to the M_8 generation. The parent cultivar 'Sparkle' and field pea cultivar 'Trapper' are well nodulated while the mutant is resistant to nodulation by 27 strains of *Rhizobium leguminosarum* tested, including 4 which nodulate the nod-resistant pea variety 'Afghanistan' (1,2). Occasionally 1-5 effective nodules are formed. Rhizobia isolated from these nodules are not infective when retested on the mutant. Segregation for nodulation in the F_1 and F_2 progeny of reciprocal crosses between the mutant and 'Sparkle' or 'Trapper' show that nodulation resistance is conditioned by a single pair of recessive alleles. This was confirmed by analysis of test crosses between F_1 plants and the mutant line. Test crosses and F_2 progeny of crosses between the mutant and 'Afghanistan' were scored for nodulation by *R. leguminosarum* strain TOM, which infects 'Afghanistan,' and by *R. leguminosarum* 128C53 which does not infect 'Afghanistan.' Segregation for nodulation indicates that there are at least two different loci controlling nodulation resistance in 'Afghanistan' (sym-2,sym-2) and the EMS derived mutant (sym-5,sym-5).

1) T.A. Lie; Ann. Appl. Biol. 88: 445 (1978).
2) F.B. Holl; Euphytica 24: 767 (1975).

ALFALFA NODULINS FROM EFFECTIVE AND INEFFECTIVE SYMBIOSES

Naomi Lang-Unnasch/Frederick Ausubel
Department of Molecular Biology
Massachusetts General Hospital
Boston, MA 02114, USA

At least 12 different nodule specific proteins (nodulins) are detected in alfalfa nodules by immune assay. The pattern of nodulins from three types of ineffective nodules differs both qualitatively and quantitatively from that of effective nodules. The nodulins range in size from leghemoglobin at about 14 kilodaltons (kd) up to 140 kd. Of these, one, a 66 kd protein, is clearly present in free living *R. meliloti* and *R. leguminosarum* as well as in the bacteroids purified from either alfalfa or pea nodules. Two other nodulins comigrate with immunoreactive bacteroid proteins in SDS polyacrylamide gel electrophoresis.

HETEROGENEITY OF GLUTAMINE SYNTHETASE POLYPEPTIDES IN *Phaseolus vulgaris* L.

Miguel Lara, Helena Porta, Jaime Padilla, Jorge Folch and Federico Sánchez. Centro de Investigación sobre Fijación de Nitrógeno. U.N.A.M. Cuernavaca, Mor. Apartado Postal 565-A. México.

Glutamine synthetase (GS) is the major enzyme for ammonia assimilation in higher plants (Miflin, Lea, 1980). Moreover multiple forms of this enzyme have been separated by ion exchange chromatography from different tissues of barley (Mann, et al. 1979), rice Guiz, et al. 1979) and bean (Lara, et al. 1983). We have characterized the differents forms of GS forms of GS from root, nodule and leaf of Phaseolus vulgaris.

Material and Methods: Glutamine synthetase from root, nodule and leaf of P. vulgaris has been purified to homogeneity based on the affinity chromatography procedure previously reported by Palacios, 1976. $NaDodSO_4$ polyacrylamide gel electrophoresis was performed in 7.5 polyacrylamide gel according to Palmiter, et al. 1971. Two dimentional gel electrophoresis was done according to O'Farrell, 1975.

Results and Discussion: On $NaDodSO_4$ polyacrylamide gel the GS purified from leaf is formed for two protein bands of 45000 and 43000 molecular weight (MW) and the root and nodule enzyme is composed for a single protein band of 43000 MW.

On two dimentional gel electrophoresis, the 45000 MW band of the leaf GS is resolved in four polypeptides with isoelectric points (IP) ranging from 5.7-6.1. The 43000MW protein band of leaf and root is composed of two polypeptides called α and β with IP of 5.8 and 6.2 respectively. The nodule GS is also formed by two polypeptides, one migrates as the β polypeptide of root and leaf and the other with IP of 6.6 (called γ) seems to be specific for nodule tissue.

References:

Guiz, C., Hirel, B., Shedlofsky, G. and Gadal, P. (1979). Plant Sci. Lett. 15, 271-277.

Lara M., Cullimare, J.V., Lea, P.J., Miflin, B.J., Johnston, A.W.B. and Lamb, J.W. (1983). Planta 157,254-258.

Mann, A.F., Fentem, P.A. and Stewart, G.R. (1979). Biochem. Biophys. Res. Commun. 88,515-521.

Miflin, B.J. and Lea, P.J. (1980). In Miflin, B.J. ed, The Biochemistry of Plants, pp. 169-202. Academic Press, New York, London.

O'Farrell, P.H. (1975). J. Biol. Chem. 250, 4007-4021.

Palacios, R. (1976). J. Biol. Chem. 251, 4787-4791.

Palmiter, R.D., Oka, T. and Schinke, R.T. (1971). J. Biol. Chem. 246, 724-734.

HOST-GENETIC CONTROL OF NITROGEN FIXATION IN THE PEA/RHIZOBIUM ASSOCIATION

T.A.LIE and P.C.J.M. TIMMERMANS
LABORATORY OF MICROBIOLOGY, AGRICULTURAL UNIVERSITY, WAGENINGEN, THE NETHERLANDS

1. INTRODUCTION

Leguminous plants indigenous in a region are often incompatible with *Rhizobium* strains originating from another region (Lie, 1981). The degree of incompatibility seems to be related to the distance between the places of origin of the two partners (Lie *et al.*, 1982). We now report on a finding that an Israeli *Rhizobium* strain induces nodules of high nitrogenase activity in pea cv. Iran, but in contrast nodules of low activity on the more distant pea cv. Afghanistan. We make use of this system to study the host-genetic control on the expression of nitrogenase in *Rhizobium*.

2. MATERIAL AND METHODS

Pea cv. Afghanistan and cv. Iran were crossed in both directions and the resultant hybrids were self-fertilized to obtain a F_2-population. The symbiotic performance of the plants was assayed under aseptic conditions in a N-free nutrient solution, and inoculated with an Israeli *Rhizobium* strain F13 (Lie, 1981). The plants were grown in a climate room at 25°C and harvested after five weeks.

3. RESULTS AND DISCUSSION

Table 1. Growth and N-fixation of pea cv. Afghanistan, cv. Iran and their reciprocal hybrids, inoculated with *Rhizobium* strain F13

	A	(AxI)	(IxA)	I
Shoot wt. (g/pl)	1.17	1.98	3.42	4.21
% N in shoot	1.57	3.12	3.24	3.96
mg N fixed	2.7	7.8	13.3	19.8
µM C_2H_4/pl/h	0.43	3.36	4.13	6.20

Table 2. Segregation of the ability to fix nitrogen in a F_2-population of a cross between pea cv. Afghanistan and cv. Iran, inoculated with *Rhizobium* strain F13 (based on shoot weight)

Parents		
cv. Afghanistan	0.50(0.16-0.77) g/pl	
cv. Iran	2.86(1.65-4.58) g/pl	
F_2-population		
Class	Found	Expected
"Afgh."type (<0.77)	23	26
Intermediate type (0.77-1.65)	27	26
"Iran" type (>1.65)	53	52

These results show that in pea cv. Afghanistan a recessive host gene prevents the full expression of nitrogenase in *Rhizobium* strain F13. Earlier Holl (1975) and Holl and LaRue (1975) suggested that a recessive gene sym 3 is present in cv. Afghanistan which confers ineffectiveness. However, this is based on experiments in non-sterile soil with a mixture of *Rhizobium* strains and the results are not reproducible by the same authors (personal communication). Other data (not shown here) indicate that the gene concerned here is not identical to sym 3. A comparison of the reciprocal hybrids clearly shows that the plant background should be taken into account when the genetic analysis is based on plant weight or total nitrogen fixed.

4. REFERENCES

Holl, F.B. (1975) Euphytica 24, 767-770
Holl, F.B. and LaRue, T.A. (1975) Proc. 1st. Int. Symp. N-fixation (eds. Newton and Newman), Wash. STate Un. Press, vol.2 p.125
Lie, T.A. (1981) Plant and Soil 61, 125-134
Lie, T.A., Timmermans, P.C.J.M. and Ladizinsky, G. (1982) Isr. J. Bot. 31, 163-167

DINITROGEN FIXATION BY PISUM SATIVUM ECOTYPE FULVUM. GENETIC CONTROL BY THE HOST PLANT.

W.J.M. LOMMEN AND D. WOUTERS.
Laboratory of Microbiology, Agricultural University, Wageningen, The Netherlands.

INTRODUCTION.

Pisum sativum ecotype fulvum lines, originating from Israel, were nodulated effectively by an Israelian *Rhizobium* strain. However, the lines differed in N_2-fixation with strains from other countries (Lie 1981).

In this investigation the genetic differences between two of these lines in N_2-fixation with a strain from the Hindukush region are studied.

MATERIAL AND METHODS.

P. sativum ecotype fulvum lines Fu 27 and Fu 62 originate from Israel, as does *R. leguminosarum* strain F13. Strain HIM originates from the Hindukush.

Seeds were germinated aseptically in the dark. When the plumula emerged, the seedlings were transplanted into perlite and N-free nutrient solution and the roots were inoculated with a suspension containing 10^7 bacteria.

C_2H_4 production was measured after incubating the whole plants for one hour in 8% C_2H_2.

Crossings between the parent lines were made in both directions.

Means were tested with the t-test at a 10% probability level. Different letters indicate a significant difference between the means.

RESULTS AND DISCUSSION.

Two fulvumlines Fu 27 and Fu 62 were able to fix nitrogen effectively with an appropriate *Rhizobium* strain, F13. Inoculated with strain HIM, the two lines differed in nitrogen fixation. Fu 27 was superior to Fu 62. The differences in ethyleneproduction are demonstrated in table 1.

Table 1. Characteristics of pea lines Fu 27 and Fu 62, 32 days after inoculation with strain F13 or strain HIM.

	strain F13 line Fu 27	strain F13 line Fu 62	strain HIM line Fu 27	strain HIM line Fu 62
C_2H_4 production in 10^{-6}moles/g fresh nodule weight.hour	19,3 a	19,4 a	8,9 b	1,4 c
C_2H_4 production in 10^{-6}moles/plant.hour	4,8 a	6,3 a	3,0 b	0,1 c

In another experiment the ethyleneproduction/plant of the parent lines and the F1 plants was measured, 21 days after inoculation with strain HIM. The F1 plants didn't differ significantly from the most effective parent line, line Fu 27. This suggests involvement of at least one dominant gene.

ACKNOWLEDGEMENTS.

This investigation is supported by the Foundation for Fundamental Biological Research (BION) which is subsidized by the Netherlands Organization for the Advancement of Pure Research (ZWO).

REFERENCE.

Lie T.A. (1981), Plant and Soil 61, 125 - 134.

EXPRESSION OF NITROGENASE REDUCTASE IN YEAST

CLAUDE V. MAINA/ALLAN YUN and ALADAR A. SZALAY
BOYCE THOMPSON INSTITUTE, CORNELL UNIVERSITY, ITHACA N.Y., 14853 U.S.A.

Our laboratory has reported (1) the table integration of the entire nitrogen-fixing gene cluster from Klebsiella pneumoniae into the yeast genome. Even though the prokaryotic DNA was stably maintained during both mitosis and meiosis no nif specific transcriptional products could be detected in the transformed yeast cells (unpublished data, this laboratory).

One of the reasons for the lack of transcription of the prokaryotic DNA might be the inability of the yeast RNA polymerase to recognize the prokaryotic promoter. Therefore, the nifH gene (coding for nitrogenase reductase) with its 5' upstream sequences was excised from the plasmid pSA30 for manipulations that would allow its expression in yeast. Its promoter region was removed by Bal 31 digestion leaving the structural gene intact. After the addition of Hind III linkers, the nifH gene was inserted into the Hind III site of pAAH5, downstream of the yeast alcohol dehydrogenase I promoter. Restriction and DNA sequence analysis was used to confirm the structure of the chimeric plasmid. pH-ADH-1 contains a 1.8 Kb fragment which carries the yeast alcohol dehydrogenase I promoter including 25 nucleotides downstream of the transcription initiation site. DNA sequence analysis showed that the fusion occurred at 32 base pairs upstream of the nifH ATG. In addition, pH-ADH-1 included the transcriptional termination signal as well as the poly-adenylation site of the yeast alcohol dehydrogenase I gene. The construct is contained on a 13.5 Kb as well as selectable markers for E.coli (amp^R) and yeast (LEU 2).

LEU 2 independent yeast transformants containing pH-ADH-1 were analyzed for the presence of nifH specific RNA and nitrogenase reductase protein. Northern analysis of total RNA showed the presence of a single nifH transcript of approximately 1.1 Kb in size. The mRNA is polyadenylated and of similar abundance as that of yeast alcohol dehydrogenase I mRNA. When glucose grown transformants were shifted to media containg 3% alcohol, the transcription of the nifH gene paralleled that of the alcohol dehydrogenase I gene, indicating that the nifH gene is transcriptionally regulated.

Analysis of protein isolated from transformants showed a polypeptide of MW 35 Kd that cross-reacted with antisera made against nitrogenase reductase. The abundance of the polypeptide is similar to that of alcohol dehydrogenase.

Similarly, fusions of the alcohol dehydrogenase I gene to the nifHDK (polycistronic) genes are presently under way and transcriptional and translational analysis of this hybrid operon will be carried out. In addition nitrogenase reductase isolated from pH-ADH-1 transformants is being used in in vitro complementation experiments to determine the biological activity of the yeast-synthesized bacterial enzyme.

REFERENCES

Zamir A, Maina CV, Fink GR, Szalay AA (1981) Proc.Natl.Acad.Sci. 78, pp. 3496-3500.
Denis CL, Ferguson J, Young ET (1983) J.Biol.Chem. 258, pp. 1165-1171.

PREDOMINANT mRNAs FROM ANU289-INDUCED NODULES OF SIRATRO AND PARASPONIA

S.E. NEWTON[1]/P.M. GRESSHOFF[2]/S.S. MOHAPATRA[2]/B.G. ROLFE[1] AND J. SHINE[1]
CENTRE FOR RECOMBINANT DNA RESEARCH[1] AND BOTANY DEPARTMENT[2], AUSTRALIAN NATIONAL UNIVERSITY, PO BOX 475 CANBERRA, A.C.T. 2601 AUSTRALIA

The slow-growing *Parasponia rhizobium* strain ANU289 forms effective nodules on the tropical legume Siratro and the non-legume *Parasponia*. We are examining the plant-specified mRNAs and proteins (nodulins) synthesized in nodules from these two different types of plants.
Analysis of *Sau* 3A-digested ds cDNA transcribed from ANU289-induced Siratro nodule poly A^+ RNA showed a number of predominant bands. The same major bands were apparent in cDNA from Siratro nodules induced by the fast-growing *Rhizobium* strain ANU240. The response of the plant to infection by a slow-growing strain and a fast grower thus appears to be similar. These were no particularly predominant species in *Sau* 3A-digested cDNA transcribed from *Parasponia* nodule poly A^+ RNA, although some faint bands were apparent.
We have constructed cDNA libraries from both Siratro and *Parasponia* poly A^+ nodule RNA, and are isolating clones which hybridize strongly to nodule-specific cDNA probes. Such clones represent predominant nodule mRNAs and we are attempting to identify and classify these clones by sequence analysis. One such clone isolated from a *Parasponia* cDNA library has some homology with the amino acid sequence of broad bean leghaemoglobin (Lb), but less homology with soybean Lb's (Table 1). It is possible that this clone may represent a *Parasponia* globin. We are currently screening a *Parasponia* genomic library with this clone. Although we have not yet isolated a Siratro Lb cDNA clone, we have isolated and purified a Siratro Lb protein (mol.wt. ≃ 16,000). In association with Dr. Frank Morgan, we have determined the amino acid sequence at the N-terminus of this Lb (Table 2). The sequence differs from that of Soybean Lbc1 by only 3 amino acids out of the first 35 N-terminal amino acids. In contrast to Soybean, however, there appears to be only one predominant Siratro Lb, since no sequence heterogeneity was observed.

TABLE 1. Amino acid sequence of a potential *Parasponia* Lb cDNA clone

	Exon 1	Exon 2
Parasponia:	AS-YFIVRGE	ILGFMKDEQTAKA-FA--KDVFINQER-KLGA
Broad bean:	YSVLFYTI--	ILQ--KAP-TAKAMFSFLKDSAGVVDSPKLGA
Soybean:	YSVVFYNS--	ILE--KAP--AKDLFSFLANGVDPTNP-KLIG
	24 30	40 50 60

Numbers represent the amino acid residue of Soybean Lb, the vertical line the junction between exons 1 and 2 in the Soybean Lb and underlined residues indicate areas of homology.

TABLE 2. N-terminal amino acid sequence of Siratro Lb

Siratro:	GAFTEKQEALVNSSYEAFKANIPQYSAVFYTSILE
Soybean:	GAFTEKQEALVSSSFEAFKANIPQYSVVFYTSILE

Underlined residues indicate sequence differences.

STRUCTURAL ORGANIZATION AND GENETIC CHARACTERIZATION OF THE nif REGION IN THE SLOW-GROWING RHIZOBIA

A.C. YUN/R. MURASKOWSKY/M.N. JAGADISH/J.D.NOTI/R.P. LEGOCKI/
A. KOERMENDY/A.A. SZALAY
Boyce Thompson Institute, Cornell University, Ithaca, NY 14853 USA

The nifK, nifD and nifH genes of the slow-growing cowpea Rhizobium IRc78 were cloned separately as HindIII fragments of respective sizes 21kb, 9.6kb and 5.1kb based on homology to hybridization probes, nifKD and nifDH of Klebsiella pneumoniae (R.G. Hadley, A.A. Szalay, unpublished). In addition, a gene bank of cowpea Rhizobium IRc78 DNA constructed in the wide host range cosmid vector, pSUP106, was screened with the previously isolated nifK, D and H genes of IRc78. Results show that nifK and nifD hybridize to a common 4.7kb EcoRI fragment, and are thus linked. However, clones containing nifK and D do not contain nifH sequences. Mapping data indicate that the nifH gene is located at least 16kb away from nifK and D.

The cloned cowpea Rhizobium IRc78 nifD and nifH genes were sequenced and an extensive DNA sequence homology was found on the order of 70% when compared with Anabaena (P.J. Lammers, R. Haselkorn 1983; M. Mevarech et al. 1980) and K. pneumoniae (K.F. Scott et al. 1981; V. Sundaresan, F.M. Ausubel 1981). The structural genes, nifD (1473bp) and nifH cowpea Rhizobium IRc78, along with their 5' - upstream regions are presented in figures. The nifD and nifH 5' - upstream sequences show no homology or consensus to each other. When compared with the nifH promoter of the fast-growing Rhizobium, R. meliloti (G. Ditta et al. 1983) and with the heptameric consensus sequence found in R. meliloti and K. pneumoniae nifH promoter region (D.W. Ow et al. 1983), no significant homology was detected.

However, the nifH 5' - upstream region from stem Rhizobium BTAil was shown to be almost identical (98% homologous) to cowpea Rhizobium nifH. This promoter region was fused to the lacZ gene of E. coli and returned into the stem Rhizobium chromosome by gene replacement. Transformed rhizobia showed significant β-galactosidase activity on X-gal/minimal medium plates (blue colonies) in response to anaerobic conditions, while both transformed and non-transformed rhizobia inhibited by nitrate show no β-galactosidase activity (white colonies).

The 21kb nifK hybridizing region was cloned into a mobilization vector, pSUP201, which replicates only in E. coli. Using Tn5 mutagenesis in E. coli, 45 clones with Tn5 insertions in the nifK gene and in the downstream flanking region were obtained. Tn5 inserted fragments were used to replace their genomic counterparts in IRc78 by double crossover exchange between the regions of homology. Host plants infected with IRc78::Tn5 have shown that insertion in the nifK gene and 9kb downstream of nifK resulted in reduced fixation phenotypes on all host plants. Hybridization patterns of DNA obtained from bacteroids and from Rhizobium in liquid culture using a labeled Tn5 fragment were identical, indicating the presence of Tn5 in its original site of insertion.

REFERENCES

Ditta G, Better M, Corbin D, Barran L, Ruiz-Argueso T and Helinski DR (1983) UCLA Symposia on Molecular and Cellular Biology, New Series.
Lammers PJ and Haselkorn R (1983) Proc.Natl.Acad.Sci.USA 80:4723-4727.
Mevarech M, Rice D and Haselkorn R (1980) Proc.Natl.Acad.Sci.USA 77:6476-6480.
Ow DW, Sundaresan V, Rothstein, DM, Brown, SE and Ausubel FM (1983) Proc. Natl.Sci.USA 80:2524-2528.
Scott KF, Rolfe BG and Shine J (1981) J. Mol. Appl. Gen. 1:71-81.
Sundaresan V and Ausubel FM (1981) J. Biol. Chem. 256:2808-2812.

BACTERIAL GENETICS

POSTER DISCUSSION 9A

ORGANIZATION OF *RHIZOBIUM* SYMBIOTIC GENES

POSTER DISCUSSION 9B

RHIZOBIUM PLASMIDS

POSTER DISCUSSION 9C

NIF GENETICS OF FREE-LIVING DIAZOTROPHS

ADVANCES IN THE GENETICS OF FREE-LIVING AND SYMBIOTIC NITROGEN FIXING BACTERIA

ALFRED PÜHLER/M.O. AGUILAR/M. HYNES/P. MÜLLER/W. KLIPP/U. PRIEFER/
R. SIMON/G. WEBER
LEHRSTUHL FÜR GENETIK, FAKULTÄT FÜR BIOLOGIE, UNIVERSITÄT BIELEFELD,
POSTFACH 8640, D-4800 BIELEFELD 1, FRG

1. INTRODUCTION

It is evident that the knowledge about biological nitrogen fixation was enormously increased when genetic systems in free-living and symbiotic nitrogen fixing bacteria were developed and used for the analysis of their nif (nitrogen fixation) genes. Due to its close relationship to Eschericia coli, Klebsiella pneumoniae, a facultatively anaerobic procaryotic microorganism, was initially selected to analyze nif genes (Streicher et al., 1971; Dixon, Postgate, 1971). For other Gram-negative nitrogen fixing microorganisms, genetic systems were recently developed and are now used to compile information in the field of nif genetics. To review all the information available would certainly exceed the scope of this paper. Therefore, we concentrate on three different nitrogen fixing species which are currently being analyzed in our laboratory: the facultative anaerobe Klebsiella pneumoniae, the phototrophic Rhodopseudomonas capsulata and the symbiotic nitrogen fixing Rhizobium meliloti. For these three species, the genetic techniques developed will be outlined. It is our special intention to show that breakthroughs are often achieved following the development of new methodology.

2. TRANSCRIPTIONAL ANALYSIS OF THE nif GENE REGION OF Klebsiella pneumoniae

Recently introduced recombinant DNA techniques helped a lot to understand the Klebsiella nif region. In particular, we constructed hybrid plasmids in E.coli which contain the chromosomal nif gene cluster of K.pneumoniae (Pühler et al., 1979a; Pühler et al., 1979b; Pühler, Klipp, 1981). Since E.coli cells harboring these plasmids are able to fix atmospheric nitrogen, it can be assumed that all essential nif genes are located on the cloned fragment. In order to analyze the fine structure of the Klebsiella nif gene region we developed a complex system including transcription of nif genes from constitutive plasmid promoters, identification of nif gene products in minicells of E.coli, and location of nif coding regions by Tn5 insertions (Pühler, Klipp, 1981; Pühler, Klipp, 1983). The resulting coding region map of the Klebsiella nif region has been published and contains the location of 15 Klebsiella nif genes (Pühler, Klipp, 1981; Pühler, Klipp, 1983; Pühler et al., 1982). The most interesting result of this work was that 2 new nif genes, nifX and nifY, could be identified and that the molecular weight of the proteins encoded by the above mentioned 15 nif genes could be determined. An improved coding region map of the Klebsiella nif region is now presented in Figure 1. In this map we have added the nifB gene and we have changed the length and the orientation of the nifF coding region. The establishment of these changes resulted from studying the transcriptional regulation of the Klebsiella nif genes in E.coli.

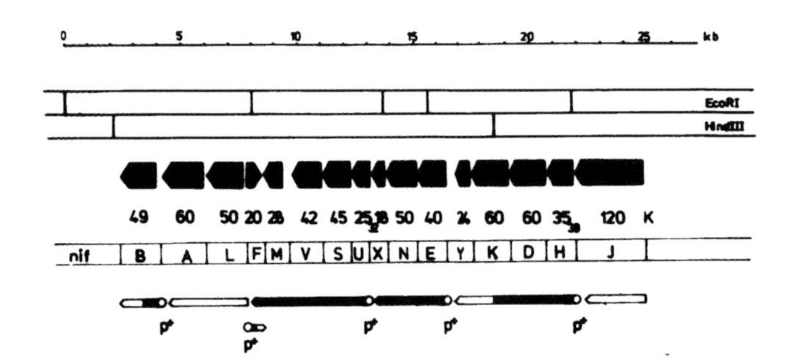

FIGURE 1. The nif gene cluster of Klebsiella pneumoniae.
The restriction map for various enzymes of the Klebsiella nif region is presented. In relation to this restriction map the coding regions for 16 nif genes are drawn. The molecular weight of the nif gene products determined by the minicell technique are indicated below the coding regions. The operon structure of the nif region is shown by arrows. Black parts of the arrows indicate DNA regions for which we were able to measure transcription starting from nifA gene product dependent promoters (p^+).

Until recently, the analysis of an EcoRI fragment of the Klebsiella nif region carrying the genes nifB, nifA and nifL (see Figure 1) only led to the identification of the nifA and the nifL gene products (nifAgp and nifLgp). This was explained by the assumption that transcription starting from the P_{Cm} promoter is terminated beyond nifA by a transcriptional terminator and that the contiguous nif promoter is not functional (Figure 2a). This assumption could be confirmed by further experiments schematically shown in Figure 2b. It is known that nifA and nifL are regulatory genes (Kennedy et al., 1981; Dixon et al., 1981). The nifAgp is believed to represent an activator necessary for the transcription of Klebsiella nif operons. In contrast, nifLgp seems to inhibit the action of the activator if conditions for nitrogen fixation such as low oxygen pressure or absence of fixed nitrogen are not fulfilled. This explanation was tested by the following experiment (Figure 2b). We deleted the nifL gene and could now express the nifB coding region in E.coli minicells. Evidently the nifAgp was able to activate the nif promoter, resulting in the expression of nifB. This activation is abolished at elevated temperature since the nifAgp is temperature sensitive (Zhu, Brill, 1981)

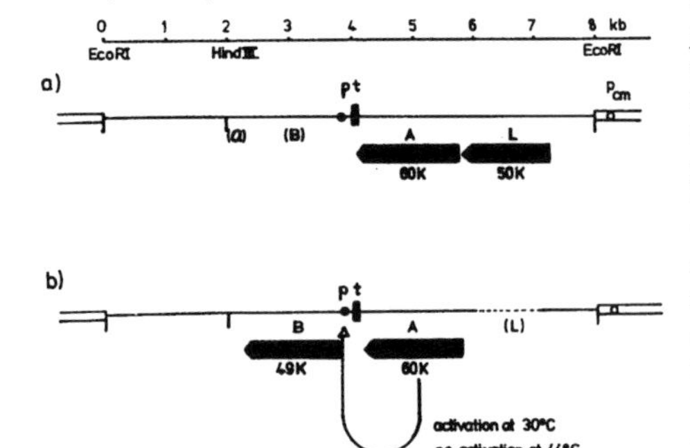

FIGURE 2. Expression of the Klebsiella pneumoniae nifB gene in minicells of Escherichia coli.
a) An EcoRI fragment of the Klebsiella nif region carrying the genes nifB, nifA and nifL was expressed in the E.coli minicell system. Polypeptides of molecular weight 50K and 60K were identified and their coding regions were mapped. b) Following deletion of the nifL gene, analysis in the minicell system showed polypeptides of 49K and 60K. The coding region of the 49K polypeptide is located on a fragment were nifB mutations map.

In order to identify nifAgp dependent nif promoters, we made use of an operon fusion technique which is explained in more detail in Figure 7. By in vitro techniques, a promoterless tetracycline resistance gene (Tc) was fused to different restriction fragments of the Klebsiella nif region. In Figure 3 an EcoRI fragment of the Klebsiella nif region carrying the genes nifM, nifV, nifS, nifU and nifX was used for such a fusion experiment. This EcoRI fragment carries a specific nif promoter since, following nifAgp activation, E.coli cells become tetracycline resistant. It is of special interest that such a nif promoter (p^+) can be mapped by a simple procedure. Insertion of Tn5 transposons into already known nif coding regions results in a tetracycline sensitive phenotype if the Tn5 insertion is located between Tc and p^+. In such a way a nifAgp dependent promoter p^+ was found next to nifU (Figure 3). We employed this technique and could identify nifAgp dependent

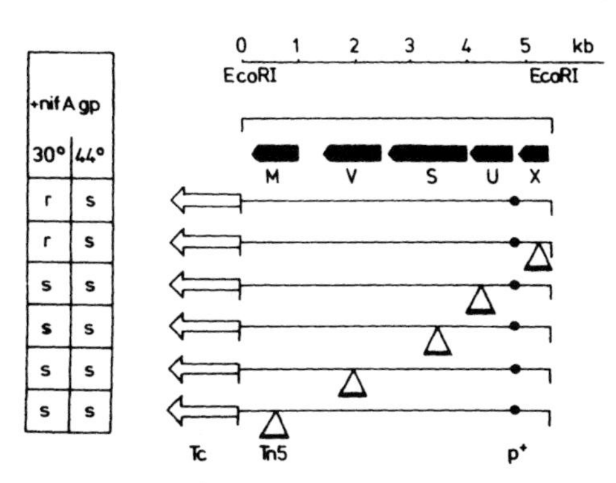

FIGURE 3. Mapping of a nifA gene product dependent promoter (p^+) next to nifU.
An EcoRI fragment carrying the genes nifX, nifU, nifS, nifV and nifM was fused to a promoterless tetracycline resistance gene (Tc). Tn5 insertions into nifM, nifV, nifS, nifU and nifX were isolated. The resulting plasmids were tested to see if the Tc gene can be transcribed from a nifAgp dependent promoter. The results are listed (r: resistant; s: sensitive).

promoters (p^+) next to nifB, nifU, nifE and nifH. In Figure 1 these p^+ promoters are indicated. In addition, the operon structure is shown as it could be determined by our experiments. The nif operons nifp^+B, nifp^+F, nifp^+USVM, nifp^+ENX and nifp^+HDKY are all transcribed from nifAgp dependent promoters. The following remarks are necessary. In our Tc fusion experiments we did not test whether nifY belongs to the nifp^+HDK operon, but earlier minicell experiments support this assumption. In addition, we could not demonstrate with our Tc fusion experiments whether nifJ is transcribed by a nifAgp dependent promoter. The transcriptional organization of the Klebsiella nif region was confirmed by Beynon et al. (in press) who detected the same nifAgp dependent promoters. and showed that they have a characteristic structure of 26 base pairs.

It should be mentioned that Sibold et al. (1983) published a molecular weight of 51.500 for the nifBgp which is in good agreement with our result. In addition, we would like to add that we did not find a gene product for nifQ. Upstream nifB we found a transcriptional terminator (Klipp, unpublished), leaving little or no space for nifQ.

Of special interest is the nifF transcription unit. We first reported a nifF gene which is transcribed and translated in the same direction as all the other Klebsiella nif genes. In our former minicell experiments we determined a gene product of 10.000 molecular weight. Now we were able to demonstrate that a nifAgp dependent promoter p^+ transcribes, in the opposite direction, a DNA region where nifF mutations are located. Recent minicell experiments revealed that this specific promoter perhaps transcribes a gene specifying a gene product of 20.000 molecular weight (Klipp, unpublished). We now assume that this gene is identical to nifF. Our finding is supported by other molecular weight determinations (Roberts et al., 1978; Nieva-Gomez et al., 198o) and also by sequencing data (Beynon et al., in press).

This short section demonstrates that the organization of the Klebsiella nif region is already very well known. But it should be mentioned that a lot of further aspects were not included in this paper, e.g. the finding of a further nifAgp dependent promoter between nifH and nifJ reading in the opposite direction compared to the promoter of the other nifp$^+$HDKY operon (Shen et al., 1983) or the transcriptional regulation of the nifpLA operon which is discussed by other authors in this volume. It is also of importance to note that in addition to the nifAgp, the glnFgp is necessary for the activation of the Klebsiella nif promoters (Sundaresan et al., in press; Merrick, 1983).

3. INDUCTION AND MAPPING OF nif MUTATIONS IN Rhodopseudomonas capsulata

The analysis of K.pneumoniae is closely related to E.coli. Thus, a lot of genetic techniques developed for E.coli could be directly applied to K.pneumoniae. As soon as other nitrogen fixing microorganisms became interesting, a lot of basic work was necessary to develop appropriate genetic systems. For Gram-negative bacteria this development was rather successful: transposon mutagenesis systems (Beringer et al., 1978; Van Vliet et al., 1978) and cloning vehicles (Bagdasarian et al., 1979; Bagdasarian et al., 1981; Ditta et al., 198o) are now available. In this section, we are going to introduce a special transposon mutagenesis system which we developed for Gram-negative nitrogen fixing bacteria. The system, which is schematically shown in Figure 4, consists of two components: special E.coli donor strains and mobilizable derivatives of E.coli vector plasmids (Mob-vectors). The donor strains (called mobilizing strains) carry the transfer genes of the broad host range plasmid RP4 (Datta et al., 1971) integrated into their chromosomes. They accept any Gram-negative bacterium as a recipient for conjugative DNA transfer. The vector plasmids are provided with the RP4 specific recognition site for mobilization (Mob-site). One example (pSUP2021) is given in Figure 4A. This plasmid is a derivative of pBR325 (Bolivar et al., 1977) in which

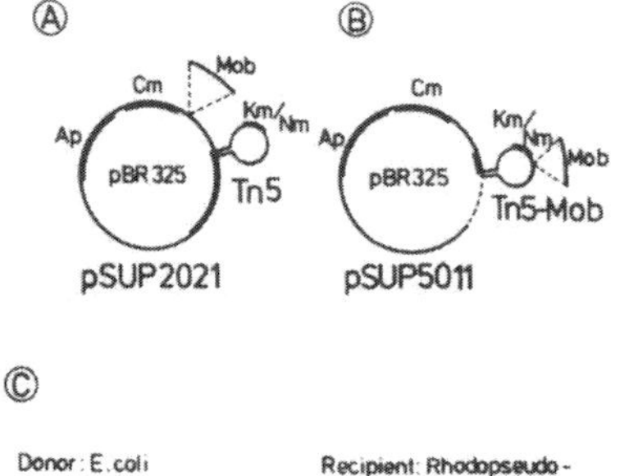

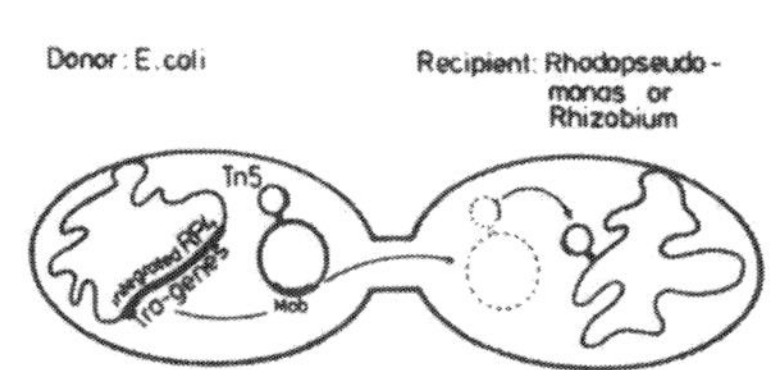

FIGURE 4. A broad host range mobilization system for transposon mutagenesis in Gram-negative bacteria.

(A) For description of Mob-vector pSUP2021 carrying Tn5 see text in section 3.

(B) For description of Mob-vector pSUP5011 carrying Tn5-Mob see text in section 5.

(C) Experimental use of Mob-vectors for transposon mutagenesis.

Abbreviations: Ap: ampicillin resistance; Cm: chloramphenicol resistance; Tc: tetracycline resistance; Km: kanamycin resistance; Nm: neomycin resistance; E: EcoRI; B: BamH1; Mob: mobilization; Tra: transfer genes.

a fragment carrying the RP4 Mob-site was integrated by in vitro techniques. In addition, this plasmid was loaded with transposon Tn5. The construction of mobilizing strains and Mob-vectors is described in several papers (Simon et al., in press a; Simon et al., in press b). Therefore, only the use of this system is shown in Figure 4C. The mobilizing E.coli donor strain containing the Mob-vector, e.g. pSUP2021, is mated with the Gram-negative recipient strain, e.g. Rhodopseudomonas or Rhizobium. The integrated RP4 plasmid of the donor strain provides the transfer functions which are necessary to form the conjugation bridge and to initiate DNA transfer. Usually, E.coli vector plasmids are non-transmissible. But since Mob-vectors carry the RP4 Mob-site they can be mobilized by plasmid RP4 and passively transferred with high frequency from the E.coli donor to any other Gram-negative recipient. It is of importance that E.coli vector plasmids can usually not replicate outside E.coli. Therefore, selection for neomycin or kanamycin resistant transconjugants leads to the isolation of clones that have received Tn5 via transposition. The result of such a mating experiment is a collection of Tn5 induced mutant clones. It should be pointed out that this system is practically applicable to all Gram-negative species.

We have employed this broad host range Tn5 mutagenesis system for the induction of Nif^- mutants of the phototrophic Rh.capsulata strain B10. Nif^- mutants were identified by their failure to grow anaerobically on nitrogen free medium. The corresponding nif mutations were mapped as follows: By R' formation with a kanamycin sensitive R68.45 plasmid Rh.capsulata DNA fragments carrying transposon Tn5 were transferred to E.coli and further analyzed. In addition, by hybridization with Rhodopseudomonas nif DNA, a cosmid clone could be identified which carries a large DNA fragment of about 20 kb necessary for nitrogen fixation. In Figure 5 the genetic data are summarized. Of 24 independently isolated Rhodopseudomonas nif::Tn5 mutants, 13 map in this 20 kb nif region. Some of them are indicated in Figure 5a. In order to confirm that this 20 kb region codes predominantly for nif functions we applied a fragment specific mutagenesis procedure. For this purpose restriction fragments of the 20 kb region were cloned into plasmid pSUP202, a pSUP2021 plasmid lacking Tn5 (Figure 4). Subsequently, the inserted fragments were mutagenized with Tn5 in E.coli. Following the conjugative transfer of the hybrid plasmids to Rh.capsulata, DNA fragments carrying Tn5 were integrated into the genome by homologous recombination. In such a way, 15

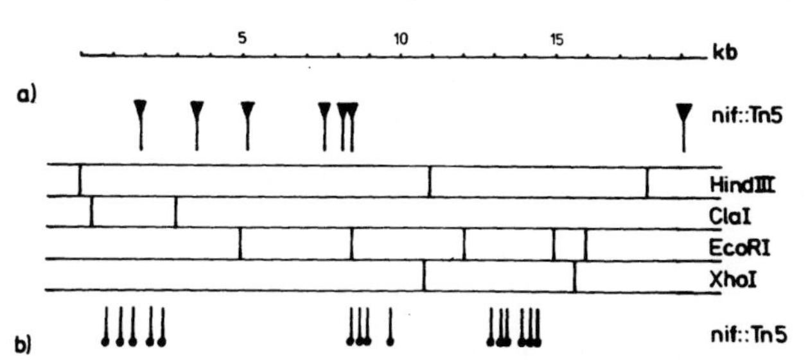

FIGURE 5. The nif mutation cluster of Rhodopseudomonas capsulata B10. Using the broad host range mobilization system general and fragment specific Tn5-induced nif mutations were isolated and located in relation to the restriction map (Figure 5a: general Tn5-induced nif mutations; Figure 5b: fragment specific Tn5-induced nif mutations).

further Tn5-induced nif mutations could be identified (Figure 5b), demonstrating that this 20 kb Rh.capsulata DNA region really contains a cluster of nif genes. In addition, by a special procedure called recombination mutagenesis it could be shown that the nif genes of this region are organized in operons (Klipp, Pühler, this volume).

Evidently, general and fragment specific transposon mutagenesis are powerful systems for the identification, mapping and cloning of genes. In particular, the recently developed broad host range mutagenesis system is very helpful for the genetic analysis of Gram-negative bacteria.

4. MAPPING AND EXPRESSION OF nif AND fix GENES OF Rhizobium meliloti

The identification of R.meliloti nif genes was facilitated by the finding that K.pneumoniae DNA coding for the structural genes of the nitrogenase hybridizes to different Rhizobium species (Ruvkun, Ausubel, 198o). In the case of R.meliloti the hybridizing EcoRI fragment was cloned in E.coli plasmid vectors (Ruvkun, Ausubel, 1981). Further studies using fragment specific mutagenesis in E.coli and transfer of the mutations to the R.meliloti genome confirmed that the cloned fragment carries nif specific genes (Ruvkun et al., 198o; Ruvkun et al., 1982). We analyzed the cloned R.meliloti fragment by the minicell technique, already mentioned in section 2, and were able to express the R.meliloti nifH gene in minicells of E.coli and to locate the nifH coding region (Weber, Pühler, 1982). In the meantime we extended our investigations to adjacent fragments known to be involved in symbiotic nitrogen fixation by R.meliloti. In Figure 6 the collected data from our laboratory are presented. In addition, to the nifH coding region, we were

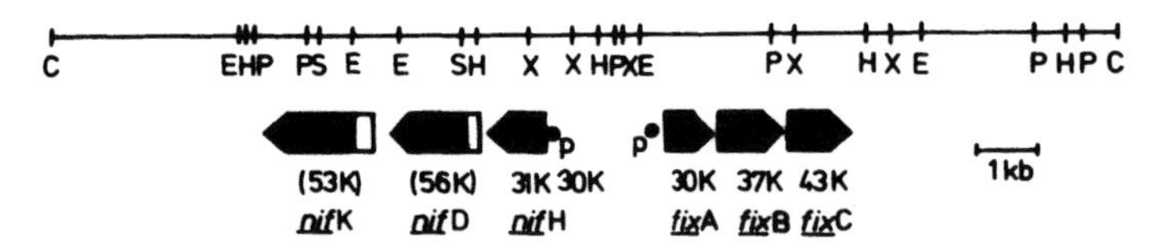

FIGURE 6. A preliminary map of Rhizobium meliloti nif and fix genes and their promoters.
In relation to the restriction map the coding regions of several R.meliloti nif and fix genes are shown. The arrows indicate the direction of transcription and translation. nifK and nifD are presented as hybrid genes (see text). The indicated promoters (p) were identified in E.coli. The nifA as well as the fixA promoter need the K.pneumoniae nifAgp for activation.
Abbreviations: C: ClaI; E: EcoRI; H: HindIII; P: PstI; S: SalI; X: XhoI.

also able to express the R.meliloti nifD and nifK genes in minicells of E.coli (Pühler et al., in press), but for these nif genes, satisfactory results were obtained only after gene fusion experiments. We constructed hybrid genes consisting of parts of the nifD and the nifK coding regions fused to the Cm gene of plasmid pACYC184 (Chang, Cohen, 1978). The molecular weights of the corresponding gene products could be determined. They were 53.000 for hybrid nifKgp and 56.000 for the hybrid nifDgp. As shown in Figure 6, the Cm gene part of these hybrid genes is in both cases the minor one. In addition to these R.meliloti nif genes, we were able to locate three further coding regions next to the nifKDH fragment. For this fragment, it was shown that Tn5 insertions result in a Fix$^-$ phenotype (Ruvkun et al., 1982). By a series of experiments we were able to demonstrate that this fragment codes for three genes which we call fixA, fixB and fixC. These genes are expressed in E.coli minicells and characterized by their molecular weights of 30.000 (fixAgp) and 37.000 (fixBgp) and 43.000 (fixCgp). The specific functions of these fix genes in the symbiotic nitrogen fixation process have not yet been determined.

The organization of the structural genes for nitrogenase in R.meliloti shows a striking similarity to that of K.pneumoniae. We therefore asked the question whether the transcriptional regulation of these genes shares common features. The experimental set-up to solve this question was identical to the previously reported operon fusion technique with K.pneumoniae nif promoters. We fused a promoterless tetracycline resistance gene (Tc) to R.meliloti DNA fragments carrying putative symbiotic promoters. Figure 7a shows such a plasmid (pGW5), which carries the Tc gene under the control of the R.meliloti nifH promoter. Surprisingly, this nif promoter can be activated by the Klebsiella nifAgp which is provided in trans from another plasmid, called pWK131. The pWK131 plasmid expresses the Klebsiella nifA gene constitutively. In addition, plasmid pWK130 is of importance for regulation studies since this plasmid carries the Klebsiella nifA and nifL genes under the control of a constitutive promoter. Both plasmids are presented in Figure 7b. For the actual experiment we constructed an E.coli strain, carrying the plasmids pGW5 and pWK131. This E.coli strain became tetracycline resistant only at low temperature (30°) demonstrating that the temperature sensitive nifAgp activates the R.meliloti nifH promoter. In contrast, an

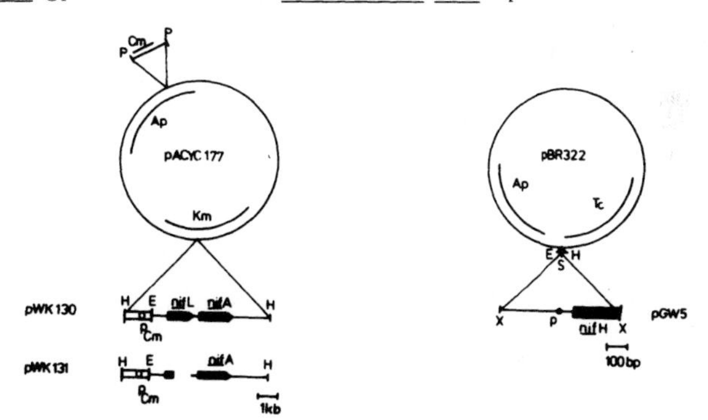

FIGURE 7. Plasmids constructed for the activation of the Rhizobium meliloti nifH promoter by the Klebsiella pneumoniae nifAgp in Escherichia coli. In Figure 7a the restriction map of plasmid pGW5 is shown. An XhoI fragment of R.meliloti DNA carrying the R.meliloti nifH promoter was cloned in the SalI site of a modified pBR322 plasmid (Pühler et al., in press).
In Figure 7b the restriction maps of the plasmids pWK130 and pWK131 are presented. The construction of these plasmids is obvious from the drawing: The EcoRI/HindIII fragment carrying the K.pneumoniae nifL and nifA genes is transcribed from the constitutive p_{Cm} promoter (Pühler et al., in press). pWK131 differs from pWK130 by a partial deletion of the K.pneumoniae insert removing most of the nifL gene.
Abbreviations: Ap: ampicillin resistance; Cm: chloramphenicol resistance; Km: kanamycin resistance; E: EcoRI; H: HindIII; P: PstI; S: SalI.

E.coli strain harboring pGW5 and pWK130 did not show tetracycline resistance at all. Evidently, the Klebsiella nifLgp abolishes the action of the activator protein nifAgp. We applied the experimental technique described above to several R.meliloti DNA fragments carrying the nif and fix genes outlined in Figure 6, and were able to map two nifAgp dependent promoters, one next to nifH and the other to fixA. Both promoters are reading in opposite directions. Our findings could be confirmed: Sundaresan et al.(1983) identified the nifH promoter, whereas Corbin et al. (1983) found the nifH as well as the fixA promoter. Sundaresan et al. (1983) sequenced the R.meliloti nifH promoter and found some homology to the K.pneumoniae nifH promoter. The activation of the R.meliloti nifH and fixA promoters by the K.pneumoniae nifAgp evokes the question whether R.meliloti harbors a "nifA like" gene. Hybridization with a Klebsiella nifA probe shows that a DNA region next to fixC possesses a certain degree of homology (data not shown). Szeto et al. (in press) reported that a Tn5 insert in this region has a regulatory effect on the expression of the structural nifH, nifD and nifK genes. It is of interest to note that, for Rhizobium leguminosarum, a "nifA like" gene has also been postulated (Downie et al., 1983).

5. ANALYSIS OF Tn5 INDUCED NODULATION MUTANTS OF Rhizobium meliloti

The Tn5 mutagenesis system for Gram-negative bacteria introduced in a previous section was also employed to analyze the nodulation process of Medicago sativa (alfalfa) by R.meliloti strain 2011. The results obtained are extensively described by Müller et al. (this volume). Therefore, we are going to mention only two mutants, 1142 and 2526, (Aguilar et al., this volume) which are completely nodulation defective. These mutants could be complemented by the hybrid plasmid pRmSL26 (Long et al., 1982) to a fully effective symbiosis. It is known that pRmSL26 carries genes for early infection. Therefore, we suppose that our R.meliloti mutants 1142 and 2526 are also blocked in one of the first nodulation steps. Surprisingly, these mutants could be also complemented by the R.leguminosarum Sym plasmid pIJ1008 (Brewin et al., 1982) indicating that the blocked nodulation step is similar in both species. We therefore tested the DNA homology of two different Rhizobium nod regions. For the experiment, we used an EcoRI fragment of the R.trifolii nod region known to carry nod genes (Rolfe, personal communication) and an EcoRI fragment of plasmid pRmSL26 caryying R.meliloti nod genes. DNAs of both fragments were prepared and their homology was tested by an electron microscopic heteroduplex experiment. Without discussing experimental details and showing actual micrographs, the collected data are presented in Figure 8. We were

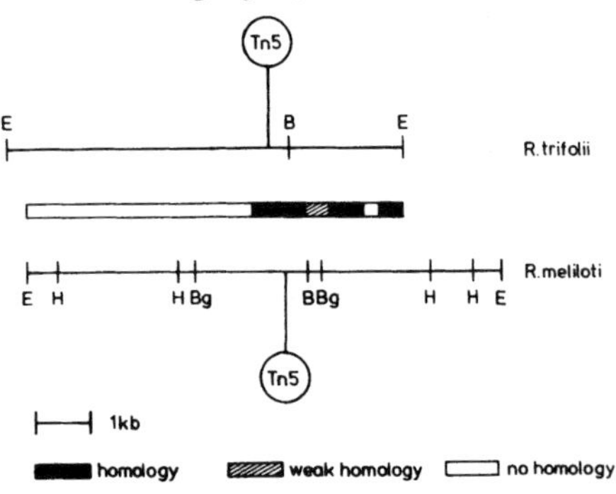

FIGURE 8. DNA homology between an R.meliloti and an R.trifolii nod fragment as revealed by an electron microscopic experiment.
An EcoRI fragment of pRT851 (Rolfe, personal communication) carrying Tn5 (R.trifolii nod fragment) and an EcoRI fragment of pRmSL26 (R.meliloti nod fragment) were used. For details see text.
Abbreviations: B: BamH1; Bg: BglII; E: EcoRI; H: HindIII.

able to define homology regions of no homology, weak homology and strong homology between the nod fragments of R.meliloti and R.trifolii. As shown in Figure 8, the R.trifolii nod fragment carries a Tn5 insert, which results in a Nod$^-$ phenotype (Scott et al., 1982). In the R.meliloti nod fragment we also isolated a Tn5 insert causing a Nod$^-$ phenotype (Aguilar et al., this volume). In the heteroduplex experiment, both Tn5 inserts could be located very near to each other. Further nod::Tn5 mutations which also map in the homology region defined in Figure 8 were isolated by Long et al. (personal communication). From these experiments one can conclude that different Rhizobium species (R.trifolii, R.leguminosarum and R.meliloti) possess nodulation genes which were evidently conserved during evolution. To analyze these genes and their functions in more detail is a future task.

6. *Rhizobium meliloti* MEGAPLASMIDS

From a series of studies, it is known that R.meliloti nif, fix and nod genes map on a so-called Megaplasmid (Banfalvi et al., 1981; Rosenberg et al., 1981). In order to analyze thse symbiotic genes encoded by the Megaplasmid in more detail, we employed the Tn5-Mob transposon (Figure 4) for conjugational transfer experiments. This Tn5-Mob transposon has been provided with the RP4 Mob-fragment by an in vitro insertion (Simon et al., in press a). Tn5-Mob shows the following characteristics: It transposes with a frequency similar to the wildtype Tn5 and provides a Mob site to its host replicon. Thus, host replicons carrying Tn5-Mob can be mobilized by plasmid RP4. This special transposon is widely employable and was, for example, used for incompatibility studies of different Rhizobium plasmids (O'Connell et al., this volume). We inserted Tn5-Mob into the R.meliloti Megaplasmid and transferred this plasmid to Agrobacterium tumefaciens, which subsequently became able to induce ineffective nodules on alfalfa. This demonstrates that the nodulation ability is encoded by this plasmid. Nodules induced by A.tumefaciens carrying the R.meliloti Megaplasmid were also studied by other groups (Wong et al., in press; Wilson et al., in press). Recently, Simon et al. (in press a) could demonstrate that two strains of R.meliloti carry a second Megaplasmid, which we call Mega 2 in contrast to the nif and nod carrying Mega 1. Unpublished results from our laboratory indicate that Mega 2 most probably does not encode nodulation functions. This was achieved by constructing A.tumefaciens strains carrying Mega 1 and/or Mega 2. A.tumefaciens was able to induce ineffective nodules only when Mega 1 was present. By a special Eckhardt gel technique (Eckhardt, 1978) we were able to band the R.meliloti Mega 1 and Mega 2 plasmids in different positions in an agarose gel demonstrating that Mega 2 is larger than Mega 1. In addition, we found a faint band above Mega 2 which perhaps represents a further Megaplasmid (Mega 3) or the R.meliloti chromosome (unpublished results).

7. ACKNOWLEDGEMENTS

We wish to thank our colleagues who allowed us to cite unpublished results. In addition, we have to mention two of our students, Monika Labes and Petra Grönger, who contributed scientifically to this paper. We also acknowledge the help of our technicians and secretary. This work was supported by Agrigenetics Corporation, Bundesministerium für Forschung und Technologie and Deutsche Forschungsgemeinschaft.

8. REFERENCES

Bagdasarian M, Bagdasarian MM, Coleman S and Timmis KN (1979) In Timmis KN and Pühler A, eds, Plasmids of Medical, Environmental and Commercial Importance, p. 411, Elsevier/North Holland Biomedical Press, Amsterdam New York.
Bagdasarian M, Lurz R, Rücker B, Franklin FCH, Bagdasarian MM, Frey J and Timmis KN (1981) Gene 16, 237.
Banfalvi Z, Sakanyan V, Koncz C, Kiss A, Dusha I and Kondorosi A (1981) Mol. Gen. Genet. 184, 318.
Beringer JE, Beynon JL, Buchanan-Wollaston AV and Johnston AWB (1978) Nature 276, 633.
Beynon J, Cannon M, Buchanan-Wollaston V and Cannon F (in press) Cell.
Bolivar F, Rodriguez R, Greene PJ, Betlach MC, Heyneker HL and Boyer HW (1977) Gene 2, 95.
Brewin NJ, Wood EA, Johnston AWB, Dibb NJ and Hombrecher G (1982) J. Gen. Microbiol. 128, 1817.
Chang ACY and Cohen SN (1978) J. Bact. 134, 1141.
Corbin D, Barran L and Ditta G (1983) Proc. Natl. Acad. Sci. USA 8o, 3oo5.
Datta N, Hedges RW, Shaw EJ, Sykes EJ and Richmond MH (1971) J. Bact. 1o8, 1244.
Ditta G, Stanfield S, Corbin D and Helinski DR (198o) Proc. Natl. Acad. Sci. USA 77, 7347.
Dixon RA and Postgate JR (1971) Nature 234, 47.
Dixon R, Kennedy C and Merrick (1981) In Glover SW and Hopwood DA, eds, Genetics as a Tool in Microbiology, p. 161, Cambridge University Press.
Downie JA, Ma QS, Knight CD, Hombrecher G and Johnston WB (1983) EMBO Journal 2.
Eckhardt T (1978) Plasmid 1, 584.
Kennedy C, Cannon F, Cannon M, Dixon R, Hill S, Jensen J, Kumar S, McLean P, Merrick M, Robson R and Postgate J (1981) In Gibson AH and Newton WE, eds, Current Perspectives in Nitrogen Fixation, p. 146, Austr. Ac. Sc., Canberra.
Long SR, Buikema WJ and Ausubel FM (1982) Nature 298, 485.
Merrick M (1983) EMBO Journal 2, 39.
Nieva-Gomez D, Roberts GP, Klevickis S and Brill WJ (198o) Proc. Natl. Acad. Sci USA 77, 2555.
Pühler A, Burkardt HJ and Klipp W (1979a) Molec. Gen. Genet. 176, 17.
Pühler A, Burkardt HJ and Klipp W (1979b) In Timmis K and Pühler A, eds, Plasmids of Medical, Environmental and Commercial Importance, p. 435, Elsevier/North Holland Biomedical Press, Amsterdam New York.
Pühler A and Klipp W (1981) In Bothe H and Trebst A, eds, Biology of Inorganic Nitrogen and Sulfur, p. 275, Springer-Verlag, Berlin Heidelberg New York.
Pühler A, Klipp W and Weber G (1982) In Proceedings of the IVth Intern. Symp. on Genetics of Industrial Microorganisms, p. 32o, Tokyo.
Pühler A and Klipp W (1983) In Müller A and Newton WE, eds, Nitrogen Fixation, p. 111, Plenum Press, New York and London.
Pühler A, Klipp W and Weber G (in press) In Pühler A, ed, Molecular Genetics of the Bacteria Plant Interaction, p. 69, Springer-Verlag, Berlin Heidelberg New York.
Roberts GP, MacNeil T, MacNeil D and Brill WJ (1978) Proc. J. Bacteriol. 136, 267.
Rosenberg C, Boistard P, Dénarié J and Casse-Delbart F (1981) Mol. Gen. Genet. 184, 326.

Ruvkun GB and Ausubel FM (198o) Proc. Natl. Acad. Sci. USA 77, 191.
Ruvkun GB, Long SR, Meade HM and Ausubel FM (198o) Cold Spring Harbor Symposia on Quant. Biol. 45, 492.
Ruvkun GB and Ausubel FM (1981) Nature 289, 85.
Ruvkun GB, Sundaresan V and Ausubel FM (1982) Cell 29, 551.
Scott KF, Hughes JE, Gresshoff PM, Beringer JE, Rolfe BG and Shine J (1982) J. Mol. Appl. Gen. 1, 315.
Shen S, Xue Z, Kong Q and Wu Q (1983) Nucl. Acids Res. 11, 4241.
Sibold L, Quiviger B, Charpin N, Paquelin A and Elmerich C (1983) Biochimie 1, 53.
Simon R, Priefer U and Pühler A (in press a) In Pühler A, ed, Molecular Genetics of the Bacteria Plant Interaction, Springer-Verlag, Berlin Heidelberg New York.
Simon R, Priefer U and Pühler A (in press b) Biotechnology.
Sundaresan V, Jones JDG, Ow DW and Ausubel FM (1983) Nature 3o1, 728.
Sundaresan V, Ow DW and Ausubel FM (in press) Proc. Natl. Acad. Sci. USA.
Szeto WW, Zimmerman JL and Ausubel FM (in press) In Pühler A, ed, Molecular Genetics of the Bacteria Plant Interaction, p. 64, Springer-Verlag, Berlin Heidelberg New York.
Streicher S, Gurney E and Valentine RC (1971) Proc. Natl. Acad. Sci. USA 68, 1174.
Vliet F Van, Silva B, Montagu M Van and Schell J (1978) Plasmid 1, 446.
Weber G and Pühler A (1982) Plant Mol. Biol. 1, 3o5.
Wilson KJ, Hirsch AM, Jones JDG and Ausubel FM (in press) In Ahmad F, Downey K, Schultz J and Voellmy RW, eds, Advances in Gene Technology: Molecular Genetics of Plants and Animals, Academic Press, New York.
Wong CH, Pankhurst CE, Kondorosi A and Broughton WJ (in press) J. Cell Biol.
Zhu J and Brill WJ (1981) J. Bact. 145, 1116.

MOLECULAR CLONING AND ORGANISATION OF GENES INVOLVED IN SYMBIOTIC NITROGEN FIXATION IN DIFFERENT RHIZOBIUM SPECIES

JOHN SHINE/PETER R. SCHOFIELD/JEREMY J. WEINMAN/FLORENCE FELLOWS/JANE BADENOCH-JONES/NIGEL MORRISON/KIERAN F. SCOTT/PETER M. GRESSHOFF/JOHN M. WATSON and BARRY G. ROLFE
CENTRE FOR RECOMBINANT DNA RESEARCH and DEPARTMENTS OF GENETICS AND BOTANY, AUSTRALIAN NATIONAL UNIVERSITY, CANBERRA, AUSTRALIA

1 INTRODUCTION

The symbiotic association between plants and bacteria of the genus Rhizobium is the result of a complex interaction between the bacterium and its host, requiring the expression of both bacterial and plant genes in a tightly co-ordinated manner. Bacteria bind to the emerging plant root hairs and invade the root tissue through the formation of an infection thread. The plant responds to this infection by the development of a highly differentiated root nodule. These nodules are the site of synthesis of the bacterial enzyme complex nitrogenase, which reduces atmospheric nitrogen to ammonia. The fixed nitrogen is then exported into the plant tissue and assimilated by plant-derived enzymes.

Rhizobium species are defined on the basis of their host range with particular legumes being nodulated only by certain Rhizobium strains. Thus R.trifolii nodulates clovers, R.leguminosarum-peas, R.meliloti-lucerne and R.japonicum-soybeans. In contrast to the narrow specificity shown by these Rhizobium species on temperate legumes, another large group known as the 'cowpea miscellany' contain those rhizobia which have a broad host range and can usually infect a variety of tropical legumes. This distinction between the two groups based on host range is paralleled by their different growth rates in the laboratory. The slow-growing rhizobia comprise most of the cowpea group of bacteria, whilst the narrow host range bacteria generally fall into the class of fast-growing rhizobia.

Although most Rhizobium symbioses are confined to leguminous plants, strains of Rhizobium have been isolated which are able to fix N_2 in symbiosis with members of the non-legume Parasponia genus. These strains belong to the slow-growing group of rhizobia and thus also form an effective symbiosis with a diverse group of tropical legumes, although fast-growing strains with the same broad host range have also recently been isolated (Trinick and Galbraith, 1980).

The use of molecular cloning techniques, coupled with specific transposon mutagenesis, has permitted the isolation and study of bacterial genes involved in each of the many steps leading to effective symbiotic nitrogen fixation (Ruvkun et al., 1980; Buchanan-Wollaston et al., 1980; Scott et al, 1982). In this report we discuss the results of such studies on the structure and organisation of genes involved in symbiotic nitrogen fixation in different strains of Rhizobium.

2. The nif/nod region of R.trifolii ANU843

Rhizobium trifolii ANU843 is a typical fast-growing, narrow host range Rhizobium which effectively nodulates clover plants. We have previously shown that this strain contains at least five large plasmids which range in size from about 180 Kb to greater than 500 Kb (Djordjevic et al., 1983). Hybridisation analysis of the separated plasmids, and of heat-cured strains missing particular plasmids, has demonstrated that both nodulation (nod) and nitrogen-fixation (nif) genes are located on the smallest (180 Kb) plasmid (Djordjevic et al., 1983; Schofield et al., 1983).

In order to identify and characterise the symbiotic genes of strain ANU843 in more detail, the region of the sym plasmid containing Nod⁻ Tn5 insertions and the nif HDK genes was analysed by extensive restriction endonuclease mapping (Fig. 1). The nif region was identified by using nifH- and nifD-specific DNA sequences, isolated from another R.trifolii strain (Scott et al., 1983), as hybridisation probes. The nod genes are identified by the point of insertion of Tn5 in different Tn5-induced Nod⁻ mutants. This approach clearly demonstrated the presence of at least two separate nod genes (or operons), the limits of which at present are defined by the positions of flanking Nod⁺, Tn5 insertions. The number of separate nod 'genes' may increase as the resolution in this region is improved by the characterisation of further Tn5 insertions. At present it appears likely that there are at least three nod 'genes' since the 7.2 Kb Eco R1 fragment which carries a complete copy of the two known nod regions does not carry enough information to restore nodulation to the sym plasmid-cured R.trifolii strain ANU845 (see below and Schofield et al. 1983).

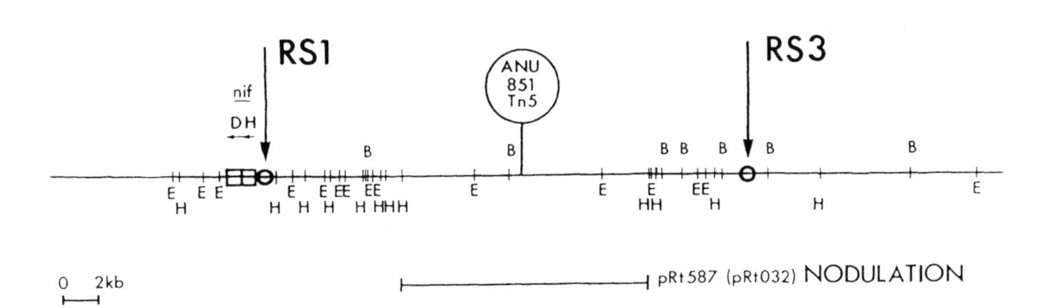

FIGURE 1. Molecular map of the nodulation and nitrogen fixation gene region of the Sym plasmid of R.trifolii strain ANU843. Restriction enzyme sites: B=BamI; E=EcoRI; H=HindIII. RS1 and RS3 are repeated sequences.

An analysis of the nodulation functions encoded on the sym plasmid was carried out by first subcloning fragments of the symbiotic region into broad host range, multi-copy plasmid vectors such as pKT240 (Bagdasarian and Timmis, personal communication). The recombinant plasmids were then mobilized from E.coli into R.trifolii ANU845 (the sym plasmid minus strain derived from ANU843) using a derivative of RP4 as the mobilising plasmid. Such recombinants were then assayed for the restoration of nodulation ability to ANU845. Using this approach, the 14 Kb Hind III fragment shown in Fig. 1 was found to contain all of the Sym plasmid-encoded information necessary for the formation of nodules (Fig. 2). Since neither ANU845 or the cloned 14 Kb Hind III fragment carry any nitrogenase structural genes, the resultant nodules were ineffective and were less well-developed than those induced by the wild-type ANU843 (Fig. 2).

The cloned 14 Kb fragment was also mobilised into Agrobacterium tumefaciens A136 and into a 'cowpea' Rhizobium strain ANU240. Although neither of these two

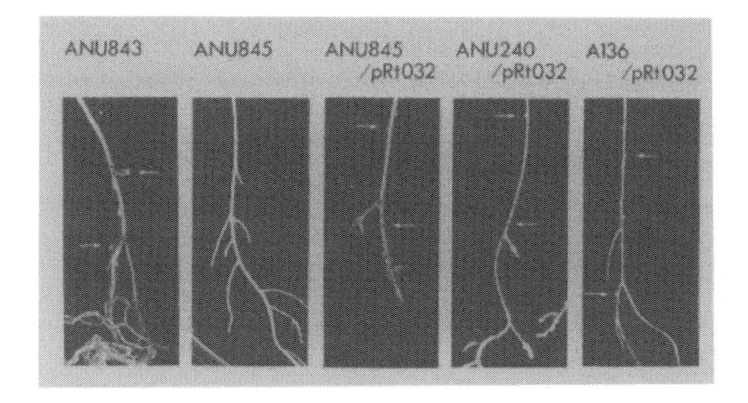

FIGURE 2. Nodules induced on the roots of white clover (Trifolium repens) plants by various derivatives of Rhizobium and Agrobacterium tumefaciens. Roots were photographed five weeks after inoculation. Nodules indicated by arrows. pRT032 is a pKT240 recombinant containing the 14 Kb HindIII fragment.

bacteria are normally capable of nodulating clover, both recombinants were found to induce nodules on white clover seedlings (Fig. 2).

We have previously shown in another R.trifolii strain (ANU329) that the nifH coding sequence is preceeded by a sequence which is repeated approx. five times in the genome (Scott et al., 1983). In strain ANU843 this sequence has now been shown to immediately precede the nifH coding region extending some 200 bp to the 5'-side of nifH (Fig. 1). Another copy of the repeat is located some 10 Kb to the other side of the nod region (RS3 in Fig. 1). At present it would appear that the repeated sequence represents a nif-specific symbiotic promoter/regulatory sequence since: a) the copy preceeding nifH contains the nifH promoter and presumptive regulatory regions; b) the sequence is repeated approximately five times; c) all copies of the repeat are restricted to the Sym plasmid (Fig. 3), and d) the promoter regions of the nod genes do not appear to contain copies of the repeated sequence.
At present we do not know if such a repeated sequence is a general feature of the Sym plasmid in other Rhizobia.

3. The Nif genes of Parasponia Rhizobium ANU289

Parasponia Rhizobium ANU289 is a slow-growing Rhizobium strain which shows similarities to the cowpea group of rhizobia but is also capable of the effective nodulation of the non-legume Parasponia. Unlike fast-growing strains, nitrogen fixation in ANU289 is inducible in vitro. This property makes the strain useful for studies on the regulation of nif gene expression in culture, as well as in the symbiotic state.

Hybridisation analysis of ANU289 DNA using as probes the previously cloned Klebsiella nifH and nifD sequences (Scott et al., 1981) demonstrated that there

is only a single copy of each of the nif structural genes in this strain. Furthermore, such an analysis suggests that perhaps, unlike the situation in Klebsiella (Scott et al., 1981) and Rhizobium trifolii (Scott et al., 1983), the nifH and nifDK genes may not be contiguous in the genome. Molecular cloning of the nifH,D,K genes and their characterisation by direct DNA sequence analysis has now demonstrated that, although in all other nitrogen fixing organisms examined to date the nifH and nifD genes are linked on the same operon, in this slow-growing Rhizobium strains these two genes are separated by at least 13 Kb and are encoded on separate operons (Scott et al., 1983a).

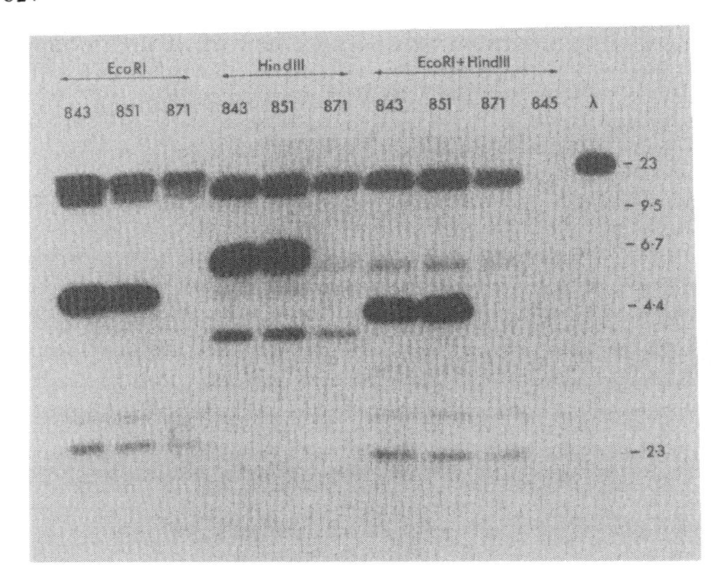

FIGURE 3. Demonstration of a repeated sequence on the Sym plasmid of Rhizobium trifolii strain ANU843. Total DNA was isolated from each of the ANU strains and digested with EcoRI and/or HindIII. After electrophoresis in an agarose gel, the separated fragments were transferred to nitrocellulose and hybridized with ^{32}P-labelled RS1. The resultant autoradiograph is shown. On the right are shown HindIII restriction fragments of phage lambda DNA with sizes in kb. Strain 851 carries a Tn5 insertion in the Nod region. 871 is a derivative of 851 in which the Sym plasmid has sustained a 45 kb deletion.

If the synthesis of nitrogenase is controlled by a general regulatory mechanism in this strain, as in Klebsiella, the co-ordinate expression of nifH and nifDK would require the action of a common regulatory effector at both the nifH and nifDK transcription initiation signals. In this case, the two promoter sequences might be expected to share common structural features corresponding to the effector recognition sites, although no regions of extensive homology are present in the primary sequence of the nifH and nifDK promoters. S1 nuclease mapping and primer extension experiments have demonstrated a large difference in the size of the 5'-untranslated regions of the two operons. The nifH region is approx. 150 nucleotides long whilst only approx. 35 nucleotides are present before the initiator codon in the nifDK transcript (Weinman et al., 1984).

4. The nif/nod region of Rhizobium strain ANU240

The Rhizobium strain ANU240 is a fast-growing Rhizobium capable of nitrogen fixation on a wide variety of tropical legumes normally nodulated by slow-growing strains. It also nodulates the non-legume Parasponia, but the symbiosis is ineffective.

Examination of the genes involved in both nodulation and nitrogen fixation suggest that ANU240 is more closely related to other fast-growing strains than it is to the slow-growers with which it shares a similar broad host range. Like most fast-growing strains, the nif and nod genes appear to be closely linked on a Sym plasmid (Morrison et al., 1984). Furthermore, at least two distinct nod 'genes' have been identified which share significant sequence homology to two of the genes found on the R.trifolii 14 Kb Hind III fragment (Fig. 1). The functional relationship of these gene products has also been demonstrated by the ability of the ANU240 sequences to complement Nod$^-$ mutations in R.trifolii and vice versa (Rolfe et al., 1983). A more extensive analysis of the 2-3 nod 'genes' in different fast-growing Rhizobium has confirmed the general observations of structural conservation (by specific hybridisation) and functional similarities (by complementation ability) (Rolfe and Shine, 1983).

Although analysis of the nif genes of ANU240 clearly demonstrates that nif H,D and K are closely linked in the genome, there is some evidence that,unlike R.trifolii or ANU289, there may be two copies of the nif structural genes.

5. REFERENCES

Buchanan-Wollastron AV, Beringer JE, Brewin NJ, Hirsch PR and Johnston AWB (1980) Mol.Gen.Genet. 178, 185-190.
Djordjevic, MA, Zurkowski W, Shine J and Rolfe BG (1983) J.Bact. (in press).
Morrison N, Chen HC, Bassam B, Plazinski J, Ridge R, Shine J and Rolfe BG (1984) - these Proceedings.
Rolfe BG and Shine J (1983) in Plant Gene Research Verma, DP Ed., Springer-Verlag, in press.
Ruvkun GB, Long SR, Meade HR and Ausubel FM (1980) Cold Spring Harbor Symp. Quant.Biol. 45, 492-499.
Schofield, PR, Djordjevic MA, Rolfe BG, Shine J and Watson JM (1983) Mol.Gen. Genet. (in press).
Scott KF, Rolfe BG and Shine J (1981) J.Mol.Appl.Genet. 1, 71-81.
Scott, KF, Hughes JE, Gresshoff PM, Beringer JE, Rolfe BG and Shine J (1982) J.Mol.Appl.Genet 1, 315-326.
Scott KF, Rolfe BG and Shine J (1983) DNA 2, 149-155.
Scott KF, Rolfe BG and Shine J. (1983a) DNA 2, 141-148.
Trinick MJ and Galbraith J (1980) New Phytol. 86, 17-26.
Weinman J, Fellows F, Gresshoff P, Shine J and Scott K (1984) these Proceedings.

REGULATION OF THE NITROGEN FIXATION (*nif*) GENES OF *Klebsiella pneumoniae* AND *Rhizobium meliloti*: ROLE OF NITROGEN REGULATION (*ntr*) GENES

F.J. DE BRUIJN*/V. SUNDARESAN+/W.W. SZETO/D.W. OW°/F.M. AUSUBEL
DEPARTMENT OF MOLECULAR BIOLOGY, MASSACHUSETTS GENERAL HOSPITAL, BOSTON, MASSACHUSETTS 02114 U.S.A.
* PRESENT ADDRESS: MAX PLANCK INSTITUT FUR ZUCHTUNGSFORSCHUNG, ERWIN BAUR INSTITUT, 5000 KOLN 30 (VOGELSANG), WEST GERMANY
+ PRESENT ADDRESS: OXFORD TRACT, DEPARTMENT OF GENETICS, UNIVERSITY OF CALIFORNIA, BERKELEY, CALIFORNIA 94720 U.S.A.
° PRESENT ADDRESS: SHANGHAI INSTITUTE OF PLANT PHYSIOLOGY, 300 FENG LIN ROAD, SHANGHAI, 200032, PEOPLE'S REPUBLIC OF CHINA

1. INTRODUCTION

In this article we describe experiments which demonstrate that there are surprising similarities in the regulation of *nif* genes in *Klebsiella pneumoniae* and *Rhizobium meliloti*.

K. pneumoniae nif genes are regulated on two distinct levels. The first level involves a centralized nitrogen regulation system (the *ntr* system), mediated by the products of *glnF*(*ntrA*), *glnL*(*ntrB*) and *glnG*(*ntrC*) (Leonardo, Goldberg, 1980; de Bruijn, Ausubel, 1981,1983; Espin et al., 1981, 1982; Ow, Ausubel, 1983; Merrick, 1982, 1983; Drummond et al., 1983). The second level of *nif* regulation involves the specific control of *nif* operons (except the *nifLA* operon) by the *nifA* and *nifL* products (Roberts, Brill, 1980; Dixon et al., 1980; Hill et al., 1981; Merrick et al., 1982; Buchanan-Wollaston et al., 1981a,b). Under conditions of nitrogen starvation, the products of *glnF*(*ntrA*) and *glnG*(*ntrC*) interact to activate the *nifLA* promoter (Ow, Ausubel, 1983; Merrick 1983). The *nifA* product in turn activates transcription of all other *nif* transcription units (Roberts, Brill, 1980; Dixon et al., 1980; Buchanan-Wollaston et al., 1981a). The *nifL* product mediates both ammonia and oxygen repression of *nif* operons once the *nif* system has been derepressed (Hill et al., 1981; Merrick et al., 1982; Buchanan-Wollaston et al., 1981a).

Recent results from our laboratory (Ow, Ausubel, 1983) and from Merrick (1983) have shown that the *nifA* and *glnG*(*ntrC*) genes and gene products are closely related. This conclusion is based on the observations that *nifA* can substitute for *glnG*(*ntrC*) in the activation of *ntr* controlled genes (Ow, Ausubel, 1983) and that *nifA* activation requires *glnF*(*ntrA*), in direct analogy with *glnG*(*ntrC*) mediated activation (Ow, Ausubel, 1983; Sundaresan et al., 1983b; Merrick, 1983). In addition, DNA homology exists between the *nifA* and *glnG*(*ntrC*) genes (F. de Bruijn, unpublished observation), further supporting the hypothesis that *nifA* and *glnG*(*ntrC*) have a common evolutionary origin.

The ability of *nifA* product to substitute for *glnG*(*ntrC*) product in the activation of the *nifL*, *glnA*, *aut*, and *put* genes, led to the discovery of a consensus sequence, TTTTGCA, located in the promoter regions of *nifA*/*glnG*(*ntrC*) activated genes (Ow et al., 1983; Sundaresan et al., 1983a). In the case of the *K. pneumoniae nifL* gene, where the start site of transcription was determined, the TTTTGCA consensus sequence was found to be located at the -10 to -15 region. Moreover, reflecting the fact that *glnG*(*ntrC*) cannot activate the *K. pneumoniae nifHDKY* operon, this promoter region contains the sequence CCCTGCA at the -10 to -15 region. Independently of these studies, Beynon et al. (1983) found a subset of the

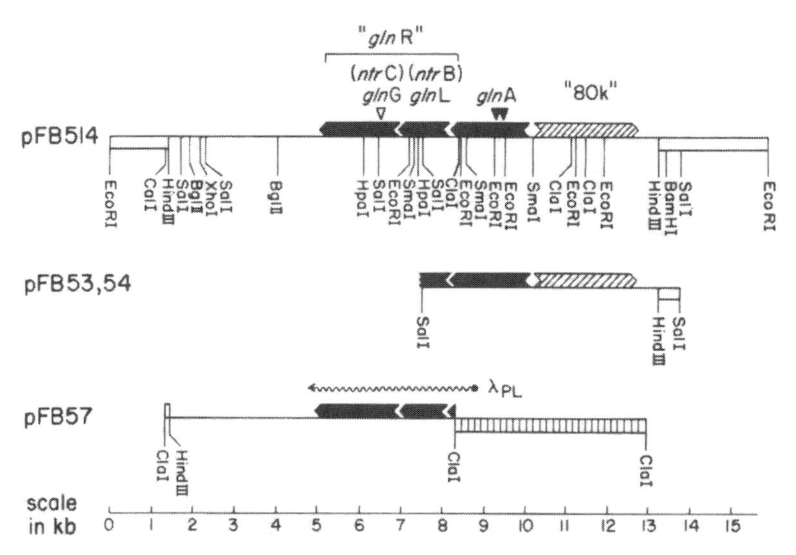

FIGURE 1. Correlated physical and genetic map of the K. pneumoniae glnALG(glnA-ntrBC) region (previously designated glnAR) carried by plasmids pFB514, 53, 54 (de Bruijn, Ausubel, 1981) and pFB57. The open boxes represent pACYC184 (Chang, Cohen, 1978) and vertically crosshatched boxes represent pλ8 (Reed, 1981). Filled in arrows represent the approximate locations of the coding sequences for the glnG(ntrC), glnL(ntrB) and glnA proteins and the crosshatched arrows the position of the "80K" protein. The wavy arrow shows the direction of transcription originating at the λP_L promoter, responsible for glnLG(ntrBC) expression on plasmid pFB57.

heptameric sequence, TTGCA, to be a consensus sequence in the -10 to -15 region of seven K. pneumoniae nif operons.

2. IDENTIFICATION OF ntr GENE PRODUCTS

We previously described the cloning of the K. pneumoniae glnALG (glnA-ntrBC) region (de Bruijn, Ausubel, 1981). We have extended this analysis by carrying out a detailed restriction endonuclease cleavage analysis of this region (see Fig. 1) and an analysis of the polypeptides synthesized by this region in Maxicells (Sankar et al., 1979)(Fig. 2). The glnA gene product (GS; 58 Kd), and a prominent polypeptide of 80 Kd, encoded by the region immediately upstream of the glnA gene, were identified (see Fig. 2, lane 4). However, no polypeptides corresponding to the glnL(ntrB) or glnG(ntrC) gene products were observed.

In order to amplify the levels of glnL(ntrB) and glnG(ntrC) products, we constructed plasmid pFB57 (Fig. 1) containing the glnLG(ntrBC) region fused to the strong λP_L promoter of plasmid pλ8 (Reed, 1982). When this gene fusion was analyzed in Maxicells, the glnL(ntrB) and glnG(ntrC) products were identified as 36 Kd and 56 Kd polypeptides, respectively (Fig. 2, lane 1), in reasonable agreement with those obtained by Espin et al. (1982) for K. pneumoniae and by Magasanik (1982) for E. coli.

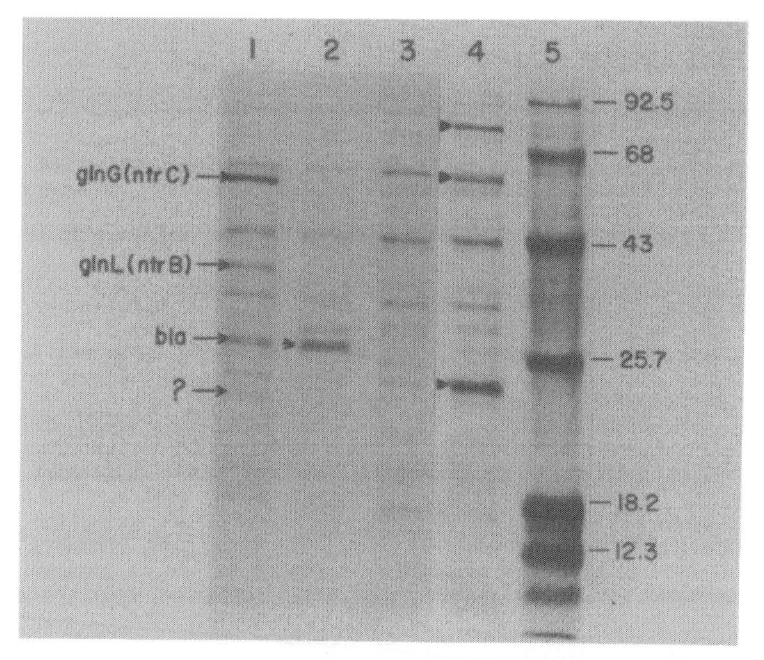

FIGURE 2. Identification of the glnA, glnL(ntrB), glnG(ntrC) and "80K" gene products. Polypeptides synthesized by plasmids pFB57, pλ8 (vector), and pFB514 in the Maxicell system (Sankar et al., 1979) and fractionated by SDS-Page (12.5%; Laemmli et al., 1970). The Maxicell extracts were prepared as described by de Bruijn and Ausubel (1983). The migration positions of glnG(ntrC), glnL(ntrB) and bla (β-lactamase; product of the Ap^r gene of pλ8) are indicated in lane 1, in addition to a minor polypeptide of unknown origin (de Bruijn, 1983); their approximate molecular weights are 58K (kilodaltons), 36K, 26K and 21K respectively, as determined by comparison with the molecular weight standards shown in lane 5. The polypeptides synthesized by Maxicells harboring pλ8 (vector) or devoid of plasmid DNA and shown in lanes 2 and 3 respectively. Lane 4 shows the polypeptides synthesized by plasmid pFB514 (see Figure 1). The migration positions of the "80K" protein, GS (glnA protein; glutamine synthetase) and the cat protein (Cm^r gene product) are indicated by arrow heads; their approximate molecular weights are 80K, 56K and 22.5K respectively.

In addition to our work on glnG(ntrC) and glnL(ntrB), we have also cloned and characterized the *K. pneumoniae* glnF(ntrA) gene. The cloned gene was subjected to Tn5 mutagenesis, a correlated physical-genetic map was constructed, and the glnF product was identified in Maxicells as an 84 Kd protein (de Bruijn, Ausubel, 1983).

3. REGULATION OF *Rhizobium meliloti* nif GENES BY ntr GENES

Relatively little is known about the organization and regulation of nif genes in the symbiotic diazotrophs such as *R. meliloti*, the endosymbiont of alfalfa (*Medicago sativa*). However, a number of observations from our laboratory suggest that both nifA and ntr-like regulation of nif genes has been conserved between the free-living and symbiotic diazotrophs.

In R. meliloti, a cluster of symbiotic genes, including the nitrogenase structural genes (nifHDK), has been identified and mapped by Tn5 mutagenesis (Ruvkun, Ausubel, 1981; Ruvkun et al., 1982; Buikema et al., 1983; Corbin et al., 1982, 1983). Recently, we screened Tn5 insertions in symbiotic genes for ones which had a pleiotropic effect on nifHDK expression and identified a presumptive regulatory gene, 5 kb upstream of the nifHDK operon, required for nifHDK transcription (Zimmerman et al., 1983; Szeto et al., submitted). This presumptive regulatory gene hybridized to an E. coli glnG(ntrC) probe (Szeto, et al., submitted). These results indicate that the nifHDK operon in R. meliloti is regulated by a ntr or nifA-like gene. A nifA-like gene has also been described by Downie et al. (1983) in R. leguminosarum.

The presence of nifA and/or glnG(ntrC)-like regulatory genes in R. meliloti is further substantiated by a series of experiments utilizing a fusion of the R. meliloti nifH promoter to the lacZ gene (Sundaresan et al., 1983a,b). Experiments carried out in wild-type, glnG(ntrC)⁻, and glnF(ntrA)⁻ strains of E. coli and K. pneumoniae showed that activation of the R. meliloti nifH promoter required the nifA or glnG(ntrC) products in conjunction with the glnF(ntrA) product. In contrast, the K. pneumoniae nifH promoter could only be activated by the K. pneumoniae nifA product (Sundaresan et al., 1983a,b; de Bruijn, Ausubel, 1983). These observations are consistent with the finding that the R. meliloti nifH promoter contained the TTTTGCA consensus sequence in the -10 to -15 region (Sundaresan et al., 1983a; Ow et al., 1983).

4. MOLECULAR CLONING OF THE R. meliloti glnA GENE

As illustrated in Fig. 1, the glnG(ntrC) gene is contained in a complex glnALG(glnA-ntrBC) operon in enteric species such as K. pneumoniae. Because of this structure, polar mutations in glnA (the structural gene for glutamine synthetase) have a Ntr⁻ phenotype. In Rhizobium species, GlnA⁻ (GSI⁻) mutants can also exhibit a Nif⁻ phenotype (Kondorosi et al., 1977; Ludwig, 1980). On the basis of these observations, we examined the R. meliloti glnA(GSI) region for the possible presence of a (linked) glnG(ntrC)-like gene. We cloned the R. meliloti glnA(GSI) gene (plasmid pFB616) in the mobilizable cosmid vector pLAFR1 (Friedman et al., 1982) by complementing a GlnA⁻ strain of E. coli deleted for the entire glnALG region. The glnA gene on pFB616 was located on a 10 kb BamHI fragment by Tn5 transposon mutagenesis and the BamHI fragment was subcloned in pACYC184, resulting in plasmid pFB6162 which directed the synthesis of a 56 Kd polypeptide in Maxicells (Fig. 3). The cloned R. meliloti glnA gene produced a heat stable GS activity (D. Keister, personal communication), suggesting that it encodes the structural gene for R. meliloti GSI (Darrow et al., 1981).

A convenient assay for the presence of a glnG(ntrC)-like gene linked to the cloned R. meliloti nifA gene utilized a plasmid containing the K. pneumoniae nifLA promoter fused to lacZ (pOB86). Since activation of the nifLA promoter is dependent on the glnG(ntrC) gene product, no β-galactosidase activity was observed when plasmid pOB86 was introduced into a glnG⁻(ntrC⁻) strain of E. coli. However, when a second plasmid carrying the cloned K. pneumoniae glnALG(glnA-ntrBC) region was also

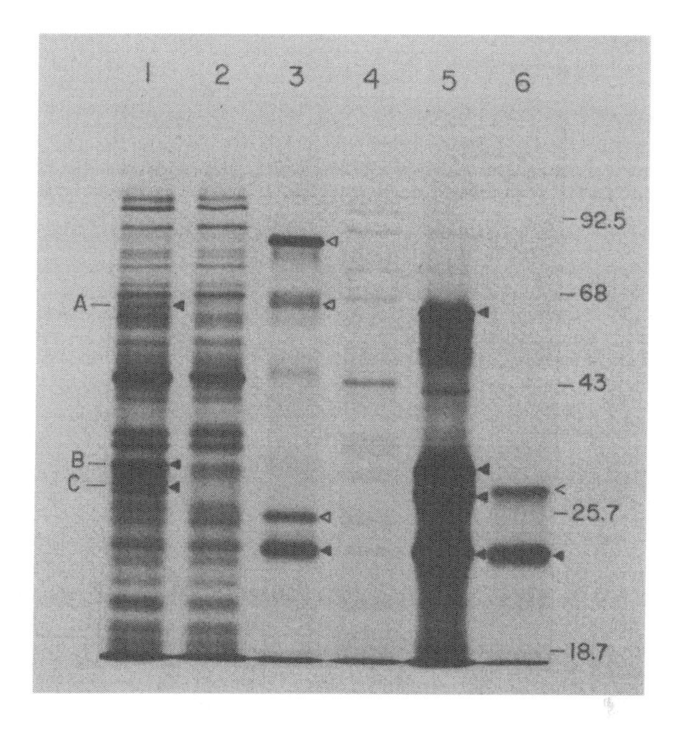

FIGURE 3. Identification of the polypeptides encoded by the cloned R. meliloti glnA (GSI) region. Lane 1 shows the polypeptides encoded by plasmid pFB616. The positions of three polypeptides, which are not synthesized by the vector plasmid (pLAFR1, lane 2) are indicated by black triangles. The molecular weights of these polypeptides are 57K(A) (GSI?), 31K(B) and 26K(C). The origin of the B and C polypeptides is unclear. Polypeptides of identical molecular weights are also synthesized by plasmid pFB6162 carrying 10 kb of cloned R. meliloti DNA, including the glnA(GSI) gene (lane 5, indicated by the top three black triangles). The bottom triangle in lane 5 represents the cat (Cm^r gene) product, also synthesized by the pACYC184 vector of pFB6162 (lane 6, black triangle). Lane 3 shows the polypeptides synthesized by plasmid pFB574 (see Figure 1) and the position of the K. pneumoniae glnA gene product (GS) is indicated by the second open triangle from the top (58K). Lane 4 shows the polypeptides synthesized by Maxicell strain CSR603 (Sankar et al., 1979), harboring no plasmid. The molecular weights of the size standards used are indicated at the extreme right of this Figure.

introduced, high levels of β-galactosidase were produced. In contrast, no β-galactosidase activity was observed when the R. meliloti glnA region was introduced, suggesting the absence of a R. meliloti glnG(ntrC)-like gene closely linked to glnA(GSI). On the other hand, due to the heterologous nature of the experimental system used, no absolute conclusions can be drawn from this experiment. Regarding the question of whether GSI itself plays any direct role in the regulation of R. meliloti nif genes, recent evidence suggests that the introduction of an insertion mutation into the structural gene for R. meliloti glnA(GSI) does not affect the expression of nif genes (J. Somerville and M. Kahn, personal communication).

5. CIRCUITRY OF nifHDK REGULATION IN R. meliloti

Although the results reported above suggest that R. meliloti has a nifA-like gene which activates nifHDK expression, the activation of the R. meliloti nifHDK promoter appears to be different in the free-living and symbiotic state. A plasmid carrying the R. meliloti nifH promoter fused to lacZ (Sundaresan et al., 1983a) was introduced into wild type R. meliloti and into a R. meliloti nifA-like mutant. In bacteroids, β-galactosidase levels were 100X higher than background in the wild type strain and equal to background in the nifA-like mutant strain. In contrast, in nitrogen starved free-living cells, β-galactosidase levels were 10X higher than background, irrespective of the mutation in the nifA-like gene (V. Sundaresan, unpublished observations; Szeto, et al., submitted). These results suggest that the R. meliloti nifHDK promoter is activated directly by the central nitrogen control system (ntr) in free-living cells, whereas in bacteroids activation depends on the nifA-like gene product. Thus, in contrast to K. pneumoniae, in R. meliloti the nifA-like gene may be under the control of a central symbiotic regulatory sytem instead of being under ntr control.

6. CONCLUSIONS

In spite of the varied physiological conditions under which free-living and symbiotic diazotrophs carry out nitrogen fixation, the mechanisms which regulate nif gene expression appear to have been conserved in evolution. This conclusion is based on the following observations:

1) The R. meliloti nifHDK promoter is activated by glnG(ntrC) + glnF(ntrA) or nifA + glnF(ntrA) in E. coli and K. pneumoniae (Sundaresan et al., 1983a,b; de Bruijn, Ausubel, 1983).

2) A nifA-like gene has been identified phenotypically in R. meliloti which hybridizes to the E. coli glnG(ntrC) gene (Zimmerman et al., 1983; Szeto et al., submitted).

3) The R. meliloti nifHDK promoter region contains the TTTTGCA consensus sequence at -10 to -15, previously identified as characteristic of glnG(ntrC) + glnF(ntrA) and/or nifA + glnF(ntrA) activated promoters (Ow et al., 1983).

Our results also suggest that the ntr system may play a direct role in the activation of the nifHDK operon in free-living R. meliloti, in contrast to K. pneumoniae where the ntr system only activates the nifLA operon. In addition, the ntr pathway may be "overridden" in nodules by a symbiotic control pathway which operates through the "nifA-like" gene (V. Sundaresan, unpublished observations; Szeto et al., submitted). Finally, our evidence suggests that the putative R. meliloti glnG(ntrC) gene is probably not closely linked to or part of the same operon as the structural gene for GSI (glnA).

7. REFERENCES

Beynon JL, Cannon MC, Buchanan-Wollaston V, and Cannon FC (1983) Cell, in press.
Buchanan-Wollaston V, Cannon MC, Beynon JL and Cannon FC (1981a) Mol. Gen. Genet. 184, 102-106.
Buchanan-Wollaston V, Cannon MC, Beynon JL and Cannon FC (1981b) Nature

294, 776-778.
Buikema WB, Long SR, Brown SE, van den Bos RC, Earl CD and Ausubel FM (1983) J. Mol. Appl. Genet., in press.
Chang ACY and Cohen SN (1978) J. Bacteriol. 134, 1141-1156.
Corbin D, Ditta, G and Helinski, DR (1982) J. Bacteriol. 149, 221-228.
Corbin D, Barran, L and Ditta, G (1983) Proc. Natl. Acad. Sci. USA 80, 3005-3009.
Darrow RA, Crist D, Evans WR, Jones BL, Keister, DI and Knotts RR (1981) In Gibson AH and Newton WE eds, Current Perspectives in Nitrogen Fixation, pp. 182-185, Austr. Ac. Sc., Canberra.
de Bruijn FJ (1983) Ph.D. Thesis, Harvard University.
de Bruijn FJ and Ausubel FM (1981) Mol. Gen. Genet. 183, 289-297.
de Bruijn FJ and Ausubel FM (1983) Mol. Gen. Genet., in press.
Dixon R, Eady RR, Espin G, Hill S, Iaccarino M, Kahn D and Merrick M (1980) Nature 286, 128-132.
Downie JA, Ma Q-S, Knight CD, Hombrecher G and Johnston AWB (1983) EMBO Journal 2, 947-952.
Drummond M, Clements J, Merrick M and Dixon R (1983) Nature 301, 302-307.
Espin G, Alvarez-Morales A and Merrick M (1981) Mol. Gen. Genet. 184, 213-217.
Espin G, Alvarez-Morales A, Cannon F, Dixon R, and Merrick M (1982) Mol. Gen. Genet. 186, 518-524.
Friedman AM, Long SR, Brown SE, Buikema WJ and Ausubel FM (1982) Gene 18, 289-296.
Hill S, Kennedy C, Kavanagh E, Goldberg RB and Hanau R (1981) Nature 290, 424-426.
Kondorosi A, Svab Z, Kiss GB and Dixon RA (1977) Mol. Gen. Genet. 151, 221-226.
Laemmli UK (1970) Nature 227, 680-685.
Leonardo JM and Goldberg RB (1980) J. Bacteriol. 142, 99-110.
Ludwig RA (1980) Proc. Natl. Acad. Sci. USA 77, 5817-5821.
Magasanik B (1982) Ann. Rev. Genet. 16, 135-168.
Merrick M (1982) Nature 297, 362-363.
Merrick M (1983) EMBO Journal 2, 39-44.
Merrick M, Hill S, Hennecke H, Hahn M, Dixon R and Kennedy C (1982) Mol. Gen. Genet. 185, 75-81.
Ow DW and Ausubel FM (1983) Nature 301, 307-313.
Ow DW, Sundaresan V, Rothstein D, Brown SE and Ausubel FM (1983) Proc. Natl. Acad. Sci. USA 80, 2524-2528.
Reed RR (1981) Cell 25, 713-719.
Roberts GP and Brill WJ (1980) J. Bacteriol. 144, 210-216.
Ruvkun GB and Ausubel FM (1981) Nature 289, 85-88.
Ruvkun GB, Sundaresan V and Ausubel FM (1982) Cell 29, 551-559.
Sankar A, Hack AM and Rupp WD (1979) J. Bacteriol. 137, 692-693.
Sundaresan V, Jones JDG, Ow DW and Ausubel FM (1983a) Nature 301, 728-732.
Sundaresan V, Ow DW and Ausubel FM (1983b) Proc. Natl. Acad. Sci. USA 80, 4030-4034.
Zimmerman JL, Szeto WW and Ausubel FM (1983) J. Bacteriol., in press.

8. ACKNOWLEDGEMENTS

This work was supported by National Science Foundation grants PCM-8104193 and PCM-8104492 awarded to FMA and by funds provided by Hoechst AG.

TRANSCRIPTIONAL CONTROL OF THE NIF REGULON IN KLEBSIELLA PNEUMONIAE

R.A. DIXON, A. ALVAREZ-MORALES, J. CLEMENTS+, M. DRUMMOND, M. MERRICK, J.R. POSTGATE
ARC UNIT OF NITROGEN FIXATION, UNIVERSITY OF SUSSEX, BRIGHTON BN1 9RQ, UK.
+ Present Address: DEPARTMENT OF GENETICS, UNIVERSITY OF LEICESTER, LEICESTER LE1 7RH, UK.

1. INTRODUCTION

Expression of the seventeen nitrogen fixation (nif) genes in Klebsiella pneumoniae is regulated in response to both nitrogen source and oxygen tension, so that nitrogenase synthesis is severely repressed by the presence of ammonia, certain amino acids or by dissolved oxygen in the growth medium. Nitrogen control of nif transcription is maintained at two levels; firstly by the ntr system which exerts a general control on nitrogen metabolism and secondly by regulatory proteins encoded by two nif genes, nifA and nifL. The nifA product is required for transcriptional activation of the nif operons (Dixon et al, 1980; Buchanan-Wollaston et al, 1981a) whereas the nifL product has been implicated in repression of nif transcription in response to fixed nitrogen and oxygen (Buchanan-Wollaston et al, 1981b; Hill et al, 1981; Merrick et al, 1982).

The nifL and nifA genes which form a single operon nifLA are themselves regulated by the ntr system, which comprises three genes ntrA (glnF) ntrB (glnL) and ntrC (glnG). These genes have been identified in several enteric bacteria (Pahel et al, 1982; MacNeil et al, 1982; McFarland et al, 1981; Rothman et al, 1982; de Bruijn, Ausubel 1981; Espin et al, 1982); in all cases the ntrBC (glnLG) genes are linked to glnA, the structural gene for glutamine synthetase whereas ntrA (glnF) is unlinked. The ntrBC (glnLG) genes are part of a complex operon in E. coli and can be transcribed either from their own promoter, P_2, or by readthrough transcription from the stronger glnA promoter P_1 (Pahel et al, 1982; Guterman et al, 1982; MacNeil et al, 1982). Recent evidence suggests that ntrA (glnF) and ntrC (glnG) are required for transcriptional activation of a number of operons under ntr control, including nifLA, whereas repression of transcription of ntr-controlled operons requires both the ntrB (glnL) and ntrC (glnG) products (see Fig. 1).

Current studies in our laboratory are directed towards analysing the functional relationships between the ntr and nifLA gene products in Klebsiella pneumoniae and determining their role in transcriptional regulation. Our data, in agreement with those from other laboratories, indicate close similarities between the ntrBC (glnLG) and nifLA systems and suggest a common role for ntrA in transcriptional activation mediated by both ntrC (glnG) and nifA. Comparative analysis of promoters subject to ntrC (glnG) or nifA control allow us to propose mechanisms for transcriptional activation and repression at these promoters.

2. ACTIVATION OF nif TRANSCRIPTION BY THE ntrC and nifA GENE PRODUCTS

Cloning of the glnA, ntrBC operon from K. pneumoniae (de Bruijn, Ausubel, 1982; Espin et al, 1982) has established that this operon has a similar genetic structure to that in E. coli and that the ntrB and ntrC gene products have molecular weights almost identical to the corresponding E. coli gene products (Espin et al, 1982; McFarland et al, 1981). The glnA and ntrB genes are not required for activation of nif transcription. Plasmids which express ntrC product from a constitutive promoter activate nitrogenase synthesis in a glnA ntrBC deletion background, indicating that ntrC is the sole gene in this regulon which is required for nif transcriptional activation (Espin et al, 1982).

The nifLA promoter is the major target for ntr-mediated activation of nif transcription. Strains lacking ntrC product fail to activate transcription from the nifLA promoter (Drummond et al, 1983; Ow, Ausubel 1983) and hence fail to synthesize the nifA product which is required for transcriptional activation at the promoters of the other seven nif operons (Dixon et al, 1980; Roberts, Brill 1980). Nitrogenase synthesis in wild-type strains therefore requires both ntrC and nifA as transcriptional activators and derepression occurs via a regulatory cascade (Fig. 1).

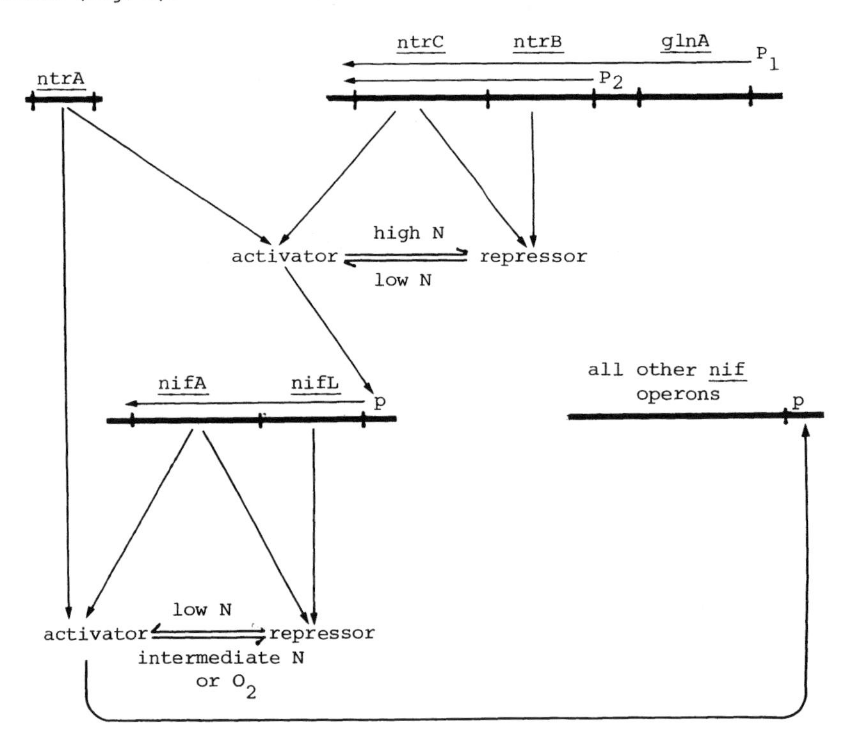

Fig. 1. Current model for nif regulation in K. pneumoniae

It is possible to obviate the requirement for ntrC product in nif derepression by constructing strains in which the nifA product is expressed constitutively. In such strains nifA transcription is uncoupled from ntr control and the ntrC product is no longer required for transcriptional activation (Buchanan-Wollaston et al, 1981a; Sibold, Elmerich 1983). By cloning the nifA gene into plasmids which allow constitutive nifA expression it has been shown that nifA can auto-activate its own transcription at the nifLA promoter in the absence of ntrBC (Drummond et al, 1983) and that the nifA product can substitute for the ntrC product at other ntr controlled promoters including glnA, hutUH and genes for arginine and proline utilisation (Merrick, 1983; Ow, Ausubel 1983). The ntrC and nifA gene products also have common physical properties; they have similar molecular weights and both are basic proteins with almost identical pI's (Merrick, 1983). However, in *K. pneumoniae* the ntrC product cannot substitute for nifA product at other nif promoters such as nifHDKY which can only be activated by nifA. Hence the ntrC has retained some transcriptional specificity. Nevertheless the functional similarity between these regulatory proteins suggests that the two genes may have a common evolutionary origin.

2.1 Requirement for ntrA (glnF) product

The ntrA gene is known to have a central role in the regulation of nitrogen assimilation, since mutations in this gene have a pleiotropic effect on nitrogen source utilization. It has recently been shown that ntrA is required for transcriptional activation by both ntrC and nifA, even in strains which produce high levels of the ntrC or nifA gene products (Merrick, 1983; Ow, Ausubel 1983). This common requirement for ntrA again suggests that ntrC and nifA share functional homology and it has been suggested that ntrA may encode a low molecular weight co-regulator of transcription, analogous to the role of cAMP in CRP-mediated activation. However, the ntrA product does not appear to respond to the nitrogen status of the cell, since full activation is observed in strains grown on ammonia when the ntrC or nifA genes are transcribed from constitutive promoters. The ntrA gene from *K. pneumoniae* has been cloned (de Bruijn, Ausubel, personal communication; Merrick, unpublished) and the product has been identified as a polypeptide of 75-85,000 daltons. The function of this gene product has yet to be elucidated.

3. ROLE OF THE ntrB AND nifL PRODUCTS IN nif REPRESSION

Repression of nif transcription, like activation, occurs at two different levels; the nifL operon mediates specific repression of nitrogenase synthesis in response to oxygen and intermediate levels of fixed nitrogen, whereas the ntrBC system ensures that nitrogen-regulated operons such as nifLA are not activated when cells are grown on ammonia. Functional similarities between these two systems again emerge since the first gene in each operon is required to mediate repression. However, the nifL gene product is apparently far more sensitive to fixed nitrogen concentration than is ntrB and also responds to the oxygen tension in the growth medium. Hence a specific repression of nitrogenase synthesis can be maintained by nifL without repressing other operons under ntr control.

There is no evidence to date to suggest that nifL is a DNA-binding protein which directly represses nif transcription and 'repression' may in fact result from a modulation of the activity of the nifA gene product. Using plasmids which constitutively synthesize nifA and nifL products, we have shown that the nifL product mediates repression of its own synthesis in strains carrying an ntrBC deletion (Dixon et al, 1983). This effect is not observed in a ntr^+ strain which may indicate that nifL exerts repression by inactivating nifA product rather than by directly repressing transcription. The ntrB product may have an analogous role in inactivating transcriptional activation by ntrC in response to ammonia at ntr-controlled promoters. The P_1 promoter of the glnA, ntrBC regulon has a basal constitutive activity in the absence of ntrBC and it is therefore possible to examine repression at this promoter in the absence of a transcriptional activator. We have found that the ntrB product alone has no regulatory effect at this promoter and that regulation of this promoter in response to the nitrogen status of the cell is dependent on the presence of both the ntrB and ntrC products (see Section 5).

4. THE nifLA PROMOTER

The promoter of the nif-specific regulatory operon nifLA, performs a central role in nif regulation since this promoter is subject to positive control by the ntr system and is also regulated autogenously by the nifL and nifA gene products. The nucleotide sequence of this promoter shows little homology with the consensus sequence for prokaryotic promoters (Drummond et al, 1983; Ow et al, 1983), which may reflect the requirement for two regulatory effectors (either ntrC plus ntrA or nifA plus ntrA) for transcriptional activation.

Transcription is initiated at the same nucleotide in vivo whether the promoter is activated by ntrC or by nifA (Drummond et al, 1983), indicating that these proteins activate transcription by similar mechanisms, possibly by recognising the same site in the DNA sequence. In order to define regions in the promoter which may be important for transcriptional activation, we constructed a series of deletions which progressively removed sequences upstream of the transcription start (Drummond et al, 1983). Full promoter activity requires sequences 150 base-pairs upstream of the transcriptional initiation start and the promoter activity progressively declines as the upstream sequence is deleted. However some activation by ntrC and nifA is still observed in deletions which remove nucleotides upstream of -28, which indicates that the -35 region does not determine transcriptional specificity. In addition the retention of positive control in deletions removing the -35 region precludes binding of the activator protein to an upstream target as the sole mechanism of activating transcription from this promoter. Any recognition sites for ntrC- or nifA-mediated activation must therefore be downstream of the -35 region in this promoter.

5. THE P_1 AND P_2 PROMOTERS OF THE glnA ntrBC REGULON

To gain a better understanding of how the ntr system regulates nif expression we have constructed translational fusions of the lacZ gene to both glnA and ntrB in order to monitor expression from the P_1 and P_2 promoters. We have also constructed plasmids which express either the ntrB product or the ntrC product, or both products, from constitutive promoters, so that we can assess the effect of individual gene products *in trans* to the P_1 or P_2 fusions. We have also analysed the effect of nifA on expression from these two promoters. (Full experimental details are given in the poster summary by Alvarez-Morales et al in this volume.) Our results indicate that ntrC product alone represses transcription from both the P_1 and P_2 promoters, whereas it activates transcription from the nifLA promoter. Normal regulation of P_1 i.e. activation and repression in response to nitrogen source concentration only occurs when both ntrB and ntrC are present whereas ntrB alone has no regulatory effect on either P_1 or P_2. Hence, ntrB is required in addition to ntrC to activate the P_1 promoter (Table 1).

TABLE 1. Comparison of the glnA ntrBC and nifLA promoters

Plasmid genotype *in trans*	Level of expression[a] from:					
	P_1		P_2[b]		P_{nifLA}	
	-N[c]	+N[d]	-N	+N	-N	+N
None	+	+	++	++	-	-
ntrB[c]	+	+	++	++	-	-
ntrC[c]	-	-	-	-	+++	+++
ntrBC[c]	+++	-	+	+	+++	-
nifA[c]	+++	+	++	++	+++	+++

[a] expression was measured by assaying β-galactosidase activities from ntrB::lac and nifL::lac fusions and glutamine synthetase activity as a measure of P_1 expression.

[b] P_2 expression was measured in the absence of transcriptional readthrough from P_1.

[c] -N indicates strains were grown on limiting nitrogen (i.e. 200 μg glutamine ml^{-1}).

[d] +N indicates strains were grown on excess nitrogen (i.e. glutamine (2 mg ml^{-1}) plus $(NH_4)_2SO_4$ (2 mg ml^{-1}).)

In contrast, nifA alone is unable to repress transcription from either the P_1 or P_2 promoters and activates the P_1 promoter in the absence of ntrB. Therefore, although ntrC and nifA share functional homology as transcriptional activators, the nifA product cannot substitute for ntrC as a repressor at either P_1 or P_2. Our data suggest that ntrB and ntrC function as a complex to modulate transcription in response to nitrogen source availability.

6. SEQUENCE COMPARISONS

Ow et al (1983) have compared the sequences of six promoters which are subject to activation by ntrC or nifA and have found a heptameric consensus sequence TTTTGCA, which is located in the -10 to -15 region of those promoters in which the transcription start has been determined. Beynon et al (1983) have also compared the sequences of seven nif promoters and have identified two regions of homology, a five base pair homology TTGCA located between -10 and -14 and a four base pair homology CTGG situated around -26. This second region of homology at -26 is less well conserved in ntr-controlled promoters. Promoters which are solely subject to nifA-mediated activation, such as the K. pneumoniae nifH promoter have less homology in the -15 region, retaining only the consensus TGCA, whereas the Rhizobium meliloti nifH promoter which can be activated either by nifA or ntrC contains the complete sequence TTTTGCA (Fig. 2). The additional T residues at the 5' end of the consensus sequence may therefore be essential for ntrC mediated control (Sundaresan et al, 1983a,b).

Fig. 2. Nucleotide sequences of ntr and nifA-activatable promoters

R. meliloti -30 -20 -10 +1

nifH[a] 5'-TCAGACGGCTGGCACGACT[TTTGCA]CGATCAGCCCTG-3'

K. pneumoniae -30 -20 -10 +1

nifH[a] 5'-CAGGCACGGCTGGTATGTTC[CCTGCA]CTTCTCTGCTG-3'

-30 -20 -10 +1

nifL[a,b,c] 5'-ACGCCGATAAGGGCGCACGG[TTTGCA]TGGTTATCACC-3'

A

glnA P_1[d] 5'-ATTCATTTTGGTGCAGCCCT[TTTGCA]CGATGATGTGC-3'

Data from [a]Ow et al, 1983; [b]Drummond et al, 1983; [c]Dixon et al, 1983; [d]Dixon R. (unpublished results). Numbering is relative to +1, the transcriptional start site. The conserved hexamer which is common to ntr controlled promoters is boxed. The arrow below the box in the nifL promoter shows the location of a strong 'down' mutation resulting from a G to A transition at -13.

We have recently sequenced the P_1 promoter of the glnA ntrBC regulon from K. pneumoniae; this promoter also contains the TTTTGCA heptamer (Fig. 2) and is homologous to the E. coli glnA promoter in the vicinity of the consensus sequence. We have also demonstrated the functional importance of this conserved sequence in the nifLA promoter by mutating the G residue in the TTTGCA box to an A at -13 (Fig. 2). This change results in a strong'down' promoter mutation and the mutant promoter is no longer subject to activation by either nifA or ntrC.

7. CONCLUSIONS

Promoters subject to ntr- or nifA-mediated activation represent a sub-set of promoters which share a unique conservation of sequence homology yet show little homology with the consensus sequence for other prokaryotic promoters (Rosenberg, Court, 1979; Siebenlist et al, 1980; Hawley, McClure, 1983). This may suggest that it is necessary to modify the transcription initiation specificity of RNA polymerase in order to recognise these promoters. The ntrA product is a possible candidate for such a role. This model would predict that ntrA acts as a specific sigma factor for nitrogen-controlled promoters and allows selective recognition of these promoters by RNA polymerase holoenzyme. Activation of transcription would also require either the ntrC or nifA products and changes in the conserved sequence would allow for specificity of activation (as observed in the K. pneumoniae nifH promoter).

Our data suggest that the ntr- or nifA-activatable promoters may fall into three separate classes:

(1) Promoters such as P_1 glnA show a basal level of activity in the absence of transcriptional activators but in wild-type cells are subject to both positive and negative regulation. These promoters are repressed by ntrC product alone but require ntrA, ntrB and ntrC for transcriptional activation. In order to maintain such complex transcriptional patterns, three structural features are required (as predicted from sequence data for the Klebsiella pneumoniae P_1 promoter): (a) a canonical promoter sequence which can be recognised by RNA polymerase in the absence of ntr gene products (b) an ntr-activatable promoter sequence containing the conserved hexamer TTTGCA and (c) a recognition sequence for the repressor form of the ntrC product. We therefore predict that there are two transcription initiation sites in the P_1 promoter, one ntr-dependent, the other ntr-independent, and that repression of both transcripts can be maintained by the ntrC product. The role of ntrB would be to prevent repression by ntrC in the absence of fixed nitrogen and hence facilitate activation of the ntr-dependent promoter.

(2) Promoters such as P_2, the internal promoter of the glnA ntrBC regulon, show activity in the absence of ntr gene products and are presumably recognised by RNA polymerase without a requirement for additional factors. Such promoters are solely subject to negative control by ntrC and presumably contain a recognition site for the repressor form of the ntrC product.

(3) Promoters such as P_{nifLA} and P_{nifH} have an absolute requirement for transcriptional activation and presumably are not recognised by RNA polymerase in the absence of ntrA and the appropriate activator. We would predict that such promoters are solely subject to positive control and do not contain a recognition site for the repressor form of ntrC. Such promoters can be activated in the absence of ntrB (or nifL) products. We predict that 'repression' at such promoters would be mediated by the latter gene products via inactivation of ntrC or nifA activity.

8. REFERENCES

Beynon J, Cannon M, Buchanan-Wollaston V, and Cannon F (1983) Cell (in press).

de Bruijn FJ and Ausubel FM (1981) Mol.Gen.Genet. 183, 289-297.

Buchanan-Wollaston V, Cannon MC, Beynon JL and Cannon FC (1981a) Nature 294, 776-778.

Buchanan-Wollaston V, Cannon MC, and Cannon FC (1981b) Mol.Gen.Genet. 184, 102-106.

Dixon R, Eady RR, Espin G, Hill S, Iaccarino M, Kahn D, and Merrick M (1980) Nature 286, 128-132.

Dixon R, Alvarez-Morales A, Clements J, Drummond M, Filser M, and Merrick M (1983) In Ahmad F, Downey K, Schultz J and Voellmy RW, eds, Advances in Gene Technology: Molecular Genetics of Plants and Animals, 15th Miami Winter Symposium (in press) Academic Press.

Drummond M, Clements J, Merrick M and Dixon R (1983) Nature 301, 302-307.

Espin G, Alvarez-Morales A, Cannon F, Dixon R, and Merrick M (1982) Mol.Gen.Genet. 147, 189-198.
Guterman SK, Roberts G and Tyler B (1982) J.Bacteriol. 150, 1314-1321.
Hawley DK and McClure W (1983) Nucleic Acids Res. 11, 2237-2255.
Hill S, Kennedy C, Kavanagh E, Goldberg RB and Hanau R (1981) Nature 290, 424-426.
MacNeil T, MacNeil D and Tyler B (1982) J.Bacteriol. 150, 1302-1313.
Merrick M (1983) EMBO J. 2, 39-44.
Merrick M, Hill S, Hennecke H, Hahn M, Dixon R and Kennedy C (1982) Mol.Gen.Genet. 185, 75-81.
McFarland N, McCarter L, Artz S and Kustu S (1981) Proc.Natl.Acad.Sci. USA, 78, 2135-2139.
Ow DW and Ausubel FM (1983) Nature 301, 307-313.
Ow DW, Sundaresan V, Rothstein DM, Brown SE and Ausubel F (1983) Proc. Natl.Acad.Sci.USA, 80, 2524-2528.
Pahel G, Rothstein DM and Magasanik B (1982) J.Bacteriol. 150, 202-213.
Roberts GP and Brill WJ (1980) J.Bacteriol. 144, 210-216.
Rosenberg M and Court D (1979) A.Rev.Genet. 13, 319-353.
Rothman N, Rothstein DM, Foor F and Magasanik B (1982) J.Bacteriol. 150, 221-230.
Sibold L and Elmerich C (1982) EMBO J. 1, 1551-1558.
Siebenlist U, Simpson RB and Gilbert W (1980) Cell 20, 269-281.
Sundaresan V, Jones JDG, Ow DW and Ausubel F (1983a) Nature 301, 728-732.
Sundaresan V, Ow DW and Ausubel F (1983b) Proc.Natl.Acad.Sci.USA (in press).

ASPECTS OF GENETICS OF AZOTOBACTERS

R. ROBSON, R. JONES, C.K. KENNEDY, M. DRUMMOND, J. RAMOS, P.R. WOODLEY,
C. WHEELER, J. CHESSHYRE, J. POSTGATE
ARC UNIT OF NITROGEN FIXATION, UNIVERSITY OF SUSSEX, BRIGHTON BN1 9RQ UK.

1. INTRODUCTION

There have been a number of studies on the genetics of N_2-fixation in Azotobacter. Early reports describe the isolation and biochemical characterization of several classes of mutants defective in N_2-fixation or normal regulation of the genes (nif) responsible (Wyss, Wyss 1950; Green et al, 1953; Fisher, Brill 1969; Sorger, Trofimenkof, 1970; Shah et al, 1973; Gordon, Brill, 1972; Shah et al, 1974). Some of the mutations were mapped by transformation to provide an approximate linkage map (Bishop, Brill, 1977). Further progress was hampered by a lack of good genetic systems for this genus. Also, Nif^- mutants aside, the wide variety of mutants useful for such work were and remain elusive (see Roberts, Brill, 1981), a problem attributed to a high level of genetic redundancy in the form of an unusually high chromosome copy number (Sadoff et al, 1979).

Developments in recombinant DNA technology have removed some technical barriers to the study of genetics in Azotobacter. The study of nif genetics in this genus should reveal those features important for aerobic N_2-fixation and add to our knowledge of the variation in nif genes among the diverse genera of diazotrophs. The work should also throw light on the possibility of an alternative nitrogenase proposed for A. vinelandii (Bishop et al, 1980).

Here we describe studies of several aspects of the genetics of A. chroococcum in which we have investigated the genomic complexity, isolated and characterized biochemically mutants affected in N_2-fixation, prepared gene banks from which clones bearing nif genes have been isolated and physically mapped. We have also investigated the degree of similarity between the mechanisms of regulation of nif genes in K. pneumoniae and Azotobacter.

I. Genome of A. chroococcum

(a) Plasmids

The finding of nif genes on large plasmids in Rhizobia (Johnston et al, 1978; Nuti et al, 1979) stimulated us to look for plasmids in A. chroococcum. Fig.1(b) shows the plasmids present in strain (MCB-1) a non-gummy variant NCIB8003. Five plasmids were detected, ranging in size from 7 to 200 Md. Fig.1(c-k) shows that plasmids detected in recent isolates of this species from local soils exhibit variability in number and size.

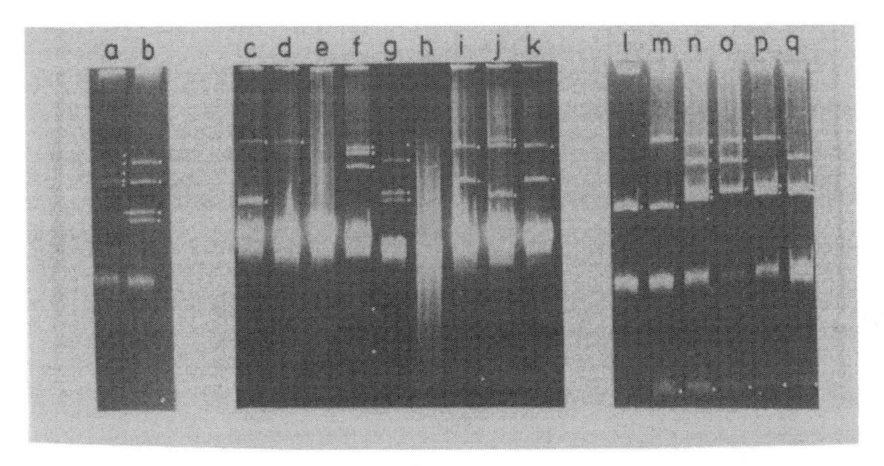

Fig.1. Plasmids in A. chroococcum. (a) Rhizobium leguminosarum T3 (plasmids 85, 90, 130, 190, > 200 Md: J. Beynon, personal communication). (b) A. chroococcum MCB1. Recent soil isolates of A. chroococcum: (e), RR2; (f), RR1; (h), CW8; (i), CW9; (j), CW6; (k), CW1 with (c), MCD1(RP4); (d), MCD1 and (g), MCC1 for comparison. Cured derivatives of MCC1: (m) MCD1(RP4), (o) MCC2, (p) MCC3, (q) MCC4, with (n) MCC1 and Escherichia coli J53 (RP4) for comparison.

To determine the function of these plasmids, we isolated derivatives of strain MCC-1 (a str^r nal^r derivative of MCB-1) 'cured' of various plasmids except the smallest (pAC1) see Fig.1(l-q). Comparisons of the cured derivatives revealed no obvious phenotypic differences. Attributes compared included: sensitivity to a wide range of commonly available antibiotics, heavy metals and antimetabolites; carbon source utilization pattern; bacteriophage or bacteriocin production. All strains fixed N_2 and exhibited normal uptake hydrogenase (Hup) activity. No plasmid showed DNA homology to pSA-30 (Cannon et al, 1979).

(b) Genome size

We sized the genome of A. chroococcum using the 2-D-agarose gel electrophoresis technique of Yee and Inouye (1982) (see Fig.2). We estimated the genome sizes for two strains MCD-1 and CW8 (which contained no large plasmid) and values of 1918 and 1784 kilobase pairs were obtained. These values suggest that the genome of this species is smaller than that of E. coli. E. coli was used as a control (Fig.2b)

The DNA content per cell was estimated for both A. chroococcum strains and mean values of 5.13 and 3.97×10^{-14} g/cell were obtained. We conclude that each strain contains on average 20 to 25 chromosome copies.

Our data essentially confirm the polyploidy shown for A. vinelandii (Sadoff et al, 1979).

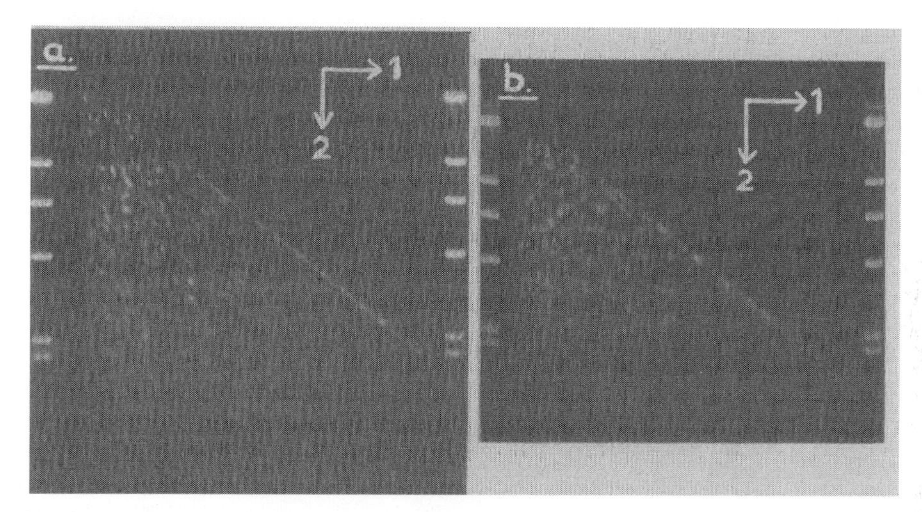

Fig.2. Genome analysis of A. chroococcum. Genomic DNA was digested sequentially with two different "6-base" restriction endonucleases. The second digest was performed after the enzyme had been diffused into the agarose lane containing the electrophoretically separated fragments from the first digest. The double digest was then resolved by electrophoresis after embedding the lane in agarose at 90° to the original orientation. DNA standards were λ Hind III fragments. (a) A.chroococcum CW8 1 EcoRI 2 Bst EII; (b) Escherichia coli Jc 1553 1 EcoRI 2 Bst EII.

2. ISOLATION AND CHARACTERISATION OF Nif^- MUTANTS OF A. CHROOCOCCUM MCD-1

Nowadays, mutagenesis by transposon insertion offers a number of advantages. Several attempts to isolate such mutants of A. chroococcum MCD-1 using Tn5 or Tn7 introduced using 'suicide' vectors such as pJB4JI (Beringer et al, 1978) or RP4::Mu::Tn7 (Van Vliet et al, 1978) have not been successful. We conclude that neither transposon transposes at a useful frequency from these vectors and as yet we know of no demonstration of transposition mutagenesis in this genus.

Nif^- mutants were isolated after NTG treatment and enriched with carbenicillin and cycloserine. They were subdivided into two classes, those which were consistently Nif^- and those which were variably Nif^- depending on the method of culture. The latter isolates were unusually O_2-sensitive: they grew on N_2 with sucrose as C-source only in semisolid media, like microaerobic N_2-fixers, or on the surface of agar plates when the pO_2 was reduced to 0.02 atm. We call this phenotype Fos^- (Fixation, oxygen sensitive).

(a) Nif^- isolates

Presumptive Nif^- isolates were characterised essentially according to published methods (Shah et al, 1973). We were particularly interested to examine strains apparently defective in regulation since we hoped to identify nif specific polypeptides other than those for nitrogenase components. 2-dimensional separations of pulse-labelled polypeptides synthesised under N-starvation conditions were compared for wild-type and Nif⁻ regulatory mutants (see Fig.3). We observed 16 commonly missing from the mutants, a number close to that seen for K. pneumoniae (Roberts, Brill, 1980). Some polypeptides were made at high rates in the mutants but not the parent: their significance is not known. All the mutants used in this study were revertible and were complemented by K. pneumoniae nifA (Kennedy, Robson, 1983) see later.

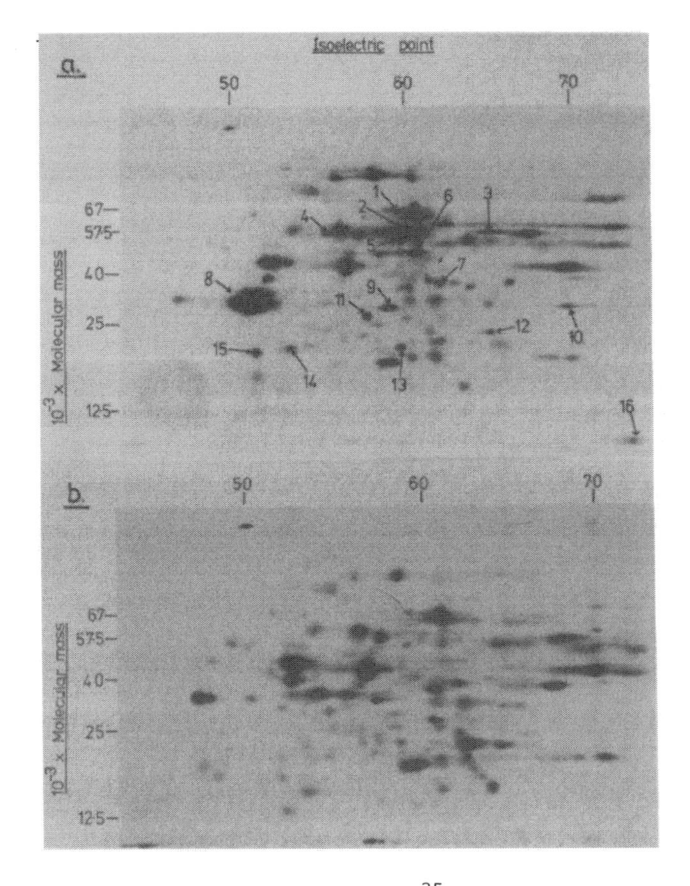

Fig.3. 2-Dimensional analysis of ^{35}S pulse-labelled polypeptides made under N-starvation conditions by A. chroococcum MCD-1(a) and its Nif⁻ regulatory mutant MCD 1008(b). Spots present in MCD1 but not MCD 1008 are arrowed. Identity of most spots unknown except for sub-units of nitrogenase MoFe protein (2 and 5), Fe protein (8) and flavodoxin (15).

Fos$^-$ mutants. Several properties are common to all these mutants. Firstly, the uncharacteristic O_2-sensitivity was observed in aerated cultures when sugars (usually sucrose) were supplied as the carbon source, though when NH_4^+ is provided as N-source growth rates were surprisingly faster even than the parent strain, indicating no general inability to metabolize the sugar. In poorly aerated cultures, growth with N_2 on sugars was observed but only after a prolonged lag. Secondly, metabolizable carboxylic acids e.g. TCA cycle intermediates, acetate or pyruvate or ethanol "correct" the O_2-sensitivity and abolish the lag phase. Thirdly, citrate (at approx. 3mM), normally a growth inhibitor for A. chroococcum MCD-1, does not inhibit growth of these isolates. Citrate resistance can be used to select further Fos$^-$ strains.

In many respects the mutants behave as Ca^{2+}-starved cultures described before (Jakobsons et al, 1962) and it is interesting to note that in Fos$^-$ mutants increasing the medium Ca^{2+} also overcomes the O_2-sensitivity. The nature of the Ca^{2+}-requirement for N_2-fixation in Azotobacters is not known. Two mechanisms are probably responsible for the capability of Azotobacters to fix N_2 in air, conformational and respiratory protection (see Robson, Postgate, 1980). Many reports have examined the role of the electron transport chain in respiratory protection. Whilst it is clear that the high respiratory rates observed in Azotobacter (Phillips, Johnson 1961) demand a highly active respiratory chain, it must also follow that intermediary metabolism must be suitably regulated to provide a flux of oxidisable substrates (see Haaker, Veeger, 1976). Since we detect no obvious differences in the properties of the electron transport chain in these mutants it is likely that the Fos$^-$ mutants are deficient in respiratory protection due to a subtle defect(s) in intermediary metabolism.

3. CLONING AND ORGANIZATION OF SOME nif GENES FROM A. CHROOCOCCUM MCD-1

We have examined the DNA homology between those K. pneumoniae nif genes borne on plasmid pSA-30 (Cannon et al, 1979) and genomic endonuclease digests of A. chroococcum MCD-1 and A. vinelandii UWr. We confirm the results reported for A. vinelandii (Ruvkun, Ausubel, 1980) and using their nomenclature for subfragments of the pSA-30 nif insert find for A. chroococcum several fragments hybridizing to the A3 fragment (nifH and part of nifD), more than could be expected were the two genes simply contiguous as in K. pneumoniae (Fig.4). We did observe weak hybridization to the A1 fragment for both species. Also we have compared the hybridization patterns for the two species using as DNA probes inserts from pLB1 and pLB3 which are sequences cloned from A. vinelandii which themselves hybridize to various pSA-30 fragments (Rizzo, Bishop, 1983; Bishop, Bott, 1983). All the data demonstrate species differences (Fig.4).

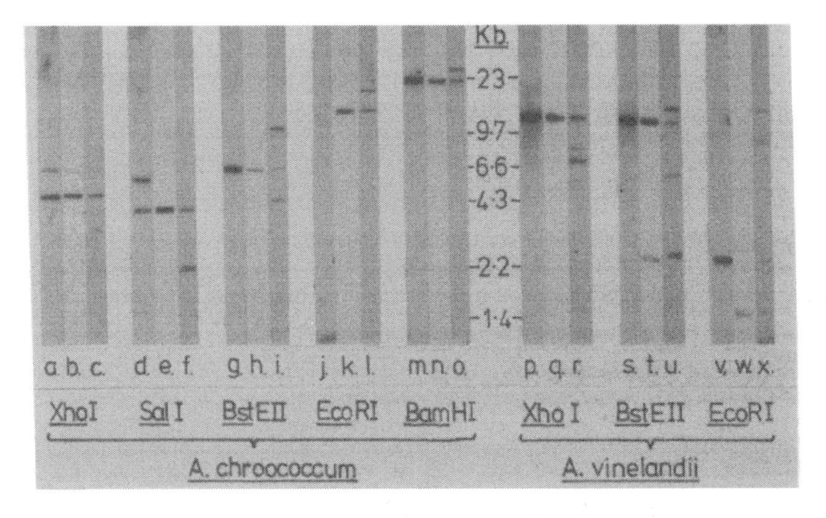

Fig. 4. Hybridization of various nif sequences to Southern blots of restriction digests of genomic DNA of Azotobacters. DNA probes used: (a), pLB1 insert (EcoR1 fragment from A. vinelandii (A.v) with homology to pSA-30 fragments A1, A2); (b), pLB3 insert (EcoR1 fragment from A. vinelandii with homology to pSA-30 fragments A2, A3) (c), pSA-30 A3 fragment. Other groups of three lanes have similar order of probes. (pSA-30 fragments as in Ruvkun, Ausubel, 1980).

In order to clone nif genes from A. chroococcum a gene bank in E. coli was constructed using the 5.1 Kb cosmid pTBE (Grosveld et al, 1982). We cloned into the BamHI site 45±10 Kb fragments from partial Sau3A digests of genomic DNA. Clones bearing nif genes were selected after colony hybridization using pLB3 or pSA-30 inserts as probes. The physical map of one such clone pACB1 is shown in Fig.5 which also indicates regions of DNA homology to several probes. Comparisons of the hybridization patterns for the several probes indicates that nifH,D,K are tightly clustered, and that the two A. vinelandii cloned fragments are contiguous in A. chroococcum probably sharing two commonly located EcoR1 sites. Also we found homology to a fragment bearing K. pneumoniae nif(M),V,(S) located between 14 and 24 Kb from nifK. We did not find homology between the K. pneumoniae nif(B),A,(L) fragment from pMC71a (Buchanan-Wollaston et al, 1981) and the cloned fragment or genomic digests.

A cluster of genes coding for nitrogenase components in Azotobacter wwould be consistent with the early map of mutants of A. vinelandii (Bishop, Brill, 1977) and the recent determination of the transcript size for this organism (Krol et al, 1982). However, this data is at variance with the hybridization patterns for total genomic digests of Azotobacter. One explanation is that one of the nitrogenase structural genes is duplicated though we cannot rule out the possibility that the intragenic region between nifH and nifJ gives homology. The available data suggest that a second sequence like nifH, but not nifD or nifK, is present in the genome. Recently, it has been reported that a second nifH-like gene may be present in Anabaena 7120 (Rice et al, 1982).

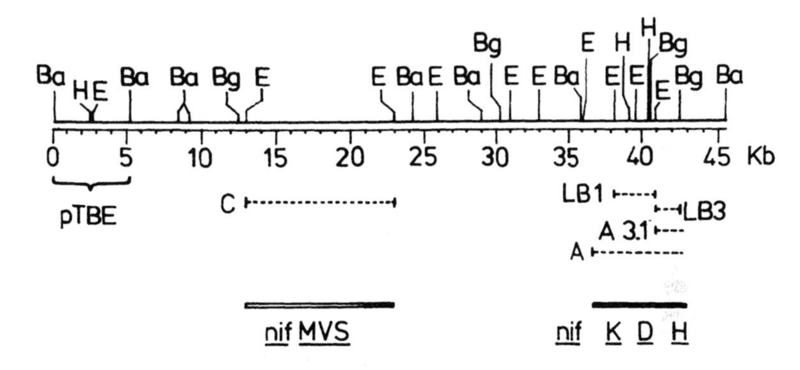

Fig.5. Physical map of pAcBl and regions of homology to nif genes. Dotted lines show max. limits of regions with homology to the following DNA sequences. A, K. pneumoniae (Kp) nif HDKY(E) from pSA30 (Cannon et al, 1979). A 3.1, Kp nif(H) 0.6 Kb EcoRl/BglII fragment from pSA30. C, Kp nif(M)V(S) from pWFl (W. Filler, R.A. Dixon unpublished). LBl and LB3. Cloned sequences from A. vinelandii (Bishop, Bott, 1983; Rizzo, Bishop, 1983) bearing homology respectively to Kp Al, A2 and A2,A3 fragments from pSA30 (as defined by Ruvkun, Ausubel, 1980). Ba, BamHI; Bg, Bgl II; E, EcoRl; H, Hind III.

4. REGULATION OF NIF IN AZOTOBACTER

Environmental factors such as N-availability or O_2-concentration affect expression of nif genes in Azotobacter as in K. pneumoniae (see Robson et al, 1983). Their mechanisms of nif regulation can be compared by introducing into Azotobacter wide host range recombinant plasmids carrying well characterized K. pneumoniae (Kp) nif regulatory genes or promoters (Buchanan-Wollaston et al, 1981; Drummond et al, 1983). This approach led to restoration and constitutivity of nif expression in Azotobacter Nif$^-$ regulatory mutants e.g. A. vinelandii UWl (Shah et al, 1973) containing Kp nifA (nif activator, Buchanan-Wollaston et al, 1981) on plasmid pCKl. This implied the presence of a gene functionally analogous to nifA in Azotobacter (Kennedy, Robson, 1983).

By contrast, introduction of multiple copies of the Kp nifL gene (nif repressor, Merrick et al, 1982) on RSF1010 (pMD132) had no effect on N_2-fixation in A. vinelandii, though it did negate the complementation of A. vinelandii UWl by nifA expressed from pCK3 (a construct of pRK290 carrying Kp nifA). These data show that Kp nifL behaves as an antagonist of nifA activity in the Azotobacter background despite having no detectable effect on expression of the host nif genes. The failure of Kp nifL to affect expression of Nif in Azotobacter shows there is a difference between the nif specific activator(s) in these two genera.

Since Azotobacters are naturally Lac$^-$ (Thompson, Skerman, 1979) and have no detectable β-galactosidase activity, we could examine them directly for activation of various Kp nif promoters fused to the lacZ gene. We showed that the Kp nifH promoter carried on plasmid pMD61 was not expressed in A. vinelandii UW and only very weakly after the additional introduction of Kp nifA on pCK3. It also did not reduce Nif expression, suggesting that the Kp nifH promoter did not compete effectively for the Azotobacter transcriptional apparatus though it is not possible to say whether this is due to differences in nif-specific activator, lack of an ntrA-like gene product as required in Klebsiella pneumoniae (Merrick, 1983; Ow, Ausubel, 1983) or differences in RNA polymerase per se. However, Kp nifL (on pMD23) and nifF (on pMD21) were strongly expressed in wild-type strains (A. vinelandii UW, A. chroococcum MCD-1) and respective regulatory mutants (UW1, MCD 1007) even in the presence of NH_4^+. This further suggests that the Azotobacter transcriptional system differs from that of Klebsiella, resembling the ntrC-activated Kp system in its ability to activate nifL and nifF, but not nifH (see Dixon et al, this book).

Acknowledgements. We would like to thank F. Grosveld MRC, Mill Hill, London for assistance in preparation of gene banks and P. Bishop for supplying plasmids prior to their publication, R. Humphrey and G. Saunders for technical assistance and to all those who assisted in preparing this manuscript.

REFERENCES

Beringer JE, Beynon JL, Buchanan-Wollaston AV and Johnston AWB (1978) Nature 276, 633-634.
Bishop PE and Bott KF (1983) Abstr.Ann.Meet.Am.Soc.Bacteriol. 188
Bishop PE and Brill WJ (1977) J.Bacteriol. 130, 954-956.
Bishop PE, Jarlenski DML and Hetherington DR (1980) Proc.Natl.Acad.Sci. USA 77, 7342-7346.
Buchanan-Wollaston V, Cannon MC, Beynon JL and Cannon FC (1981) Nature 294, 776-778.
Cannon FC, Riedel GE and Ausubel FM (1979) Mol.Gen.Genet. 174, 59-66.
Drummond M, Clements J, Merrick M and Dixon R (1983) Nature 301, 302-307.
Fisher RH and Brill WJ (1969) Biochim.Biophys.Acta, 184, 99-105.
Green M, Alexander M and Wilson PW (1953) J.Bacteriol. 66, 623-624.
Gordon JK and Brill WJ (1972) Proc.Natl.Acad.Sci.USA, 69, 3501-3503.
Grosveld FG, Lund T, Murray EJ, Mellor AL, Dahl HHM and Flavell RA (1982) Nucl.Acid.Res. 10, 6715-6732.
Haaker H and Veeger C (1976) Eur.J.Biochem. 63, 499-507.
Jakobsons A, Zell EA and Wilson PW (1962) Archiv.für Mikrobiol. 41, 1-10.
Johnston AWB, Beynon JL, Buchanan-Wollaston AV, Setchell SM, Hirsch PR and Beringer JE (1978) Nature 276, 634-636.
Kennedy CK and Robson RL (1983) Nature 301, 626-628.
Krol AJM, Hontelez JGJ, Roozendaal B and Van Kammen A (1982) Nucl.Acid. Res. 10, 4147-4157.
Merrick M (1983) EMBO J. 2, 39-44.
Merrick M, Hill S, Hennecke H, Hahn M, Dixon R and Kennedy C (1982) Mol.Gen.Genet. 185, 75-81.
Nuti MP, Lepidi AA, Prakash RK, Schilperoort RA and Cannon FC (1979) Nature 282, 533-535.
Ow DW and Ausubel FM (1983) Nature 301, 307-313.
Phillips DH and Johnson MJ (1961) J.Biochem.Microbiol.Technol.Eng. 3, 277-309.
Rice D, Mazur BJ and Haselkorn R (1982) J.Biol.Chem. 257, 13157-13163.
Rizzo TM and Bishop PE (1983) Abst.Ann.Meet.Am.Soc.Bacteriol. 188.
Roberts GP and Brill WJ (1980) J.Bacteriol. 144, 210-216.

Roberts GP and Brill WJ (1981) Ann.Rev.Microbiol. 35, 207-235.
Robson RL and Postgate JR (1980) Ann.Rev.Microbiol. 34, 183-207.
Robson RL, Kennedy CK and Postgate JR (1983) Canad.J.Microbiol. (in the press).
Ruvkun GG and Ausubel FM (1980) Proc.Natl.Acad.Sci.USA, 77, 191-195.
Sadoff HL, Shimei B and Ellis S (1979) J.Bacteriol. 138, 871-877.
Shah VK, Davis LC, Gordon JK, Orme-Johnson WH and Brill WJ (1973) Biochim.Biophys.Acta, 292, 246-255.
Shah VK, Davis LC, Stieghorst M and Brill WJ (1974) J.Bacteriol. 117, 917-919.
Sorger GJ and Trofimenkoff D (1970) Proc.Natl.Acad.Sci.USA, 65,74-80
Thompson JP and Skerman VBD (1979) Azotobacteraceae: The Taxonomy and Ecology of the Aerobic Nitrogen-Fixing Bacteria. Academic Press Inc (London) Ltd.
Van Vliet F, Silva B, Van Montagu M and Schell J (1978) Plasmid 1, 446-455
Wyss O and Wyss MB (1950) J.Bacteriol. 59, 287-291
Yee T and Inouye M (1982) J.Mol.Biol. 54, 181-196

ORGANIZATION AND TRANSCRIPTION OF NITROGENASE GENES IN THE CYANOBACTERIUM ANABAENA

ROBERT HASELKORN/PETER J. LAMMERS/DOUGLAS RICE/STEVEN J. ROBINSON
DEPARTMENT OF BIOPHYSICS AND THEORETICAL BIOLOGY, UNIV. OF CHICAGO, CHICAGO, IL. 60637 USA

The genetic approach to the elucidation of biochemical pathways has been consistently powerful since its introduction by Beadle and Tatum nearly forty years ago. The ability to obtain mutants blocked in each step of a pathway makes it possible to accumulate intermediates, to analyze individual steps in detail, and to study regulation of the pathway. In the case of nitrogen fixation, application of the methods of genetic analysis developed for Escherichia coli to its nitrogen-fixing cousin Klebsiella pneumoniae has made the latter the best understood azotroph. In contrast, the absence of a laboratory gene transfer system has seriously hampered comparable studies of the cyanobacteria. However, we have made some progress in determining the organization and regulation of the nitrogen fixation (nif) genes in the cyanobacterium Anabaena by exploiting recombinant DNA methods to mitigate the absence of a gene transfer system (Mazur et al., 1980; Rice et al., 1982).

We cloned the genes coding for the structural components of nitrogenase from Anabaena in the following way: total DNA from Anabaena was digested with several restriction enzymes, fractionated by electrophoresis, blotted, and hybridized with a DNA probe that contains the nifHDK genes of Klebsiella. That probe identified homologous sequences in a number of small HindIII fragments and several large EcoRI fragments. The latter were isolated from suitable recombinant DNA libraries prepared in lambda vectors and the nif genes in these fragments were mapped by examination of heteroduplex DNA molecules in the electron microscope and by Southern hybridization, in each case comparing the Anabaena DNA fragment to its Klebsiella counterpart. The results of these experiments are summarized in Figure 1 (Rice et al., 1982). The nif gene region of the Klebsiella chromosome is shown in the upper part of the Figure (Ausubel, Cannon, 1980). Genes H, D and K code for the nitrogenase complex and genes F and J code for electron carriers. Other of the genes are involved in synthesis of the molybdenum cofactor or maturation of the nitrogenase complex. Finally, genes A and L are regulatory, in the following way: gene A codes for an activator and gene L for a repressor of transcription of the other nif genes (Ow, Ausubel, 1983; Drummond et al., 1983). The nif transcription units, defined by polarity of insertion mutations, are shown by horizontal arrows. Initiation of each of these transcripts requires a functional nifA gene product. If nifA is missing, or shut down, the other nif genes cannot be turned on.

The organization of the nif genes in Anabaena is clearly different. The most striking difference is that the nifK and nifD genes, which code for the β-subunit and α-subunit of dinitrogenase, respectively, are separated by 11 kbp in Anabaena. In Klebsiella, these two genes are adjacent and, based on polar effects of insertion mutations, they are transcribed as a unit. We will present several lines of evidence below to show that nifK and nifD are transcribed independently in Anabaena.

The only nif gene, other than nif HDK, sufficiently conserved between Klebsiella and Anabaena to serve as a hybridization probe, is nifV (or nifS). These genes are required for an unspecified step in the maturation of dinitrogenase. In Klebsiella they are located in a large operon to the left of nif HDK; in Anabaena, a 600 bp region of homology to nifV (or nifS) is located to the right of nifH, oriented in the same direction as in Klebsiella.

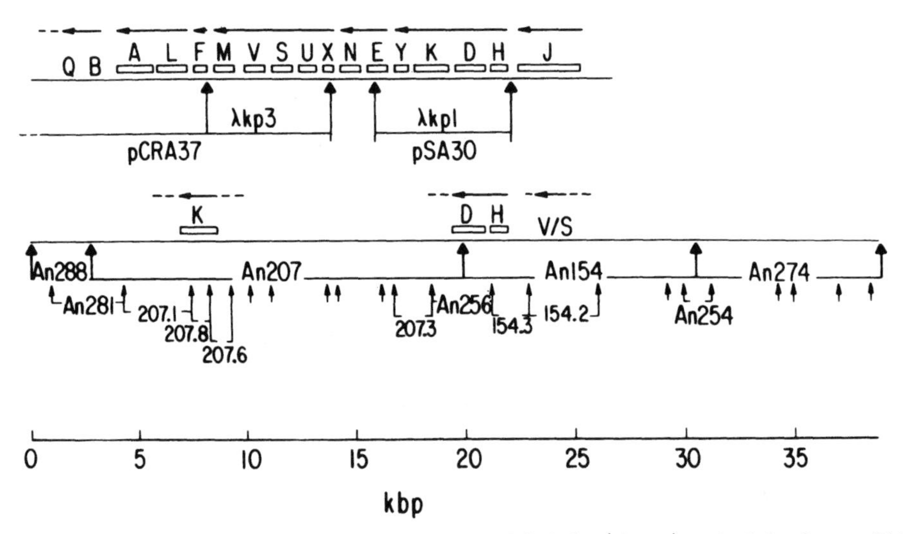

FIGURE 1. Physical map of the nif genes of Klebsiella (above) and of Anabaena 7120 (below). Sites for EcoRI and HindIII are shown as thick and thin arrowheads, respectively. Horizontal arrows indicate direction and extent of transcription units; dashed lines mean that the ends of the units are not yet defined. Boxes indicate gene product sizes where they are known. Numbers below each map (pCRA37, An288, etc.) refer to recombinant clones containing the indicated fragment. The numbered HindIII fragments of Anabaena DNA are the ones transcribed during nitrogenase induction.

All of the HindIII fragments in Figure 1 were subcloned and some of them have been sequenced. The first of these was the Anabaena nifH gene, coding for nitrogenase reductase, which was found entirely within the HindIII fragment in pAn154.3 (Mevarech et al., 1980). That same fragment also contained the start of the α subunit of nitrogenase (nifD), 115 bp downstream from the nifH gene terminator. The nifD sequence was completed by determining the sequence of pAn256 from the HindIII site leftward to a point 500 bp beyond the EcoRI site that divides An207 from An154 (Lammers, Haselkorn, 1983). The nifK gene was located by sequencing fragments 207.6, 207.8 and 207.1; the gene codes for the β subunit of nitrogenase (Mazur, Chui, 1982).

The complete amino acid sequences of the components of the Anabaena nitrogenase complex, determined by translating the nucleotide sequences of the genes, are shown in Figure 2. The nifH genes from Klebsiella and from Rhizobium meliloti have been sequenced as well; the nifH protein from Clostridium was also sequenced, along with all the cysteine-containing peptides of the protein from Azotobacter. From comparison of the five sequences thus available it could be concluded that the segments underlined in the figure are highly conserved and probably involved in critical aspects of the protein's function. For the nifD product, the Anabaena sequence is the only one completed, although partial information from Klebsiella, Clostridium, and Azotobacter is available. For the nifK product, only the complete Anabaena sequence and the Azotobacter cysteine-peptide sequences can be compared. The underlined segments in Figure 2 are placed according to those comparisons and indicate conserved regions.

nifH

MTDENIRQEAFYGKGGIGKSTTSQNTLAAMAGMGQRIMIVGCDPKADSTR (10–50)

LMLHSKAQTTVLHLAAERGAVEDLELHEVMLTGFRGVKCVESGGPEPGVG (60–100)

CAGRGIITAINFLEENGAYQDLDFVSYDVLGDVVCGGFAMPIREGKAQEI (110–150)

YIVTSGEMMAMYAANNIARGILKYAHSGGVRLGGLICNSRKVDREDELIM (160–200)

NLAERLNTQMIHFVPRDNIVQHAELRRMTVNEYAPDSNQGQEYRALAKKI (210–250)

NNDKLTIPTPMEMDELEALKIEYGLLDDDTKHSEIIGKPAEATNRSCRN (260–290)

nifD

MTPPENKNLVDENKELIQEVLKAYPEKSRKKREKHLNVHEENKSDCGVKS (10–50)

NIKSVPGVMTARGCAYAGSKGVVWGPIKDMIHISHGPVGCGYWSWSGRRN (60–100)

YYVGVTGINSFGTMHFTSDFQERDIVFGGDKKLTKLIEELDVLFPLNRGV (110–150)

SIQSECPIGSIGDDIEAVAKKTSKQIGKPVVPLRCEGFRGVSQSLGHHIA (160–200)

NDAIRDWIFPEYDKLKKETRLDFEPSPYDVALIGDYNIGGDAWASRMLLE (210–250)

EMGLRVVAQWSGDGTLNELIQGPAAKLVLIHCYRSMNYICRSLEEQYGMP (260–300)

WMEFNFFGPTKIAASLREIAAKFDSKIQENAEKVIAKYTPVMNAVLDKYR (310–350)

PRLEGNTVMLYVGGLRPRHVVPAFEDLGIKVVGTGYEFAHNDDYKRTTHY (360–400)

IDNATIIYDDVTAYEFEEFVKAKKPDLIASGIKEKYVFQKMGLPFRQMHS (410–450)

WDYSELGDGVQMSDEVRFFCEGRKKSLFLA (460–480)

FIGURE 2. Complete amino acid sequences for the three proteins of the nitrogenase complex of Anabaena 7120, determined by translation of the gene sequences. Data for nifH are from Mevarech, et al. (1980); for nifD from Lammers, Haselkorn (1983); for nifK from Mazur, Chui (1982). Sequences are shown in the one-letter amino acid code. Underlined sequences are conserved in at least one other bacterial species. Cysteine residues likely to be important in ligand binding are boxed.

nifK

```
                  10                  20                  30                  40                  50
M P Q N P E R T D H V D L F K Q P E Y T E L F E N K R K N F E G A H P P E E V E R V S E W T K S W D

                  60                  70                  80                  90                 100
Y R E K N F A R E A L T V N P A K G [C] Q P V G A M F A A L G F E G T L P F V Q G S Q G [C] V A Y F R T

                 110                 120                 130                 140                 150
H L S R H Y K E P [C] S A V S S S M T S D A A V F G G L N N M I E G M Q V S Y Q L Y K P K M I A V C T

                 160                 170                 180                 190                 200
T [C] M A E V I G D D L G A F I T N S K N A G S I P Q D F P V P F A H T P S F V G S H I T G Y D N M M

                 210                 220                 230                 240                 250
K G I L S N L T E G K K K A T S N G K I N F I P G F D T Y V G N N R E L K R M M G V M G V D Y T I L

                 260                 270                 280                 290                 300
S D S S D Y F D S P N M G E Y E M Y P S G T K L E D A A D S I N A K A T V A L Q A Y T T P K T R E Y

                 310                 320                 330                 340                 350
I K T Q W K Q E T Q V L R P F G V K G T D E F L T A V S E L T G K A I P E E L E I E R G R L V D A I

                 360                 370                 380                 390                 400
T D S V A W I H G K K F A I Y G D P D L I I S I T S F L L E M G A E P V H I L [C] N N G D D T F K K E

                 410                 420                 430                 440                 450
M E A I L A A S P F G K E A K V W I Q K D L W H F R S L L F T E P V D F F I G N S Y G K Y L W R D T

                 460                 470                 480                 490                 500
S I P M V R I G Y P L F D R H H L H R Y S T L G Y Q G G L N I L N W V V N T L L D E M D R S T N I T

                 510
G K T D I S F D L I R
```

The Anabaena nifK and nifD genes are of comparable size but their nucleotide sequences have no detectable regions of homology to each other. Similarly, the amino acid sequences show no obvious evolutionary relationship. However, a plot of the secondary structures of the α- and β-subunits, based on the amino acid sequences, shows that there are regions of structural homology roughly in the area of the first three conserved cysteines in each subunit (Lammers, Haselkorn, 1983). Low resolution structural homology between α- and β-subunits of Clostridium dinitrogenase has been inferred from X-ray crystallographic data (Yamane et al., 1982).

Hybridization of Klebsiella nif DNA to total Anabaena DNA produced one additional unexpected result. In the EcoRI digest of total Anabaena DNA, a third band of ~19 kbp was detected in addition to the 17 and 10.5 kbp bands mapped and shown in Figure 1. This band was recovered from a lambda Charon 4A library and shown to contain a second sequence related to the Anabaena nifH gene. The restriction map of this gene differs from the one shown in Figure 1 and there does not appear to be a nifD gene adjacent to it. However, determination of the nucleotide sequence of the region of homology shows an open reading frame nearly identical in amino acid sequence to the product of the original nifH gene (S. J. Robinson, unpublished). It will be of some interest to determine whether this gene is transcribed when nitrogenase is induced.

This brings us to the question of the regulation of nif gene expression which, as will be seen, is at the level of transcription. The lower portion of Figure 1 shows the distribution of HindIII sites along the 39 kbp of cloned Anabaena DNA containing the known nif genes (other than the extra nifH gene). Each of the HindIII fragments was cloned separately into plasmids and used for a quick test of transcription, as follows: total RNA was extracted from Anabaena cells grown on ammonia or from cells induced anaerobically for nitrogenase. Each RNA preparation was bound to filters and

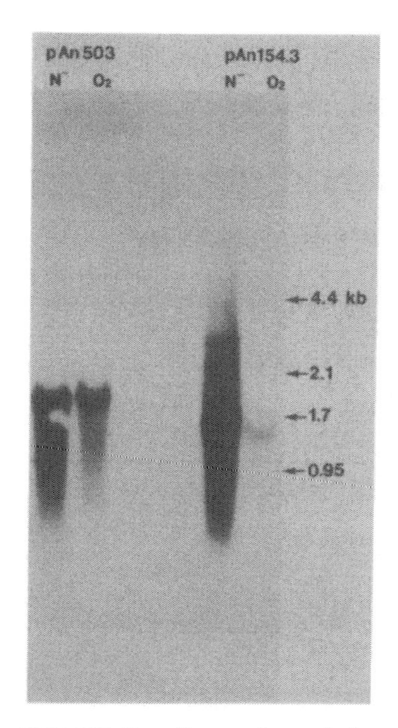

FIGURE 3. Transcripts of the glnA and nifH genes in Anabaena. RNA was prepared from cells induced anaerobically for nitrogenase (N^-) or induced and then gassed with O_2 for 45 min. Each preparation was fractionated by electrophoresis, transferred to nitrocellulose paper, and probed with radioactive pAn503, which contains the glnA gene (Fisher et al., 1981), or pAn154.3, which contains the nifH gene and the 5' end of nifD. Ammonia-grown cells have no detectable nif messenger RNA; the induced cells have several sizes of nifH mRNA; the O_2-treated cells have lost their nif mRNA. GlnA mRNA is present in ammonia-grown cells at roughly the same size as shown in the figure. Unlike nif mRNA, the gln mRNA is stable in O_2.

then each pair of filters (N^+ RNA, N^- RNA) was annealed with nick-translated plasmid. Those plasmids bound by N^- RNA, but not by N^+ RNA, were judged to have contained sequences transcribed at elevated levels during nitrogenase induction (Rice et al., 1982). Every such fragment is numbered in the lowest part of Figure 1. Fragments without a numerical designation are not detectably transcribed. All of the fragments known to contain nif genes are transcribed. In addition, the fragment to the left of nifD and the entire region to the left of nifK are transcribed. These may contain more nif genes.

The qualitative transcription studies were extended in several ways. One was to examine the size and stability of the messenger RNA corresponding to the cloned genes. So far this examination has been done only for the nifH gene, using pAn154.3 as the probe, with the result shown in Figure 3 (S. J. Robinson, unpublished results). Here, two RNA preparations are compared. One is from cells induced anaerobically for nitrogenase and the second is from induced cells to which O_2 has been added for 45 min. RNA prepared from cells grown on ammonia contains no detectable nif mRNA. It can be seen that nitrogenase induction results in the synthesis of several size classes of nifH messenger RNA. The largest of these is sufficiently long to code for both nifH and nifD proteins, while the more abundant 1.4 kb message is too small

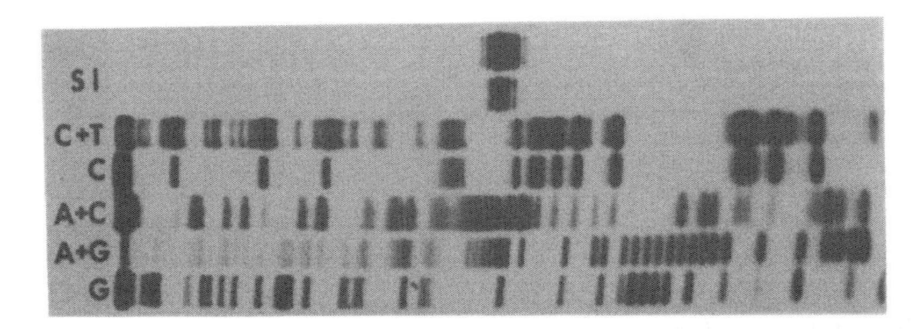

FIGURE 4. Determination of the start site for transcription of nifH messenger RNA in Anabaena. The outside lanes contain end-labeled DNA fragments that had been annealed with RNA from cells induced for nitrogenase and then digested with nuclease S1. The interior lanes contain the same end-labeled DNA fragments subjected to the Gilbert-Maxam sequencing reactions. Comparison of the S1 lanes with the sequence ladder permits exact assignment of the start of the message.

```
                  -180                              -160
TAACACCCAAAAGAACTTTCACAACTACATAACGAACCCATCATGAAC

      -140               ▽      -120                    -100
ACTAATTCTACTGGTTTTTCTGTGGAGCGATCGCCCCCTCTTCGGCGACTG
                                  Sau3a

                    -80                           -60
TTCTACATAACCCCTCACAGCCATAGCTCAAACAGGCGTGAGATCCAAAC
                                                 Sau3a

      -40                        -20                     1
ACAAAGACCGACCAACTAACCAACCAATTGCAGGAAAAGAGAACA ATG

ACT GAC GAA AAC ATT AGA CAG ATA GCT TTC etc.
```

FIGURE 5. Nucleotide sequence of the DNA fragment containing the origin of transcription of nifH messenger RNA in Anabaena. The arrow shows the origin determined by S1 nuclease protection and primer extension. Underlined sequences are found in comparable positions of other Anabaena nif-related transcripts.

---

to code for both, and perhaps terminates within the nifD gene. When O_2 is admitted, both species of messenger RNA disappear. This result is observed because the induction was done anaerobically (argon plus DCMU); conditions under which heterocyst differentiation is incomplete, nitrogenase is induced in all the cells and is completely oxygen-sensitive. This result is of interest because it provides one of the reasons that nitrogenase proteins appear late in heterocyst differentiation: nif gene transcription does not occur until anaerobiosis is established. Oxygen repression of nif gene transcription is also a feature of nif gene regulation in Klebsiella, where the responsible gene is nifL (Hill et al., 1981).

The other two lanes in Fig. 3 contain the same RNA preparations probed with pAn503, which contains the Anabaena gene coding for glutamine synthetase, glnA. The enzyme is present in cells growing on ammonia or fixing nitrogen, so this experiment serves as a control of the quality of the RNA prepared from O_2-treated cells, to demonstrate that the absence of a band in the nifH gene-probed lane was not due to general degradation of the RNA. That part of the experiment is satisfactory. But we have subsequently learned that the glnA gene is transcribed from a promoter, under N^- conditions, that is very similar in nucleotide sequence to the nifH gene promoter (Tumer et al., 1983). Since very similar promoters are used for one gene that is not transcribed under N^-, O_2^+ conditions and for another gene that is, there must be other regulatory sequences that distinguish between the two.

The cloned Anabaena nif gene fragments and the knowledge of their sequences has made possible a deeper analysis of transcription based on determination of the promoter sequences for several genes. This analysis proceeded in two ways. In the first, a suitable DNA fragment which contains the suspected promoter was labeled at the 5' end, annealed with messenger RNA from cells induced for nitrogenase, and then digested with nuclease S1. The precise size of the DNA fragment protected from nuclease digestion by the messenger RNA was determined by comparison with a sequencing ladder on a urea gel. In the alternate procedure, a DNA fragment corresponding to a sequence within the gene was labeled at the 5' end, annealed with messenger RNA, and then extended back to the transcript start using AMV reverse transcriptase and deoxynucleoside triphosphates. Again, the product size was measured accurately, on a sequencing gel. Ideally the two methods should agree and should indicate, within one or two nucleotides, the start site of transcription. Data for the determination of the start of Anabaena nifH messenger RNA by S1 nuclease protection are shown in Figure 4. The promoter region of the nifH gene sequence is shown in Figure 5. The underlined residues in the -35 and -10 region are of considerable interest: they correspond to very poor promoter sequences in E. coli. This result is perfectly consistent with a requirement for specific transcriptional activation of the nif genes, as is the case for Klebsiella.

This promoter is not transcribed in E. coli. This is known because no nifH product is seen in E. coli cells containing the cloned Anabaena nifH gene unless there is a strong lambda promoter or plasmid promoter suitably placed upstream of the Anabaena nifH gene. Moreover, the nifH DNA fragment is not transcribed in vitro, either by Anabaena or E. coli RNA polymerase (Catherine Richaud, personal communication), while both polymerases recognize conventional E. coli-like promoters on other cloned Anabaena DNA fragments (Nilgun Tumer, personal communication). All of these observations suggest that Anabaena nif genes must be activated for transcription or are transcribed by a modified RNA polymerase.

REFERENCES

Ausubel FM and Cannon FC (1980) Cold Spring Harbor Symp Quant Biol 45, 4723-4727.
Drummond M, Clements J, Merrick M and Dixon R (1983) Nature 301, 302-307.
Fisher R, Tuli R and Haselkorn R (1981) Proc. Natl. Acad. Sci. USA 78, 3393-3397.
Lammers PJ and Haselkorn R (1983) Proc. Natl. Acad. Sci. USA 80, 4723-4727.
Mazur BJ and Chui, C-F (1982) Proc. Natl. Acad. Sci. USA 79, 6782-6786.
Mazur BJ, Rice D and Haselkorn R (1980) Proc. Natl. Acad. Sci. USA 77, 186-190.
Mevarech M, Rice D and Haselkorn R (1980) Proc. Natl. Acad. Sci. USA 77, 6476-6480.
Ow DW and Ausubel FM (1983) Nature 301, 307-313.
Rice D, Mazur BJ and Haselkorn R (1982) J. Biol. Chem. 257, 13157-13163.
Tumer NE, Robinson SJ and Haselkorn R (1983) Nature, in press.
Yamane T, Weininger MS, Mortenson, LE and Rossmann MG (1982) J. Biol. Chem. 257, 8042-8048.

THE EXPRESSION OF SYM-PLASMIDS AND TI PLASMIDS IN RHIZOBIA AND AGROBACTERIA

P.J.J. HOOYKAAS[+], A.A.N. VAN BRUSSEL, R.J.M. VAN VEEN[+], C.A. WIJFFELMAN
DEPARTMENT OF PLANT MOLECULAR BIOLOGY
+ BIOCHEMICAL LAB, WASSENAARSEWEG 64, 2333 AL LEIDEN, THE NETHERLANDS
++BOTANICAL LAB, NONNENSTEEG 3, 2311 VJ LEIDEN, THE NETHERLANDS

1. INTRODUCTION

Agrobacteria and rhizobia belong to the bacterial family of Rhizobiaceae.The classification of bacteria belonging to this family into species is based on phytopathogenic and symbiotic properties. In the genus Agrobacterium three species can be distinguished viz A.tumefaciens, the causative agent of the crown gall disease, A.rhizogenes,which induces hairy root in plants, and A.radiobacter, which is non-pathogenic. The genus Rhizobium contains species such as R.trifolii, which forms root nodules on clovers, R.phaseoli, which nodulates beans, and R.leguminosarum, which induces nodulation on peas, lathyrus and vetches.
For Agrobacterium it has been shown that the phytopathogenic properties are determined by genes on large (120-150 Mdal.) plasmids. In the case of A.tumefaciens the plasmid has been called tumor-inducing (Ti)-plasmid (Van Larebeke et al, 1974; 1975; Watson et al, 1975), in the case of A.rhizogenes root-inducing (Ri)-plasmid (Albinger, Beiderbeck, 1977; White, Nester, 1980). Agrobacteria genetically engineer plant cells. During crown gall induction as well as during hairy root induction a piece of genetic material is transferred from the agrobacteria to the plant cells, and subsequently is integrated somewhere into the nuclear DNA of the plant cells. The transferred DNA (T-DNA) corresponds to a portion of the Ti plasmid in the case of crown gall induction (e.g.Chilton et al, 1977; Ooms et al, 1982), and to a portion of the Ri plasmid in the case of hairy root induction (Chilton et al, 1982; Spano et al, 1982; White et al, 1982). Crown gall tissues have two novel properties as compared to normal plant tissues, viz they can grow in the absence of phytohormones (Braun, 1958) and they contain enzymes that are responsible for the synthesis of unusual compounds (opines) such as octopine and nopaline (Otten et al, 1977; Goldmann, 1977). Genes that are located in the T-DNA are responsible for the novel characteristics of the crown gall cell (Ooms et al, 1981; Garfinkel et al, 1981; Schröder et al, 1981; Murai, Kemp, 1982). The molecular genetics of crown gall tumorigenesis has recently been reviewed (Hooykaas, Schilperoort, 1983). For Rhizobium it has been shown that large (100->300 Mdal.) plasmids are present in all strains sofar examined (Nuti et al, 1977; Casse et al, 1979). Some of these large plasmids have a region of DNA-homology with the Klebsiella pneumoniae structural nif genes (Nuti et al, 1979; Hombrecher et al, 1981; Prakash et al, 1981), indicating that the structural nif genes in Rhizobium are located on large plasmids. These same large plasmids also contain genes which determine nodulation ability and host specificity for nodulation in R.leguminosarum, R.phaseoli and R.trifolii and have therefore been named Sym(biotic) plasmids (Johnston et al, 1978; Hombrecher et al, 1981; Hooykaas et al, 1981; 1982a; Lamb et al, 1982).

2. RESULTS AND DISCUSSION

Since the phytopathogenic and symbiotic properties in Agrobacterium and Rhizobium are largely determined by genes on plasmids, we investigated

Figure 1. Tumors induced by Agrobacterium tumefaciens on kalanchoë.

the effect of the introduction of Ti plasmids into rhizobia or of Sym plasmids into agrobacteria.

2.1. Transfer of Ti and Ri plasmids to rhizobia: The Ti plasmids are conjugative plasmids, but transfer only occurs if a specific opine is added to the conjugation medium. For octopine Ti plasmids octopine, ornopine and lysopine can act as aphrodisiacs (Genetello et al, 1977; Kerr et al, 1977; Hooykaas et al, 1979). The Ti plasmids can also be transferred to new hosts via mobilization by an R plasmid (Hooykaas et al, 1980a). In such mobilization experiments relatively stable R::Ti cointegrates may be obtained (Holsters et al, 1978; Klapwijk et al, 1978). We succeeded in transferring the octopine plasmid into R.trifolii by mobilization with RP4 (Hooykaas et al, 1977) as well as by inducing the Ti transfer genes by the addition of octopine to the medium (Hooykaas, 1979). The ability to utilize octopine as a nitrogen source was used for the selection of octopine Ti plasmid carrying transconjugants. A Tra^c nopaline Ti plasmid and a Ri plasmid were also introduced into this same new host. The R.trifolii transconjugants with a Ti or an Ri plasmid had the same growth characteristics as the parental strains, and they still nodulated effectively (only) their proper clover hosts (Table 1). In agreement the R.trifolii Sym plasmid was still present in the transconjugants, which shows that this plasmid belongs to a different incompatibility group than Ti and Ri plasmids. Importantly, the R.trifolii Ti$^+$ transconjugants were able to induce crown gall on dicotyledonous plants (e.g. tomato, tobacco), whereas Ri$^+$ transconjugants induce hairy root on the same plant species. Opines were detected in the tumors and roots incited by these strains, but opines were not detected in root nodules induced by the same strains. This indicates that T-DNA transfer does not occur in the root nodules. Vice versa fixation of nitrogen did not take place in the tumors induced by these R.trifolii strains, showing that the nif genes are not activated when the bacteria are present in a tumor. Similar results were obtained if an RP4::Ti cointegrate was transferred into an R.leguminosarum strain. After transfer the recipient became able to form crown gall on dicotyledonous plants. However, the presence of this Ti plasmid interfered

TABLE 1. Properties of R.trifolii Ti$^+$ and Ri$^+$ transconjugants

	LPR 5001	LPR 5011	LPR 5055	LPR 5079
Plasmid introduced	-	pTiB6	pTiT37	pRi 1855
Sensitivity to:				
rhizobiophages	+	+	+	+
agrobacteriophages	-	-	-	-
Nodulation of clovers	+	+	+	+
Nitrogen fixation	+	+	+	+
Tumor induction	-	+	+	roots
Opines in tumors		octopine	nopaline	agropine
Opine catabolism		octopine	nopaline	not tested

LPR 5001, LPR 5055 LPR 5079 are directly derived from LPR 5001.

with proper nodulation on the pea: the nodules remained small and no nitrogen was fixed (Hooykaas, 1979). No interference occurred with nodulation of some other hosts (Vicia hirsuta,Vicia sativa) belonging to the same cross-inoculation group.
Different types of Ti plasmids were similary introduced into R.meliloti strains. The R.meliloti Ti$^+$ transconjugants had unaltered growth characteristics as compared to the parental strains and they nodulated effectively alfalfa. In contrast to the R.trifolii and R.leguminosarum Ti$^+$ transconjugants the R.meliloti Ti$^+$ transconjugants were not able to induce crown gall (Hooykaas, 1979). The reason for this may be that R.meliloti cannot attach to the cell walls of dicotyledonous plants (Draper et al, 1983; Krens, 1983). This is a chromosomally determined property in Agrobacterium (Douglas et al, 1982), which must also be present in R.leguminosarum and R.trifolii (see above).

2.2. Transfer of Sym plasmids to agrobacteria: In order to be able to detect the transfer of Sym plasmids to new hosts these plasmids must first be tagged, for instance with transposon Tn5 (Beringer et al., 1978). Some Sym plasmids are conjugative, but others seem to be non-conjugative. However, transfer of the latter can be brought about by the introduction into the cells of a mobilizing R plasmid (Hooykaas et al, 1982b). First, a conjugative 230Mdal. R.trifolii Sym plasmid was introduced into a Ti$^-$ A.tumefaciens strain (Hooykaas et al., 1981). The properties of the transconjugants are summarized in Table 2.

TABLE 2. Properties of A.tumefaciens pSym$^+$ transconjugants

	LBA 288	LBA 2712	LBA 2702	LBA 2709
Plasmid introduced	-	pSym 1	pSym 5	pSym 9
Sensitivity to:				
rhizobiophages	-	-	-	-
agrobacteriophages	+	+	+	+
Tumor induction	-	-	-	-
Nodulation of:				
vetches	-	+	-	-
clovers	-	-	+	-
beans	-	-	-	+
Nitrogen fixation	-	-	-	-

LBA 2712, LBA 2702 and LBA 2709 are directly derived from LBA 288.

The growth characteristics of the agrobacteria were not altered by the entrance of this Sym plasmid. Interestingly, the transconjugants were able to induce root nodules on clover species in contrast to the parental strains. However, nodules appeared somewhat later than after infection with R.trifolii,and no nitrogen was fixed in these nodules. Studies of the nodules with the electron microscope showed the presence of numerous infection threads, and, within the plant cells, agrobacteria surrounded by a peribacteroid membrane. Apparently, agrobacteria with a R.trifolii Sym plasmid are released from the infection threads into the plant cells. No evidence for conversion of these agrobacteria into bacteroids was obtained. Results similar to those described above were obtained when agrobacteria with Sym plasmids from R.leguminosarum or R.phaseoli were tested for nodulation of plants of the proper cross-inoculation-group (Hooykaas et al, 1982a; Van Brussel et al., 1982). Agrobacteria with a R.leguminosarum Sym plasmid had a smaller host range for nodulation than rhizobia with the same Sym plasmid (Van Brussel et al., 1982). Resident Ti or Ri plasmids in agrobacteria with a Sym plasmid did not cause an interference in the expression of the nod genes on the Sym plasmid and vice versa the Sym plasmid sym genes did not interfere in the expression of the Ti plasmid vir- and onc-genes. Genes for medium bacteriocin production and for small bacteriocin sensitivity, which are located on the R.leguminosarum Sym plasmid pRL1, were properly expressed in Agrobacterium (Wijffelman et al., 1983).

3. CONCLUDING REMARKS

From our work it appears that plasmids largely determine the phytopathogenic properties of agrobacteria. However, also factors encoded by other parts of the genome are important. R.meliloti does not become oncogenic after receiving a Ti plasmid. Other rhizobia with Ti plasmids, induce tumors, but the tumors induced are generally somewhat smaller than those induced by agrobacteria. Sym plasmids largely determine the symbiotic properties of rhizobia, but the introduction of Sym plasmids into agrobacteria converts these into nodulating, but not into nitrogen fixing bacteria although the nif genes are present on the plasmids. The lack of bacteroids in agrobacterial root nodules indicates that some factors that are not encoded by the Sym plasmids are necessary for a complete nodule development. A determining factor might be the presence of a bacteroid-state adaptable cell wall. However, agrobacteria differ in more aspects from rhizobia. For instance in contrast to rhizobia agrobacteria contain enzymes which catalyze the production of H_2O_2 from amino acids (Clark, 1972). The octopine and nopaline Ti plasmids form a group of related plasmids of a similar size (100 - 150 Mdal.) and belong to the inc Rh-1 incompatibility group (Hooykaas et al, 1980b). The leucinopine Ti plasmids, although related to the inc Rh-1 plasmids on the basis of DNA homology, belong to a different inc-group (inc Rh-2) and the Ri plasmids form a third group (inc-Rh-3) (Hooykaas, 1979; Costantino et al, 1980). The sym plasmids form a very heterogeneous group of plasmids enormously varying in size (100-> 300 Mdal.) and in incompatibility properties. Sofar no incompatibility with Ti or Ri plasmids has been observed, but some DNA homology has been detected between Sym plasmids and Ti plasmids (Prakash, Schilperoort, 1982). The significance of this finding is not clear yet. Rhizobium Sym plasmids cannot bring about the transfer of the T-region from the Ti plasmid to plant cells (Hoekema, pers. communication).

4. REFERENCES

Albinger G and Beiderbeck R (1977) Phytopath.Z. 90, 306-310
Beringer JE, Beynon JL, Buchanan-Wollaston AV and Beringer JE (1978) Nature 276, 633-634
Braun AC (1958) Proc. Natl. Acad. Sci. USA 44, 344-349
Casse F, Boucher C, Julliot JS, Michel M and Dénarié J (1979) J. gen. Microbiol. 113, 229-242
Chilton MD, Drummond MH, Merlo DJ, Sciaky D, Montagu AL, Gordon MP and Nester EW (1977) Cell 11, 263-271
Chilton MD, Tepfer DA, Petit A, David C and Tempé J (1982) Nature 295, 432-434
Clark AG (1972) Phytopath.Z. 74, 74-83
Constantino P, Hooykaas PJJ, Den Dulk-Ras H and Schilperoort RA (1980) Gene 11, 79-87
Douglas CJ, Halperin W and Nester EW (1982) J. Bacteriol. 152, 1265-1275
Draper J, Mackenzie IA, Davey MR and Freeman JP (1983) Plant Sc. Lett. 29, 227-236
Garfinkel DJ, Simpson RB, Ream LW, White FF, Gordon MP and Nester EW (1981) Cell 27, 143-153
Genetello C, Van Larebeke N, Holsters M, De Picker A, Van Montagu M and Schell J (1977) Nature 265, 561-563
Goldman A (1977) Plant Sci. Lett. 10, 49-58

Holsters M, Silva A, Genetello C, Engler G, Van Vliet F, De Block M, Villarroel R, Van Montagu M and Schell J (1978) Plasmid 1, 456-467
Hombrecher G, Brewin NJ and Johnston AWB (1981) Mol.Gen. Genet. 182, 133-136
Hooykaas PJJ, Klapwijk PM, Nuti MP, Schilperoort RA and Rörsch A (1977) J. gen. Microbiol. 98, 477-484
Hooykaas PJJ (1979) The role of plasmid determined functions in the interactions of Rhizobiaceae with plant cells. A genetic approach. Thesis, Leiden, The Netherlands
Hooykaas PJJ, Roobol C and Schilperoort RA (1979) J.gen. Microbiol. 110,99-109
Hooykaas PJJ, Den Dulk-Ras H and Schilperoort RA (1980a) Plasmid 4, 64-75
Hooykaas PJJ, Den Dulk-Ras H, Ooms G and Schilperoort RA (1980b). J. Bacteriol. 143, 1295-1306
Hooykaas PJJ, Van Brussel AAN, Den Dulk-Ras H, Van Slogteren GMA and Schilperoort RA(1981) Nature 291, 351-353
Hooykaas PJJ, Snijdewint FGM and Schilperoort RA (1982a) Plasmid 8, 73-82
Hooykaas PJJ, Den Dulk-Ras H and Schilperoort RA (1982b) Plasmid 8, 94-96
Hooykaas PJJ and Schilperoort RA (1983) Adv. Genet. 22,
Johnston AWB, Beynon JL, Buchanan-Wollaston AV, Setchell SM, Hirsch PR and Beringer JE (1978) Neture 276, 634-636
Kerr A, Manigault P and Tempé J (1977) Nature 265, 560-561
Klapwijk PM, Schleuderman T and Schilperoort RA (1978) J. Bacteriol. 136, 775-785
Krens FA (1983) Studies on transformations of tobacco leaf protoplasts: Ti plasmid DNA transformation and cucoltivation with Agrobacterium tumefaciens. Thesis, Leiden, The Netherlands
Lamb JW, Hombrecher G and Johnston AWB (1982) Mol. Gen. Genet. 186, 449-452
Murai N and Kemp JD (1982) Proc.Natl.Acad. Sci. USA 79, 86-90
Nuti MP, Ledeboer AM, Lepidi AA and Schilperoort RA (1977) J.gen. Microbiol. 100, 241-248
Nuti MP, Lepidi AA, Prakash RK, Schilperoort RA and Cannon FC (1979) Nature 282, 533-535
Ooms G, Hooykaas PJJ, Moolenaar G and Schilperoort RA (1981) Gene 14,33-50
Ooms G, Bakker A, Molendijk L, Wullems GJ, Gordon MP, Nester EW and Schilperoort RA (1982) Cell 30, 589-597
Otten LABM, Vreugdenhil D and Schilperoort RA (1977) Biochim.Biophys.Acta 485, 268-277
Prakash RK, Schilperoort RA and Nuti MP (1981) J. Bacteriol. 145, 1129-1136
Prakash RK and Schilperoort RA (1982) J. Bacteriol. 149, 1129-1134
Schröder J, Schröder G, Huisman H, Schilperoort RA and Schell J (1981) FEBS Lett. 129, 166-168
Spano L, Pompni M, Costantino P, Van Slogteren GMS and Tempé J (1982) Plant Mol. Biol. 1, 291-300
Van Brussel AAN, Tak T, Wetselaar A, Pees E and Wijffelman CA (1982) Plant Sc. Lett. 27, 317-315
Van Larebeke N, Engler G, Holsters M, Van den Elsacker S, Zaenen I, Schilperoort RA and Schell J (1974) Nature 252, 169-170
Van Larebeke N, Genetello C, Schell J, Schilperoort RA, Hermans AK, Hernalsteens JP and Van Montagu M (1975) Nature 255, 742-743
Watson B, Currier TC, Gordon MP, Chilton MD and Nester EW (1975) J. Bact. 123, 255-264
White FF and Nester EW (1980) J. Bacteriol. 141, 1134-1141
White FF, Ghidossi G, Gordon MP and Nester EW (1982) Proc. Natl. Acad. Sci. USA 79, 3193-3197
Wijffelman CA, Pees E, Van Brussel AAN and Hooykaas PJJ (1983) Mol.Gen.Genet.

POSTER DISCUSSION 9A ORGANIZATION OF RHIZOBIUM SYMBIOTIC GENES

HAUKE HENNECKE, Mikrobiologisches Institut, ETH Zurich, CH-8092 Zurich, Switzerland

A total of 36 posters provided the basis for this discussion, whereby the reports concerning nodulation genes were excluded; these were discussed in the session on Rhizobium plasmids (9B).
The major emphasis was on the organization of Rhizobium genes coding for functions in nitrogen fixation (nif, or more general fix), in particular those for the proteins of the nitrogenase enzyme complex (nifH for component 2 and nifD/nifK for the two different polypeptides of component 1). From the excellent work presented in the posters it is now clear that the structural organization of the nitrogenase genes both in the free-living nitrogen fixing microorganisms as well as in the symbiotic rhizobia is more versatile and complex as one may have expected initially. The nifHDK operon structure was demonstrated in several fast-growing rhizobia (R.meliloti; R.trifolii, R.leguminosarum and most likely also in R.phaseoli and R.sp.Sesbania). In slow growing rhizobia, however, there are two operons, nifH and nifDK (R.japonicum, R.sp.Parasponia, R.sp.Cowpea). Linkage between the two operons has not yet been demonstrated, i.e. the distance in kilobase pairs between them is not known. In slow-growing rhizobia each of the nifH and nifDK operons seem to be present in one copy per genome, whereas some of the nif genes in fast growers are reiterated: at least nifH homologous DNA appears to be present in more than one copy in R.phaseoli and R.sp.Sesbania. Four such nifH copies were found in R.phaseoli, and partial nucleotide sequences revealed almost complete identity among them. From published work it was known that, in fast growers, the nif genes are located on indigenous large plasmids. In a few R.phaseoli strains reiterated nifH sequences were reported here to be distributed on two plasmids. The question that now remains to be answered is whether or not these additional genes are expressed in the nitrogen fixing symbiosis, and if yes, whether their gene products are functionally identical or not.
In the search for other nif or fix genes it became evident that these genes are not as tightly linked to the nifH, D, and K genes as known for the nif clusters of the free-living Klebsiella pneumoniae. Further groups of fix genes were detected in R.leguminosarum and R.meliloti but little is known what functions they determine. In this connection it was of interest to see that fragments of R.sp. Sesbania were cloned which apparently bear homology to the K.pneumoniae nifJ and nifNE genes. It was suggested that a more intensive search for Rhizobium fragments hybridizing to defined K.pneumoniae nif fragments could lead to the isolation of further analogous Rhizobium genes. One such gene could be the nifA gene coding for the protein which activates other nif genes. There is some evidence by different experimental approaches that an analogous gene might exist in R.meliloti, R.leguminosarum, and R.trifolii. Yet, convincing evidence for the functioning of such a system needs to be worked out. The question was raised whether or not a nifA/ntr-type regulatory system would operate in Rhizobium. In the near future a number of groups will certainly attempt to answer this question. One of the incentives towards this goal is obviously the finding that the characteristic CTGG-8bp-TTGCA sequence involved in this regulation was found in promoters of some rhizobial nitrogenase genes (e.g. in R.meliloti and R.japonicum).
A respectable amound of DNA sequence information concerning Rhizobium nif genes is now accumulating. This symposium has added the R.phaseoli and R.sp.Cowpea nifH sequences to the previously published nifH sequences from R.meliloti, R.trifolii and R.sp.Parasponia. Furthermore, the first complete rhizobial nifD sequences were exhibited (from R.japonicum and R.sp. Parasponia). In all

extensive interspecies homologies were found both on the nucleotide sequence level, and even more so on the level of the predicted amino acid sequence. Unfortunately, no amino acid sequencing is being done, e.g. of the NH_2-termini of purified nitrogenase polypeptides in order to make sure that the predicted N-terminal amino acids are indeed present in mature nitrogenases. With more and more sequence data becoming available one might soon be able to propose a concept on the evolution of rhizobial nitrogenase genes.
A great interest in the audience was provoked by the finding of repeated sequence in several Rhizobium strains. Particularly intriguing are those of R.meliloti and R.trifolii as some of them provide the promoter regions of nif genes. A repeat sequence was also found in the vicinity of the R.japonicum nifK operon, but this one does not carry the promoter region and it looks more like an IS-type element. At present one can only speculate on the origin and functional involvement of the repeated sequences with regard to symbiotic nitrogen fixation. The field is open for studies on perhaps a more dynamic view of the Rhizobium genome including genome rearrangements as a possible means of switching on and off symbiotic genes.

MOLECULAR ANALYSIS OF *Rhizobium meliloti* GENES INVOLVED IN NODULATION

O. MARIO AGUILAR/P. MÜLLER/P. GRÖNGER/A. PÜHLER
LEHRSTUHL FÜR GENETIK, FAKULTÄT FÜR BIOLOGIE, UNIVERSITÄT BIELEFELD,
POSTFACH 8640, D-4800 BIELEFELD 1, FRG

We have isolated *R.meliloti* Tn*5*-induced mutants which failed to induce nodules on *Medicago sativa* (alfalfa). Using the mobilizable strain *E.coli* S-17-1, we transferred the plasmid pRmSL26 into these Nod$^-$ mutants. Transconjugants possessing pRmSL26 were tested for nodulation (Table I). For a further characterization of the positively complemented mutants, we subcloned different fragments from pRmSL26 into the vector pSUP1o4. Two hybrids were constructed (pMA10, pMA11), and then conjugated into Nod$^-$ mutants. In one case (2526) pMA10 restored the ability to form active nodules (Table I). Furthermore, we analysed this region in homogenotization experiments using the 5.3 kb *Pst*I fragment cloned into the vector pSUP202. The resultant hybrid, pMA22, was mutagenized with Tn*5* and then mobilized into wild type *R.meliloti* 2011. Transconjugants with phenotype Nm$^+$ Tc$^-$ Cm$^-$, were isolated and tested on plants. One Tn*5* insert showed Nod$^-$ phenotype (Figure I, A). From these results, we assumed that the region which complemented the Nod$^-$ mutant 2526 should be located in the right side of the cloned fragment in pMA10. The fact that mutant 1142 were complemented by the whole pRmSL26 but not by pMA10 or pMA11, indicates that regions other than that cloned in pMA10 contain essential genes for early steps of nodulation. In addition, in heteroduplex experiments between a *R.trifolii* sequence cloned in pRt8002, and the 8.7 kb *Eco*RI fragment from pRmSL26 we observed a homologous region extending 3.0 kb into the right side of pMA10. Also by cloning the same subfragments in *E.coli* vector plasmids we studied their expression in *E.coli* minicells. With the 8.7 kb *Eco*RI fragment (Figure I, B) we detected a 52 K fusion polypeptide with part of the CAT gene. The *R.meliloti* coding region extends 1.1 kb from the left and its termination agreed when compared with results from the 2.2 *Hin*dIII fragment. Our results indicate that only the *R.meliloti* DNA adjacent to the external promoter is expressed in *E.coli* minicells as a fusion product. Therefore, transcription initiation from *R.meliloti* promoter should not occur.

TABLE I

FIGURE 1

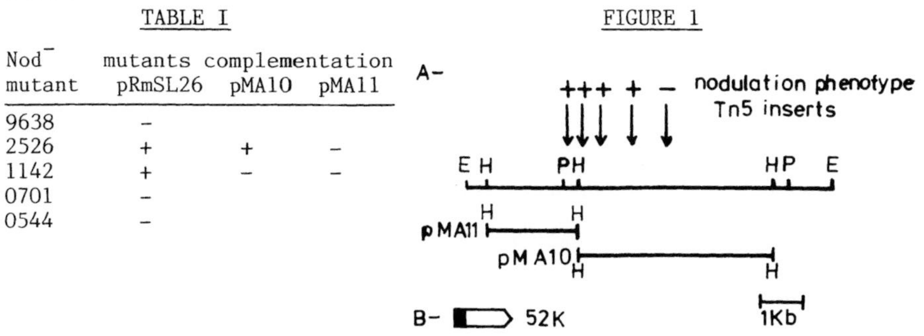

Nod$^-$ mutant	mutants complementation pRmSL26	pMA10	pMA11
9638	-		
2526	+	+	-
1142	+	-	-
0701	-		
0544	-		

O.M.A. is fellowship from Consejo Nacional de Investigaciones Cientificas y Tecnicas de la Republica Argentina.

IDENTIFICATION OF PLASMIDS CARRYING SYMBIOTIC GENES IN FAST-GROWING R. JAPONICUM USING DNA HYBRIDIZATION AND Tn5 MUTAGENESIS

EDWARD R. APPELBAUM/ERIC JOHANSEN/NICOLE CHARTRAIN
Agrigenetics Advanced Research Laboratory, 5649 East Buckeye Road, Madison, Wisconsin 53716 USA.

Fast-growing rhizobia that nodulate soybeans (PRC strains) have recently been described (1). Strain USDA191 forms nodules with nitrogenase activity on many soybean cultivars, while the others form effective nodules only on cultivar Peking (1,4; T. McLoughlin, unpublished). In five PRC strains, a plasmid was found to have homology to a cloned DNA fragment containing the *nifD* and *nifH* genes of *R. meliloti* (3). In this study, we have used both DNA hybridization and Tn5 mutagenesis to determine whether plasmids in USDA191 and other PRC strains carry genes for nodulation and nitrogen fixation.

Ten PRC strains (obtained from D. Weber) were examined for plasmids using an in-well lysis and gel electrophoresis technique. Nine strains contained a plasmid of at least 450Mdal (megaplasmid) in addition to 1-3 smaller plasmids. USDA257 contained only one smaller plasmid (260Mdal) and no megaplasmid. The plasmids of all 10 strains were transferred from gels to filters and hybridized to heterologous ^{32}P-labeled probes. In strains USDA191, USDA192, USDA193, USDA201, USDA206, USDA217, and USDA257, a single plasmid of 200-260Mdal hybridized to both pEA105 (a probe containing *K. pneumoniae* genes *nifMVSUXNEYKDHJ*; E.R.A., unpublished) and pRMSL26 (a probe that contains a 20Kb segment of *R. meliloti* megaplasmid DNA that includes *nod* genes; ref. 2). The vector portions of pEA105 and pRMSL26 did not hybridize to PRC plasmids. These results suggest that *nif* genes and *nod* genes may be linked on a single plasmid in these seven strains. pEA105 and pRMSL26 hybridized to different plasmids or to no plasmid in USDA194, USDA205 and USDA208.

Tn5 mutagenesis was carried out by transferring a suicide vector (pSUP1011; R. Simon and A. Puhler, personal communication) from *E. coli* to a streptomycin resistant derivative of USDA191. The mutagenized cells were examined for symbiotic effectiveness by inoculation onto soybeans (*Glycine max* cv. Peking). Mutants were identified by plant appearance and by the absence of nitrogenase activity. Four Tn5-induced symbiotic mutants were examined in blot hybridization experiments using pEA105, pRMSL26, and a plasmid containing Tn5 sequences as probes. One Fix$^-$ mutant contained a single copy of Tn5 that was inserted into a 4.2Kb *EcoRI* fragment of the 200Mdal plasmid. The 4.2Kb fragment hybridized to pEA105 and therefore contains one or more *nif* genes. A Nod$^-$ mutant contained a deletion that reduced the plasmid size from 200Mdal to 175Mdal and removed two pEA105-hybridizable and several pRMSL26-hybridizable *EcoRI* fragments. Tn5 and other pSUP1011 sequences were substituted for the deleted DNA. These results indicate that genes required for both nitrogen fixation and nodulation are carried by the 200Mdal plasmid in USDA191. In two other independent Fix$^-$ mutants, Tn5 was in the chromosome and there was no alteration of the pEA105- and pRMSL26-hybridizable fragments. The possibility remains that some symbiotic genes in USDA191 may be chromosomal.

1. Keyser HH *et al.* (1982) Science 215, 1631-1632.
2. Long SR *et al.* (1982) Nature 298, 485-488.
3. Masterson *et al.* (1982) J. Bacteriol. 152, 928-931.
4. Yelton MM *et al.* (1983) J. Gen. Microbiol. 129, 1537-1541.

NITROGENASE GENES IN THE FAST-GROWING BROAD-HOST RANGE *RHIZOBIUM* STRAIN ANU240

J. Badenoch-Jones, C. Bates, K. Scott, B. Rolfe and J. Shine
Centre for Recombinant DNA Research and Department of Genetics, Research School of Biological Sciences, Australian National University, P.O. Box 475, Canberra City, A.C.T., Australia.

The *Rhizobium* strain ANU240 (a derivative of strain NGR234 (Trinick, 1980) is a fast-growing strain capable of nitrogen fixation on a wide range of tropical legumes normally nodulated by slow-growing strains. It also nodulates the non-legume, *Parasponia*, but the symbiosis is ineffective (Trinick and Galbraith, 1980).

The nitrogenase structural genes, *nifH*, *nifD* and *nifK*, have been isolated by molecular cloning and appear to be continuous in the genome, as has been found in other fast-growing strains, but in contrast to the slow-growing *Parasponia Rhizobium* sp. ANU289 (which fixes nitrogen on *Parasponia*) in which the *nifH* and *nifD* genes are unlinked (Scott et al., 1983). This was demonstrated by hybridization of *nifH*-, *nifD*- and *nifK*-specific probes to total genomic digests.

There is evidence that there may be two copies of the *nif* structural genes per genome. Genomic DNA, when digested with a variety of restriction endonucleases, yielded at least two fragments which hybridized to the *nif* structural genes of *K. pneumoniae* or *R. trifolii*. Detailed analysis of phage λ clones, however, indicated that each clone either carried only one of these fragments (*EcoRI* and *HindIII* digests) or that the two fragments were conserved between the two loci, so that each copy of the *nif* structural genes yielded the same two fragments (*BamHI* digests).

DNA sequence analysis of the first 54 base pairs from the 5' end of the coding region of one of the *nifH* genes of strain ANU240 demonstrated considerable sequence homology between this strain and *R. trifolii* strain ANU329 and *Parasponia Rhizobium* sp. ANU289. There was some homology, but considerably less, between these strains for the 27 base pairs upstream from the start of the *nifH* coding region.

REFERENCES

Scott KF, Rolfe BG and Shine J (1983) DNA 2, 147-153.
Trinick MJ (1980) J. Appl. Bact. 49, 39-53.
Trinick MJ and Galbraith J (1980) New Phytol. 86, 17-26.

THE PSYM PLASMID OF *RHIZOBIUM MELILOTI*. I. PHYSICAL MAP OF A 290 KB BLOCK OF PSYM AND SEARCH FOR SYMBIOTICALLY DEFECTIVE MUTANTS ON THE PSYM OUTSIDE OF THE *NOD-NIF* REGION

J. BATUT/M. GHERARDI/E. TERZAGHI* AND T. HUGUET
LABORATOIRE DE BIOLOGIE MOLÉCULAIRE DES RELATIONS PLANTES-MIROROGANISMES, CNRS-INRA, B.P. 12, 31320-CASTANET TOLOSAN, FRANCE
*PRESENT ADDRESS: MASSEY UNIVERSITY, PALMERSTON NORTH, NEW ZEALAND

Rhizobium meliloti strains are characterized by the presence of a megaplasmid greater than 700 kb which carries the nitrogenase genes and genes controlling alfalfa nodulation (1,2). pGMI42 contains a 290 kb fragment of pSym of *Rhizobium meliloti* 2011 cloned *in vivo* in the broad host range plasmid RP4 (3). We have verified that the cloned fragment of pGMI42 is colinear with the corresponding region of pSym. Physical mapping of this pSym derivative was done by cloning the products of an incomplete *Hin*dIII digestion into pBR322. A total of 916 hybrid clones were screened and multiple enzyme digests of selected hybrid plasmids allowed us to establish the physical map of pGMI42 in respect to *Hin*dIII, *Eco*RI, *Sma*I, *Hpa*I and *Xba*I sites. The 290 kb pSym insert starts from 5 kb to the right of *nifKDH* and extends to 275 kb to the left of *nifKDH* and includes the *nod* region previously described (4). Some regions of pGMI42 appear to be significantly transcribed in mature bacteroids (see David *et al*, this volume). Tn5 mutagenesis directed towards some of these transcribed regions has yielded mutants defective in nitrogen fixation. The linkage between the Tn5 insertion and the Fix phenotype is currently under investigation.

(1) Banfalvi *et al*, (1981) Molec. Gen. Genet. 184:334-339
(2) Rosenberg *et al*, (1981) Molec. Gen. Genet. 184:326-333
(3) Dusha *et al*, (1983) in Molecular Gentics of the Bacteria Plant Interaction, A. Pühler Ed., Springer Verlag, in press
(4) Long *et al*, (1982) Nature 298:485-488

A GENETIC ANALYSIS OF THE FAST-GROWING *RHIZOBIUM* STRAIN IHP100

G.L.Bender, J.Plazinski, J.M. Watson, J.E.Olsson, P.J.Dart, and B.G.Rolfe.

Department of Genetics, Research School of Biological Sciences, The Australian National University, Canberra, Australia 2601.

Rhizobium strain ORS571 is the only fast-growing strain shown conclusively to reduce acetylene in the free-living state (Elmerich et al,1982). We report here on the characterization of *Rhizobium* strain IHP100, a fast-growing "cowpea" strain also capable of *in vitro* acetylene reduction at a rate comparable to that of slow-growing "cowpea" *Rhizobium* strain CB756 but significantly lower than for strain ORS571 under the same conditions.

Using four different methods for plasmid visualization (Eckhardt 1978; Kado, Liu 1981; Simons et al. 1981; Plazinski et al. 1983) strain IHP100 was shown to harbour four plasmids with approximate molecular weights of 500, 450, 300 and 115. Probes containing *nod* genes from fast-growing "cowpea" strain NGR234 and *R.trifolii* strain 843 hybridized to the 300 Mdal plasmid. The same result was obtained using *nif* probes from *Klebsiella pneumoniae*, strain 843 and slow-growing *Parasponia Rhizobium* strain ANU289. In contrast, use of the same probes for strains CB756 and ORS571 revealed a chromosomal location for *nod* and *nif* genes. Repeated attempts to visualize endogenous plasmids in these strains were unsuccessful despite the fact that the presence of the plasmid pRP_1::Tn501 could be detected using the same procedure.

The plasmid location of *nod* and *nif* genes, in combination with the well known advantages in applying genetic techniques to fast-growing rhizobia, make strain IHP100 potentially useful for a genetic analysis of *nif* gene expression both *in vitro* and in symbiosis.

References

Eckhardt T. (1978) Plasmid. 1, 584-588
Elmerich C, Dreyfus B.L. Reysset G. and Aubert J.P. (1982) The EMBO Journal 1, 499-505
Kado C.I. and Liu S.T. (1981) J. Bacteriol. 145, 1365-1373
Plazinski J. Dart P.J. and Rolfe B.G. (1983) J.Bacteriol. 155, (in press)
Simons R. Weber G. Arnold W. and Pühler A (1981) Proceedings of the 8th North American *Rhizobium* Conference, pp61-90. University of Manitoba, Canada

COMPARISON AND EXPRESSION OF RHIZOBIUM MELILOTI SYMBIOTIC PROMOTERS

Marc Better, Barbara Lewis, David Corbin, Gary Ditta, and Donald R. Helinski
Department of Biology, University of California at San Diego, B-22, La Jolla, CA 92093, USA

The ability of rhizobia to induce nodulation and to fix nitrogen requires the specific expression of certain bacterial genes. One set of these genes common to all effective *Rhizobium* species encodes the subunits of nitrogenase. The nitrogenase structural genes from a number of species, including *R. meliloti*, have been cloned and characterized. In *R. meliloti*, the *nifHDK* operon is found closely linked to other genes essential for nitrogen fixation. Within this region, the 5' ends of two transcription units have been described. Promoter 1 (P1) controls the expression of the *nifHDK* operon and promoter 2 (P2) controls the expression of a second transcription unit which is synthesized in a direction opposite to the nitrogenase transcript.

We have extended our previous studies, and determined precisely the transcription start sites for these two *nif* transcription units by S1 nuclease protection and Maxam-Gilbert DNA sequencing. Strikingly, the DNA sequences upstream from these two RNA start-points share extensive sequence homology, extending upstream from the RNA start points for 160 bp with about 85% homology. The DNA sequences 5' to 3' of a portion of P1 and P2, aligned to maximize the homology between them are shown below. (Arrows indicate RNA starts; Homologous regions are underlined.)

```
P1  GTCCGTAGCCCTTGTCGGCTTAGCGACACGAGTTGTTCGCTCAACCATCTGGTCAATTTCCAGATCTAACTA
P2  G CCCTCGCCTCTGTCGGCCCCTCGACA GA TTGTTCCTTCAAGCATGCGGCCAATTTCCCGATCTAACTA
                                                                           ↓

    TCTGAAAGAAAGCCGAGTAGTTTTATTTCAGACGG          CTGGCACGACTTTTGCACGATCAGCCCT
    TTTGAAAGAAAGCA ATTAGCATTATTTCAGTCACCTCTGCGACCTGGCACGACTTTTGCACGATCATCCCC
                                                                           ↑
```

Hybridization experiments using a 284 bp DNA fragment containing P2 to probe the remainder of the *nif* cluster identified a DNA segment located between P1 and P2 with promoter homology. This region (P3) contains extensive (greater than 160 bp) DNA sequence homology to P1 and P2, and is also specifically transcribed in root nodules. Such extensive regions of conserved DNA sequences adjacent to symbiotically-regulated transcription units suggests that transcription may be coordinately regulated.

When used as a hybridization probe to total *R. meliloti* 102F34 DNA digested with *XhoI*, the P2 promoter hybridized to 6 fragments. Three correspond to the *nif* DNA fragments containing P1, P2 and P3, while three are present elsewhere in the genome. DNA sequence analysis of one of the latter fragments, designated P4, revealed a sequence very similar to P1, P2 and P3.

We have examined the hybridization of the *R. meliloti* P2 promoter to genomic DNA from *R. trifolii*, *R. leguminosarum*, *R. japonicum*, and *R. phaseoli*. In each case, one or more bands hybridized to the promoter probe. Hybridization to identical blots with the *K. pneumoniae nifHDK* hybridization probe pSA30 identifies some of these same bands to contain the nitrogenase genes; this is found for *R. meliloti*, *R. japonicum* and *R. trifolii*. This result suggests an extensive conservation among rhizobia of regulatory sequences concerned with symbiotically-controlled transcription of *nif* genes.

TRANSPOSON MUTAGENESIS AND MOLECULAR CLONING OF SYMBIOTIC GENES FROM *RHIZOBIUM LOTI*

K.Y. CHUA[+], B.D.W. JARVIS[+], C.E. PANKHURST[++], D.B. SCOTT[++]
+ DEPARTMENT OF MICROBIOLOGY AND GENETICS, MASSEY UNIVERSITY, PALMERSTON NORTH, NEW ZEALAND
++ APPLIED BIOCHEMISTRY DIVISION, DSIR, PALMERSTON NORTH, NEW ZEALAND

INTRODUCTION

Strains of rhizobia that nodulate *Lotus* species include both fast (*R. loti*) and slow growers. The fast growers have less than 5% DNA homology with the slow growers. The aim of this work is to isolate the nodulation genes from *R. loti* strain NZP2037 and to compare at a molecular level the homology of these genes with the slow growing group.

RESULTS

The strategy used to isolate the Nod gene region from *R. loti* NZP2037 was to isolate a Tn5 induced Nod$^-$ mutant and to then clone the EcoRI fragment carrying the Tn5. The Tn5 mutagenesis was carried out by transferring pSUP1011 (Simon, unpublished) from *E. coli* SM10 to a Str^R derivative (PN184) of *R. loti* strain NZP2037. 1060 Neo^R transconjugants were single colony purified and tested on *Lotus pedunculatus*. Thirteen symbiotically defective mutants were isolated, 12 that were Nod$^+$Fix$^-$ and one (PN233) that was Nod$^-$Fix$^-$. The Fix$^-$ mutants were characterised by light and electron microscopy and found to fall into three classes: (i) those that formed normal bacteroid containing nodules but did not fix N_2 (ii) those that formed infection threads but failed to release bacteria into the plant cell, and (iii) tumour-like nodules which did not contain intercellular bacteria. Insertion of Tn5 in the chromosome of these mutants was confirmed by Southern hybridisation of EcoRI cut total DNA to [^{32}P]-labelled pKan2 (Scott et al, 1982). Three of the 13 mutants were also shown to contain vector (pACYC184) sequences.

The Nod$^-$ mutant contained Tn5 on a 15kb EcoRI fragment and this was subsequently cloned into pBR328. This plasmid, pPN301, was used to isolate the wild type genes from a pLAFRI (Friedman et al, 1982) gene library to *R. loti* NZP2037 by colony hybridisation. The two pLAFRI clones isolated both complemented the Nod$^-$ mutation in PN233. pPN301 was also found to hybridise to the *Lotus* slow growing strain CC814S.

REFERENCES

Scott KF, Hughes JE, Gresshoff PM, Beringer JE, Rolfe BG and Shine J (1982) J. Mol. Appl. Gen. 1, 315-326.
Friedman AM, Long SR, Brown SE, Buikema WJ and Ausubel FM (1982) Gene 18, 289-296.

THE pSYM PLASMID OF RHIZOBIUM MELILOTI. II. TRANSCRIPTION STUDIES

Michel David, Odile Domergue and Daniel Kahn
Laboratoire de Biologie Moléculaire, I.N.R.A., B.P.12, F31320 Castanet-Tolosan

The establishment of a functional nitrogen-fixing symbiosis between Rhizobium and a legume involves the differenciation of cells from both partners into specialized states -the bacteroids and the nodule cells respectively. We are interested in the biochemical transformations occurring between the free-living and the endosymbiotic forms of the lucerne symbiont Rhizobium meliloti. Here we report studies on genetic expression in bacteroids and free-living bacteria.

R. meliloti pSym megaplasmid carries some nif and nod genes (Banfalvi et al., Rosenberg et al., 1981). Transfer of R. meliloti pSym to Agrobacterium tumefaciens confers A. tumefaciens the ability to specifically nodulate lucerne (Kondorosi et al., 1982). Thus, pSym carries important functions relevant to symbiosis. Part of pSym, cloned in pGMI42, has been mapped by Batut et al. (this volume). We have used pGMI42 subclones to probe transcripts from this region of pSym.

MATERIALS AND METHODS

R. meliloti strain GMI56 was grown exponentially on Vincent's minimal medium containing 0.2% sucrose. GMI56 bacteroids were isolated from 2-month old lucerne nodules by differential centrifugation. RNA was extracted and end-labelled with ^{32}P. Known HindIII restriction fragments from pGMI42 were probed by hybridization with labelled RNA.

RESULTS AND DISCUSSION

The figure shows a HindIII restriction map of pGMI42 and indicates pSym regions expressed in bacteroids. In addition to the nif-nod region (Ruvkun et al., Long et al.,1982) other regions of pSym are actively transcribed in bacteroids. Moreover two regions of pSym, cloned in pGMI42, are actively transcribed in free-living bacteria growing exponentially on minimal medium. Current genetic investigation of these previously unidentified regions should shed light on their functions (Batut et al., this volume).

Our results agree with previous studies showing that a 17 kb region containing nif is specifically transcribed in R. meliloti bacteroids (Corbin et al., 1982).

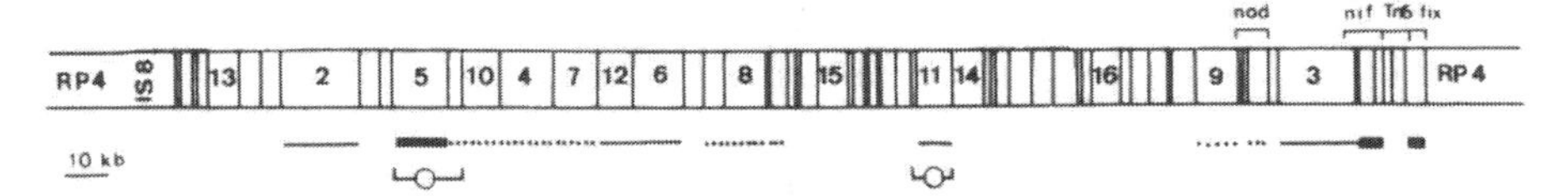

Preliminary transcription map of pGMI42.
HindIII fragments that are transcribed at high (▬), medium (——) and low (·······) levels in bacteroids are indicated. Regions transcribed in free-living bacteria are shown (└─O─┘). nod, nif, fix and Tn5 insertion n°20 are mapped as published (Long et al., Ruvkun et al., 1982).

REFERENCES

Banfalvi Z, Sakanyan V, Koncz C, Kiss A, Dusha I and Kondorosi A (1981) Molec. Gen. Genet. 184, 334-339.

Corbin D, Ditta G and Helinski DR (1982) J. Bacteriol. 149, 221-228.

Kondorosi A, Kondorosi E, Pankhurst CE, Broughton WJ and Banfalvi Z (1982) Molec. Gen. Genet. 188, 433-439.

Long SR, Buikema WJ and Ausubel FM (1982) Nature 298, 485-488.

Rosenberg C, Boistard P, Dénarié J and Casse-Delbart F (1981) Molec. Gen. Genet. 184, 326-333.

Ruvkun GB, Sundaresan V and Ausubel FM (1982) Cell 29, 551-559.

24K PROTEIN - A GENETIC MARKER FOR THE SYMBIOTIC REGION OF RHIZOBIUM LEGUMINOSARUM

N.J. DIBB, G.M. SORENSEN, J.A. DOWNIE, G. HOMBRECHER, AND N.J. BREWIN
DEPARTMENT OF GENETICS, JOHN INNES INSTITUTE, COLNEY LANE, NORWICH NR4 7UH, U.K.

When grown on agar slants, all strains of Rhizobium leguminosarum so far examined synthesised large amounts of a polypeptide, molecular weight 24kdal, for which the structural gene is located on the symbiotic plasmid. In the symbiotic plasmid pRL1JI, it maps between determinants for nodulation and the structural genes for nitrogenase. Although characteristic of R. leguminosarum, the gene is inessential for symbiotic N_2 fixation and has no known phenotype. The gene product, which can be detected by a simple antibody staining reaction (figure 1), provides a convenient genetic marker flanking the symbiotic region. Using this method, cosmids carrying cloned symbiotic genes have been isolated from gene banks and transfer of natural symbiotic plasmids between strains has been monitored. In addition, the structural gene for 24K protein has been used as a target for site-directed reversed mutagenesis with Tn5 to provide a selectable marker at a convenient place on natural symbiotic plasmids.

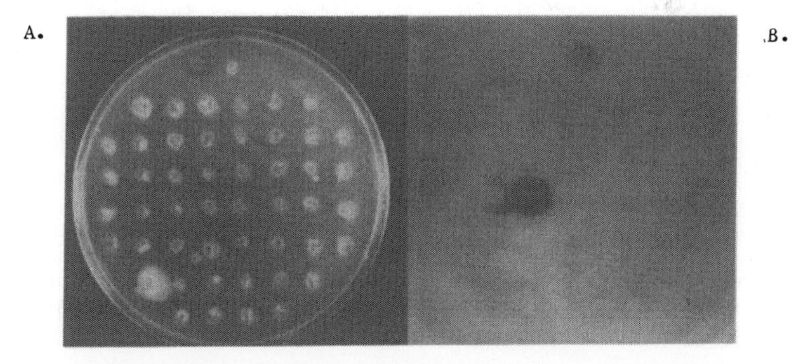

Figure 1. **Immunochemical identification of colonies containing cloned nod DNA from R. leguminosarum, using antibody against a flanking gene product (24 kdal polypeptide).** A gene bank was constructed in E. coli with DNA from R. leguminosarum B151 pRL5JI (see Hombrecher et al. this volume). This was mated with strain B151 (cured of its symbiotic plasmid, and hence $24K^-$, Nod^-). Colonies (A) were replica-plated to Whatman filter papers, lysed with 2% SDS, blotted to nitrocellulose sheets (B), and then treated first with rabbit antiserum to purified 24 kdal polypeptide and secondly with peroxidase-coupled goat anti-rabbit Igg. The sheets were then stained for peroxidase activity by the method of Hawkes et al. (Anal. Biochem. 119, 142-147, 1982).

THE NODULATION GENES OF RHIZOBIUM LEGUMINOSARUM

J. A. DOWNIE, Q-S. MA, B. WELLS, C. D. KNIGHT, G. HOMBRECHER,
A. W. B. JOHNSTON
JOHN INNES INSTITUTE, COLNEY LANE, NORWICH NR4 7UH, NORFOLK, U.K.

In the *R. leguminosarum* plasmid pRL1JI, Tn5-induced *nod* alleles were located between two clusters of *fix* alleles (Downie *et al*. 1983a). DNA from one *fix* region was homologous to the *K. pneumonia nifHD* genes and *nifA* DNA hybridized to the other.

Four recombinant plasmids which together spanned this 50kb symbiotic region of pRL1JI were isolated from an *R. leguminosarum* clone bank. Two of these plasmids, pIJ1085 and pIJ1089 could each complement strains of *R. leguminosarum* and *R. phaseoli* that were cured of their nodulation plasmids and the resultant pea nodules contained bacteroids. The two plasmids overlap by only 10kb and so this must be the limit of the pRL1JI DNA that is required for nodulation and the determination of host range.

Following Tn5 mutagenesis of pIJ1085 and pIJ1089 several more mutants defective in nodulation were isolated. All the mutations were located within the 10kb overlap region of pIJ1085 and pIJ1089. One group of mutants which completely failed to nodulate peas and did not deform root hairs were due to insertions within a 1kb region. Other mutants were delayed in nodulation and of these some formed only few nodules; however all these mutants retained the ability to deform pea root hairs.

A 6.6kb *EcoR1* fragment from the *nod* region, subcloned into a wide host-range plasmid vector was transferred to strains of *Rhizobium* cured of their symbiotic plasmid. The transconjugants induced root hair deformation on peas but failed to nodulate. Moreover, strains of *E. coli* containing this cloned 6.6kb fragment also induced root hair curling on peas.

A plasmid containing this fragment was used as a template in an *E. coli*-based *in vitro* transcription/translation system. At least three polypeptide products encoded by the inserted DNA were identified, some of which were absent when DNA fragments containing the Tn5 *nod* alleles were used as the template.

REFERENCES

Downie JA, Ma Q-S, Knight CD, Hombrecher G and Johnston AWB (1983a) EMBO J. 2, 947-952.
Downie, JA, Hombrecher G, Ma Q-S, Knight CD, Wells B and Johnston AWB (1983b) Mol. gen. Genet. 190, 359-365.

SYM PLASMIDS IN RHIZOBIUM JAPONICUM

Kim S. Engwall, Nancy DuTeau, and Alan G. Atherly. Dept. of Genetics, Iowa State Uniersity, Ames, IA.

Rhizobium japonicum strains PRC191 and PRC193 are fast-growing strains that were isolated in the People's Republic of China (H. Keyser et al., 1982). Recently it was shown that these strains have nif and nod genes on large plasmids (R. Masterson et al., 1982, and unpublished). We have introduced Tn5 into the Sym plasmids of PRC191 and PRC193. The Sym plasmids are not self-transmissible. Therefore, we introduced a helper plasmid, pRL180 (P.J.J. Hooykaas et al., 1982), into these strains to promote plasmid mobilization to other Rhizobia and Agrobacterium strains.

The Sym plasmids from PRC191 and PRC193 were introduced by conjugation into Agrobacterium tumefaciens LBA288 and A136, R. japonicum 3Ilb110, R. meliloti 102F28, and R. leguminosarum 128653. Kanamycin resistant transconjugants of these recipients contained intact Sym plasmids as visualized on agarose gels, however approximately ten percent of the transconjugants of A. tumefaciens LBA288 contained large deletions of the Sym plasmids. Transfer of kanamycin resistance to R. trifolii and R. phaseoli occurred, but intact Sym plasmids could not be detected by agarose gel electrophoresis. Preliminary observations indicate the PRC191 Sym plasmid may be unstable in R. leguminosarum and R. meliloti, however, the PRC193 Sym plasmid was transferred into A. tumefaciens LBA288 from R. leguminosarum 128653 (pPRC193::Tn5) and stably maintained.

Transconjugants of A. tumefaciens LBA288 and A136 containing the Sym plasmid from PRC191 and PRC193 were tested for nodulation and nitrogen fixation functions. Transconjugants containing non-deleted Sym plasmids formed small, white ineffective nodules on soybeans. One transconjugant, carrying about a 50 megadalton deletion of the 194 megadalton plasmid from PRC191 did not form nodules. Deletions in large Sym plasmids may yield valuable information about symbiotic properties and are currently being examined.

REFERENCES

Hooykaas PJJ, den Dulk-Ras H, and Schilperoort RA (1982) Plasmid 8, 94-95.

Keyser HH, Bohlool BB, Hu TS, and Weber DF (1982) Science 215, 1631-1632.

Masterson RV, Russell PR, and Atherly AG (1982) J. Bacteriol. 152, 928-931.

A CLONING VECTOR DERIVED FROM AN INDIGENOUS SMALL PLASMID OF *RHIZOBIUM MELILOTI* - CONSTRUCTION AND MOBILIZATION INTO THE *RHIZOBIACEAE*.

J.P. FACON/M. BECHET/J.B. GUILLAUME

LABORATOIRE DE MICROBIOLOGIE, UNIVERSITE DES SCIENCES ET TECHNIQUES DE LILLE, F - 59655 VILLENEUVE D'ASCQ CEDEX, FRANCE.

Detection and isolation of a small (MW ~ 7.3 kb) cryptic plasmid from a strain of *Rhizobium meliloti* (M19S) opened up the possibility of using it as a cloning vector for studies on bacterial genes involved in symbiotic nitrogen fixation. Restriction endonuclease analyses revealed that pRme 19a contains five cleavage sites for ClaI, two for BclI and an unique site for SalI. The latter allowed the *in vitro* insertion of pRme 19a into the narrow-host-range cloning vector pBR322 (1). The resulting ampicillin-resistance-conferring hybrid plasmid (pBRRM 1) was introduced by transformation into *Escherichia coli* strain C600 where it replicated stably and was amplifiable by chloramphenicol. pBRRM 1 was then found to exhibit two unexpected properties :

- it was mobilized efficiently by pRK2013 (2) in triparental matings,
- it coded for a kanamycine resistance.

Both latter features which were linked to the presence of an additional DNA sequence (MW ~ 1 kb) in the pBR322 part, enabled the transfer and selection of pBRRM 1 (MW ~ 12.2 kb) into a variety of *Rhizobium* and *Agrobacterium* strains where it maintained without selective pressure thanks to the pRme 19a replication functions.

Analyses of pBRRM 1 DNA isolated from *E. coli* C600 (*dam*$^+$ - methylation of GATC sequences by *E. coli dam* methylase), from *E. coli dam*$^-$ derivatives and from *R. meliloti* strain M19S showed that *dam* methylation is absent from at least *R. meliloti* strain M19S.

This is the first report describing a potential shuttle vector that carries a *Rhizobium* replicon. At the moment, pBRRM1 cloning properties and possible improvments are under investigation.

1. Bolivar F *et al.* (1977) Gene 2, 95-113.
2. Ditta G *et al.* (1980) Proc. Natl. Acad. Sci. USA 77, 7347-7351.

SYMBIOSIS-SPECIFIC GENETIC LOCI OF RHIZOBIUM PHASEOLI.

L. FERNANDEZ, A. SANCHEZ, J. LEEMANS, AND K.D. NOEL+
Centro de Investigación sobre Fijación de Nitrógeno, Cuernavaca, Morelos, Mexico and +Biology Dept., Marquette University, Milwaukee, Wisconsin, U.S.A.

Genetic analysis of *R. phaseoli* relevant to its symbiosis with *Phaseolus* beans has been limited to the large indigenous self-transmissible plasmids which carry homology to *Klebsiella nif* (Quinto, et al, 1982) and determine at least some nodulation and nitrogen fixation abilities (Lamb, et al, 1982). Toward a more complete accounting of symbiosis-specific (*sym*) genes and to see if these plasmids merit exclusive attention, *R. phaseoli* was mutated by random insertion of Tn5 with "suicide" plasmid pJB4JI (Beringer, et al, 1978).

RESULTS

About 2% of the Tn5-carrying transconjugants grew normally on rich and minimal media but showed symbiotic defects upon extensive re-testing with plants. Eleven such Sym mutants of strain CFN42 *str* were studied. One did not initiate visible nodule tissue (Nod^-). Eight led to small or very small nodule-like growths (Nod). Two formed inactive nodules that during the first two weeks were pink and normal in size (Fix^-).

Plasmids were analyzed by Eckhardt-type agarose gel electrophoresis. The Nod^- mutant was missing plasmid pCFN42d, the plasmid with homology to *nif*. The others had normal plasmid content. In three cases, including the two Fix^- mutants, Tn5 was present in plasmid pCFN42d. Tn5 hybridization indicated the other eight insertions to be in the chromosome.

These mutants (*sym str* Km) were crossed with a Nod^+ Fix^+ CFN42 *rif* strain. Rif^r Str^s Km^r transconjugants arose at $4x10^{-6}$ per donor having Tn5 in pCFN42d and $2x10^{-7}$ per donor having Tn5 in the chromosome. Use of Km^s R68.45 enhanced chromosomal transfer five-fold but did not affect pCFN42d transfer. Transconjugants were tested for symbiotic defects. Co-transfer of *sym* Tn5 was 98-100% from seven mutants, including one Fix^-, 75% and 87% from 2 mutants with Tn5 in pCFN42d, and 8-14% from two mutants.

Total DNA of the mutants having Tn5 insertions on pCFN42 (one Nod, two Fix^-) was digested with EcoR1 and separated on agarose gels. Each had a single Tn5-containing fragment of 10-15 kb. The three fragments of strain CFN42 with *nif* homology were unchanged.

CONCLUSIONS

A range of symbiotic defects can be generated by Tn5 mutagenesis of *R. phaseoli*. Plasmid pCFN42d (approximate MW $170x10^6$) carries genes essential for nodule initiation, development, and nitrogen fixation. More mutations were in the chromosome; all affected nodule development. The Tn5 insertion and the *sym* mutation may be one and the same in seven cases, but in four instances definitely are not, including a Nod^- mutant lacking pCFN42d.

REFERENCES

Beringer JE, Beynon JL, Buchanan-Wollaston AV, and Johnston AWB (1978) Nature 276, 633-635.
Lamb JW, Hombrecher G, and Johnston AWB (1982) Mol. Gen. Genet. 186, 449-452.
Quinto CH, de la Vega H, Flores M, Fernandez L, Ballado T, Soberon G, and Palacios R (1982) Nature 299, 724-726.

IN RHIZOBIUM JAPONICUM THE NIFH AND NIFDK GENES ARE SEPARATED

MARTIN FUHRMANN, KLAUS KALUZA, AND HAUKE HENNECKE
MIKROBIOLOGISCHES INSTITUT, EIDGENÖSSISCHE TECHNISCHE HOCHSCHULE, ETH-ZENTRUM, CH-8092 ZÜRICH, SWITZERLAND

I. INTRODUCTION

We have shown previously that a cloned HindIII restriction fragment from *Rhizobium japonicum* 3Ilb 110 contains the nitrogenase structural genes *nifD* and *nifK* (Fig. 1). However, we have not obtained evidence for the existence of the gene for nitrogenase reductase (*nifH*) adjacent to the *nifDK* cluster.

II. RESULTS

By interspecies hybridization, a 12.1 kb *PstI* fragment with a *nifH*-homologous region was identified. This fragment was cloned into vector pHE3 yielding the hybrid plasmid pRJ7000; it is not linked to the *nifDK* region. A restriction map of pRJ7000 was determined; specific hybridization to *K.pneumoniae nifH* fragments allowed an exact localization of the *nifH*-homologous region (Fig. 1).

A partial DNA sequence was determined (72 bp); it revealed extensive homology to all other published *nifH* sequences at the amino acid level.

The recloning of subfragments of pRJ7000 into plasmid vectors carrying strong promoters facilitated the expression of *Rhizobium*-specific genes in *E.coli* minicells. Transposon insertion mutants allowed exact mapping of coding regions. The *nifH* gene product (the nitrogenase reductase) was thus found to have a molecular weight of 33,000.

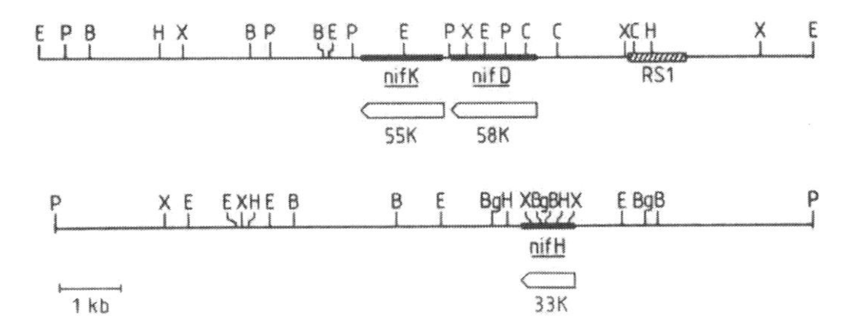

Fig. 1. The cloned *R.japonicum* DNA fragments containing the *nifDK* and *nifH* genes.
A repeated DNA sequence (RS1) with at least 8 copies per genome was found; one of these copies is located approximately 1.8 kb upstream of the *nifDK* genes. No such sequence was found in the vicinity of *nifH*.

LOCALIZED TN5 MUTAGENESIS IN RHIZOBIUM JAPONICUM: CONSTRUCTION OF DEFINED NIF$^-$ MUTANTS

MATTHIAS HAHN, DANIEL STUDER and HAUKE HENNECKE
MIKROBIOLOGISCHES INSTITUT, EIDGENÖSSISCHE TECHNISCHE HOCHSCHULE, ETH-ZENTRUM, CH-8092 ZÜRICH, SWITZERLAND

1. INTRODUCTION

Two regions of the R.japonicum genome containing nifDK and nifH have been cloned in our laboratory. To determine the Fix phenotype of mutations within nif and the surrounding DNA we have used a procedure for the localized Tn5 mutagenesis.

2. METHODS

R.japonicum nif DNA was cloned into the mobilizable plasmids pSUP101, 201, or 202 (Simon et al., 1983) and mutagenized in E.coli with Tn5. These plasmids were transferred from E.coli strains Sm10 or 17-1 into R.japonicum 110spc4 in which they cannot replicate. Kan/neo-resistant R.japonicum colonies were obtained at a frequency of 10^{-7} to $5x10^{-5}$. By colony hybridization or screening for tet^R (pSUP202) vector-containing strains (with cointegrations) were sorted out. The remaining strains were tested by Southern blot hybridization for the correct position of Tn5 at the homologous site (true marker replacement mutants).

3. RESULTS

The Tn5 mutants were tested for their symbiotic phenotype (Fig. 1). NifD::Tn5 and nifK::Tn5 mutants were Nod$^+$Fix$^-$, all other insertions on either side of nifDK were Nod$^+$Fix$^+$. Thus the nifDK operon is confined to about 2.8 kb length. In strain H1, which is also Fix$^-$, Tn5 is located in or near the beginning of the nifH coding region. R.japonicum Tn5 mutants were found to be stable: reisolates from soybean root nodules had retained Tn5 in the original location. The ultrastructure of nodules infected by Fix$^-$ strain A3 was investigated and compaired to nodules infected by the wild-type. We have obtained preliminary evidence for a massive accumulation of poly-β-hydroxybutyric acid in the mutant bacteroids, and for a premature degeneration of the peribacteroid membrane.

4. REFERENCE

Simon et al. (1983) in: Pühler (ed): Molecular genetics of the bacteria-plant interaction. Springer, Heidelberg, in press.

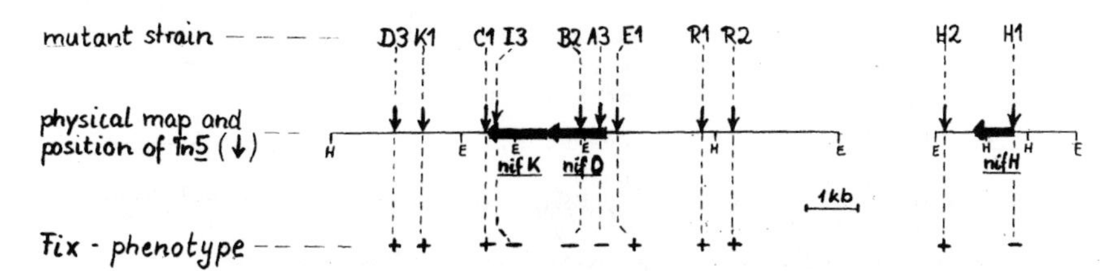

Fig. 1: Location and phenotype of Tn5-mutations

CHARACTERIZATION OF R. JAPONICUM HUP GENE COSMIDS

RICHARD A. HAUGLAND/MICHAEL A. CANTRELL/JON S. BEATY/F. JOE HANUS/STERLING A. RUSSELL AND HAROLD J. EVANS
LABORATORY FOR NITROGEN FIXATION RESEARCH
OREGON STATE UNIVERSITY, CORVALLIS, OREGON 97331 USA

We have reported the isolation of a number of recombinant cosmids from a gene bank of the Hup^+ *R. japonicum* strain 122DES DNA which return hydrogen uptake (Hup) activity to Hup^- mutants of *R. japonicum* (Cantrell et al, 1983). We now describe experiments conducted to determine the physical location and genetic organization of Hup determinants within cosmid sequences and the capability of these cosmids to return hydrogenase activity to additional Hup^- *Rhizobium* strains. We also describe work in progress which is directed toward the use of subcloned DNA fragments from these cosmids to localize *hup* gene point mutations and isolate additional DNA flanking the presently isolated sequences.

Materials and Methods: Restriction maps of *hup*-gene cosmids were constructed using restriction endonuclease digestions and hybridization of ^{32}P-labeled DNA fragments to Southern blots. Site-specific Tn5 mutagenesis (Ruvkun, Ausubel, 1981) was performed using the cosmid pHU1. Bacterial colonies were screened for hydrogenase activity by methylene blue reduction (Haugland et al, 1983).

Results and Discussion: Physical mapping of the *R. japonicum* insert DNA of pHU1 has shown the EcoRI fragments to be in the order of 0.6 kb, 12.9 kb, 2.9 kb, 2.3 kb, 1.6 kb, and 5.0 kb. The fragment order has been shown to correspond to the order in the genome of strain 122DES.

Five out of six *R. japonicum* Hup^- mutants tested showed a Hup^+ phenotype when pHU1 was introduced into them by conjugation. Three of the mutations occurring in these strains are in genes completely contained within pHU1 while the other two must be in transcriptional units which are at least partially contained within pHU1. These results indicate that the majority of the Hup specific DNA in *R. japonicum* strain 122DES may have been isolated.

The analysis of 14 Tn5 insertions in the 122DESnal genome indicates that Hup-specific sequences occur in a region spanning at least 16 kb in the right hand portion of the pHU1 insert. Ten of these Tn5 insertion strains have been physically analyzed and shown to have arisen from marker exchange.

Complementation analysis was performed by introducing pHU1 into 6 of the Hup^- 122DESnal::Tn5 strains. All of the Tn5 insertions except the right-most insertion are apparently in transcriptional units totally contained in pHU1.

We have found that pBR325 derivatives can be transferred to *R. japonicum* strain PJ18nal if the derivatives contain a region of homology with the *R. japonicum* genome. Transfer of pBR325 derivatives containing various *hup*-region DNA fragments has shown that the site of the PJ18nal *hup* lesion must be on DNA corresponding to either the 2.9 or 2.3 kb EcoRI fragments of pHU1. If the pBR325 derivatives are maintained in strain PJ18nal as a result of a single homologous recombination event, we would suggest that the PJ18nal *hup* mutation is transdominant and located on a transcriptional unit which has one boundary in the 2.9 or the 2.3 kb DNA fragment. Strain 122DES *hup* genes therefore appear to be organized in at least three transcriptional units.

Cantrell, M.A., Haugland, R.A., and Evans, H.J. (1983) Proc. Natl. Acad. Sci. USA 80, 181-185.
Haugland, R.A., Hanus, F.J., Cantrell, M.A., and Evans, H.J. (1983) Appl. Environ. Microbiol 45, 892-897.
Ruvkun, G.B. and Ausubel, F.M. (1981) Nature 289, 85-88.

CLONING THE GENES THAT ALLOW A *RHIZOBIUM LEGUMINOSARUM* STRAIN TO NODULATE AFGHAN PEAS

G. HOMBRECHER, J.A. DOWNIE, R. GOTZ, A.W.B. JOHNSTON, AND N.J. BREWIN
DEPARTMENT OF GENETICS, JOHN INNES INSTITUTE, COLNEY LANE, NORWICH NR4 7UH, U.K.

Unlike most strains of *R. leguminosarum*, the Turkish strain TOM is able to nodulate pea *cv* Afghanistan, in addition to commercial varieties of peas(Winarno, Lie, 1979). The extended host range of TOM is a property of its symbiotic plasmid, pRL5JI (Brewin *et al*. 1980). A gene bank was constructed in the cosmid vector pLAFR1. Following transfer to a strain of *R. leguminosarum* that had been cured of its symbiotic plasmid, several clones were isolated that were each able to nodulate both cv. Afghanistan and commercial pea varieties. One of these clones, pIJ1095, was subjected to mutagenesis with transposon Tn5. From one hundred Tn5 insertions screened, 11 affected the nodulation ability and all these were mapped to a region of only 6kb. The mutants fell into three classes on the basis of their map positions and phenotypes (table 1).

Table 1 Nodulation phenotype of strains containing derivatives of the *nod* clone pIJ1095 after Tn5 mutagenesis when tested in wild-type *R. leguminosarum* strain B264 (normally unable to nodulate cv Afghanistan) and in the Nod$^-$ strain 8401 (cured of its symbiotic plasmid). Wis = commercial pea variety (Wisconsin Perfection); Afg = cv Afghanistan.

Mutant class	No. of mutants	Strain 8401		Strain B264	
		Wis	Afg	Wis	Afg
without pIJ1095	-	-	-	+	-
wild type pIJ1095	89	+	+	+	+
I	7	-	-	+	+
II*	3	±	+	+	+
III	1	-	-	+	-

*In the case of 8401 derivatives containing Class II derivatives of pIJ1095 the mean number of nodules per plant was reduced from 40 to 7 on Wisconsin Perfection, but nodulation of *cv* Afghanistan was normal.

REFERENCES

Brewin, N.J. *et al*. 1980. J.Gen.Microbiol. 120, 413-420.
Winarno, R., Lie, T.A., 1979. Plant and Soil 51, 135-142.

EXPRESSION OF SYM-PLASMID GENES IN BACTEROIDS OF RHIZOBIUM LEGUMINOSARUM

J.G.J. Hontelez, P. Mol, C. van Dun, R. Schetgens, A. van Kammen and R.C. van den Bos.
Department of Molecular Biology, Agricultural University, De Dreijen 11, 6703 BC WAGENINGEN, The Netherlands

DNA-RNA hybridization techniques were used to detect genes, involved in symbiotic nitrogen fixation by bacteroids of Rhizobium leguminosarum, strain PRE, in root nodules of Pisum sativum. Strain PRE contains two plasmids with molecular weights of approximately 230 x 10^6 and more than 500 x 10^6. The smaller one, containing the sym-genes (1), was isolated from bacteria and after partial digestion, fragments were cloned into the EcoRI-site of phage λ-EMBL 3. Plaques of this plasmid gene library were screened by hybridization with ^{32}P-labeled RNA, purified from actively nitrogen fixing bacteroids, isolated from 17 days old plants. After selection of the positive plaques (5% of the library), the recombinant DNAs were isolated, EcoRI-digested and Southern blots were hybridized again with ^{32}P-RNA from bacteroids. Study of the physical map of the cloned fragments revealed, that many overlap. Thus two regions with more or less strongly expressed stretches of sym-plasmid DNA were found.
One of these regions, made up of five overlapping clones, represents a sym-plasmid fragment of about 60 kb, and contains the nif KDH-cluster (see poster-paper of Schetgens et al.). From hybridization with ^{32}P-RNA from bacteroids these latter genes, coding for the components I (DK) and II (H) of the nitrogenase complex, appeared to be the most strongly expressed genes on the sym-plasmid, in accordance to the fact, that about 20% of the total protein in bacteroids in this stage of growth consist of the nitrogenase components. In this region less strongly expressed sym-plasmid fragments were found, upstream the nif KDH-cluster, indicating that there are other sym-genes present.
Another strongly expressed sym-plasmid fragment was found in the second region (clone λRleH6). So far we have not yet succeeded in mapping this clone on the sym-plasmid and finding its position, relative to nif KDH. Using DNA-probes derived from the K.pneumoniae clone pGR102 (2) (nif QBALFM), we found λRleH6 to bear homology with nif AL as well as nif F. Hybridizations were carried out in 5 x SSC, at a temperature of 65°C. The nif F homology was located in the strongly expressed fragment of λRleH6. Thus we found evidence for a gene, coding for a protein with probably the same function as flavodoxin in K.pneumoniae, i.e. an electrondonor to nitrogenase.
In order to further characterize more of these in bacteroids expressed genes we are using several approaches. One is the study of the expression in a minicell-producing E.coli strain DS410. Selected fragments of sym-plasmid DNA are therefore being subcloned into the EcoRI-site of the expression vector pACYC184 and transformed into DS410. So far we found five different proteins to be synthesized in this system encoded by the nif AL and nif F region. These proteins are being further analysed. The second approach we make is the hybridization of selected sym-plasmid DNA-fragments with ^{32}P-RNA, purified from bacteroids in various stages of nodule development. Thus we hope to gain information about sequential order of expression of the various genes and their role in the regulation of nitrogen fixation.

References: 1. Krol, A. et al., Nucleic Acids Res. vol. 8, nr. 19, (1980).
2. Riedel, G.E. et al., P.N.A.S. 76, (1979).

PHYSICAL AND GENETIC ANALYSIS OF RHIZOBIUM MELILOTI 1021 NODULATION REGION

T. JACOBS, T. EGELHOFF, A. HIRSCH* AND S.R. LONG
STANFORD UNIVERSITY, STANFORD CA 94305 U.S.A.
*WELLESLEY COLLEGE, WELLESLEY MA 02181 U.S.A.

1. INTRODUCTION

Rhizobium meliloti stimulates the development of nitrogen fixing root nodules on its host plant, alfalfa (lucerne). A bacterial DNA region required for nodulation was identified genetically through mutations (1). This region was cloned as plasmid pRmSL26, and by comparison with other clones was mapped to the R. meliloti megaplasmid (2).

2. EXPERIMENTAL PROCEDURE AND RESULTS

A. Function. This region appears to be required for invasion of the host plant, since mutations produce a non-nodulation phenotype. The mutants fail to curl root hairs, an early stage of infection (3). Clone pRmSL26 restores function to these mutants. It does not alter the host range of wild type strains of R. trifolii or R. leguminosarum. If conjugated into Hac^- R. trifolii mutants ANU851 and ANU453 (4), pRmSL26 restores them to nodulation, but only on their original host plants. Cloned fragments from this region can confer upon Agrobacterium tumefaciens the ability to induce nodules on alfalfa roots. This has been found with large clones including both nod and nif regions (5,6) and with small subclones from pRmSL26 (7). These nodules have normal meristem development, but are not normal in terms of root hair curling or host range, raising the question of whether the invasion mechanism is the same as that of normal Rhizobium. This region does appear to be sufficient to induce organogenesis in host plants, and it may include functions shared by fast-growing Rhizobium strains. If it controls host range, it does not do so as a simple dominant function. It may be that host range is influenced by both positive and negative (e.g. avirulence) functions.

B. Map. The DNA inserted in clone pRmSL26 was subcloned and mapped with restriction enzymes. Transposon Tn5 insertions were obtained in a 8.7 kb EcoR1 fragment from pRmSL26, and these were homogenotized into R. meliloti and tested for nodulation phenotype. Out of 30 Tn5 insertions in this fragment, 9 were Nod^-. These are located on the right side of the 8.7 kb fragment. Complementation studies with the mutants and with subclones indicate the possible presence of three transcription units. Preliminary sequence analysis of the rightmost transcript region reveals a potential open reading frame and ribosome binding site. Expression studies will need to be done to test whether this represents a gene, however.

1. Long, S.R. et al. (1981) Genetic engineering, ed. Panopoulos. pp. 129-143. Praeger, New York.
2. Long, S.R. et al. (1982) Nature 289:485-488.
3. Hirsch, A. et al (1982) J. Bacteriol. 151:411-419.
4. Djordevic, M.A. et al (1983) J. Bacteriol., in press.
5. Wilson, K. et al (1983) Miami Winter Symposium, in press.
6. Hirsch, A. and Ausubel, F. (1983) J. Bacteriol., submitted.
7. Hirsch, A. and Long, S.R. (1983) manuscript in preparation.

DNA SEQUENCE ANALYSIS OF *NIF* GENES FROM *R.JAPONICUM*: COMPLETE NUCLEOTIDE SEQUENCE OF THE *NIFD* GENE

KLAUS KALUZA AND HAUKE HENNECKE
MIKROBIOLOGISCHES INSTITUT, EIDGENÖSSISCHE TECHNISCHE HOCHSCHULE, ETH-ZENTRUM, CH-8092 ZÜRICH, SWITZERLAND

1. INTRODUCTION

In the slow-growing *Rhizobium japonicum* the *nif* structural genes are organized in two independent operons.

We have sequenced the complete *nifD* gene and part of the *nifK* gene, encoding both subunits of nitrogenase component 1.

2. METHODS

DNA was sequenced by both the chemical cleavage method and the chain termination method using M13 derivatives mp8, 9, 10, 11 for subcloning experiments.

3. RESULTS

The *nifD* region in between a *BglII* site and 11 nucleotides beyond a *SmaI* site bears an open reading frame of 512 codons. The deduced protein has a molecular weight of 57,647 daltons and contains 8 cysteine residues. This is in good agreement with the molecular weight of 58,000, determined for the *nifD* gene product in our *E.coli* minicell experiments.

The amino acid sequence was compared to part of the *nifD* gene from *K.pneumoniae* (Scott et al., 1981). Highly conserved regions of amino acids, particularly around the cysteine residues, were found.

The promoter region indicates homology on the DNA sequence level with other *nif* promoters. The sequence TTGCA was detected, which appears to be a consensus recognition site for the *glnG/nifA* mediated regulation (Ow et al., 1983).

Preliminary results of S1 nuclease mapping experiments with extracted total RNA from free-living *R.japonicum* cultures and from soybean nodule bacteroids show two presumptive start sites for transcription.

Downstream of the *nifD* gene there is a Shine-Dalgarno sequence, which is followed by an open reading frame. This and other partial sequences share homology on the amino acid level to the *Anabaena* sp. 7120 *nifK* gene product (Mazur, Chui, 1982).

4. REFERENCES

Scott KF, Rolfe BG, Shine J (1981) J.Mol.Appl.Genet. 1, 71-81.
Ow DW, Sundaresan V, Rothstein DW, Brown SE, Ausubel FM (1983) Proc. Natl.Acad.Sci.USA 80, 2524-2528.
Mazur BJ, Chui CF (1982) Proc.Natl.Acad.Sci.USA 79, 6782-6786.

INTRODUCTION OF TRANSPOSON TN5 INTO RHIZOBIUM ASTRAGALI

Li Fudi, Cao Yanzhen, Li Reya, and Chen Hua Kui
Huazhong Agricultural College, Wuhan, Peoples' Republic of China

Plasmid attached transposon Tn5 was used as mutagenic factor. Rhizobium astragali strains were used as recipients. Conjugational crossing over was studied.

Materials and Methods. The following donors were used: E. coli 1830 (pJB4JI) (Beringer et al 1978), E. coli RL29 (p R64::Tn5), E. coli RL30 (pR136::Tn5) and E. coli RL32 (pR199::Tn5) (Li, Beringer, 1981). Recipients used were Rhizobium astragali Ra6 $Nod^+Nal^rSm^r$ and Ra18Nod^+Nal^r (this laboratory). Modified Vidaver's medium (NBY) was used for E. coli, TY medium (Beringer, 1974) for rhizobia. Selective medium was Y (Beringer 1974) plus supplements (FY) with kanamycin to a final concentration of 50 $ugml^{-1}$ and nalidixic acid of 20 $ugml^{-1}$. The mating of bacteria was made as described by Beringer and Hopwood (1976).

Results and Discussion. Frequencies of mating for kanamycin resistant marker transfer are listed in the Table. Spontaneous mutation of recipient strains were lower than 10^{-9}. The Table indicated that transposon Tn5 was definitely transfered into R. astragali from E. coli by conjugation. Two series of transconjugants were obtained, i.e., Nod^+Km^r and Nod^-Km^r.

E. coli donor	Km^r transfer frequency	
	Ra6	Ra18
1830	1.3×10^{-6}	3×10^{-8}
RL29	4.3×10^{-7}	
RL30	5.3×10^{-7}	1×10^{-7}
RL32	5.6×10^{-7}	4.4×10^{-7}

They were prepared to conduct two kinds of experimentation. Using nodulation defective mutant (Nod^-Km^r) of R. astragali to analyse the steps of establishment of symbiosis is now under way. Mutation by transposons is a handy method both for analysis and reconstruction. It is also useful to study a number of questions encountered in the applied rhizobiology, such as the cause and mechanism of lossing nodulation ability, which is a fairly common phenomenon in laboratory. By using stably linked Nod^+Km^r mutant the loss of nodulation ability can be tested with ease on agar plate containing kanamycin in place of the relative time-consuming method of plant test.

References.

Beringer JE (1974) J. Gen. Microbiol. 84, 188-198
Beringer JE, Hopwood DA (1976) Nature 264, 291-293
Beringer JE et al. (1978) Nature 276, 633-634
Li F, Beringer JE (1981) Rothamsted Exp. St. Ann. Re. (1981), 215-216
Vidaver AK (1967) Appl. Microbiol. 15, 1523-1524.

CLONING OF RHIZOBIUM TRIFOLII NODULATION GENES

Z. LORKIEWICZ/A. SKORUPSKA/M. DERYLO/M. GLOWACKA AND D. GOSZCZNSKI
INSTITUT OF MICROBIOLOGY, M. CURIE-SKLODOWSKA UNIVERSITY, AKADEMICKA 19, 20-033 LUBLIN, POLAND

Cloning of nodulation genes of *Rhizobium meliloti*, was recently described (Long et al. 1982).
We have carried out experiments on cloning of *R. trifolii* nod genes by complementation of Nod$^-$ mutants. EcoRI partial digests of DNA from nodulation-proficient strain *R. trifolii* 24 (Zurkowski, Lorkiewicz, 1979) were used as a source of nod genes, where plasmid pRK290 as the cloning vehicle (Ditta et al. 1980).
375 Tc^r transformants of *E. coli* HB101 were isolated and conjugated with four different *R. trifolii* Nod$^-$ recipients in tri-parental mating using pRK2013 as a helper plasmid. 400 colonies of Tc^r transconjugants from each conjugation were collected and examined for clover nodulation. Non-nodulating mutant *R. trifolii* 24AR defective in root adhesion, after acquisition of the *in vitro* constructed plasmid p200, produced nodules of normal size. Besides that *R. trifolii* 24AR (p200) showed alterations in morphology of colonies, phage sensitivit and increased expolysaccharide production. It was further demonstrated that the hybrid plasmid p200 differed from the vector pRK290 by an insert of 20kb and three supplementary EcoRI sites. Plasmid p200 was also transferred to *R. meliloti* L5-30 Nod$^-$ mutants. Both tested mutants RM152 and RM6873 nodulated alfalfa after p200 introduction. Nod$^-$ strains 2406, 24 nod3 and 24K, deficient in hair curling, produced only small abortive nodules when infected with *in vitro* constructed plasmids p65 and p160.
Plasmid p200 complements mutation in hair adhesion but it does not contain sequences required for complementation of mutations in hair curling. This plasmid harbors probably a common nod gene(s) for different Rhizobium species.

REFERENCES

Ditta G, Stanfield S, Corbin D and Helinski DR (1980) Proc. Natl. Acad. Sci. USA 77, 7347-7351.
Long SR, Buikema WJ and Ausubel FM (1982) Nature 298, 485-488.
Zurkowski W and Lorkiewicz Z (1979) Arch. Microbiol. 123, 195-201.

CLONING AND CHARACTERIZATION OF A NODULATION LOCUS FROM *RHIZOBIUM PARASPONIUM*

D.J.MARVEL/[+]G. KULDAU/*A.M. HIRSCH/*J. PARK/[≠]J.G. TORREY/[+]AND F.M. AUSUBEL[≠]
+DEPARTMENT OF ORGANISMAL AND EVOLUTIONARY BIOLOGY, HARVARD UNIVERSITY
*DEPARTMENT OF BIOLOGICAL SCIENCES, WELLESLEY COLLEGE
≠DEPARTMENT OF MOLECULAR BIOLOGY, MASSACHUSETTS GENERAL HOSPITAL

INTRODUCTION

A clone bank of DNA from a slow growing, broad host range strain of *Rhizobium parasponium* (Rp501) has been constructed in the mobilizable vector pLAFR1. This clone bank was mated *en masse* into *R. meliloti* Hac⁻Nod⁻ strains 1027 and 1126. Exconjugants of the mating were used to inoculate alfalfa seedlings. After five weeks, symbiotic phenotypes were scored.

METHODS AND MATERIALS

The vector chosen for these experiments, pLAFR1, is a 26kb long cosmid which confers tetracycline resistance. Cloning and packaging of recombinant plasmids *in vitro* in lambda phage heads was performed as described in Friedman et al (1982). Plasmids packaged in phage heads were used to infect *E.coli* strain HB101. The bank in HB101 contains about two thousand members.

The strategy for selection of clones conferring nodulation capacity on Nod⁻ *R. meliloti* strains 1027 and 1126 was that used by Long et al (1982). Triparental matings of the clone bank into Rm1027 and Rm1126 were performed as described in Ruvkun and Ausubel (1981). Tetracycline resistant exconjugants were inoculated onto alfalfa plants grown asceptically in test tubes on nitrogen free nutrient agar slants.

RESULTS

Twenty-three out of twenty-nine alfalfa plants inoculated with Rm1126 exconjugants were nodulated after five weeks, while twenty-eight out of thirty-eight plants inoculated with Rm1027 exconjugants were nodulated. Nod⁻ parental strains did not nodulate plants, nor did Rp501. Exconjugants which nodulated plants also induced root hair deformations.

Plants inoculated with Rm1126 exconjugants reduced acetylene at ninety-two percent of wild type levels, while those inoculated with Rm1027 exconjugants showed forty percent of wild type levels.

Light microscope examinations show that nodules elicited by infection with Rm1126 exconjugants are not significantly different from wild type induced alfalfa nodules in overall stucture and internal anatomy.

REFERENCES

Friedman, A.M., Long,S.R., Brown,S.E., Buikema,W.J., and Ausubel F,M, (1982) Gene 18, 289-292.
Long,S.R., Buikema, W.J., and Ausubel,F.M. (1982) Nature, 298, 485-488.
Ruvkun,G.B., and Ausubel, F.M. (1981) Nature 289 85-88.

Tn5 MUTAGENESIS AND SYMBIOTIC MUTANTS OF A FAST-GROWING COWPEA RHIZOBIUM

NIGEL MORRISON/HAN CAI CHEN/BRANT BASSAM/ JACEK PLAZINSKI/ROBERT RIDGE and BARRY ROLFE
DEPARTMENT OF GENETICS, R.S.B.S., THE AUSTRALIAN NATIONAL UNIVERSITY, CANBERRA CITY 2601, AUSTRALIA

INTRODUCTION

The fast-growing cowpea *Rhizobium* strain NGR234 is able to nodulate a wide range of tropical legumes as well as the non-legume *Parasponia* (1). Nodulation and structural nitrogenase genes are located on a large Sym-plasmid in this strain (2). A cryptic megaplasmid of greater than 450 Mdals also exists in NGR234.

Transposon mutagenesis: The vector pSUP1011 was used to generate random mutations in NGR234. Plant screening resulted in the isolation of a non-nodulating (Nod^-) mutant also unable to induce root-hair curling (Hac^-). This mutant had Tn5 on the Sym-plasmid. Other mutants defective in nodulation had Tn5 insertions which did not map on the Sym-plasmid. These mutants were Hac^+Nod^- or blocked at an early nodulation stage.

Mobilisation of the NGR234 Sym-plasmid to other bacteria: pSUP1011 cointegrated randomly in the NGR234 genome at a rate comparable to true Tn5 transposition. This phenomenon was exploited to isolate a cointegrate between pSUP1011 and the NGR234 Sym-plasmid. Since pSUP1011 carries a mobilisation site (*mob*) this cointegrate molecule (called pNM4AN) can be mobilized in *trans* by the plasmid pJB3JI. pNM4AN could transfer to classic fast-growing *R.trifolii*, *R.leguminosarum*, *R.meliloti* and *Agrobacterium tumefaciens* strains at a rate of about 10^{-2}. *R.meliloti* ZB157 is a Hac^- mutant and is Nod^- on lucerne or siratro. Strain ZB157 (pNM4AN) was able to nodulate siratro in a normal fashion producing Nod^+ Fix^+ nodules. Microscopy demonstrated a normal nodule morphology. *R.leguminosarum* 6015 is a Nod-nif deletion mutant and cannot nodulate peas or siratro. Strain 6015 (pNM4AN) was Nod^+Fix^- on siratro but was still unable to nodulate peas. *R.trifolii* strain ANU1064 is a heat-cured Nod^- mutant and cannot nodulate clovers. Strain ANU1064 (pNM4AN) was Nod^+Fix^- on siratro but still could not nodulate clovers. *A.tumefaciens* strain C58 carrying pNM4AN was able to form distorted root nodules on siratro. These structures were devoid of bacteria. pNM4AN was stably maintained in all strains. These results show that both nodulation and nitrogen fixation functions for symbiosis with cowpea type plants can be expressed in classic fast-growing *Rhizobium* strains. Since these transconjugant strains could not nodulate their normal hosts there must be genes on the Sym-plasmid which control the host specificity of this broad host range cowpea *Rhizobium*. All attempts to introduce this plasmid into slow-growing cowpea *Rhizobium* strains were unsuccessful.

Cloning of the Hac region: Wild-type DNA sequences corresponding to the Sym-plasmid Hac^- mutant, mentioned above, were cloned from a λ charon -28 gene library of strain NGR234. We verified that the correct gene region was cloned by introducing the cloned fragment, on a mobilisable vehicle, to the Hac^- mutant. The mutant lesion was corrected immediately leading to a Nod^+Fix^+ response on siratro. The cloned fragment was also capable of correcting certain *R.trifolii* Hac^- mutants. Other Hac genes were mapped on this fragment by using specific probes derived from *R.trifolii* Hac genes.

1. Trinick, MJ (1980) J.Appl.Bact. 49, 39-53.
2. Morrison, NA *et al*. (1983) J.Bacter. 153, 527-531.

Tn5 INDUCED SYMBIOTIC MUTANTS OF *Rhizobium meliloti* 2011

PETER MÜLLER/REINHARD SIMON/ALFRED PÜHLER
LEHRSTUHL FÜR GENETIK, FAKULTÄT FÜR BIOLOGIE, UNIVERSITÄT BIELEFELD,
POSTFACH 8640, D-4800 BIELEFELD 1, FRG

Mutants of *Rhizobium meliloti* strain 2011 were induced by general Tn5 transposition using pBR325-Mob-Tn5 as a mobilising vector and inoculated on sterilized seedlings of *Medicago sativa* (alfalfa). After 3 or 4 weeks plants were examined for nodulation and also tested by the acetylene reduction assay. Of about 2400 mutants screened by this procedure, 23 showed defects or significant morphological and physiological changes in their symbiotic properties. In total, nine mutants failed to induce nodule formation. Among these there were five auxotrophic mutants (two for leucine, two for valine/isoleucine and one mutant with two Tn5 insertions but unknown auxotrophic markers). Further auxotrophic mutants (arginine, adenine and cysteine) induced small nodules which sometimes were tumorlike in their appearance. These mutants did not fix nitrogen. Among the ineffective mutants, there were three prototrophic ones with an altered colony type.
Eight mutants of particular interest were characterized using both classical genetic manipulations and recombinant DNA techniques:

1) pIJ1008, a symbiotic plasmid from *R.leguminosarum*, and pRmSL26, a cosmid harboring *R.meliloti nod* genes, as well as subclones of this cosmid were used in complementation analysis. Two Tn5 induced mutants could be complemented by pIJ1008 and pRmSL26 to a fully effective symbiosis while all others could not. One of these two mutants could be complemented additionally by pMA10, which contains a subcloned 4.6 kb *Hin*dIII-fragment of pRmSL26.
2) Cosmid-banks of four different strains (Nod^- and Nif^-) were obtained by ligating partially digested total DNA to the vector p205, *in vitro* packaging in λ heads and infection of *E.coli* K802.
3) A neomycin sensitive mutant of R68.45 was used for R-prime formation. The R-primes contained the Tn5 mutagenised regions of five original *R.meliloti* Nod^--mutants.

Cosmid-clones and subclones of R-primes were transformed into *E.coli* 17.1 and transferred back to *R.meliloti* 2011 wild type by bacterial conjugation, resulting in a single crossover with the vector integrated into the genome (Nm^R and Tc^R) or a double crossover with a concomitant loss of the vector (Nm^R and Tc^S). The ratio of double to single crossover was about 5 %. Exconjugants were tested on plants. The phenotypes of the primary mutants (obtained by general Tn5 mutagenesis) and the secondary mutants (which arose from site specific Tn5 insertions) were expected to be identical or at least similar. However, it could be demonstrated, that one Nod^--mutant was not induced by Tn5 but by an unknown effect. One Fix^--mutant was not able to reduce acetylene after homogenotisation and infection of plants. Thus it could be proved that the strategy outlined above can be effectively applied. Hybridizing blots of Eckhardt-gels with a ^{32}P-labelled fragment of Tn5, it could be shown that the mutants which could be complemented by pRmSL26 are carrying Tn5 within the megaplasmid. A mixed inoculum of six different Nod^--mutants did not result in nodule formation, whereas a mixed inoculum of different Nod^--mutants in combination with Fix^--mutants led to formation of fully effective nodules on alfalfa plants.

NITROGEN FIXATION IN A TROPICAL *RHIZOBIUM* ASSOCIATED WITH *SESBANIA ROSTRATA*.

F. NOREL / A. KUSH / P. DENEFLE / N. CHARPIN / C. ELMERICH
INSTITUT PASTEUR - P.O. BOX 75724 PARIS (CEDEX 15) FRANCE.

1. INTRODUCTION

The *Rhizobium* strain 571, which is associated with the tropical legume *Sesbania rostrata* (Dreyfus, Dommergues 1981) has the property to fix nitrogen in the free living state (Elmerich et al. 1982, Dreyfus et al. 1983). This property prompted us to study biochemistry and genetics of nitrogen fixation using the same methodology as with other free living diazotrophs.

2. NITROGENASE PURIFICATION

Nitrogenase components were purified from cells grown under conditions of nitrogen fixation in a 15 liter fermentor, with a constant tension of dissolved oxygen adjusted at 1 %. Crude extract obtained in Tris-HCl 100 mM, pH 8.0 had a specific activity of 65 nmoles ethylene /min/mg protein. Component 1 (SA 750 U/mg) was made up of two subunits of 56.000 and 59.000 daltons and component 2 (SA 1700 U/mg) was made up of a single subunit of 36.000 daltons.

3. *IN VIVO* SWITCH OFF OF NITROGENASE ACTIVITY BY AMMONIA.

Nitrogen fixing cultures of strain 571 were found to be subjected to *in vivo* nitrogenase activity switch off by ammonia. As in photosynthetic bacteria, no nitrogenase activity was detected in crude extracts of ammonia inhibited cultures, but it was restored after addition of pure component 2.

4. ISOLATION OF MUTANTS DEVOID OF NITROGENASE ACTIVITY.

Mutants impaired in nitrogenase activity in the free living state were isolated after EMS mutagenesis. Among them, strain 5740 was genetically and biochemically characterized as impaired in nitrogenase component 1 activity (Elmerich et al. 1982). All the mutants examined had a Nod$^+$ phenotype and most of them were also Fix$^-$ *in planta*.

5. PROTEIN PRODUCT SYNTHESIS SPECIFIC FOR NITROGEN FIXATION

Comparison of polypeptides synthesized under conditions of nitrogen fixation or ammonia assimilation was performed by one and two dimension gel electrophoresis. Six polypeptides were identified, three of them corresponded to nitrogenase subunits. The six polypeptides were detected in strain 5740. None of them was found in strain 5751 a Nif$^-$ Fix$^-$ Nod$^+$ mutant which was likely to be a regulatory mutant.

6. HOMOLOGY WITH *KLEBSIELLA PNEUMONIAE NIF* GENES AND CLONING OF *NIF* DNA.

Homology between total DNA and probes carrying *nif* genes of *K. pneumoniae* was detected for *nifHDK* (Elmerich et al. 1982), *nifNE* and *nifJ*. DNA fragments homologous to *nifHDK* and *nifJ* were cloned in the λL47-1 vector. Our data are consistent with the existence of several copies of *nifJ* in strain 571.

7. REFERENCES

Dreyfus BL and Dommergues Y (1981) In Gibson AH and Newton WE, eds, Current perspectives in Nitrogen Fixation, pp. 471, Austr. Ac. Sc. Canberra.
Dreyfus BL, Elmerich C and Dommergues Y (1983) Enviromn. Appl. Microbiol. 45, 711-713. Elmerich C, Dreyfus BL, Reysset G and Aubert JP (1982) EMBO J. 1, 499-503.

CLONING GENE(S) INVOLVED IN c-AMP SYNTHESIS IN *RHIZOBIUM*

F. O'Gara, B. Kiely, M. O'Regan and A. McGetrick

Department of Microbiology, University College, Cork, Ireland.

The formation of nitrogen fixing root nodules is a complex process and involves the differential expression of several rhizobial genes. However, the nature of the regulatory element(s) involved in controlling gene expression in root nodule bacteria are not well understood. In other gram negative bacteria cycic 3'5' - adenosine monophosphate (c-AMP) is involved in regulating gene expression. Although c-AMP has been detected in *Rhizobium* (Lim, Shanmugan, 1979), its role in these oxidative bacteria is unclear. To help elucidate the nature of c-AMP in these agronomically important bacteria, we have isolated a cloned DNA sequence encoding gene(s) for adenyl cyclase from *R. meliloti*.

Materials and Methods

The strategy adopted for identifying gene(s) involved in c-AMP synthesis in *R. meliloti* was to complement *E. coli* mutants deleted in the adenyl cyclase (*cya*) locus with DNA fragments from a *R. meliloti* gene bank (Ditta et al., 1980). Enzyme activities (β-galactosidase and adenyl cyclase), c-AMP levels, and DNA analyses were performed following standard published procedures.

Results and Discussion

Recombinant plasmids from the *R. meliloti* F34 gene bank capable of complementing *E. coli* Δ *cya* mutants were identified by selecting for lac^+ clones. The cloned DNA only weakly complemented Δ *cya* mutants. The c-AMP concentration detected in the complemented clones was approximately 15% of the level in the wild type *E. coli* strain. The level of adenyl cyclase expressed was also significantly lower than that detected in the wild type cya^+ strain. The complemented Δ *cya* clones were also capable of inducing a number of c-AMP dependent carbon utilisation operons. A common 5.4 Kb *Bgl* II DNA fragment, with internal *Eco* RI, *Bam* HI, and *Pst* I sites, and inserted in different orientations was identified in the complementing recombinant plasmids. The identification of this DNA sequence will be a useful tool in elucidating the role of c-AMP in these agronomically important bacteria.

References:

Ditta, G., Stanfield, S., Corbin, D., Helinski, D.R. (1980). Proc. Natl. Acad. Sci. USA 77: 7347-7351.

Lim, S.T., Shanmugan, K.T. (1979). Biochim. Biophys. Acta. 584: 479-492.

ORGANIZATION OF NITROGEN FIXATION GENE SEQUENCES IN Rhizobium phaseoli.

C. QUINTO, M. FLORES, H. DE LA VEGA, R. AZPIROZ, E. MARTINEZ, M.A. CEVALLOS, E. CALVA, M.L. GIRARD, J. LEEMANS AND R. PALACIOS.
CENTRO DE INVESTIGACION SOBRE FIJACION DE NITROGENO, UNIVERSIDAD NACIONAL AUTONOMA DE MEXICO. AP. POSTAL 565-A, CUERNAVACA, MOR. MEXICO.

1. INTRODUCTION

Nitrogen fixation gene sequences are reiterated in R. phaseoli. Three different DNA regions containing such sequences were isolated from a genome library of strain CFN-42 (C.Quinto et al., 1982). Here we report on the organization of nitrogenase reductase gene (H) sequences.

2. RESULTS AND DISCUSSION

The nucleotide sequence of gene H from one of the cloned R. phaseoli nif regions was determined. This gene (H-a) shares 92%; 88% and 69% homology with the aminoacid sequence of R. meliloti (A.Torok, A.Kondorosi,1981), R. trifolii (K. Scott et al., 1983) and K. pneumoniae (K.Scott et al., 1981), respectively.

A 300 bp fragment of the coding region of this gene was used as hybridization probe against total DNA of different R. phaseoli strains. DNA was digested with EcoRl restriction endonuclease, separated by agarose-gel electrophoresis, blotted into nitrocellulose membranes and hybridized under stringent conditions. All strains screened showed at least four hybridization bands indicating gene H sequence reiterations.

In CFN-42 four of the nif regions (a, b, c and d) are present in a Sym plasmid while a fifth region is present in other plasmids (J. Leemans et al, these Proceedings). The genome of CFN-42 was cloned in phage λ 1059. The four different nif regions of the Sym plasmid were isolated as part of aproximately 20 Kb inserts. Preliminary analysis shows no-overlap thus suggesting that the different nif regions are not tightly clustered.

A more detailed analysis of region a by hybridization between non overlapping subclones of gene H-a suggests the presence of at least a fragment of a second gene H.

A survey for the presence of gene H reiterations in fast growing strains, isolated from different legumes, was initiated. This complex organization of nif genes appears to be a general feature in strains that infect and fix nitrogen in legumes belonging to the genus Phaseolus.

3. REFERENCES

- Quinto, C., De la Vega, H., Flores, M., Fernández, L., Ballado, T., Soberón, G. and Palacios, R. (1982). Nature, 299 , 724-726.
- Scott, K., Rolfe, B.G. and Shine, J. (1981). J. Mol. Appl. Genet. 1, 71-81.
- Scott, K., Rolfe, B.G. and Shine, J. (1983). DNA 2, 147-153.
- Torok, A. and Kondorosi, A. (1981). Nucl. Ac. Res. 9, 5711-5723

CONSTRUCTION OF A GENE BANK OF HUP$^+$ RHIZOBIUM JAPONICUM STRAIN CB1003 IN A BROAD HOST RANGE CLONING COSMID VECTOR

TOMÁS RUIZ-ARGÜESO
DEPARTMENT OF MICROBIOLOGY, E.T.S. INGENIEROS AGRÓNOMOS, MADRID-3, SPAIN

In the symbiotic association with soybeans some strains of *Rhizobium japonicum* induce a H_2-oxidation system that is responsible for recycling the H_2 generated by nitrogenase. The presence of an active H_2-uptake system in *R. japonicum* seems to be beneficial for the soybean symbiosis and apparently results in a greater plant productivity (1,2). Since the H_2-uptake phenotype (Hup$^+$) is sparcely distributed among the species of *Rhizobium*, it should be useful the transfer of the Hup character to agricultural important strains of *Rhizobium* that lack an efficient H_2-recycling system.

In order to identify and isolate genes of *Rhizobium* involved in the oxidation of H_2 in legume nodules ("hup" genes), a gene bank of the Hup$^+$ *R.japonicum* strain CB1003 (3) has been constructed in *Escherichia coli*. The broad host range cosmid pLAFR1 (4) was chosen as cloning vector because it can be mobilized into and stably replicates in *Rhizobium*. A partial Eco R1 digest of total *R. japonicum* DNA was size fractionated by preparative gel electrophoresis and fragments 10 to 35 kb were ligated to Eco R1-cleaved pLAFR1. The ligated mixture was packaged "in vitro" into phages heads and used to infect *E. coli* HB101. Tetracyline-resistant transductants were generated at a frequency of 6.4×10^4 per ug of vector DNA. The resulting gene bank contained approximately 10,000 clones. The analysis of 30 clones chosen at random showed that 96% harboured recombinant plasmids. The size of the insert DNA ranged between 9.5 and 33.0 kb and averaged 24 kb.

Based on theoretical calculations (5) the number of clones needed to give a probability of 99 % that a particular gene of 1 kb is contained in the above gene bank is about 750, assuming a size of the genome of *Rhizobium* Of 2.5×10^9 daltons and a random distribution of cutting sites. Clones containing the hup genes can be identified by mating the gene bank "en masse" to *R. japonicum* Hup$^-$ mutants and selection of exconjugants complemented for the Hup$^+$ phenotype.

(1) D.W. Emerich et al. J. Bacteriol. 137, 153 (1979)
(2) J.E. Lepo et al. J. Bacteriol. 146, 614 (1981)
(3) T. Ruiz-Argüeso et al. Arch. Microbiol. 128, 275 (1981)
(4) A.M. Friedman et al. Gene 18, 289 (1982)
(5) L. Clarke et al. Proc.Natl. Acad. Sci.USA 72, 4361 (1975)

EXPRESSION OF SYMBIOTIC SPECIFIC SEQUENCES IN THE Rhizobium phaseoli-Phaseolus vulgaris ASSOCIATION.

F. Sánchez, A. Ayala, R. Basurto, R. Palacios, H. de la Vega and C. Quinto.

Centro de Investigación sobre Fijación de Nitrógeno.
Universidad Nacional Autónoma de México.
Apdo. Postal No. 565-A. Cuernavaca, Morelos. México.

Bacteria of the genus Rhizobium interact with the roots of legumes and induce the formation of nitrogen fixing nodules. During the symbiosis there seems to be a coordinated expression of bacterial and plant genetic information Vanden Bos et al., 1982 and Fuller et al., 1983. We are interested in the isolation and characterization of bacterial nodule specific sequences that are expressed during the Rhizobium phaseoli-Phaseolus vulgaris symbiosis.

Material and Methods: Total RNA from nodules was extracted according to Chirgwin et al., 1979. Library screening was made according to Grunstein, M., Hogness, D.S., 1975. Plasmids profiles of Rhizobia strains were made according to Hirsch, P.R., 1980 and Eckardt, T., 1978.

Results and Discussion: Most of the major transcripts of nodules induced by CFN42 (wild type) strain are derived from the Sym plasmid (p42d). A CFN42 genome library was screened with ^{32}P-RNA from mature nodules. 20 strong hybridizing clones were obtained. Most of the clones hybridized with the CFN42 Sym plasmid. However there are also clones derived from the chromosome and from other plasmids. This results suggest that not only genes contained in the Sym plasmid participate in the establishment of the symbiotic state. By Southern blot analysis we have shown that adjacent regions to nitrogen fixation genes as well as some symbiotic specific sequences are reiterated in the plasmid and in the chromosome. In the present work we conclude that reiteration is not a particular characteristic of nif genes as previously reported by Quinto, C., et al., 1982., but also pertains to other symbiotic sequences.

References

1. Chirgwin, J.M. et al. 1979. Biochemistry 24: 5294-5299.
2. Eckhardt, T. 1978. Plasmid 1: 584-588.
3. Fuller, F. et al. 1983. Proc. Natl. Acad. Sci. USA 80: 2594-2598.
4. Grunstein, M., Hogness, D.S. 1975. Proc. Natl. Acad. Sci. USA. 72: 3961-
5. Hirsch, P.R. et al. 1980. J. Gen. Microbiol. 120: 403-412.
6. Quinto, C. et al. 1982. Nature 299: 724-726.
7. Van den Bos et al. 1982. Molecular Genetics of the bacteria-plant interaction. 1st. International Symp. G.F.R.

IDENTIFICATION AND ANALYSIS OF THE EXPRESSION OF RHIZOBIUM LEGUMINOSARUM PRE SYMBIOTIC GENES

Th.M.P. Schetgens, G. Bakkeren, C. van Dun, J.G.J. Hontelez, A. van Kammen and R.C. van den Bos. Department of Molecular Biology, Agricultural University, De Dreijen 11, 6703 BC WAGENINGEN, The Netherlands.

We studied the organization and expression of R.leguminosarum PRE genes involved in nodulation (nod) and nitrogen fixation (fix) during symbiosis with Pisum sativum. DNA fragments bearing relevant coding units were selected by hybridization of sym-plasmid clonebanks with specific R.meliloti or K.pneumoniae probes. Three clusters of genes could be identified by this approach.

At first the structural nitrogenase operon was discovered on account of homology with a R.meliloti-nif HD probe (pRmR2; ref. 1). Transposon Tn5 mutations were introduced into this structural nif-region of the R.leguminosarum PRE genome according to Ruvkun and Ausubel (2). These PRE::Tn5 strains were inoculated on pea and analyzed by isolating bacteroid proteins from induced nodules, and by subsequent incubation of Western blots with antisera specific against components I and II of nitrogenase and with ^{125}I protein A to visualize immune complexes. Two different Tn5 insertions within nif D-homologous sequences caused a $nod^{+}fix^{-}$ phenotype and resulted in inhibition of CIα (nif D) and CIβ (nif K)-synthesis, suggesting linkage between these genes; concomittant decline of CII-production could be due to a lower translational efficiency of shortened messengers or to dependence of nif H expression on the presence of intact CI-product. The influence of these Tn5-mutations at the bacterial level on host plant gene expression (nodulin-synthesis) was also investigated, see symposium paper T. Bisseling et al. (S33).

Alternatively, the chromosomal organization of nif structural genes was studied by expression in E.coli minicells (3) of different cloned fragments, containing the nif DH-homologous region and deletion derivatives thereoff as well as an adjacent part probably bearing nif K coding sequences. Nif-gene products were identified by immunoprecipitation of ^{35}S-labeled proteins with nitrogenase-specific antisera and subsequent gel electrophoresis. Assignment of functions to protein bands therefore was not exclusively based upon molecular weight measurements. All three structural nif HDK genes were effectively expressed in E.coli minicells. This expression resulted from translation of hybrid mRNAs, transcribed from a strong vector promoter. Protein synthesis was initiated on Rhizobium-specific ribosome binding sites. In the case of nif H also weak endogenous promoter activity was observed (as deduced from weak CII-synthesis from the reverse insert orientation). Theoretically the nif K-encoded protein could represent a fusion product of fortuitously the same molecular weight as intact CIβ because the position of gene K within the concerned subclone must be very close to the vector promoter and also to nif D (4). Thus a contiguous organization of the structural nif-genes was found

Isolation of a second cluster i.e. putative nodulation genes, at this stage of the experiments is only based upon determined DNA-homology with a R.meliloti nod-region (pRmSL26; ref. 5).

Finally a region of 7.45 kb bearing well-defined homology with nif AL regulatory genes (activator-repressor for nitrogenase synthesis) and with nif F a flavodoxin-encoding gene was detected (probe pGR102; ref. 6), see poster abstract J.G.J. Hontelez et al. (9A-22). Linkage between the three clusters described here on the sym-plasmid has not yet been ascertained.

1. GB Ruvkun, FM Ausubel, 1980, P.N.A.S. 77, 191-195.
2. GB Ruvkun, FM Ausubel, 1981, Nature 289, 85-88.
3. J Reeve, 1979, Methods Enzymology 68, 493-503.
4. AJM Krol et al, 1982, Nucleic Acid Res. 10, 4147-4157.
5. SR Long et al, 1982, Nature 298, 485-488.
6. GE Riedel et al, 1979, P.N.A.S. 76, 2866-2870.

RHIZOBIUM TRIFOLII NODULATION AND NITROGEN FIXATION GENES - A GENETIC AND MOLECULAR CHARACTERIZATION

Peter R.Schofield, Michael A.Djordjevic, Robert W.Ridge, Barry G.Rolfe, John Shine and John M.Watson.

Centre for Recombinant DNA Research and Department of Genetics, R.S.B.S. Australian National University, G.P.O.Box 4, Canberra City,2601,Australia

Bacteria of the genus *Rhizobium* interact symbiotically with leguminous plants, forming root nodules in which atmospheric nitrogen is fixed as ammonia. For fast-growing *Rhizobium* species, this interaction is highly specific: *R.trifolii* only nodulates clover (*Trifolium*).

Transposon mutagenesis of *R.trifolii* strain ANU843 produced a mutant that was defective in root hair curling (Hac$^-$ phenotype), an early step in the nodulation process. Recombinant plasmids containing overlapping sequences were obtained from this region. The nitrogenase structural genes *nif*HD were identified and mapped some 16kb from the Tn5-induced nodulation defective gene.

A 14kb *Hin*dIII fragment spanning the point of the Hac$^-$::Tn5 insertion was cloned in a broad-host-range vector and the recombinant (pRt032) introduced into a Sym plasmid-cured Nod$^-$ derivative of ANU843 (ANU845). Nodulation functions were restored. Transfer of pRt032 to *Agrobacterium tumefaciens* and to an unrelated "cowpea" *Rhizobium* conferred on these strains the ability to nodulate clover plants. Thus the 14kb fragment carries the genetic determinants for nodule induction and development, as well as for clover host specificity. A 7.2kb subclone of this fragment does not induce nodule development.

Tn5 mutagenesis of the cloned 14kb region and homologous marker exchange allowed the identification of two further nodulation-defective mutants. Another Tn5 insertion reduced the nitrogen-fixing capacity of nodules induced by this mutant. Merodiploid strains of these Nod$^-$ mutants were able to complement each other demonstrating that at least two transcriptional units are involved in clover nodulation. DNA sequence analysis of one of these nodulation gene regions revealed two potential coding sequences. A transcription terminator was also located adjacent to this region.

A sequence located in the 300bp upstream of the *nif*H gene has been shown to be repeated at least five times exclusively on the Sym plasmid of all *R.trifolii* strains examined. The sequence also appears specific for *R.trifolii*. DNA sequence analysis of independent copies of this *R.trifolii*-specific repeated sequence shows that the repeat is at least 150bp long and is highly conserved. The exclusive Sym plasmid location of these sequences suggests that they play an important role in the symbiosis.

PHYSICAL MAPPING OF THE NIF AND NOD REGIONS OF A *RHIZOBIUM TRIFOLII* SYM::R68.45 COINTEGRATE PLASMID

D.B. SCOTT[+], C.B. COURT[+], C.W. RONSON[++]
+ APPLIED BIOCHEMISTRY DIVISION, DSIR, PALMERSTON NORTH, NEW ZEALAND
++ GRASSLANDS DIVISION, DSIR, PALMERSTON NORTH, NEW ZEALAND

INTRODUCTION

Previous work has shown that the *Rhizobium trifolii* symbiotic plasmid, pRtr514a, readily forms stable cointegrates with the broad host range IncP group plasmid R68.45 (Scott, Ronson, 1982). One such cointegrate, pPN1, has been transferred into *E. coli*, *Pseudomonas aeruginosa* and a range of fast growing *Rhizobium* strains (Ronson, Scott, 1983). We now report on the cloning and physical mapping of the Nif and Nod gene regions of pPN1.

RESULTS

Using a modification of the method of Kado and Liu (1981) pPN1 was purified from *E. coli* strain HB101 and 10-30kb Sau3A fragments of the plasmid were cloned into the BamH1 site of pBR328. Clones from the Nif and Nod gene regions were isolated from this library by colony hybridisation using [^{32}P]-labelled pW602 and pW587 DNA probes from *R. trifolii* strain SU843 (Watson et al, unpublished). Using such clones a physical map of 45kb in the Nif and Nod region was constructed for EcoRI and HindIII. Using the 0.7kb EcoRI-BglII fragment from pSA30 (Ruvkin, Ausubel, 1981), the 4.5kb BamH1-EcoRI fragment from pSA30 (Ruvkun, Ausubel, 1981) and the 3.0kb SalI fragment from pMC71A (Buchanan-Wollaston et al, 1981), DNA regions homologous to nifH, nifDK and nifA have been located. As with *R. trifolii* strains SU329 (Scott et al, 1983) and SU843 (Watson et al, unpublished) a symbiotic plasmid specific repetitive sequence was found adjacent to nifH. Using the 0.9kb ClaI fragment of pW564 (Watson et al, unpublished) as a probe for the repetitive sequence, four other copies of this sequence were found on pPN1.

REFERENCES

Scott DB and Ronson CW (1982) J. Bacteriol. 151, 36-43.
Ronson CW and Scott DB (1983) In Proceedings of the Ist International Symposium on the Molecular Genetics of the Bacteria-Plant Interaction (in press).
Kado CI and Liu S-T (1981) J. Bacteriol 145, 1365-1373.
Ruvkun GB and Ausubel FM (1981) Nature 289, 85-88.
Buchanan-Wollaston V, Cannon MC, Beynon JL and Cannon FC (1981) Nature 294, 776-778.
Scott KF, Rolfe BG and Shine J (1983) DNA 2, 149-155.

ACKNOWLEDGEMENT

D.B. Scott thanks Drs J. Shine and B.G. Rolfe in whose laboratories part of this work was carried out.

THE pSYM PLASMID OF RHIZOBIUM MELILOTI. III. EXPRESSION OF pSYM FUNCTIONS IN AGROBACTERIUM TUMEFACIENS

G. Truchet[1], C. Rosenberg[2], J. Vasse[1], J.S. Julliot[2], S. Camut[2] and J. Dénarié[2]
Laboratoire de cytologie et de Biologie Cellulaire, Faculté des Sciences Marseille-Luminy, 13288 MARSEILLE Cedex 2, France[1]
Laboratoire de Biologie Moléculaire des Relations Plantes-Microorganismes, C.N.R.S.-I.N.R.A., B.P. 12 - 31320 CASTANET-TOLOSAN, France[2]

The pSym megaplasmid of R. meliloti 2011, mobilized by RP4, or pGMI42 wich carries a 290 Kb pSym fragment including nitrogenase and nod genes (described by Batut et al. in Poster I of this series), was introduced into Agrobacterium tumefaciens. The resulting transconjugants induced root deformations specifically on the homologous hosts Medicago sativa and Melilotus alba and not on the heterologous hosts Trifolium pratense and T. repens. The root deformations were shown to be genuine nodules by physiological and cytological studies. Thus, host specificity nodulation genes are located on the pSym megaplasmid.

Host nodulation specificity did not seem to require recognition at the root-hair level since no infection threads could be detected in the root hairs. Cytological observations indicated that bacteria penetrated only the superficial layers of the host root tissue by an atypical infection process. The submeristematic zone and the central tissue of the nodules were shown to be bacteria-free. Thus nodule organogenesis was probably triggered from a distance by the bacteria.

Agrobacterium transconjugants carrying the pSym induce the formation of more numerous and larger nodules than those carrying the RP4-prime pGMI42, suggesting that some genes controlling nodule organogenesis are located in a pSym region(s) outside that which has been cloned into pGMI42.

A GENE BANK OF RHIZOBIUM LEGUMINOSARUM AND CLONING OF THE REC GENE

E. VINCZE[+], A. LAMBERTI, A. RICCIO and M. IACCARINO
INTERNATIONAL INSTITUTE OF GENETICS AND BIOPHYSICS, CNR, VIA MARCONI 10, I-80125 NAPLES, ITALY

We prepared a gene bank of a rif^R derivative of R.leguminosarum 1001. A DNA preparation (containing both plasmid and chromosomal DNA) was partially digested with Sau3A1, centrifuged on a saccharose gradient and the approximately 30 kb fraction was used after phosphatase treatment to insert fragments into the BamHI site of the pMMB34 broad host range cosmid. According to the equation $P=1-(1-f)^N$ (Clarke, Carbon 1979), 777 clones were needed to reach 99% probability of the presence of any R.leguminosarum gene in the bank. In fact the average insert size was 26.6 kb and the R.leguminosarum genome size was estimated 4500 kb and therefore f=26.6/4500. 2000 of the 30.000 independent cosmids were transducted into E.coli HB101 and used for further analysis. The following evidence for gene bank representativity was found: a) the size of fragments of randomly chosen clones was different; b) nif genes were found by hybridization to ^{32}P-labeled bacterial and/or bacteroid RNA; d) 3 rif^R clones were obtained; e) 4 clones with a Rec^+ phenotype were obtained by screening for UV^R or nitrofurantoin^R.
The Rec^+ phenotype was transferable by conjugation and transformation. We proved by hybridization that in R.leguminosarum the rec gene is localized on a 7 kb EcoRI-SstI fragment. We recloned a 2.7 kb subfragment still containing the rec gene into plasmid p101, a 6.6 kb derivative of pACYC184 which can be transferred by conjugation. We are trying to inactivate the rec gene in vitro and in vivo in order to obtain rec^- derivatives of R.leguminosarum by recombination.

REFERENCE

Clarke L and Carbon J (1979) In Wu R ed, Methods in Enzymology, Vol. 68, pp. 396-408. Academic, New York.

ACKNOWLEDGEMENTS

We thank Prof. M. Bagdasarian and Dr. R. Simon who kindly sent us plasmids pMMB334 and p101 before publication.

[+]Institute of Genetics, Biological Research Center, p.o.box 521, Szeged, Hungary.

ORGANIZATION AND PRIMARY STRUCTURE OF NITROGENASE GENES IN THE *PARASPONIA RHIZOBIUM* STRAIN ANU289

JEREMY WEINMAN[+,++]/FLORENCE FELLOWS[+]/PETER GRESSHOFF[++]/JOHN SHINE[+] and KIERAN SCOTT[+]

[+]CENTRE FOR RECOMBINANT DNA RESEARCH, R.S.B.S. and [++]DEPARTMENT OF BOTANY, FACULTY OF SCIENCE, AUSTRALIAN NATIONAL UNIVERSITY, G.P.O. BOX 4, CANBERRA CITY, 2601, AUSTRALIA

The *Parasponia Rhizobium* strain ANU289 effectively nodulates a wide range of tropical legumes and also the non-legume *Parasponia* (Trinick and Galbraith, 1980). In addition, nitrogenase activity in this strain can be induced *in vitro* (Mohapatra *et al*., 1982).
The structural genes for nitrogenase have been isolated from a lambda genomic library of ANU289 DNA by hybridization to the cloned nitrogenase structural genes of *Klebsiella pneumoniae* (Scott *et al*., 1983). Restriction analysis of the lambda clones carrying these genes has shown that the gene encoding the Fe-protein (*nifH*) is separated from the genes encoding the MoFe-protein (*nifD* and *nifK*) by at least 10 Kb of DNA. This arrangement is in contrast to that seen previously in free-living nitrogen-fixing bacteria and fast-growing *Rhizobium* strains in which *nifH,D,K* are contigious. The gene for *nifK* lies immediately to the 3'side of the *nifD* gene suggesting that the two genes are encoded on the same operon.
Analysis of DNA sequence data has allowed the prediction of the amino acid sequence for the nitrogenase sub-units. The *nifH* gene product shows strong conservation to the amino acid sequences of all other known Fe-proteins. Further, it contains 5 cysteine residues which are conserved in all of these sequences. The *nifD* gene product shows homology (34%) to the amino acid sequence of *Clostridium*. In particular, 5 of the cysteine residues present are conserved when compared to other species (Hase *et al*., 1984). The predicted amino acid sequence for the *nifK* gene product shows considerable homology (44%) to the amino acid sequence obtained for *Anabaena* (Mazur *et al*., 1982). Again, 5 cysteine residues are conserved between these species.
The 5'ends of the *nifH* and *nifDK* transcripts have been mapped by S1-nuclease and primer extension experiments. The *nifH* gene is preceeded by a leader sequence of 155 base pairs while the *nifDK* coding region is preceeded by a 37 base pair leader sequence.
The sequences preceeding both transcripts show regions of homology as indicated below.

```
                                                  ↓
nifH     TAAGCGCGGACAGTGTTGGCATGGCGATTGCTGTTGAGTTGCAGCAACAC
nifDK    TGGCTCGCGCCGCGCTAAACATGCTCGTTGCAGTCTTGTTCAAGAAGCTG
                                                  ↑
```

It is interesting to note the presence of regions at about -10 which are similar to the sequence recently proposed as a "Pribnow box" for nitrogen starved conditions (Beynon *et al*., 1983). These regions of homology 5' to the two transcripts may play a role in the coordinate regulation of these transcripts.

REFERENCES

Beynon J, Cannon F, Buchanon-Wollaston V and Cannon F (1983) Cell. in press.
Hase T, Wakabayashi S, Nakano K, Zumft WG and Matsubara H (1984) this meeting.
Mazur BJ and Chiu CF (1982) PNAS USA 79, 6782-6786.
Mohapatra S, Bender G, Shine J, Rolfe B and Gresshoff P (1982) Arch.Microbiol. 134, 12.
Scott K, Rolfe B and Shine J (1983) DNA 2, 141-148.
Trinick MJ and Galbraith J (1980) New Phytol. 86, 17-26.

POSTER DISCUSSIONS 9B RHIZOBIUM PLASMIDS

A.A. SZALAY, Boyce Thompson Institute, Cornell University, Ithaca, N.Y. 14853, U.S.A.

The rapid development in this field is reflected by the number of posters which exceed 40 in the Rhizobium plasmid area. Plasmids carrying symbiotic (nod and fix) functions are identified in fast-growing rhizobia such as R.meliloti on a megaplasmid, R.leguminosarum, R.phaseoli, R.trifolii on large plasmids, and Rhizobium strain ANU240, NGR234, wide host range, fast-growing cowpea strain. Rhizobium strain IHP100, stem nodulating rhizobium also on large plasmids. Similarly sym plasmids identified in fast-growing R.japonicum strain in several laboratories. In contrast, the presence of plants carrying nif and nod plasmids has not yet been documented in slow-growing rhizobia such as R.japonicum strain 110, R.cowpea strain IRC78, and R.parasponia strain ANU289. A. Kondorosi summarized structural and genetic organization of nod and fix functions in R.meliloti megaplasmids. He and his collaborators documented that in addition to the functionally characterized common nod region, a separate region responsible for host specificity is also present on the megaplasmid. S. Long discussed the genetic characterization of the common nod region by establishing complementation groups. G. Truchet and his collaborators found that in addition to the identified symbiotic region based on nodule-specific RNA hybridization, additional symbiotic functions are localized on the megaplasmid. A. Downie and A. Johnston eloquently documented the presence of the common nod function of R.leguminosarum on a 10 kb plasmid DNA fragment including the root hair curling (hac) region. The authors suggest the presence of more than one gene involved in the nodulation function. R. Palacios presented the identification of symbiotic functions on plasmids of R.phaseoli using R.plasmids. The characterization of host specificity and nodulation genes of R.trifolii were discussed by P. Schofield. A 14 kb DNA fragment restored nodulation functions in a sym plasmid cured strain of R.trifolii and upon transfer enabled Agrobacterium and an unrelated cowpea Rhizobium strain to nodulate clover plants. These findings strongly indicate the presence of both nod and host specificity genes in this segment of the plasmid. A. Morrison characterized the plasmid coded symbiotic functions of cowpea Rhizobium strain NGR234. Constructing mobilizable cointegrates Morrison demonstrated the expression of nod functions and root hair curling capacity of this strain in fast-growing rhizobia using interspecific complementation. A. Atherly and R. Applebaum documented the presence of sym plasmids in fast-growing R.japonicum by hybridization and mutant complementation. Marvel and her colleagues reported functional complementation of the common nod region of R.meliloti by cosmid clones obtained from slow-growing R.parasponia.

In spite of the fact that some of the nod functions can now be complemented among different Rhizobium species, very little is known about the problems of incompatibility. J. Beringer and B. Rolfe discussed two forms of incompatibility among Rhizobium, the plasmid incompatibility at the DNA replication level and functional incompatibility. Many of the participants agreed that the problem of incompatibility should receive much more attention in the near future.

For further gene transfer and mutant complementation to exist, the construction of rec-deficient recipient rhizobium strains and the development of broad host range vectors using rhizobium specific replication origins which are stabily maintained in rhizobia are of general interest.

In summary, concerning the nod functions the following picture has emerged. The common nod region seems to be conserved among different rhizobia and

therefore the *R.meliloti* *nod* region can be used as a hybridization probe.
Secondly, the nodulation function and host specificity are located on separate regions of the sym plasmid and are presently being physically and genetically characterized.
Thirdly, it appears that the *nod* region may contain as many as 3 to 5 genes.

EFFECTS OF DIFFERENT SYMBIOTIC PLASMIDS ON THE COMPETITIVENESS OF *RHIZOBIUM LEGUMINOSARUM* STRAINS

N.J. BREWIN
JOHN INNES INSTITUTE, COLNEY LANE, NORWICH NR4 7UH, U.K.

Four different symbiotic plasmids from *Rhizobium leguminosarum* were introduced into three different recipient strains that lacked plasmid-linked symbiotic determinants. The twelve synthetic strains so constructed were each tested for competitiveness on peas when coinoculated with a standard reference strain (726). Strains were distinguished on the basis of antibiotic resistances. An analysis of variance showed that the recipient strain and the introduced symbiotic plasmid contributed about equally to competitiveness in forming root nodules on pea plants, but the recipient strain alone contributed to competitiveness for growth on the legume root surface (see Brewin *et al.*, 1983a).

This distribution of effects between the symbiotic plasmid and elsewhere was also observed when nitrogen fixation efficiency was examined (DeJong *et al.*, 1981, 1982). Taken together, these data suggest that for fast-growing strains of *Rhizobium* it should be possible to 'breed' for improved *Rhizobium* strains simply by exchanging or recombining natural symbiotic plasmids from strains of diverse origin, and screening the progeny for improved symbiotic performance (Brewin *et al.*, 1980a, b, c; Brewin, 1982, 1983b).

Symbiotic plasmids are sometimes self-transmissible by conjugation and transfer between strains occurs spontaneously. Alternatively, they may be mobilised by recombination with transmissible plasmids (Brewin *et al.* 1980a, 1982), or after insertion of transposon Tn5-mob (Simon *et al.* 1983). Likewise, a variety of methods exist which allow the detection of transfer (or loss) of symbiotic plasmids in *Rhizobium* strains: these include direct selection for nodulation ability on plants (Brewin *et al.*, 1980a, b), immunochemical screening for a plasmid-determined gene product (Dibb *et al.*, this volume), colony hybridisation to cloned *nod* DNA, and the introduction of transposons as selectable markers (Beringer *et al.*, 1978, Ruvkun, Ausubel, 1981).

REFERENCES

Beringer J.E. *et al.* (1978) Nature 276, 633-634.
Brewin N.J. *et al.* (1980a) J.Gen.Microbiol 116, 261-270.
Brewin N.J. *et al.* (1980b) J.Gen.Microbiol 120, 413-420.
Brewin N.J. *et al.* (1980c) Nature 288, 77-79.
Brewin N.J. *et al.* (1982) J.Gen.Microbiol 128, 1817-1827.
Brewin N.J. *et al.* (1983a) J.Gen.Microbiol. In press.
Brewin N.J. *et al.* (1983b) In Puhler A. (Ed.) Molecular Genetics of the Bacteria-Plant Interaction. Springer-Verlag, Berlin. In press.
DeJong T.M. *et al.* (1981) J.Gen.Microbiol 124, 1-7.
DeJong T.M. *et al.* (1982) J.Gen.Microbiol 128, 1829-1838.
Ruvkun G., Ausubel, F.M. (1981) Nature 289, 85-88
Simon R. *et al.* (1983) In Puhler A. (Ed.) Molecular Genetics of the Bacteria-Plant Interaction. Springer-Verlag, Berlin. In press.

METHIONINE SULFOXIMINE SENSITIVE (MS^S) MUTATIONS THAT IMPAIR NITROGEN FIXATION IN Rhizobium phaseoli.

Guadalupe Espín, Enrique Morett and Soledad Moreno
Centro de Investigación sobre Fijación de Nitrógeno
U.N.A.M. Apdo. Postal 565-A Cuernavaca, Mor. MEXICO

INTRODUCTION

In Klebsiella pneumoniae, the expression of glutamine synthetase (GS) and nitrogenase is controlled by a regulatory system, including the genes ntrA, ntrB, and ntrC (Merrick 1983). To investigate whether in R. phaseoli nitrogen fixation is controlled by an analogous system, we isolated mutants in which the synthesis and or regulation of the enzymes GS has been altered.

RESULTS AND DISCUSSION

ISOLATION AND GENETIC CHARACTERIZATION OF MUTANTS. R. phaseoli CFN2030, a stp^r derivative of CFN42 (Quinto et al 1982) was mutagenized with transposon Tn5, Km^r derivaties sensitive to 5µg/ml of MS, (an specific inhibitor of GS activity) were isolated. All mutants contain a single copy of Tn5 on the chromosome or on plasmid E in CFN2013 and CFN2026 (CFN2030 harbours 6 plasmid A to F). Genetic linkage data showed that MS^s mutations are caused by insertion of Tn5.

GS ACTIVITIES AND SYMBIOTIC PROPERTIES OF MS^s MUTANTS. The rhizobeacea posses two forms of GS: GSI and GSII (Fuchs, Keisther 1980). GS activities and the ability to fix nitrogen were determined for all mutants. We identified four phenotypic classes: Class I mutants, (e.g. CFN2011), have GSI, GSII, and fix nitrogen undistinguisable to those of the wild type. Class II mutants (e.g. CFN2012 and CFN2017), lack GSII activity, but retain GSI and nitrogenase activities similar to the wild type. Class III mutants like CFN2013 and CFN2026, have wild type levels of GSI and GSII, do nodulate, but are unable to fix nitrogen. Class IV mutants like CFN2014 and CFN2018, do nodulate, but lost the ability to reduce nitrogen; GSI and GSII activities are present at a reduced level compared to CFN2030.

Our data suggest that in R. phaseoli, loss of GSII activity does not affect nitrogen fixation. This agrees with previously reported mutants in which loss of GSII activity does not result in a Fix phenotype (Ludwig 1980).

Plasmid D, the sym plasmid of CFN42, carries genes for nodulation and the gene nifH (Leemans et al, Quinto et al, these proceedings). However CFN2013 and CFN2026 carry mutations on plasmid E that impair nitrogen fixation. Thus, not all genes for nitrogen fixation are on a single plasmid.

In CFN2018 and CFN2014, both the synthesis of glutamine and nitrogen fixation are altered. This suggest that these mutations are in genes involved in regulation of nitrogen fixation, similar to the ntr genes from K. pneumoniae.

REFERENCES

Fuchs R L and Keister D L (1980) J. Bacteriol. 141, 996-998
Ludwig R A (1980) PNAS 77, 5817-5821
Merrick M J (1983) EMBO J 2, 39-44
Quinto C et al (1982) Nature 299, 724-726

ANALYSIS OF SYM-PLASMID-EXPRESSION BY IMMUNOAFFINITY CHROMATOGRAPHY

S. HIGASHI, T. UCHIUMI AND M. ABE
DEPARTMENT OF BIOLOGY, FACULTY OF SCIENCE, KAGOSHIMA UNIVERSITY, KAGOSHIMA 890, JAPAN.

Studies of nodulation using several sizes of the Sym-plasmid are one approach to elucidate the nodulation mechanism of *Rhizobium*. We have attempted to clarify expression of the Sym-plasmid in strains of *R. trifolii* 4S and mutants causing quasi-nodulation (Fix^-).

MATERIALS AND METHODS Strains. *Rhizobium trifolii* 4S (Inf^+, Nod^+, Fix^+; 350, 280 and 210 Md plasmids) and A1 (a mutant derived from st. 4S spontaneously, Inf^-, Nod^-; 350 and 280 Md plasmids) were used. Immunoaffinity-chromatography. IgG of the supernatant of A1-cell homogenate, which was coupled to CNBr Sepharose 4B, was used (abb. A1-IgG-ligand column). Plasmids. Plasmids were extracted by Casse's method (2). Nod^+ elimination. The elimination of Nod^+ was performed by Zurkowski's method (3). Nodulation test. It was described in the previous paper (1).

RESULTS AND DISCUSSION The high temperature treatment of st. 4S resulted in loss of nodulation ability in about 50% of colonies. These colonies (H-series) simultaneously eliminated the 210 Md plasmid. This result suggests a relationship between nodulation and the expression of the 210 Md plasmid. A transformant ($A1^{4S}$) was obtained by transformation with st. 4S (as donor) and st. A1 (as recipient). Strain $A1^{4S}$ has three plasmids, 350, 280 and 62 Md, and formed some quasi-nodules (Fix^-). It can therefore be assumed that the genes of quasi-nodule formation are located on the 62 Md plasmid (included in the 210 Md plasmid ?).
We examined an approach to analyze the materials involved in nodulation by *Rhizobium* using the A1-IgG-ligand column. The electrophoretic patterns of peak 1s, which did not bind the column, of the strains revealed different patterns and staining intensities. When a mixture of the 4S-peak 1 of the 4S-cell homogenate and A1-cells was inoculated on white clover, quasi-nodules formed (Fig. 1). The nodules did not form when plants were inoculated with only A1-cells or 4S-peak 1. Furthermore, the normal infection threads were not observed in these nodules by light, SEM and TEM. Some reisolated colonies (Qn-series) from quasi-nodules had quasi-nodulation ability for several generations. Strain Qn1 also has three plasmids, 350, 280 and 62 Md, the same as st. $A1^{4S}$. However, we have no clear evidence concerning the origin of the 62 Md plasmid in st. Qn1. Diphenylamine positive material was included in 4S-peak 1 (6.25 μg/mg DW), however, it was not detected on agarose gel electrophoresis (when sample volume was 10 mg/track) by ethydium bromide staining. In all cases, the 62 Md plasmid must be important in quasi-nodule formation (Fig. 2). These results suggest that the 4S-peak 1 includes key factor(s) for quasi-nodule formation.
Fig. 1 Quasi-nodules., Fig. 2 Detection of plasmids in st. 4S, in its Nod^- derivatives and in quasi-nodule formants.

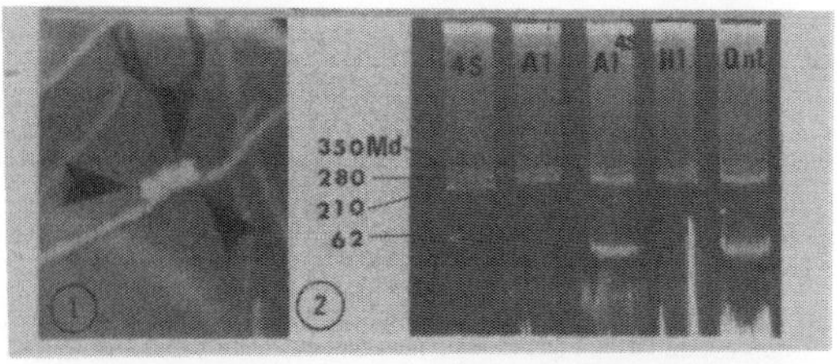

REFERENCES 1) Abe M. et al. (1982) Plant, Soil 64, 315-324. 2) Casse F et al. (1979) J. Gen. Microbiol. 113, 229-242. 3) Zurkowski W, Lorkiewicz Z (1978) Genet. Res. Camb. 32, 311-314.

GENERAL ORGANIZATION OF Rhizobium phaseoli nif PLASMIDS

J. Leemans, G. Soberón, M.A.Cevallos, L. Fernández, M.A. Pardo, H. de la Vega, M. Flores, C. Quinto and R. Palacios.
CENTRO DE INVESTIGACION SOBRE FIJACION DE NITROGENO, UNIVERSIDAD NACIONAL AUTONOMA DE MEXICO. AP. POSTAL 565-A, CUERNAVACA, MOR. MEXICO

1. INTRODUCTION

In *R. phaseoli*, as in many other rhizobia, sequences homologous to *Klebsiella pneumoniae nif* genes are plasmid located. In *R. meliloti*, *R. trifolii*, and *R. leguminosarum*, *sym* plasmids have been identified whose genes are directly involved in nodulation and N2-fixation. We report on the symbiotic and physical properties of *R. phaseoli* plasmids that contain *nif* genes.

2. RESULTS AND DISCUSSION

We used a 300 bp coding region fragment from the *nif* H-a gene of *R. phaseoli* CFN-42 (Quinto et al, these proceedings) as probe to hybridize with plasmid patterns from 8 different *R. phaseoli* isolates. In all strains a plasmid of about 250 Kb hybridized and in 4 of them a plasmid of about 140 Kb also hybridized. We propose that plasmids containing *nif* genes should be called "*nif* plasmids". CFN-42 contains 6 plasmids (a to f, size from 90 to 350 Md), 2 of them (p42-a and p42-d) share homology to the *nif* H-a gene. Hybridization experiments with purified p42-a and p42-d showed that p42-d harbors 4 *nif* regions, whereas p42-a contains a fifth region (see also Quinto et al, these proceedings).

CFN-2001, a CFN-42 derivative obtained after treatment at 37°C for 7 days, had lost both *nif* plasmids along with nodulation capacities. Tn5-*mob* insertions were isolated in CFN-42 at a frecuency of 2.10^{-5} (Simon et al, 1983). 5 out of 6 Tn5*mob* marked plasmids could be mobilized into *Agrobacterium tumefaciencs* C58Cl (Van Larebeke et al 1974). p42-a and p42-d were mobilized into CFN-2001. Transconjugants of 42-a remain - nodulation deficient whereas CFN-2001 (p42-d) nodulates and fixes nitrogen, undistinguishable from the wild-type CFN-42. *Agrobacterium* C58Cl containing p42-d nodulates bean plants but does not fix nitrogen. Thus the concept of a *sym* plasmid appears applicable to p42-d of CFN-42. We can not assign yet any symbiotic function to p42-a. Using C58Cl that contains (p42-d::Tn5*mob*) and (pJB3) (Brewin et al, 1980) in a cross to *E. coli* HB101, kanamycin resistant R-prime plasmids were isolated at 2.10^{-7}/recipient. Their analysis shows that *nod* and *nif* genes including the 4 *nif* regions, are clustered on p42-d, since a functional ($Nod^{+}Fix^{+}$) R prime was isolated that contains 50% of p42-d.

The isolation of such R prime plasmids in *E. coli* facilitates their genetic analysis via transposon mutagenesis and allows to study the expression of the *R. phaseoli* symbiotic genes in other hosts.

3. REFERENCES

- N.J. Brewin et al (1980). J. Gen. Microbiol. 120, 413-420.
- R. Simon et al (1983). In A. Puhler ed. Proceedings of the Bielefeld Symposium Springer Verlag, Berlin.
- Van Larebeke et al (1974). Nature 252: 169-170.

CHARACTERIZATION OF A PLEIOMORPHIC MUTANT OF *RHIZOBIUM MELILOTI*

MROZ C./ COURTOIS B./HORNEZ J.P./DERIEUX J.C.

LABORATOIRE DE MICROBIOLOGIE, UNIVERSITE DES SCIENCES ET TECHNIQUES DE LILLE, F - 59655 VILLENEUVE D'ASCQ CEDEX, France.

Cells of rhizobia are regular rod-shaped in culture, but within effective legume nodules, they are larger and very pleiomorphic. These large cells fix nitrogen and are called bacteroids. Cells, which morphologically resemble bacteroids are occasionally found in culture, especially after nutrient enrichments (2) (4) (5) (6).
Following N-Methyl-N'-Nitro-N-Nitrosoguanidine mutagenesis, mutant 1N8-25 of *Rhizobium meliloti* was isolated, which showed an important number of pleiomorphic cells after plating on Iswaran medium. The original strain $M5N_1$ did not give any distorted cells under identical conditions.
First we have checking the mutant origin by alfalfa nodulation and specific phage lysis tests. Two similar plasmid bands were observed for strain M5N1 and its mutant by agarose gel electrophoresis according to an Eckhardt's (1) modified method. Secondly we tried to define the influence of culture conditions. (a) The mutant pleiomorphism (80-100 % of distorted cells) appeared after plating on Iswaran medium, not in liquid Iswaran medium. This medium contained Yeast Extract (2 g/l) and Potassium Gluconate (1,5 g/l). These two nutrient factors were shown to be involved in the pleiomorphism : only the mutant was distorted in presence of gluconate (b) and was more sensitive than $M5N_1$ to the deforming effect of Yeast Extract (c) as previously described by Skinner (5).
Mutant 1N8-25 was a still nodulating *R. meliloti* that would have some parietal deficiences. The latter didn't damage the phages receptor sites. This pleiomorphism property was not linked to an alteration of the plasmid content. The abnormally-shaped cells of the mutant were obtained after plating on Iswaran medium because this one contained Potassium Gluconate and Yeast Extract. Further studies of such mutants should allow to approach the symbiotic state of *Rhizobium*, since cells capable of fixing N_2 in culture are also pleiomorphic (3) (7).

REFERENCES

1. Eckhardt T (1978) Plasmid 1, 584-588.
2. Jordan DC, Coulter WH (1965) Can. J. Microbiol. 11, 709-720.
3. Pankhurst CE, Craig AS (1978) J. Gen. Microbiol. 106, 207-219.
4. Sato K et al. (1982) Agric. Biol. Chem. 46, 501-505.
5. Skinner FA et al. (1977) J. of Appl. Bact. 43, 287-297.
6. Urban JE (1979) Appl. Environ. Microbiol. 38, 1173-1178.
7. Van Brussel AAN et al. (1979) Can. J. Microbiol. 25, 352-361.

TRANSFER OF SYMBIOTIC AND CONJUGATIVE PLASMIDS TO AND FROM FAST-GROWING RHIZOBIUM JAPONICUM.*

KENNETH D. NADLER
MICHIGAN STATE UNIVERSITY, EAST LANSING, MI 48824-1312 USA.

Symbiotic and conjugative plasmids have been transferred to fast-growing strains of Rhizobium japonicum. The self-transmissible plasmid pJB5J1 (Johnston et al 1978) encoding genes for effective pea symbiosis and kanamycin resistance (Tn5) was conjugated from effective R. leguminosarum to R. japonicum PRC205 rif. Kanamycin resistant transconjugants resemble PRC 205 rif in all vegetative propertives; none induced root nodules on pea plants. Two transconjugant classes were distinguished by their symbiotic behavior on 'Peking' soybeans and by their plasmid profiles. Class I does not induce nodules on soybeans; electrophoretograms of lysates reveal the presence of a 134 Mdal plasmid co-migrating with pJB5J1 in addition to the resident plasmids on PRC 205 rif. Class II induces effective root nodules on 'Peking' soybeans and contains plasmid pAL1 (molec. wt. ca. 110 Mdal.) in addition to the resident PRC 205rif plasmids. Presumably pAL1 arose by deletion of a 20 Mdal. segment of pJB5J1 containing genes which suppress soybean nodulation in fast-growing R. japonicum. Both transconjugant classes donate kn^r at high frequency (ca. 10^{-3}) to non-nodulating R. leguminosarum 6015; class I donates a plasmid which co-migrates with pJB5J1 whereas Class II donates a plsmid which apparently co-migrates with pAL1. The resulting 6015(pJB5J1) nodulate peas whereas the resulting 6015(pAL1) do not nodulate peas; neither nodulates 'Peking' soybeans.

The conjugative R plasmid 468.45 was transferred from R. meliloti to PRC205rif, selecting for either tc^r or kn^r. The resulting transconjugants resemble PRC 205rif in vegetative and symbiotic properties; all further donate tc^r, kn^r and ap^r at high frequency to E. coli HB101 and to various species of Rhizobium. Several such donor strains transfer (at 10^{-6}) cys 24^+ to auxotrophic strains of R. meliloti; the resulting prototrophic recombinants then donate cys 24^+ and other markers to R. meliloti strains at similar frequencies (ca. 10^{-6}).

*This work was supported by a grant from the USDA Competitive Research Grant Program (#59-2261-1-1-723-0) and from the Michigan Soybean Committee.

PLASMID INTERACTIONS IN Rhizobium; INCOMPATIBILITY BETWEEN SYMBIOTIC PLASMIDS

MICHAEL O'CONNELL[+]/D. DOWLING[++]/J. NEILAN[++]/R. SIMON[+]/L.K. DUNICAN[++]/
A. PÜHLER[+]

[+] LEHRSTUHL FÜR GENETIK, FAKULTÄT FÜR BIOLOGIE, UNIVERSITÄT BIELEFELD, POSTFACH 8640, D-4800 BIELEFELD 1, FRG

[++] DEPT. OF MICROBIOLOGY, UNIVERSITY COLLEGE GALWAY, IRELAND

Plasmid profiles were determined for twenty four strains of R.trifolii. All the strains harboured at least three plasmids and as many as eight were present in some strains. Southern hybridisation with cloned R.meliloti nif genes was used to identify indigenous Nif plasmids. Each strain carried a single Nif plasmid and these plasmids varied in size from 112 Mdal to at least 300 Mdal.

A 150 Mdal Nif plasmid identified in strain G1008 was labelled with Tn5 mob, a recombinant transposon carrying the origin of transfer of RP4. This facilitated the mobilisation of the Nif plasmid to R.leguminosarum 6015 which then acquired the ability to effectively nodulate clover, indicating the presence of both nod and nif genes on the mobilised plasmid. During subsequent mobilisation of this sym plasmid a 55 Mdal deleted derivative was isolated. This plasmid harbours all the genes necessary for effective nodulation and is termed pUG507.

Tn5 mob has been used to label and facilitate mobilisation of numerous plasmids. The introduction of an asymbiotic plasmid from strain G1067 to strain G1006 produced an alteration in the plasmid profile of the recipient. The sym plasmid of G1006 was no longer observed and a larger plasmid, presumably a cointegrate, was generated. Subsequent mobilisation resulted in breakdown of the cointegrate and the formation of a novel 140 Mdal sym plasmid termed pUG506.

Mobilisation using Tn5 mob has facilitated the investigation of plasmid incompatibility. The sym plasmid of G1008 exhibits incompatibility with the sym plasmid of R.trifolii 1024 and with the novel sym plasmid pUG506. In addition pUG506 was found to be incompatible with the sym plasmid of R.phaseoli 8002. All these plasmids were thus allocated to the same incompatibility group, termed IncRtr1. An asymbiotic plasmid from R.trifolii G1015 could also be assigned to this group. Furthermore, the sym plasmid of R.trifolii G1027 was found to be incompatible with pRL6J1, a sym plasmid from R.leguminosarum, and these were allocated to a second incompatibility group, termed IncRtr2, since pRL6J1 is compatible with plasmids from the IncRtr1 group.

A MODEL SYSTEM TO DETECT RHIZOBIUM PROMOTOR ACTIVITY

R.J.H.Okker, C.W.Wijffelman, and R.A.Schilperoort
Dept. Plant Molecular Biology, University of Leiden,
Wassenaarseweg 64, Leiden, the Netherlands.

The genes of fast growing Rhizobia that code for the nodulation properties (Nod genes) are only expressed in bacterial cells that are involved in symbiosis. Only a small number of cells is involved in the formation of each nodule. A special approach is therefore needed to study the expression of rhizobial Nod genes during the development of symbiosis.This paper describes the development of a method to determine the time of onset of each of the Nod promotors. The method uses bacteriophage Mu d (Ap^R, lac) (Casadaban, Cohen, 1979) as an indicator of promotor activity and a specific histochemical staining to detect the Mu-lac coded β-galactosidase (b-gal.).
Bacteriophage Mu-lac can insert in any DNA sequence.The internal genes of Mu are replaced by the structural genes of the lactose operon of E.coli, but the promotor and operator are absent. The lac genes can be transcribed only from a promotor upstreams of the inserted phage. The phage is recognizable by the Tn3 derived ampicillin resistance.
Before using Mu-lac in Rhizobium, two questions must be settled first. 1) Are plasmids with Mu-lac insertions stable in Rhizobium? Plasmids with insertions of wildtype Mu are highly unstable in Rhizobium. 2)Is the Mu-lac coded b-gal. detectable in the symbiotic situation?
To test the stability, Mu-lac was inserted in the Km marker of the broad host range plasmid R702. Two Tra regions are at each side of the insertion at ca. 8 kb. Any deletion emerging from Mu-lac larger than that distance will cause a loss of transfer properties.The plasmid-pMP1- was crossed to R.leguminosarum LPR1105 and R.trifolii LPR5045. The frequency of transfer was 3-5 x 10^{-1}. Ten isolates of Rhizobium pMP1 were crossed with E.coli recA (Mu A,B). pMP1 was transferred from all isolates with the same frequency to E.coli as the control plasmid R702.(1.2 -3.6 x 10^{-2}).The genetic evidence for intactness of the plasmid in Rhizobium was confirmed by gelelectrophoresis of plasmids isolated from Rhizobium.Within the range of detection (4 kb) pMP1 from Rh. did not differ from pMP1 from E.coli. Rhizobium pMP1 was grown for several generations without selection pressure for pMP1. All isolates had still an intact plasmid.Plasmids with Mu-lac are stable in Rhizobium.

Rhizobium and the roots of their host plants contain a b-gal.This background of b-gal. makes it impossible to detect Mu-lac induced b-gal. in the symbiotic situation. However, the Rhizobium and plant b-gal's could be selectively inactivated by incubation at 56°C during 12 min. Thereafter the roots were incubated 2-4 h. at 37°C in the presence of 5 μg.ml^{-1} X-gal. The Mu-lac induced b-gal. could be detected specifically on places were Rhizobium pMP1 cells were present.

REFERENCE:Casadaban M and Cohen S (79) Proc Natl Ac. Sc 76, 4530-4533

THE BIOLOGICAL CHARACTERIZATION OF SESBANIA ROSTRATA INFECTION BY RHIZOBIUM SPECIES

JANE E. OLSSON/B.G. ROLFE/J. SHINE/J. PLAZINSKI/M. NAYUDU/R. RIDGE/P. DART
GENETICS DEPARTMENT AND CENTRE FOR RECOMBINANT DNA RESEARCH, AUSTRALIAN NATIONAL UNIVERSITY, CANBERRA AUSTRALIA.

ABSTRACT

Certain fast growing strains of Rhizobia, for example strain ORS-571 and strain WE7, are able to induce both stem and root nodules on the tropical legume *Sesbania rostrata*.
Interest in this species centres around its ability to produce spherical nodules which exhibit a very high acetylene reduction ability (Dreyfus and Dommergues 1981). Thus it is of value to study the biological characteristics of the *S. rostrata-Rhizobium* symbiosis as it represents a broadening of the infection and plant association characteristics of *Rhizobium* strains. Both the mode of infection and the location of nodule formation differs from the "normal" root hair entry process displayed by other "classical" *Rhizobium* species.

ASSAY PROCEDURES

Due to a shortage of seeds, it was necessary to develop an alternative method for growing *S. rostrata* material; a "stem assay" which involved growing surface sterile stem cuttings (approx 5-6 cm long) in either tubes or plates containing a nitrogen free nutrient medium at approx 23-26°C. With this procedure it was possible to induce both root and stem nodules in the presence of strains ORS-571 and WE7 rhizobia. It also provided a rapid assay for the screening of bacterial nodulation mutants and produced large quantities of nodule material for microscopic examination.

THE INFECTION PROCESS

Infection threads were observed in both stem and root nodules. The site of invasion in stems appears to be the point between the epidermis and the emerging adventitious root. This contrasts with previous suggestions (Dreyfus and Dommergues 1981) that lenticels are the site of invasion. In root nodulation however, lateral root initials are invaded by rhizobia.

THE INVASION HYPOTHESIS

Delayed innoculation experiments suggest that in the development of both stem and root nodules the rhizobia seem to "take over" the meristematic zones. The dedifferentiation of root cap cells into nodule cortex cells seems to occur such that a nodule results. This observation is consistent with Nutman's Focus Hypothesis (1956) in which rhizobia penetrate the root and produce nodules only at the points of incipient meristematic activity (i.e at the focus of lateral or adventitious root formation).

PLASMID PROFILES AND HYBRIDIZATION STUDIES

No plasmids were observed in strain ORS-571 rhizobia, however in strain WE7, two plasmids (approx 300 and 500 Mdal in size) have been found. Hybridization studies using the nodulation region of *R. trifolii* ANU843 strain indicate that in strains ORS-571 this region is chromosomally located.

(B.L. Dreyfus & Y.R. Dommergues (1981) FEMS Microbiol.Letters 10,313-17)

[Nutman's hypotheses summarised in : The Biology of Nitrogen Fixation (1974) ed. A. Quispel. Nth. Holland Pub. Co. pp445-47.]

SIMILARITIES BETWEEN HIGHLY TRANSMISSABLE PLASMIDS OF RHIZOBIUM AND AGROBACTERIUM

E. PEES, C.A. WIJFFELMAN, A.A.N. VAN BRUSSEL, P.J.J. HOOYKAAS AND W.J.E. PRIEM
DEPARTMENT OF PLANT MOLECULAR BIOLOGY,UNIVERSITY OF LEIDEN,
NONNENSTEEG 3, 2311 VJ LEIDEN, THE NETHERLANDS

Most of the fast-growing Rhizobium strains produce a similar small molecular weight bacteriocin (Hirsch, 1979). R.leguminosarum strain 248, containing the highly self-transmissable plasmid pRL1JI, is non-small bacteriocin producing strain and is sensitive to small bacteriocin. By introducing pRL1JI in a small bacteriocin producing strain the production of small bacteriocin is stopped by a function located on this plasmid that represses small bacteriocin synthesis (Brewin et al., 1980).
We investigated 50 independent field isolates of R.leguminosarum, two R.trifolium strains, and two R.phaseoli strains on their ability to produce small bacteriocin and on the presence of a self-transmissable plasmid. Most of these strains produced small bacteriocin. Four strains did not and only in these four strains the presence of a self-transmissable plasmid could be demonstrated. The strains were R.leguminosarum RBL4 and RBL16, R. trifolii RCR5 and R.phaseoli RBL822. The presence of a self-transmissable plasmid in the strains was confirmed by gelelectrophoresis of transconjugant strains.
So there seems to be a strong correlation between the absence of small bacteriocin production and the presence of a self-transmissable plasmid in fast-growing Rhizobium strains.
All small producing strains are insensitive to small bacteriocin. The non-small producing strains RBL16, RDR5 and RBL822 are also small-insensitive, whereas 248 and RBL4 are sensitive to small bacteriocin. After introduction of the self-transmissable plasmids in the small bacteriocin producing strain 897 (transfer frequences 10^{-3} - 10^{-4}) the synthesis of small bacteriocin stops. So on all transmissable plasmids a gene (Rps) is located which is responsible for the repression of the synthesis (or excretion) of small bacteriocin. Transconjugants containing the plasmids from 248 or RBL4 were in addition sensitive to small bacteriocin. These two plasmids contain also a gene (Sbs) rendering the cells sensitive to small bacteriocin. It was demonstrated by W. Priem (these Proceedings) that Tra, Rps and Sbs are genetically coupled on plasmid pRL1JI.
Recently comparable data were described for Agrobacterium tumefaciens. The nopaline Ti plasmid pTiC58 is transmissable to other Agrobacteria with a very low frequency (10^{-7}). Mutants of pTiC58 were isolated which transfer constitutively (10^{-1}). These Tra^c plasmids cause a constitutive uptake of agrocinopine A and a greater sensitivity to the bacteriocin agrocin 84. Moreover, the Tra^c plasmids prevent the excretion of agrocin 84 (Ellis et al, 1982). There is a striking resemblance between these functions on the Tra^c nopaline Ti plasmids and the functions of Sbs and Rps on the Tra^c plasmids pRL1JI and pAB4 of RBL4. The similarity may even extends further in that the Sbs function may be a transport system for opines associated with the root-nodule symbiosis.

REFERENCES

Brewin NJ, Beringer JE, Johnston AWB (1980) J. Gen. Microbiol. 120, 413-420
Ellis JG, Murphy PJ, Kerr A (1982) Mol. Gen. Genet. 186, 275- 281
Hirsch PJ (1979) J. Gen. Microbiol. 113, 219-228

SELECTION OF CURED STRAINS AND MUTANTS OF THE SYM-PLASMID pRL1JI BY SMALL BACTERIOCIN

W.J.E. PRIEM AND C.A. WIJFFELMAN
DEPARTMENT OF PLANT MOLECULAR BIOLOGY
STATE UNIVERSITY OF LEIDEN, NONNENSTEEG 3,
2311 VJ LEIDEN, THE NETHERLANDS

The Rhizobium leguminosarum plasmid pRL1JI carries the genes for nodulation and nitrogen fixation. Other known functions on this pRL1 plasmid are those for medium bacteriocin production (Mep), transfer (Tra) and repression production small bacteriocin (Rps) (for review, see Beringer, 1980). Nearly all fast growing Rhizobium strains are insensitive to small bacteriocin. Introduction of pRL1 makes these strains sensitive. This sensitivity is due to the presence of a gene located on pRL1, which is called Sbs (Wijffelman et al, in press). By plating fast growing Rhizobium strains, harbouring pRL1 marked with Tn5(Km^r) or Tn 1831(Sp^r), on agar plates supplemented with small bacteriocin, mutants arise which are insensitive to small bacteriocin (frequency $< 10^{-4}$). These Sbs^- mutants fall in two categories: 1. strains cured of pRL1, 2. mutants in the Sbs gene.
The frequency of appearance of cured strains among the Sbs^- mutants is high, and varies per selection between 10 and 70%. In these cured strains all above mentioned plasmid functions are absent. The frequency of cured strains among the Sbs^- mutants is independent of the site of the transposon in pRL1 and also independent of the bacterial background. It is even possible to cure strains harbouring a pRL1 without any transposon insertion. In that case, however, the frequency is much lower.
The second category of mutants is not cured of pRL1 but is mutated in the Sbs gene. Analysis of these mutants for above mentioned genes on pRL1 gives 6 classes of mutants (see Table).

		Mep	Tra	Sbs	Rps	Nod
class	1	+	+	-	+	+
	2	+	+	-	-	+
	3	+	-	-	-	+
	4	+	-	-	+	+
	5	-	-	-	+	+
	6	-	-	-	-	+

Class 2 up to 6 mutants are not only mutated in the Sbs but also in one or more other known functions. These classes are most probably deletions. On base of the occurrence of the several classes, the sequence of the genes will be Mep Tra Sbs Rps.
Class 6 mutants in which 4 plasmids functions are absent, have no detectable deletion in pRL1. For this reason the genes Mep, Tra, Sbs and Rps must be located as a cluster on pRl1.

REFERENCES

Beringer JR, et al, (1980) Heredity 45, 161-186
Wijffelman CA et al. MGG, in press.

RNA POLYMERASE OF RHIZOBIUM JAPONICUM: COMPARISON OF THE ENZYME FROM AEROBIC CELLS, NITROGENASE DEREPRESSED CELLS, AND RIFAMPICIN RESISTANT CELLS

BRIGITTE REGENSBURGER AND HAUKE HENNECKE
MIKROBIOLOGISCHES INSTITUT, EIDGENÖSSISCHE TECHNISCHE HOCHSCHULE, ETH-ZENTRUM, CH-8092 ZÜRICH, SWITZERLAND

1. INTRODUCTION

Previous observations have suggested, that the ability of certain *Rhizobium* strains to develop nitrogenase activity in culture (and analogously in bacteroids) may be dependent upon a structurally distinctive RNA polymerase. (1) The expression of nitrogen fixation was selectively more sensitive to rifampicin inhibition than cellular growth (Werner, 1978; Pankhurst et al., 1981). (2) Slow-growing *Rhizobium* strains able to derepress nitrogenase activity in culture were more resistant to rifampicin than strains unable to express nitrogenase (Pankhurst et al., 1982). (3) Several rifampicin-resistant mutants of both slow- and fast-growing *Rhizobium* strains formed ineffective nodules (Pankhurst, 1977; Pain, 1979). In this work we have used purified RNA polymerase from *R.japonicum* to test, (i) if modifications of RNA polymerase occur under nitrogen fixing conditions, and (ii) to detect strains with mutationally altered RNA polymerases that could have a selective influence on nitrogen fixation.

2. RESULTS

RNA polymerase from *R.japonicum* was purified. The subunit structure was found to be $\beta\beta'\alpha_2\sigma$ (molecular weights: (β/β') 150,000 each, (σ) 96,000, (α) 40,000). The enzyme is Mg^{2+}-dependent, rifampicin sensitive, and has optimal activity at pH 8 to 10 and at 35 to 40 ^oC, which is higher than the optimal growth temperature of *R.japonicum* (28 ^oC).

RNA polymerase subunits from nitrogen fixing (microaerobic) *R.japonicum* cultures migrated in two dimensional O'Farrell gels at the same positions as the subunits of the purified RNA polymerase, which was obtained from cells grown aerobically. This suggests that no chemical modification of RNA polymerase occurs when cells undergo this metabolic differentiation.

Spontaneous rifampicin resistant mutants were selected; part of them possessed resistant RNA polymerase due to modified (more acidic) β-subunits, part of them were resistant presumably due to altered uptake properties. Both types of mutants formed effective nodules and expressed nitrogenase in free living culture.

In conclusion, the results suggest that RNA polymerase itself has not a selective regulatory function for the expression of genes involved in symbiotic nitrogen fixation.

3. REFERENCES

Pain AN (1979) Appl.Bacteriol. 47, 53-64.
Pankhurst CE (1977) Can.J.Microbiol. 23, 1026-1033.
Pankhurst CE, Scott DB, Ronson CW, White DWR (1981) In Gibson AH, Newton WE, eds, Current Perspectives in Nitrogen Fixation, p.447, Austr.Ac.Sc., Canberra.
Pankhurst CE, Scott DB, Ronson CW (1982) FEMS Microbiol.Lett. 15, 137-139.
Werner D (1978) Z.Naturforsch. 33, 859-862.

CONSERVATION OF ROOT HAIR CURLING (HAC) FUNCTIONS IN FAST-GROWING RHIZOBIA

B.G.Rolfe, R.W.Ridge, N.A.Morrison, P.R.Schofield, B.J.Bassam, J.M.Watson, J.Shine and M.A.Djordjevic.

Genetics Department, Research School of Biological Sciences, Australian National University, Canberra, Australia, 2601.

ABSTRACT

Endogenous *Rhizobium* plasmids (called Sym plasmids) are thought to encode genes which affect nodulation and host-specificity. Conjugation experiments with various Sym plasmids from *R.leguminosarum*, *R.trifolii* and *R. meliloti*, have resulted in the transfer of host specific nodulation capacity (1,2). Nod genes located on various Sym plasmids have been isolated using molecular cloning techniques and mapping data suggests that these genes are restricted to a small region of DNA (some 4 to 5kb) (3,4,5).

METHODS AND RESULTS

To test whether Sym plasmid Nod genes are involved in host specificity, complementation tests with heterologous Sym plasmids were conducted. Plasmids encoding early nodulation information from a wide range of bacteria were transferred to several transposon-induced Nod$^-$ mutants to determine whether the heterologous nodulation information could restore the ability of the mutant to nodulate the original plant host. Clover nodulation could be restored to three *R.trifolii* Nod$^-$ (Hac) strains by the transfer of either a *R.leguminosarum* Sym plasmid (pJB5JI), a *R.meliloti* plasmid (pRMSL26), or by a plasmid (pNM4AN) from a fast growing *Rhizobium* strain NGR234 (6) that nodulates various cowpea legumes and the non-legume *Parasponia*. While clover nodulation by *R.trifolii* transconjugants possessing either pJB5JI or pRMSL26 was normal, the frequency of nodule formation was low in *R.trifolii* transconjugant strains possessing pNM4AN. Lucerne nodulation could be restored to two *R.meliloti* Nod (Hac$^-$) mutants by clover (pBRIAN or pRT032) or pea Sym plasmids. However, *R.meliloti* transconjugants possessing plasmid pNM4AN could not nodulate lucerne. Nodulation of the tropical legume siratro could be restored to a Tn5-induced Nod$^-$ derivative of NGR234 by clover and pea Sym plasmids although the frequency of nodule formation was low. In contrast, transfer of pRMSL26 to this Nod$^-$ mutant did not result in restoration of siratro nodulation (over 100 plants tested). Positive hybridization between DNA fragments from the clover nodulation region was detected to the Sym plasmid Nod region of several fast-growing *Rhizobium* strains including *R.leguminosarum*, *R.meliloti* strains, and to the NGR234 strain. These results indicate that at least some early nodulation functions in a broad spectrum of fast-growing *Rhizobium* strains are conserved.

REFERENCES

1) Brewin N.J. *et al.*, 1980. J. Gen Microbiol. 12: 413-420
2) Djordjevic M.A. *et al.*, 1983. J. Bacteriol. in press
3) Long S.R. *et al.*, 1982. Nature 289: 485-488
4) Downie J.H. *et al.*, 1983. EMBO 2: 947-952
5) Schofield P.R. *et al.*, 1983 M.G.G. in press
6) Morrison N.A. *et al.*, 1983. J. Bacteriol 153: 527-531

TRANSFER OF HOST-RANGE PLASMIDS BETWEEN *Rhizobium leguminosarum* AND FAST GROWING RHIZOBIA THAT NODULATE SOYBEANS.

J.E. RUIZ-SAINZ*/ J.E. BERINGER+/ R. JIMENEZ-DIAZ*
* Department of Microbiology, Faculty of Biology, University of Seville, Avda. Reina Mercedes s/n, SEVILLE, Spain.
+ Soil Microbiology Department, Rothamsted Experimental Station, HARPENDEN, Hertfordshire, AL5 2JQ, England.

Fast growing strains of *Rhizobium* that nodulate soybeans have been isolated in China (Keyser *et al.*, 1982). They formed effective nitrogen-fixing associations with wild soybeans (*Glycine soja*) and other unbred soybean cultivars as *Glycine max* cv. Peking and Malayan, but were largely ineffective as nitrogen-fixing symbionts with american soybean cultivars. We report here on preliminary experiments designed to study the genetic control of host range determination in these bacteria.

The *R. leguminosarum* host-range plasmid pJB5JI has been transferred by membrane crosses from *R. leguminosarum* to the fast growing strains USDA 192 and USDA 194 (kindly provided by H. Keyser, USDA, Beltsville) at frequencies of 10^{-5} and 10^{-2} respectively. Selection for inheritance of pJB5JI was for kanamycin resistance. Clones of USDA 192 and USDA 194 carrying pJB5JI were tested for their ability to nodulate pea plants as described by Beringer (1974). Six weeks after inoculation, a few 'bumps' were occasionally obtained, but no nodules were formed. All transconjugants of USDA 192 and USDA 194 carrying pJB5JI were effective on soybeans.

Agarose gel electrophoresis (as described by Rosenberg *et al.*, 1982) showed that USDA 192 harbours two plasmids. The clones of USDA 192 carrying pJB5JI present an extra band corresponding in MW to pJB5JI. One of the tested clones, carrying pJB5JI (USDA 192-5), had lost its smallest plasmid, probably as a consequence of the adquisition of pJB5JI. USDA 194 has two plasmid bands with a MW for the smallest one, very similar to that of pJB5JI (130 Mdal) and consequently clones of USDA 194 carrying pJB5JI showed only two bands.

Transconjugants of both strains carrying pJB5JI were able to transfer pJB5JI back to a non-nodulating *R. leguminosarum* strain which regained the ability to nodulate and fix nitrogen in peas.

Beringer JE (1974) J. Gen. Microbiol. 84, 188-198.
Keyser HH *et al* (1982) Science 215, 1631-1632.
Rosenberg C *et al* (1982) J. Bacteriol. 150, 402-406

LOCALIZATION OF NIF- AND HUP-SPECIFIC SEQUENCES ON THE PLASMIDS OF NEWLY ISOLATED STRAINS OF *RHIZOBIUM LEGUMINOSARUM*

BRIGITTE L. SEIFERT/H.V. TICHY/LOUISE M. NELSON[+]/M.A. CANTRELL[++]/R.A. HAUGLAND[++]/ W. LOTZ
INSTITUT FÜR MIKROBIOLOGIE UND BIOCHEMIE, UNIV. ERLANGEN-NÜRNBERG, D-8520 ERLANGEN, FRG
+ PLANT BIOTECHNOLOGY INSTITUTE, NATIONAL RESEARCH COUNCIL CANADA, SASKATOON S7N OW9, SASK., CANADA.
++LABORATORY FOR NITROGEN FIXATION, OREGON STATE UNIVERSITY, CORVALLIS, OREGON 97331, USA

Thirtyfive strains of *R.leguminosarum* were isolated from pea plants grown at two different locations. Location "B" was a garden bed in which pea plants had been grown annually for at least 4 years. Location "W" was a meadow in which pea plants had not been grown for a number of years. The plasmid content of the isolated strains has been analyzed previously (Tichy, Lotz 1981).

The plasmid pRmR2 (Ruvkun, Ausubel 1980) carrying the structural *nif* genes of *R.meliloti* was used as a probe in Southern hybridization. For this purpose the plasmids were separated on gels using the method of Eckardt (1978). Although the strains have up to six plasmids of different size, only one plasmid per strain was found to hybridize with the *R.meliloti* part of pRmR2. The size of the *nif* plasmids varied from 170 to 350 Mdal. The *nif* sequences were never found on the largest plasmids ($>$ 500 Mdal; Tichy, Lotz 1981) seen in these strains.

Four of the isolated strains have been tested for the expression of uptake hydrogenase (*Hup*) activity. Strains B5 and B10 were found to be *Hup*$^+$, whereas strains W5 and W12 were *Hup*$^-$. Using cloned *hup* DNA from *Rhizobium japonicum* (Cantrell et al. 1983) as a probe, we found specific hybridization with the *nif* plasmid of the *Hup*$^+$ *R.leguminosarum* strains B5 and B10. No hybridization could be detected with the DNA of the *Hup*$^-$ strains *R.leguminosarum* 300, PRE, W5 and W12. Our 9 tested strains isolated from location "W", including strains W5 and W12, did not hybridize with the *hup* probe. In contrast, 15 of the 17 strains tested from location "B" were found to hybridize with *R.japonicum hup* DNA. In each of the 15 strains it was the respective *nif* plasmid which also hybridized with the *hup* probe.

References

Cantrell MA, Haugland RA, Evans HJ (1983) Proc. Natl. Acad. Sci. 80, 181-185.
Eckardt T (1978) Plasmid 1, 584-588.
Ruvkun GB, Ausubel FM (1980), Proc. Natl. Acad. Sci. 77, 191-195.
Tichy HV, Lotz W (1981) FEMS Microbiol. Letters 10, 203-207.

RNA POLYMERASE AND SYMBIOTIC PROPERTIES OF SPONTANEOUS RIFAMPICIN RESISTANT MUTANTS OF RHIZOBIUM LEGUMINOSARUM STRAINS PRE AND 300

W. SELBITSCHKA AND W. LOTZ
INSTITUT FÜR MIKROBIOLOGIE UND BIOCHEMIE
UNIVERSITÄT ERLANGEN-NÜRNBERG
D-8520 ERLANGEN, FEDERAL REPUBLIC OF GERMANY

Mutations leading to rifampicin resistance may affect the symbiotic properties of Rhizobium (Pain, 1979; Pankhurst, 1977). It has been discussed that the rhizobial RNA polymerase may have a specific regulatory function in the development of nitrogen fixing bacteroids (Pain, 1979; Beringer et al., 1980). We isolated 100 spontaneous Rif^r mutants each of R.leguminosarum strains 300 and PRE and examined them for nodulation (Nod) and symbiotic nitrogen fixation (Fix) with garden pea plants. The mutants were grouped into the following categories:

I. Nod^+Fix^+: formation of N_2-fixing nodules 3 weeks after inoculation. Generation time of 170 min; this corresponds to the generation time of the respective wild type, (189 of the 200 mutants).

II. $Nod^+Fix^{(+)}$: formation of N_2-fixing nodules 4 weeks after inoculation. Generation time of 200 min, (10 of the 200 mutants).

III. Nod^+Fix^-: formation of ineffective nodules. Generation time of 240 min, (1 of the 200 mutants); this mutant was isolated from R.leguminosarum strain PRE.

The RNA polymerase of a strain from each of the three mutant groups has been isolated using methods described earlier (Lotz et al., 1981) and was shown by in vitro transcription assays to be rifampicin resistant.
We postulate that the mapped rif locus of R.leguminosarum strain 300 (Beringer et al., 1978) is located in the structural gene (rpo B) for the ß subunit of the rhizobial RNA polymerase. In strain 300 rif and str genes are closely linked (Beringer et al., 1978). Therefore, $rif^s str^r$ markers were co-transduced into two mutant strains of group II by the generally transducing phage RL 38 (Buchanan-Wollaston, 1979). Seven independent $rif^s str^r$ transductants of each recipient were tested for nodulation and symbiotic nitrogen fixation. The acetylene reduction of the transductants corresponded to that of the wild type (3 weeks after inoculation).
In the case of our $Nod^+Fix^{(+)}$ (group II) mutants it is possible that effectivity is delayed due to a "slow down" of a number of physiological functions, which may also result in the observed prolonged generation time of the vegetative cells. In the case of the one Nod^+Fix^- (group III) mutant we cannot exclude the possibility that the rif^r mutation may interfere with a symbiosis-specific regulatory function of the rhizobial RNA polymerase.

References:
Beringer JE, Brewin NJ and Johnston AWB (1980) Heredity 45, 161-186.
Beringer JE, Hoggan SA and Johnston AWB (1978) J. Gen. Microbiol. 104, 201-207.
Buchanan-Wollaston V (1979) J. Gen. Microbiol. 112, 135-142.
Lotz W, Fees H, Wohlleben W and Burkhardt HJ (1981) J. Gen. Microbiol. 125, 301-309.
Pain AN (1979) J. appl. Bacteriol. 47, 53-64.
Pankhurst CE (1977) Can. J. Microbiol. 23, 1026-1033.

STUDY OF SYM MUTANTS OF *RHIZOBIUM MELILOTI*

N. TORO/J. OLIVARES
DEPARTAMENTO DE MICROBIOLOGIA, ESTACION EXPERIMENTAL DEL ZAIDIN, CSIC, GRANADA, SPAIN

1. INTRODUCTION

In *R. meliloti*, a very large plasmid of MW>300 Mdal, called pSym, carries some genes with the symbiotic characteristics Nod and Fix. In addition to pSym, practically all strains have other large plasmids, pRme, with unknown implications in symbiosis.
We have analyzed a large collection of Sym mutants of *R. meliloti* obtained by heat treatment.

2. MATERIAL AND METHODS

R. meliloti strains GR4 and L5-30 were used to obtain Sym mutants. In overnight curing experiments, cultures were diluted to 10^7-10^8 cells mL^{-1} and incubated at different temperatures. Single colonies were tested with *Medicago sativa*. Matings were performed as usual. R68.45 pMO60 or pMO61 were used as vectors. Plasmids were identified according to Rosenberg et al. (1981). Plasmid DNA was isolated by the procedure of Ish-Horowicz, Burke (1981). DNA was labeled according to Rigby et al. (1977) and hybridization probes were done as described by Forrai et al. (1982).

3. RESULTS AND DISCUSSION

We have found that *R. meliloti* GR4 harbours four indigenous plasmids, two pRme (pRmeGR4a of a MW of 100 and pRmeGR4b of 115) and two larger ones of a MW higher than 400 (pRmeGR4c and pRmeGR4d). In wild type cells, only one of these two larger plasmids is detectable, but in some of Sym mutants, two bands are clearly differentiated. Using pID1 and pSA30 labeled DNA, *nif* genes have been detected in one of these megaplasmids. In addition, a possible cointegrate of pRmeGR4a and b can be observed. Sym mutants have been obtained by heat treatment, after finding the optimal temperatures of 38.5° and 37° for GR4 and L5-30, respectively. Although some deletions or even losses of plasmids a and b have been obtained, we have not been able to relate this fact to any phenotypic characteristic.
In mating experiments, pMO61 appeared as the most efficient vector for the transfer of Sym characters.

4. REFERENCES

Forrai T, Vincze E, Bânfalvi Z, Kiss GB, Rhandhawe GS and Kondorosi A (1982) J. Bacteriol. 153, 635-643.
Ish-Horowicz D and Burke JF (1981) Nucleic Acids Res. 9, 2980-2998.
Rigby PWJ, Dieckmann M, Rhodes C and Berg P (1977) J. Mol. Biol. 113, 237-255.
Rosenberg C, Casse-Delbart F, David M, Dusha I and Boucher C (1982) J. Bacteriol. 150, 402-406.

FUNCTIONAL ANALYSIS OF Tn 5 - INSERTION NODULATION MUTANTS OF pRL1JI, A RHIZOBIUM LEGUMINOSARUM SYM-PLASMID

A.A.N. VAN BRUSSEL /E. PEES /C.A. WIJFFELMAN /T. TAK
DEPARTMENT OF PLANT MOLECULAR BIOLOGY, UNIVERSITY OF LEIDEN,
NONNENSTEEG 3, 2311 VJ LEIDEN, THE NETHERLANDS

The presence of Sym-plasmids in Rhizobium facilitates the isolation of mutants affected in genes essential for root nodule morphogenesis, but unimportant for normal growth of Rhizobium in culture. These mutants will provide us with a better insight into the factors involved in and the mechanism of root nodule morphogenesis.

Tn 5 insertion mutants in the Sym-plasmid pRL1JI were isolated by transposon mutagenesis with pJB4JI (Beringer, et al., 1978) and transferred to the Sym-plasmid cured strain LPR 5039 (Hooykaas et al., 1981). 1100 pRL1JI transposon mutants in LPR 5039 were investigated for their nodulation and nitrogen fixation properties on Vicia sativa nigra (Van Brussel, 1982). We found 27 mutants affected in nodulation (Nod^-), whereas 43 mutants still nodulated, but did not reduce acetylene (Fix^-). The remaining mutants (1030) formed effective root nodules.

Two categories of Nod^- mutants were found viz. 15 mutants with a normal plasmid size of 135 Mdal and 12 mutants with a deletion of about 40 Mdal (MW: 100 Mdal). The pRL1JI character of the deletion mutants was concluded from the presence of pRL1JI markers: medium bacteriocin production (Mep) and immunity (Mei), small bacteriocin sensitivity (Sbs) and repression of small bacteriocin excretion (Rps).

Three classes were differentiated in the 10 investigated non-deletion Nod^- mutants. Class 1 mutants ($Hac^+Tsr^+Nod^-$), which still induce specific root hair curling (Hac) and thick short roots (Tsr) on Vicia sativa nigra, was found 4 times. One mutant, class 2 ($Hac^+Tsr^-Nod^-$), only induces Hac and not Tsr; while 5 mutants, class 3 ($Hac^-Tsr^-Nod^-$) induced neither Hac, nor Tsr. The 12 deletion mutants have a class 3 phenotype.

The following arguments are in favour of a physiological and genetical coupling of Tsr and Nod: 1) The Tsr marker has been found in all nodulating strains of R.leguminosarum and R.trifolii and not in Sym-plasmid cured strains. 2) Introduction of Sym-plasmids in cured strains and Agrobacterium made these strains Tsr inducing. 3) All Tsr^- mutants did not nodulate. 4) 50% of the Nod^- mutants were Tsr^-. 5) Tsr is coupled genetically to Nod by co-transduction.

Tsr was visible on some Vicia sativa strains and represents a teratological reaction to a substance with a still unknown function in root nodule morphogenesis.

On the transduction map of pRL1JI (Pees, unpublished) the Nod-area is located in between two areas needed for nitrogen fixation: the Fix I and the Fix II areas. Tsr and Hac map both in this Nod area and assuming that they are in the same operon, the probable gene order is Hac $\rightarrow$ Tsr.

References

Beringer JE, Beynon JL, Buchanan-Wollaston AV and Johnston AWB (1978) Nature 276, 633-634.

Brussel AAN van, Tak T, Wetselaar A, Pees E and Wijffelman CA (1982) Plant Science Letters 27, 317-325.

Hooykaas PJJ, van Brussel AAN, den Dulk-Ras H, van Slogteren GMS and Schilperoort RA (1981) Nature 291, 351-353.

TRANSFER OF THE SYM-PLASMID pRL1JI TO AGROBACTERIUM BY THE FORMATION OF CO-INTEGRATES

C.A. WIJFFELMAN /C.H.M. MULDERS /A.A.N. VAN BRUSSEL AND E. PEES
DEPARTMENT OF PLANT MOLECULAR BIOLOGY, UNIVERSITY OF LEIDEN,
NONNENSTEEG 3, 2311 VJ LEIDEN, THE NETHERLANDS

The sym-plasmid pRL1 can be transferred to *Agrobacterium tumefaciens*. Exconjugants of *A.tumefaciens* harbouring pRL1::Tn5 produce medium bacteriocin and are able to nodulate on *Vicia sativa* (Van Brussel *et al.*, 1982). Transfer of the self-transmissable plasmid pRL1 to *A.tumefaciens* was much lower than to Rhizobium. Transfer of pRL1 only occurred when 1062 pRL1 was the donor strain and transfer frequencies were very low e.g. 10^{-6} (table 1, lines 1,2). Using *A.tumefaciens* pRL1::Tn5 as a donor the plasmid could be retransferred to 1062 and 5039 with equal frequencies (10^{-4}). When now these two strains, harbouring pRL1::Tn5-A (the A stands for the transfer via *A.tumefaciens*) were used as donors pRL1::Tn5-A was transferred to *A.tumefaciens* in both cases with a rather high frequency (10^{-4}). The resulting transconjugants *A.tumefaciens* pRL1::Tn5-A produced medium bacteriocin and nodulated *V.sativa*. The extra plasmid in these exconjugants was larger than the original pRL1 viz. 180 Mdal in stead of 135 Mdal. This large plasmid appeared to be a cointegrate between pRL1::Tn5 and pRL8, a 100 Mdal plasmid of 1062 (Johnston *et al.*, 1982), because it was incompatible with pRL8. So pRL1 can only be transferred to *A.tumefaciens* as a cointegrate. This explains the low transfer frequencies when 1062 pRL1 is the donorstrain and the normal transfer frequencies when the cointegrate has been established.

table 1 Transfer frequency of pRL1::Tn5 to *A.tumefaciens*C58

donor	acceptor	transfer frequency
1062 pRL1	A.tumefaciens 202	10^{-6}
5039 pRL1	A.tumefaciens 202	10^{-8}
1062 pRL1-A	A.tumefaciens 202	10^{-4}
5039 pRL1-A	A.tumefaciens 202	10^{-4}
5039 pRL1	A.tumefaciens 202 pRL8	10^{-5}

Only in one other case we were able to demonstrate the formation of a cointegrate between pRL1::Tn5 and another plasmid. This plasmid R 180 is an derivative of RP4, and has a MW of 35 Mdal. The resulting cointegrate (160 Mdal) is transferred to *A.tumefaciens* with a very high frequency (10^{-1}).
The presence of pRL1 in *A.tumefaciens* only as a cointegrate can be due to the fact that transfer of the plasmid to *A.tumefaciens* only occurs in a cointegrate form or that pRL1 can only be maintained in *A.tumefaciens* as a cointegrate. To discriminate between these alternatives 5039 pRL1::Tn5 was used as a donor in crosses to *A.tumefaciens* and *A.tumefaciens* pRL8. Only in the latter case (table 1, line 5) transfer of pRL1 was observed indicating that transfer of pRL1 to *A.tumefaciens* in a non-cointegrated form is possible, but that for the maintenance in *A.tumefaciens* a cointegrated form is necessary.

Van Brussel, AAN, Tak, T, Pees, E, Wetselaar, A, Wijffelman, CA (1982) Plant Sci. Lett. 27, 317-325.
Johnston AWB, Hombrecher, G, Brewin, NJ and Cooper, MC (1982) J. of Gen. Microbiol. 128, 85-93.

EXPRESSION OF NITROGENASE POLYPEPTIDES IN YEAST

A. Zilberstein, J. Berman, D. Salomon, D. Holland, A. Hochman*, R. Bitoun and A. Zamir
Dept. of Biochemistry, The Weizmann Institute of Science, Rehovot 76100, Israel,
*Dept. of Biochemistry, Tel-Aviv University, Tel-Aviv 69978, Israel

To investigate the expression of nitrogen fixation functions in eukaryotes we transferred nif genes coding for nitrogenase components from Klebsiella pneumoniae into the yeast Saccharomyces cerevisiae. NifH,D and K (encoding respectively the subunit of dinitrogenase reductase and the α and β subunits of dinitrogenase) were not expressed in yeast when introduced as part of the intact nif cluster. Therefore, individual genes were joined to yeast promoter sequences and introduced into yeast in autonomously replicating plasmids. To facilitate the screening of nif gene expression in yeast, the genes were first fused in-phase to a truncated lacZ sequence and subsequently reconstituted.
NifH expression: a translational nifH'-'lacZ fusion was constructed in pMC 1403. The hybrid gene was fused to the yeast ura3 promoter in plasmid pRB45-48 (Rose et al. (1981) Proc. Natl. Acad. Sci. 78, 246). To allow proper alignment, the two components were first resected with Bal 31 nuclease. Plasmid pJB3, isolated from transformed yeast expressing β galactosidase activity, contained a tribride gene consisting of the promoter and first 15 codons of ura3 followed by the 5th codon of nifH, fused in turn to lacZ. Subsequently, the lacZ portion was substituted with the remainder of nifH to give pJB21. Yeast transformed with pJB3 or pJB21 expressed low but significant levels of polypeptides crossreacting with antibodies against Kp2 of approximately 140 and 35 kd, respectively.
NifD expression: an in-phase nifD'-'lacZ fusion was constructed in pMC 1403. The hybrid gene was introduced between the promoter and terminator sequences of the yeast adhl gene in plasmid pAAR6 (Ammerer and Hall). Transformed yeast exhibited significant β galactosidase activity. Reconstituted nifD specified in yeast an anti-Kpl crossreacting polypeptide of 55-60 kd.

In the course of these studies we have also (1) characterized mutations allowing nifA-independent expression from the nifHDK promoter in E. coli and in K. pneumoniae (Bitoun et al., Proc. Natl. Acad. Sci. in press); (2) shown that in K. pneumoniae nifD and nifK are separated by an intercistronic region of approximately 400 bp and (3) provided evidence suggesting that the modified dinitrogenase reductase specified by ura3'-'nifH may retain some biological activity.

Acknowledgements: We thank R.R. Eady for anti Kpl and Kp2, D. Botstein and G. Ammerer for plasmids, H. Dixon, W.J. Brill for bacterial strains and G.R. Fink for yeast strains.

POSTER DISCUSSION 9C NIF GENETICS OF FREE-LIVING DIAZOTROPHS

C. ELMERICH, Institut Pasteur, Unité de Physiologie Cellulaire, 75015 Paris, France

Most of what is known about the genetics of nitrogen fixation was discovered in Klebsiella pneumoniae which may be regarded as a model system. Thus the information obtained in these studies provides the basis for a molecular approach of the development of genetics in other diazotrophs. In the poster session 9C which was denoted to this topic, a third of the contributions dealt with this model system and the others with free-living diazotrophs i.e. Azospirillum, Azotobacter, Enterobacter, blue green-algae and photosynthesis bacteria. Development of genetics of these micro-organisms is an important objective since a large number of them can be associated with, or were isolated from, the roots of non-leguminous plants.
In this report I wish to give a short overview of the information presented or discussed and the reader is invited to look for detail in the poster abstracts and papers of this volume, and in the recent literature.

1. News on the organization of the nif clusters of K.pneumoniae.
A revised map of the niftranscriptional units was presented (Beynon et al, Cannon et al) which showed that nif F and possibly nif J were transcribed in the opposite direction to the other transcriptional units. Nif X is likely to belong to the nif EN transcript and nif M was found to be transcribed from two promoters (Puehler, this volume). These were the nif V promoter and a promoter preceeding nif M In addition, the sequence of the nif promoters was established, showing two regions of conserved sequence which differ from the 'consensus promoter' of the enteric bacteria.

2. Physical organization of the nif genes in the free-living diazotrophs.
Using homology with K.pneumoniae nif probes th. physical organization, as well as the localization of the corresponding nif HDK genes was examined in various diazotrophs. In most cases there was not indication that the nif genes were located on plasmids. However, an exception to this rule is strain AVY5 of A.vinelandii which carries a large plasmid, containing genes homologous to nif HDK and to genes involved in benzoate degradation (Yano et al, Robson this volume). Another example of plasmid location was reported for a strain of Enterobacter agglomerans isolated from the wheat rhizosphere (Singh and Klingmüller). Differences in the organisation of the nif HDK cluster have been encountered:) one copy of the nif HDK cluster, as is usually observed in Enterbacteraceae, (Burns and Reeve, Uozumi et al) even when the genes are plasmid borne (Yano et al). ii) the existence of an extra nif H copy as shown in some cyanobacteria (Haselkorn this volume) and Azotobacter (Robson this volume); iii) the existence of several copies of the nif HDK cluster only one being functional, as demonstrated for Rhodopseudomonas capsulata (Scolnik et al). In addition, in this organism a large number of mutants were isolated (Klipp and Puehler, Wall and Love, Willison and Vignais). The nif H, D, and K cisterons which are organized in a single transcriptional unit, were identified by genetic complementation (Scolnik et al). iv) the existence of remnant genes was detected by hybridization in several non-nitrogen-fixing Klebsiella strains (Burns and Reeve).

3. Regulation: is nif A ubiquitous?
New data concerning the model of regulation of nif gene repression in K.pneumoniae were presented (Zhu Jia-bi and Shen San-chiun, Alvarez-Morales et al, Buchanan-Wallaston and Cannon, De Bruijn and Ausubel). The role of the nif gene and of the nif LA genes is reviewed in detail (Dixon this volume). It is

notable that cloning of the nif A gene of K.pneumoniae, as well as the identification of its product was reported (De Bruijn and Ausubel). In Azotobacter and Azospirillum the data presented (Drummond and Kennedy, Zhu Jai-bi and Shen San-chiun) suggest the involvement of a nif A-like function, since complementation of some nif mutants with the nif A gene of K.pneumoniae was reported. Moreover, constitutive repression of the nif A production resulted in nitrogenase activity presence of NH_4^+ ions. It is likely that in these bacteria the system of regulation of the nif genes is similar to K.pneumoniae and that nif genes are also involved. However, no nif A-type of activation could be demonstrated in Rhodopseudomonas capsulata nor in cyanobacteria. In the case of cyanobacteria homology with the E.coli nif gene was detected in a strain of Anabaena (Machray et al) but studies developed with the Anabaena strain 7120 lead to a model of regulation independent of ntr, nif A mediated control (Tumer et al, Haselkorn this volume).

4. Topics to be developed.
The structure of the nif cluster of K.pneumoniae as well as the regulation of nif gene expression is now well understood. However, very little is known on the functions of the nif genes in biochemical terms. This is due to the lack of an assay for most of the nif products which makes purification of an active product problematical. In this respect several posters (Chance and Orme-Johnson, McLean et al, Hansen et al) dealt with a methodology to study nif functions in using nif clones that higher produce nif polypeptides. This approach looks very promising not only to study nif gene functions in the model system but also to provide a strategy to identify nif functions in other diazotrophs. Intergenetic complementation was tested in various diazotrophs when plasmid pRD1 was constructed. It looks as if more sophisticated tools can be constructed to find out if specific genes can be expressed in the background of other species The first one of this methodology was the transfer of the K.pneumoniae nif A gene into Azotobacter (see Robson this volume) and now the use of nif-lac fusions appears to be an efficient alternative approach (Drummond and Kennedy). In addition, this methodology could allow the construction of hyper-performing strain in respect of nitrogen fixation activity, constitutivity, oxygen tolerance. Another exciting topic that should be understood is the reiteration of some nif DNA sequences in nitrogen fixation species and also the existence of remnant genes in some non-nitrogen fixing, but related bacteria. This raises the question of the role of these sequences, which may represent silent copies of nif DNA but which also could be expressed in some physiological conditions that have yet to be elucidated.
There is also a need to develop specific tools for genetic analysis, applicable to non-enteric diazotrophs. (Puehler et al, De Vries and Ludwig).

Finally an important field, and in my view, the most important, is to develop studies of the genetic and molecular basis of associative symbiosis with plants. Since a large number of the free-living diazotrophs should be now considered as rhizospheric-diazotrophs.

REGULATION OF THE P_1 AND P_2 PROMOTERS OF THE *KLEBSIELLA PNEUMONIAE* *glnA ntrBC* REGULON

A. ALVAREZ-MORALES, R. DIXON and M. MERRICK
ARC UNIT OF NITROGEN FIXATION, UNIVERSITY OF SUSSEX, BRIGHTON BN1 9RQ, UK.

As an extension of our studies on nitrogen-regulated promoters, such as *nifLA* in *K. pneumoniae*, we have examined regulation of two other nitrogen-regulated promoters, P_1 and P_2, in the P_1 *glnA* P_2 *ntrBC* regulon. We constructed two translational *ntrB*::*lacZ* fusions carrying either $P_1 + P_2$ (pAM123) or P_2 alone (pAM125). With pAM123 glutamine synthetase activity is a measure of P_1 expression and β-galactosidase activity is a measure of $P_1 + P_2$ expression; whilst with pAM125 β-galactosidase is a measure of P_2 expression alone (see Fig. 1).

Fig. 1.

P_1 glnA P_2 *ntrB*::*lacZ* pAM123

P_2 *ntrB*::*lacZ* pAM125

P_1 and P_2 expression were analysed in conditions of limiting or excess nitrogen and with *ntrB*, *ntrC*, *ntrB+C* or *nifA* in *trans* (see Fig. 2).

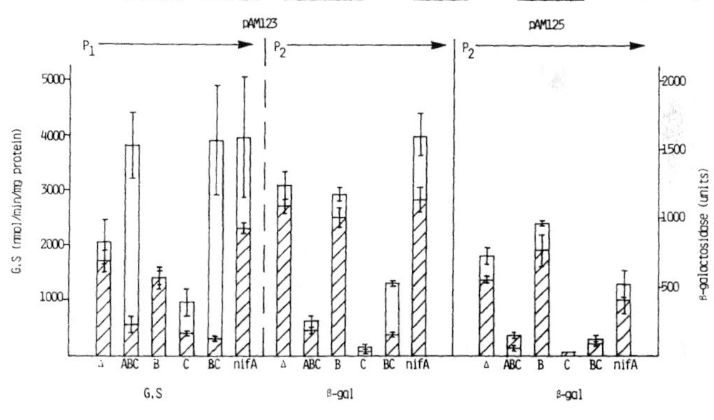

These data show that i) the P_1 promoter is both positively and negatively regulated. In the absence of *ntrB*, the *ntrC* product represses P_1 transcription but with *ntrB* and *ntrC* together positive control of P_1 is observed. The *ntrB* product alone has no regulatory effect and we believe the role of *ntrB* product is to 'modulate' *ntrC* activity; ii) the P_2 promoter is negatively controlled by *ntrC* product and shows no positive regulation; iii) *ntrBC* expression is from P_1 under N-deficiency and from P_2 under N-excess; iv) whilst the *nifA* product can substitute the positive regulatory functions of *ntrC* product at certain promoters including *nifLA* it cannot substitute as a negative regulator at P_1 or P_2.

Comparison of the modes of regulation observed at P_1 and P_2 with that observed for *nifLA* suggests a general model for expression of nitrogen-regulated promoters which is discussed in the paper of Dixon et al. (this volume).

THE NIF PROMOTERS OF KLEBSIELLA PNEUMONIAE HAVE A CHARACTERISTIC PRIMARY STRUCTURE

JIM BEYNON/MAURA CANNON/VICKI BUCHANAN-WOLLASTON AND FRANK CANNON
BIOTECHNICA INTERNATIONAL, INC., 85 BOLTON STREET, CAMBRIDGE, MASSACHUSETTS 02140, U.S.A.

We present a revised transcription map of the *nif* gene cluster of *Klebsiella pneumoniae* and compare the sequences of seven *nif* promoters. The following revisions based on nuclease S1 mapping studies are made to the transcription map of the *nif* gene cluster: (a) there is a promoter preceeding *nifU* but not *nifX*; (b) *nifF* is transcribed in the opposite direction to that previously proposed; and (c) *nifM* is transcribed from two promoters.
Comparison of the *nif* promoters has revealed a characteristic 26 base pair structure situated between bases -1 and -26 relative to transcription initiation This structure contains two conserved regions of homology, one of which our results suggest confers activator specificity on the *nif* promoters, and the other being equivalent to a "Pribnow box" for promoters expressed under nitrogen starved conditions.

THE CLONING AND CHARACTERIZATION OF THE GLNF(NTRA) GENE OF KLEBSIELLA PNEUMONIAE: ROLE OF GLNF(NTRA) IN THE REGULATION OF THE NITROGEN FIXATION (NIF) AND OTHER NITROGEN ASSIMILATION GENES.

FRANS J. DE BRUIJN* AND FREDERICK M. AUSUBEL

DEPARTMENT OF CELLULAR AND DEVELOPMENTAL BIOLOGY, HARVARD UNIVERSITY AND DEPARTMENT OF MOLECULAR BIOLOGY, MASS. GENERAL HOSPITAL, BOSTON, USA
* PRESENT ADDRESS: MAX PLANCK INSTITUT, ABT. SCHELL, 5000 KOLN 30 FRG

Introduction.

The K. pneumoniae nif gene cluster is regulated on at least two different levels, in response to fluctuating levels of ammonia in the cell. The first level is part of the general nitrogen regulation (ntr) system, by which enteric bacteria regulate a number of nitrogen assimilation genes such as glnA (glutamine synthetase), nif, and the hut, put and aut genes (Magasanik, 1982). The ntr system includes the glnF(ntrA), glnL(ntrB) and glnG(ntrC) genes. As in E. coli, the K. pneumoniae glnF(ntrA)+glnG(ntrC) gene products act as co-activators of ntr controlled genes, while the glnL(ntrB)+ glnG(ntrC) gene products act as negative co-regulators (Leonardo, Goldberg 1980; de Bruijn, Ausubel 1981; Espin et al. 1981,1982; Ow, Ausubel 1983; Merrick 1983).

In order to continue our study of ntr mediated nif regulation, we have cloned the K. pneumoniae glnF(ntrA) gene, characterized it and identified its protein product.

Results and Discussion.

We cloned the glnF(ntrA) gene of K. pneumoniae by complementation of the GlnA$^-$Ntr$^-$ phenotype of a glnF$^-$(ntrA$^-$) strain of E. coli. A plasmid was isolated, carrying a 2.0 kb ClaI insert, capable of complementing all previously identified phenotypic defects associated with glnF$^-$(ntrA$^-$) strains of E. coli and K. pneumoniae (GlnA$^-$, Hut$^-$, Put$^-$, Aut and Nif$^-$). Plasmid pFB71 was mutagenized with transposon Tn5 to delimit the glnF(ntrA) structural gene and the glnF(ntrA) gene product was identified in maxicells as an 80,000 dalton polypeptide.

Plasmid pFB71 also restores activation of the K. pneumoniae and Rhizobium meliloti nifHDK promoters in glnF$^-$(ntrA$^-$) strains, as determined by measuring ß-galactosidase production by plasmids carrying fusions of these promoters to the lacZ gene (Sundaresan et al. 1983).

In conclusion, we report here the cloning and characterization of the K. pneumoniae glnF(ntrA) gene, the identification of its protein product and the confirmation of the hypothesis that the glnF(ntrA) gene encodes a trans-acting co-activator of ntr controlled genes, including the nif genes of K. pneumoniae and the nifHDK operon of R. meliloti. The latter finding suggests the presence of a ntr-like control system in this strictly symbiotic diazotroph. The results presented here are discussed in more detain in de Bruijn et al. (this volume).

de Bruijn FJ and Ausubel FM (1981) Mol. Gen. Genet. 183, 289-297.
Espin G, Alvarez-Morales A and Merrick M (1981) Mol. Gen. Genet.184,213.
Espin G, Alvarez-Morales A, Cannon F, Dixon R and Merrick M (1982) Mol. Gen. Genet. 186, 518-524.
Leonardo JM and Goldberg RB (1980) J. Bacteriol. 142, 99-110.
Magasanik B (1982) Ann. Rev. Genet. 16, 135-168.
Merrick M (1983) EMBO Journal 2, 39-44.
Ow DW and Ausubel FM (1983) Nature 301, 307-313.
Sundaresan V, Ow DW and Ausubel FM (1983) Proc. Natl. Acad. Sci.80,4030.

REGULATION OF *nif* TRANSCRIPTION IN *KLEBSIELLA PNEUMONIAE*

VICKI BUCHANAN-WOLLASTON AND FRANK CANNON
BIOTECHNICA INTERNATIONAL, INC., 85 BOLTON STREET, CAMBRIDGE, MASSACHUSETTS 02140, U.S.A.

Transcription of the *nif* operons in *K. pneumoniae* is activated by the *nif*A protein and repressed by the *nif*L protein. The *nif*L protein is present during *nif* derepression and only becomes a repressor in the presence of fixed nitrogen or oxygen. We have shown that the *nif*L protein can repress *nif*A activated transcription of promoters normally activated by *ntr*C such as the *nif*LA promoter and the promoters of operons for histidine and proline utilization, while it has no repressive effect on these promoters when they are activated by the *ntr*C gene product. This suggests that the *nif*L protein represses transcription by inactivating the *nif*A protein.

The close similarity between the *nif*A and *ntr*C proteins may suggest a common ancestral gene but the physiological significance of the activation by *nif*A protein of promoters normally activated by *ntr*C is not clear. We have shown that, in the presence of all the other *nif* promoters, *nif*A activation of the *nif*LA promoter is not detectable. This suggests that the *nif*A protein has a low affinity for the *nif*L promoter and autoactivation of this promoter is probably negligible in a wild type *ntr nif* strain.

HOMOLOGY STUDIES OF THE NIF REGION IN DIFFERENT WILD STRAINS OF KLEBSIELLA SPECIES

A.T.H. BURNS/E.C.R. REEVE
DEPARTMENT OF GENETICS, UNIVERSITY OF EDINBURGH,
EDINBURGH EH9 3JN, SCOTLAND.

We have examined nif DNA homology among a number of wild Klebsiella strains, using as probes the plasmids pSA30 (which contains the nitrogenase structural genes from Klebsiella M5a1) and pCRA37 (which covers most of the rest of the nif region)(Cannon et al., 1977; Cannon et al., 1979). Bacterial DNA was digested with EcoRI restriction enzyme, and standard methods were used to obtain the restriction patterns of the DNA fragments which hybridised with each probe.

Four major categories of Nif^+ Klebsiella strains could be clearly distinguished:

1) Strains which gave very similar restriction patterns to M5a1: i.e. they were identical for the pSA30 probe and shared only minor variations for the non-structural nif genes (pCRA37 probe). Thus three strains gave bands of 5 and 3.2 kb in place of the 8.3 kb fragment for M5a1, and in two other strains this 8.3 kb band appeared to be replaced, respectively, by a 10.4 kb fragment and by fragments of 7.6 and 3 kb.
2) Strains showing minor restriction site differences for pSA30 and much greater variation from the M5a1 pattern for pCRA37. Greater divergence seems to have occurred within the non-structural nif genes than within the structural genes in these strains. This would imply less stringent selection on the non-structural genes, consistent with the findings of Ruvkun & Ausubel (1980) for intergenic nif DNA homology.
3) Strains sharing a paucity of EcoRI sites within their nif genes, since both probes detected major bands of 18-20 kb.
4) Strains showing a much increased number of EcoRI sites, leading to complex restriction site patterns which differ strikingly from that of M5a1.

There is clearly a considerable degree of restriction site variation among Klebsiella strains, within the nif region, despite the strong selection pressure which must be assumed for its conservation.

In addition, tests on a number of naturally Nif^- Klebsiellas detected several which gave consistent homology to both the pSA30 and pCRA37 probes, but no similarity to the restriction site patterns of any of the Nif^+ strains tested. It may be assumed that this homology reflects the presence of remnants of an active nif system in these currently Nif^- strains.

References

Cannon, F.C., Riedel, G.E. and Ausubel, F.M., (1977), Proc. Natl. Acad. Sci. U.S.A. 74, 2963-2967.
Cannon, F.C., Riedel, G.E., and Ausubel, F.M. (1979), Mol. Gen. Genet. 174, 59-66.
Ruvkun, G.B. and Ausubel, F.M., (1980), Proc. Natl. Acad. Sci. U.S.A. 77, 191-195.

KLEBSIELLA PNEUMONIAE NIFF AND NIFJ

MAURA CANNON*/JANET DEISTUNG/SUSAN HILL AND FRANK CANNON*
ARC UNIT OF NITROGEN FIXATION, UNIVERSITY OF SUSSEX, BRIGHTON BN1 9RQ, U.K.
*PRESENT ADDRESS: BIOTECHNICA INTERNATIONAL, INC., 85 BOLTON STREET, CAMBRIDGE, MASSACHUSETTS 02140, U.S.A.

Previous studies have shown that *nifF* and *nifJ* are involved in electron transport to nitrogenase in *Klebsiella pneumoniae*. Genetic and physical experiments determined the locations of *nifF* and *nifJ* in the *nif* gene cluster. However, the direction of transcription and operon organization had not been determined. In this study we show (1) that *nifF* and *nifJ* are transcribed in the opposite direction to that of the other *nif* operons; (2) that *nifF* and *nifJ* are the only monocistronic operons in the *nif* gene cluster; and (3) the transcription start and termination of *nifF* in the DNA sequence and confirmatory evidence that this transcript is that of *nifF*.

CLONING OF THE *E. coli* lamB GENE WITH CONSTITUTIVE EXPRESSION.

GERT E. DE VRIES AND ROBERT A. LUDWIG
UNIVERSITY OF CALIFORNIA SANTA CRUZ USA

E. coli HB101 was mutagenized by λ::Tn5. After segregation, λ-resistant mutants were selected and screened for a non-reverting Mal^- phenotype, yielding strain ECG1. ECG1 could be complemented to Mal^+, λ^s by plasmid pH7 carrying wild type $malT^+$. However in genomic hybridizations using cloned malT as a probe, Tn5 in ECG1 was not inserted in malT, nor were deletions observed in this region. P1-transduction and F-mediated conjugation suggest instead that the Tn5 induced mutation is tightly linked to the malB region. ECG1 displays a leaky phenotype for lamB: *in vivo* packaged selectable cosmids infect ECG1 at a frequency 10^{-5} that of wild-type.

10^{10} ECG1 cells were infected with λ::Tn5132 (Tc^r). Only one out of eight derivatives obtained displayed a λ^s phenotype and segregated Tc^r after maintainance on Km only (ECG10). Electrophoresis revealed a lamB protein in the ECG10 membrane fraction indistinguishable from that of wild-type. However, maltose or glucose did not regulate lamB levels in ECG10.

λ^r-Derivatives were selected from ECG10. Infection of a λ-packaged cosmid could not be detected in one such derivative (ECG17) This strain was used as host to receive cloned DNA of ECG10, using BglII to restrict genomic DNA and pLAFR-BAM as a vector. Tc^r-ECG17 transformants carrying putative lamB regions were screened by Ap^r after infection with *in vivo* packaged compatible cosmids. The smallest plasmid so obtained (pTROY9) containing the malK-lamB region, showed that IS3 was inserted in malK in ECG10. Sequencing of a HindIII-SalI fragment proximal to lamB, confirmed the presence of IS3 in orientation II inserted at the distal end of malK. IS3 contains a functional outward promoter. When pTROY9 was cojugated to *Klebsiella pneumoniae*, λ-infection gave rise to plaque formation indistinguishable from *E. coli*. In *Salmonella typhimurium*/pTROY9, λ did not proliferate but cosmids could infect and λ::Tn phages could transposon-mutagenise.

USE OF CLONED NIF REGULATORY ELEMENTS FROM K.PNEUMONIAE TO EXAMINE NIF REGULATION IN AZOTOBACTER AND THE MODE OF ACTION OF THE NIFL GENE PRODUCT

MARTIN DRUMMOND AND CHRISTINA KENNEDY
ARC UNIT OF NITROGEN FIXATION, UNIVERSITY OF SUSSEX, BN1 9RQ, UK

1. INTRODUCTION

We have constructed wide host range plasmids *in vitro* carrying the *K.pneumoniae* (Kp) regulatory genes *nifA* and *nifL* or *nif* promoters fused to *lacA*, the gene for β-galactosidase. We reported recently that constitutively-produced *nifA* activates *nif* gene expression in two species of *Azotobacter* (1).

2. METHODS

A.vinelandii is naturally Lac$^-$ and can be assayed for *E.coli lacZ* expression by conventional assays of β-galactosidase. Plasmids derived from RSF1010 (2) were: pMD23 (Kp *nifL::lacZ* fragment (3) inserted at the EcoRI site of RSF1010); pMD21 (Kp *nifF::lacZ* (3) inserted at EcoRI); pMD61 (Kp *nifH:: lacZ* (4) inserted at PstI); pMD132 (Kp *nifL* inserted at EcoRI). Plasmid pCK3 is a cointegrate of pMC71a (4) and pRK290 (5).

3. RESULTS AND DISCUSSION

The Kp *nifL* gene on pMD132 had no effect on nitrogen fixation in wild-type did, however, prevent correction of the Nif$^-$ phenotype of regulatory mutant UW1 by Kp *nifA*. This suggests that *nifL* product inactivates *nifA* product rather than binding to DNA regulatory sequences. Also, the *A.vinelandii* activator probably differs from *nifA* in its interaction with *nifL* product.
The Kp *nifH* promoter on pMD61 was not expressed in *A.vinelandii* and only weakly after introduction of constitutively-produced *nifA* on pCK3. These results indicate that the transcriptional apparatus of *A.vinelandii* differs from that of *K.pneumoniae* and that *nif*-specific activators recognize somewhat different activator sites. This is borne out by the failure of multiple copies of Kp *nifH* promoter to prevent *nif* expression in *A.vinelandii* by activator titration as it does in *K.pneumoniae* (6,7).
In contrast, the Kp *nifL* and *nifF* promoters on pMD23 and pMD21 were strongly expressed in both wild type *A.vinelandii* and the mutant UW1; expression was not repressed by NH_4^+. Recognition by *Azotobacter* of the Kp *nifL* and *nifF* promoters on pMD 23 and pMD21 were strongly expressed in both wild type *A.vinelandii* and the mutant UW1; expression was not repressed by NH_4^+. Recognition by *Azotobacter* of the Kp *nifL* and *nifF* promoters but not of the *nifH* promoter may indicate the presence of an activator with properties resembling those of *ntrC* in *K.pneumoniae* (3,8,9).

REFERENCES

1. Kennedy CK and Robson RL (1983) Nature 301 626
2. Barth PT In Plasmids of Medical, Environmental and Commercial Importance (eds Timmis and Pühler) Elsevier 1979 399.
3. Drummond MH Clements J Merrick M and Dixon RAD (1983) Nature 301 302
4. Buchanan-Wollaston V Cannon MC Beynon JL and Cannon FC (1981) Nature 294 776.
5. Ditta G Stanfield S Corbin D and Helinski DR (1980) Proc.Natl.Acad.Sci. USA 77 7347.
6. Buchanon-Wollaston V Cannon MC and Cannon FC (1981) Mol.Gen.Genet. 184, 102
7. Riedel GR Brown SE and Ausubel FM (1983) J.Bact. 153 45
8. Merrick M (1983) EMBO J. 2 39.
9. Ow D and Ausubel F (1983) Nature 301 307.

OVERPRODUCTION OF NIF KD AND NIF SUX FROM KLEBSIELLA PNEUMONIAE

F.B. HANSEN, P.A. MCLEAN AND W.H. ORME-JOHNSON
DEPARTMENT OF CHEMISTRY, MASSACHUSETTS INSTITUTE OF TECHNOLOGY
CAMBRIDGE, MA 02139 USA

The molybdenum-iron protein and the iron protein from *Klebsiella* pneumoniae catalyze the reduction of dinitrogen to ammonia. The MoFe-protein, believed to harbour the site(s) for dinitrogen binding, contains 2 bound molybdenum, 30-32 irons and 28 acid-labile sulfurs. These metals are arranged in 3 types of clusters: 2 M-centers each consisting of 8 irons, 8 acid-labile sulfurs and 1 molybdenum atom. This FeMo-cofactor can be extracted into N-methyl formamide; 4[4Fe-4S] clusters called P-clusters; and 2 residual irons termed the S-cluster(s). A total of 16 genes have been implicated in regulation and synthesis of both introgenase proteins. The genes coding for the structural polypeptides of iron- and molybdenum-iron proteins are contained in one operon (*nif* HKDY). The *nif* H promoter is not only regulated by the activator *nif* A and the repressor *nif* L but also requires a fully matured MoFe-protein and/or Fe-protein for maximal expression. The products of 4-5 genes (*nif* B, N. E, V and C) are involved in insertion of the FeMo-cofactor into the MoFe-protein. Whether the remaining metal clusters, which are believed to be inserted prior to the FeMo-cofactor, are matured by an enzymatic process remains to be determined. There are several genes, notably *nif* SUX, whose functions are yet obscure; some of these may participate in P-cluster and S-cluster insertion.

We have initiated attempts to determine whether or not P-clusters and S-clusters are inserted enzymatically and whether *nif* SUX play a part in this. To get production of the structural polypeptides of the MoFe-protein independent of the complex regulation of the *nif* H promoter, plasmids were constructed in which the strong *tac* promoter was inserted into the *nif* H gene. This results in very high expression of the *nif* KD polypeptides. We have furthermore constructed compatible plasmids that express high levels of *nif* SUX. We present details on the construction of these plasmids and their protein-labelling patterns.

ISOLATION AND CHARACTERIZATION OF nif MUTANTS IN Rhodopseudomonas capsulata

WERNER KLIPP/A. PÜHLER
LEHRSTUHL FÜR GENETIK, FAKULTÄT FÜR BIOLOGIE, UNIVERSITÄT BIELEFELD,
POSTFACH 8640, D-4800 BIELEFELD 1, FRG

Transposon Tn5 was introduced into *Rhodopseudomonas capsulata* strain B10 by using *in vitro* constructed plasmids that are mobilizable by broad host range plasmid RP4 but do not replicate outside *E.coli*. Mutants deficient in nitrogen fixation were identified by retesting Tn5 containing *Rh.capsulata* strains for their ability to grow on N_2 gas as sole source of nitrogen.

A Km^S derivative of R68.45 was used to construct R' plasmids carrying part of the *Rh.capsulata* chromosome including the nif::Tn5 mutant gene. Cloning of these Tn5 containing DNA fragments in *E.coli* vector plasmids allow restriction analysis of the mutated gene regions.

One of these hybrid plasmids was used as a probe to identify from a *Rh.capsulata* cosmid gene bank clones containing nif genes. The restriction map of such a cosmid was compared to the isolated nif::Tn5 DNA fragments of different nif mutants. It could be shown that out of 24 so far analysed nif::Tn5 mutants, 13 map within a region of 20 kb.

For the identification of further nif genes appropriate DNA fragments were cloned into mobilizable vector plasmids, mutagenized by Tn5 in *E.coli* and subsequently the Tn5 mutated gene was introduced into the *Rh.capsulata* chromosome by homologous recombination. By this site specific mutagenesis it could be shown that also regions between already mapped nif::Tn5 mutants encode nif genes. This indicates that at least part of the genes necessary for nitrogen fixation are clustered in the chromosome.

The integration of plasmids with cloned DNA fragments carrying only a central part of a transcription unit into the chromosome by single cross-over recombination, results in the inactivation of this gene region. By this recombination mutagenesis transcription units that should contain more than one gene were identified.

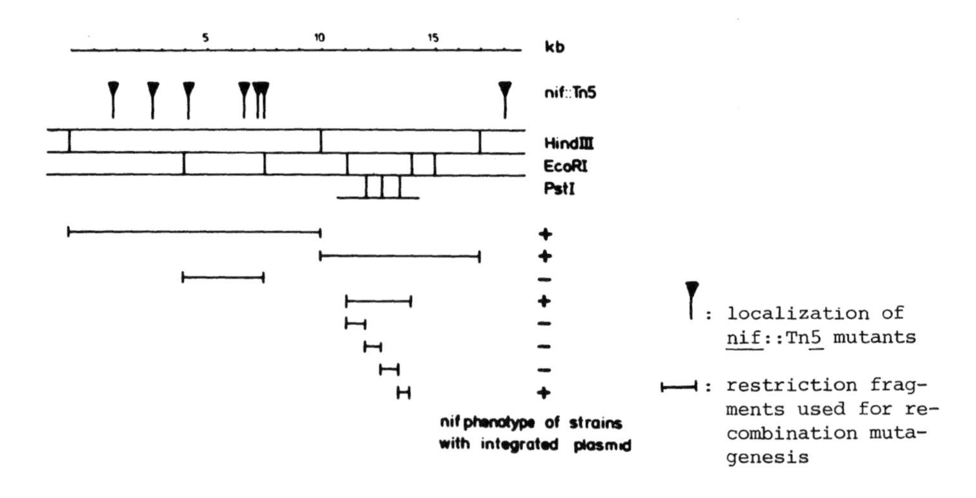

CONSTRUCTION OF PLASMIDS THAT OVERPRODUCE nif REGULATORY PROTEINS

PAUL V. LEMLEY, FINN B. HANSEN, MARK CHANCE, PAUL McLEAN, W.H. ORME-JOHNSON
Department of Chemistry, Massachusetts Institute of Technology,
77 Massachusetts Avenue, Cambridge, Massachusetts 02139 U.S.A.

The functions and interactions of the nif regulatory proteins, nif A and nif L are not well understood. Our objective is to purify, study and further characterize these proteins. The expression of these proteins in vivo, however, is controlled by the glutamine synthetase regulatory (gln) system. We sought a way to produce nif A and nif L in a gln-free background since the regulatory properties of both types of proteins (e.g. gln G and nif A) have been shown to be similar. This could present problems later in assays and purification.

Using the tac promoter, we have constructed plasmids which overproduce nif A and nif L proteins as well as nif A without nif L and nif L without nif A. By using the tac promoter the expression of these regulatory proteins is controlled by the lac repressor and can be induced by the addition of isopropyl-β-D-thio-galactoside (IPTG). This, then, allows expression of nif A and nif L independent of the glutamine synthetase regulatory system.

We present the construction and protein labelling properties of the following plasmids:

pVL1 was used to construct plasmids which overproduce nif A and nif L (pVL2) and nif L and a truncated nif A polypeptide (pVL3). When radioactively pulse-labelled, pVL1 produces a truncated nif L^1 protein (from the start codon to the SAL I site in nif L) under + IPTG conditions.

pVL2, when radioactively pulse-labelled, overproduces nif A and nif L under + IPTG conditions.

pVL3, when radioactively pulse-labelled, overproduces nif L and a truncated nif A (from its start to SST I) polypeptide under + IPTG conditions.

Finally, pVL4 was constructed from pVL2 by removing the center of nif L (~700 base pairs). When radioactive pulse labelling is done, the nif A protein is overproduced under + IPTG conditions.

Following the purification of nif A and nif L, further experimentation will be done on the binding of nif A and nif L to DNA and/or to each other as well as experiments with varying conditions of temperature, NH_4^+, O_2, etc.

NITROGEN-REGULATORY GENES IN NITROGEN-FIXING CYANOBACTERIA

G.C. MACHRAY, C.E. HAGAN, M. BOXER AND W.D.P. STEWART
ARC RESEARCH GROUP ON CYANOBACTERIA AND DEPARTMENT OF BIOLOGICAL SCIENCES, UNIVERSITY OF DUNDEE, DUNDEE DD1 4HN, U.K.

The genetic control of nitrogen-fixation and assimilation in photosynthetic nitrogen-fixing cyanobacteria is of particular interest. Nitrogen-regulatory gene systems controlling these processes have been found in nitrogen-fixing bacteria - in *K. pneumoniae* two regulatory genes, *ntr* B and *ntr* C, lie adjacent to the structural gene (*gln* A) for glutamine synthetase (Espin et al. 1982), a key enzyme involved in ammonia assimilation. These regulatory genes, in concert with a third unlinked regulatory gene *ntr* A, constitute a control gene system by which *gln* A gene transcription is activated or repressed in response to levels of fixed nitrogen (Merrick 1983). This system also exerts control over other operons involved in nitrogen assimilation and, importantly, it regulates transcription of the *nif* gene cluster and hence nitrogen fixation. We report results of our attempts to determine if an equivalent gene system exists in nitrogen-fixing cyanobacteria.

Southern blot hybridisation analyses using heterologous probes have been used to identify and clone structural genes such as *gln* A (Fisher et al. 1981) and *nif* K, D and H (Rice et al. 1982) from cyanobacteria. We have used the plasmid pMM12 (kindly supplied by Dr M. Merrick) which contains nitrogen-regulatory gene sequences of *K. pneumoniae* to probe Southern blots of restriction enzyme digests of DNA from the nitrogen-fixing cyanobacterium *Anabaena* 27893. pMM12 contains a small portion of *ntr* B, the entire *ntr* C gene and DNA down-stream from *ntr* C. It hybridises to a single 8 kb large fragment of *Anabaena* 27893 DNA digested with Hind III. Hind III/Bgl II and Hind III/EcoR1 double digests of *Anabaena* 27893 DNA yield fragments of 7.5 and 4.8 kb respectively which hybridise to pMM12. In addition we have used pMM12 as a probe to screen a library of EcoR1 fragments of *Anabaena* 27893 DNA in the phage vector Charon 4A. One positive clone, λCh4A-GM124, which contains approximately 16 kb of *Anabaena* 27893 DNA was obtained.

To exclude the possibility that non-*ntr* sequences of *Klebsiella* DNA cloned in pMM12 were responsible for the above result we subcloned from pMM12 an EcoR1-Pvul fragment which contains only *ntr* gene sequences into pBR325. The resultant plasmid, pGC121, was nick-translated and hybridised to a Southern blot of λCh4A-GM124 DNA. Positive hybridisation to the *Anabaena* DNA insert of this recombinant phage indicated DNA sequence homology with the *ntr* genes of *Klebsiella*. Thus this clone contains putative cyanobacterial *ntr* gene sequences. DNA sequence analysis and complementation studies using *ntr* gene mutants of *E. coli* and *K. pneumoniae* will be used to compare the properties of the *Anabaena* DNA cloned in λCh4A-GM124 with those of the *ntr* B and C genes of *K. pneumoniae*.

REFERENCES

Espin G, Alvarez-Morales A, Cannon F, Dixon R and Merrick M (1982) Mol. Gen. Genet. 186, 518-524.
Fisher R, Tuli R and Haselkorn R (1981) Proc. Natl. Acad. Sci. USA 78, 3393-339
Merrick MJ (1983) EMBO J. 2, 39-44.
Rice D, Mazur BJ and Haselkorn R (1982) J. Biol. Chem. 257, 13157-13163.
Southern EM (1975) J. Mol. Biol. 98, 503.

CONSTRUCTION OF RECOMBINANT PLASMIDS WHICH OVERPRODUCE THE NIF B, Q, N, E, & X GENE PRODUCTS

PAUL A. McLEAN/FINN B. HANSEN/W. H. ORME-JOHNSON
Department of Chemistry, Massachusetts Institute of Technology, Cambridge, MA 02139 USA.

1. INTRODUCTION

The aim of this work is to determine the structure and synthetic pathway of the iron-molybdenum cofactor (FeMo-co) of nitrogenase. In order to study the functions of the genes implicated in synthesis and/or insertion of FeMo-co (*nif* B, *nif* E, *nif* N, nif N, nif V and possibly *nif* Q and *nif* X)[1,2,3], we are currently constructing recombinant plasmids which overproduce the proteins coded for by these genes and, subsequently, will purify these gene products. Overproduction is necessary because there are no known assays for these gene products, and so their purification will have to be monitored by polyacrylamide gel electro phoresis, therefore requiring easy visibility of protein bands.

2. METHODS

The method of cloning uses the procedure developed by Guarente *et al*[4] for the high level expression of genes for which there is no assay in *E. coli*. There are basically five steps to this procedure:

1) The gene to be overproduced is fused *in vitro* to the β-galactosidase (*lac* Z) gene. The *lac* Z gene lacks its 8 N-terminal amino acids including its initiator methionine codon and ribosome binding (Shine-Dalgarno) sequence. Beta-galactosidase activity is only obtained when the N-terminal of another protein is fused in phase with the β-galactosidase sequence.
2) The phase at the junction of the *nif* and *lac* Z is corrected (if necessary).
3) The plasmid is cut at a restriction site upstream of the N-terminal of the fused *nif* protein. The ends of DNA generated by the restriction enzyme are digested by exonuclease Bal 31 generating a population of deletions, some of which end close to the N-terminal start of the *nif*-*lac* gene fusion.
4) A strong promoter, in our case the *tac* promoter[5], is positioned at the deletion end points, in some cases generating clones which produce high levels of beta-galactosidase.
5) The *lac* Z portion is removed and the original *nif* gene sequence restored, thus giving a plasmid which will overproduce *nif* gene products.

3. RESULTS

We have made in phase fusions to *nif* B and to *nif* E and have obtained beta-galactosidase activity of approx. 1500 units after Bal 31 digestion in one particular experiment. (Wild-type β-galactosidase at 1% of cell protein has 1000 units of activity.) However, when the *nif* BQ and *nif* ENX operons were subsequently regenerated, no detectable levels of the *nif* proteins were seen. We are currently repeating the Bal 31 digestion to position the promoter more precisely to obtain greater expression.

REFERENCES

1) Roberts *et al.* (1978) *J. Bacteriol.* *136*, 267-279
2) Hawkes & Smith (1983) *Biochem. J.* *209*, 43-50
3) Hawkes, McLean & Smith (submitted to Biochem. J.)
4) Guarente *et al.* (1980) *Cell* *20*, 543-553
5) DeBoer *et al.* (1983) *Proc. Natl. Acad. Sci.(USA)* *80*, 21-25

GENOME ORGANISATION IN AZOTOBACTER VINELANDII: CONSTRUCTION OF A LIBRARY OF GENES

MEETHA MEDHORA, SUHAS H. PHADNIS and H.K.DAS
SCHOOL OF ENVIRONMENTAL SCIENCES
JAWAHARLAL NEHRU UNIVERSITY
NEW DELHI-110067 (INDIA)

Genetic manipulations with *Azotobacter vinelandii* have been relatively unsuccessful. This may be due to an unusual genome organisation in laboratory cultured cells of this microbe. Sadoff et al. (1979) have suggested that these cells may contain 40 copies of chromosomes.

We have made a library of genes of *Azotobacter* in the cloning vehicle pHC79. The recombinant cosmids constructed were introduced by transduction into *E. coli leuB*$^-$ cells. On screening the library for Leu$^+$ transductants, we obtained a higher frequency of leucine prototrophe, than expected from a cloned genome the size of *Azotobacter*. The results shows that the *leuB* locus must be present in about 40 copies in cultured *Azotobacter* cells as has been suggested by Sadoff et al., (1979). Cosmids isolated from independently obtained Leu$^+$ transductants shared significant sequence homology. Multicopies of *Azotobacter* DNA could therefore be structurally similar.

We also screened the library by *in situ* colony hybridization for recombinant cosmids harbouring *Azotobacter* DNA sequences that can hybridize with the *Klebsiella pneumoniae nif KDH* fragment. Four putative *nif* sequence containing cosmids have been isolated. Restriction analysis and Southern hybridization of these cosmids, with labelled *Klebsiella nif* DNA have been carried out.

REFERENCES

Sadoff HL, Shimei B and Ellis S (1979) J.Bacteriol. 138, 871-877

ENHANCED LEVELS OF NIF EXPRESSION IN KLEBSIELLA PNEUMONIAE WITHOUT ACCUMULATION OF PPGPP.

M.B. NAIR AND R.R. EADY
UNIT OF NITROGEN FIXATION, UNIVERSITY OF SUSSEX BRIGHTON BN1 9RQ UK.

ppGpp has been postulated to play an essential role in the regulation of *nif* expression in *K.pneumoniae*, since relaxed mutants unable to synthesize ppGpp do not derepress nitrogenase when starved of N (Riesenberg *et al*. 1982). However, in this organism *nif* expression is not always repressed in the presence of fixed N (Shanmugam and Morandi 1976). We have investigated the correlation between ppGpp levels and the extent of *nif* expression following NH_4^+ shift-down in the presence and absence of glutamine.

Derepression of suspensions of different population densities in the absence of glutamine, showed an optimum for maximum expression of nitrogenase activity. The presence of glutamine in the derepression medium resulted in a marked change in the pattern of derepression.

Table 1

Cell density Eel 650 nm	protein $\mu g\ ml^{-1}$	Specific activity of nitrogenase nmol C_2H_4 min^{-1} mg protein^{-1} - glutamine	+ glutamine
17	158	0.15	7.9
27	270	0.95	0.02
35	379	2.7	0.03
45	538	1.8	0.02
62	905	1.1	0.04

Glutamine was 600 $\mu g\ ml^{-1}$; specific activity is after 5 hr derepression.

At the lowest cell density tested, glutamine enhanced nitrogenase activity 52-fold, but produced almost complete repression at the optimum cell density for unsupplemented suspensions.

In order to determine whether the lack of activity was due to repression of *nif* expression, we investigated the effect of glutamine on the rates of transcription from the *nif* L and *nif* H promotors using *nif::lac* fusions (Dixon *et al* 1980) in merodiploid (Nif^+) strains. In contrast to nitrogenase activity, in the absence of glutamine the level of expression from both promotors was independent of the cell density.

The presence of glutamine in the derepression medium repressed transcription of *nif* H at the highest cell density, but a graded release from repression occurred as the cell density was decreased. At the lowest density expression was stimulated 7-fold compared with suspensions without glutamine. Similar data were obtained with *nif* L except that repression was not complete.

ppGpp levels were measured in suspensions of these *nif lac* fusions in the presence and absence of glutamine, under conditions where *nif* expression was enhanced or repressed. The pattern was independent of cell density. In the absence of glutamine, ppGpp rapidly accumulated on NH_4^+-shift down to a maximum of 1100 pmol mg protein^{-1} and then declined to a plateau level. In contrast, in the presence of glutamine, ppGpp remained at the basal level observed in NH_4^+ repressed cultures, even under conditions where *nif* expression and activity was stimulated 7-8 fold. These data make ppGpp an unlikely candidate for regulating *nif* expression following NH_4-shift down in *K.pneumoniae*.

References

1. Dixon, R. *et al*. (1980) Nature 286: 128-132.
2. Riesenberg, D. *et al*. (1982) Mol.Genet. 185: 198-203.
3. Shanmugan, K.T. and Morandi, C. Biochim.Biophys.Acta (1976) 437: 322-332.

HOMOLOGY BETWEEN *KLEBSIELLA PNEUMONIAE NIF* GENES AND *AZOSPIRILLUM* DNA.

S.K. NAIR / C. ELMERICH
INSTITUT PASTEUR - P.O. BOX 75724 PARIS (CEDEX 15) FRANCE.

1. INTRODUCTION

In order to clone *Azospirillum nif* DNA and to initiate genetic analysis of nitrogen fixation in this bacterium, we performed hybridization experiments between *Azospirillum* total DNA and a series of *K. pneumoniae nif* probes covering the entire *nif* cluster.

2. STRAINS, PLASMIDS AND METHODOLOGY

Total DNA was purified from strains *A. brasilense* 7000 (ATCC29145) and *A. lipoferum* Br17 (ATCC29709) according to Quiviger et al (1982). The *nif* plasmids used as probes were : pPC936 : *nifQBALF'*, *Hind*III-*Xho*I fragment ; pMC71A : *nifA* (Buchanan-Wollaston et al. 1981) ; pPC937 : *nifFMVSUX*, *Xho*I-*Eco*RI fragment (this work) ; pGR113 : *nifNE* and pSA30 : *nifYKDH* (from Riedel, Cannon and Ausubel) pPC880 : *nifJ*, *Eco*RI-*Bgl*II fragment (this work). Total DNA was restricted *Bam*H1 (B) ; *Bgl*II (Bg) ; *Hind*III (H) and *Eco*RI (R). Hybridization were performed on Southern blots under non stringent conditions.

3. HOMOLOGY WITH *K. PNEUMONIAE NIF* PROBES

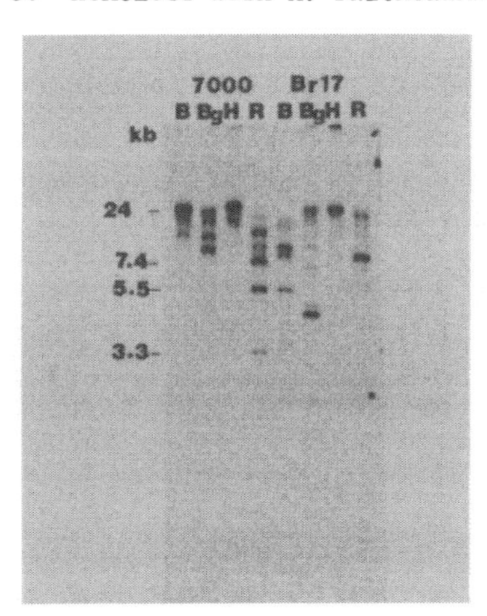

Homology with pSA30 was previously reported as well as the cloning of a *nifHDK* cluster from strain 7000 (Quiviger et al. 1982) Using the *Sal*I fragment that carries the *nifA* gene purified from pMC71A hybridization was detected with strains 7000 and Br17 DNAs (Figure 1). It is likely that some of the bands visible in Figure 1 are due to partial digestions. The 7.4 kb *Eco*RI fragment from strain 7000 was cloned in a bacteriophage vector, and hybridization with the *K. pneumoniae nifA* probe was confirmed. A faint homology was detected between pPC880 that carries *nifJ* and strain Br17 DNA, but no hybridization was found with strain 7000. No homology was observed with pPC937 and pGR113.

4. CONCLUSION

It is still too early to conclude that *Azospirillum* spp contain genes whose functions are analogous to *K. pneumoniae nifA* and *nifJ*, even though our results favor this hypothesis. In addition further experiments are necessary to determine whether the *nif* genes are clustered or not.

5. REFERENCES

Buchanan Wollaston V, Cannon MC and Cannon FC (1981) Nature 294, 776-778
Quiviger B, Franche C, Lutfalla G, Rice D, Haselkorn R and Elmerich C (1982) Biochimie 64, 495-502.

MULTIPLE COPIES OF GENES CODING FOR NITROGENASE IN THE PHOTOTROPHIC BACTERIUM RHODOPSEUDOMONAS CAPSULATA

PABLO A. SCOLNIK/PETER AVTGES/ROBERT HASELKORN
DEPT. OF BIOPHYSICS AND THEORETICAL BIOLOGY, UNIVERSITY OF CHICAGO, CHICAGO, IL. 60637 USA

R. capsulata is a phototrophic bacterium capable of fixing nitrogen anaerobically or microaerobically. Heterologous probes from Klebsiella pneumoniae and Anabaena were used to identify the genes for the structural components of nitrogenase (nifHDK) in R. capsulata. Multiple copies of all three genes were found. We cloned, from cosmid libraries, two HindIII fragments (11.8 kbp and 4.8 kbp) containing sequences homologous to the three structural genes. The 11.8 kbp fragment (copy I) was subcloned into the mobilizable vector pRK292, and the resulting plasmid was conjugated into Nif⁻ point mutants. Complementation was observed with four mutants: J56, J57, J602 and PA1. This fragment was then mutagenized with Tn5, and the resulting plasmids were used in complementation and marker rescue experiments. The results indicate that in R. capsulata the nifHDK genes have the same organization and polarity as in K. pneumoniae. We then designed a method to make insertions and deletions at specific sites using the gene transfer agent, a defective phage which transduces 4.6 kbp linear fragments of DNA. Insertions and deletions in the 11.8 kbp (copy I) region generate a Nif$^-$ phenotype, confirming the results of the mutant complementation analyses, which indicate that, under laboratory conditions, only the copy I genes are active. However, at a low frequency, Nif$^+$ pseudorevertants can be isolated. Some of these pseudorevertants retain the mutagenic insertion in the copy I gene, indicating that the extra copies can be activated to produce functional nitrogenase.

The physiological meaning of the nif gene extra copies is still unclear. We are considering two possibilities: i) the extra copies are pseudogenes or ii) they are activated under growth conditions different from the ones used in this work. The first model implies that the extra copies are permanently silent, and that the pseudorevertants are mutants which create a promoter sequence in front of the coding region(s) of one or more extra copies, or they "repair" the DNA sequence to produce active gene products, or both. The second model predicts that the nitrogenase coded by the extra gene copies has properties (ammonia switch-off, substrate affinity, cofactor requirements) different from copy I nitrogenase and that the cell can selectively express different copies of the nif genes according to environmental cues. DNA sequencing and transcriptional studies should allow us to distinguish between these possibilities.

LOCATION OF NITROGEN FIXATION (nif) GENES ON INDIGENOUS PLASMIDS OF ENTEROBACTER AGGLOMERANS

M. SINGH and W. KLINGMÜLLER

Lehrstuhl für Genetik, Universität Bayreuth, D-8580 Bayreuth, FRG

INTRODUCTION

Several recent reports have shown that the nitrogenase structural genes in Rhizobia are located on large plasmids (Nuti et al, 1979; Banfalvi et al, 1981; Rosenberg et al, 1981). Although large plasmids have also been detected in free-living, N_2-fixing bacteria (Singh, Wenzel, 1982), Elmerich, Franche, 1982), there has been no report of plasmids containing nif-genes in this group. We have recently identified large plasmids in N_2-fixing strains of E. agglomerans (Singh et al, 1983) and present evidence that at least the nif-structural genes are located on these plasmids.

MATERIALS AND METHODS

Crude lysates of plasmid DNAs or the restriction fragments of purified plasmids were electrophoresed in agarose gels and used for Southern hybridisation (Southern, 1975). DNA fragments were isolated from gels by electroelution.

RESULTS AND DISCUSSION

One plasmid (70-100 Mdal) from each of the five strains analysed was found to hybridise strongly to the ^{32}P-labelled Eco RI fragment of pSA30 containing nif HDKY genes of K. pneumoniae. When we purified the 78 Mdal plasmid pEA3 (Singh et al, 1983) and used it as labelled probe, it hybridised strongly to the 6.2 kb Eco RI fragment and to both the HindIII fragments of pSA30. Labelled pEA3 also hybridised to 4 out of five PstI fragments generated by Eco RI PstI double digestion of pSA30 indicating that pEA3 shares considerable homology with the nif HDKY of K. pneumoniae. Digestion of pEA3 with Eco RI produces 18 fragments, while 17 fragments are generated by HindIII and 13 by Bgl II. Our results indicate that the nif-structural genes are clustered on one HindIII fragment of pEA3. We have cloned nif HDK from pEA3 in order to study the organisation of nif-genes in E. agglomerans and to compare it with that of the other N_2-fixing bacteria.

REFERENCES

Banfalvi Z, Sakanyan V, Koncz C, Kiss A, Dusha I, Kondorosi A (1981) Mol. Gen. Genet. 184, 318-325.
Elmerich C, Franche C (1982) In Klingmüller W, ed, Azospirillum Genetics Physiology Ecology, pp 9-17, EXS 42, Birkhäuser, Basel.
Nuti MP, Lepidi AA, Prakash RK, Schilperoort RA, Cannon FC (1979) Nature 282, 533-535.
Rosenberg C, Boistard P, Denarie J, Casse-Delbart F (1981) Mol. Gen. Genet. 184, 326-333.
Singh M and Wenzel W (1982) In Klingmüller W, ed, Azospirillum Genetics Physiology Ecology, pp 44-51, EXS 42, Birkhäuser, Basel.
Singh M, Kleeberger A, Klingmüller W (1983) Mol. Gen. Genet. 190, 373-378.
Southern M (1975) J. Mol. Biol. 98, 503-517.

STUDIES OF MUTANTS OF N_2-FIXING HETEROCYSTOUS CYANOBACTERIA

D.W. SPENCE/W.D.P. STEWART
A.R.C. RESEARCH GROUP ON CYANOBACTERIA AND DEPARTMENT OF BIOLOGICAL SCIENCES, UNIVERSITY OF DUNDEE, DUNDEE DD1 4HN, U.K.

Mutants of the N_2-fixing cyanobacterium *Anabaena* PCC 7120 have been isolated following exposure to NTG. Class 1 mutants do not produce heterocysts, show GS biosynthetic activity but show no nitrogenase activity (*het*$^-$ *gln*$^+$ *nif*$^-$). Class 2 mutants do not produce heterocysts and show GS activity but fix N_2 anaerobically only (*het*$^-$*gln*$^+$*nif*$^+$). Class 3 mutants differentiate heterocysts, have GS activity but do not fix N_2 (*het*$^+$*gln*$^+$*nif*$^-$). Class 4 mutants make heterocysts and fix N_2 and have GS activity but require exogenous glutamine for growth.

Under N_2/CO_2 (99.96/0.04, v/v) *het*$^-$*gln*$^+$*nif*$^+$ mutants show approximately 10% of the wild type nitrogenase activity; this activity is inhibited by up to 90% by DCMU (10 μm). SDS-polyacrylamide gel electrophoretic analysis suggests that the *het*$^-$*gln*$^+$*nif*$^+$ mutants (Class 2) synthesize nitrogenase structural polypeptides. The *het*$^+$*gln*$^+$*nif*$^-$ mutants and the Class 4 mutants differentiate heterocysts with similar frequency and spacing to the wild type on -N medium. The *het*$^+$*gln*$^+$*nif*$^-$ mutants do not show nitrogenase activity. Two of the *het*$^+$*gln*$^+$*nif*$^-$ mutants tested show normal or higher GS activity relative to the wild type. The Class 4 mutants show reduced nitrogenase activity (approximately 50% of the wild type) despite having a heterocyst spacing pattern similar to the wild type. Two of the Class 4 mutants tested show 65% and 100% respectively of the wild type GS activity. The altered GS activities in the Class 3 and Class 4 mutants are due to altered levels of glutamine synthetase protein as evidenced by rocket immunoelectrophoresis using antibodies against glutamine synthetase from *Anabaena cylindrica*.

We conclude that under N_2/CO_2 nitrogenase activity is synthesized in vegetative cells (Class 2 mutants); (2) photosynthetic electron transport is required for sustained nitrogenase activity; (3) The heterocyst spacing pattern may not be determined by NH_4^+; (4) There may be a relationship between nitrogenase synthesis and GS activity since no mutant was isolated which had full nitrogenase activity and low GS activity. Mutants were isolated, however, which had low nitrogenase activity but full glutamine synthetase activity.

TRANSCRIPTION FROM DIFFERENT PROMOTERS MAINTAINS EXPRESSION OF THE GLUTAMINE SYNTHETASE STRUCTURAL GENE IN ANABAENA GROWING ON AMMONIA OR FIXING NITROGEN

NILGUN EREKEN TUMER/STEVEN J. ROBINSON/ROBERT HASELKORN
DEPARTMENT OF BIOPHYSICS AND THEORETICAL BIOLOGY, UNIVERSITY OF CHICAGO, CHICAGO, IL. 60637

We have been interested in the regulatory circuits controlling transcription of the genes for nitrogen fixation and assimilation in cyanobacteria (blue-green algae). Glutamine synthetase (coded by the *glnA* gene) is a key enzyme in nitrogen assimilation in all organisms. In the heterocyst-forming *Anabaena* 7120, nitrogen reduced to ammonia by nitrogenase is incorporated into glutamine by glutamine synthetase. The glutamine is then transported into the vegetative cells to serve in the biosynthesis of various metabolites. Glutamine synthetase is needed whether *Anabaena* cells are growing on ammonia or fixing nitrogen; thus the enzyme is present in the vegetative cells as well as in the heterocysts.

Several of the genes for nitrogen fixation in *Anabaena* have been cloned and sequenced and the origins of their transcripts have been determined by S1 nuclease mapping and primer extension. Examination of the DNA sequences in the -10 and -35 regions of these genes revealed a consensus for *Anabaena nif* gene promoters that differs both from the *nif* gene promoters of *Klebsiella* and from the conventional promoters of *E. coli* and *B. subtilis*. Transcripts corresponding to these genes are found in *Anabaena* only under conditions under which nitrogenase is derepressed. Since glutamine synthetase is present in *Anabaena* under both nitrogenase repressing and derepressing conditions, it was of interest to determine the origins of the *glnA* transcripts produced under both conditions.

In order to determine the promoter sequences, *Anabaena glnA* was sequenced by the procedure of Maxam and Gilbert. RNA was isolated from cells grown in the presence (N^+) or absence (N^-) of ammonia, the two RNA preparations were then used for S1 nuclease protection experiments. When *Anabaena* cells are grown on ammonia as nitrogen source, major transcripts of *glnA* are found beginning 267 (RNA_{IV}), 155 (RNA_{II}) and 93 (RNA_I) nucleotides before the coding sequence. The -35 and -10 sequences for RNA_{II} are TTGTGC and TAATAT, respectively. These sequences are similar to *E. coli* promoters; they are used by the *Anabaena glnA* gene only when ammonia is present in the growth medium. Under nitrogen-fixing conditions, RNA_I is the major *glnA* transcript in *Anabaena*. The -35 and -10 sequences of RNA_I are CAAAAC and TCTAC, respectively. These sequences are very different from *E. coli* promoter sequences. However, they are very similar to the -35 and -10 regions of the *Anabaena nifH* promoter. Primer extension experiments agree with S1-mapping and confirm that RNA_{II} is present only under N^+ conditions, but that RNA_I is present in both N^+ and N^- conditions. Additional longer transcripts (RNA_{III} and RNA_{IV}) appear under both conditions.

The *glnA* transcripts appear in both ammonia-grown and ammonia-starved *Anabaena* cells. However, S1 mapping and primer extension show that these transcripts are initiated from different promoters. If RNA polymerase is modified for abundant transcription of *nif* genes and for cessation of transcription of genes coding for proteins involved in CO_2 fixation and O_2 evolution, the *glnA* gene has to be transcribed by both forms of the enzyme. Multiple promoters would be a reasonable and simple solution to the problem of ensuring continued expression of *glnA* in the presence or absence of ammonia.

CLONING AND EXPRESSION IN E. coli OF THE WHOLE nif GENES OF Klebsiella oxytoca, A NITROGEN FIXER IN THE RHIZOSPHERE OF RICE.

T. UOZUMI, P. L. WANG, S. K. KOH, K. S. CHUNG AND T. BEPPU
DEPARTMENT OF AGRICULTURAL CHEMISTRY, THE UNIVERSITY OF TOKYO,
BUNKYO-KU, TOKYO, JAPAN

K. oxytoca NG13 is a nitrogen fixer isolated from rhizosphere of rice in Japan. We found that NG13 has a nif gene cluster similar to that of K. pneumoniae M5a1, by Southern hybridization. We cloned nifQ-K(16.2 Kb) and nifDHJ -(22.0 Kb) fragments produced by HindIII cutting, on pBR322 in E. coli. And, these 2 fragments were put together on pBR322 to make a recombinant plasmid pNOW25(42.6 Kb) which contained the whole nif genes of K. oxytoca and some non-nif genes flanking on the right side of nifJ. The non-nif DNA fragment (11.5 Kb) was cut out by XhoI, and a smaller plasmid pNOK31(31.1 Kb) was reconstructed, which contained the whole nif genes.

E. coli K060 (C strain) containing pNOW25 or pNOK31 can grow on a N-free medium and reduce acetylene. The acetylene reduction activity of K060(pNOK31) was 100 nmole/day per 5 ml of semisolid N-free medium, whereas the activities of K060(pNOW25) and K060(pRD1) were 11-26 nmole and 270-450 nmole, respectively. Under the same conditions, K. pneumoniae M5a1 and K. oxytoca NG13 showed activities of 580-2300 nmole and 2400 nmole, respectively. Thus, the expressed activity of the nif system of K. oxytoca is rather low in E. coli even if the nif genes are cloned on a multicopy plasmid.

Restriction maps of the cloned nif genes of NG13 were made up. As for Hind III, EcoRI and BamHI, the restriction sites of NG13 nif genes are the same as those of K. pneumoniae M5a1. But, there are considerable differnces as for PstI, SalI and BglII, differing in 8 sites among 18 sites, 1 among 8, and 1 among 3, respectively.

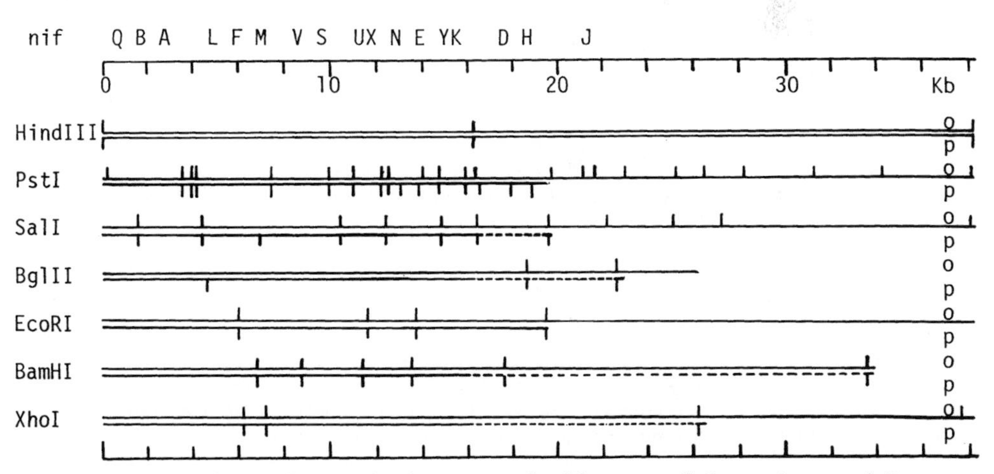

Figure 1. Comparison of restriction maps of nif genes of K. oxytoca and K. pneumoniae. Upper line (o): K. oxytoca NG13, lower line (p): K. pneumoniae M5a1. Restriction sites on solid line were checked on agarose gels by our hands. Those on dotted line were cited from literature(Ausubel FM et al.(1980) in Stewart WDP and Gallon JR, eds, Nitrogen Fixation, pp.395-421, Academic Press, London).

SPONTANEOUS NIF$^-$ MUTANTS OF RHODOPSEUDOMONAS CAPSULATA

JUDY D. WALL AND JEAN LOVE
BIOCHEMISTRY DEPARTMENT, UNIV. OF MISSOURI, COLUMBIA, MO 65211 U.S.A.

1. INTRODUCTION

The suggestion has been made that nitrogenase mediated H_2 evolution from the photosynthetic bacteria might function in energy charge maintenance or redox balance (Gest, 1972; Hillmer, Gest, 1977). For experimental purposes, we assumed that H_2 production did not serve an essential function and was deleterious, such that cells having lost this capability should have a growth advantage. Therefore wild-type *Rhodopseudomonas capsulata* cells were serially subcultured photosynthetically in minimal medium with glutamate as N source so that nitrogenase was always derepressed and H_2 was being evolved. Spontaneous Nif$^-$ (nitrogen fixation) mutants, no longer producing H_2, were strongly selected in this growth mode.

2. MATERIALS AND METHODS

R. capsulata B100 (Hillmer, Gest, 1977) was the wild-type parent of the mutants. Photosynthetic growth for H_2 production was in completely filled 17 ml screw-capped tubes incubated at 30-33°C, illuminated with three 60w Lumiline incandescent lamps (6000 lux).

3. RESULTS AND DISCUSSION

In a wild-type culture (<0.1% Nif$^-$ mutants), Nif$^-$ mutants had become 2-5% of the population after 3-4 daily subcultures in medium derepressing for nitrogenase and 95% in 10 days. All of the six independently isolated mutants harbored mutations which were tightly linked when mapped via the species-specific gene transfer agent (Yen, Marrs, 1976).

The doubling time of the Nif$^-$ mutants was not different from that of the wild type when ammonium was N but was significantly more rapid (3.4 ± 0.10 h vs 4.3 ± 0.30 h) when glutamate was N. Biochemical analyses showed that the mutants lacked detectable nitrogenase activity in whole cells or extracts. In fact, all three of the nitrogenase specific polypeptides were greatly reduced or absent from the mutants when analyzed by SDS PAGE. Of the other major enzymes involved in nitrogen metabolism (Johansson,Gest 1976), glutamine synthetase was elevated about 2 fold, glutamate synthase was not altered, and glutamate dehydrogenase was still undetectable.

If the original hypothesis were correct, the detrimental effect of H_2 production should be amplified by lowering the rate of ATP synthesis. Thus at low light intensity, the growth rate differential on glutamate should be increased. It was not. Some other, unknown factor may become limiting at low light or the production of H_2 may be useful under these conditions.

4. REFERENCES

Gest H (1972) Adv. Microbiol. Physiol. 7, 243-282.
Hillmer P and Gest H (1977) J. Bacteriol. 129, 724-731.
Johansson BC and Gest H (1976) J. Bacteriol. 128, 683-688.
Yen H-C and Marrs B (1976) J. Bacteriol. 126, 619-629.

5. ACKNOWLEDGEMENTS

SERI Contract XK-2-02103-01 and NSF grant PCM-7904223 provided support.

PMYL, A LARGE PLASMID OF AZOTOBACTER VINELANDII AVY5, HAS DNA SEQUENCES HYBRIDIZABLE TO KPNIF AND PPXYL

KEIJI YANO/MIYO ANAZAWA/FUKASHI MURAI/MASAO FUKUDA
DEPARTMENT OF AGRICULTURAL CHEMISTRY, THE UNIVERSITY OF TOKYO,
1-1-1, YAYOI, BUNKYO, TOKYO 113, JAPAN

1. INTRODUCTION

Nitrogen fixation is an anaerobic and energy-consuming process. It must have evolved very early. The structural genes of "nitrogenase", at least a part, have been highly conserved among evolutionarily distant nitrogen fixers irrespective of the responses to oxygen.

Azotobacter vinelandii is a free-living, absolute aerobe and has between 10 and 40 times as much DNA per cell as does E. coli (Sadoff, H.L. et al. 1979), probably multigenomic. It capable of not only nitrogen fixation but benzoate assimilation which requires oxygen. A possible mechanism to acquire these ambivalent abilities is the acquisition of a plasmid or plasmids.

We have reported previously (Yano,K. et al.1983) that A.vinelandii AVY5 harbors three indigenous plasmids, pMY1 (120Md), pMY2 (5.2Md), pMY3 (1.2Md). The largest pMY1 functions both nitrogen fixation and benzoate assimilation and has DNA hybridizable to KpNif and PpXyl, respectively.

2. MATERIALS AND METHODS

A. vinelandii AVY5 was grown in L broth or Burk's medium. E. coli GM4/pSA30, E.coli FMA185/pCM1, and E.coli 20SO/pTS1 were used as plasmid amplifiers. Plasmid isolation and other experimental procedures were described elsewhere (Yano, K. et al. 1983). LY (large yellow) stands for wild type, whereas SW (small white) for deficient in nitrogen fixation and benzoate assimilation. SWs were obtained by mitomycin C treatment of LY or appeared spontaneously.

3. RESULTS

Cells at the state of non-nitrogen fixation were necessary to detect plasmid bands on agarose gel electrophoresis. LY harbored 3 plasmids while SWs showed polymorphic profile in which pMY1 disappeared or deleted. pMY1 was apt to disintegrate (resolve?) to smaller bands. ^{32}P-Labelled probes prepared from pCM1, pSA30 and pTS1 consisting of Klebsiella nifQ~K, nifYKDH and Pseudomonas XylE (catechol-2,3-dioxygenase), respectively, were used in Southern hybridization. KpnifQ~K showed no significant results. KpnifYKDH clearly hybridized with total DNA from AVY5 LY but not with total DNA from SW on "dot hybridization" after depurination with acetic acid. After HindIII digestion, plasmids preparation eliminated l-DNA by alkaline treatment hybridized definitely while l-DNA, mainly chromosomal DNA fragments, faintly. Observed hybrid bands ascribed to pMY1 and also to pMY2 with KpnifKDH. HindIII digested pMY1 apparently hybridized with PpxylE. pMY1 was undoubtedly a composite plasmid since minor plasmid bands appeared in not only SWs but LY, too. Restriction enzyme analyses showed that pMY1 has MW of at least 119Md and pMY3 is suitable for a plasmid vector material.

4. References

Sadoff HL, Shimel B and Ellis S (1979) J. Bact. 138, 871-877.
Yano K et al. (1983) In Ikeda Y and Beppu T, eds, Genetics of Industrial Microorganisms, 1982, pp. 328-337, Kodansha Ltd, Tokyo.

EFFECT OF nifA PRODUCT ON DEREPRESSION OF nif GENES IN KLEBSIELLA PNEUMONIAE

Zhu Jia-bi, Yu Quo-chiao, Wang Li-wen, Shen Szi-shih and Shen San-chiun

Laboratory of Molecular Genetics, Shanghai Institute of Plant Physiology, Academia Sinica, Shanghai, China

Genetical and biochemical evidences have been presented that nifA gene plays a positive role on the regulation of nitrogen fixation in K. pneumoniae[1][2]. Our present work is concerned with the mechanism of NH_4^+ and oxygen[3] regulation of nif genes with special reference to the role of nifA product on these regulations.

A recombinant plasmid pST1021 which carries nifA gene under the promoter of tetracycline resistant gene was constructed. When this nifA carried plasmid was introduced into a glnG mutant, the Nif$^-$ phenotype of this gln mutant was restored to be Nif$^+$. Furthermore, if this recombinant plasmid was introduced into the wild type strain and the glnG mutant, derepression of nitrogenase synthesis in ammonia was readily demonstrated in both strains. The nif transcription was examined with the extracts of these bacterial cells which had been pulse labeled with ^{3}H-leucine and were then subjected to two dimensional gel electrophoresis. The results showed that both glnG mutant and the wild type strain which harbour the pST1021 plasmid[4] produce the notable nif polypeptides in ammonium. The role of nifA product on derepression of nif genes under oxygen was also investigated. We adopted the method of fusing nif promoters to lacZ gene with phage Mud (plac), thus lacZ is expressed under the control of nif promoters in such nif-lac fusions. We found that all aerobically grown nif-lac fusion strains when received the nifA gene carried plasmid pST1021 synthesized B-galactosidase constitutively, escaping the repression from oxygen.

The pnenomenon of derepression of nif genes by the product of nifA has been also observed in Enterobacter cloacae, nitrogen fixing rhizospheric bacteria of rice plant. To test whether a DNA sequence homologous to nifA of K. pneumoniae is present in E. cloacae and in other nitrogen fixing bacteria, the ^{32}P-labeled DNA fragments containing nif'BAL' from the plasmid pST1021 and the ^{32}P-labeled fragment containing nif'KDHJ' from the plasmid pJC363 were used as probes to hybridize the Southern blots of restricting-endonuclease digested DNAs from 3 different species of nitrogen fixing bacteria, Enterobacter cloacae, Rhizobium astragali and Alcaligenes faecalis. It was found that as expected, DNAs from all 3 nitrogen fixing species hybridized to the K. pneumoniae DNA containing nif'KDHJ' and that only the DNA from E. cloacae hybridized both the DNA sequence containing nif'KDHJ' and that containing nif'BAL' of K. Pneumoniae. It indicates that the regulatory nif gene is less conserved than the nitrogenase structural genes among nif bacteria during evolution.

1. Buchanan-Wollaston, V., Cannon, M. C. and Cannon, F. C. Molec. gen. Genet. 184, 102-106 (1981).
2. Kong, Q. T., Wu, Q. L., Jia, L. B., M. Syvanen, E. C. C. Lin, and S. C. Shen, Scientia Sinica 25, 1062-1070 (1982).
3. St. John, R. T., V. K. Shah, and W. J. Brill, J. Bacterial 119, 266-269 (1974).
4. Zhu J. B., Yu Q. C., Wang L. W., and S. C. Shen, Scientia Sinica (in the press).

AUTHOR INDEX